PREFACE

Congratulations on taking the first step toward learning to becoming an Aviation Maintenance Technician. The *A&P Technician General Textbook* contains the answers to many of the questions you may have as you begin your training program. It is based on the "study/review" concept of learning. This means detailed material is presented in an uncomplicated way, then important points are summarized through the use of bold type and illustrations. The textbook incorporates many design features that will help you get the most out of your study and review efforts. These include:

Illustrations — Illustrations are carefully planned to complement and expand upon concepts introduced in the text. The use of bold in the accompanying caption flag them as items that warrant your attention during both initial study and review.

Bold Type — Important new terms in the text are printed in bold type, then defined.

Federal Aviation Regulations — Appropriate FARs are presented in the textbook. Furthermore, the workbook offers several exercises designed to test your understanding of pertinent regulations.

This textbook is the key element in the training materials. Although it can be studied alone, there are several other components which we recommend to make your training as complete as possible. These include the *A&P Technician General Workbook* and *Study Guide*, as well as *AC 43.13-1B/2A* and *FAR Handbook for Aviation Maintenance Technicians*. When used together, these various elements provide an ideal framework for you and your instructor as you prepare for the FAA computerized and practical tests.

The A&P Technician General course is the first segment of your training as an aviation maintenance technician. The General section introduces you the basic concepts, terms, and procedures that serve as the building blocks for the more complex material you will encounter later on in your training.

TABLE OF CONTENTS

A&P TECHNICIAN
GENERAL
TEXTBOOK

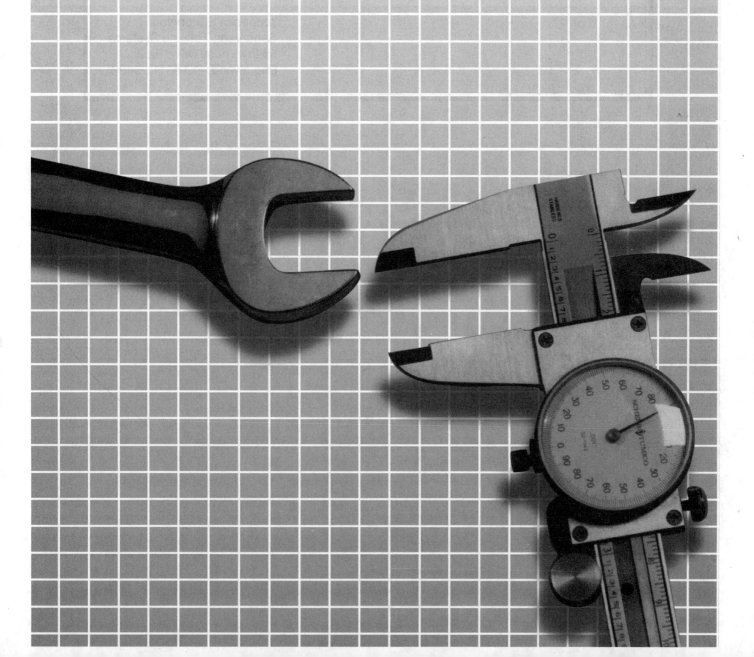

Support Materials

Look for these support materials to complement your A&P Technician General Textbook:

- **A&P Technician General Workbook**
- **A&P Technician General Study Guide**
- **Federal Aviation Regulations**
- **AC 43.13-1B/2A Acceptable Methods, Techniques and Practices/Aircraft Alterations**
- **Aircraft Technical Dictionary**
- **Standard Aviation Maintenance Handbook**

These items are among the wide variety of Jeppesen reference materials available through your authorized Jeppesen Dealer. If there is no Jeppesen Dealer in your area, you can contact us directly:

Jeppesen Sanderson
Sanderson Training Systems
55 Inverness Drive East
Englewood, CO 80112-5498

JS312690-004

MATHEMATICS

INTRODUCTION

Mathematics is the basic language of science and technology. It is an exact language that has a vocabulary and meaning for every term. Since math follows definite rules and behaves in the same way every time, scientists and engineers use it as their basic tool. Long before any metal is cut for a new aircraft design, there are literally millions of mathematical computations made. Aviation maintenance technicians perform their duties with the aid of many different tools. Like the wrench or screwdriver, mathematics is an essential tool in the repair and fabrication of replacement parts. With this in mind, you can see why you must be able to use this important tool.

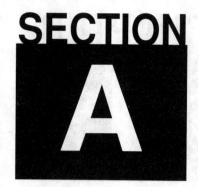

SECTION A

ARITHMETIC

Just as studying a new language begins with learning basic words, the study of mathematics begins with arithmetic, its most basic branch. Arithmetic uses real, non-negative numbers, which are also known as counting numbers, and consists of only four operations, addition, subtraction, multiplication, and division. While you have been using arithmetic since childhood, a review of its terms and operations will make learning the more difficult mathematical concepts much easier.

NUMBER SYSTEMS

Numbers are a large part of everyone's life, and you are constantly bombarded with figures. Yet little attention is paid to the basic structure of the numbering system. In daily life, most people typically use a "base ten" or **decimal system**. However, another numbering system that is used in computer calculations is the **binary**, or "base two" system.

THE DECIMAL SYSTEM

The decimal system is based on ten whole numbers, often called **integers**, from zero to nine. Above the number nine the digits are reused in various combinations to represent larger numbers. This is accomplished by arranging the numbers in columns based on a multiple of ten. With the addition of a negative (-) sign, numbers smaller than zero are indicated.

To describe quantities that fall between whole numbers, fractions are used. **Common fractions** are used when the space between two integers is divided into equal segments such as fourths. When the space between integers is divided into ten segments, **decimal fractions** are typically used.

THE BINARY SYSTEM

Because the only real option in an electrical circuit is ON or OFF, a number system based on only two digits is used to create electronic calculations. The base two, or binary system, only utilizes the digits zero and one. For example, when a circuit is ON a one is represented, and when a circuit is OFF a zero is indicated. By converting these ON or OFF messages to represent numbers found in the decimal system, a computer can perform complex tasks.

To build a binary number system that corresponds to the decimal system, begin with one switch. When this switch is in the OFF position, a zero is indicated. When it is in the ON position, a one is represented. Because these are the only possibilities for a single switch, additional switches must be added to represent larger quantities. For example, a second switch represents the quantity 2. When the first switch is OFF and the second switch is ON, the quantity 2 is indicated. When both the first and second switch are ON, the 1 and 2 are added to indicate the quantity 3. This procedure of adding switches continues with each switch value doubling as you progress. For example, the first 10 values in the binary system are 1, 2, 4, 8, 16, 32, 64, 128, 256, and 512. [Figure 1-1]

WHOLE NUMBERS

While integers are useful in communicating a given quantity, you must be able to manipulate them to discover their full power. There are four fundamental mathematical operations with which you must be familiar. They are addition, subtraction, multiplication, and division.

ADDITION

The process of finding the total of two or more numbers is called addition. This operation is indicated by the plus (+) symbol. When numbers are combined by addition, the resulting total is called the **sum**.

When adding whole numbers whose total is more than nine, it is necessary to arrange the numbers in columns so that the last digit of each number is in the same column. The ones column contains the values zero through nine, the tens column contains multiples of ten, up to ninety, and the hundreds column consists of multiples of one hundred.

Example:

hundreds	tens	ones
	^{1}7	8
^{1}2	4	3
+ 4	6	2
7	8	3

DECIMAL NUMBER	BINARY NUMBERS								BINARY NUMBER
	128	64	32	16	8	4	2	1	
0	0	0	0	0	0	0	0	1	1
2	0	0	0	0	0	0	1	0	10
3	0	0	0	0	0	0	1	1	11
4	0	0	0	0	0	1	0	0	100
5	0	0	0	0	0	1	0	1	101
27	0	0	0	1	1	0	1	1	11011
48	0	0	1	1	0	0	0	0	110000
92	0	1	0	1	1	1	0	0	1011100
117	0	1	1	1	0	1	0	1	1110101
168	1	0	1	0	1	0	0	0	10101000

Figure 1-1. This binary conversion chart illustrates how a decimal number is converted to a binary number. For example, the binary equivalent to 48 is 110000.

To check addition problems, add the figures again in the same manner, or in reverse order from bottom to top. It makes no difference in what sequence the numbers are combined.

SUBTRACTION

The process of finding the **difference** between two numbers is known as subtraction and is indicated by the minus (-) sign. Subtraction is accomplished by taking the quantity of one number away from another number. The number which is subtracted is known as the **subtrahend**, and the number from which the quantity is taken is known as the **minuend**.

To find the difference of two numbers, arrange them in the same manner used for addition. With the minuend on top and the subtrahend on the bottom, align the vertical columns so the last digits are in the same column. Beginning at the right, subtract the subtrahend from the minuend. Repeat this for each column.

Example:

hundreds	tens	ones
7	$^5\cancel{8}$	$^1 4$
-4	3	6
3	2	8

To check a subtraction problem, you may add the difference to the subtrahend to find the minuend.

MULTIPLICATION

Multiplication is a special form of repetitive addition. When a given number is added to itself a specified number of times, the process is called multiplication. The sum of 4 + 4 + 4 + 4 + 4 = 20 is expressed by multiplication as 4 × 5 = 20. The numbers 4 and 5 are called **factors** and the answer, 20, represents the **product**. The number multiplied (4) is called the **multiplicand**, and the **multiplier** represents the number of times the multiplicand is added to itself. Multiplication is typically indicated by an (×), (•), or in certain equations, by the lack of any other operation sign.

One important fact to remember when multiplying is that the order in which numbers are multiplied does not change the product.

Example:

$$\begin{array}{r} 3 \\ \times 4 \\ \hline 12 \end{array} \quad \text{or} \quad \begin{array}{r} 4 \\ \times 3 \\ \hline 12 \end{array}$$

Like addition and subtraction, when multiplying large numbers it is important they be aligned vertically. Regardless of the number of digits in the multiplicand or the multiplier, the multiplicand should be written on top, and the multiplier beneath it. When multiplying numbers greater than nine, multiply each digit in the multiplicand by each digit in the multiplier. Once all multiplicands are used as a multiplier, the products of each multiplication operation are added to arrive at a total product.

Example:

532	Multiplicand
× 24	Multiplier
2128	First partial product
1064	Second partial product
12,768	

DIVISION

Just as subtraction is the reverse of addition, division is the reverse of multiplication. Division is a means of finding out how many times a number is contained in another number. The number divided is called the **dividend**, the number you are dividing by is the **divisor**, and the result is the **quotient**. With some division problems, the quotient may include a **remainder**. A remainder represents that portion of the dividend that cannot be divided by the divisor.

Division is indicated by the use of the division sign (÷) with the dividend to the left and the divisor to the right of the sign, or a ⌐ with the dividend inside the sign and the divisor to the left. Division also is indicated in fractional form. For example, in the fraction $^3/_4$, the 3 is the dividend and the 4 is the divisor. When division is carried out, the quotient is .75.

The process of dividing large quantities is performed by breaking the problem down into a series of operations, each resulting in a single digit quotient. This is best illustrated by example.

Example:

$$
\begin{array}{r}
52 \\
8\overline{)416} \\
\underline{40} \\
16 \\
\underline{16} \\
0
\end{array}
$$

To check a division problem for accuracy, multiply the quotient by the divisor and add the remainder (if any). If the operation is carried out properly, the result equals the dividend.

SIGNED NUMBERS

If zero is used as a starting point, all numbers larger than zero have a positive value, and those smaller than zero have a negative value. This is illustrated by constructing a **number line**. [Figure 1-2].

ADDING SIGNED NUMBERS

When adding two or more numbers with the same sign, ignore the sign and find the sum of the values and then place the common sign in front of the answer. In other words, adding two or more positive numbers always results in a positive sum, whereas adding two or more negative numbers results in a negative sum.

When adding a positive and negative number, find the difference between the two numbers and apply the sign (+ or -) of the larger number. In other words, adding a negative number is the same as subtracting a positive number. The result of adding or subtracting signed numbers is called the **algebraic sum** of those numbers.

Add 25 + (-15)

25		25	
+ (-15)	or	− 15	
10		10	

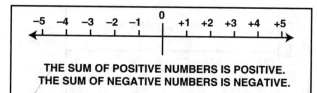

THE SUM OF POSITIVE NUMBERS IS POSITIVE.
THE SUM OF NEGATIVE NUMBERS IS NEGATIVE.

Figure 1-2. When creating a number line, negative values are identified with a minus sign (-), and positive values are identified by the plus (+) sign or by the absence of a sign.

SUBTRACTING SIGNED NUMBERS

When subtracting numbers with different signs, change the operation sign to plus and change the sign of the subtrahend. Once this is done, proceed as you do in addition. For example, +3 - -4 is the same as +3 + +4. It makes no difference if the subtrahend is larger than the minuend, since the operation is done as though the two quantities are added.

Example:

Subtract 48 from -216.

Step 1: Set up the subtraction problem.

-216 - 48

Step 2: Change the operation sign to a plus sign and change the sign of the subtrahend. Now add.

-216 + -48 = -264

MULTIPLYING SIGNED NUMBERS

Multiplication of signed numbers is accomplished in the same manner as multiplication of any other number. However, after multiplying, the product must be given a sign. There are three rules to follow when determining a product's sign.

1. The product of two positive numbers is always positive.
2. The product of two negative numbers is always positive.
3. The product of a positive and a negative number is always negative.

Example:

6 × 2 = 12	-6 × -2 = 12
(-6) × (-2) = 12	(-6) × (2) = -12

DIVIDING SIGNED NUMBERS

Like multiplying signed numbers, division of signed numbers is accomplished in the same manner as dividing any other number. The sign of the quotient is determined using rules identical to those used in multiplication.

1. The quotient of two positive numbers is always positive.
2. The quotient of two negative numbers is always positive.
3. The quotient of a positive and a negative number is always negative.

Example:

$$12 \div 3 = 4 \qquad 12 \div (-3) = -4$$
$$(-12) \div (-3) = 4 \qquad (-12) \div 3 = -4$$

COMMON FRACTIONS

A common fraction represents a portion or part of a quantity. For example, if a number is divided into three equal parts, each part is one-third ($^1/_3$) of the number. A fraction consists of two numbers, one above and one below a line, or **fraction bar**. The fraction bar indicates division of the top number, or **numerator**, by the bottom number, or **denominator**. For example, the fraction $^3/_4$ indicates that three is divided by four to find the decimal equivalent of .75.

When a fraction's numerator is smaller than the denominator, the fraction is called a **proper fraction**. A proper fraction is always less than 1. If the numerator is larger than the denominator, the fraction is called an **improper fraction**. In this situation the fraction is greater than 1. If the numerator and denominator are identical, the fraction is equal to 1.

A **mixed number** is the combination of a whole number and a proper fraction. Mixed numbers are expressed as $1\ ^5/_8$ and $29\ ^9/_{16}$ and are typically used in place of improper fractions.

The numerator and denominator of a fraction can be changed without changing the fraction's value. One way this is done is by multiplying the numerator and denominator by the same number.

Example:

$$\frac{3}{8} \times \frac{3}{3} = \frac{9}{24}$$

A fraction's value also remains the same if both the numerator and denominator are divided by the same number. This type of operation allows you to simplify, or reduce, large fractions to their smallest terms.

Example:

$$\frac{3}{9} \div \frac{3}{3} = \frac{1}{3}$$

$$\frac{21875}{100000} \div \frac{25}{25} = \frac{875}{4000} \div \frac{25}{25} = \frac{35}{160} \div \frac{5}{5} = \frac{7}{32}$$

REDUCING FRACTIONS

It is generally considered good practice to reduce fractions to their lowest terms. The simplest reductions occur when the denominator is divisible by the numerator. If the denominator is not evenly divided by the numerator, you must find a number by which the numerator and denominator are divided evenly. Here are a few tips to help in the selection of divisors:

1. If both numbers are even, divide by 2.
2. If both numbers end in 0 or 5, divide by 5.
3. If both numbers end in 0, divide by 10.

Example:

Reduce $^{15}/_{45}$ to its lowest terms.

Step 1: Divide both the numerator and denominator by 5.

$$\frac{15}{45} \div \frac{5}{5} = \frac{3}{9}$$

Step 2: Reduce further by dividing both terms by 3.

$$\frac{15}{45} = \frac{3}{9} \div \frac{3}{3} = \frac{1}{3}$$

When neither the numerator or denominator can be divided evenly, the fraction is reduced to its lowest terms.

LEAST COMMON DENOMINATOR

You cannot add or subtract common fractions without first converting all of the denominators into identical units. This process is known as finding the least common denominator (LCD). For example, the quickest way to find the least common denominator for $^1/_3$ and $^1/_2$ is to multiply the two denominators ($3 \times 2 = 6$). To determine the numerators, multiply the numerator by the same number used to obtain the LCD.

Example:

$$\frac{1}{3} \times \frac{2}{2} = \frac{2}{6}$$

$$\frac{1}{2} \times \frac{3}{3} = \frac{3}{6}$$

ADDING COMMON FRACTIONS

As mentioned earlier, you cannot add common fractions without first determining the least common denominator. However, once this is done, you only need to add the numerators to arrive at a sum. This answer is then reduced to its lowest terms.

Example:

Add $\frac{1}{12} + \frac{2}{6} + \frac{1}{3}$

Step 1: Rewrite using the least common denominator.

$$\frac{1}{12} + \frac{4}{12} + \frac{4}{12}$$

Step 2: Add the numerators and reduce to lowest terms, if possible.

$$\frac{1}{12} + \frac{4}{12} + \frac{4}{12} = \frac{9}{12} \div \frac{3}{3} = \frac{3}{4}$$

SUBTRACTING COMMON FRACTIONS

Subtracting fractions also requires an LCD to be determined. Once this is accomplished, subtract the numerators, express the difference over the LCD, and reduce the answer to its lowest terms.

Example:

Subtract $^2/_8$ from $^1/_3$

Step 1: Rewrite using the least common denominator.

$$\frac{8}{24} - \frac{6}{24}$$

Step 2: Subtract the numerators and reduce to lowest terms.

$$\frac{8}{24} - \frac{6}{24} = \frac{2}{24} \div \frac{2}{2} = \frac{1}{12}$$

MIXED NUMBERS

Mixed numbers contain both whole numbers and proper fractions. Before adding or subtracting mixed numbers, you must convert them to improper fractions. To convert a mixed number to an improper fraction, multiply the whole number by the denominator and add the product to the numerator. The sum of these two numbers becomes the numerator.

Example:

Convert 3 $^3/_4$ to an improper fraction.

$$3\frac{3}{4} = \frac{(4 \times 3) + 3}{4} = \frac{15}{4}$$

ADDING MIXED NUMBERS

When adding mixed numbers, either to other mixed numbers or to proper fractions, you must convert the mixed numbers to improper fractions. Once accomplished, determine the least common denominator and add in the same manner as with proper fractions.

When adding improper fractions, the sum is usually another improper fraction. When faced with an improper fraction in an answer, you should convert it to a mixed number. To do this, divide the numerator by the denominator to determine the whole number. If there is a remainder, leave it in fractional form.

Example:

Add the following:

$$2\frac{2}{3} + 3\frac{1}{4} + 5\frac{1}{2}$$

Step 1: Convert each to an improper fraction.

$$2\frac{2}{3} = \frac{(2 \times 3) + 2}{3} = \frac{8}{3}$$

$$3\frac{1}{4} = \frac{(3 \times 4) + 1}{4} = \frac{13}{4}$$

$$5\frac{1}{2} = \frac{(5 \times 2) + 1}{2} = \frac{11}{2}$$

Step 2: Find the LCD and add.

$$\frac{32}{12} + \frac{39}{12} + \frac{66}{12} = \frac{137}{12}$$

Step 3: Convert the improper fraction to a mixed number.

$$\frac{137}{12} = 11\,\frac{5}{12}$$

SUBTRACTING MIXED NUMBERS

To subtract a mixed number from another mixed number or proper fraction, begin by converting the mixed number to an improper fraction. Once converted, find the LCD and perform the subtraction. To complete the problem, convert the resulting improper fraction into a mixed number.

Example:

Subtract $2\,{}^{1}/_{6}$ from $5\,{}^{2}/_{3}$.

Step 1: Convert to improper fractions.

$$5\,\frac{2}{3} = \frac{(5 \times 3) + 2}{3} = \frac{17}{3}$$

$$2\,\frac{1}{6} = \frac{(2 \times 6) + 1}{6} = \frac{13}{6}$$

Step 2: Find the LCD and subtract.

$$\frac{34}{6} - \frac{13}{6} = \frac{21}{6}$$

Step 3: Convert to a mixed number.

$$\frac{21}{6} = 3\,\frac{3}{6} = 3\,\frac{1}{2}$$

MULTIPLYING FRACTIONS

Multiplication of fractions is performed by multiplying the numerators of each fraction to form the product numerator, and multiplying the individual denominators to form the product denominator. The resulting fraction is then reduced to its lowest terms.

Example:

Multiply the following: ${}^{8}/_{32} \times {}^{5}/_{8} \times {}^{4}/_{16}$

Step 1: Multiply the numerators and the denominators.

$$\frac{8}{32} \times \frac{5}{8} \times \frac{4}{16} = \frac{160}{4096}$$

Step 2: Reduce to lowest terms.

$$\frac{160}{4096} \div \frac{32}{32} = \frac{5}{128}$$

SIMPLIFY FRACTIONS FOR MULTIPLICATION

It was mentioned earlier that the value of a fraction does not change when you perform the same operation (multiplication or division) on both the numerator and denominator. You can use this principle to simplify the multiplication of fractions. For example, ${}^{8}/_{32} \times {}^{5}/_{8} \times {}^{4}/_{16}$ is equivalent to

$$\frac{8 \times 5 \times 4}{32 \times 8 \times 16}$$

Notice that there is an 8 in the numerator and denominator. Since these are equivalent values, they can be removed from the equation. Furthermore, the 16 in the denominator is divisible by the 4 in the numerator. Therefore, when both are divided by 4, the 4 in the numerator reduces to 1 and the 16 reduces to 4.

Example:

Simplify by cancellation, then multiply:

$$\frac{8}{32} \times \frac{5}{8} \times \frac{4}{16}$$

Step 1: Simplify

$$\frac{\overset{1}{\cancel{8}}}{32} \times \frac{5}{\underset{1}{\cancel{8}}} \times \frac{4}{16}$$

$$\frac{1}{32} \times \frac{5}{1} \times \frac{\overset{1}{\cancel{4}}}{\underset{4}{\cancel{16}}}$$

$$\frac{1}{32} \times \frac{5}{1} \times \frac{1}{4}$$

Step 2: Multiply and reduce, if possible.

$$\frac{1}{32} \times \frac{5}{1} \times \frac{1}{4} = \frac{5}{128}$$

DIVIDING FRACTIONS

Division of common fractions is accomplished by inverting, or turning over, the divisor and then multiplying. However, it is important that you invert the divisor only and not the dividend. Once the divisor is inverted, multiply the numerators to obtain a new numerator, multiply the denominators to obtain a new denominator, and reduce the quotient to its lowest terms.

Example:

Divide $^2/_3$ by $^1/_4$.

Step 1: Invert the divisor and multiply.

$$\frac{2}{3} \div \frac{1}{4} = \frac{2}{3} \times \frac{4}{1}$$

Step 2: Multiply and simplify the product.

$$\frac{2}{3} \times \frac{4}{1} = \frac{8}{3} = 2\frac{2}{3}$$

DECIMALS

Working with fractions is typically time consuming and complex. One way you can eliminate fractions in complex equations is by replacing them with **decimal fractions** or decimals. A common fraction is converted to a decimal fraction by dividing the numerator by the denominator. For example, $^3/_4$ is converted to a decimal by dividing the 3 by the 4. The decimal equivalent of $^3/_4$ is .75. Improper fractions are converted to decimals in the same manner. However, whole numbers appear to the left of the decimal point.

In a decimal, each digit represents a multiple of ten. The first digit represents tenths, the second hundredths, the third thousandths.

Example:

.5	is read as five tenths
.05	is read as five hundredths
.005	is read as five thousandths

When writing decimals, the number of zeros to the right of the decimal does not affect the value as long as no other number except zero appears. In other words, numerically, 2.5, 2.50, and 2.5000 are the same.

ADDING DECIMALS

The addition of decimals is done in the same manner as the addition of whole numbers. However, care must be taken to correctly align the decimal points vertically.

Example:

Add the following: 25.78 + 5.4 + 0.237

Step 1: Rewrite with the decimal points aligned, and add.

```
   25.78
    5.4
 + 0.237
  31.417
```

Once everything is added, the decimal point in the answer is placed directly below the other decimal points.

SUBTRACTING DECIMALS

Like adding, subtracting decimals is done in the same manner as with whole numbers. Again, it is important that you keep the decimal points aligned.

Example:

If you have 325.25 pounds of ballast on board and remove 30.75 pounds, how much ballast remains?

```
  325.25
 - 30.75
  294.50
```

MULTIPLYING DECIMALS

When multiplying decimals, ignore the decimal points and multiply the resulting whole numbers. Once the product is calculated, count the number of digits to the right of the decimal point in both the multiplier and multiplicand. This number represents the number of places from the left the decimal point is placed in the product.

Example:

```
  26.757        3 decimal places
 × .32          2 decimal places
  53514
 80271
 856224
 8.56224        Count 5 decimal places
                to the left of the 4
```

DIVIDING DECIMALS

When dividing decimals, the operation is carried out in the same manner as division of whole numbers. However, to ensure accurate placement of the decimal point in the quotient, two rules apply:

1. When the divisor is a whole number, the decimal point in the quotient aligns vertically with the decimal in the dividend when doing long division.
2. When the divisor is a decimal fraction, it should first be converted to a whole number by moving the decimal point to the right. However, when the decimal in the divisor is moved, the decimal in the dividend must also move in the same direction and the same number of spaces.

Example:

Divide 37.26 by 2.7

Step 1: Move the decimal in the divisor to the right to convert it to a whole number.

$$27.\overline{)37.26}$$

Step 2: Move the decimal in the dividend the same number of places to the right.

$$27\overline{)372.6}$$

Step 3: Divide.

```
          13.8
    27)372.6
        27
        102
         81
        216
```

CONVERTING DECIMALS TO FRACTIONS

Although decimals are typically easier to work with, there are times when the use of a fraction is more practical. For example, when measuring something, most scales are in fractional increments. For this reason it is important that you know how to convert a decimal number into a fraction. For example, .125 is read as 125 thousandths, which is written as 125/1000. This fraction is then reduced to its lowest terms.

Example:

Convert 0.625 into a common fraction.

Step 1: Rewrite as a fraction.

$$0.625 = \frac{625}{1000}$$

Step 2: Reduce to lowest terms.

$$\frac{625}{1000} \div \frac{25}{25} = \frac{25}{40} \div \frac{5}{5} = \frac{5}{8}$$

ROUNDING DECIMALS

Because decimal numbers can often be carried out an unreasonable number of places, they are usually limited to a workable size. This process of retaining a certain number of digits and discarding the rest is known as **rounding**. In other words, the retained number is an approximation of the computed number.

Rounding is accomplished by viewing the digit immediately to the right of the last retained digit. If this number is 5 or greater, increase the last retained digit to the next highest value. When the number to the right of the last retained digit is less than 5, leave the last retained digit unchanged. For example, when rounding 3.167 to 2 decimal places, the 7 determines what is done to the 6, which is the last retained digit. Since 7 is greater than 5, the rounded number is 3.17.

PERCENTAGE

Percentages are special fractions whose denominator is 100. The decimal fraction 0.33 is the same as $^{33}/_{100}$ and is equivalent to 33 percent or 33%. You can convert common fractions to percentages by first converting them to decimal fractions, and then multiplying by 100. For example, 5/8 expressed as a decimal is 0.625, and is converted to a percentage by moving the decimal right two places, becoming 62.5%.

To find the percentage of a number, multiply the number by the decimal equivalent of the percentage. For example, to find 10% of 200, begin by converting 10% to its decimal equivalent which is .10. Now multiply 200 by .10 to arrive at a value of 20.

If you want to find the percentage one number is of another, you must divide the first number by the second and multiply the quotient by 100. For instance, let's say an engine develops 85 horsepower of a possible 125 horsepower. What percentage of the total power available is developed? To solve this, divide 85 by 125 and multiply the quotient by 100.

Example:

$$85 \div 125 = .68 \times 100 = 68\% \text{ power is developed.}$$

Another way percentages are used is to determine a number when only a portion of the number is known. For example, if 4,180 rpm is 38% of the maximum speed, what is the maximum speed? To determine this, you must divide the known quantity, 4,180 rpm, by the decimal equivalent of the percentage.

Example:

$$4,180 \div .38 = 11,000 \text{ rpm maximum}$$

A common mistake made on this type of problem is multiplying by the percentage instead of dividing. One way to avoid making this error is to look at the problem and determine what exactly is being asked. In the problem above, if 4,180 rpm is 38% of the maximum, then the maximum rpm must be greater than 4,180. The only way to get an answer that meets this criterion is to divide by .38.

RATIO AND PROPORTION

A ratio provides a means of comparing one number to another. For example, if an engine turns at 4,000 rpm and the propeller turns at 2,400 rpm, the ratio of the two speeds is 4,000 to 2,400, or 5 to 3, when reduced to lowest terms. This relationship can also be expressed as $^5/_3$ or 5:3.

The use of ratios is common in aviation. One ratio you must be familiar with is compression ratio, which is the ratio of cylinder displacement when the piston is at bottom center to the cylinder displacement when the piston is at top center. For example, if the volume of a cylinder with the piston at bottom center is 96 cubic inches and the volume with the piston at top center is 12 cubic inches, the compression ratio is 96:12 or 8:1 when simplified.

Another typical ratio is that of different gear sizes. For example, the gear ratio of a drive gear with 15 teeth to a driven gear with 45 teeth is 15:45 or 1:3 when reduced. This means that for every one tooth on the drive gear there are three teeth on the driven gear. However, when working with gears, the ratio

of teeth is opposite the ratio of revolutions. In other words, since the drive gear has one third as many teeth as the driven gear, the drive gear must complete three revolutions to turn the driven gear one revolution. This results in a revolution ratio of 3:1, which is opposite the ratio of teeth.

A proportion is a statement of equality between two or more ratios and represents a convenient way to solve problems involving ratios. For example, if an engine has a reduction gear ratio between the crankshaft and the propeller of 3:2, and the engine is turning 2,700 rpm, what is the speed of the propeller? In this problem, let "x" represent the unknown value, which in this case is the speed of the propeller. Next, set up a proportional statement using the fractional form, $^3/_2 = 2700/x$. To solve this equation, cross multiply to arrive at the equation $3x = 2 \times 2,700$, or 5,400. To solve for (x), divide 5,400 by 3. The speed of the propeller is 1,800 rpm.

$$\frac{3}{2} = \frac{Engine\ Speed}{Propeller\ Speed}$$

$$\frac{3}{2} = \frac{2700}{x}$$

$$3x = 5,400$$

$$x = 1,800 \text{ rpm}$$

This same proportion may also be expressed as 3:2 = 2,700 : x. The first and last terms of the proportion are called the **extremes**, and the second and third terms are called the **means**. In any proportion, the product of the extremes is equal to the product of the means. In this example, multiply the extremes to get 3x, and multiply the means to get $2 \times 2,700$, or 5,400. This results in the identical equation derived earlier; 3x = 5,400.

$$3:2 = \text{engine speed : propeller speed}$$

$$3:2 = 2,700 : x$$

$$3x = 2 : 2,700$$

$$3x = 5,400$$

$$x = 1,800 \text{ rpm}$$

POWERS AND ROOTS

When a number is multiplied by itself, it is said to be raised to a given power. For example, $6 \times 6 = 36$; therefore, $6^2 = 36$. The number of times a **base number** is multiplied by itself is expressed as an **exponent** and is written to the right and slightly above the base number. A positive exponent indicates how many times a number is multiplied by itself.

Example:

3^2 is read "3 squared" or "3 to the second power." Its value is found by multiplying 3 by itself.

$$3 \times 3 = 9$$

2^3 is read "2 cubed" or "2 to the third power." Its value is found by multiplying 2 by itself 3 times.

$$2 \times 2 \times 2 = 8$$

A negative exponent implies division or fraction of a number. It indicates the inverse, or reciprocal of the number with its exponent made positive.

Example:

2^{-3} is read "2 to the negative third power." The inverse, or reciprocal of 2^{-3} with its exponent made positive is

$$\frac{1}{2^3} = \frac{1}{2 \times 2 \times 2} = \frac{1}{8}$$

Any number, except zero, that is raised to the zero power equals 1. When a number is written without an exponent, the value of the exponent is assumed to be 1. Furthermore, if the exponent does not have a sign (+ or -) preceding it, the exponent is assumed to be positive.

The root of a number is that value which, when multiplied by itself a certain number of times, produces that number. For example, 4 is a root of 16 because when multiplied by itself, the product is 16. However, 4 is also a root of 64 because $4 \times 4 \times 4 = 64$. The symbol used to indicate a root is the **radical** sign ($\sqrt{\ }$) placed over the number. If only the radical sign appears over a number, it indicates you are to extract the **square root** of the number under the sign. The square root of a number is the root of that number, when multiplied by itself, equals that number. When asked to extract a root other than a square root, an **index number** is placed outside the radical sign. For example, the cube root of 64 is expressed as

$$\sqrt[3]{64}$$

SCIENTIFIC NOTATION

Many engineering and scientific calculations involve very large or very small numbers. To ease

manipulation and decrease the possibility for error, scientific notation is used. Scientific notation is based on multiplying a number by a power of ten. Therefore, you must understand how to use exponents. [Figure 1-3]

Positive Powers of Ten	Negative Powers of Ten
$10^0 = 1$	$10^{-1} = 0.1$
$10^1 = 10$	$10^{-2} = 0.01$
$10^2 = 100$	$10^{-3} = 0.001$
$10^3 = 1,000$	$10^{-4} = 0.0001$
$10^4 = 10,000$	$10^{-5} = 0.00001$
$10^5 = 100,000$	$10^{-6} = 0.000001$
$10^6 = 1,000,000$	

Figure 1-3. This table illustrates a portion of both the positive and negative powers of ten.

When using scientific notation, multiply the number you want to change by a power of ten equal to the number of places you want to move the decimal point. The net result does not change the value of the number, only the way it is written.

Example:

$$2,540,000 = 2.54 \times 10^6$$

As you can see, the decimal point was moved six places; therefore, the resulting number must be multiplied by a power of ten equal to 1,000,000, which is 10^6.

If the number you are working with is smaller than 1, and you want to move the decimal point to get a number between 1 and 10, count the number of places you want to move the decimal point and multiply the number by a power of ten. For example, 0.000004 is equal to 4.0×10^{-6}. Since the decimal point was moved 6 places to the right, you must multiply the number by 0.000001, which is 10^{-6}.

$$0.000004 = 4.0 \times 10^{-6}$$

MULTIPLYING BY SCIENTIFIC NOTATION

Multiplication of very large or very small numbers is often made easier when using scientific notation. To begin, convert each of the numbers being multiplied to scientific notation. Once this is done, the product is found by multiplying the numbers and finding the algebraic sum of the exponents.

Example:

Multiply $0.275 \times 30,000.0$ using scientific notation.

Step 1: Convert to scientific notation.

$$0.275 = 2.75 \times 10^{-1}$$

$$30,000 = 3.0 \times 10^{4}$$

Step 2: Multiply the numbers and add the exponents.

$$(2.75 \times 10^{-1}) \times (3.0 \times 10^{4}) = 8.25 \times 10^{3}$$

DIVISION BY SCIENTIFIC NOTATION

Division using scientific notation is performed in a manner similar to multiplication. Begin by converting the numbers to their scientific notation equivalents. Perform the division operation as you normally would, and find the power of ten by subtracting the exponents.

Example:

Divide 5,280 by 0.25 using scientific notation.

Step 1: Convert to scientific notation.

$$5,280 = 5.28 \times 10^{3}$$

$$0.25 = 2.5 \times 10^{-1}$$

Step 2: Divide the numbers and subtract the exponents.

$$(5.28 \times 10^{3}) \div (2.5 \times 10^{-1}) = 2.112 \times 10^{4}$$

Remember, when multiplying or dividing using scientific notation, you must calculate the algebraic sum of the exponents. Pay attention to the signs of the exponents, and observe the rules for adding and subtracting signed numbers.

ALGEBRA

Algebra is a form of arithmetic that uses letters or symbols to represent numbers in equations and formulas. For example, if an airplane cruises at 200 knots, how long will it take to fly 600 nautical miles? To solve this problem, an equation is set up with the unknown variable of time represented by the letter "T." The equation is 200 kts. × T = 600 n.m. Through algebra, you calculate the time (T) required of 3 hours. While some forms of algebra are extremely complex, others are fairly simple and straightforward. This section introduces you to the basic algebra you need to know to perform your duties as an aviation maintenance technician.

EQUATIONS

One way to express a math problem is to write it out in words. For example, "What is 24 divided by 3?" This is written in an algebraic sentence in the form 24 ÷ 3 = x. In this example, "x" represents the unknown quantity, or **variable**, you are solving for. The expression 24 ÷ 3 = x is called an **equation**. The purpose of the equation is to identify two equal quantities. Typically, once you get a math problem set up in an equation, the problem is fairly easy to solve. For example, if asked to determine what quantity, when added to 23, results in 48, your first step should be to set up an equation. The equation used to solve this problem is 23 + x = 48. To find the value of "x," subtract 23 from both sides of the equation. The equation now reads x = 48 - 23. Once simplified, the equation reads x = 25.

ALGEBRAIC RULES

There are some basic rules you must use to simplify and solve algebraic equations. First, consider fractions. As discussed earlier, when working with fractions, the numerator and denominator can be changed without changing the fraction's value as long as you do the same operation to both. This is often useful in reducing or combining fractions. For example, to reduce $^{18}/_{45}$ to its lowest terms, divide both the numerator and denominator by 9.

Example:

$$\frac{18}{45} \div \frac{9}{9} = \frac{2}{5}$$

This same principle also is used to simplify fractions and cancel out units such as gallons, miles, or foot-pounds. For example, in the given equation

$$\frac{60 \text{ Miles}}{\text{Hour}} \times \frac{1}{2} \text{ Hour} = x$$

Since "Hour" is in each element, it cancels. Furthermore, since the 60 in the numerator is divisible by the 2 in the denominator, both figures reduce. Once complete, you are left with the formula

$$\frac{30 \text{ Miles}}{1} = x$$

which is equivalent to x = 30 miles.

In another example, determine the number of revolutions a gear completes in 30 seconds when the gear turns at 100 revolutions per minute (rpm).

Example:

Step 1: Convert the word problem to an equation.

$$\frac{100 \text{ Revolutions}}{\text{Minute}} \times \frac{1}{2} \text{ Minute} = x$$

Step 2: Cancel the like terms and reduce where appropriate.

$$\frac{\overset{50}{\cancel{100}} \text{ Revolutions}}{\text{Minute}} \times \frac{1}{\underset{1}{\cancel{2}}} = \frac{50 \text{ Revolutions}}{1}$$

Therefore, x = 50 revolutions.

It is important to keep all labels in an equation. If this is not done, it may be difficult to determine the appropriate label for the answer.

Another important rule you must follow when solving algebraic equations is to never perform an operation to one side of an equation without performing the identical operation to the other side. In other words, you can add, subtract, multiply, or divide on one side of an equation as long as you do the same thing to the other side. For example, when solving the equation $x + 16 = 30$, 16 is subtracted from both sides of the equation.

Example:

$$x + 16 = 30$$

Subtract 16 from both sides to solve for x.

$$x + 16 - 16 - = 30 - 16$$

$$x = 14$$

SOLVING FOR A VARIABLE

Most of the algebra you do in everyday life requires you to solve for a variable. For example, suppose you want to determine your car's gas mileage. You filled the gas tank and drove 270 miles, then added 9 gallons to the tank. How many miles per gallon did the car get? To begin, build an equation and let "x" = miles per gallon.

$$x = \frac{\text{Miles driven}}{\text{Gallons used}} = \frac{270 \text{ Miles}}{9 \text{ Gallons}} = 30 \text{ miles/gallon}$$

Now, suppose you are planning a trip and want to know how far you can drive without stopping for gas. Your owner's manual says the car has a fuel capacity of 17.9 gallons. Using the same formula used to calculate miles per gallon, the problem reads

$$30 \text{ miles / gallon} = \frac{x \text{ miles}}{17.9 \text{ gallons}}$$

Multiply both sides by 17.9 gallons.

$$30 \text{ mi./gal.} \times 17.9 \text{ gal.} = \frac{x \text{ miles}}{17.9 \text{ gal.}} \times 17.9 \text{ gal.}$$

The 17.9 gallons cancels out on the right side of the

equation and the label gallons cancels out on the left. Do the multiplication to solve for x.

$$30 \text{ miles} \times 17.9 = x \text{ miles}$$

$$537 = x \text{ miles}$$

The car's range is 537 miles.

USE OF PARENTHESES

In algebra, **parentheses** indicate an operation that must be carried out before any other operation. For example, in the expression $10 \times (8 + 7)$, the 8 and 7 must be added first. When using parentheses, the absence of an operation sign between a number and a parenthetical statement indicates multiplication. For example, 8(3-2) is the same as $8 \times (3 - 2)$. Furthermore, if a negative sign (-) precedes the parentheses, it is the same as multiplying -1 by each of the quantities within the parentheses.

Example:

$$-(6 + 4 - 8) = -2$$

or

$$-6 - 4 + 8 = -2$$

ORDER OF OPERATIONS

When solving complex equations, the only way you can arrive at the correct answer is if you follow the correct order of operations. For example, when solving the equation $4 \times 3 + 2 \times 5$, it is possible to arrive at several different answers by doing the math operations in different orders. You could, for example, multiply before you add, add before you multiply, or work the equation from left to right. However, only one of these results in the correct answer. The proper order for performing mathematical operations is as follows.

1. **P**arentheses: Operations contained in parentheses are always done first.

2. **E**xponents: Once all operations within parentheses are complete, exponent operations are done.

3. **M**ultiplication and **D**ivision: The operations of multiplication and division are performed from left to right after exponents.

4. **A**ddition and **S**ubtraction: Once the operations of multiplication and division are done, you may add and subtract from left to right.

A memory aid used to remember the proper order of operations is the mnemonic "**P**lease **E**xcuse **M**y **D**ear **A**unt **S**ally."

Apply the correct order of operations to solving this equation.

$$x = \frac{(12 + 6)\,(2)^2 + 9\,(14)}{\sqrt{16}}$$

Do the operations in parentheses contained in the numerator.

$$x = \frac{(18)\,(4) + 9\,(14)}{\sqrt{16}}$$

Now multiply from left to right.

$$\frac{72 + 126}{\sqrt{16}}$$

Add the terms in the numerator.

$$\frac{198}{\sqrt{16}}$$

Calculate the square root in the denominator.

$$\frac{198}{4}$$

Reduce the fraction to a decimal number.

$$\frac{198}{4} = 49.5$$

Notice that you now have a fraction, and that you did not do the division for the fraction after multiplication. When an equation is presented in the form of a fraction, complete all operations in the numerator and denominator before you reduce the fraction. The same is true for operations within a square root sign. Do the operations within the square root sign in the proper order before extracting the root.

If you perform mathematical operations in the proper order, calculations typically go smoother and you obtain the right answer. The importance of proper order becomes clear when you begin performing more complex equations.

COMPLEX EQUATIONS

The algebraic rules presented in this section are not only useful for answering test questions, they also allow you to use the complex formulas frequently found in the study of electricity and weight and balance computations. Work the following equation.

$$x = \sqrt{2.246^2 + (.75 - 1.22)^2}$$

Perform the calculations in parentheses first.

$$x = \sqrt{2.246^2 + (-.47)^2}$$

Exponent work is done next.

$$x = \sqrt{5.0445 + .2209}$$

Add the elements within the square root sign.

$$x = \sqrt{5.2654}$$

Extract the root.

$$x = \sqrt{5.2654} = 2.2946$$

While it may seem complex, a problem of this type is fairly simple to solve when you observe the basic rules of signed numbers and follow the proper order of operations.

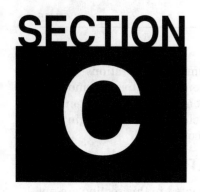
GEOMETRY AND TRIGONOMETRY

Geometry is the measurement of dimensions, areas, and volumes of geometric shapes, and is quite useful in aviation maintenance. In fact, it is geometry that allows you to calculate the displacement of a cylinder, determine the volume of a fuel tank, and calculate the surface area of a wing. On the other hand, trigonometry allows you to determine unknown lengths and angles of a triangle. In addition to aiding you when fabricating sheet metal, trigonometry plays a large part in the theory of alternating current.

COMPUTING AREA

The **area** of a surface is two dimensional and is expressed in square units. An area that is square and measures one inch on each side is called a square inch. This same relationship holds true for other units of measure such as square feet, square yards, square miles, and square meters. The area of a figure is equal to the number of square units the figure contains.

THE RECTANGLE

As you know, a rectangle is a four-sided plane. It is distinguished by having opposite sides of equal length, and four angles each equal to 90 degrees. The area (A) of a rectangle is found by multiplying its length (L) by its width (W), or A = L × W. However, before the mathematical operation can be carried out, both measurements must reflect the same unit of measure. For example, given a sheet of aluminum that is 48 inches wide by 12 feet long, you must convert either the width to feet or the length to inches. By converting the width of 48 inches to feet, the area of the sheet of aluminum is calculated to be 48 square feet (12 ft. × 4 ft. = 48 sq. ft.). If you later find that you need the area in square inches rather than square feet, multiply 48 square feet by 144 which is the number of square inches in a square foot. The result is 6,912 square inches.

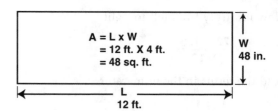

THE SQUARE

A square is a symmetrical plane in which all four sides are of equal length. The same formula used for a rectangle is used to find the area of a square. However, since all sides of a square are of equal length, the formula is sometimes expressed as the square of the sides or:

$$A = S^2$$

THE TRIANGLE

The triangle is a three-sided figure consisting of three angles whose combined measurement equals 180 degrees. Three basic types of triangles you should be familiar with are: the **scalene triangle**, which consists of three unequal angles and sides, the **equilateral triangle**, which has equal sides and equal angles, and the **isosceles triangle**, which has two equal angles.

Triangles are further classified by the measurement of one angle. For example, a **right triangle** is one that has one angle measuring 90 degrees. In an **obtuse triangle**, one angle is greater than 90 degrees, while in an **acute triangle** all angles are less than 90 degrees.

There are several terms associated with triangles. For instance, the **base** of a triangle is the side the triangle rests or stands on. Depending on a triangle's orientation, any side may be the base. The **vertex** is a common endpoint, or the point where the sides of the triangle meet. The **altitude** of a triangle is the **height** of the vertex above the base.

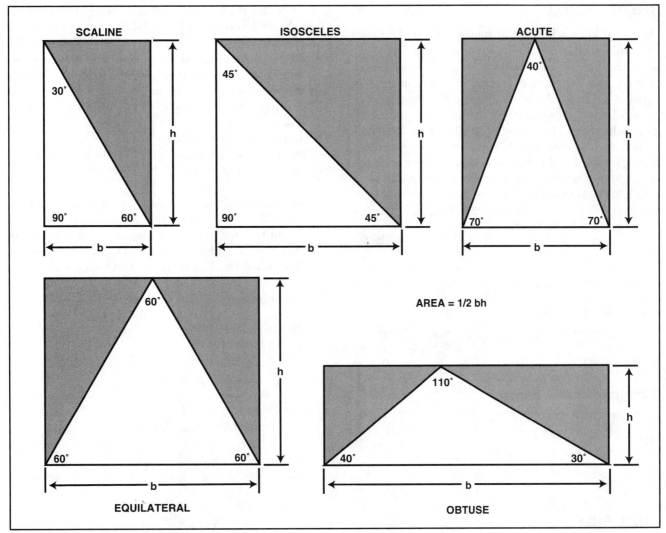

Figure 1-4. The shaded portion of each rectangle is equal in area to the triangle that is not shaded. The area of a triangle is calculated with the formula A = $^1/_2$ bh.

If a triangle is set in a rectangle and the triangle's base and height are equal to two of the rectangle's sides, the area of the triangle is exactly one-half that of the rectangle. Therefore, the formula for calculating the area of a triangle is one-half the base times the height, or $^1/_2$ bh. [Figure 1-4]

Find the area of a triangle whose base is 6 inches and height is 15 inches.

Step 1: Insert given values into the formula.

$$A = \frac{1}{2} 6 \times 15$$

Step 2: Perform multiplication.

$$A = 45 \text{ square inches}$$

THE PARALLELOGRAM

The parallelogram, like the rectangle, has opposite sides that are parallel and equal in length. However, the corner angles of a parallelogram are some measurement other than 90 degrees. The area of a parallelogram is calculated by multiplying the length by the height (A = l × h). The height is measured perpendicular to the length, similar to the way the altitude of a triangle is determined.

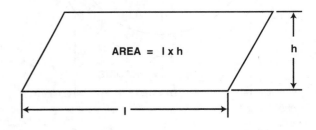

THE TRAPEZOID

A trapezoid is a four-sided figure that has one set of parallel sides. If you lay two trapezoids side by side so the top and bottom sides form straight lines, a parallelogram is formed with a base that is equal to the combined length of the trapezoid's parallel sides. As discussed earlier, the area of a parallelogram is found by multiplying the length, which in this case equals the sum of the parallel sides, by the height. However, because the area of a single trapezoid is one-half that of the parallelogram, the trapezoid's area is equal to one-half the product of the base times the height. This is expressed with the formula:

$$\text{AREA} = \tfrac{1}{2}(b_1 + b_2)\,h$$

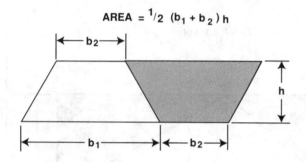

THE CIRCLE

A circle is a closed figure bounded by a single curved line. Every point on the line forming a circle is an equal distance from the center. The distance

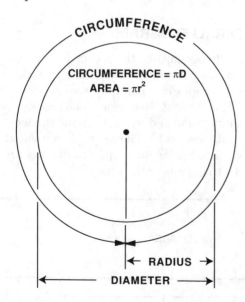

CIRCUMFERENCE

CIRCUMFERENCE = πD
AREA = πr²

RADIUS
DIAMETER

from the center to the line forming the circle is called the **radius**, and the distance around the circle is known as the **circumference**. The **diameter** of a circle is represented by a line that touches two points on the outside of the circle and passes through the circle's center. The circumference has a definite relationship with the diameter. This relationship is represented by the Greek letter pi (π), and is equal to 3.1416. The ratio of the circumference to the diameter of a circle is always pi. Regardless of the size of the circle, pi is a constant.

The circumference of a circle is found by multiplying pi times the diameter, and the area is calculated by multiplying pi times the square of the radius. For example, if a circle has a diameter of 10 inches, determine the circumference and area.

Example:

Circumference = $\pi\,D$

$$C = 3.1416 \times 10$$

$$C = 31.416 \text{ inches}$$

Area = πr^2

$$A = 3.1416 \times 5^2$$

$$A = 3.1416 \times 25$$

$$A = 78.54 \text{ square inches}$$

COMPUTING VOLUME

Solids are objects with three dimensions: length, width, and height. Having the ability to calculate volume enables you to determine the capacity of a fuel tank or reservoir, figure the capacity of a cargo area, or calculate the displacement of a cylinder. Volumes are calculated in cubic units, such as cubic inches, cubic feet, and cubic centimeters. However, volumes are easily converted to useful terms such as gallons. For example, to convert cubic inches to gallons, divide the total number of cubic inches by 231. If converting cubic feet to gallons, remember that 1 cubic foot holds 7.5 gallons.

VOLUME OF A RECTANGLE

The volume of a rectangular solid is found by multiplying the dimensions of length, width, and height. When calculating volume, it is important that all measurements be in like terms. The formula for determining the volume of a rectangular solid is:

$$V = L \times W \times H$$

Where:

V = volume
L = length
W = width
H = height

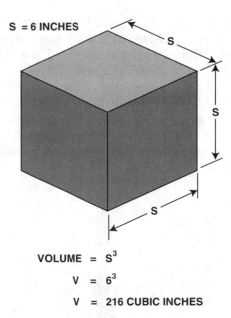

S = 6 INCHES

VOLUME = S^3

V = 6^3

V = 216 CUBIC INCHES

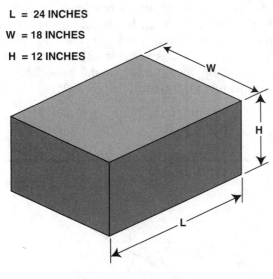

L = 24 INCHES
W = 18 INCHES
H = 12 INCHES

VOLUME = L x W x H

V = 24 x 18 x 12

V = 5,184 CUBIC INCHES

VOLUME OF A CYLINDER

A **cylinder** is a solid with circular ends and parallel sides. Its volume is found by multiplying the area of one end by the cylinder's height. The formula is expressed as:

$$\text{Volume} = \pi r^2 H$$

D = 8 inches
H = 8 inches

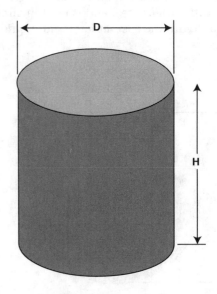

VOLUME = $\pi r^2 H$

V = 3.1416 x 4^2 x 8

V = 402.1248

VOLUME OF A CUBE

A **cube** is a solid with equal sides. Since all dimensions of a cube are identical, its area is calculated by multiplying one dimension by itself three times.

As a technician, you can use this formula to calculate piston displacement. Piston displacement is defined as the volume of air displaced by a piston as it moves from bottom center to top center. For example, one cylinder of a four-cylinder aircraft engine has a bore, or diameter, of four inches and the piston has a stroke of six inches. Stroke is defined as the distance the head of a piston travels from bottom center to top center. What is the total cylinder displacement?

Where:

Bore = 4 inches
Stroke = 6 inches

$$\text{Volume} = \pi r^2 H$$
$$\text{Volume} = 3.1416 \times 2^2 \times 6$$
$$\text{Volume} = 75.4 \text{ cubic inches}$$

Once you know the volume of one cylinder, you can calculate the engine's total displacement. The total piston displacement is defined as the total volume displaced by all the pistons during one crankshaft revolution. To calculate the displacement of an entire engine, multiply the volume of one cylinder by the number of cylinders on the engine.

VOLUME OF A SPHERE

A **sphere** is any round body having a surface on which all points are an equal distance from the center of the sphere. A sphere has the greatest volume for its surface area, and is used in aircraft systems for hydraulic accumulators and liquid oxygen converters.

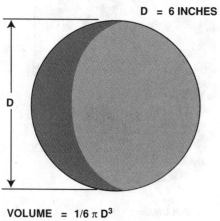

D = 6 INCHES

$$\text{VOLUME} = 1/6 \, \pi \, D^3$$
$$V = .1667 \times 3.1416 \times 216$$
$$V = 113.1 \text{ CUBIC INCHES}$$

The volume of a sphere is determined by multiplying the cube of the diameter by a factor which is $1/6$ pi, or 0.5236. If you want to find the volume of a sphere that is 6 inches in diameter, you must first cube the diameter and multiply the resulting value by 0.5236. For example, calculate the volume of a sphere with a diameter of 6 inches.

TRIGONOMETRIC FUNCTIONS

Trigonometry basically deals with the relationships that exist within a right triangle and is commonly used in the shop for sheet metal layout. Because trigonometry is a based on the ratio of the sides of a right triangle to one another, you must be familiar with how these ratios are derived. Figure 1-5 illustrates a right triangle with the sides and angles labeled for identification. Angle C is the right angle (90°). For this explanation, angle A is the angle for which you are setting up the relationships. Side c is the hypotenuse, which, by definition, is the side opposite the right angle. Side a is the side opposite angle A, and side b is the side adjacent, or next to, angle A. Using these labels, examine the three relationships that exist within this triangle.

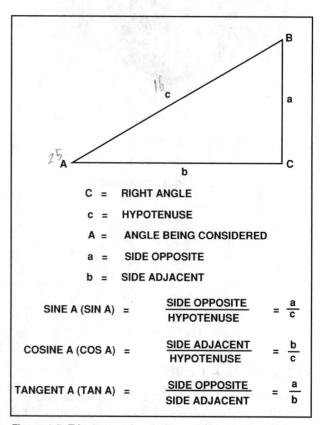

C =	RIGHT ANGLE	
c =	HYPOTENUSE	
A =	ANGLE BEING CONSIDERED	
a =	SIDE OPPOSITE	
b =	SIDE ADJACENT	
SINE A (SIN A) =	$\dfrac{\text{SIDE OPPOSITE}}{\text{HYPOTENUSE}}$	= $\dfrac{a}{c}$
COSINE A (COS A) =	$\dfrac{\text{SIDE ADJACENT}}{\text{HYPOTENUSE}}$	= $\dfrac{b}{c}$
TANGENT A (TAN A) =	$\dfrac{\text{SIDE OPPOSITE}}{\text{SIDE ADJACENT}}$	= $\dfrac{a}{b}$

Figure 1-5. Trigonometric relations of a right triangle.

Because trigonometric relationships are constant for a given angle, they are often times presented in a **Table of Trigonometric Functions**. Trigonomic tables usually only list the angles to 45 degrees. [Figure 1-6]

By referring back to figure 1-5, you see that the **Sine** (Sin) of angle A is the ratio of the length of the side opposite the angle to the length of the hypotenuse. For any degree of angle A, this ratio is constant, regardless of the size of the triangle. In the trig table, the sine of 30°, which is written Sin 30°, is 0.500. This means that the side opposite a 30° angle is 50%, or one-half the length of the hypotenuse. For a 45° angle, the ratio is 0.7071, indicating the side opposite the 45° angle is 0.7071 times the length of the hypotenuse.

The **Cosine** (Cos) of an angle is the ratio of the length of the side adjacent to the angle, to the length of the hypotenuse. Cosine is found on the same table used to find the sine of the angle. The cosine of 30° is 0.8660 and tells you that the length of the side adjacent to the 30° angle is 0.8660 times the length of the hypotenuse.

The third ratio appearing on the Trigonometric Function Table is **Tangent** (Tan). This is the ratio of the length of the side opposite the angle to the length of the side adjacent the angle.

THE METRIC SYSTEM

The metric system is the dominant language of measurement in use today. However, in the United States, the customary units of measurement include the English units of inches, feet, ounces, and pounds. While these units have proved functional for many years, they are cumbersome to convert.

The **meter** as a standard was developed in France and adopted by the National Assembly of France in the late 18th century. The United States government legalized the use of the metric system in 1866, but it was not until the Metric Bill of 1975 that its use became common. We are still in a time of slow conversion to the metric system and in many cases both Metric and English units appear together on packaging, speedometers, and thermometers.

The metric system is built on decimal units. Each basic unit is divided or multiplied by ten as many times as necessary to get a convenient size. Each of

TRIGONOMETRIC FUNCTIONS

DEG	SIN	COS	TAN		
0	.0	1.000	.0	+	90
1	.0175	.9999	.0175	57.29	89
2	.0349	.9994	.0349	28.64	88
3	.0523	.9986	.0524	19.08	87
4	.0698	.9976	.0699	14.30	86
5	.0872	.9962	.0875	11.43	85
6	.1045	.9945	.1051	9.514	84
7	.1219	.9926	.1228	8.144	83
8	.1392	.9903	.1405	7.115	82
9	.1564	.9877	.1584	6.314	81
10	.1737	.9848	.1763	5.671	80
11	.1908	.9816	.1944	5.145	79
12	.2079	.9782	.2126	4.705	78
13	.2250	.9744	.2309	4.331	77
14	.2419	.9703	.2493	4.011	76
15	.2588	.9659	.2680	3.732	75
16	.2756	.9613	.2868	3.487	74
17	.2924	.9563	.3057	3.271	73
18	.3090	.9511	.3249	3.078	72
19	.3256	.9455	.3443	2.904	71
20	.3420	.9397	.3640	2.747	70
21	.3584	.9336	.3839	2.605	69
22	.3746	.9272	.4040	2.475	68
23	.3907	.9205	.4245	2.356	67
24	.4067	.9136	.4452	2.246	66
25	.4226	.9063	.4663	2.145	65
26	.4384	.8988	.4877	2.050	64
27	.4540	.8910	.5095	1.963	63
28	.4695	.8830	.5317	1.881	62
29	.4848	.8746	.5543	1.804	61
30	.5000	.8660	.5774	1.732	60
31	.5150	.8572	.6009	1.664	59
32	.5299	.8481	.6249	1.600	58
33	.5446	.8387	.6494	1.540	57
34	.5592	.8290	.6745	1.483	56
35	.5736	.8192	.7002	1.428	55
36	.5878	.8090	.7265	1.376	54
37	.6018	.7986	.7536	1.327	53
38	.6157	.7880	.7813	1.280	52
39	.6293	.7772	.8098	1.235	51
40	.6428	.7660	.8391	1.192	50
41	.6561	.7547	.8693	1.150	49
42	.6691	.7431	.9004	1.111	48
43	.6820	.7314	.9325	1.072	47
44	.6947	.7193	.9657	1.036	46
45	.7071	.7071	1.0000	1.000	45
	COS	SIN		TAN	DEG

Figure 1-6. In this table, the numbers in the left-hand column go from 0° to 45° whereas the numbers in the right-hand column begin at 45° and continue to 90°. The names of the columns at the bottom are opposite the names at the top of the same column. When reading values for angles less than 45°, use the columns labeled on the top. When the angle is greater than 45°, use the column headings at the bottom of the chart.

the multiples has a definite prefix, symbol, and name. As a technician, you must be familiar with each of them. [Figure 1-7]

Number	Prefix	Symbol	Scientific Notation
1,000,000,000,000	tera	t	1×10^{12}
1,000,000,000	giga	g	1×10^{9}
1,000,000	mega	M	1×10^{6}
1,000	kilo	k	1×10^{3}
100	hecto	h	1×10^{2}
10	deka	dk	1×10^{1}
0.1	deci	d	1×10^{-1}
0.01	centi	c	1×10^{-2}
0.001	milli	m	1×10^{-3}
0.000,001	micro	u	1×10^{-6}
0.000,000,001	nano	n	1×10^{-9}
0.000,000,000,001	pico	p	1×10^{-12}

Figure 1-7. Listed are the common prefixes, symbols, and multiples for basic metric quantities.

There are six base units in the metric system. The unit of length is called the **meter**, and is approximately 39 inches. The metric unit of mass, or weight, is the **gram**. The unit of time is the **second**. The unit of electrical current is the **ampere**. The unit of temperature is the degree **Celsius**, formerly called degree **Centigrade**. The unit of luminous intensity is the **candela**. All other units of measurement in the **International System of Units**, which is now called **SI**, are derived from these six.

Conversion from English to Metric systems is accomplished in a variety of ways. Many hand-held electronic calculators now have specific function keys for making these conversions. There are also several charts that supply conversion factors. [Figure 1-8]

MATHEMATICAL HARDWARE

While there is no substitute for a basic understanding of mathematical principles and proficiency in performing mathematical operations, certain mathematical hardware offers increases in both calculating speed and accuracy. This hardware often takes the form of detailed function tables, or the electronic calculator. Either one is easily obtainable, portable, and easy to use.

MATHEMATICAL TABLES

The **Decimal Equivalent chart** (Figure 1-9) and the **Function of Numbers chart** (Figure 1-10) are presented for the convenience of making common calculations. As you become familiar with the information presented on these charts you will begin to see the advantages of "ready-made" computations.

ELECTRONIC CALCULATORS

The advent of the inexpensive hand-held electronic calculator has changed forever the way mathematical operations are performed. While you still need a basic understanding of mathematical operations and logic, the calculator typically helps increase speed and reduce errors.

There is no "right" calculator for the aviation maintenance technician. However, there are calculators with function keys for many of the operations discussed in the chapter. The selection of a calculator should be based on its anticipated use.

To Convert From	To	Multiply By
acres	sq ft	4.356 x 10⁴
atmospheres	cm Hg at 0°C	76.00
	in. Hg at 0°C	29.92
	lb/sq in	14.696
	bars	1.013
bars	lb/sq in	14.5
Btu	ft-lb	778.26
	kilowatt-hr	2.931×10^{-4}
	joules	1055
Btu/sec	watts	1055
centimeters	in	0.3937
cm/second	Ft/sec	3.281×10^{-2}
circular mils	sq in	7.854×10^{-7}
cu centimeters	cu in	6.102×10^{-2}
	U.S. gal	2.642×10^{-4}
cu ft	cu cm	2.832×10^{4}
	U.S. gal	7.481
	liters	28.32
cu ft H₂0	lb	62.428
cu in.	cu cm	16.39
	liters	1.639×10^{-2}
	U.S. gal	4.329×10^{-3}
cu meters	U.S. gal	264.2
feet	meters	3.048×10^{-1}
ft/min	mph	1.136×10^{-2}
	km/hr	1.829×10^{-2}
ft/sec	mph	.6818
	cm/sec	30.48
	knots	.5925
knots	nautical mph	1.0
	ft/sec	1.688
	mph	1.151
liters	cu cm	10³
	cu in.	61.03
	U.S. gal	2.642×10^{-1}
meters	in.	39.37
	ft.	3.281
meter-kilogram	ft-lb	7.233
meter/sec	ft/sec	3.281
microns	in.	3.937×10^{-5}
miles [stat.]	ft	5280
	km	1.609
mph	ft/sec	1.467
	km/hr	1.609
	knots	8.690×10^{-1}
millibars	in. Hg at 0°C	2.953×10^{-2}
nautical miles [naumiles]	ft	6076.1
	miles [stat.]	1.151
	m	1852

To Convert From	To	Multiply By
ft-lb/min	hp	3.030×10^{-5}
ft-lb/sec	hp	1.818×10^{-3}
	kilowatts	1.356×10
fluid oz	dram	8
gal, Imperial	cu in.	277.4
	U.S. gal	1.201
	Liters	4.546
gal, U.S. dry	U.S. gal, liquid	1.164
gal, U.S. liquid	cu in.	231.0
grams	oz avdp	3.527×10^{-2}
	lb avdp	2.205×10^{-3}
grams/cm	lb/ft	6.721×10^{-2}
horsepower	ft-lb/min	33,000
	ft-lb/sec	550
	m-kg/sec	76.04
	kilowatts	7.457×10^{-1}
	Btu/sec	7.068×10^{-1}
horsepower, metric	hp	9.863×10^{-1}
in.	cm	2.540
in. water at 4 °C	in. Hg at 0 °C	7.355×10^{-2}
kilograms	lb	2.205
	oz	35.27
kilometers	ft	3.281×10^{3}
	miles	6.214×10^{-1}
	nautical miles	5.400×10^{-1}
km/hr	ft/sec	9.113×10^{-1}
	knots	5.396×10^{-1}
	mph	6.214×10^{-1}
kilowatts	Btu/sec	9.480×10^{-1}
	hp	1.341
ounces. fluid	cu in.	1.805
lb/cu in.	grams/cu cm	27.68
lb/sq in.	in. Hg at 0°C	2.036
radians/sec	deg/sec	57.30
	rev/min	9.549
revolutions	radians	6.283 [2π]
rev/min	radians/sec	1.047×10^{-1}
slug	lb	32.174
sq cm	sq in.	1.550×10^{-1}
sq ft	sq cm	929.0
sq in.	sq cm	6.452
sq meters	sq ft	10.76
sq miles	sq km	2.590
watts	Btu/sec	9.481×10^{-4}

Figure 1-8. Metric conversion factors.

Figure 1-9. Decimal Equivalent Chart.

Fractions	Inches Decimals	MM	Fractions	Inches Decimals	MM	Fractions	Inches Decimals	MM	Fractions	Inches Decimals	MM	Fractions	Inches Decimals	MM
-	.0004	.01	39/64	.609	15.478	-	1.5748	40.	-	2.7953	71.	4-1/16	4.062	103.188
-	.004	.10	5/8	.625	15.875	1-19/32	1.594	40.481	2-13/16	2.8125	71.4376	4-1/8	4.125	104.775
-	.01	.25	-	.6299	16.	-	1.6142	41.	-	2.8346	72.	-	4.1338	105.
1/64	.0156	.397	41/64	.6406	16.272	1-5/8	1.625	41.275	2-27/32	2.844	72.2314	4-3/16	4.1875	106.363
-	.0197	.50	-	.6496	16.5	-	1.6535	42.	-	2.8740	73.	4-1/4	4.250	107.950
-	.0295	.75	21/32	.656	16.669	1-21/32	1.6562	42.069	2-7/8	2.875	73.025	4-5/16	4.312	109.538
1/32	.03125	.794	-	.6693	17.	1-11/16	1.6875	42.863	2-29/32	2.9062	73.819	-	4.3307	110.
-	.0394	1.	43/64	.672	17.066	-	1.6929	43.	-	2.9134	74.	4-3/8	4.375	111.125
3/64	.0469	1.191	11/16	.6875	17.463	1-23/32	1.719	43.656	2-15/16	2.9375	74.813	4-7/16	4.438	112.713
-	.059	1.5	45/64	.703	17.859	-	1.7323	44.	-	2.9527	75.	4-1/2	4.500	114.300
1/16	.062	1.588	-	.7087	18.	1-3/4	1.750	44.450	2-31/32	2.969	75.406	-	4.5275	115.
5/64	.0781	1.984	23/32	.719	18.256	-	1.7717	45.	-	2.9921	76.	4-9/16	4.562	115.888
-	.0787	2.	-	.7283	18.5	1-25/32	1.781	45.244	3	3.000	76.200	4-5/8	4.625	117.475
3/32	.094	2.381	47/64	.734	18.653	-	1.8110	46.	3-1/32	3.0312	76.994	4-11/16	4.6875	119.063
-	.0984	2.5	-	.7480	19.	1-13/16	1.8125	46.038	-	3.0315	77.	-	4.7244	120.
7/64	.109	2.778	3/4	.750	19.050	1-27/32	1.844	46.831	3-1/16	3.062	77.788	4-3/4	4.750	120.650
-	.1181	3.	49/64	.7656	19.447	-	1.8504	47.	-	3.0709	78.	4-13/16	4.8125	122.238
1/8	.125	3.175	25/32	.781	19.844	1-7/8	1.875	47.625	3-3/32	3.094	78.581	4-7/8	4.875	123.825
-	.1378	3.5	-	.7874	20.	-	1.8898	48.	-	3.1102	79.	-	4.9212	125.
9/64	.141	3.572	51/64	.797	20.241	1-29/32	1.9062	48.419	3-1/8	3.125	79.375	4-15/16	4.9375	125.413
5/32	.156	3.969	13/16	.8125	20.638	-	1.9291	49.	-	3.1496	80.	5	5.000	127.
-	.1575	4.	-	.8268	21.	1-15/16	1.9375	49.213	3-5/32	3.156	80.169	-	5.1181	130.
11/64	.172	4.366	53/64	.828	21.034	-	1.9685	50.	3-3/16	3.1875	80.963	5-1/4	5.250	133.350
-	.177	4.5	27/32	.844	21.431	1-31/32	1.969	50.006	-	3.1890	81.	5-1/2	5.500	139.700
3/16	.1875	4.763	55/64	.859	21.828	2	2.000	50.800	3-7/32	3.219	81.756	-	5.518	140.
-	.1969	5.	-	.8661	22.	-	2.0079	51.	-	3.2283	82.	5-3/4	5.750	146.050
13/64	.203	5.159	7/8	.875	22.225	2-1/32	2.03125	51.594	3-1/4	3.250	82.550	-	5.9055	150.
-	.2165	5.5	57/64	.8906	22.622	-	2.0472	52.	-	3.2677	83.	6	6.000	152.400
7/32	.219	5.556	-	.9055	23.	2-1/16	2.062	52.388	3-9/32	3.281	83.344	6-1/4	6.250	158.750
15/64	.234	5.953	29/32	.9062	23.019	-	2.0868	53.	-	3.3071	84.	-	6.2992	160.
-	.2362	6.	59/64	.922	23.416	2-3/32	2.094	53.181	3-5/16	3.312	84.1377	6-1/2	6.500	165.100
1/4	.250	6.350	15/16	.9375	23.813	2-1/8	2.125	53.975	3-11/32	3.344	84.9314	-	6.6929	170.
-	.2559	6.5	-	.9449	24.	-	2.126	54.	-	3.3464	85.	6-3/4	6.750	171.450
17/64	.2656	6.747	61/64	.953	24.209	2-5/32	2.156	54.769	3-3/8	3.375	85.725	7	7.000	177.800
-	.2756	7.	31/32	.969	24.606	-	2.165	55.	-	3.3858	86.	-	7.0866	180.
9/32	.281	7.144	-	.9843	25.	2-3/16	2.1875	55.563	3-13/32	3.406	86.519	-	7.4803	190.
-	.2953	7.5	63/64	.9844	25.003	-	2.2047	56.	-	3.4252	87.	7-1/2	7.500	190.500
19/64	.297	7.541	1	1.000	25.400	2-7/32	2.219	56.356	3-7/16	3.438	87.313	-	7.8740	200.
5/16	.312	7.938	-	1.0236	26.	-	2.244	57.	-	3.4646	88.	8	8.000	203.200
-	.315	8.	-	1.0312	26.194	2-1/4	2.250	57.150	3-15/32	3.469	88.106	-	8.2677	210.
21/64	.328	8.334	1-1/16	1.062	26.988	2-9/32	2.281	57.944	3-1/2	3.500	88.900	8-1/2	8.500	215.900
-	.335	8.5	-	1.063	27.	-	2.2835	58.	-	3.5039	89.	-	8.6614	220.
11/32	.344	8.731	1-3/32	1.094	27.781	2-5/16	2.312	58.738	3-17/32	3.531	89.694	9	9.000	228.600
-	.3543	9.	-	1.1024	28.	-	2.3228	59.	-	3.5433	90.	-	9.0551	230.
23/64	.359	9.128	1-1/8	1.125	28.575	2-11/32	2.344	59.531	3-9/16	3.562	90.4877	-	9.4488	240.
-	.374	9.5	-	1.1417	29.	-	2.3622	60.	-	3.5827	91.	9-1/2	9.500	241.300
3/8	.375	9.525	1-5/32	1.156	29.369	2-3/8	2.375	60.325	3-19/32	3.594	91.281	-	9.8425	250.
25/64	.391	9.922	-	1.1811	30.	-	2.4016	61.	-	3.622	92.	10	10.000	254.001
-	.3937	10.	1-3/16	1.1875	30.163	2-13/32	2.406	61.119	3-5/8	3.625	92.075	-	10.2362	260.
13/32	.406	10.319	1-7/32	1.219	30.956	2-7/16	2.438	61.913	3-21/32	3.656	92.869	-	10.6299	270.
-	.413	10.5	-	1.2205	31.	-	2.4409	62.	-	3.6614	93.	11	11.000	279.401
27/64	.422	10.716	1-1/4	1.250	31.750	2-15/32	2.469	62.706	3-11/16	3.6875	93.663	-	11.0236	280.
-	.4331	11.	-	1.2658	32.	-	2.4803	63.	-	3.7008	94.	-	11.4173	290.
7/16	.438	11.113	1-9/32	1.281	32.544	2-1/2	2.500	63.500	3-23/32	3.719	94.456	-	11.8110	300.
29/64	.453	11.509	-	1.2992	33.	-	2.5197	64.	-	3.7401	95.	12	12.000	304.801
15/32	.469	11.906	1-5/16	1.312	33.338	2-17/32	2.531	64.294	3-3/4	3.750	95.250	13	13.000	330.201
-	.4724	12.	-	1.3386	34.	-	2.559	65.	-	3.7795	96.	-	13.7795	350.
31/64	.484	12.303	1-11/32	1.344	34.131	2-9/16	2.562	65.088	3-25/32	3.781	96.044	14	14.000	355.601
-	.492	12.5	1-3/8	1.375	34.925	2-19/32	2.594	65.881	3-13/16	3.8125	96.838	15	15.000	381.001
1/2	.500	12.700	-	1.3779	35.	-	2.5984	66.	-	3.8189	97.	-	15.7480	400.
-	.5118	13.	1-13/32	1.406	35.719	2-5/8	2.625	66.675	3-27/32	3.844	97.631	16	16.000	406.401
33/64	.5156	13.097	-	1.4173	36.	-	2.638	67.	-	3.8583	98.	17	17.000	431.801
17/32	.531	13.494	1-7/16	1.438	36.513	2-21/32	2.656	67.469	3-7/8	3.875	98.425	-	17.7165	450.
35/64	.547	13.891	-	1.4567	37.	-	2.6772	68.	-	3.8976	99.	18	18.000	457.201
-	.5512	14.	1-15/32	1.469	37.306	2-11/16	2.6875	68.263	3-29/32	3.9062	99.219	19	19.000	482.601
9/16	.563	14.288	-	1.4961	38.	-	2.7165	69.	-	3.9370	100.	-	19.6350	500.
-	.571	14.5	1-1/2	1.500	38.100	2-23/32	2.719	69.056	3-15/16	3.9375	100.013	20	20.000	508.001
37/64	.578	14.684	1-17/32	1.531	38.894	2-3/4	2.750	69.850	3-31/32	3.969	100.806			
-	.5906	15.	-	1.5354	39.	-	2.7559	70.	-	3.9764	101.			
19/32	.594	15.081	1-9/16	1.562	39.688	2-25/32	2.781	70.6439	4	4.000	101.600			

No.	Square	Cube	Square Root	Cube Root	Circumference	Area
1	1	1	1.000	1.0000	3.1416	0.7854
2	4	8	1.4142	1.2599	6.2832	3.1416
3	9	27	1.7321	1.4422	9.4248	7.0686
4	16	64	2.0000	1.5874	12.5664	12.5664
5	25	125	2.2361	1.7100	15.7080	19.635
6	36	216	2.4495	1.8171	18.850	28.274
7	49	343	2.6458	1.9129	21.991	38.485
8	64	512	2.8284	2.0000	25.133	50.266
9	81	729	3.0000	2.0801	28.274	63.617
10	100	1,000	3.1623	2.1544	31.416	78.540
11	121	1,331	3.3166	2.2240	34.558	95.033
12	144	1,728	3.4641	2.2894	37.699	113.10
13	169	2,197	3.6056	2.3513	40.841	132.73
14	196	2,744	3.7417	2.4101	43.982	153.94
15	225	3,375	3.8730	2.4662	47.124	176.71
16	256	4,096	4.0000	2.5198	50.265	201.06
17	289	4,913	4.1231	2.5713	53.407	226.98
18	324	5,832	4.2426	2.6207	56.549	254.47
19	361	6,859	4.3589	2.6684	59.690	283.53
20	400	8,000	4.4721	2.7144	62.832	314.16
21	441	9,261	4.5826	2.7589	65.973	346.36
22	484	10,648	4.6904	2.8020	69.115	380.13
23	529	12,167	4.7958	2.8439	72.257	415.48
24	576	13,824	4.8990	2.8845	75.398	452.39
25	625	15,625	5.0000	2.9240	78.540	490.87
26	676	17,576	5.0990	2.9625	81.681	530.93
27	729	19,683	5.1962	3.0000	84.823	572.56
28	784	21,952	5.2915	3.0366	87.695	615.75
29	841	24,389	5.3852	3.0723	91.106	660.52
30	900	27,000	5.4772	3.1072	94.248	706.86
31	961	29,791	5.5678	3.1414	97.389	754.77
32	1,024	32,768	5.6569	3.1748	100.53	804.25
33	1,089	35,937	5.7446	3.2075	103.67	855.30
34	1,156	39,304	5.8310	3.2396	106.81	907.92
35	1,225	42,875	5.9161	3.2717	109.96	962.11
36	1,296	46,656	6.0000	3.3019	113.10	1,017.88
37	1,369	50,653	6.0828	3.3322	116.24	1,075.21
38	1,444	54,872	6.1644	3.3620	119.38	1,134.11
39	1,521	59,319	6.2450	3.3912	122.52	1,194.59
40	1,600	64,000	6.3246	3.4200	125.66	1,256.64
41	1,681	68,921	6.4031	3.4482	128.81	1,320.25
42	1,764	74,088	6.4807	3.4760	131.95	1,385.44
43	1,849	79,507	6.5574	3.5034	135.09	1,452.20
44	1,936	85,184	6.6332	3.5303	138.23	1,520.53
45	2,025	91,125	6.7082	3.5569	141.37	1,590.43
46	2,116	97,336	6.7823	3.5830	144.51	1,661.90
47	2,209	103,823	6.8557	3.6088	147.65	1,734.94
48	2,304	110,592	6.9282	3.6342	150.80	1,809.56
49	2,401	117,649	7.0000	3.6593	153.94	1,885.74
50	2,500	125,000	7.0711	3.6840	157.08	1,963.50

Figure 1-10. Function of Numbers Chart.

No.	Square	Cube	Square Root	Cube Root	Circumference	Area
51	2,601	132,651	7.1414	3.7084	160.22	2,042.82
52	2,704	140,608	7.2111	3.7325	163.36	2,123.72
53	2,809	148,877	7.2801	3.7563	166.50	2,206.18
54	2,916	157,464	7.3485	3.7798	169.65	2,290.22
55	3,025	166,375	7.4162	3.8030	172.79	2,375.83
56	3,136	175,616	7.4833	3.8259	175.93	2,463.01
57	3,249	185,193	7.5498	3.8485	179.07	2,551.76
58	3,364	195,112	7.6158	3.8709	182.21	2,642.08
59	3,481	205,379	7.6811	3.8930	185.35	2,733.97
60	3,600	216,000	7.7460	3.9149	188.50	2,827.43
61	3,721	226,981	7.8102	3.9365	191.64	2,922.47
62	3,844	238,328	7.8740	3.9579	194.78	3,019.07
63	3,969	250,047	7.9373	3.9791	197.92	3,117.25
64	4,096	262,144	8.0000	4.0000	201.06	3,126.99
65	4,225	274,625	8.0623	4.0207	204.20	3,381.31
66	4,356	287,496	8.1240	4.0412	207.34	3,421.19
67	4,489	300,763	8.1854	4.0615	210.49	3,525.65
68	4,624	314,432	8.2462	4.0817	213.63	3,631.68
69	4,761	328,509	8.3066	4.1016	216.77	3,739.28
70	4,900	343,000	8.3666	4.1213	219.91	3,848.45
71	5,041	357,911	8.4261	4.1408	233.05	3,959.19
72	5,184	373,248	8.4853	4.1602	226.19	4,071.50
73	5,329	389,017	8.5440	4.1793	229.34	4,185.39
74	5,476	405,224	8.6023	4.1983	232.48	4,300.84
75	5,625	421,875	8.6603	4.2172	235.62	4,417.86
76	5,776	438,976	8.7178	4.2358	238.76	4,536.46
77	5,929	456,533	8.7750	4.2543	241.90	4,656.63
78	6,084	474,552	8.8318	4.2727	245.05	4,778.36
79	6,214	493,039	8.8882	4.2908	248.19	4,901.67
80	6,400	512,000	8.9443	4.3089	251.33	5,026.55
81	6,561	531,441	9.0000	4.3267	254.47	5,153.00
82	6,724	551,368	9.0554	4.3445	257.61	5,281.02
83	6,889	571,787	9.1104	4.3621	260.75	5,410.61
84	7,056	592,704	9.1652	4.3795	263.89	5,541.77
85	7,225	614,125	9.2195	4.3968	267.04	5,674.50
86	7,396	636,056	9.2376	4.4140	270.18	5,808.80
87	7,569	638,503	9.3274	4.4310	273.32	5,944.68
88	7,744	681,472	9.3808	4.4480	276.46	6,082.12
89	7,921	704,969	9.4340	4.4647	279.60	6,221.14
90	8,100	729,000	9.4868	4.4814	282.74	6,361.73
91	8,281	753,571	9.5394	4.4979	285.88	6,503.88
92	8,464	778,688	9.5917	4.5144	289.03	6,647.61
93	8,649	804,357	9.6437	4.5307	292.17	6,792.91
94	8,836	830,584	9.6954	4.5468	295.31	6,939.81
95	9,025	857,375	9.7468	4.5629	298.45	7,088.22
96	9,216	884,736	9.7980	4.5789	301.59	7,283.23
97	9,409	912,673	9.8489	4.5947	304.73	7,389.81
98	9,604	941,192	9.8995	4.6104	307.88	7,542.96
99	9,801	970,299	9.9499	4.6261	311.02	7,697.69
100	10,000	1,000,000	10.0000	4.6416	314.16	7,853.98

Figure 1-10. Function of Numbers Chart.

CHAPTER 2

PHYSICS

INTRODUCTION

As an aviation maintenance technician, you must have a basic knowledge of physics, and the laws that govern the behavior of the materials with which you work. Physics is the science that deals with matter and energy and their interactions. Physics operates with absolutes whose properties and values behave in the same way every time. Not only do these absolutes make flight possible, but they also allow engineers and technicians to design, build, and maintain aircraft.

MATTER AND ENERGY

By definition, **matter** is anything that occupies space and has mass. Therefore, the air, water, and food you need to live, as well as the aircraft you maintain, are all forms of matter. The Law of Conservation states that matter cannot be created or destroyed. You can, however, change the characteristics of matter. When matter changes state, **energy**, which is the ability of matter to do work, can be extracted. For example, as coal is heated it changes from a solid to a combustible gas which produces heat energy.

CHEMICAL NATURE OF MATTER

In order to better understand the characteristics of matter it is typically broken down to smaller units. The smallest unit that can exist is the **atom**. The three subatomic particles that form atoms are **protons**, **neutrons**, and **electrons**. The positively charged protons and neutrally charged neutrons coexist in an atom's nucleus. However, the negatively charged electrons

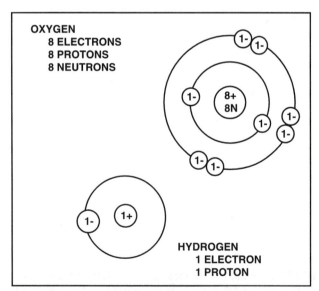

Figure 2-1. The hydrogen atom has one proton, no neutrons, and one electron. The oxygen atom, on the other hand, has eight protons, eight neutrons, and eight electrons.

orbit around the **nucleus** in orderly rings, or shells. The hydrogen atom is the simplest atom. It has one proton in the nucleus, no neutrons, and one electron. A more complex atom is the oxygen atom. An oxygen atom contains eight protons and eight neutrons in the nucleus, and has eight electrons orbiting around the nucleus. [Figure 2-1]

There are currently 109 known **elements** or atoms. Each have an identifiable number of protons, neutrons and electrons. In addition, every atom has its own **atomic number**, as well as its own **atomic mass**. [Figure 2-2]

Generally, when atoms bond together they form a **molecule**. However, there are a few molecules that exist as single atoms. Two examples that you will most likely use in aircraft maintenance are helium and argon. All other molecules are made up of two or more atoms. For example, water (H_2O) is made up of two atoms of hydrogen and one atom of oxygen.

When atoms bond together to form a molecule they share electrons. In the example of H_2O, the oxygen atom has six electrons in its outer, or **valence shell**. However, there is room for eight electrons. Therefore, one oxygen atom can combine with two hydrogen atoms by sharing the single electron from each hydrogen atom. [Figure 2-3]

PHYSICAL NATURE OF MATTER

Matter is composed of several molecules. The molecule is the smallest unit of a substance that exhibits the physical and chemical properties of the substance. Furthermore, all molecules of a particular substance are exactly alike and unique to that substance.

Matter may exist in one of three physical states, **solid**, **liquid**, and **gaseous**. All matter exists in one of these states. A physical state refers to the

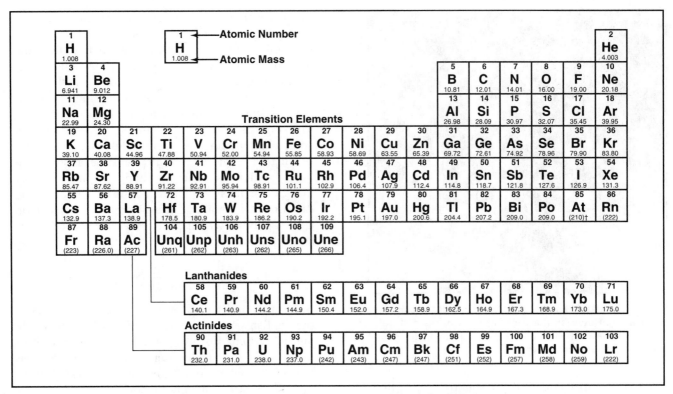

Figure 2-2. This chart contains each of the known elements and their corresponding atomic numbers and atomic masses.

physical condition of a compound and has no affect on a compound's chemical structure. In other words, ice, water, and steam are all H_2O and the same type of matter appears in all of these states.

All atoms and molecules in matter are constantly in motion. This motion is caused by the heat energy in the material. The degree of motion determines the physical state of matter.

SOLID

A solid has a definite volume and shape, and is independent of its container. For example, a rock that is put into a jar does not reshape itself to form to the jar. In a solid there is very little heat energy and, therefore, the molecules or atoms cannot move very far from their relative position. For this reason a solid is incompressible.

LIQUID

When heat energy is added to solid matter, the molecular movement increases. This causes the molecules to overcome their rigid shape. When a material changes from a solid to a liquid, the material's volume does not significantly change. However, the material conforms to the shape of the container its held in. An example of this is a melting ice cube.

Liquids are also considered incompressible. Although the molecules of a liquid are farther apart than those of a solid, they are still not far enough apart to make compressing possible.

In a liquid, the molecules still partially bond together. This bonding force is known as surface tension and prevents liquids from expanding and spreading out in all directions. Surface tension is

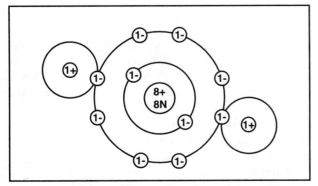

Figure 2-3. A molecule of water (H_2O) is formed when two atoms of hydrogen join one atom of oxygen.

Figure 2-4. A liquid conforms to the shape of the container it is held in. However, the cohesive force of the molecules forms a surface tension that allows the liquid to extend slightly above the container.

evident when a container is slightly over filled. [Figure 2-4]

GAS

As heat energy is continually added to a material, the molecular movement increases further until the liquid reaches a point where surface tension can no longer hold the molecules down. At this point the molecules escape as gas or vapor. The amount of heat required to change a liquid to a gas varies with different liquids and the amount of pressure a liquid is under. For example, at a pressure that is lower than atmospheric, water boils at a temperature less than 212°F. Therefore, the boiling point of a liquid is said to vary directly with pressure.

Gases differ from solids and liquids in the fact that they have neither a definite shape nor volume. Chemically, the molecules in a gas are exactly the same as they were in their solid or liquid state. However, because the molecules in a gas are spread out, gasses are compressible.

WEIGHT AND MASS

Contrary to popular belief, the weight and mass of a material are not the same. Weight is the force with which gravity attracts a mass. However, it's more important to note that the force of gravity varies with the distance between a body and the center of the earth. In other words, the farther away an object is from the center of the earth, the less it weighs.

The mass of an object is described as the amount of matter in an object and is constant regardless of its location. For example, an astronaut has the same mass on earth as when in space. However, an astronaut's weight is much less when in space than it is on earth. Another definition sometimes used for mass is the measurement of an object's resistance to change its state of rest or motion. This is seen by comparing the force required to move a jet as compared to a single engine airplane. Because the jet has a greater resistance to change, it has a greater mass. The mass of an object may be found by dividing the weight of the object by the acceleration of gravity, which is 32.2 feet per second every second an object falls.

$$\text{Mass} = \frac{\text{Weight}}{\text{Acceleration due to gravity}}$$

Both mass and weight are measured in pounds in the English system and in grams or kilograms in the metric system. However, another common unit of measure for mass is the slug. A **slug** is a unit of mass that is equivalent to approximately 32.175 pounds under standard atmospheric conditions.

DENSITY

The density of a substance is its weight per unit volume. The density of solids and liquids varies with temperature. However, the density of a gas varies with temperature and pressure. To find the density of a substance, divide the weight of the substance by its volume. This results in a weight per unit volume.

$$\text{Density} = \frac{\text{Weight}}{\text{Volume}}$$

For example, the liquid which fills a certain container weighs 1,497.6 pounds. The container is 4 feet long, 3 feet wide, and 2 feet deep. Therefore, its volume is 24 cubic feet (4 ft. × 3 ft. × 2 ft.). Based on this, the liquid's density is 62.4 lbs./ft³.

$$62.4 \text{ pounds per cubic foot} = \frac{1,497.6}{24 \text{ ft}^3}$$

Because the density of solids and liquids vary with temperature, a standard temperature of 4°C is used when measuring the density of each. Although temperature changes do not change the weight of a substance, they do change the volume of a substance through thermal expansion or contraction. This changes a substance's weight per unit volume.

As mentioned earlier, when measuring the density of a gas, temperature and pressure must be considered. Standard conditions for the measurement of gas density is established at 0°C and a pressure of 29.92 inches of mercury which is the average pressure of the atmosphere at sea level.

SPECIFIC GRAVITY

It is often necessary to compare the density of one substance with that of another. For this reason, a standard is needed from which all other materials can be compared. The standard when comparing the densities of all liquids and solids is water at 4°C. The standard for gases is air.

In physics the word "specific" refers to a ratio. Therefore, specific gravity is calculated by comparing the weight of a definite volume of substance with the weight of an equal volume of water. The following formulas are used to find specific gravity (sp. gr.) of liquids and solids:

$$\text{sp. gr.} = \frac{\text{Weight of a substance}}{\text{Weight of equal volume of water}}$$

$$\text{sp.gr.} = \frac{\text{Density of a substance}}{\text{Density of water}}$$

The same formulas are used to find the density of gases by substituting air for water. Specific gravity is not expressed in units, but as a pure number. For example, if a certain hydraulic liquid has a specific gravity of 0.8, 1 cubic foot of the liquid weighs 0.8 times as much as 1 cubic foot of water. Specific gravity is independent of the size of the sample under consideration and varies only with the substance the sample is made of. [Figure 2-5]

A device called a hydrometer is used to measure the specific gravity of liquids. This device has a tubular-shaped glass float contained in a larger glass tube. The float is weighted and has a vertically graduated scale. The scale is read at the surface of the liquid in which the float is immersed. A reading of 1000 is shown when the float is immersed in pure water. When filled with a liquid having a density greater than pure water, the float rises and indicates a greater specific gravity. For liquids of lesser density, the float sinks below 1000. [Figure 2-6]

LIQUID		SOLID	
Gasoline	0.72	Ice	0.917
Jet fuel JP-4	0.785	Aluminum	2.7
Jet fuel JP-5	0.871	Titanium	4.4
Alcohol (ethyl)	0.789	Zinc	7.1
Kerosene	0.82	Iron	7.9
Lubricating oil	0.89	Brass	8.4
Synthetic oil	0.928	Copper	8.9
Water	1.000	Lead	11.4
Sulfuric acid	1.84	Gold	19.3
Mercury	13.6	Platinum	21.5

GASES	
Hydrogen	0.0695
Helium	0.138
Acetylene	0.898
Nitrogen	0.967
Air	1.000
Oxygen	1.105
Carbon dioxide	1.528

Figure 2-5. This table includes the specific gravity of common substances. The standard for liquids and solids is water whereas the standard for gases is air. Both have a specific gravity of 1.

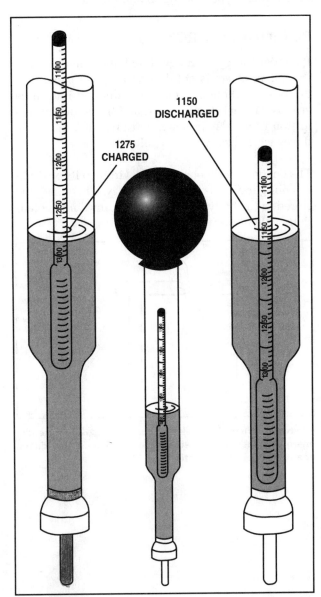

Figure 2-6. The specific gravity of a liquid is measured with a hydrometer.

An example of specific gravity that holds interest for an aviation maintenance technician, is the electrolyte of a lead-acid battery. When a battery is discharged, the calibrated float immersed in the electrolyte indicates approximately 1150. The indication of a charged battery is between 1275 and 1300. Since specific gravity is based on the density of the electrolyte, temperature is a consideration. Therefore, battery electrolyte is measured at 80 degrees Fahrenheit. If electrolyte is at a different temperature, a correction must be applied.

ENERGY

Energy, in its practical form, is the capacity of an object to perform work. It is classified into two rather broad types, potential and kinetic.

POTENTIAL ENERGY

Potential energy is energy stored in a material. Even though an object is not doing work, it is capable of doing work. Potential energy is divided into three groups: (1) that due to position, (2) that due to distortion of an elastic body, and (3) that which produces work through chemical action.

The energy a body possesses by virtue of its position or configuration is potential energy. This energy is stored in the body which retains it, until it is potentially able to release it. [Figure 2-7]

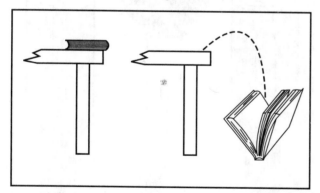

Figure 2-7. When the book is at rest, it possesses potential energy because of its position; but as it falls, it has kinetic energy because of its motion.

The amount of potential energy possessed by an elevated weight is computed using the following formula:

$$\text{Potential Energy} = \text{Weight} \times \text{Height}$$

When a spring is compressed or stretched from its normal condition, it possesses potential energy which may be released when the spring is allowed to return to its at-rest condition.

Chemical energy is stored in an aircraft battery and is there to produce mechanical work when the starter switch is pressed. Electrical energy can also be stored in a capacitor to produce light when a strobe light tube is fired.

KINETIC ENERGY

When potential energy is released and causes motion, it is changed to kinetic energy. Kinetic energy is known as "energy of motion."

When you raise a hammer above your head, the hammer gains potential energy because of its position, or height. As you bring the hammer down, the potential energy stored in the hammer, plus the kinetic energy your muscles put into the hammer, gives it a great deal of kinetic energy. When the hammer strikes a nail, work is done forcing the nail into a piece of wood. However, not all the kinetic energy does work. Some of the energy is dissipated as heat caused by the friction between the nail and wood.

UNITS OF ENERGY

In order to better understand energy, you must recognize the units with which it is expressed. The most common unit of measure of mechanical energy is the horsepower and is equivalent to 33,000 foot-pounds of work done in one minute. In the metric system the measure of mechanical energy is the Joule. For electrical energy the typical unit of measure is the watt. These units are used extensively in the study of machines and electricity.

WORK, POWER, FORCE, AND MOTION

Work, power, force, and motion are important concepts of physics. As an aircraft maintenance technician, you must understand these concepts and be able to use the associated formulas to fully comprehend simple machines like the lever, pulley, or gear.

WORK

If a force is applied to an object and the object moves, work is done. The amount of work done is directly proportional to the force applied and the distance the object moves. In mathematical terms, work is defined as the product of force times distance. For example, if an engine weighing 400 pounds is lifted 10 feet, 4,000 foot-pounds of work is done. This is expressed by the equation:

Work = Force (F) × Distance (D)

Work = 400 lbs. × 10 ft.

Work = 4,000 foot/pounds

If a force is applied to an object and the object does not move, no work is done. By the same token, no work is done if an object moves with no force applied to it.

In the English system, work is typically measured in **foot-pounds**. One foot-pound is equal to one pound of force applied to an object through the distance of one foot. To convert foot-pounds to **inch-pounds**, multiply the number of foot-pounds by the number of inches in a foot, or 12 inches. One foot-pound is equivalent to 12 inch-pounds.

In the metric system, the unit of work is the **joule**. One joule is the work done by a force of one **newton** acting through a distance of one meter. One pound is equal to 4.448 newtons. When using metric units of measure, the formula for computing work remains the same algebraically, and only the units of measure change.

$$W \text{ (Joules)} = F \text{ (Newtons)} \times D \text{ (meters)}$$

POWER

When determining the amount of work done, the time required to do the work is not considered. Power, on the other hand, does take time into consideration. For example, a low powered motor can be geared to lift a large weight if time is not a factor. However, if it is important to lift the weight quickly, more power is required. Power is calculated with the formula:

$$\text{Power} = \frac{\text{Force} \times \text{Distance}}{\text{Time}}$$

Power is defined as the time-rate of doing work. In the English system, power is expressed in foot-pounds per second, whereas the unit of power in the metric system is joules per second.

Another unit of measure for power is the **horsepower**. Horsepower was first used by James Watt to compare the performance of his steam engine with a typical English dray horse. One horsepower is the amount of power required to do 33,000 foot-pounds of work in one minute or 550 foot-pounds of work in one second. Therefore, the formula used to calculate horsepower is:

$$\text{Horsepower} = \frac{\text{Force} \times \text{Distance}}{33,000 \text{ ft lbs} \times \text{Time}}$$

Example:

Find the horsepower required to raise a 12,000 pound airplane six feet in one-half minute.

Given:

Aircraft Weight = 12,000 Pounds
Height = 6 Feet
Time = 1/2 Minute

$$\text{Horsepower} = \frac{12,000 \times 6}{33,000 \times 1/2}$$

$$\text{Horsepower} = \frac{72,000}{16,500}$$

$$\text{Horsepower} = \frac{720}{165}$$

$$\text{Horsepower} = 4.36$$

The electrical unit of measure for horsepower is the **watt**. One horsepower is equal to 746 watts. Therefore, in the example above, 3,252.56 watts of electrical power is required to operate the 4.36 horsepower motor.

As mentioned earlier, the metric system utilizes joules per second to measure power. To convert joules per second to watts use the relationship:

$$1 \text{ watt} = \frac{1 \text{ joule}}{1 \text{ second}}$$

Based on this relationship, a motor with a power output of 5,000 watts is capable of doing 5,000 joules of work per second. Since one kilowatt (kw) is equal to 1,000 watts, the above motor has a power output of 5 kw.

Using a 5 kw motor, how much time does it take to hoist a 12,000 pound aircraft 30 meters? To determine this, you must rewrite the formula for calculating power as:

$$t = \frac{\text{Force} \times \text{Distance}}{\text{Power}}$$

Now substitute the given values.

Force = 53,376 newtons (12,000 lbs. × 4.448 N)
Distance = 30 meters
Power = 5 kw

$$t = \frac{53,376 \times 30}{5,000}$$

Time = 320.256 seconds

FORCE

You now know that work is the product of a force applied to an object, times the distance the object moves. However, many practical machines use a **mechanical advantage** to change the amount of force required to move an object. Some of the simplest mechanical advantage devices used are: the **lever**, the **inclined plane**, the **pulley**, and **gears**.

Mechanical advantage is calculated by dividing the weight, or resistance (R) of an object by the effort (E) used to move the object. This is seen in the formula:

$$\text{Mechanical Advantage (MA)} = \frac{R}{E}$$

A mechanical advantage of 4 indicates that for every 1 pound of force applied, you are able to move 4 pounds of resistance.

LEVERS

A lever is a device used to gain a mechanical advantage. In its most basic form, the lever is a seesaw that has a weight at each end. The weight on one end of the seesaw tends to rotate the board counterclockwise while the weight on the other end tends to rotate the board clockwise. Each weight produces a **moment** or turning force. The moment of an object is calculated by multiplying the object's weight by the distance the object is from the balance point or **fulcrum**.

A lever is balanced when the algebraic sum of the moments is zero. In other words, a 10-pound weight located six feet to the left of a fulcrum has a moment of negative 60 foot-pounds while a 10-pound weight located six feet to the right of a fulcrum has a moment of positive 60 foot-pounds. Since the sum of the moments is zero, the lever is balanced. [Figure 2-8]

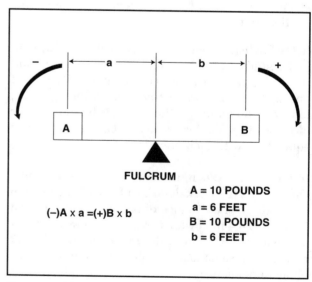

Figure 2-8. In a balanced lever, the sum of the moments is zero.

FIRST-CLASS LEVER

Figure 2-9 illustrates a practical application of a first-class lever. The end of a four-foot bar is placed under a 100-pound weight, so the fulcrum is one-half foot from the weight's center of gravity. This leaves three and one half feet between the weight and the point at which the force, or effort is applied. When effort (E) is applied, it acts in the direction opposite the weight's movement. To calculate the amount of effort required to lift the weight, you must calculate the moments on each side of the fulcrum. This is done using the formula:

$$\frac{L}{l} = \frac{R}{E}$$

Where:

L = length of effort arm
l = length of resistance arm
R = resistance
E = effort force

$$\frac{3.5}{.5} = \frac{100}{E}$$

$$3.5 \text{ x } E = 50$$

$$E = 14.28$$

Although less effort is required to lift the 100-pound weight, a lever does not reduce the amount of work done. Remember, work is the product of force and distance; therefore, when you examine the ratio of the distances moved on either side of the fulcrum,

you notice that the effort arm must move 21 inches to move the resistance arm 3 inches. The work done on each side is the same.

$$3 \text{ in. } \times 100 \text{ lbs. } = 21 \text{ in. } \times 14.28 \text{ lbs.}$$

$$300 \text{ in.-lbs. } = 300 \text{ in.-lbs.}$$

SECOND-CLASS LEVER

Unlike the first-class lever, the second-class lever has the fulcrum at one end of the lever and effort is applied to the opposite end. The resistance, or weight, is typically placed near the fulcrum between the two ends. The most common second-class lever is the wheelbarrow. When using a wheelbarrow, the lever, or handle, is used to gain mechanical advantage to reduce the effort required to carry a load. For example, if a wheelbarrow has 200 pounds of weight concentrated 12 inches from the wheel axle and effort is applied 48 inches from the axle, only 50 pounds of effort is needed to lift the weight. You calculate this by using the same relationship derived for a first-class lever. [Figure 2-10]

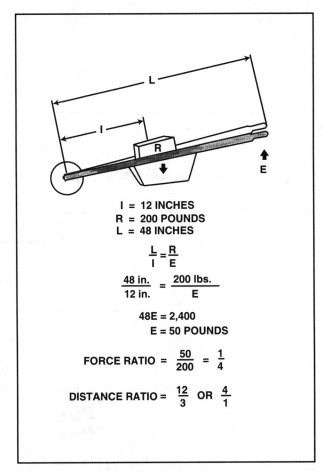

I = 12 INCHES
R = 200 POUNDS
L = 48 INCHES

$$\frac{L}{I} = \frac{R}{E}$$

$$\frac{48 \text{ in.}}{12 \text{ in.}} = \frac{200 \text{ lbs.}}{E}$$

$$48E = 2,400$$
$$E = 50 \text{ POUNDS}$$

FORCE RATIO = $\frac{50}{200}$ = $\frac{1}{4}$

DISTANCE RATIO = $\frac{12}{3}$ OR $\frac{4}{1}$

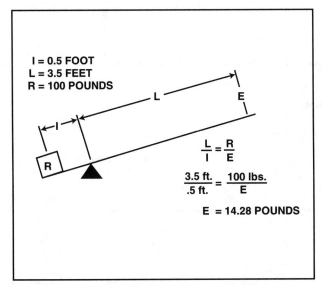

I = 0.5 FOOT
L = 3.5 FEET
R = 100 POUNDS

$$\frac{L}{I} = \frac{R}{E}$$

$$\frac{3.5 \text{ ft.}}{.5 \text{ ft.}} = \frac{100 \text{ lbs.}}{E}$$

E = 14.28 POUNDS

Figure 2-9. The effort required to lift a weight using a first-class lever is determined by balancing the moments on each end of the fulcrum.

Figure 2-10. Using the second-class lever illustrated, 50 pounds of effort is required to lift 200 pounds of resistance.

The mechanical advantage gained using a second-class lever is the same as that gained when using a first-class lever. The only difference is that the resistance and effort on the second-class lever move in the same direction. [Figure 2-11]

THIRD-CLASS LEVER

In aviation, the third-class lever is primarily used to move a resistance a greater distance than the effort applied. This is accomplished by applying the effort between the fulcrum and the resistance. However, when doing this, a greater effort is required to produce movement. An example of a third-class lever is a landing gear retracting mechanism. The effort required to retract the landing gear is applied near the fulcrum while the resistance is at the opposite end of the lever. [Figure 2-12]

INCLINED PLANES

Another way to gain mechanical advantage is through the inclined plane. An inclined plane achieves an advantage by allowing a large resistance to be moved by a small effort over a long distance. The amount of effort required is calculated through the formula:

$$\frac{L}{l} = \frac{R}{E}$$

Where:
L = length of the ramp
l = height of the ramp
R = weight or resistance of the object moved
E = effort required to raise or lower the object

Determine the amount of effort required to roll a 500 pound barrel up a 12 foot inclined plane to a platform that is 4 feet high.

Example:

Given:

L = 12 feet
l = 4 feet
R = 500 pounds

Solve for effort

$$\frac{12}{4} = \frac{500}{E}$$

$$12 \times E = 2000$$

$$E = 166.7 \text{ lbs.}$$

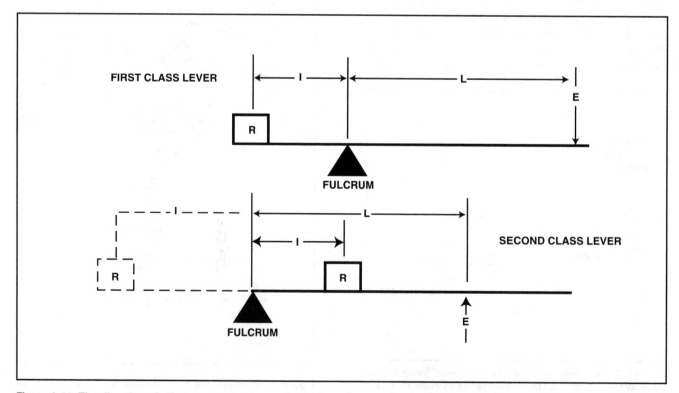

Figure 2-11. The direction of effort applied to a second-class lever is opposite that applied to a first-class lever.

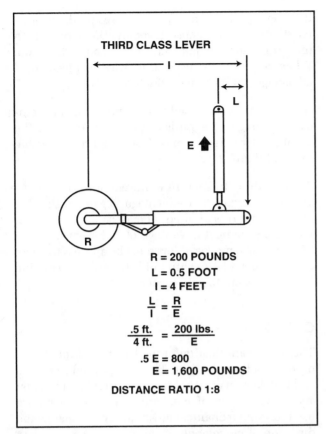

Figure 2-12. To raise a 200-pound wheel, 1,600 pounds of effort are required. However, the ratio of movement between the point where effort is applied and the resistance is 1:8. This means the effort moves six inches to raise the wheel four feet.

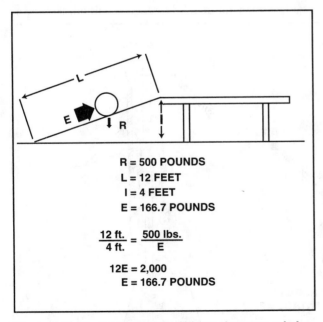

Figure 2-13. The inclined plane follows the same relationship used with levers. The mechanical advantage gained is the ratio of the length of the ramp to the height of the platform, or in this example 3:1. Therefore, only 166.7 pounds of effort are required to raise the barrel 4 feet.

By using an inclined plane, 500 pounds of resistance is moved by an effort of 166.7 pounds. [Figure 2-13]

The wedge is a special application of the inclined plane, and is actually two inclined planes set back-to-back. By driving a wedge full-length into a material, the material is forced apart a distance equal to the width of the broad end of the wedge. The greatest mechanical advantage exists in long, slim wedges.

PULLEYS

Pulleys are another type of simple machine that allow you to gain mechanical advantage. A single fixed pulley is identical to a first-class lever. The fulcrum is the center of the pulley and the arms that extend outward from the fulcrum are identical in length. Therefore, the mechanical advantage of a single fixed pulley is 1. When using a pulley in this fashion, the effort required to raise an object is equal to the object's weight. [Figure 2-14]

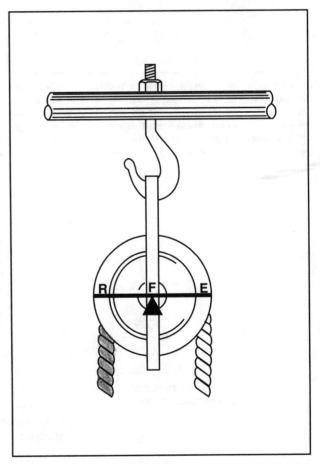

Figure 2-14. With the center of the pulley acting as the fulcrum, the two arms, RF and FE, extending outward are the same length. Therefore, the effort required to lift the object equals the resistance.

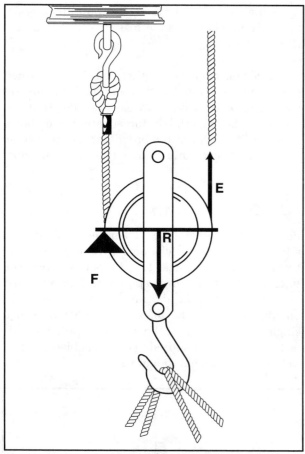

Figure 2-15. In a single movable pulley, the effort acts upward on arm "EF," which is the diameter of the pulley. The resistance acts downward on arm "FR," which is the radius of the pulley. Since the effort arm is twice the length of the resistance arm, the mechanical advantage is 2.

If a single pulley is not fixed, it takes on the characteristics of a second class lever. In other words, both the effort and weight act in the same direction. When a pulley is used this way, a mechanical advantage of 2 is gained. [Figure 2-15]

A common method used to determine the mechanical advantage of a pulley system is to count the number of ropes that move or support a movable pulley. [Figure 2-16]

Another thing to keep in mind when using pulleys is that as mechanical advantage is gained, the distance the effort is applied increases. In other words, with a mechanical advantage of 2, for every 1 foot the resistance moves, effort must be applied to 2 feet of rope. This relationship holds true wherever using a pulley system to gain mechanical advantage.

GEARS

There is no application of the basic machine that is used more than the gear. The gear is used in clocks and watches, in automobiles and aircraft, and in just about every type of mechanical device. Gears are used to gain mechanical advantage, or to change the direction of movement.

To gain a mechanical advantage when using gears, the number of teeth on either the drive gear or driven gear is varied. For example, if both the drive gear and driven gear have the identical number of teeth, no mechanical advantage is gained. However,

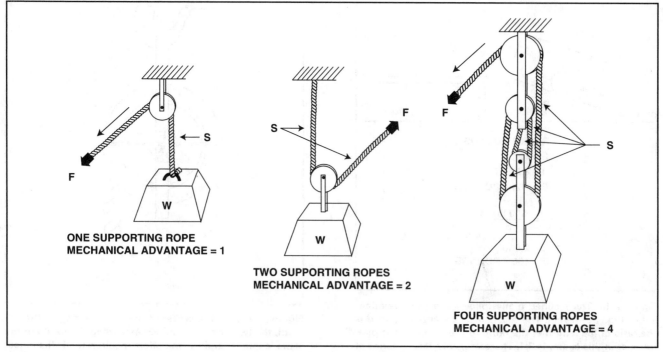

Figure 2-16. The mechanical advantage of a pulley system is equal to the number of ropes that move or support a movable pulley.

if a drive gear has 50 teeth and a driven gear has 100 teeth a mechanical advantage of 2 is gained. In other words, the amount of power required to turn the drive gear is reduced by half.

Another thing to keep in mind is that the revolution ratio between two gears is opposite the ratio of their teeth. Using the earlier example of a drive gear with 50 teeth, and a driven gear with 100 teeth, the gear ratio is 1:2. However, for every one turn of the drive gear the driven gear makes one-half turn. This results in a revolution ratio of 2:1.

There are many types of gears in use. **Spur gears** have their teeth cut straight across their circumfer-

ence and are used to connect two parallel shafts. When both gears have external teeth, the shafts turn in opposite directions. If it is necessary for both shafts to turn in the same direction, one gear must have internal teeth. [Figure 2-17]

If a drive shaft and driven shaft are not parallel to each other, **beveled gears** are used. However, because the teeth on beveled gears are external, the rotational direction of each shaft is opposite. Tail rotor gear boxes on helicopters typically use beveled gears. [Figure 2-18]

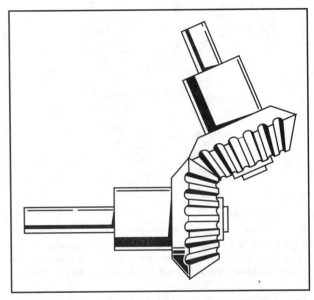

Figure 2-18. With beveled gears, the angle between the drive shaft and driven shaft is typically 90 degrees. However, the angle can be any value less than 180 degrees.

When an extreme amount of mechanical advantage is needed, a **worm gear** is used. A worm gear uses a spiral ridge around a shaft for the drive gear with the shafts usually at right angles to each other. One complete rotation of the drive shaft moves the driven gear one tooth. [Figure 2-19]

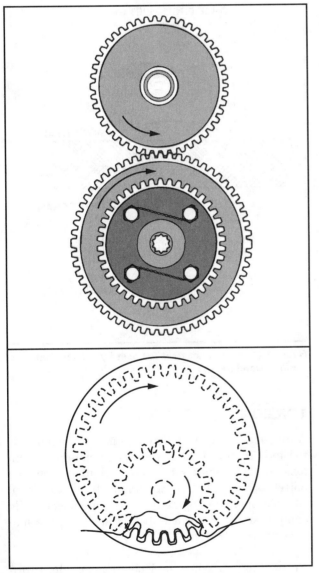

Figure 2-17. External spur gears provide a mechanical advantage and reverse the rotational direction of the drive shafts. However, a spur gear system with internal teeth provides a mechanical advantage without reversing the rotational direction.

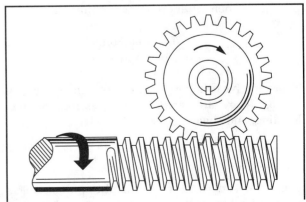

Figure 2-19. A worm gear system provides a very high mechanical advantage.

Planetary gear systems are typically used to reduce the propeller shaft speed on more powerful aircraft engines. This allows the engine to turn at a higher rpm and develop more power. In a planetary gear system, the propeller mounts on a spider-like cage that holds the planetary gears. These planetary gears rotate around a fixed central sun gear. [Figure 2-20]

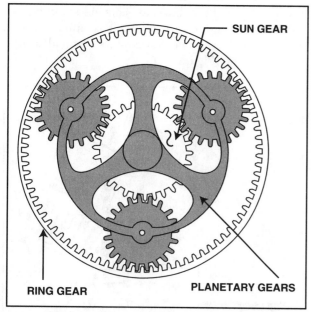

Figure 2-20. In a planetary gear system, the propeller mounts to a cage that holds the planetary gears. As the ring gear turns, the planetary gears rotate around a fixed sun gear.

In some planetary gear systems, the sun gear is the drive gear and the ring gear is fixed in the nose section of the engine. In this situation, the planetary gears act as simple idler gears in the system.

STRESS

When an external force acts on a body, it is opposed by an internal force called stress. The English measure for stress is pounds per square foot, or pounds per square inch. Stress is shown as the ratio:

$$\text{Stress} = \frac{\text{External Force}}{\text{Area of Applied Force}}$$

There are five different types of stress in mechanical bodies. They are **tension**, **compression**, **torsion**, **bending**, and **shear**.

TENSION

Tension describes forces that tend to pull an object apart. Flexible steel cable used in aircraft control systems is an example of a component that is designed to withstand tension loads. Steel cable is easily bent and has little opposition to other types of stress; however, when subjected to a purely tensional load it performs exceptionally well.

COMPRESSION

Compression is the resistance to an external force that tries to push an object together. Aircraft rivets are driven with a compressive force. When compression stresses are applied to a rivet, the rivet shank expands until it fills the hole and forms a butt to hold materials together. [Figure 2-21]

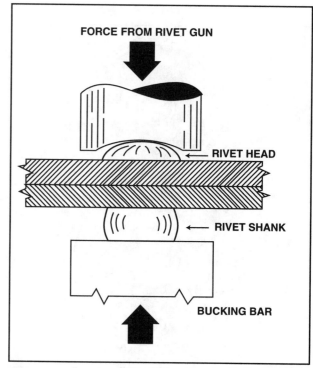

Figure 2-21. An aircraft rivet is upset by the application of compression stresses.

TORSION

A torsional stress is applied to a material when it is twisted. Torsion is actually a combination of both tension and compression. For example, when an object is subjected to torsional stress, tensional stresses operate diagonally across the object while compression stresses act at right angles to the tension stresses. [Figure 2-22]

An engine crankshaft is a component whose primary stress is torsion. The pistons pushing down on the connecting rods rotate the crankshaft against the opposition caused by the propeller. The resulting stresses attempt to twist the crankshaft.

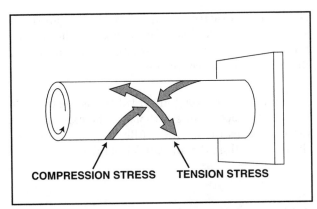

Figure 2-22. Torsional stresses are a combination of tension and compression stresses.

BENDING

In flight, the force of lift tries to bend an aircraft's wings upward. When this happens the skin on top of the wing is subjected to a compression force, while the skin below the wing is pulled by a tension force. When the aircraft is on the ground the force of gravity reverses the stresses. In this case, the top of the wing is submitted to tension stress while the lower skin experiences compression stress. [Figure 2-23]

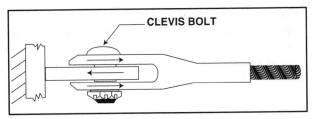

Figure 2-24. When the control cable moves, the forces created attempt to slice the bolt apart, or shear it.

STRAIN

As stated earlier, stress is a force within an object that opposes an applied external force. Strain is the deformation of an object that is caused by stress.

Hooke's law states that if strain does not exceed the elastic limit of a body, it is directly proportional to the applied stress. This fact allows beams and springs to be used as measuring devices. For example, as force is applied to a hand torque wrench, its deformation, or bending, is directly proportional to the strain it is subjected to. Therefore, the amount of torque deflection can be measured and used as an indication of the amount of stress applied to a bolt. [Figure 2-25]

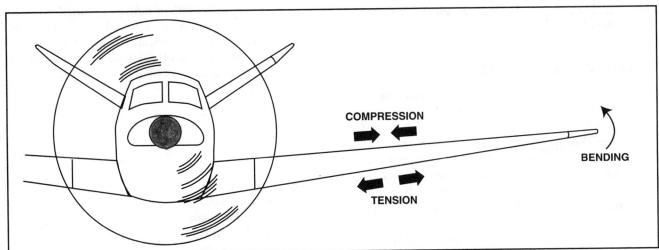

Figure 2-23. Like torsional stress, bending stresses are a combination of tension and compression.

SHEAR

A shear stress tries to slice a body apart. A clevis bolt in an aircraft control system is designed to withstand shear loads. Clevis bolts are made of a high-strength steel and are fitted with a thin nut that is held in place by a cotter pin. Whenever a control cable moves, shear forces are applied to the bolt. However, when no force is present, the clevis bolt is free to turn in its hole. [Figure 2-24].

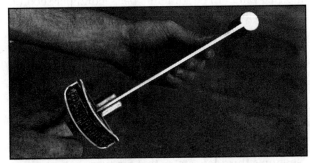

Figure 2-25. The deflection, or strain, in the beam of this torque wrench is directly proportional to the stress applied to the bolt being tightened.

When making a repair to an aircraft structure, it is important that you do not make abrupt changes in the cross-sectional area of a structural member. Abrupt changes concentrate stresses and can lead to structural failure.

MOTION

An English physicist by the name of Sir Isaac Newton proposed three laws of motion that explain the effect of force on matter. These laws are commonly referred to as **Newton's Laws of Motion.**

NEWTON'S FIRST LAW

Newton's first law of motion explains the effect of inertia on a body. It states that a body at rest tends to remain at rest, and a body in motion tends to remain in uniform motion (straight line), unless acted on by some outside force. Simply stated, an object at rest remains at rest unless acted on by a force. By the same token, an object in motion on a frictionless surface continues in a straight line at the same speed indefinitely. However, you know this does not happen because every object in motion encounters friction.

NEWTON'S SECOND LAW

Newton's second law states that the acceleration produced in a mass by the addition of a given force is directly proportional to the force, and inversely proportional to the mass. When all forces acting on a body are in balance, the object remains at a constant velocity. However, if one force exceeds the other, the velocity of the object changes. Newton's second law is expressed by the formula:

Force = mass × acceleration (F = ma).

An increase in velocity with time is measured in the metric system in centimeters per second per second. In the English system it is measured in feet per second per second. This is an important relationship when working with the acceleration of gravity. For example, if a body is allowed to fall freely under the effect of gravity, it accelerates uniformly at 32.17 feet per second every second it falls.

NEWTON'S THIRD LAW

Newton's third law states that for every action, there is an equal and opposite reaction. When a gun is fired, expanding gases force a bullet out of the bar- rel and exert exactly the same force back against the shooter. This is felt as a familiar kick. The magnitudes of both forces are exactly equal, however, their directions are opposite.

An application of Newton's third law is the turbojet engine. The action in a turbojet is the exhaust as it rapidly leaves the engine while the reaction is the thrust propelling the aircraft forward.

SPEED AND VELOCITY

Speed and velocity are often used interchangeably, however, they are actually quite different. Speed is simply a rate of motion, or the distance an object moves in a given time. It is usually expressed in terms like miles per hour, feet per second, kilometers per hour, or knots. Speed does not take into consideration any direction. Velocity, on the other hand, is the rate of motion in a given direction, and is expressed in terms like five hundred feet per minute downward, or 300 knots eastward.

An increase in the rate of motion is called acceleration and a decrease is called deceleration. Both acceleration and deceleration are measured in terms such as feet per second per second, or meters per second per second. Acceleration is calculated using the following formula:

$$A = \frac{V_f - V_i}{t}$$

Where:

A = Acceleration
V_f = the final velocity
V_i = the initial velocity
t = the elapsed time

VECTORS

A vector quantity is a mathematical expression having both magnitude and direction. Velocity is a vector quantity because it has both of these characteristics. Since all vector quantities have magnitude and direction, they can be added to each other. One of the simplest ways to add vectors is to draw them to scale. For example, vectors A and B have a known velocity and direction. To add these two vectors begin by drawing vector A to the correct length and direction. Then, place the tail of vector B at the head of vector A and draw it to scale. Once both vectors are laid out, draw the resultant vector from the starting point to the head of the last vector. The resulting vector has the same velocity as the sum of the two vectors laid out. [Figure 2-26]

CIRCULAR MOTION

When an object moves in a uniformly curved path at a uniform rate, its velocity changes because of its constant change in direction. If you tie a weight onto a string and swing it around your head, it follows a circular path. The force that pulls the spinning object away from the center of its rotation is called **centrifugal** force. The equal and opposite force required to hold the weight in a circular path is called **centripetal** force. [Figure 2-27]

Centripetal force is directly proportional to the mass of the object in motion and inversely proportional to the size of the circle in which the object travels.

Thus, if the mass of the object is doubled, the pull on the string must double to maintain the circular path. By the same token, if the radius of the string is cut in half and the speed remains constant, the pull on the string must increase. The reason for this is that as the radius decreases, the string must pull the object from its linear path more rapidly. Using the same reasoning, the pull on the string must increase if the object is swung more rapidly in its orbit. Centripetal force is thus directly proportional to the velocity of the object. The formula used to calculate centripetal force is:

$$CP = \frac{MV^2}{R}$$

When working with most grinding wheels, notice that they are rated for a maximum rpm. The reason for this is, if too much centrifugal force builds up, the binding materials within the wheel can not hold the wheel together.

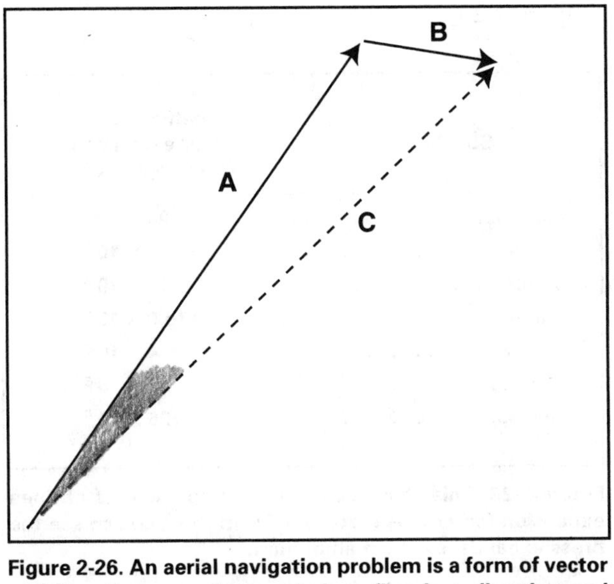

Figure 2-26. An aerial navigation problem is a form of vector addition. For example, an airplane flies in a direction and speed represented by vector A. However, the wind, represented by vector B, blows in a given direction and speed. Therefore, the airplane actually follows the ground track illustrated by vector C whose magnitude and direction are the sum of vectors A and B.

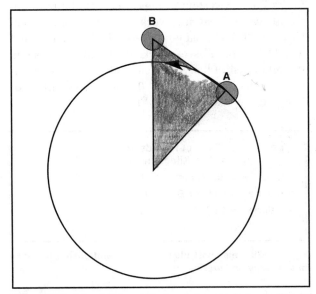

Figure 2-27. A body in motion tends to remain in motion in a straight line. When the body is forced into a curved path, centrifugal force tends to pull the body away from the center of the curve. The force used to counter centrifugal force is centripetal force and is provided by the string or wire attached to the body.

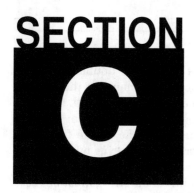

SECTION C

GAS AND FLUID MECHANICS

HEAT

Heat is one of the most useful forms of energy because of its direct relationship with work. For example, to slow an airplane after touchdown, the brakes are applied. When this is done, the kinetic energy of the moving aircraft is changed into heat energy by brake pad friction against the disks. This slows the wheels and produces additional friction between the wheels and the runway which, finally, slows the airplane.

In the English system of measurement, the relationship between heat and work is the British Thermal Unit, or **BTU**, of heat energy. One BTU is equivalent to 778 foot-pounds of work. In the metric system, the relationship between heat and work is the calorie, which is equivalent to 0.42 meter-kilogram, or 3.09 foot-pounds of work. [Figure 2-28]

1 Btu	= 778 Foot-Pounds or 107.59 Kilogram-Meters.
1 Btu	= 252 Calories or 1,055 Joules.
1 Calorie	= 0.00397 Btu or 4.186 Joules.
1 Calorie	= 3.088 Foot-Pounds or

Figure 2-28. This chart illustrates the relationship between heat energy and work.

LATENT HEAT

When a pan of water is left on a stove and heat energy is continually added, the water temperature rises to a peak and remains constant. Beyond this point the water changes its state from a liquid to a vapor. The heat that causes the change of state is called the heat of vaporization, or latent heat. This latent heat remains in the water as long as it is in its vapor form. It requires 539.6 calories of heat energy to change one gram of liquid water into water vapor.

DIMENSIONAL CHANGES

When heat energy is added to an object, the molecular movement within the object increases and the object expands in physical dimension. This expansion change is most noticeable in gases. In fact, the expansion of gases is so great that the expansion energy produced can be harnessed and used to do work.

Solids and liquids expand far less than gases. Therefore, you can determine the amount of expansion by multiplying a material's original dimension by its **coefficient of linear expansion**. Every material has its own coefficient of linear expansion. [Figure 2-29]

Substance	Coefficient of linear expansion (per degree C)
Aluminum	24×10^{-6}
Brass	19×10^{-6}
Copper	17×10^{-6}
Glass	4 to 9×10^{-6}
Quartz	0.4×10^{-6}
Steel	11×10^{-6}
Zinc	26×10^{-6}

Figure 2-29. This chart illustrates the coefficients of linear expansion for various materials. From this you can see that brass expands less than aluminum.

To estimate the expansion of an object, it is necessary to know three things: 1) the object's length, 2) the rise in temperature the object is subjected to, and 3) the object's coefficient of expansion. This relationship is expressed by the equation:

$$\text{Expansion} = kL \, (t_2 - t_1).$$

Where:

k = the coefficient of expansion
L = the length of the object
t_2 = ending temperature
t_1 = beginning temperature

If a steel rod measures exactly 9 feet at 21 degrees C, what is its length at 55 degrees C? The value of "k" for steel is 11×10^{-6}.

$$e = kL(t_2 - t_1)$$

$$e = (11 \times 10^{-6}) \times 9 \times (55 - 21)$$

$$e = 0.000011 \times 9 \times 34$$

$$e = 0.003366$$

When this amount is added to the rod's original length, it makes the rod 9.00336 feet long.

Although the increase in length is relatively small, if the rod were placed where it could not expand freely, a tremendous amount of force could result from thermal expansion. Thus, thermal expansion is taken into consideration when designing airframes, powerplants, or related equipment.

SPECIFIC HEAT

Different materials require different amounts of heat energy to change their temperature. The heat required to raise the temperature of one gram of a material one degree Celsius is known as material's specific heat. The specific heat of water is 1 while the specific heat of most other materials is less than 1. [Figure 2-30]

Due to the high specific heat of water, oceans and large lakes serve as temperature stabilizers. Land

MATERIAL	SPECIFIC HEAT
Lead	0.031
Mercury	0.033
Brass	0.094
Copper	0.095
Iron or Steel	0.113
Glass	0.195
Alcohol	0.547
Aluminum	0.712
Water	1.000

Figure 2-30. From this chart you can see that it requires about three percent as much heat energy to raise the temperature of lead one degree Celsius as it does to raise the temperature of an equal weight of water.

surfaces, on the other hand, have a much lower specific heat; therefore, temperatures can vary significantly throughout the course of a day.

TRANSFER OF HEAT

There are three methods by which heat is transferred from one location to another or from one substance to another. These three methods are conduction, convection, and radiation.

CONDUCTION

Conduction requires physical contact between a body having a high level of heat energy and a body having a lower level of heat energy. When a cold object touches a hot object, the violent action of the molecules in the hot material speed up the molecules in other material. This action spreads until the heat is equalized throughout both bodies.

A good example of heat transfer by conduction is the way excess heat is removed from an aircraft engine cylinder. When gasoline burns inside a cylinder, it releases a great deal of heat. This heat passes to the outside of the cylinder head by conduction and into the fins surrounding the head. The heat is then conducted into the air as it flows through the fins. [Figure 2-31]

Various metals have different rates of conduction. In some cases, the ability of a metal to conduct heat is a major factor in choosing one metal over another. It is interesting to note that the thermal conductivity

Figure 2-31. Heat is removed from an aircraft engine cylinder by conduction.

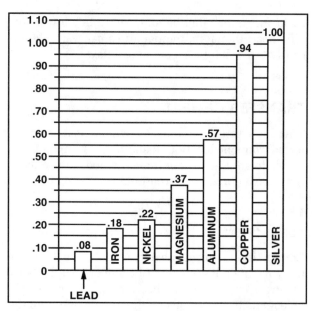

Figure 2-32. Various metals conduct heat at different rates.

of a certain metal has no relationship to its coefficient of thermal expansion. [Figure 2-32]

Liquids are poor conductors of heat. For example, when ice and water are placed in a test tube together and heat is applied, the ice does not melt rapidly, even though the water at the top boils. This is because pure water does not conduct enough heat to melt the ice. [Figure 2-33]

Gases are even worse conductors of heat than liquids. This is why it is possible to stand close to a stove without being burned. Because the molecules are farther apart in gases than in solids, the gases are much poorer conductors of heat.

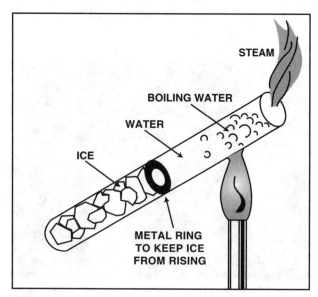

Figure 2-33. Water is a poor conductor of heat.

Insulators are materials that prevent the transfer of heat. A wooden handle on a pot or a soldering iron serves as a heat insulator. Certain material, such as finely spun glass, is a particularly poor heat conductor and, therefore, is used in many types of insulation.

CONVECTION

Convection is the process by which heat is transferred by movement of a heated fluid. For example, when heat is absorbed by a free-moving fluid, the fluid closest to the heat source expands and its density decreases. The less dense fluid rises and forces the more dense fluid downward. A pan of water on a stove is heated in this way. The water at the bottom of the pan heats by conduction and rises. Once this occurs, the cooler water moves toward the bottom of the pan. [Figure 2-34]

A similar phenomenon takes place in a house that is heated by a hot air furnace. As warm air is forced through a house, the cool air sinks and is drawn to the furnace. This type of heating is known as forced convection because a blower is used in place of natural convection.

Transfer of heat by convection is often hastened by using a ventilating fan to move the air surrounding a hot object. For example, a hot object cools faster when cool air passes over the object's surface.

When the circulation of gas or liquid is not rapid enough to remove sufficient heat, fans or pumps are used to accelerate the motion of the cooling material. In some installations, pumps are used to circulate water or oil to help cool large equipment. In airborne installations, electric fans and blowers are used to aid convection.

Figure 2-34. Water in the pan is heated by natural convection.

RADIATION

The third way heat is transferred is through radiation. Radiation is the only form of energy transfer that does not require the presence of matter. For example, the heat you feel from an open fire is not transferred by convection because the hot air over the fire rises. Furthermore, the heat is not transferred through conduction because the conductivity of air is poor and the cooler air moving toward the fire overcomes the transfer of heat outward. Therefore, there must be some way for heat to travel across space other than by conduction and convection.

The term "radiation" refers to the continual emission of energy from the surface of all bodies. This energy is known as **radiant energy**. Sunlight is a form of radiant heat energy. This form of heat transfer explains why it is warm in front of a window where the sun is shining when the outside temperature is very low.

TEMPERATURE

Heat is a form of energy that causes molecular agitation within a material. The amount of agitation is measured in terms of temperature. Therefore, temperature is a measure of the kinetic energy of molecules.

In establishing a temperature scale, two conditions are chosen as a reference. These conditions are the points at which pure water freezes and boils. In the Fahrenheit system, water freezes at 32°F and boils at 212°F. The difference between these two points is divided into 180 equal divisions called "degrees."

A more convenient division is made on the **Centigrade** scale. It is divided into 100 graduated increments with the freezing point of water represented by 0°C and the boiling point 100°C. The Centigrade scale was renamed the Celsius scale, after the Swedish astronomer Anders Celsius who first described the centigrade scale in 1742.

To convert Celsius to Fahrenheit, remember that 100 degrees Celsius represents the same temperature difference as 180 degrees Fahrenheit. Therefore, you must multiply the Celsius temperature by 1.8 or 9/5, and add 32 degrees.

$$°F = (1.8°C) + 32$$

or

$$°F = 9/5 \ °C + 32$$

To convert Fahrenheit to Celsius, first subtract 32° from the Fahrenheit temperature and then divide by 1.8, or multiply by 5/9.

$$°C = (°F - 32) \div 1.8$$

or

$$°C = 5/9 \ (°F - 32)$$

In 1802, the French chemist and physicist Joseph Louis Gay Lussac found that when you increase the temperature of a gas one degree Celsius, it expands 1/273 of its original volume. Based on this, he reasoned that if a gas were cooled, its volume would decrease by the same amount. Therefore, if the temperature were decreased to 273 degrees below zero, a gas' volume would decrease to zero and there would be no more molecular activity. The point at which there is no molecular activity is termed **absolute zero**. On the Celsius scale absolute zero is –273°C. On the Fahrenheit scale it is –460°F.

Many of the gas laws relating to heat are based on a condition of absolute zero. To facilitate working with these terms, two absolute temperature scales are used. They are the **Kelvin** scale, which is based on the Celsius scale, and the **Rankine** scale, which is based on the Fahrenheit scale. On the Kelvin scale, water freezes at 273 degrees Kelvin and boils at 373 degrees Kelvin. However, on the Rankine scale, water freezes at 492 degrees Rankine and boils at 672 degrees Rankine. [Figure 2-35]

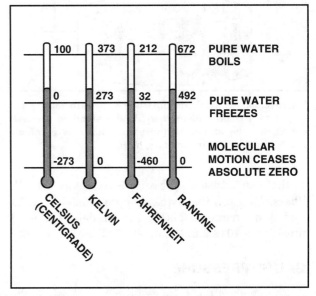

Figure 2-35. The four temperature scales commonly used are the Celsius, Kelvin, Fahrenheit, and Rankine. Notice that pure water freezes at 0 degrees Celsius, 273 degrees Kelvin, 32 degrees Fahrenheit, and 492 degrees Rankine.

To convert Celsius to Kelvin, add 273 to the Celsius temperature.

$$°K = °C + 273$$

To convert Fahrenheit to Rankine, add 460 to the Fahrenheit temperature.

$$°R = °F + 460$$

PRESSURE

Pressure is the measurement of a force exerted on a given area and is usually expressed in pounds per square inch, or **psi**. Atmospheric pressure on the other hand is measured in inches or millimeters of mercury and represents the height that a column of mercury rises in a glass tube when exposed to the atmosphere. Standard atmospheric pressure at sea level is equivalent to 29.92 inches of mercury, or 14.7 pounds per square inch at 59°F. [Figure 2-36]

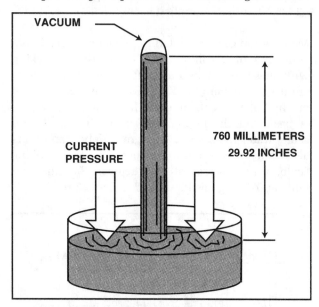

Figure 2-36. Atmospheric pressure is determined by the height of a column of mercury. Under standard sea-level conditions, the atmospheric pressure supports a column 29.92 inches, or 760 millimeters, high.

In the metric system, pressure is expressed in **millibars**. One millibar equals approximately .02953 inches of mercury. Therefore, standard sea level pressure is 1013.2 millibars at 15°C.

GAUGE PRESSURE

As the name implies, gauge pressure, or **psig** is the pressure read directly from a gauge, and represents the pressure in excess of barometric pressure. You can see this when you open the valve on an oxygen

bottle. When doing this, the oxygen rushes out until the pressure inside the cylinder equals that of the atmosphere, or 14.7 psi. However, once the pressures equalize, the gauge reads zero. [Figure 2-37]

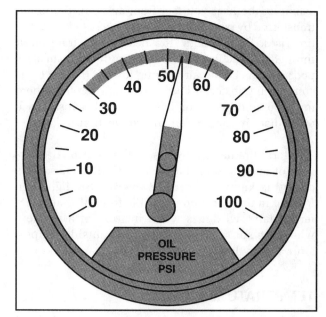

Figure 2-37. An engine oil pressure gauge measures gauge pressure.

ABSOLUTE PRESSURE

When pressure is referenced from zero pressure rather than from atmospheric pressure, it is known as absolute pressure, or **psia**. By definition, absolute pressure is gauge pressure plus atmospheric pressure. In a laboratory, absolute pressure is measured by a mercury barometer. In aircraft, the absolute pressure within an engine's induction system is indicated on a manifold pressure gauge. However, when the engine is not running, the gauge reads the existing atmospheric pressure. [Figure 2-38]

DIFFERENTIAL PRESSURE

Differential pressure, or **psid** is nothing more than the difference between any two pressures. One of the most familiar differential pressure gauges in an aircraft is the airspeed indicator. This instrument measures the difference between ram air pressure that enters the pitot tube and static air pressure. [Figure 2-39]

GAS LAWS

Gases and liquids are both fluids that are used to transmit forces. However, gases differ from liquids in that gases are compressible, while liquids are

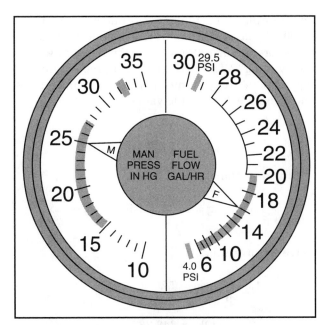

Figure 2-38. An aircraft manifold pressure gauge measures absolute pressure when the engine is running.

considered to be incompressible. The volume of a gas is affected by temperature and pressure. The degree to which temperature and pressure affect volume is defined in the gas laws of Boyle and Charles.

BOYLE'S LAW

In 1660, the British physicist Robert Boyle discovered that when you change the volume of a confined

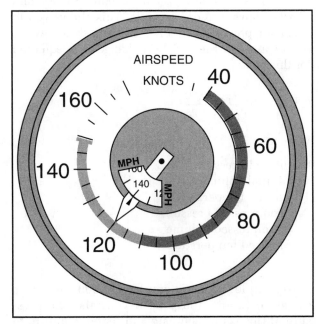

Figure 2-39. An airspeed indicator is actually a differential pressure gauge that measures the difference between impact air pressure and static air pressure.

gas that is held at a constant temperature, the gas pressure also changes. For example, using Boyle's Law, if the temperature of a gas is held constant, and the volume decreases, the pressure increases. [Figure 2-40]

Volume and pressure are said to be inversely related and the relationship is expressed in the formula:

$$\frac{V_1}{V_2} = \frac{P_2}{P_1}$$

Where:

V_1 = initial volume
V_2 = compressed volume
P_1 = initial pressure
P_2 = compressed pressure

Example:

If six cubic feet of oxygen is pressurized to 1,200 psig, what will the pressure be if the oxygen is allowed to expand to 10 cubic feet?

$$\frac{6}{10} = \frac{P_2}{1,200}$$

$$10P_2 = 7,200$$

$$P_2 = 720 \text{ psig}$$

Since the volume increased, the pressure must decrease. The new pressure is 720 psig.

CHARLES' LAW

Jacques Charles found that all gases expand and contract in direct proportion to any change in

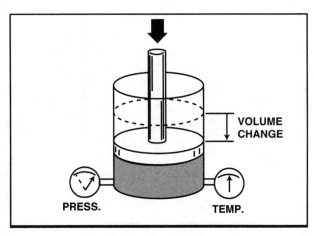

Figure 2-40. If the temperature of a gas remains constant, the pressure of the gas increases as volume decreases.

absolute temperature. This relationship, known as Charles' Law, is illustrated in figure 2-41 and is represented by the equation:

$$\frac{V_1}{V_2} = \frac{T_1}{T_2}$$

Where:

V_1 = initial volume
V_2 = revised volume
T_1 = initial temperature
T_2 = revised temperature

Charles' Law also states that if the volume of a gas is held constant, the pressure increases and decreases in direct proportion to changes in absolute temperature. This relationship is represented by the equation:

$$\frac{P_1}{P_2} = \frac{T_1}{T_2}$$

Where:

P_1 = initial pressure
P_2 = compressed pressure
T_1 = initial temperature
T_2 = revised temperature

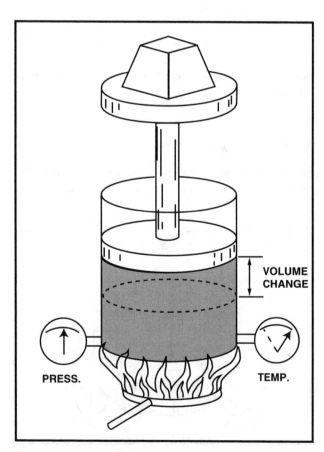

Figure 2-41. When the pressure of a gas remains constant, a gas' volume increases as temperature increases.

For example, if a cylinder of gas under a pressure of 1,500 psig at 60°F is left in the sun and heats up to a temperature of 100°F, what is the pressure within the cylinder? When working a gas problem involving temperature, you must convert all temperatures and pressures to absolute values.

60°F = 520° Rankine

100°F = 560° Rankine

1,500 psig = 1,514.7 psia

Solve for pressure.

$$\frac{1,514.7 \text{ psia}}{P_2} = \frac{520°\text{R}}{560°\text{R}}$$

$$520P_2 = 848,232$$

$$P_2 = 1,631.22 \text{ psia}$$

To convert this value to a gauge pressure, subtract the atmospheric pressure.

1,631.22 psia − 14.7 = 1,616.52 psig

GENERAL GAS LAW

The general gas law combines both Boyle's and Charles' laws into one formula. This allows you to calculate pressure, volume, or temperature when one or more of the variables change. The equation for this law is:

$$\frac{P_1 V_1}{T_1} = \frac{P_2 V_2}{T_2}$$

Where:

P_1 = initial pressure
V_1 = initial volume
T_1 = initial temperature
P_2 = compressed pressure
V_2 = compressed volume
T_2 = revised temperature

For example, let's say you have 1,000 liters of a gas at 30°C, and 100 psig. If you raise the pressure of the gas to 300 psig and its temperature to 150°C, you can find its volume by using the general gas law.

Example:

Remember, when using any of the gas law formulas, you must convert both temperature and pressure to absolute values.

P_1 = 100 psig = 114.7 psia
P_2 = 300 psig = 314.7 psia
V_1 = 1,000 liters
V_2 = 508.8 liters
T_1 = 30°C = 303°K
T_2 = 150°C = 423°K

$$\frac{P_1 V_1}{T_1} = \frac{P_2 V_2}{T_2}$$

$$\frac{114.7 \times 1,000}{303} = \frac{314.7 \times V_2}{423}$$

$$95,354.1 \; V_2 = 48,518,100$$

$$V_2 = 508.82 \text{ liters}$$

DALTON'S LAW

When a mixture of two or more gases which do not combine chemically is placed in a container, each gas expands to fill the container. However, as discussed earlier, when a gas expands its pressure decreases; therefore, each gas exerts a **partial pressure**. According to Dalton's Law, the total pressure exerted on the container is equal to the sum of the partial pressures.

Partial pressure is important when working with the aircraft environmental systems. For example, at about 10,000 feet, the partial pressure of oxygen drops and is no longer capable of forcing its way into a person's blood. Because of this, either cabin pressurization or supplemental oxygen is needed at high altitudes.

FLUID MECHANICS

A **fluid** is any substance that flows or conforms to the outline of a container. Both liquids and gases are fluids that follow many of the same rules. However, for all practical purposes, liquids are considered incompressible, while gases are compressible.

Much of the science of flight is based on the principles of fluid mechanics. For example, the air that supports an aircraft in flight and the liquid that flows in hydraulic systems both transmit force through fluid mechanics.

FLUID PRESSURE

The pressure exerted by a column of liquid is determined by the height of the column and is not affected by the volume of the liquid. For example, pure water weighs 62.4 pounds per cubic foot, which is equivalent to 0.0361 pounds per cubic inch. If you stack 1,000 cubic inches of water vertically in a column with a base of one square inch, the column would extend 1,000 inches high and would weigh 36.1 pounds. There would also be a pressure, or force per unit area of 36.1 pounds per square inch at the bottom of this column. The amount of fluid has no effect on the pressure at the bottom of the column. [Figure 2-42]

Gasoline has a specific gravity of 0.72, which means its weight is 72% that of water, or 0.026 pounds per cubic inch. Therefore, a column with a base of 1 square inch and 1,000 inches high results in a pressure of 26.0 pounds per square inch. To achieve a pressure of 1 psi, the column must be 38.5 inches high.

A practical example of how you use this information is when you are adjusting the float level on a carburetor. Assume the carburetor specifications require an application of 3 psi to the inlet of the carburetor. As stated earlier, a column of water 1,000 inches high exerts a pressure of 36.1. Therefore, a column 83.1 inches height results in a pressure of 3 psi. However, since you are working on a carburetor, gasoline is used instead of water. As stated earlier,

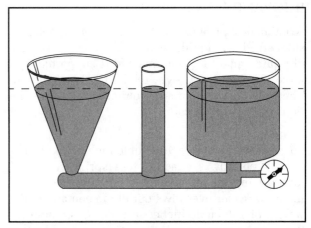

Figure 2-42. The pressure exerted by a column of liquid is determined by the height of the column and is not affected by the volume of the liquid.

the specific gravity of gasoline is 0.72. Therefore, to determine the height of a column of gasoline required to produce 3 psi, divide the height of a column of water required to produce 3 psi by 0.72. The column of gasoline must be 115.4 inches high (83.1 in. ÷ 0.72 = 115.4 in.).

With this known, connect a small container of gasoline to the carburetor by a piece of small diameter tubing. Suspend the container so the level of the liquid is 115.4 inches above the inlet valve of the carburetor. This applies the correct pressure at the inlet. [Figure 2-43]

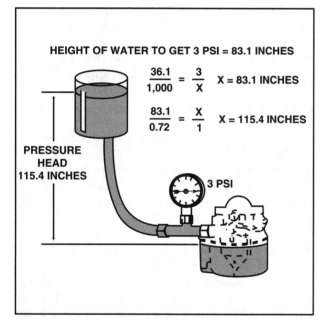

HEIGHT OF WATER TO GET 3 PSI = 83.1 INCHES

$$\frac{36.1}{1,000} = \frac{3}{X} \qquad X = 83.1 \text{ INCHES}$$

$$\frac{83.1}{0.72} = \frac{X}{1} \qquad X = 115.4 \text{ INCHES}$$

PRESSURE HEAD 115.4 INCHES

3 PSI

Figure 2-43. A given pressure head of a liquid may be produced by raising a container of the liquid to a specified height.

BUOYANCY

Archimedes' principle states that when an object is submerged in a liquid, the object displaces a volume of liquid equal to its volume and is supported by a force equal to the weight of the liquid displaced. The force that supports the object is known as the liquid's **buoyant force**.

For example, when a 100 cubic inch block weighing 9.7 pounds is attached to a spring scale and lowered into a full container of water, 100 cubic inches of water overflows out of the container. The weight of 100 cubic inches of water is 3.6 pounds; therefore, the buoyant force acting on the block is 3.6 pounds and the spring scale reads 6.1 pounds. [Figure 2-44]

If the object immersed has a specific gravity that is less than the liquid, the object displaces its own weight of the liquid and floats. The effect of buoyancy is not only present in liquids, but also in gases. Hot air balloons are able to rise because they are filled with heated air that is less dense than the air they displace.

PASCAL'S LAW

Pascal's Law explains that when pressure is applied to a confined liquid, the liquid exerts an equal pressure at right angles to the container that encloses it. For example, assume a cylinder is filled with a liquid and fitted with a one square inch piston. When a force of one pound is applied to the piston, the resulting pressure of the confined liquid is one pound per square inch everywhere in the container. [Figure 2-45]

You can find the amount of force (F) produced by a hydraulic piston by multiplying the area (A) of the piston by the pressure (P) exerted by the fluid. This is expressed in the formula F = A × P. For example, when 800 psi of fluid pressure is supplied to a cylinder with a piston area of 10 square inches, 8,000 pounds of force is generated. To determine

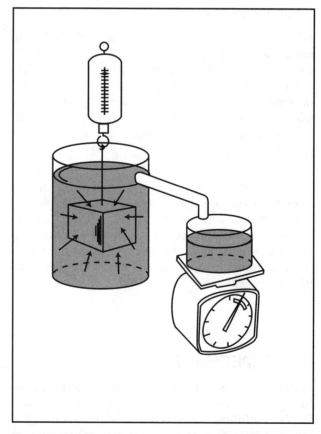

Figure 2-44. A body immersed in a fluid is buoyed up by a force equal to the weight of the fluid it displaces.

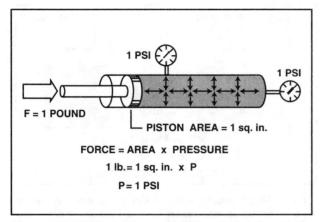

Figure 2-45. The pressure produced in a hydraulic cylinder acts at right angles to the cylinder.

the area needed to produce a given amount of pressure, divide the force produced by the pressure applied. [Figure 2-46]

Since the shape of a container has no effect on pressure, connecting one cylinder to a larger cylinder results in a gain in mechanical advantage. For example, a cylinder with a 1 square inch piston is connected to a cylinder with a 10 square inch piston. When 1 pound of force is applied to the smaller piston, the resulting pressure inside both cylinders is 1 psi. This means that the piston in the larger cylinder has an area of 10 square inches, and one pound of pressure acts on every square inch of the piston, the resulting force applied to the larger piston is 10 pounds. [Figure 2-47]

When gaining mechanical advantage this way it is important to note that the pistons do not move the same distance. In the previous example, when the small piston moves inward 5 inches, it displaces 5 cubic inches of fluid. When this is spread out over the 10 square inches of the larger piston, the larger piston only moves one-half inch. [Figure 2-48]

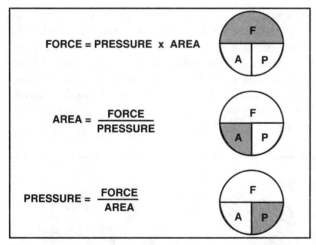

Figure 2-46. The relationship between force, pressure, and area.

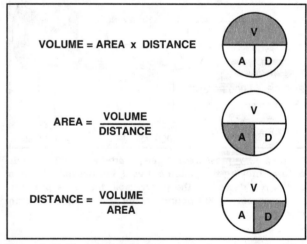

Figure 2-48. The relationship between volume, area, and distance.

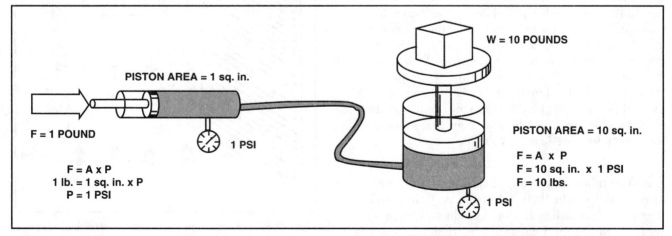

Figure 2-47. A mechanical advantage may be obtained in a hydraulic system by using a piston with a small area to force fluid into a cylinder with a larger piston. For example, when applying a force of 1 pound to a 1 square inch piston, you push upward against the 10 square inch piston with a force of 10 pounds.

DIFFERENTIAL AREAS

As stated earlier, the amount of force produced by a piston is calculated by multiplying the area of the piston by the pressure applied to the fluid. However, consider a piston that is subjected to pressure on both ends. For example, a cylinder has a piston with a surface area of four square inches on one side, while the other side is connected to a one square inch rod. When 1,000 psi of pressure is applied to the side of the piston without the rod, 4,000 pounds of force is produced (4 sq. in. × 1,000 psi = 4,000 lbs.). However, when the same amount of pressure is applied to the opposite side, the fluid acts on only three square inches. Therefore, only 3,000 pounds of force is produced (3 sq. in. × 1,000 psi = 3,000 lbs.). [Figure 2-49]

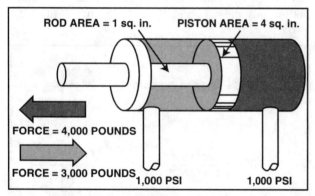

Figure 2-49. A pressure of 1,000 psi produces 4,000 pounds of force moving the piston outward, but because the area of the rod decreases the piston area, the same pressure produces only 3,000 pounds of force moving the piston inward.

BERNOULLI'S PRINCIPLE

The Swiss mathematician and physicist Daniel Bernoulli developed a principle that explains the relationship between potential and kinetic energy in a fluid. As discussed earlier, all matter contains potential energy and/or kinetic energy. In a fluid, the potential energy is that caused by the pressure of the fluid, while the kinetic energy is that caused by the fluid's movement. Although you cannot create or destroy energy, it is possible to exchange potential energy for kinetic energy or vice versa.

A **venturi tube** is a specially shaped tube that is narrower in the middle than at the ends. As fluid enters the tube, it is traveling at a known velocity and pressure. When the fluid enters the restriction, it must speed up, or increase its kinetic energy. However, when the kinetic energy increases, the potential

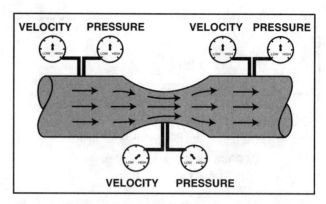

Figure 2-50. Bernoulli's principle states that when energy is neither added to nor taken from a fluid in motion, the potential energy, or pressure decreases when the kinetic energy, or velocity increases.

energy decreases. Then, as the fluid continues through the tube, both velocity and pressure return to their original values. [Figure 2-50]

Bernoulli's principle is used in a carburetor to draw gasoline from the float bowl. The fuel discharge nozzle is in the narrowest part of the venturi, and as air is drawn through the venturi, the air pressure past the nozzle decreases. Once this occurs, the atmospheric pressure in the float bowl becomes strong enough to force fuel out of the nozzle. [Figure 2-51]

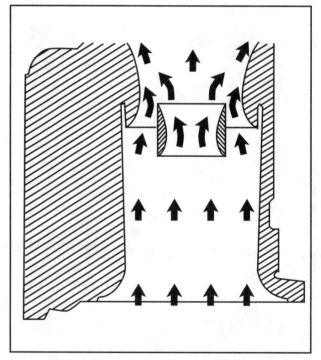

Figure 2-51. The venturi in a carburetor is designed to speed up the flow of air entering the engine. This increase in velocity lowers the pressure so fuel can be forced out of the carburetor and into the engine.

SOUND

In the study of physics as it applies to practical aviation maintenance, we are interested in sound. Not so much for what we hear, but as a study of the way it travels either through the air or through a solid medium.

SOUND INTENSITY

Sound intensity is determined by the amplitude of the sound wave. The higher the amplitude, the louder the sound. Fortunately, the human ear does not register changes in sound in a uniform, or linear scale. Instead, we hear in a logarithmic fashion, and we are able to detect faint changes in low-level sound; such as, the ticking of a watch, but not tremendous differences in sound energy when the intensity is very high, like when the afterburner on a jet fighter lights up.

Sound intensity is measured in **decibels** (db). A decibel is the ratio of one sound to another. One db is the smallest change in sound intensity the human

ear can detect. Approximately 60 db is a comfortable level for conversation and background music. On the other hand, loud music peaks at more than 100 db. Sound that reaches approximately 130 db can cause pain. [Figure 2-52]

FREQUENCY

Frequency is the number of vibrations completed per second. It is expressed in cycles per second, or hertz. The frequency of a sound determines its pitch. The higher the frequency, the higher the pitch.

SOUND PROPAGATION

There are three things necessary for the transmission and reception of sound. They are a source, a medium for carrying the sound, and a detector. Sound is any vibration of an elastic medium in the frequency range that affects our auditory senses. This range varies among different types of animals and even among different humans. However, the

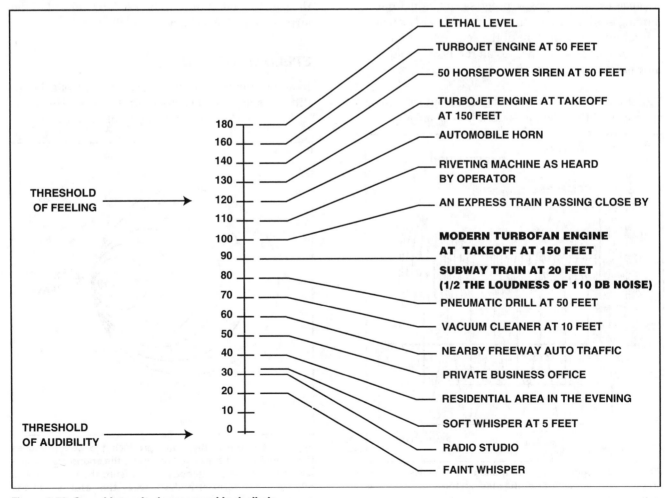

Figure 2-52. Sound intensity is measured in decibels.

range is generally considered to be between about 20 and 20,000 hertz (Hz).

Sound waves differ from water waves in that they are made of a series of compressions and expansions or rarefactions. When the tine of the tuning fork vibrates, it moves back and forth away from the fork. With each movement in one direction the surrounding air compresses and expands, setting up a series of pressure changes. These pressure changes, while extremely slight, are enough to vibrate the eardrum. [Figure 2-53]

DOPPLER EFFECT

When a source of sound moves, the frequency ahead of the source is higher than the frequency behind it. This change in frequency is called the Doppler effect and it accounts for the difference in sound as an airplane passes over you. As an airplane approaches, the sound becomes both louder and more highly pitched. Then, as it passes, the intensity drops off and the frequency decreases. This is because as sound waves progress outward from a moving source, the source moves forward causing the sound waves in front of the source to compress. Conversely, the sound waves aft rarefy as the source moves away. [Figure 2-54]

Vibrations cause waves which radiate out from their source at the speed of sound. When an airplane

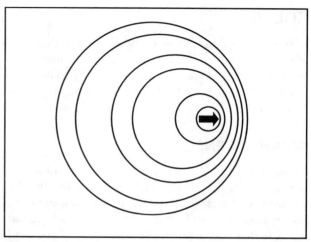

Figure 2-54. When a body that is producing sound moves through the air, the sound waves progress out ahead of the body at a higher frequency.

moves through the air it creates vibrations, or sound waves which move outward. If an airplane flies slower than the speed of sound, the sound waves move out ahead of the airplane. However, if an airplane flies at the speed of sound, the sound waves cannot travel ahead of the aircraft. In this situation, the sound waves build up immediately ahead of the airplane forming a **shock wave**. [Figure 2-55]

SPEED OF SOUND

In any uniform medium, under given physical conditions, sound travels at a definite speed. However, in some substances, the velocity of sound is higher

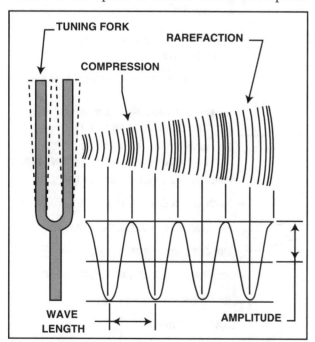

Figure 2-53. Sound waves are made up of a series of compression and rarefactions of the medium through which the sound passes.

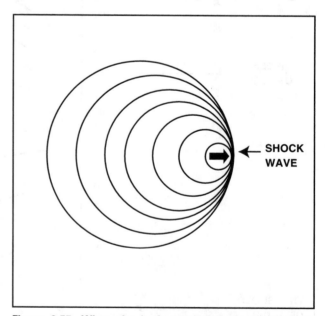

Figure 2-55. When the body producing a sound moves through the air at the speed of sound, the sound waves cannot move ahead of the body. As a result, the sound waves pile up in the form of a shock wave that attaches to the body.

than in others. As a general rule, the denser the medium, the faster sound travels. This is why sound travels much faster in water than in air.

The speed of sound varies directly with the temperature of the air. For example, at sea level, in standard atmospheric conditions, sound travels at 661.2 knots, or about 340.4 meters per second. However, as altitude increases, the temperature decreases causing a decrease in the speed of sound. This decrease continues until slightly above 36,000 feet. There, the temperature stabilizes at approximately −56.5°C, and the speed of sound remains at 573.3 knots. Around 67,000 feet the temperature begins to increase and therefore, the speed of sound begins to increase slightly.

MACH NUMBER

When studying aircraft that fly at supersonic speeds, it is customary to represent aircraft speed in relation to the speed of sound. The term Mach number is used to represent the ratio of the speed of the airplane to the speed of sound in the same atmospheric conditions. When the airplane travels at exactly the speed of sound, it is said to be traveling at Mach one. Two times the speed of sound is Mach two. If the airplane is traveling at three-fourths the speed of sound, it is traveling at Mach .75

RESONANCE

The natural frequency of a given object is the frequency where that object vibrates naturally, or without an external force. If two objects have the same natural frequency and are set next to each other, when one of them vibrates, it can transfer its wave energy to the other object making it vibrate. This transfer of energy is known as resonance.

Because resonance induces vibration, it can exert destructive forces on an aircraft. For example, it is possible to have portions of an aircraft, such as the propeller, vibrate in resonance at certain engine speeds. If the vibrations are strong enough they can create stresses in the aircraft that could lead to structural failure.

SECTION D

AERODYNAMICS

Airplanes and helicopters are able to fly only when they generate an amount of lift that equals their own weight. To understand exactly how lift is produced you must first be familiar with the air itself.

THE ATMOSPHERE

The earth's atmosphere is composed of about 78 percent nitrogen and 21 percent oxygen. The remaining 1 percent is made up of several other gases, primarily argon and carbon dioxide. The percentage of these gases remain constant regardless of their altitude. The atmosphere also contains some water vapor, but the amount varies from almost zero to about five percent by volume.

In order to have a reference for all aerodynamic computations, the International Civil Aeronautics Organization (ICAO) has agreed upon a **standard atmosphere**. The pressures, temperatures, and densities in a standard atmosphere serve as a reference only. When all aerodynamic computations are related to this standard, a meaningful comparison of flight test data between aircraft can be made.

Standard atmospheric conditions are based on a sea level temperature of 15°C or 59°F, and a barometric pressure of 29.92 inches of mercury or 1013.2 millibars. Pressure and temperature normally decrease with increases in altitude. The standard pressure lapse rate for each 1,000 feet of altitude change is approximately 1.00 inch of mercury, whereas the standard temperature lapse rate per 1,000-foot change in altitude is 2°C or 3.5°F.

PRESSURE

Everything on the earth's surface is under pressure due to the weight of the atmosphere. The amount of pressure applied equals the weight of a column of air one square inch in cross sectional area that extends to the top of the atmosphere. There are three systems commonly used in aviation for measuring pressure. They are pounds per square inch, inches of mercury, and millibars.

POUNDS PER SQUARE INCH (PSI)

The force in pounds that the air exerts on each square inch of area is expressed in pounds per square inch or psi. At sea level, the air exerts a pressure of 14.69 psi. Approximately one-half of the air in the atmosphere is below 18,000 feet. Therefore, the pressure at 18,000 feet is about 7.34 psi, or half of that at sea level. Most pressures expressed in pounds per square inch generally do not account for atmospheric pressure and, therefore, indicate the pressure in excess of atmospheric pressure. These pressures are generally read from a gauge and are referred to as a gauge pressure (psig). Fuel and oil pressure are both measured in gauge pressure and indicate the amount a pump raises the pressure of the liquid above atmospheric pressure.

INCHES OF MERCURY

When a tube is filled with mercury and inverted in a bowl, the mercury drops in the tube until the atmospheric pressure exerted on the mercury in the bowl equals the weight of the mercury in the tube. Standard atmospheric pressure can support a column of mercury that is 29.92 inches tall. [Figure 2-56]

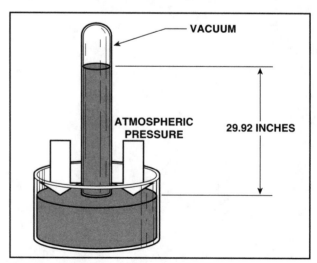

Figure 2-56. Under standard atmospheric conditions, the atmosphere can support a column of mercury 29.92 inches tall.

Figure 2-57. An altimeter measures absolute pressure and displays the result in feet above sea level.

Pressure that is referenced from zero pressure is called **absolute pressure** and is usually measured in inches of mercury. Absolute pressure in the induction system of a piston engine forces the fuel-air mixture into the cylinders. Thus, the pilot reads absolute pressure in inches of mercury on the manifold pressure gauge.

Another instrument that displays absolute pressure is an altimeter. However, the scale is marked in feet rather than in units of pressure. When the barometric scale in an altimeter is set to the current barometric pressure, the altimeter displays its height above sea level. The barometric scale used in most altimeters in the United States is calibrated in inches of mercury. [Figure 2-57]

The metric unit of measure for barometric pressure is millibars. One **millibar** is approximately equivalent to .0295 in. Hg., and, therefore, standard sea level pressure is equivalent to 1013.2 millibars. Some altimeters have their barometric scale calibrated in millibars.

TEMPERATURE

As it relates to the study of aerodynamics, there are two temperature scales you must be familiar with. They are the Fahrenheit scale and the Celsius scale. The Celsius scale has 100 divisions between the freezing and boiling points of pure water. Water freezes at zero degrees C and boils at 100 degrees C. The Fahrenheit scale is also based on the freezing and boiling points of water. However, water freezes at 32 degrees F and boils at 212 degrees F. The standard temperature for all aerodynamic computations is 15°C or 59°F.

AIR DENSITY

By use of the general gas laws studied earlier, you can derive that for a particular gas, pressure and temperature determine density. Since standard pressures and temperatures are associated with each altitude, the density of the air at these standard temperatures and pressures is also considered standard. Therefore, a particular atmospheric density is associated with each altitude. This gives rise to the expression **density altitude**. A density altitude of 10,000 feet is the altitude at which the density is the same as that considered standard for 10,000 feet. However, density altitude is not always the same as true altitude. For example, on a day when the atmospheric pressure is higher than standard and the temperature is lower than standard, the standard air density at 10,000 feet might occur at 12,000 feet. In this case, at a true altitude of 12,000 feet, the air has the same density as standard air at 10,000 feet. Therefore, density altitude is a calculated altitude obtained by correcting pressure altitude for temperature.

The water content of the air has a slight effect on the density of the air. It should be remembered that humid air at a given temperature and pressure is lighter than dry air at the same temperature and pressure.

HUMIDITY

As you know, the air is seldom completely dry. It contains water vapor in the form of fog or water vapor. Fog consists of minute droplets of water held in suspension by the air. Clouds are composed of fog.

As a result of evaporation, the atmosphere always contains some moisture in the form of water vapor. The moisture in the air is often referred to as humidity.

Fog and humidity both affect the performance of an aircraft. For example, since humid air is less dense than dry air, the allowable takeoff gross weight of an aircraft is generally reduced when operating in humid conditions. Secondly, since water vapor is incombustible, its presence in the atmosphere results in a loss of engine power output. The reason for this is that as the mixture of water vapor and air is drawn through the carburetor, fuel is metered into it as though it were all air. Since water vapor does not burn, the effective fuel/air ratio is enriched and the engine operates as though it were on an excessively rich mixture. The resulting horsepower loss under humid conditions is therefore attributed to the loss in volumetric efficiency due to displaced air, and the incomplete combustion due to an excessively rich fuel/air mixture. The power loss on a piston engine can be as high as 12 percent, while a turbine seldom loses more than two or three percent.

Absolute Humidity

Absolute humidity refers to the actual amount of water vapor in a mixture of air and water. The amount of water vapor the air can hold varies with air temperature. The higher the air temperature the more water vapor the air can hold.

Relative Humidity

Relative humidity is the ratio between the amount of moisture in the air to the amount that would be present if the air were saturated. For example, a relative humidity of 75 percent means that the air is holding 75 percent of the total water vapor it is capable of holding. Relative humidity has a dramatic effect on airplane performance because of its effect on air density. In equal volumes, water vapor weighs 62 percent as much as air. This means that in high humidity conditions the density of the air is less than that of dry air.

Dewpoint

Dewpoint is the temperature at which air reaches a state where it can hold no more water. When the dewpoint is reached, the air contains 100 percent of the moisture it can hold at that temperature, and is said to be **saturated**. If the temperature drops below the dewpoint, condensation occurs.

AIRPLANES

One of the first things you are likely to notice when you begin working on aircraft is the wide variety of airplane styles and designs. Although, at first glance, you may think that airplanes look quite different from one another, you will find that their major components are quite similar.

THE FUSELAGE

The fuselage serves several functions. Besides being a common attachment point for the other major components, it houses a cockpit where a flight crew operates the aircraft, and a cabin area for passengers or cargo. If the aircraft is meant to carry passengers, the cabin is typically equipped with seats, galleys,

and lavatories. In this configuration, cargo and baggage are placed in a dedicated cargo area or pit in the lower part of the fuselage.

THE WING

When air flows around the wings of an airplane, it generates a force called "lift" that helps the airplane fly. Wings are contoured to take maximum advantage of this force.

The wings have two types of control surfaces attached to the rear, or trailing, edges. They are referred to as ailerons and flaps. **Ailerons** extend from about the midpoint of each wing outward to the tip. They move in opposite directions — when one aileron goes up, the other goes down. **Flaps** extend outward from the fuselage to the midpoint of each wing. They always move in the same direction. If one flap is down, the other is down. [Figure 2-58]

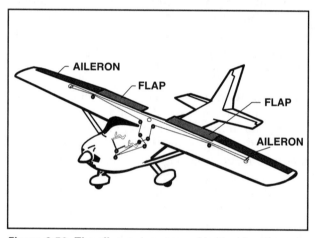

Figure 2-58. The ailerons are moved by turning the control wheel in the cockpit. When the wheel is turned to the left, the left aileron moves up and the right moves down. Turning the wheel to the right has the opposite effect. The flaps operate using a switch or handle located in the cockpit.

Some larger aircraft employ **spoilers** on the top of the wing to disrupt the airflow and reduce lift. Spoilers are used both as a flight control to aid the ailerons in rolling the aircraft, or as air brakes to slow the aircraft during landing and rollout.

THE EMPENNAGE

The empennage consists of the **vertical stabilizer**, or fin, and the horizontal stabilizer. These two surfaces are stationary and act like the feathers on an arrow to steady the airplane and help you maintain a straight path through the air. [Figure 2-59]

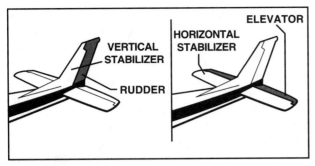

Figure 2-59. Besides the two fixed components, the empennage has two important movable surfaces called the rudder and the elevator.

The **rudder** is attached to the back of the vertical stabilizer to move the airplane's nose left and right. Actually, the rudder is used in combination with the ailerons during flight to initiate a turn.

The **elevator** is attached to the back of the horizontal stabilizer. During flight, the elevator is used to move the nose up and down to direct the airplane to the desired altitude, or height above the ground.

STABILATOR

Some empennage designs vary from the type of horizontal stabilizer just discussed. They have a one-piece horizontal stabilizer that pivots up and down from a central hinge point. This type of design, called a stabilator, requires no elevator. The stabilator is moved using the control wheel, just as you would the elevator. An **antiservo tab** is mounted at the back of the stabilator, to provide a control "feel" similar to what you experience with an elevator. Without the antiservo tab, control forces from the stabilator would be very light and a pilot could "over control" the airplane. [Figure 2-60]

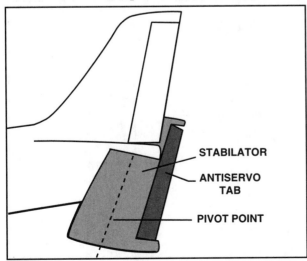

Figure 2-60. The stabilator pivots up and down as the control wheel is moved. The antiservo tab moves in the same direction as the trailing edge of the stabilator.

Some aircraft merge two flight controls into one system. For example, some Beechcraft Bonanza models have **ruddervators** that combine rudders and elevators into one control surface. Depending on control inputs, the ruddervators move in the same direction for pitch control, and in opposite directions for yaw control. Still other aircraft combine spoilers and ailerons into **spoilerons** to provide roll. Furthermore, some delta-wing aircraft merge elevator and aileron functions through control surfaces called **elevons**.

TRIM CONTROLS

Trim tabs are small movable portions of the trailing edge of the control surface. These movable trim tabs are controlled from the cockpit and alter the control surface camber to create an aerodynamic force that deflects the control surface. Trim tabs can be installed on any of the primary control surfaces. If only one tab is used, it is normally on the elevator. This is used to adjust the tail load so the airplane can be flown hands-off at any given airspeed. [Figure 2-61]

A **fixed trim tab** is normally a piece of sheet metal attached to the trailing edge of a control surface. This fixed tab is adjusted on the ground by bending it in the appropriate direction to eliminate flight control forces for a specific flight condition. The fixed tab is normally adjusted for zero-control forces in cruise flight. Adjustment of the tab is a trial-and-error process where the aircraft must be flown and the trim tab adjusted based on the pilot's report. The aircraft must then be flown again to see if further adjustment is necessary. Fixed tabs, normally found on light aircraft, are used to adjust rudders and ailerons.

Large aircraft obviously require larger control surfaces, and more force is required to move them. To assist in moving these larger surfaces, control tabs are mounted on the trailing edges of the control surfaces. For example, a **servo tab** is connected directly

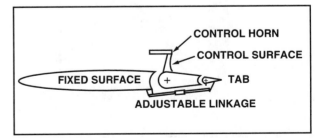

Figure 2-61. Trim tabs alter the camber of a control surface to create an aerodynamic force that deflects the control surface.

to an aircraft's flight controls, and moves in a direction opposite the desired control surface movement. As the tab deflects into the airstream, it forces the control surface in the opposite direction. **Balance tabs** receive no control inputs, but are connected to a main structure. When a control is activated, the balance tab deflects into the airstream and helps move the control surface.

FOUR FORCES OF FLIGHT

During flight, there are four forces acting on an airplane. They are lift, weight, thrust, and drag. **Lift** is the upward force created by the effect of airflow as it passes over and under the wings. It supports the airplane in flight. **Weight** opposes lift. It is caused by the downward pull of gravity. **Thrust** is the for-

ward force which propels the airplane through the air. It varies with the amount of engine power being used. Opposing thrust is **drag**, which is a backward, or retarding, force that limits the speed of the airplane. [Figure 2-62]

The arrows which show the forces acting on an airplane are often called **vectors**. The magnitude of a vector is indicated by the arrow's length, while the direction is shown by the arrow's orientation. When two or more forces act on an object at the same time, they combine to create a resultant. [Figure 2-63]

LIFT

Lift is the key aerodynamic force. It is the force that opposes weight. In straight-and-level, unaccelerated flight, when weight and lift are equal, an airplane is in a state of equilibrium. If the other aerodynamic factors remain constant, the airplane neither gains nor loses altitude.

During flight, the pressures on the upper and lower surfaces of the wing are not the same. Although several factors contribute to this difference, the shape of the wing is the principle one. The wing is designed to divide the airflow into areas of high pressure below the wing and areas of comparatively lower pressure above the wing. This pressure differential, which is created by movement of air about the wing, is the primary source of lift.

BERNOULLI'S PRINCIPLE

The basic principle of pressure differential of subsonic airflow was discovered by Daniel Bernoulli, a

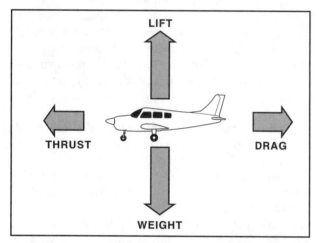

Figure 2-62. The four aerodynamic forces are in equilibrium during straight-and-level, unaccelerated flight. Lift is equal to and directly opposite weight and thrust is equal to and directly opposite drag.

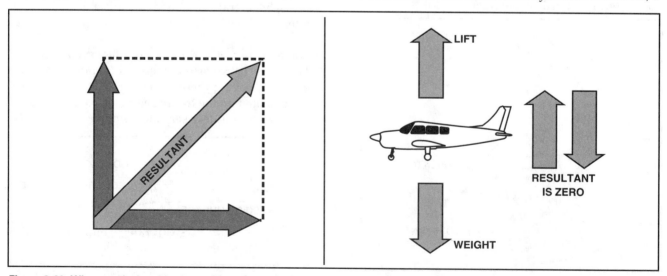

Figure 2-63. When vertical and horizontal forces are applied, as shown on the left, the resultant acts in a diagonal direction. When two opposing vertical forces are applied, as shown on the right, they tend to counteract one another. If the forces are equal in magnitude, the resultant is zero.

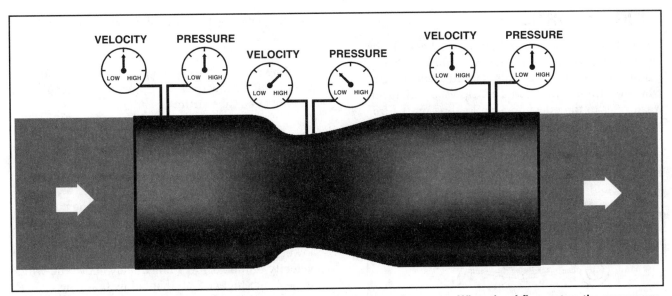

Figure 2-64. As the air enters the tube, it is traveling at a known velocity and pressure. When the airflow enters the narrow portion, the velocity increases and the pressure decreases. Then, as the airflow continues through the tube to the wider portion, both the velocity and pressure return to their original values.

Swiss physicist. **Bernoulli's Principle**, simply stated, says, "as the velocity of a fluid (air) increases, its internal pressure decreases."

One way you can visualize this principle is to imagine air flowing through a tube that is narrower in the middle than at the ends. This type of device is usually called a **venturi**. [Figure 2-64]

It is not necessary for air to pass through an enclosed tube for Bernoulli's Principle to apply. Any surface that alters airflow causes a venturi effect. For example, if the upper portion of the tube is removed, the venturi effect applies to air flowing along the lower section of the tube. Velocity above the curvature is increased and pressure is decreased. You can begin to see how the venturi effect works on a wing, or airfoil, if you picture an airfoil inset in the curved part of the tube. [Figure 2-65]

Air flowing over the top of an airfoil reaches the trailing edge in the same amount of time as air flow-

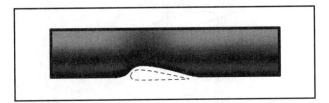

Figure 2-65. An airfoil changes the relationship between air pressure and velocity in the same way it is changed in a venturi.

ing along the relatively flat bottom. Since both the upper and lower surfaces pass through a block of air at the same speed, the air flowing over the curved upper surface travels farther. This means it must go faster, resulting in lower pressure above the airfoil and a greater pressure below. An airfoil is specially designed to produce a reaction with the air that passes over it. This difference in pressure is the main source of lift.

NEWTON'S THIRD LAW OF MOTION

The remaining lift is provided by the wing's lower surface as air striking the underside is deflected downward. According to Newton's Third Law of Motion, "for every action there is an equal and opposite reaction." Therefore, the air that is deflected downward by the wing also produces an upward (lifting) reaction.

Since air is much like water, the explanation for this source of lift may be compared to the planing effect of skis on water. The lift which supports the water skis (and the skier) is the force caused by the impact pressure and the deflection of water from the lower surfaces of the skis.

Under most flying conditions, the impact pressure and the deflection of air from the lower surface of the wing provide a comparatively small percentage of the total lift. The majority of lift is the result of the decreased pressure above the wing rather than the increased pressure below it.

WINGTIP VORTICES

Wingtip vortices are caused by the air beneath the wing rolling up and around the wingtip. This causes a spiral or vortex that trails behind each wingtip whenever lift if being produced. Upwash and downwash refer to the effect an airfoil exerts on the free airstream. **Upwash** is the deflection of the oncoming airstream upward and over the wing. **Downwash** is the downward deflection of the airstream as it passes over the wing and past the trailing edge. [Figure 2-66]

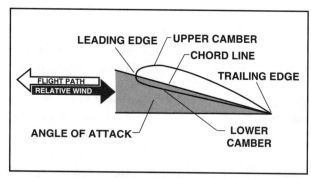

Figure 2-66. Whenever an airplane generates lift, air spills over the wingtips from the high pressure areas below the wings to the low pressure areas above them. This flow causes rapidly rotating whirlpools of air called wingtip vortices, or wake turbulence. The intensity of the turbulence depends on aircraft weight, speed, and configuration.

THE AIRFOIL

An airfoil is any surface, such as a wing or propeller, that provides aerodynamic force when it interacts with a moving stream of air. Remember, an airplane's wing generates a lifting force only when air is in motion about it. Some of the terms used to describe the wing, and the interaction of the airflow about it, are listed here. [Figure 2-67]

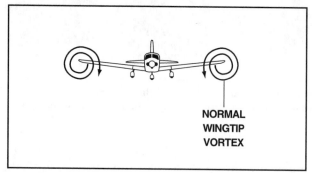

Figure 2-67. This figure shows a cross-section of a wing in straight-and-level flight. Chord and camber are terms which help to define the wing's shape, while flight path and relative wind help define the movement of the wing with respect to the surrounding air. Angle of attack is determined by the wing's chord line and the relative wind.

Leading edge — This part of the airfoil meets the airflow first.

Trailing edge — This is the portion of the airfoil where the airflow over the upper surface rejoins the lower surface airflow.

Chord line — The chord line is an imaginary straight line drawn through an airfoil from the leading edge to the trailing edge.

Camber — The camber of an airfoil is the characteristic curve of its upper and lower surfaces. The upper camber is more pronounced, while the lower camber is comparatively flat. This causes the velocity of the airflow immediately above the wing to be much higher than that below the wing.

Relative wind — This is the direction of the airflow with respect to the wing. If a wing moves forward horizontally, the relative wind moves backward horizontally. Relative wind is parallel to and opposite the flight path of the airplane.

Angle of attack — This is the angle between the chord line of the airfoil and the direction of the relative wind. It is important in the production of lift. The angle formed by the wing chord line and relative wind is called the angle of attack. [Figure 2-68]

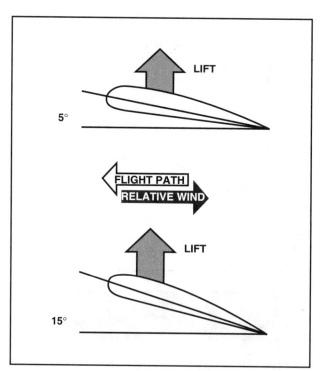

Figure 2-68. As the angle of attack increases, lift also increases. Notice that lift acts perpendicular to the relative wind, regardless of angle of attack.

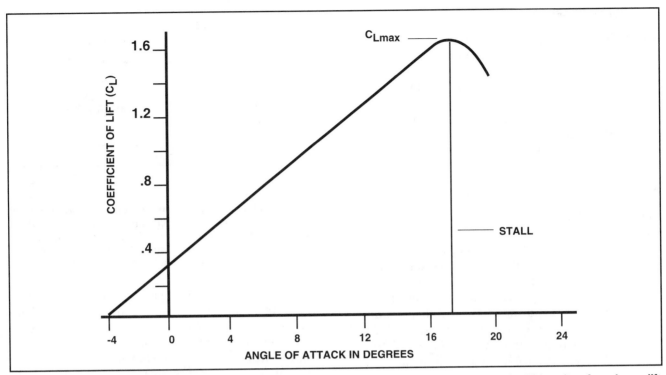

Figure 2-69. As angle of attack increases, C_L also increases. This continues to a point where C_L peaks. This point of maximum lift is called C_{Lmax}. In this example, C_{Lmax} occurs at about 17°. If the maximum lift angle is exceeded, lift decreases rapidly and the wing stalls.

CHANGING ANGLE OF ATTACK

Pilots have direct control over angle of attack. During flight at normal operating speeds, if a pilot increases the angle of attack, lift increases. The angle of attack is changed anytime the control column is moved fore or aft during flight. At the same time, the coefficient of lift is changed. The coefficient of lift (C_L) is a way to measure lift as it relates to angle of attack. C_L is determined by wind tunnel tests and is based on airfoil design and angle of attack. Every airplane has an angle of attack where maximum lift occurs. [Figure 2-69]

A **stall** is caused by the separation of airflow from the wing's upper surface. This results in a rapid decrease in lift. For a given airplane, a stall always occurs at the same angle of attack, regardless of airspeed, flight attitude, or weight. This is the stalling or critical angle of attack. It is important to remember that an airplane can stall at any airspeed, in any flight attitude, or at any weight. [Figure 2-70]

To recover from a stall, smooth airflow must be restored. The only way to do this is to decrease the angle of attack to a point below the stalling or critical angle of attack.

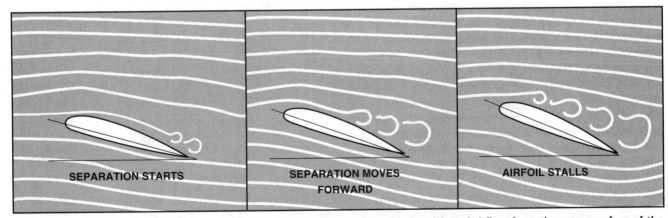

Figure 2-70. Increasing the angle of attack beyond C_{Lmax} causes progressive disruption of airflow from the upper surface of the wing. At first, the airflow begins to separate at the trailing edge. As angle of attack is further increased, the airflow separation progresses forward until the wing is fully stalled.

CHANGING AIRSPEED

The faster the wing moves through the air, the greater the lift. Actually, lift is proportional to the square of the airplane's speed. For example, at 200 knots, an airplane has four times the lift of the same airplane traveling at 100 knots if the angle of attack and other factors are constant. On the other hand, if the speed is reduced by one-half, lift is decreased to one-quarter of the previous value.

ANGLE OF ATTACK AND AIRSPEED

The relationship between angle of attack and airspeed in the production of lift is not as complex as it may seem. Angle of attack establishes the coefficient of lift for the airfoil. At the same time, lift is proportional to the square of the airplane's speed. Therefore, total lift depends on the combined effects of airspeed and angle of attack. In other words, when speed decreases, the angle of attack must increase to maintain the same amount of lift. Conversely, if the same amount of lift is maintained at a higher speed, the angle of attack must decrease.

USING FLAPS

When properly used, flaps increase the lifting efficiency of the wing and decrease stall speed. This allows the aircraft to fly at a reduced speed while maintaining sufficient control and lift for sustained flight.

The ability to fly slowly is particularly important during the approach and landing phases. For example, an approach with full flaps allows the aircraft to fly slowly and at a fairly steep descent angle without gaining airspeed. This allows for touch down at a slower speed.

In airplanes, **configuration** normally refers to the position of the landing gear and flaps. When the gear and flaps are up, an airplane is in a clean configuration. If the gear is fixed rather than retractable, the airplane is considered to be in a clean configuration when the flaps are in the up position. Flap position affects the chord line and angle of attack for that section of the wing where the flaps are attached. This causes an increase in camber for that section of the wing and greater production of lift and drag. [Figure 2-71]

There are several common types of flaps. The **plain flap** is attached to the wing by a hinge. When deflected downward, it increases the effective camber and changes the wing's chord line. Both of these

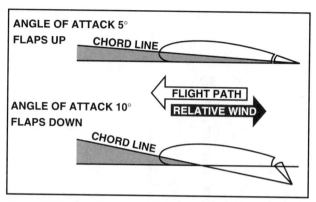

Figure 2-71. Flaps increase lift (and drag) by increasing the wing's effective camber and changing the chord line, which increases the angle of attack. In some cases, flaps also increase the area of the wing. Most flaps, when fully extended, form an angle of 35° to 40° relative to the wing.

factors increase the lifting capacity of the wing. The **split flap** is hinged only to the lower portion of the wing. This type of flap also increases lift, but it produces greater drag than the plain flap because of the turbulence it causes. The **slotted flap** is similar to the plain flap. In addition to changing the wing's camber and chord line, it also allows a portion of the higher pressure air beneath the wing to travel through a slot. This increases the velocity of the airflow over the flap and provides additional lift. Another type of flap is the **Fowler flap**. It is attached to the wing by a track and roller system. When extended, it moves rearward as well as down. This rearward motion increases the total wing area, as well as the camber and chord line. The Fowler flap is the most efficient of these systems. As you might expect, it also is the most expensive. [Figure 2-72]

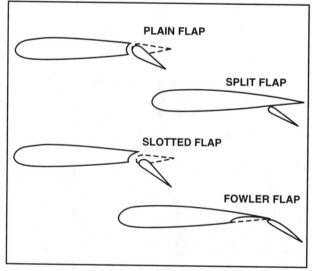

Figure 2-72. Flap types include plain, split, slotted, and Fowler. Although some flaps increase the wing's surface area, most change only the effective camber and the chord line.

LEADING EDGE DEVICES

A stall occurs when the angle of attack becomes so great that the energy in the air flowing over the wing can no longer pull air down to the surface. The boundary layer thickens and becomes turbulent, and airflow separates from the surface. This separation can be delayed until a higher angle of attack by increasing the energy of air flowing over the surface. One way to do this is by installing a **slot** in the leading edge of the wing. This slot is simply a duct for air to flow from below the wing to the top. Once there, it is directed over the surface in a high-velocity stream. Slots are typically placed ahead of the aileron to keep the outer portion of the wing flying after the root has stalled. This maintains aileron effectiveness and provides lateral control during most of the stall. [Figure 2-73]

Many high-performance aircraft utilize **slats** which are mounted on the leading edge on tracks. These extend outward and create a duct to direct high-energy air down over the surface. This delays separation until a very high angle of attack.

In many aircraft these slats are actuated by aerodynamic forces and are entirely automatic in their operation. As the angle of attack increases, the low pressure just behind the leading edge on top of the wing increases. This pulls the slat out of the wing. When the slat moves out, it ducts the air from the high-pressure area below the wing to the upper surface and increases the velocity of air in the boundary layer. When the angle of attack is lowered, air pressure on the slat moves it back into the wing where it has no effect on airflow.

Some aircraft have slats operated by either hydraulic or electric actuators. They are lowered

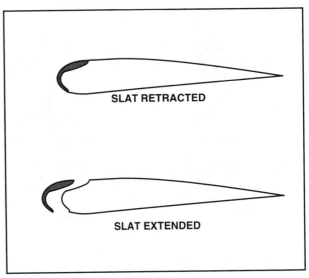

Figure 2-74. A movable slat moves out of the leading edge of the wing at high angles of attack to form a duct for high-energy air.

when the trailing edge flaps are lowered to prevent airflow from breaking away from the upper surface. Flaps used with slats are slotted and they duct high-energy air over the deflected flap sections so the air will not break away over their surface. [Figure 2-74]

On some aircraft, the leading edge of the wing deflects downward to increase camber. These **leading edge flaps** are electrically or hydraulically actuated and are used in conjunction with the trailing edge flaps.

The swept wings of large turbine-engined transports develop little lift at low speeds. To remedy this problem, leading edge devices called **Krueger flaps** are used to effectively increase a wing's camber and hence its lift. A Krueger flap is hinged to a wing's leading edge and lays flush to its lower surface when stowed. When the flap is deployed, it extends down and forward to alter the wing profile.

AIRFOIL DESIGN FACTORS

Wing design is based on the anticipated use of the airplane, cost, and other factors. The main design considerations are wing planform, camber, aspect ratio, and total wing area.

Planform refers to the shape of the airplane's wing when viewed from above or below. Each planform design has advantages and disadvantages. The rectangular wing (straight wing) is used on most light aircraft because it has a tendency to stall first at the

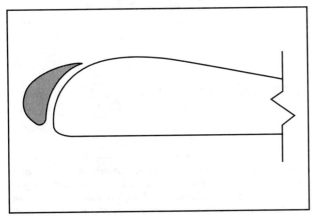

Figure 2-73. A fixed slot ducts air over the top of the wing at high angles of attack.

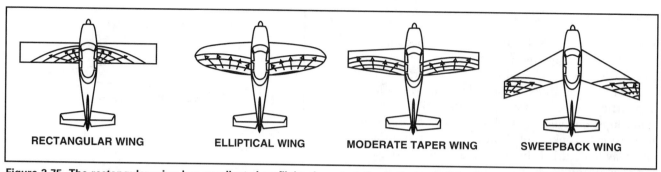

RECTANGULAR WING   ELLIPTICAL WING   MODERATE TAPER WING   SWEEPBACK WING

Figure 2-75. The rectangular wing has excellent slow-flight characteristics in that a stall begins at the wing root providing adequate stall warning and aileron effectiveness throughout the stall. The elliptical and tapered wings, on the other hand, typically stall along the entire trailing edge providing little stall warning. The sweptback and delta wings used on higher performance aircraft are efficient at high speeds, but not at low speeds.

root, providing adequate warning and aileron effectiveness. [Figure 2-75]

It is important that a wing stall at the root first so the ailerons are able to provide lateral control throughout a stall. If a wing does not have this characteristic naturally, it can be obtained by installing small triangular **stall strips** at the root of the wing's leading edge. When the angle of attack is increased and a stall occurs, the strips disturb the air enough to hasten the stall on the section of wing behind them. This loss of lift causes the nose of the airplane to drop while the outer portion of the wing is still flying and the ailerons are still effective. If a stall strip is missing from a wing's leading edge, the aircraft will consequently have asymmetrical aileron control at or near stall angles of attack. In other words, as the aircraft approaches a stall, the wing lacking its stall strip loses its aileron effectiveness first. As a result, the aircraft is left with uneven aileron control.

Camber, as noted earlier, affects the difference in the velocity of the airflow between the upper and lower surfaces of the wing. If the upper camber increases and the lower camber remains the same, the wing is said to be **asymmetrical** and the velocity differential increases. **Symmetrical** wings, on the other hand, have the same curve on the top and bottom of the wing. This type of wing relies on a positive angle of attack to generate lift and is generally used on high performance, aerobatic aircraft. [Figure 2-76]

Aspect ratio is the relationship between the wing span to the average chord, or mean chord. It is one of the primary factors in determining the three dimensional characteristics of a wing and its lift/drag characteristics. An increase in aspect ratio at a given velocity results in a decrease in drag, especially at high angles of attack, and a lower stalling speed. [Figure 2-77]

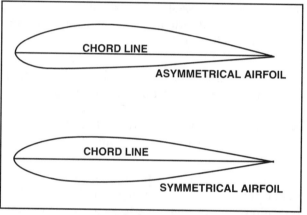

CHORD LINE

ASYMMETRICAL AIRFOIL

CHORD LINE

SYMMETRICAL AIRFOIL

Figure 2-76. The asymmetrical wing has a greater camber across the wing's upper surface than the lower surface. This type of wing is primarily used on low speed aircraft. This symmetrical wing, on the other hand, is typically used on high performance aircraft and has identical upper and lower cambers.

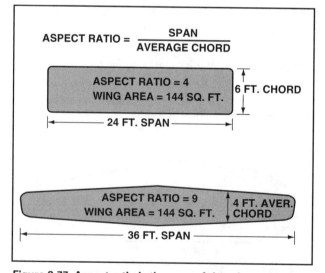

$$ASPECT\ RATIO = \frac{SPAN}{AVERAGE\ CHORD}$$

ASPECT RATIO = 4
WING AREA = 144 SQ. FT.

6 FT. CHORD

24 FT. SPAN

ASPECT RATIO = 9
WING AREA = 144 SQ. FT.

4 FT. AVER. CHORD

36 FT. SPAN

Figure 2-77. Aspect ratio is the span of the wing, wingtip to wingtip, divided by its average chord. In general, the higher the aspect ratio the higher the lifting efficiency of the wing. For example, gliders may have an aspect ratio of 20 to 30, while typical light aircraft have an aspect ratio of about seven to nine.

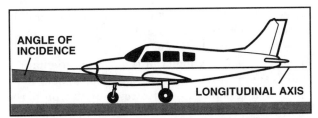

Figure 2-78. Angle of incidence refers to the angle between the wing chord line and a line parallel to the longitudinal axis of the airplane. A slight positive angle of incidence provides a positive angle of attack while the airplane is in level flight at normal cruising speed.

Wing area is the total surface area of the wings. Most wings don't produce a great amount of lift per square foot, so wing area must be sufficient to support the weight of the airplane. For example, in a light aircraft at normal operating speed, the wings produce about 10.5 pounds of lift for each square foot of wing area. This means a wing area of 200 square feet is required to support an airplane weight of 2,100 pounds during straight-and-level flight.

Once the design of the wing is determined, the wing is mounted on the airplane. Usually it is attached to the fuselage with the chord line inclined upward at a slight angle, which is called the **angle of incidence**. [Figure 2-78]

WEIGHT

The weight of the airplane is not a constant. It varies with the equipment installed, passengers, cargo, and fuel load. During the course of a flight, the total weight of the airplane decreases as fuel is consumed. Additional weight reduction may also occur during some specialized flight activities, such as crop dusting, fire fighting, or sky diving flights.

In contrast, the direction in which the force of weight acts is constant. It always acts straight down toward the center of the earth.

THRUST

Thrust is the forward-acting force which opposes drag and propels the airplane. In some airplanes, this force is provided when the engine turns the propeller. Each propeller blade is cambered like the airfoil shape of a wing. This shape, plus the angle of attack of the blades, produces reduced pressure in front of the propeller and increased pressure behind it. As is the case with the wing, this produces a reaction force in the direction of the lesser pressure. This is how the propeller produces thrust, the force which moves the airplane forward.

Jet engines produce thrust by accelerating a relatively small mass of air to a high velocity. A jet engine draws air into its intake, compresses it, then mixes it with fuel. When this mixture burns, the resulting heat expands the gas, which is expelled at high velocity from the engine's exhaust, producing thrust.

DRAG

Drag is caused by any aircraft surface that deflects or interferes with the smooth airflow around the airplane. A highly cambered, large surface area wing creates more drag (and lift) than a small, moderately cambered wing. If you increase airspeed, or angle of attack, you increase drag (and lift). Drag acts in opposition to the direction of flight, opposes the forward-acting force of thrust, and limits the forward speed of the airplane. Drag is broadly classified as either parasite or induced.

PARASITE DRAG

Parasite drag includes all drag created by the airplane, except that drag directly associated with the production of lift. It is created by the disruption of the flow of air around the airplane's surfaces. Parasite drag normally is divided into three types: form drag, skin friction drag, and interference drag.

Form drag is created by any structure which protrudes into the relative wind. The amount of drag created is related to both the size and shape of the structure. For example, a square strut creates substantially more drag than a smooth or rounded strut. Streamlining reduces form drag.

Skin friction drag is caused by the roughness of the airplane's surfaces. Even though these surfaces may appear smooth, under a microscope they may be quite rough. A thin layer of air clings to these rough surfaces and creates small eddies which contribute to drag.

Interference drag occurs when varied currents of air over an airplane meet and interact. This interaction creates additional drag. One example of this type of drag is the mixing of the air where the wing and fuselage join.

Each type of parasite drag varies with the speed of the airplane. The combined effect of all parasite drag varies proportionately to the square of the air-

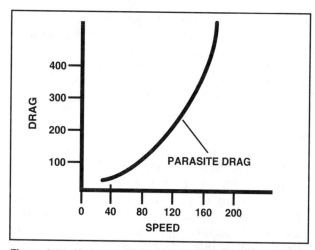

Figure 2-79. If airspeed is doubled, parasite drag increases fourfold. This is the same formula that applies to lift. Because of its rapid increase with increasing airspeed, parasite drag is predominant at high speeds.

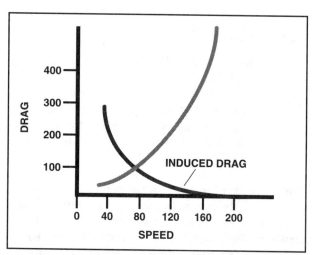

Figure 2-80. Induced drag is inversely proportional to the square of the speed. If speed is decreased by half, induced drag increases fourfold. It is the major cause of drag at reduced speeds near the stall; but, as speed increases, induced drag decreases.

speed. In other words, if airspeed is doubled, parasite drag increases by a factor of four. [Figure 2-79]

INDUCED DRAG

Induced drag is the main by-product of the production of lift. It is directly related to the angle of attack of the wing. The greater the angle, the greater the induced drag. Since the wing usually is at a low angle of attack at high speed, and a high angle of attack at low speed, the relationship of induced drag to speed also can be plotted. [Figure 2-80]

Over the past several years the **winglet** has been developed and used to reduce induced drag. As dis-

cussed earlier in this section, the high pressure air beneath the wing tends to spill over to the low pressure area above the wing, producing a strong secondary flow. If a winglet of the correct orientation and design is fitted to a wing tip, a rise in both total lift and drag is produced. However, with a properly designed winglet the amount of lift produced is greater than the additional drag, resulting in a net reduction in total drag.

TOTAL DRAG

Total drag for an airplane is the sum of parasite and induced drag. The total drag curve represents these combined forces and is plotted against airspeed. [Figure 2-81]

THREE AXES OF FLIGHT

All maneuvering flight takes place around one or more of three axes of rotation. They are called the **longitudinal**, **lateral**, and **vertical axes** of flight. The common reference point for the three axes is the aircraft's **center of gravity** (CG), which is the theoretical point where the entire weight of the airplane is considered to be concentrated. Since all three axes pass through this point, you can say that the airplane always moves about its CG, regardless of which axis is involved. The ailerons, elevator, and

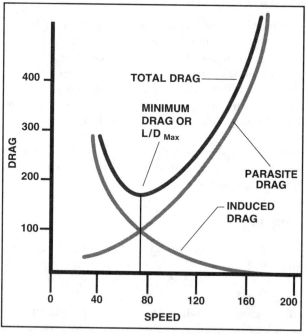

Figure 2-81. The low point on the total drag curve shows the airspeed at which drag is minimized. This is the point where the lift-to-drag ratio is greatest. It is referred to as L/D_Max. At this speed, the total lift capacity of the airplane, when compared to the total drag of the airplane, is most favorable. This is important in airplane performance.

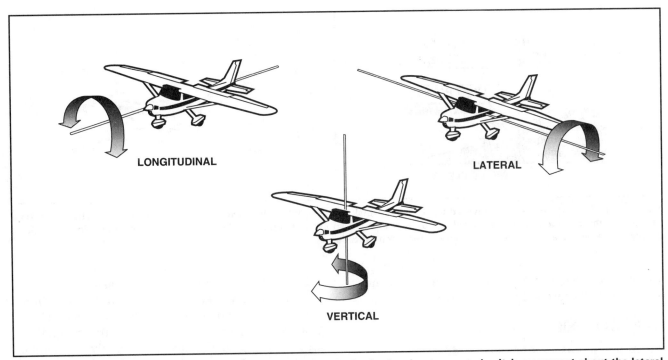

Figure 2-82. Ailerons control roll movement about the longitudinal axis; the elevator controls pitch movement about the lateral axis; and the rudder controls yaw movement about the vertical axis.

rudder create aerodynamic forces which cause the airplane to rotate about the three axes. [Figure 2-82]

LONGITUDINAL AXIS

When the ailerons are deflected, they create an immediate rolling movement about the longitudinal axis. Since the ailerons always move in opposite directions, the aerodynamic shape of each wing and its production of lift is affected differently. [Figure 2-83]

LATERAL AXIS

Pitch movement about the lateral axis is produced by the elevator or stabilator. Since the horizontal stabilizer is an airfoil, the action of the elevator (or stabilator) is quite similar to that of an aileron. Essentially, the chord line and effective camber of the stabilizer are changed by deflection of the elevator. In other words, as the elevator is deflected in one direction, the chord line changes and increases the angle of attack. This

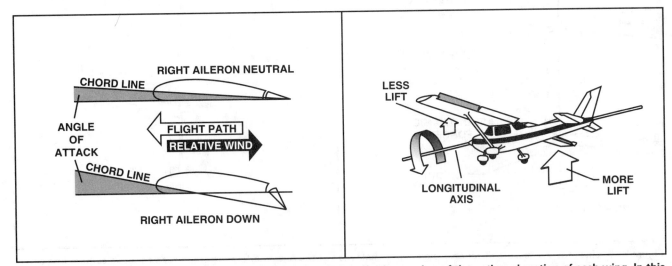

Figure 2-83. Deflected ailerons alter the chord line and change the effective camber of the outboard section of each wing. In this example, the angle of attack increases for the right wing, causing a corresponding increase in lift because of a decrease in its angle of attack. The airplane rolls to the left, because the right wing is producing more lift than the left wing.

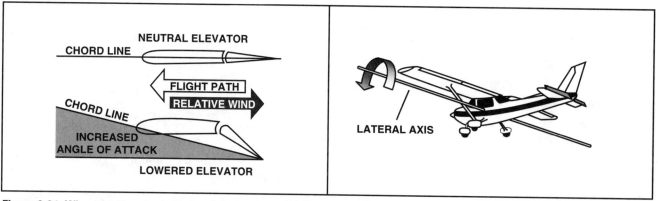

Figure 2-84. When the elevator is lowered the angle of attack of the stabilizer increases, and produces more lift. The lifting force created by the stabilizer causes the airplane to pivot forward about its lateral axis. The net result is a decrease in the angle of attack of the wings, and an overall decrease in the pitch attitude.

increased angle of attack produces more lift on one side of the tail causing it to move. [Figure 2-84]

VERTICAL AXIS

When pressure is applied to the rudder pedals, the rudder deflects into the airstream. This produces an aerodynamic force that rotates the airplane about its vertical axis. This is referred to as yawing the airplane. The rudder may be displaced either to the left or right of center, depending on which rudder pedal is depressed. [Figure 2-85]

TURNING FLIGHT

Before an airplane turns, however, it must overcome inertia, or its tendency to continue in a straight line.

The necessary turning force is created by banking the airplane so that the direction of lift is inclined. Now, one component of lift still acts vertically to oppose weight, just as it did in straight-and-level flight, while another acts horizontally. To maintain altitude, lift must be increased by increasing back pressure and, therefore, the angle of attack until the vertical component of lift equals weight. The horizontal component of lift, called **centripetal force**, is directed inward, toward the center of rotation. It is this center-seeking force which causes the airplane to turn. Centripetal force is opposed by **centrifugal force**, which acts outward from the center of rotation. When the opposing forces are balanced, the airplane maintains a constant rate of turn, without gaining or losing altitude. [Figure 2-86]

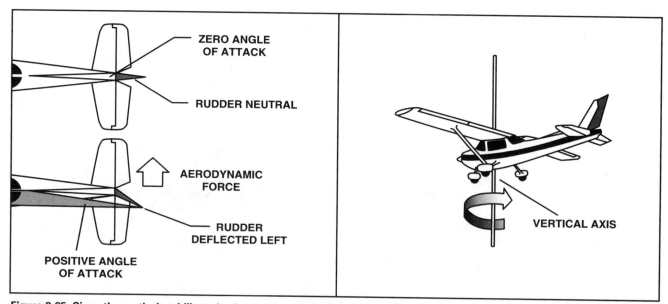

Figure 2-85. Since the vertical stabilizer also is an airfoil, deflection of the rudder alters the stabilizer's effective camber and chord line. In this case, left rudder pressure causes the rudder to move to the left. With a change in the chord line, the angle of attack is altered, generating an aerodynamic force toward the right side of the vertical fin. This causes the tail section to move to the right, and the nose of the airplane to yaw to the left.

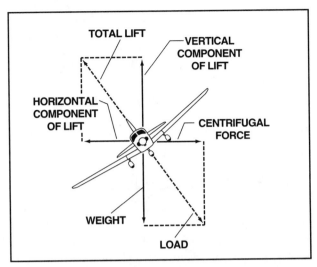

Figure 2-86. The resultant of the horizontal component (centripetal force) and the vertical component of lift is the total lift provided by the wings. The resultant of centrifugal force and weight is the total load the wings must support.

ADVERSE YAW

When an airplane rolls into a turn, the aileron on the inside of the turn is raised, and the aileron on the outside of the turn is lowered. The lowered aileron on the outside increases the angle of attack and produces more lift for that wing. Since induced drag is a by-product of lift, the outside wing also produces more drag than the inside wing. This causes a yawing tendency toward the outside of the turn, which is called adverse yaw. [Figure 2-87]

The coordinated use of aileron and rudder corrects for adverse yaw. For a turn to the left, the left rud-

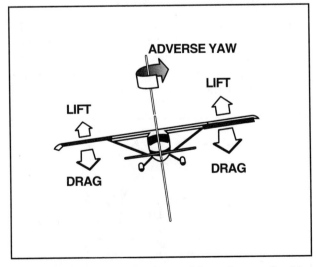

Figure 2-87. If a turn is rolled into without the use of rudder to help establish the turn, the airplane will yaw about its vertical axis opposite to the direction of the turn. Adverse yaw is caused by higher induced drag on the outside wing, which is producing more lift.

der pedal is depressed slightly. Once established in the turn, both aileron and rudder controls are neutralized.

Another way to compensate for adverse yaw is by the use of **Frise** ailerons. A Frise aileron has its hinge point located behind its leading edge. When the aileron is raised, its leading edge sticks out into the airstream below the wing. This produces enough parasitic drag to counter the aileron's induced drag.

LOAD FACTOR

So far in this discussion, you have looked at the combination of opposing forces acting on a turning airplane. Now it's time to examine load factors induced during turning flight. To better understand these forces, picture yourself on a roller coaster. As you enter a banked turn during the ride, the forces you experience are very similar to the forces which act on a turning airplane. On a roller coaster, the resultant force created by the combination of weight and centrifugal force presses you down into your seat. This pressure is an increased load factor that causes you to feel heavier in the turn than when you are on a flat portion of the track.

The increased weight you feel during a turn in a roller coaster is also experienced in an airplane. In a turning airplane, however, the increase in weight and loss of vertical lift must be compensated for or the aircraft loses altitude. This is done by increasing the angle of attack with back pressure on the control wheel. The increase in the angle of attack increases the total lift of the airplane. Keep in mind that when lift increases, drag also increases. This means thrust must be increased to maintain the original airspeed and altitude.

During turning maneuvers, weight and centrifugal force combine into a resultant which is greater than weight alone. Additional loads are imposed on the airplane, and the wings must support the additional load factor. In other words, when flying in a curved flight path, the wings must support not only the weight of the airplane and its contents, but they also must support the load imposed by centrifugal force.

Load factor is the ratio of the load supported by the airplane's wings to the actual weight of the aircraft and its contents. If the wings are supporting twice as much weight as the weight of the airplane and its contents, the load factor is two. You are probably more familiar with the term "G-forces" as a way to describe flight loads caused by aircraft maneuvering.

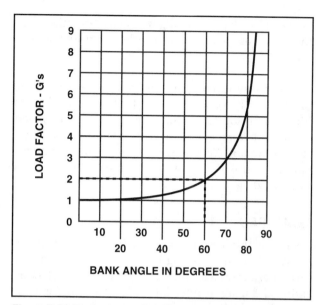

Figure 2-88. During constant altitude turns, the relationship between load factor, or G's, and bank angle is the same for all airplanes. For example, with a 60° bank, two G's are required to maintain level flight. This means the airplane's wings must support twice the weight of the airplane and its contents, although the actual weight of the airplane does not increase.

"Pulling G's" is common terminology for higher performance airplanes. For example, an acrobatic category airplane may pull three or four G's during a maneuver. An airplane in cruising flight, while not accelerating in any direction, has a load factor of one. This one-G condition means the wings are supporting only the actual weight of the airplane and its contents. [Figure 2-88]

LIMIT LOAD FACTOR

When the Federal Aviation Administration certifies an airplane, one of the criteria they look at is how much stress the airplane can withstand. The limit load factor is the number of G's an airplane can sustain, on a continuing basis, without causing permanent deformation or structural damage. In other words, the limit load factor is the amount of positive or negative G's an airframe is capable of supporting.

Most small general aviation airplanes with a gross weight of 12,500 pounds or less, and nine passenger seats or less, are certified in either the normal, utility, or acrobatic categories. A normal category airplane is certified for nonacrobatic maneuvers. The maximum limit load factor in the normal category is 3.8 positive G's, and 1.52 negative G's. In other words, the airplane's wings are designed to withstand 3.8 times the actual weight of the airplane and its contents during maneuvering flight.

In addition to those maneuvers permitted in the normal category, an airplane certified in the utility category may be used for several maneuvers requiring additional stress on the airframe. A limit of 4.4 positive G's and 1.76 negative G's is permitted in the utility category. An acrobatic category airplane may be flown in any flight attitude as long as its limit load factor does not exceed 6 positive G's or 3 negative G's.

LEFT-TURNING TENDENCIES

Propeller-driven airplanes are subject to several left-turning tendencies caused by a combination of physical and aerodynamic forces — torque, gyroscopic precession, asymmetrical thrust, and spiraling slipstream.

TORQUE

In airplanes with a single engine, the propeller rotates clockwise when viewed from the pilot's seat. Torque can be understood most easily by remembering Newton's third law of motion. The clockwise action of a spinning propeller causes a torque reaction which tends to rotate the airplane counterclockwise about its longitudinal axis. [Figure 2-89]

Generally, aircraft have design adjustments which compensate for torque while in cruising flight. For example, some airplanes have aileron trim tabs which correct for the effects of torque at various power settings.

GYROSCOPIC PRECESSION

The turning propeller of an airplane also exhibits characteristics of a gyroscope — rigidity in space and precession. The characteristic that produces a left-turning tendency is precession. **Gyroscopic precession** is the resultant reaction of an object

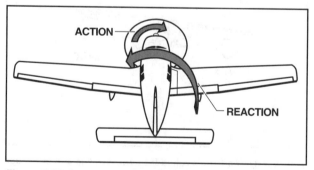

Figure 2-89. In a single-engine airplane, the greatest effect of torque occurs during takeoff or climb when the aircraft in a low-airspeed, high-power, high angle of attack flight condition.

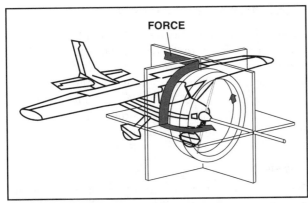

Figure 2-90. If the pitch attitude of an airplane is changed from a nose-high to a nose-level position, a force will be exerted near the top of the propeller's plane of rotation. A resultant force will then be exerted 90° ahead in the direction of rotation, which will cause the nose of the airplane to yaw to the left. This typically happens when the tail of a conventional-gear airplane is raised on the takeoff roll.

when force is applied. The reaction to a force applied to a gyro acts 90° in the direction of rotation. [Figure 2-90]

ASYMMETRICAL THRUST

When a single-engine airplane is at a high angle of attack, the descending blade of the propeller takes a greater "bite" of air than the ascending blade on the other side. The greater bite is caused by a higher angle of attack for the descending blade, compared to the ascending blade. This creates the uneven, or asymmetrical thrust, which is known as P-factor. **P-factor** makes an airplane yaw about its vertical axis to the left. [Figure 2-91]

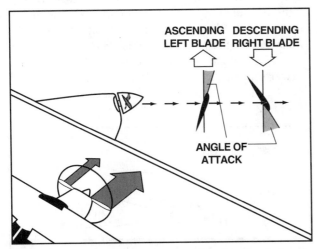

Figure 2-91. Asymmetrical thrust occurs when an airplane is flown at a high angle of attack. This causes uneven angles of attack between the ascending and descending propeller blades. Consequently, less thrust is produced from the ascending blade on the left than from the descending blade on the right. This produces a tendency for the airplane to yaw to the left.

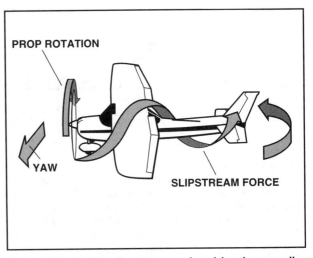

Figure 2-92. As the slipstream produced by the propeller rotation wraps around the fuselage, it strikes the left side of the vertical fin. A left-turning tendency is created as the air "pushes" the tail of the airplane to the right and yaws the nose left.

SPIRALING SLIPSTREAM

As the propeller rotates, it produces a backward flow of air, or slipstream, which wraps around the airplane. This spiraling slipstream causes a change in the airflow around the vertical stabilizer. Due to the direction of the propeller rotation, the resultant slipstream strikes the left side of the vertical fin. [Figure 2-92]

AIRCRAFT STABILITY

Although no aircraft is completely stable, all aircraft must have desirable stability and handling characteristics. This quality is essential throughout a wide range of flight conditions — during climbs, descents, turns, and at both high and low airspeeds.

Stability is a characteristic of an airplane in flight that causes it to return to a condition of equilibrium, or steady flight, after it is disturbed. For example, if you are flying a stable airplane that is disrupted while in straight-and-level flight, it has a tendency to return to the same attitude.

STATIC STABILITY

Static stability is the initial tendency that an object displays after its equilibrium is disrupted. An airplane with **positive static stability** tends to return to its original attitude after displacement. A tendency to move farther away from the original attitude following a disturbance is **negative static stability**. If

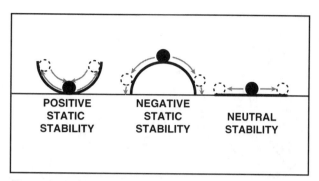

Figure 2-93. Positive static stability is the most desirable characteristic, because the airplane attempts to return to its original trimmed attitude. Almost all light aircraft exhibit this characteristic. However, static stability is highly desirable in aerobatic aircraft and some high-performance military aircraft.

an airplane tends to remain in its displaced attitude, it has **neutral static stability**. [Figure 2-93]

DYNAMIC STABILITY

Dynamic stability describes the time required for an airplane to respond to its static stability following a displacement from a condition of equilibrium. It is determined by its tendency to oscillate and damp out successive oscillations after the initial displacement. Although an airplane may be designed with positive static stability, it could have positive, negative, or neutral dynamic stability.

Assume an airplane in flight is displaced from an established attitude. If its tendency is to return to the original attitude directly, or through a series of decreasing oscillations, it exhibits **positive dynamic stability**. If you find the oscillations increasing in magnitude as time progresses, **negative dynamic stability** is exhibited. **Neutral dynamic stability** is indicated if the airplane attempts to return to its original state of equilibrium, but the oscillations neither increase nor decrease in magnitude as time passes. [Figure 2-94]

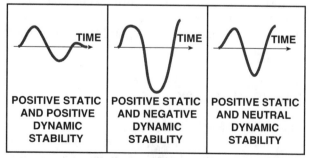

Figure 2-94. The most desirable condition is a combination of positive static stability with positive dynamic stability. In this situation, less effort is required to return the airplane to its original attitude because it "wants" to return there.

LONGITUDINAL STABILITY

When considering the overall stability of an airplane, remember that the airplane moves about three axes of rotation. Longitudinal stability involves the pitching motion or tendency of the airplane about its lateral axis. An airplane which is longitudinally stable tends to return to its trimmed angle of attack after displacement.

BALANCE

To achieve longitudinal stability, most airplanes are designed so they're slightly nose heavy. This is accomplished during the engineering and development phase by placing the center of gravity slightly forward of the center of pressure.

The **center of pressure** is a point along the wing chord line where lift is considered to be concentrated. For this reason, the center of pressure is often referred to as the **center of lift**. During flight, this point along the chord line changes position with different flight attitudes. It moves forward as the angle of attack increases and aft as the angle of attack decreases. As a result, pitching tendencies created by the position of the center of lift in relation to the CG vary. For example, with a high angle of attack and the center of lift in a forward position (closer to the CG) the nose-down pitching tendency is decreased. The position of the center of gravity in relation to the center of lift is a critical factor in longitudinal stability.

If an airplane is loaded so the CG is forward of the forward CG limit, it becomes too nose heavy. Although this tends to make the airplane seem stable, adverse side effects include longer takeoff distance and higher stalling speeds. The condition gets progressively worse as the CG moves to an extreme forward position. Eventually, stabilator (elevator) effectiveness becomes insufficient to lift the nose. [Figure 2-95]

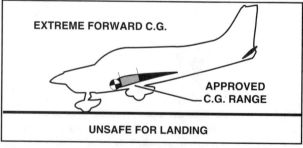

Figure 2-95. If the CG is well forward of the approved CG range, stabilator effectiveness becomes insufficient to exert the required tail-down force needed for a nose-high landing attitude. This may cause the nosewheel to strike the runway before the main gear.

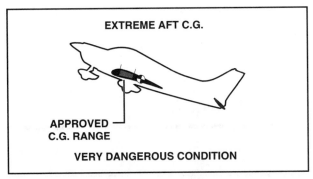

Figure 2-96. If the CG is too far aft, there is not enough sta-bilator effectiveness to raise the tail and lower the nose of the airplane. As a result, a pilot would be unable to recover from a stalled condition and a spin may develop.

A CG located aft of the approved CG range is even more dangerous than a CG that is too far forward. With an aft CG, the airplane becomes tail heavy and very unstable in pitch, regardless of speed. [Figure 2-96]

LATERAL STABILITY

Stability about an airplane's longitudinal axis, which extends nose to tail, is called lateral stability. If one wing is lower than the opposite wing, lateral stability helps return the wings to a wings-level atti-tude. This tendency to resist lateral, or roll, move-ment is aided by specific design characteristics of an airplane.

The most common design for lateral stability is known as wing dihedral. **Dihedral** is the upward angle of the airplane's wings with respect to the hor-izontal. When you look at an airplane, dihedral makes the wings appear to form a spread-out V. Dihedral usually is just a few degrees.

When an airplane with dihedral enters an uncoor-dinated roll during gusty wind conditions, one wing is elevated and the opposite wing drops. This causes an immediate side slip downward toward the low wing. The side slip makes the low wing approach the air at a higher angle of attack than the high wing. The increased angle of attack on the low

wing produces more lift for that wing and tends to lift it back to a level flight attitude. [Figure 2-97]

DIRECTIONAL STABILITY

Stability about the vertical axis is called directional stability. Essentially, an airplane in flight is much like a weather vane. You can compare the pivot point on the weather vane to the center of gravity pivot point of the airplane. The nose of the airplane corresponds to the weather vane's arrowhead, and the vertical fin on the airplane acts like the tail of the weather vane. [Figure 2-98]

High-speed jet aircraft are subject to a form of dynamic instability called **Dutch roll**, which is characterized by a coupling of directional and lat-eral oscillation. In other words, an aircraft experi-encing Dutch roll tends to wander about the roll and yaw axes. Dutch roll generally occurs when an air-craft's dihedral effect is larger than its static direc-tional stability. To prevent Dutch roll, many aircraft employ **yaw dampers** that sense uncommanded roll and yaw motion and activate appropriate flight con-trols to overcome them.

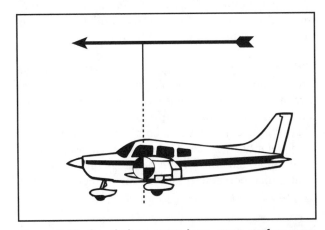

Figure 2-98. An airplane must have more surface area behind the CG than it has in front of it. The same is true for a weather vane; more surface aft of the pivot point is a basic design feature. This arrangement helps keep the air-plane aligned with the relative wind, just as the weather vane points into the wind.

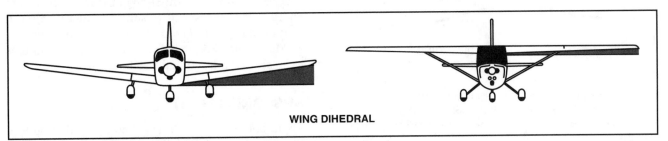

Figure 2-97. Wing dihedral contributes to lateral stability. Low-wing airplanes commonly have more dihedral than high-wing air-planes. The center of gravity is well below the wing in a high-wing airplane. Because this CG position has a stabilizing effect, very little dihedral is necessary.

HIGH-SPEED AERODYNAMICS

Advancements in technology have produced high performance airplanes that are capable of very high speed flight. The study of aerodynamics at these high flight speeds is significantly different from the study of low speed aerodynamics.

COMPRESSIBILITY EFFECTS

At low speeds, the study of aerodynamics is simplified by the fact that as air passes over a wing it experiences a relatively small change in pressure and density. This airflow is termed incompressible since the air undergoes changes in pressure without apparent changes in density. This is seen when a fluid passes through a venturi. As the fluid enters a restriction, its velocity increases and the pressure decreases. [Figure 2-99]

At high speeds the change in air pressure and density is significant. For example, as air enters a venturi at supersonic speeds, the airflow slows down, and therefore, must compress to pass through the restriction. Once a fluid compresses, its pressure and density increase. The study of high speed airflow must account for these changes in air density and must consider that the air is compressible. [Figure 2-100]

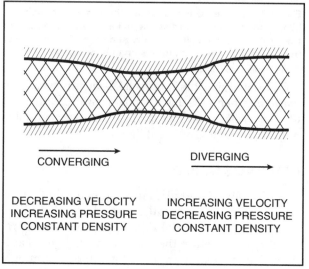

Figure 2-100. At supersonic speeds, as air converges into a venturi, its velocity decreases while its pressure and density increase.

SPEED OF SOUND

As discussed earlier in this chapter, when air is disturbed, longitudinal waves are created. This causes the air pressure to increase and decrease. These changes radiate concentrically from the point of disturbance. At standard sea level, sound waves travel at 661.7 knots. However, when the air temperature changes so does the speed of sound. [Figure 2-101]

One of the most important measurements in high-speed aerodynamics is based on the speed of sound. The Mach number is the ratio of the true airspeed of an aircraft to the speed of sound. When an aircraft flies at the speed of sound, it is said to be traveling at Mach one. Flight below the speed of sound is expressed as a fractional Mach number.

SUBSONIC FLIGHT

Flight below Mach 0.75 is called subsonic flight. In this speed range all airflow is subsonic and aircraft behave in accordance with the concepts discussed in the aerodynamics section of this chapter.

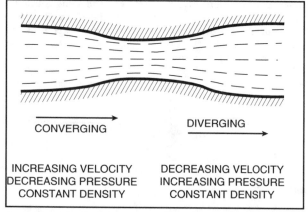

Figure 2-99. At subsonic speeds, as air converges into a venturi, its velocity increases while its pressure decreases. However, there is almost no change in air density.

ALTITUDE FT.	DENSITY RATIO σ	PRESSURE RATIO δ	TEMPER- ATURE F°	SPEED OF SOUND α KNOTS
0	1.0000	1.0000	59.00	661.7
1000	0.9711	0.9644	55.43	659.5
2000	0.9428	0.9298	51.87	65702
3000	0.9151	.08962	48.30	654.9
4000	0.8881	0.8637	44.74	652.6
5000	0.8617	0.8320	41.17	650.3
6000	0.8359	0.8014	37.60	647.9
7000	0.8106	0.7716	34.04	645.6
8000	.07860	0.7428	30.47	643.2
9000	0.7620	0.7148	26.90	640.9
10000	0.7385	0.6877	23.34	628.6
15000	0.6292	0.5643	5.51	626.7
20000	0.5328	0.4595	-12.32	614.6
25000	0.4481	0.3711	-30.15	602.0
30000	0.3741	0.2970	-47.98	589.5
35000	0.3099	0.2353	-65.82	576.6
*36089	0.2971	0.2234	-69.70	573.8
40000	0.2462	0.1851	-69.70	573.8
45000	0.1936	0.1455	-69.70	573.8
50000	0.1522	0.1145	-69.70	573.8
55000	0.1197	0.0900	-69.70	573.8
60000	0.0941	0.0708	-69.70	573.8
65000	0.0740	0.0557	-69.70	573.8
70000	0.0582	0.0428	-69.70	573.8
75000	0.0458	0.0344	-69.70	573.8
80000	0.0360	0.0271	-69.70	573.8
85000	0.0280	0.0213	-64.80	577.4
90000	0.0217	0.0168	-56.57	583.4
95000	0.0169	0.0134	-48.34	589.3
100000	0.0132	0.0107	-40.11	595.2

Figure 2-101. As the air temperature decreases, the speed of sound also decreases.

TRANSONIC FLIGHT

The most difficult realm of flight is that at transonic speed which is between Mach 0.75 and 1.20. At this speed some of the airflow passing over the wings is subsonic, and some is supersonic. At these speeds, shock waves form and an airfoil's aerodynamic center of lift shifts from 25% of the chord at subsonic speeds to 50%. This shift causes large changes in aerodynamic trim and stability. The speed at which the local flow over the top of the wing becomes sonic is known as the airfoils **critical Mach number**.

SUPERSONIC FLIGHT

Flight between Mach 1.20 and Mach 5.00 is smooth and efficient. This regime is characterized by all the airflow being supersonic. At these speeds, the shock and expansion waves attach to the wing and remain stationary. Flight speeds above Mach 5.00 are called **hypersonic flight**, and thermal limitations are encountered as well as aerodynamic problems.

SHOCK WAVES

As an aircraft flies it disturbs air, and sound waves radiate from every part of its surface. As long as the aircraft is moving slower than the speed of sound, the sound waves move out from the aircraft in all directions. However, when the aircraft approaches the speed of sound, the sound waves cannot move out ahead of the source, so they start to pile up. This forms a **compression wave** which causes a large difference in the static pressure and density of the air. [Figure 2-102]

OBLIQUE SHOCK WAVES

When a wedge-shaped airfoil passes through the air at a supersonic velocity, the air must change its direction suddenly. This causes a compression wave or shock wave to form and attach to the wing's leading edge. The supersonic airflow that passes through a compression wave slows in velocity.

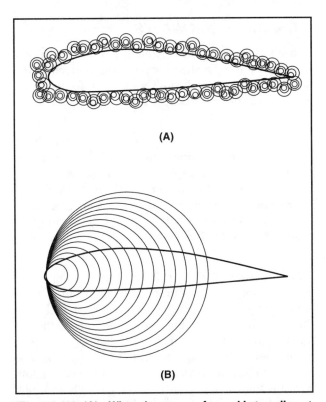

(A)

(B)

Figure 2-102. (A)—When the source of sound is traveling at subsonic speed, the sound waves propagate outward from each disturbance. (B)—When the source of sound is traveling at the speed of sound, the sound waves cannot move ahead of the source, and therefore, pile up and form a shock wave.

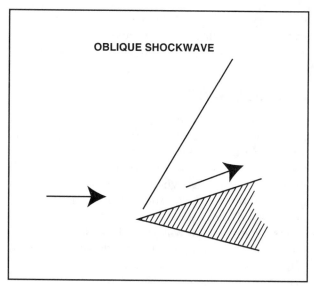

Figure 2-103 An oblique shock wave forms on the leading edge of wedge-shaped airfoils.

However, the velocity is still supersonic as it leaves the shock wave and both the static pressure and the density of the air increase. Some of the energy in the airstream is dissipated in the form of heat. [Figure 2-103]

NORMAL SHOCK WAVES

If a blunt airfoil passes through the air at a supersonic velocity, the shock wave cannot attach to the leading edge. Instead, the shock wave forms ahead of the airfoil and perpendicular to the airstream.

When the airstream passes through a normal shock wave, its direction does not change. However, the airstream does slow to a subsonic speed with a large increase in its static pressure and density. A normal shock wave forms the boundary between supersonic and subsonic airflow when there is no change in direction of air as it passes through the wave.

In addition to forming in front of the leading edge, normal shock waves also form on an airfoil in transonic flight. For example, as an airfoil is forced through the air at a high subsonic speed, the air passing over the top of the wing speeds up to a supersonic velocity and a normal shock wave forms. Once the shock wave forms, it slows the airflow beyond the wave to a subsonic speed. These shock waves form on top of the airfoil first and then on the bottom. As airspeed increases beyond the transonic range, both shock waves move aft and attach to the wing's trailing edge to form an oblique shock wave. [Figure 2-104]

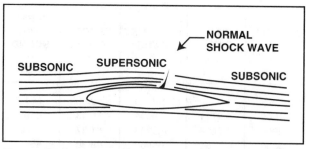

Figure 2-104. As the airflow on the top of the wing accelerates to supersonic speed, a normal shock wave forms and slows the velocity of the air to subsonic speeds on the wing's trailing edge.

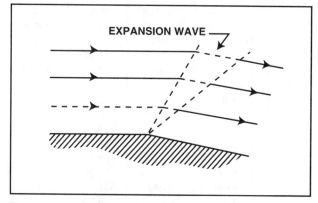

Figure 2-105. Creation of an expansion wave as supersonic airflow follows the surface.

EXPANSION WAVES

When a supersonic stream of air turns away from its direction of flow to follow the surface of an airfoil, its speed increases and both static pressure and density decrease. Since an expansion wave is not a shock wave, no energy in the airstream is lost. [Figure 2-105]

HIGH-SPEED AIRFOILS

Transonic flight displays the greatest airfoil design problems because only a portion of the airflow passing over the wing is supersonic. When an airfoil moves through the air at a speed below its critical Mach number, all of the airflow is subsonic and the pressure distributions are as you would expect. However, as flight speed exceeds the critical Mach number for an airfoil, the airflow over the top of the wing reaches supersonic velocity and a normal shock wave forms. Normally, the airflow over the top of a wing creates an area of low pressure that pulls the air to the wing's surface. However, when a shock wave forms on top of the wing, airflow passing though it slows causing the air's static pressure to increase. This destroys the area of low pressure above the wing and allows the air to separate from

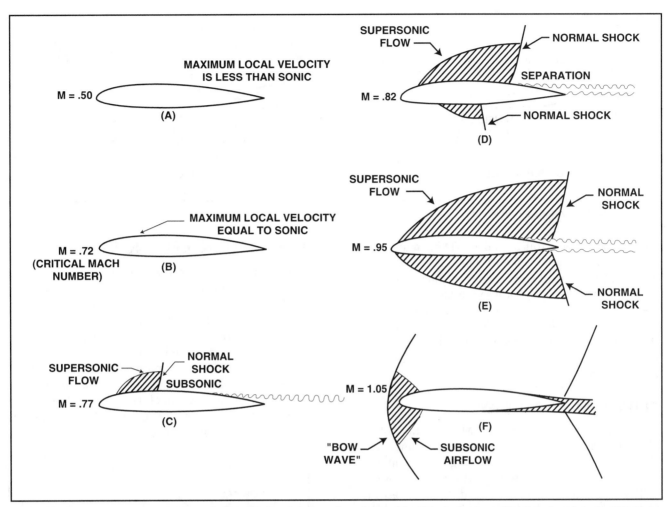

Figure 2-106. (A)—When an airfoil travels through the air below the wing's critical Mach number, all airflow is subsonic. (B)—Once the critical mach number is reached, the local flow air passing over the wing's greatest camber reaches sonic speed. (C)—As the airfoil exceeds the critical mach number a shock wave begins to form. (D)—As more of the airflow passing over the wing reaches supersonic speed, the shock wave moves aft and shock induced separation begins to occur. A second shock wave also begins to form on the lower camber. (E)—When almost all of the airflow is supersonic, both shock waves continue to move aft and attach to the trailing edge. (F)—Once the airfoil moves through the air above Mach 1, a bow wave begins to form ahead of the wing.

the surface. This **shock-induced separation** causes a loss of lift and can reduce control effectiveness. [Figure 2-106]

Because the supersonic flow is a local condition, its effects can be reduced by the use of vortex generators. A **vortex generator** is a small airfoil mounted ninety degrees to the surface of the wing. It has a low aspect ratio and produces a strong vortex or flow of controlled air that moves high energy air from the airstream into the boundary layer. This vortex delays airflow separation. To obtain maximum benefit, vortex generators are mounted in pairs so the vortices are combined. Although they actually add drag at low speed, the benefit at high speed is a good tradeoff. [Figure 2-107]

Airfoil sections designed for supersonic flight are typically a **double wedge** or **biconvex** shape with

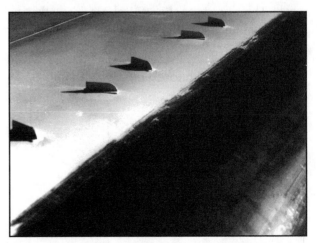

Figure 2-107. Vortex generators are typically mounted on the upper surface of high-speed aircraft wings. Their function is to bring high energy to the surface of the wing by keeping the airflow from seperating from the surface of the wing, thereby delaying shock induced separation.

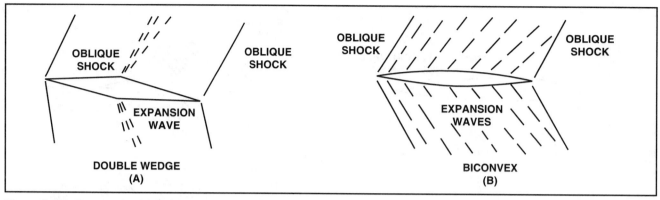

Figure 2-108. Supersonic airfoil shapes.

sharp leading and trailing edges. Their maximum thickness is at the 50% chord position. As soon as either of these airfoil sections pass through the transonic range, oblique shock waves attach to the leading and trailing edges. The expansion waves form at the point where the airflow must deflect to follow the surface. Since there are no normal shock waves on either airfoil section, there is no subsonic airflow. [Figure 2-108]

CRITICAL MACH NUMBER

As discussed earlier, the critical Mach number of an aircraft is the flight Mach number where a portion of the airflow passing over the wing travels at the speed of sound. This is the speed where compressibility effects are first encountered. In most cases it is better to have a high critical Mach number to delay the formation of a shock wave. One way the critical Mach number can be increased is by using a thin airfoil with a maximum thickness well back from the leading edge. Another design feature that increases the critical Mach number is wing sweepback. This increases the critical Mach number by effectively decreasing the thickness ratio of the wing. [Figure 2-109]

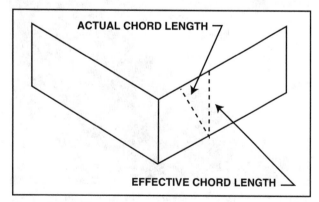

Figure 2-109. The air flowing across the wing in the line of flight travels farther than the distance perpendicular to the leading edge. This longer travel for the same thickness increases the critical Mach number.

SUPERSONIC ENGINE INLETS

The air flow into a turbine engine must be subsonic for maximum effectiveness. This holds true for aircraft that fly at supersonic speeds. One way to slow engine intake airflow is to create shock waves in front of the inlet. At low supersonic speeds, the engine inlet is designed to create a single normal shock wave at its entry point. However, at higher speeds it is necessary to have a **diffuser** in the inlet that forms an oblique shock wave and slows the air before it enters the main inlet duct where a normal shock wave is formed. [Figure 2-110]

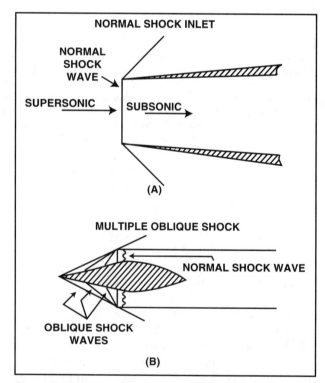

Figure 2-110. (A)—At low supersonic speed, engine inlets are designed to produce a normal shock wave to slow the airflow into the engine. (B)—However, at high supersonic speeds it is often necessary to use a diffuser that generates multiple oblique shock waves in addition to the normal shock wave to slow the intake air sufficiently.

AERODYNAMIC HEATING

One of the biggest problems of hypersonic flight is aerodynamic heating of the aircraft structure. A "shooting star" is nothing more than an extreme example of aerodynamic heating. The material overheats to the point of destruction from friction heating with the earth's atmosphere. In fact, if it were not for the special ceramic tile insulation on the structure of the space shuttle, the same thing would happen to it upon reentry. New materials and applications need to be developed before flight in the hypersonic region is practical, particularly with point to point flights as compared to outer space flights.

HELICOPTER AERODYNAMICS

Five hundred years ago, Leonardo da Vinci conceived the idea of a vehicle with a screw type thread on a vertical shaft, that if properly shaped and powered, would be able to takeoff vertically, hover, and land. However, it was not until the twentieth century that such a vehicle became a reality. Today, versatile rotary-wing aircraft, called helicopters, are vital tools for many civil and military applications. This chapter acquaints you with the fundamentals of helicopter design and structure, and with the function of each major component. You also become familiar with the dynamic nature of airflow and the dynamic and static forces acting upon a helicopter in flight. Additionally, this chapter provides a basic knowledge of the factors affecting helicopter performance.

THE HELICOPTER

There are many variations in the design and complexity of modern helicopters. Even though helicopters vary greatly in size and appearance, they share many of the same major components. [Figure 2-111]

MAIN ROTOR SYSTEMS

Main rotor systems are classified according to how the main rotor blades move relative to the rotor hub. The main classifications are: fully articulated, semirigid, and rigid.

In a **fully articulated rotor system**, the blades are hinged at the rotor hub so that each is free to move up and down and back and forth, as well as around the spanwise (pitch change) axis. These rotor systems normally have three or more blades. The blades are attached to the main hub through a vertical hinge which allows the blades to move back and forth independently as they rotate. This hinge is called a drag or lag hinge. Movement around this hinge is called dragging, lead/lag, or hunting. [Figure 2-112]

A **semirigid rotor system** uses two blades which are rigidly attached to the main rotor hub. There is no vertical drag hinge. The main rotor hub, however, is free to tilt and rock independently of the main rotor shaft, on what is know as a teetering hinge. This

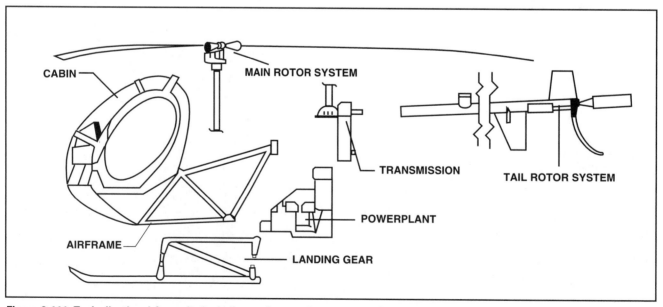

Figure 2-111. Typically, the airframe is the helicopter's central component since the cabin, powerplant, main and tail rotor system, and landing gear are all attached to it.

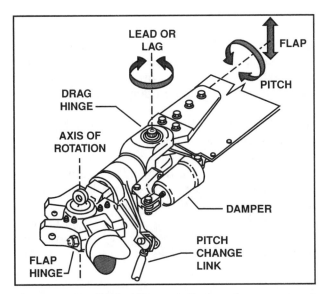

Figure 2-112. Dampers prevent excessive movement around the drag hinges. The blades can also move around their spanwise axis. This movement, called feathering, changes the pitch angle of the blade. The blade hub is also hinged to permit each blade to flap independently on the hub's horizontal axis in an up and down motion.

allows the blades to flap together. As one blade flaps down, the other flaps up. [Figure 2-113]

The **rigid rotor system** is mechanically simple, but structurally complex because operating loads must be absorbed in bending rather than through hinges. In a rigid rotor system, the blades cannot flap or drag, but can be feathered.

ANTITORQUE SYSTEMS

Most helicopters with a single main rotor system require a separate rotor to overcome torque, which is

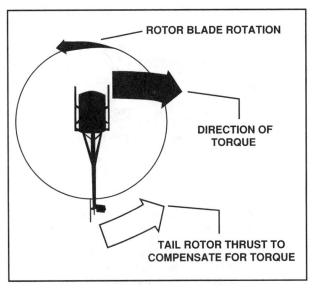

Figure 2-114. The antitorque rotor produces thrust to oppose torque, and thereby, prevent the helicopter from turning.

the tendency of the helicopter to turn in the opposite direction of main rotor rotation. The thrust of the **antitorque rotor** is varied to maintain directional control when main rotor torque changes, or to make heading changes while hovering. [Figure 2-114]

Another form of antitorque rotor is the **fenestron** or "fan-in-tail" design. This system uses a series of rotating blades shrouded within a vertical tail. Because the blades are located within a circular duct, they are less likely to come into contact with people or objects. [Figure 2-115]

The **no tail rotor** system is an alternative to the antitorque rotor. The system uses low-pressure air

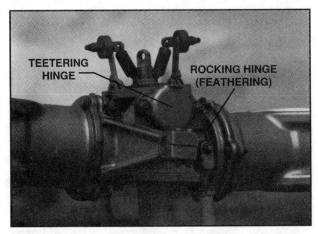

Figure 2-113. A rocking hinge, perpendicular to the teetering hinge, allows the rotor head to move in response to inputs from the cyclic pitch control. Cyclic pitch control inputs are transmitted to the rotor head through a swash plate assembly.

Figure 2-115. Compared to an unprotected tail rotor, the fenestron antitorque system provides better aerodynamics while in flight, and greatly improves safety during ground operations.

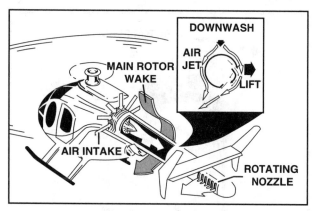

Figure 2-116. The low pressure air coming from horizontal slots located on the right side of the tailboom, allow the main rotor downwash to create a phenomenon called "coanda effect" as it flows around the tail. This creates a force on the right side of the tailboom. While in a hover, this force produces approximately two-thirds of the lift necessary to maintain directional control. The rest is created by directing the thrust from the controllable rotating nozzle.

which is forced into the tailboom by a fan mounted within the helicopter. This air is then fed through horizontal slots and a controllable rotating nozzle in the tailboom to provide antitorque and directional control. [Figure 2-116]

Tandem rotor helicopters do not require a separate antitorque rotor. In this configuration, torque from one rotor is opposed by the torque of the other so the turning tendencies cancel each other.

AERODYNAMIC FORCES

The four forces acting on a helicopter are identical to those for aircraft and include lift, weight, thrust, and drag. **Lift** is the force created by the effect of airflow as is passes around an airfoil. **Weight** opposes lift and is caused by the downward pull of gravity. **Thrust** is the force which propels the helicopter through the air. Opposing thrust is **drag**, which is the retarding force created by the movement of an object through the air. [Figure 2-117]

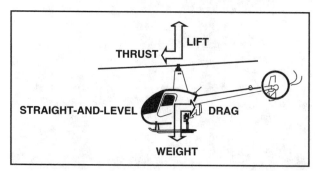

Figure 2-117. Lift opposes weight and thrust opposes drag. When a helicopter is in straight-and-level, unaccelerated flight, the opposing forces balance each other; lift equals weight and thrust equals drag.

LIFT

The principles applied to the venturi are also utilized to create lift for a helicopter. However, in lieu of a venturi, rotor blades with an airfoil shape are used to create a pressure differential in the air.

According to Bernoulli's Principle, there is an acceleration or increase in the velocity of the air as it flows around an airfoil shape; therefore, there is an acceleration of the relative wind as it flows above and below the surface of the rotor blade. As air flows over the upper surface of the rotor blade, the curvature of the airfoil causes the speed of the airflow to increase. This increase in speed over the rotor blade causes a decrease in pressure above the airfoil. [Figure 2-118]

The remaining lift is provided by the blade's lower surface as air striking the underside is deflected downward. However, even at high angles of attack, the lift generated by the impact of air on the bottom area of the rotor blade amounts to only a fraction of the lifting force needed to sustain the helicopter in flight. More than 75% of the lift is caused by the lower pressure above the rotor blade.

AERODYNAMIC TERMINOLOGY

Many aerodynamic phenomena associated with helicopter operations can best be understood in terms of their definitions. They are presented here so you can become familiar with rotorcraft control characteristics.

The **span** of a rotor blade is the distance from the rotor hub to the blade tip. The **twist** of a rotor blade refers to a changing chord line from the blade root to the blade tip. Twisting a rotor blade causes it to produce a more even amount of lift along its span. This is necessary because rotational velocity

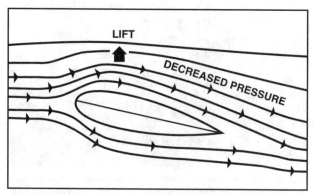

Figure 2-118. The resulting pressure differential, caused by the acceleration of airflow over the top of an airfoil creates an upward force. This force is the main source of lift.

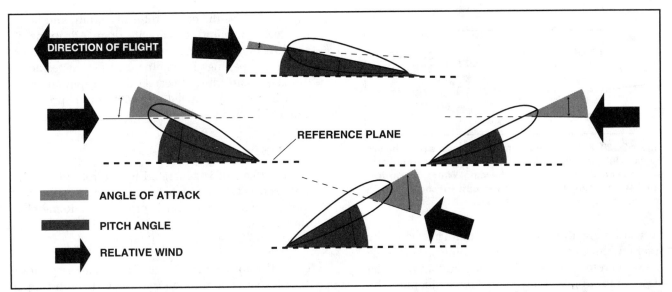

Figure 2-119. During the rotation of a rotor blade when a helicopter is in forward flight, the angle of attack is equal to the pitch angle when it is directly forward or aft of the helicopter's nose. When 90° to the right of the helicopter, the angle of attack is less than the pitch angle. When left of the helicopter, the angle of attack is greater. This assumes the rotation of the blade is from right to left.

increases toward the blade tip. The **pitch angle** is the angle between the blade chord line and a reference plane determined by the main rotor hub. The pitch angle is changed in the cockpit by pulling up on the collective.

A rotor's **angle of attack** is the angle between the rotor's chord and the relative wind, and should not be confused with the relative pitch angle. It may be greater than, less than, or the same as the pitch angle. If the pitch angle is increased, the angle of attack is increased, if the pitch angle is reduced, the angle of attack is reduced. [Figure 2-119]

Increasing the angle of attack increases lift, up to a point. As the airfoil is inclined, the air flowing over the top is diverted over a greater distance, resulting in an even greater increase in air velocity and more lift. As the angle of attack is increased further, it becomes more difficult for the air to flow smoothly across the top of the airfoil. Thus, it starts to separate from the airfoil and enters a burbling or turbulent pattern. This turbulence results in a loss of lift in the area where it is taking place.

There are several other terms you should know. One of these is **disc loading** of a helicopter, which is the ratio of weight to the total main rotor disc area. Disk loading is determined by dividing the total helicopter weight by the rotor disc area, which is the area swept by the blades of a rotor. Disc area is found by using the span of one rotor blade as the radius of a circle and then determining the area the blades encompass during a complete rotation.

The **solidity ratio** is the ratio of the total rotor blade area, which is the combined area of all the main rotor blades, to the total rotor disc area.

The **tip-path plane** is the imaginary circular plane outlined by the rotor blade tips as they make a cycle of rotation.

Gyroscopic precession is a characteristic of all rotating bodies. When a force is applied to the outside of a rotating body, parallel to its axis of rotation, the rotating body tilts 90° in the direction of rotation from the point where the force was applied.

The **axis of rotation** of a helicopter rotor is the imaginary line about which the rotor rotates. It is represented by a line drawn through the center of, and perpendicular to, the tip-path plane.

CONTROL OF LIFT

Lift is increased in two ways: by increasing the rotor speed through the air or by increasing the angle of attack. A pilot can increase or decrease rotor speed and, thereby, increase or decrease lift. However, in most helicopter operations, rotor rpm is kept constant and changes in lift can be accomplished by changing the angle of attack with the collective.

In order for a helicopter to generate lift, the rotor blades must be turning. This creates a relative wind which is opposite the direction of rotor system rotation. The rotation of the rotor system creates centrifugal force which tends to pull the blades straight

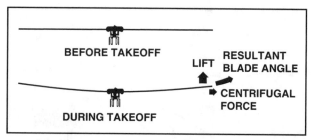

Figure 2-120. Rotor blade coning occurs as the rotor blades begin to lift the weight of the helicopter. In a semirigid rotor system, coning results in blade bending. With a fully articulated system, the blades assume an upward angle through movement about the flapping hinges.

outward from the main rotor hub. The faster the rotation, the greater the centrifugal force. This force gives the rotor blades their rigidity and, in turn, the strength to support the weight of the helicopter. The centrifugal force generated determines the maximum operating rotor rpm due to structural limitations on the main rotor system.

As the pitch angle is increased, lift increases. When this occurs, the rotor blades flex upward under the stress of the load. This upward flexing is called **coning**. Coning is caused by the combination of lift and centrifugal force. [Figure 2-120]

WEIGHT

Normally, weight is thought of as being a known, fixed value; that is, the weight of the helicopter, fuel, and occupants. To lift the helicopter off the ground vertically, the rotor system must generate enough lift to overcome or offset the total weight of the helicopter and its occupants. This is accomplished by increasing the pitch angle of the main rotor blades.

DRAG

Drag is the force which opposes helicopter movement. The total drag acts parallel to the relative wind in a direction opposite the helicopter's movement. You learned earlier that there are two main types of drag — induced and parasite. **Induced** drag is created in the process of the rotor blades developing lift. Higher angles of attack which produce more lift also increase induced drag. As airspeed increases, induced drag decreases. It is the major cause of drag at reduced airspeeds.

In addition to the induced drag caused by the development of lift, there is parasite drag due to skin friction and form. **Parasite** drag is present any time the helicopter is moving through the atmosphere and increases with airspeed. Components of the heli-

copter, such as the cabin, rotor mast, tail, and landing gear, contribute to parasite drag. Additionally, any loss of momentum by the airstream due to such things as openings for engine cooling creates additional parasite drag. Because of its rapid increase with increasing airspeed, parasite drag is predominate at high airspeeds.

THRUST

Thrust, like lift, is generated by the rotation of the main rotor system. In a helicopter, thrust can be forward, rearward, sideward, or vertical. Remember the resultant of lift and thrust determines the direction of movement of the helicopter.

The tail rotor also produces thrust. The amount of thrust is variable through the use of the anti-torque pedals and is used to control the helicopter's movement around the vertical axis.

FORCES IN FLIGHT

Once a helicopter leaves the ground, it is supported by the four aerodynamic forces. In this section, we will examine these forces as they relate to a helicopter in flight.

HOVERING

For standardization purposes, this discussion assumes a stationary hover in a no-wind condition. During hovering flight, the variable forces are directed to make the helicopter perform as needed. While hovering, the amount of main rotor thrust is changed to maintain the desired hovering altitude. This is done by changing the angle of attack of the main rotor blades and by varying power, as needed. In this case, thrust acts in the same vertical direction as lift. [Figure 2-121]

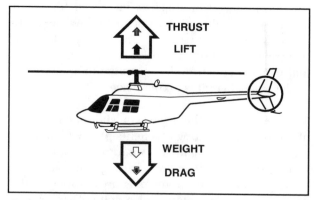

Figure 2-121. To maintain a hover at a constant altitude, enough lift and thrust must be generated to equal the weight of the helicopter and the drag produced by the rotor blades.

The weight that must be supported is the total weight of the helicopter and its occupants. If the amount of thrust is greater than the actual weight, the helicopter gains altitude; if thrust is less than weight, the helicopter loses altitude.

The drag of a hovering helicopter is mainly induced drag incurred while the blades are producing lift. There is, however, some parasite drag on the blades as they rotate through the air. Throughout the rest of this discussion, the term "drag" will include both induced and parasite drag.

An important consequence of producing thrust is torque. Remember, for every action there is an equal and opposite reaction. Therefore, as the engine turns the main rotor system in a counterclockwise direction, the helicopter fuselage tends to turn clockwise. The amount of torque is directly related to the amount of engine power being used to turn the main rotor system. Therefore, as power changes, torque changes.

To counteract this torque-induced turning tendency, an antitorque tail rotor is incorporated into most helicopter designs. The amount of thrust produced by the tail rotor is varied in relation to the amount of torque produced by the engine. As the engine supplies more power, the tail rotor must produce more thrust.

While hovering in helicopters with the main rotor rotation from right to left, the entire helicopter has a tendency to drift toward the right because of the horizontal thrust produced by the tail rotor. This drifting tendency is called **translating tendency**. [Figure 2-122]

Since the main body of the helicopter is supported by the main rotor system at a single point, it is free to swing or oscillate. This oscillating tendency, sometimes referred to as **pendular action**, can be exaggerated by rough control usage and overcontrolling.

GROUND EFFECT

When hovering near the ground, a phenomenon known as ground effect occurs. This effect usually occurs less than one rotor diameter above the surface. As rotor downwash is restricted by the surface friction, it increases the lift vector. This allows a lower rotor blade angle for the same amount of lift, which reduces induced drag. As a result, you need less power to hover near the ground.

Wind reduces ground effect as do energy absorbing ground surfaces. As wind speed or helicopter

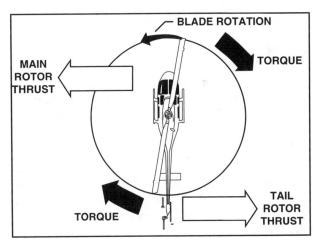

Figure 2-122. The tail rotor is designed to produce thrust in a direction opposite torque. The trust produced by the tail rotor is sufficient to move the helicopter laterally. This is normally counteracted in the design of the helicopter by slightly tilting the rotor mast opposite the direction of tail rotor thrust.

groundspeed exceeds approximately five miles per hour, ground effect becomes less effective. Tall grass and water surfaces absorb and dissipate thrust and, therefore, also reduce ground effect. Ground effect is at its maximum in a no-wind condition over a firm, smooth surface.

FORWARD FLIGHT

Up to this point, the helicopter has been considered to be in a stationary hover with no wind. Now, the aerodynamic effects of moving through the air (or hovering with a wind) will be presented.

To begin forward flight, the main rotor disc is tilted in the desired direction of travel. This diverts some thrust forward and causes the helicopter to move. [Figure 2-123]

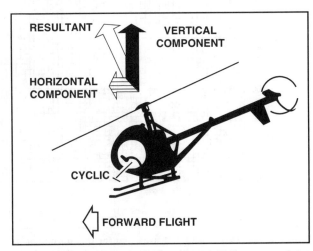

Figure 2-123. To transition into forward flight, some of the vertical thrust must be vectored horizontally. You initiate this by forward movement of the cyclic control.

As the helicopter moves forward, it begins to lose altitude because of the lift that is lost as thrust is diverted forward. However, as the helicopter begins to accelerate, the rotor system becomes more efficient due to the increased airflow through it. The result is excess power. Continued acceleration causes more airflow through the rotor disc and more excess power.

TRANSLATIONAL LIFT

When the helicopter accelerates to an airspeed of approximately 15 miles per hour, a rapid increase in excess power develops, as evidenced by a transient induced aerodynamic vibration. This building of excess power is called translational lift. The additional performance, as a result of translational lift, is sometimes called "effective translational lift." It can be present in a stationary hover if the wind speed is in excess of 15 mph. [Figure 2-124]

DISSYMMETRY OF LIFT

As the helicopter moves through the air, the relative airflow through the main rotor disc is different on the advancing side than on the retreating side. The relative wind encountered by the advancing blade is increased by the forward speed of the helicopter, while the relative wind speed acting on the retreating blade is reduced by the helicopter's forward airspeed. Therefore, as a result of the relative wind speed, the advancing blade side of the rotor disc produces more lift than the retreating blade side. This situation is defined as dissymmetry of lift. [Figure 2-125]

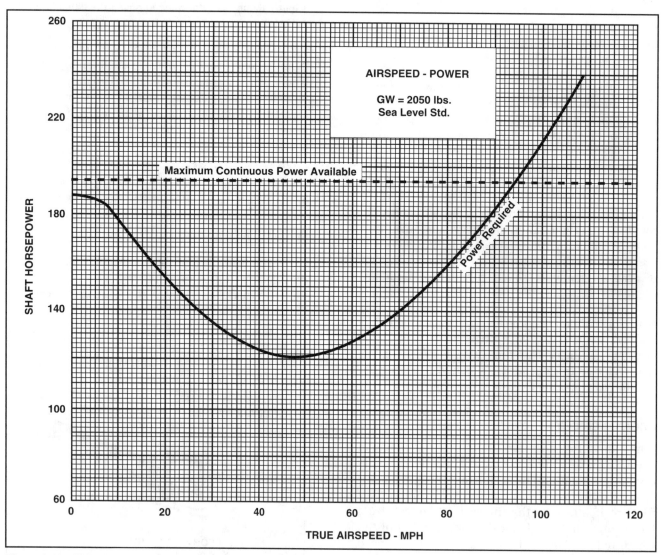

Figure 2-124. This chart illustrates that for this particular helicopter, a significant amount of horsepower is required at a stationary hover. As the airspeed increases from zero, the power required decreases. This is in part due to translational lift. The increase in power required beyond 47 mph is due to drag. The lowest part of the curve (47 mph) represents the best rate-of-climb airspeed, since this is the speed where there is the most excess power.

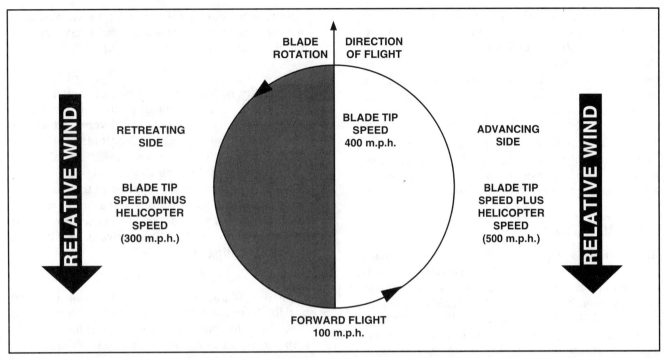

Figure 2-125. The blade tip speed of most helicopters is approximately 400 mph. If the helicopter is moving forward at 100 mph, the relative wind speed on the advancing side is 500 mph. On the retreating side, it is only 300 mph. This difference in speed causes a dissymmetry of lift.

If this condition is allowed to exist, the helicopter will roll to the left because of the difference in lift. In reality, the main rotor blades flap and feather automatically to equalize lift across the rotor disc. Fully articulated rotor systems (normally three or more blades) incorporate a horizontal hinge to allow the individual rotor blades to move (flap) up and down as they rotate. Semirigid rotor systems (two blades) utilize a teetering hinge which allows the blades to flap as a unit. When one blade flaps up, the other flaps down.

As lift increases on the advancing side of the rotor disc due to the increased relative wind speed, the rotor blade flaps up. This upward blade movement with respect to the relative wind reduces the angle of attack which, in turn, reduces lift. The opposite movement occurs on the retreating blade side. As the lift decreases because of decreased relative wind speed, the rotor blade flaps down. This downward blade movement increases the angle of attack and associated lift. The combined upward flapping (reduced lift) of the advancing blade and downward flapping (increased lift) of the retreating blade equalizes lift.

The combination of blade flapping and slow relative wind acting on the retreating blade normally limits the maximum forward speed of a helicopter. At a high forward speed, the retreating blade stalls because of a high angle of attack and slow relative wind speed. This situation is called **retreating blade stall** and, as a result of gyroscopic precession, is evidenced by a nose pitchup, vibration, and a rolling tendency — usually to the left.

CORIOLIS EFFECT

When the rotor blades flap, their center of mass moves closer or farther from the axis of rotation. As a blade flaps up, its center of mass moves closer to the axis of rotation. Coriolis effect then causes that blade to accelerate. Conversely, when a blade flaps down, it decelerates, until it reaches the neutral, or level, position — if below neutral it again accelerates. This rotor blade acceleration/deceleration is termed leading, lagging, or hunting. [Figure 2-126]

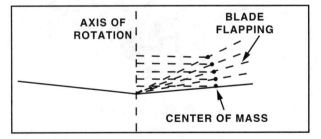

Figure 2-126. Coriolis effect causes the acceleration or deceleration of the rotor blade as a result of the blade's center of mass moving closer to or further from the axis of rotation.

A vertical hinge, sometimes called the drag hinge, is incorporated into fully articulated rotor systems to accommodate this blade acceleration/deceleration. Dampers are installed to cushion and control these forces. A semirigid system does not use these hinges, but instead uses blade bending or flexing to counteract this force.

The same principles just discussed for forward flight also apply to rearward or sideward flight, or hovering with a crosswind except that high speed and retreating blade stall are not factors.

AUTOROTATION

Autorotation is the state of flight where the main rotor system is being turned by the action of the relative wind passing up through the disc, rather than engine power. It is the means by which a helicopter is landed in the event of an engine failure. All helicopters must have this capability in order to be certificated. Autorotation is permitted mechanically because of a freewheeling unit which allows the main rotor to continue turning even if the engine is not running. [Figure 2-127]

To successfully change the downward flow of air to an upward flow during an autorotation, the pitch angle of the main rotor blades must be reduced by lowering the collective pitch. This can be compared to lowering the nose or pitch attitude of a fixed-wing aircraft in order to establish a glide.

THE THREE AXES

Like airplanes, helicopters have a vertical, lateral, and longitudinal axis that pass through the center of gravity. A helicopter yaws around the vertical axis, pitches around the lateral axis, and rotates around the longitudinal axis. [Figure 2-128]

Moving the cockpit controls causes motion about each of these axes. Although rotation about the lateral, or pitch, axis is affected by several factors, the control of the helicopter's pitch attitude is primarily a function of tilting the main rotor disc. This is accomplished by moving the cyclic pitch control. When forward pressure is applied to the cyclic, the main rotor disc tilts forward. This causes the nose of the helicopter to pitch down. When back pressure is applied, the rotor disc tilts rearward, resulting in the nose pitching up.

In forward flight, moving the cyclic toward the left or right, as desired, tilts the rotor disc sideward and rolls the helicopter about the longitudinal axis. To stop the roll, the cyclic is returned to the neutral position.

Movement about the vertical axis is referred to as yaw. The antitorque system controls the rotation around the vertical axis. By using the antitorque pedals mounted on the floor of the cabin, a pilot can increase or decrease tail rotor thrust by changing the pitch on the tail rotor blades.

Applying pressure to the right pedal decreases the pitch of the tail rotor blades and, therefore, decreases the amount of thrust produced. Applying pressure to the left pedal increases the pitch and increases the thrust produced.

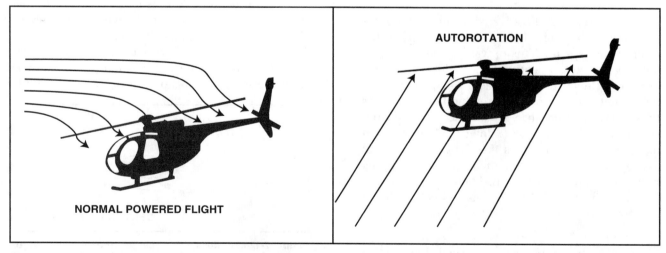

Figure 2-127. In normal powered flight, air is being drawn into the main rotor system from above and exhausted downward. During autorotation, airflow enters the rotor disc from below as the helicopter descends. This upward flow of relative wind permits the main rotor blades to rotate at their normal speed. In effect, the blades are "gliding" in their rotational plane.

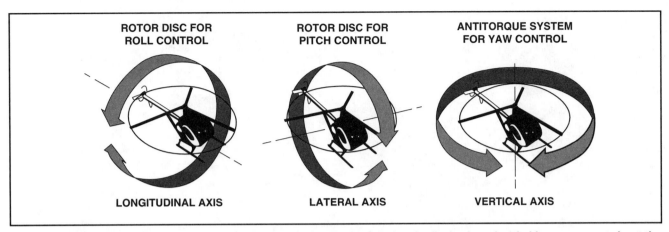

Figure 2-128. Movement of the rotor disc causes a rolling action about the longitudinal axis and a pitching movement about the lateral axis. The antitorque system controls the yawing movement around the vertical axis.

CHAPTER 3

BASIC ELECTRICITY

INTRODUCTION

As you know, electricity runs a great deal of all modern devices. Like any other means of transportation, aircraft rely on electricity for a number of functions, including, but not limited to, lighting, communication, navigation, and environmental control. In fact, some newer aircraft designs eliminate control cables and rely entirely on electrically actuated flight controls. As an aircraft maintenance technician, you will encounter electricity every day, and a solid grasp of this subject is essential.

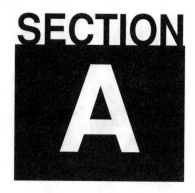
THEORY AND PRINCIPLES

Although electricity and electronics encompass a considerable body of knowledge, the basic elements are neither difficult nor exceedingly complex. Electricity is simply the flow of electrons through a conductor. It is produced by a variety of chemical and physical means, but all use the manipulation of subatomic positively and negatively charged particles. This section discusses the theory needed to understand the production and control of the flow of electrons.

DISCOVERY OF ELECTRICITY

One of the first recorded mentions of electricity was by the Greek philosopher Thales in about 500 B.C. He reported that when substances such as amber and jet were rubbed with a piece of cloth, they attracted light objects such as feathers and bits of straw. Later in the eighteenth century, it was discovered that there were two kinds of forces, or charges, caused by rubbing certain materials together. Charges of the same kind repelled each other while opposite charges attracted.

In about the middle eighteenth century, the practical mind of Benjamin Franklin found a way to prove that lightning was a form of electricity. In his famous kite experiment, he flew a kite into a thunderstorm and found that sparks jumped to the ground from a metal key attached to the wet string. Franklin made a logical assumption that whatever it was that came down the string was flowing from a high level of energy to a lower level. He assigned the term "positive" to the high energy, and "negative" to the lower level. It was not known what actually came down the string, but Franklin used a term associated with the flow of water, and said that it was "current" that flowed down the string, from positive to negative.

The assumption that electricity flowed from positive to negative was accepted until the discovery of the electron in 1897. At that time, it was discovered that electrons, or negatively charged particles, actually move through a circuit. However, there are still many textbooks in use today that speak of current as being from positive to negative.

ELECTRON THEORY

When a light bulb is connected to a source of electrical energy by solid conductors, or wires, there appears to be no movement within the conductor. However, if you could see inside the wires, you would find that they are not really solid. In fact, you would see that the wire contains far more empty space than expected. This space allows for electron flow between atoms.

THE ATOM

As discussed in Chapter 2, all of the material in the universe is composed of atoms, which are the smallest particles that can exist, either alone or in combination with other atoms. You should also recall that each atom consists of a **nucleus** containing posi-

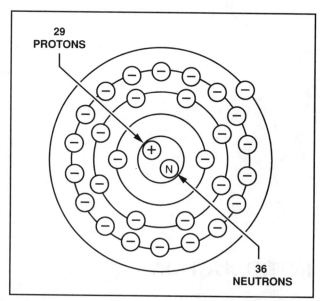

Figure 3-1. The nucleus of a copper atom consists of 29 positively charged protons and 36 electrically neutral neutrons. Spinning around the nucleus are 29 negatively charged electrons. Since the positive charges equal the number of negative charges, the atom is said to be electrically balanced.

tively charged **protons** and neutrally charged **neutrons**. Surrounding the nucleus and traveling at high speeds are **electrons**, each of which is negatively charged and weighs about 1/1845 as much as a proton.

All electrons are alike, as are all protons and all neutrons. However, the number and arrangement of these elementary building blocks determine the material that atoms make up. For example, copper has a nucleus consisting of 29 protons and 36 neutrons. Surrounding the nucleus and spinning in four rings, or "**shells**," are 29 electrons. This combination yields what is called an electrically balanced atom. In other words, there are exactly the same number of positive charges as there are negative charges. The neutrons have no electrical charge and do not affect the flow of electricity. [Figure 3-1]

All matter contains energy, and energy in an atom causes the electrons to spin around the nucleus. As the electrons spin, centrifugal force tends to pull them away from the nucleus. However, the electrostatic attraction between protons and electrons produces a force which opposes this centrifugal force and holds the electrons in a specific orbit.

The electrons spin around the nucleus in shells at a constant radius. Some atoms have up to seven shells. Each shell holds a certain number of electrons. For example, the first shell holds 2 electrons, the second shell holds 8, the third shell holds 18, and so on. The outermost shell containing at least one electron is called the **valence shell**. Likewise, the electrons in the valence shell are called **valence electrons**.

IONS

Positive electrical forces outside an atom tend to attract or rob electrons from an atom's outer ring. This results in an unbalanced electrostatic condition and leaves the atom with an electrical charge. Charged atoms are called **ions**. If an atom possesses an excess of electrons, it is said to be negatively charged, and is called a **negative ion**. On the other hand, an atom with excess protons is called a **positive ion**. For example, copper has one electron in its outer ring. When a positive force is applied to the atom, the valence electron is drawn from the atom and leaves it with more protons than electrons. The atom is now a positive ion and tries to attract an electron from a nearby balanced atom. Electrons constantly move within a material from one atom to another in a random fashion.

CONDUCTORS AND INSULATORS

Some materials have an atomic structure that easily permits the movement of electrons. These materials are referred to as **conductors**. Materials are typically good conductors if they have fewer than five electrons in their outer shells. Four excellent conductors are silver, copper, gold, and aluminum. Materials which oppose the movement of electrons are called **insulators**. Insulators typically have between five and eight valence electrons and therefore do not easily accept additional electrons. To prevent the inadvertent flow of electricity, insulators are often placed around conductors. Some common insulating materials are plastic, rubber, glass, ceramics, air (or vacuum), and oil.

ELECTRON FLOW

Consider what happens when a conductor made of copper is connected across a source of electrons. The positive terminal of the source attracts an electron from an atom in the conductor and the atom leaves the conductor. The atom which lost the electron now becomes a positive ion and pulls an electron away from the next atom. This exchange continues until the electron that left the conductor initially is replaced by one from the source's negative terminal. [Figure 3-2]

Electron movement takes place within the conductor at about the speed of light, which is approximately 186,000 miles per second. However, this does not mean that a single electron moves from one end of a conductor to the other at this speed.

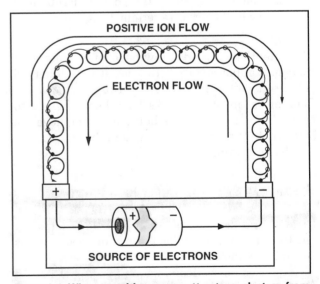

Figure 3-2. When a positive source attracts an electron from a conductor, it leaves a positive ion. This ion attracts an electron from an adjoining atom. This exchange continues through the conductor until an electron is furnished by the negative terminal to replace the one taken by the source.

Instead, an electron entering one end of the conductor almost immediately forces another electron out the other end. [Figure 3-3]

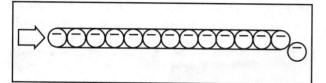

Figure 3-3. When one electron enters a conductor, it immediately forces another electron out of the opposite end.

EFFECTS OF ELECTRON FLOW

Although you cannot see the movement of electrons within a conductor, you can see and use the effects of this movement. For example, as electrons flow through a conductor they produce a magnetic field around the conductor. The greater the flow, the stronger the field. Furthermore, as electrons flow, the opposition to their flow produces heat within the conductor.

DIRECTION OF FLOW

Since the flow of electricity could not be observed, it was only natural to assume that it flowed from a high level of energy to a lower level or, in electrical terms, from "positive" to "negative." This theory worked well for years. In fact, many textbooks were written calling the flow of electrons "current flow," and assumed a flow from the positive terminal of the source to the negative terminal.

As scientists gained knowledge of the atom, it became apparent that the negatively charged electron actually moved through a circuit. Therefore, most textbooks have been revised to explain electron flow as being from the negative terminal, through the load, and back into the positive terminal.

Because electricity was thought to flow from positive to negative for so long, the theory is still discussed and is referred to as **conventional flow**. Although the conventional flow of electricity is technically incorrect, it does follow the arrow symbology used on semiconductors. The proper flow of electricity is termed **electron flow**. You may use either method for tracing flow, as long as you remain consistent. This chapter follows electron flow and uses the terms electron flow and current

interchangeably. In chapters dealing with semiconductor devices and their symbols, the flow of conventional current is used. This is because the arrows used in semiconductor symbols point in the direction of conventional current flow. [Figure 3-4]

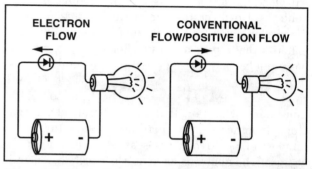

Figure 3-4. Electron flow refers to the flow of electrons from the negative terminal to the positive terminal. On the other hand, conventional current is said to travel from positive to negative. It is easier to think in terms of conventional current when working with semiconductor symbols.

UNITS OF ELECTRICAL MEASUREMENT

The electron is such a small particle that an enormous number of them are required to obtain a measurable unit. The **coulomb** is the basic unit of electrical quantity and is equivalent to 6.28 billion billion electrons. This is typically written as 6.28×10^{18}. The symbol for quantity is **Q**.

When one coulomb of electrons flows past a point in one second, there is a flow of one **ampere**, or one **amp**. It makes no difference whether you think in electron flow or conventional flow, it is all generally called current. The symbol used to represent current is **I**.

The **ohm** is the standard unit of resistance, or opposition to current flow, and is represented by an Ω. One ohm is the resistance through which a force of one volt results in a flow of one ampere.

The force that causes electrons to flow is called the **electromotive force** or **EMF**. An electromotive force is measured in **volts**. One volt represents the amount of force required to cause one amp of flow through one ohm of resistance. A number of terms are used to express electrical force. They are: voltage, voltage drop, potential, potential difference, EMF, and IR drop. These terms have slightly different meanings, but are often used interchangeably. The symbol used to represent the volt is **E**.

Typically, electricity is used to generate **power**. The standard unit of measure for electrical power is the **watt**. One watt is the amount of power dissipated when one amp of current flows under a force of one volt. The symbol for power is **P**. [Figure 3-5]

CHARACTERISTIC	SYMBOL	UNIT
Electrical Charge (Quantity)	Q	Coulomb
Electromotive Force	E or V	Volt
Current	I	Ampere or Amp
Resistance	R	Ohm
Power	P	Watt

Figure 3-5. This table illustrates a summary of electrical characteristics and their corresponding units.

METRIC PREFIXES

Many terms used in the study of electricity deal with numbers that are either extremely large or extremely small. Because of this, metric prefixes are used extensively. For example, the basic unit of capacitance is the farad, and one farad is much too large for practical use in aircraft electronics. Therefore, a typical capacitor has a capacity of 0.000,000,000,002 farad. A number such as this is awkward to work with, and its use encourages errors. A more convenient way to express this unit is to use the prefix pico, which represents 0.000,000,000,001. By doing this the capacitance is represented as two picofarads, or 2pf. [Figure 3-6]

MULTIPLIER	PREFIX	SYMBOL
1,000,000,000,000	tera	t
1,000,000,000	giga	g
1,000,000	mega	M
1,000	kilo	k
100	hecto	h
10	deka	dk
0.1	deci	d
0.01	centi	c
0.001	milli	m
0.000,001	micro	μ (mu)
0.000,000,001	nano	n
0.000,000,000,001	pico	p

Figure 3-6. The metric prefixes, pico, micro, milli, and kilo are used extensively in the study of electricity.

POWERS OF TEN

Another method used to express very small and very large numbers is through scientific notation. In this method of handling numbers, the primary number is converted into a value between one and ten by moving the decimal the appropriate number of places. For example, 0.000,000,002 is converted into 2.0 by moving the decimal to the right nine places. Since the original value is smaller than one, the number two must be multiplied by a negative power of ten. When 0.000,000,002 is converted to scientific notation, it becomes 2×10^{-9}.

Numbers larger than one are converted in exactly the same way, except they are multiplied by a positive power of ten. For example, one coulomb contains 6,280,000,000,000,000,000 electrons. This number is easier to work with when the decimal is moved to the left 18 places. The resulting value in scientific notation becomes 6.28×10^{18}.

If you recall from Chapter 1, numbers that have been converted to scientific notation are multiplied or divided by performing the required mathematical operation to the numbers, and then adding or subtracting the exponents.

Example:

$$0.0025 \times 5,000 = (2.5 \times 10^{-3}) \times (5 \times 10^{3})$$
$$= 12.5 \times 10^{0}$$
$$= 12.5$$

$$5,000,000 \div 250,000 = (5 \times 10^{6}) \div (2.5 \times 10^{5})$$
$$= 2 \times 10^{1}$$
$$= 20$$

STATIC ELECTRICITY

There are two basic types of electricity; they are **current** and **static**. In current electricity, electrons move through a circuit and perform work through the magnetic field created by their movement, or by the heat generated when forced through a resistance. Static electricity, on the other hand, serves little useful purpose. In fact, static electricity is more often a nuisance rather than a useful form of electrical energy.

Static electricity is of real concern during the fueling operation of an aircraft. For example, as an aircraft flies, friction between the air and the aircraft surface builds up a static charge. Once an aircraft lands, the static charge cannot readily dissipate, since the tires insulate the aircraft from the ground. If the first thing to contact a statically charged aircraft is a fuel nozzle in the filler neck, a spark can ignite the explosive fumes and cause a serious fire. To prevent this, you must always ground an aircraft prior to fueling. This is usually accomplished by connecting the aircraft to the fuel truck, which is in turn grounded to the earth. [Figure 3-7]

Figure 3-7. Airplanes and fuel trucks should be grounded together to neutralize any charge of static electricity before the fueling nozzle is put into the tank.

POSITIVE AND NEGATIVE CHARGES

As you know, you cannot see static electricity. However, you can observe its effects through the use of a glass rod and a pair of pith balls suspended by a string. For example, if you rub a glass rod with a piece of wool or fur, the rod picks up additional electrons, and therefore becomes negatively charged. When the rod is held close to a suspended pith ball with a neutral charge, the rod attracts the ball. However, once the ball touches the rod, the excess electrons on the rod flow to the ball and give the ball a negative charge. Whenever the rod and ball have like charges, they repel each other.

In a second example, if you rub a glass rod with a piece of silk, the rod gives up electrons and becomes positively charged. If you then hold the rod near a neutrally charged pith ball, the rod attracts the ball. However, once the ball touches the rod, the ball loses some electrons to the rod and assumes a positive charge and is repelled. [Figure 3-8]

The strength of repelling and attracting forces vary as the inverse of the square of the distance between the two charges. For example, if the distance between two objects with dissimilar charges is dou-

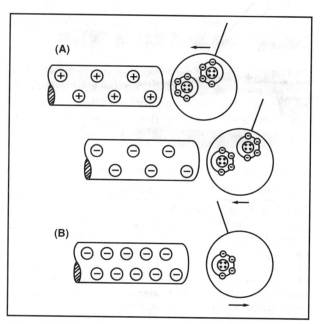

Figure 3-8. (A) — An uncharged pith ball is attracted to a rod that has either a positive or a negative charge. (B) — Once the charged rod contacts the ball, the ball assumes the same charge as the rod and is repelled.

bled, the force of attraction is reduced to one-fourth its original value. If the distance between the two objects is tripled, the force becomes one-ninth its original value. On the other hand, if the distance between two electrically charged objects is cut in half, the force between them increases by a factor of four. [Figure 3-9]

ELECTROSTATIC FIELDS

If you could see the lines of electrostatic force between two opposite charges, you would see that the lines leave one charged object and enter the other. If the charges are close together all of the lines link and the two charges form a neutral, or an uncharged, group. However, the lines of electrostatic force from like charges repel each other and tend

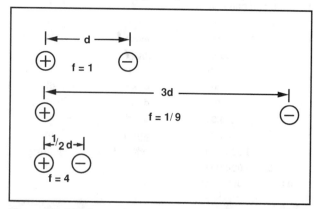

Figure 3-9. The force of repulsion or attraction varies as the inverse of the square of the distance between the charges.

to push the charges apart. Electrostatic fields are also known as **dielectric fields**. [Figure 3-10]

DISTRIBUTION OF ELECTRICAL CHARGES

When a body having a smooth or uniform surface is electrically charged, the charge distributes evenly over the entire surface. If the surface is rough or irregular in shape, the charge concentrates at points or areas having the sharpest curvature. This explains the action of **static dischargers** used on many aircraft control surfaces. As discussed earlier, when an airplane flies through the air, friction causes a static charge to build on the aircraft's surface. To help prevent an excessive charge from building, many aircraft utilize **static wicks**, or **null field dischargers**. These devices are attached to the aircraft's control surfaces and provide points where electricity can concentrate and then discharge into the air. [Figure 3-11]

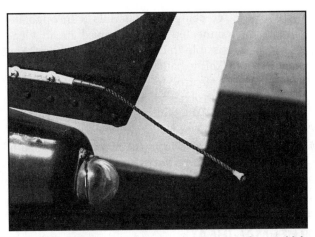
Figure 3-11. Static dischargers provide points from which static charges are dissipated into the air before a high potential builds on the control surface.

An aircraft's control surfaces are connected to the airframe structure by hinges which do not provide a good conductive path for static electricity to move to static dischargers. Because of this, several aircraft utilize **bonding straps** which provide a conductive path between the two structures. The maximum permissible resistance of a bonding strap is .003 ohms. [Figure 3-12]

SOURCES OF ELECTRICITY

As you know, energy cannot be created or destroyed. However, energy can be converted from one form to another. The conversion of chemical, thermal, pressure, light, and magnetic energy into electricity, and the exchange of electricity back into these forms, is commonplace today.

CHEMICAL

Some materials exist as positive ions having a positive charge, while others exist as negative ions and

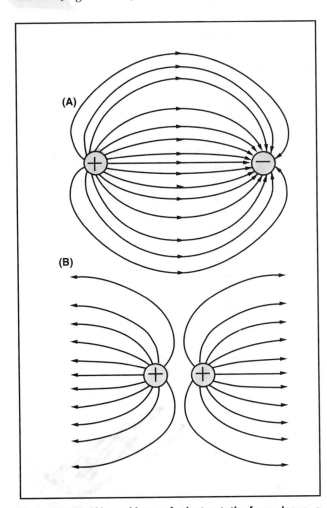
Figure 3-10. (A) — Lines of electrostatic force leave a charged body at right angles to its surface, and then spread apart. They enter an oppositely charged body at right angles to its surface. (B) — Charged bodies reject lines of electrostatic force from other bodies having the same charge.

Figure 3-12. Bonding straps provide a low-resistance path between a control surface and the aircraft structure.

carry a negative charge. If materials with opposite charges are connected and immersed in an electrolyte, it is possible to create an electron flow. For example, an alkaline battery consists of a carbon rod and paste-like electrolyte enclosed in a zinc container. The electrolyte reacts chemically with the zinc and changes the zinc to zinc chloride. As this reaction takes place, the zinc releases electrons. When you run a wire from the zinc can, through a light bulb, and back to the carbon rod, the electrons flow through the light and into the carbon rod. [Figure 3-13]

THERMAL (HEAT)

A **thermocouple** is a loop of two wires made of dissimilar metals that are joined at two places. When a temperature difference exists between the two junctions, electrical current flows. This property makes thermocouples valuable as temperature sensors in aircraft. Iron/constantan and chromel/alumel are the metal pairs most commonly used in thermocouples. A cylinder head temperature measuring system has one junction held tightly against a hot engine cylinder head by a spark plug, while the other junction is mounted in an area where the temperature remains relatively constant. [Figure 3-14]

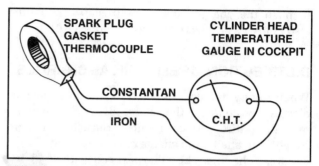

Figure 3-14. Electrons flow in a thermocouple made of dissimilar metals when there is a temperature difference between the two junctions.

PRESSURE

When a crystalline material such as quartz is bent or deformed by a mechanical force, an excess of electrons accumulates on one surface. This is known as the **piezoelectric** effect and is commonly used in crystal microphones and phonograph pickups. A piece of crystal vibrates at one natural frequency and when a crystal is excited by pulses of electrical energy, it vibrates at this frequency. As it vibrates, it produces alternating voltage that has a very specific frequency. [Figure 3-15]

LIGHT

When light strikes certain **photoemissive** materials such as selenium, light energy is absorbed. When this occurs, electrons are discharged. These electrons can be channelled through a conductor to an electrical circuit. Solar powered calculators take advantage of electrical current produced by light. [Figure 3-16]

Figure 3-13. Electrons flow between two dissimilar materials when they are connected by a conductor and immersed in an electrolyte.

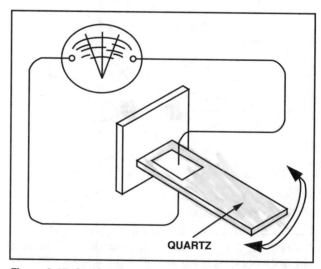

Figure 3-15. An electrical potential difference builds across the faces of certain crystalline materials when they are bent or otherwise subjected to mechanical pressure.

MAGNETISM

One of the most effective devices used to produce electricity is the magnet. By definition, a **magnet** is a body that has the ability to attract ferrous substances and produce an external magnetic field. It is these magnetic fields that are of interest in the study of electricity.

The ends of the magnet are called the **poles**, and are referred to as the north- and south-seeking poles. The north-seeking end of a magnet is labeled "N," and the opposite south-seeking end is labeled "S." These labels refer to the direction sought by the pole of the magnet.

Magnetism follows the same rules as charges of static electricity. Like poles repel each other, and the force of repulsion follows the inverse square law. This means that if the distance between the poles is doubled, the force of repulsion is reduced to one-fourth. On the other hand, the force of attraction is squared as the distance decreases. In other words, when you decrease the separation by half, the force of attraction increases four times.

As stated earlier, a magnet produces an external magnetic field. A magnetic field consists of invisible lines called **lines of magnetic flux**. Lines of flux are always complete loops that leave the north-seeking pole of the magnet at right angles to its surface and re-enter the south pole in the same fashion. Since lines of flux are polarized in the same direction, they repel each other and spread out between the poles. [Figure 3-17]

The number of lines of flux that loop through a magnet is an indication of a magnet's strength. One line of flux is called one **maxwell**. On the other hand, flux density is measured in **gauss** which represents the number of lines of flux per given area. One gauss represents a density of one maxwell per square centimeter.

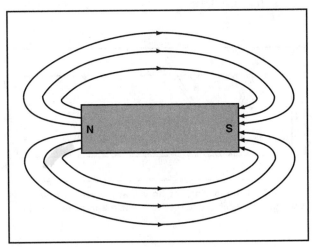

Figure 3-17. Lines of magnetic flux form complete loops, leaving the magnet at its north pole and returning at its south pole.

The inside of a piece of unmagnetized iron contains an almost infinite number of magnetic fields oriented in a random fashion. However, if a piece of iron is placed in a strong magnetic field, all of the fields, or **domains**, align themselves with the induced magnetic field. Once this occurs, the iron becomes a magnet having a north and south pole, and lines of magnetic flux. [Figure 3-18]

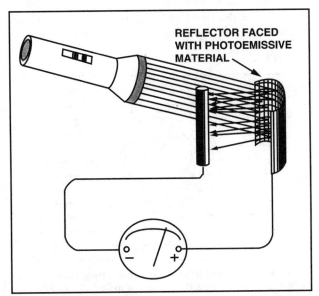

Figure 3-16. A photoemissive material emits electrons when struck by light.

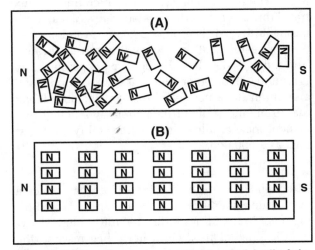

Figure 3-18.(A) — In an unmagnetized material, all of the individual magnetic fields, or domains, are arranged in a random fashion and cancel each other. (B) — When the material is magnetized, all of the domains are aligned, and the material has a north and south pole.

The domain theory of magnetism is supported by the fact that each magnet has both a north and south pole, regardless of the size of the magnet. For example, if you break a bar magnet in two, each half demonstrates the characteristics of the original magnet. If you break each of these halves in two, all of the pieces still retain magnetic properties. [Figure 3-19]

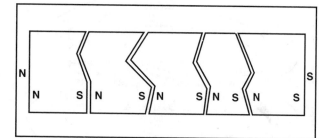

Figure 3-19. Regardless of how many pieces a magnet is broken into, each piece retains a north and a south pole.

Soft iron has a very low **retentivity**, meaning that as soon as a magnetizing force is removed, the domains lose their alignment and the iron loses its magnetism. On the other hand, materials such as hard steel and some iron alloys have very high retentivities and retain their magnetic properties long after they are magnetized. Materials with high retentivity are used as permanent magnets in aircraft magnetos, instruments, and radio speakers.

As lines of flux travel from the north pole to the south, they always follow the path of least resistance. The measure of ease with which lines of flux travel through a material is referred to as a material's **permeability**. Air is used as a reference and is given the permeability of one. Since flux travels through iron much easier than through air, its permeability is around 7,000. Other materials such as copper and aluminum have permeabilities considerably lower than iron, and some of the extremely efficient permanent magnet alloys have permeability values as high as 1,000,000.

Although lines of flux are invisible, if you place a magnet under a piece of paper and sprinkle iron filings over it, the filings form a definite pattern showing the lines of flux. On a horseshoe magnet, the flux lines pass directly between the poles of the magnet. However, if you place a piece of soft iron above the poles, the lines flow through the iron to the south pole. [Figure 3-20]

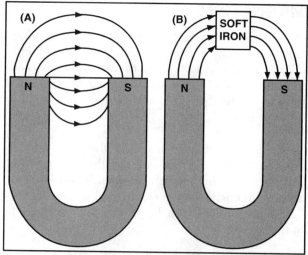

Figure 3-20. (A) — Lines of magnetic flux leave the north pole of a magnet at right angles to its surface and travel to the south pole, where they enter at right angles to its surface. (B) — The flux lines always seek the path of least resistance, even if it means traveling longer distances.

Lines of flux have the ability to pass through any material. However, if it is important that a device be protected from magnetic fields, it can be surrounded by a soft iron shield. Since flux lines travel the path of least resistance, any lines of flux flow through the iron leaving an area inside the shield with no magnetic field. [Figure 3-21]

The characteristic of flux lines to pass through a permeable material also explains the attraction of ferrous metals to a magnet. For example, since flux lines seek the path of least resistance, lines of flux want to link the poles with the shortest possible loops. Flux lines exert a strong pull on permeable metals to center them between the poles. As a piece of metal is pulled in closer, more lines of flux pass

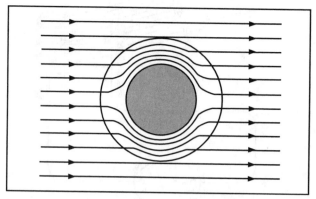

Figure 3-21. One way to shield an object from lines of magnetic flux is to enclose it in a shield made of a highly permeable material. The lines of flux flow through the shield and bypass its center.

through it and the pull becomes stronger. When the piece of metal centers, it resists any force that tries to lengthen the lines of flux. [Figure 3-22]

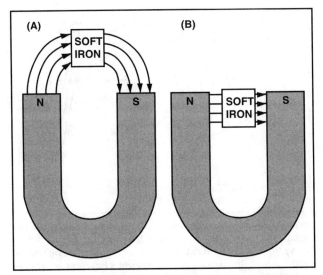

Figure 3-22. (A) — Lines of flux pass through a material having a high permeability. In their effort to keep all of the force loops as short as possible, a force is exerted on the soft iron to pull it into the center of the magnetic field. (B) — When the soft iron is centered between the poles, it resists any attempt to lengthen the lines of flux.

Almost all magnets, regardless of their retentivity, lose some of their magnetic strength when their lines of flux pass through the air. Because of this, a magnet whose strength is critical is stored with **keepers**. A keeper is a piece of soft iron that is used to link the poles and provide a highly permeable path for the flux.

MAGNETISM AND ELECTRICITY

If a conductor is moved through the lines of magnetic flux that pass between the poles of a magnet, a flow of electrons is induced in the conductor. This is called **electromagnetic induction** and is the most common form of electric power generation in use today. Most aircraft use generators or alternators to produce electricity by this method. In fact, atomic, hydro-electric, and fossil fuel powerplants produce power by the same procedure.

The amount of electricity induced depends on the rate at which the lines of flux are cut. This rate can be increased by increasing the number of flux lines, by making the magnet stronger, or by moving the conductor through the lines faster. [Figure 3-23]

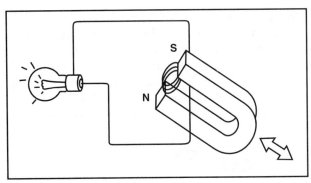

Figure 3-23. The amount of electricity generated by electromagnetic induction is determined by the rate at which the conductor cuts through lines of magnetic flux.

ELECTROMAGNETISM

Although the effects of magnetism were observed for centuries, it was not until 1819 that the relationship between electricity and magnetism was discovered. The Danish physicist Hans Christian Oersted discovered that the needle of a small compass deflected when it was held near a wire carrying electric current. This deflection was caused by the invisible magnetic field surrounding the wire.

You can see the magnetic field produced by a conductor by sprinkling iron filings on a plate that surrounds a current-carrying conductor. When this is done, the filings arrange themselves in a series of concentric circles around the conductor. The reason for this is when electrons travel through a conductor, they produce lines of flux. The greater the amount of flow, the stronger the magnetic field. [Figure 3-24]

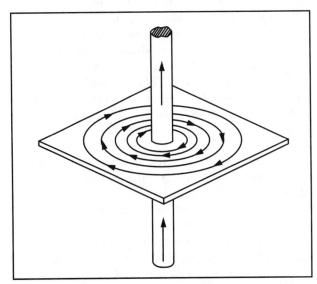

Figure 3-24. Lines of magnetic flux encircle a current-carrying conductor. These lines are relatively weak, and have no polarity.

One way to determine the direction the lines of flux travel is with the **left-hand rule**. For example, if you grasp the conductor in your left hand with your thumb pointing in the direction of electron flow, your fingers encircle the conductor in the direction of the lines of flux travel. [Figure 3-25]

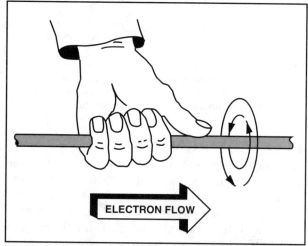

Figure 3-25. When a current-carrying conductor is grasped with the left hand, with the thumb pointing in the direction of electron flow, your fingers encircle the conductor in the same direction the flux lines travel.

Because the magnetic field around a conductor does not have any poles and is relatively weak, it does not serve a practical purpose. However, if the conductor is wound into the form of a coil, the lines of flux become concentrated and the coil attains the characteristics of a magnet. [Figure 3-26]

In an electromagnet, the lines of flux surrounding each turn of wire reinforce the flux around every other turn of wire. This results in a magnetic field that leaves the north end of the coil and enters the south end. To determine which end of an electromagnet is north and which is south, you can use the **left-hand rule for coils**. This rule states that if you grasp a coil with your left hand so your fingers wrap around the coil in the direction of electron flow, your thumb points to the coil's north pole. [Figure 3-27]

The strength of an electromagnet is determined by the number of turns in the coil, the amount of current flowing through it, and the type of material used for a core. A coil's strength or its **magnetomotive force**, is similar to the electromotive force in an electrical circuit. However, magnetomotive force is measured in **gilberts**. One gilbert is equal to .7968 ampere-turns and is symbolized by the letters "**Gb.**"

The field intensity of an electromagnet is measured in gauss just like a conventional magnet. One gauss

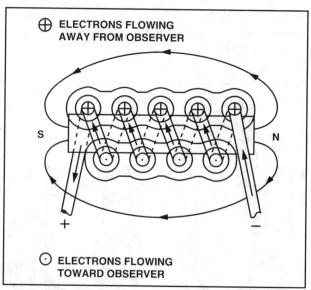

Figure 3-26. Electrons flow into the coil from the right. As the conductor passes over the top of the coil, the electrons flow away from you, as indicated by the cross representing the tail of the arrow. Below the coil, the electrons flow toward you, as indicated by the dot representing the head of the arrow. When the electron flow is away from you, the lines encircle the conductor in a counterclockwise direction. When they come toward you, the field circles the conductor clockwise.

represents a density of one maxwell per square centimeter. To increase the density of the lines of flux, a highly permeable material, such as soft iron, is used for the core.

The law for magnetic circuits states that one gilbert is the amount of magnetomotive force produced by one maxwell flowing through a magnetic circuit having one unit of reluctance. **Reluctance** is the opposition in a circuit to the flow of magnetic flux and is inversely proportional to permeability.

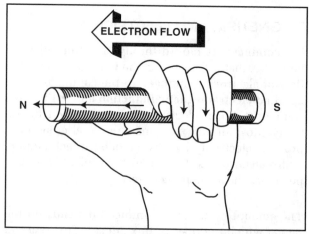

Figure 3-27. If a coil is grasped with the left hand in such a way that your fingers encircle it in the same direction as the electron flow (from negative to positive), your thumb points to the north pole of the electromagnet formed by the coil.

Simply stated, an electromagnet having one ampere-turn produces a magnetomotive force of 1.256 gilberts.

CURRENT ELECTRICITY AND OHM'S LAW

As discussed earlier, a quantity of electrons produces an electromotive force that causes electrons to flow through a circuit. However, as electrons flow through a conductor, they are met by an opposition or resistance. By assigning values to the force, flow, and opposition within a circuit, the relationship that exists between these items becomes apparent.

It was the German scientist George Simon Ohm who documented the relationship between force, flow, and opposition. **Ohm's law** is a basic statement which says the current that flows in a circuit is directly proportional to the voltage (force) that causes it, and inversely proportional to the resistance (opposition) in the circuit.

To ease the handling of these terms in formulas, voltage is represented by the letter **E**, current, or amperes, by the letter **I**, and resistance by the letter **R**. A statement of Ohm's law in the form of a formula is: Volts = Amps × Resistance, or **E = IR**. In other words, voltage equals amps times resistance. Through algebra, this same relationship is used to determine current and resistance. For example, to find current, use the formula **I = E/R**, and resistance is found by the formula **R = E/I**. [Figure 3-28]

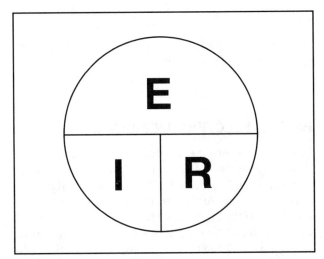

Figure 3-28. The relationship between volts, amps, and resistance is often illustrated in this way. To determine an unknown quantity, cover the unknown quantity with your thumb. The location of the uncovered letters indicates the mathematical operation to be performed. For example, to find ``R,'' you must divide volts by amps.

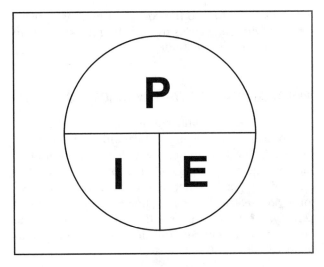

Figure 3-29. The relationship between power, current, and voltage.

Power in an electrical circuit is measured in **watts**. One watt is the amount of power used to move one amp of current under a force of one volt. Therefore, power is calculated using the formula:

Power = Amps × Volts, or **P = IE**.

The relationship between voltage, current, resistance, and power is such that if any two values are known, then the other two can be determined. For example, given a light which provides 20 ohms of resistance in a circuit carrying 3 amperes, how much power will the light dissipate? Given the resistance and current, calculate the circuit voltage using the formula E = IR. Once the voltage is known, calculate the power dissipated with the formula P = IE. [Figure 3-29]

Example:

Resistance = 20 ohms
Current = 3 amperes

Calculate the voltage.

E = IR

E = 3 × 20

E = 60 volts

Now calculate the power dissipated.

P = IE

P = 3 × 60

P = 180 watts

One easy way to find the correct formula is to use a series of divided circles representing the symbols in the formula. [Figure 3-30]

MECHANICAL POWER IN CIRCUITS

Power, as you remember from physics, is the time-rate of doing work. The practical unit of measure for power is the horsepower, which is the amount of power required to do 33,000 foot-pounds of work in one minute, or 550 foot-pounds of work in one second. In electrical terms, 1 horsepower is equal to 746 watts. Using this relationship, determine the amount of current required to raise a 1,000-pound load six feet in 30 seconds.

Example:

Given:
Hoist = 24 volts
Force = 1,000 pounds
Distance = 6 feet

Step 1: Calculate the horsepower required.

$$\text{Horsepower} = \frac{\text{Force} \times \text{Distance}}{550 \times t}$$

$$\text{Horsepower} = \frac{1,000 \text{ lbs.} \times 6 \text{ ft.}}{550 \times 30 \text{ sec.}}$$

$$\text{Horsepower} = .364$$

Step 2: Convert the horsepower to watts.

1 horsepower = 746 watts

.364 hp $\times$ 746 = 271.5 watts

Step 3: Calculate the amperes required.

$$I = \frac{P}{E}$$

$$I = \frac{271.5}{24}$$

$$I = 11.31 \text{ amps}$$

No electric motor is 100 percent efficient. Therefore, motors require more than 746 watts to produce 1 horsepower. To determine the number of watts

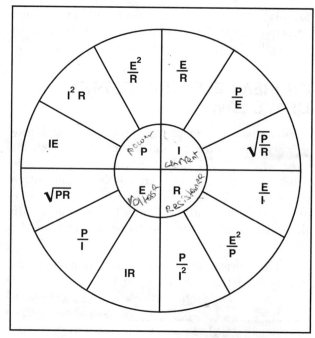

Figure 3-30. Summary of basic equations using the volt, ampere, ohm, and watt.

required to produce a given horsepower, multiply the horsepower by 746 watts and divide by the motor's efficiency. For example; how many watts will a 90 percent efficient motor require to generate 2 horsepower?

Watts required = (2 horsepower $\times$ 746 watts) $\div$ .90

= 1,492 $\div$ .90

= 1,657.8 watts

A 90 percent efficient motor requires 1,657.8 watts to produce 2 horsepower.

HEAT IN ELECTRICAL CIRCUITS

In circuits where mechanical work is not actually done, power is still an important consideration. For example, if you install a resistor in a light circuit to drop the voltage from 12 volts to 3 volts for a light bulb that requires 150 milliamps, you must find the resistance and power the resistor must dissipate. To solve this problem, first find the voltage to be dropped.

Example:

E = 12 − 3 = 9 volts

Find the resistance required to drop the voltage:

$$R = E/I = 9/0.15 = 60 \text{ ohms}$$

Find the power dissipated in the resistor:

$$P = I \times E = .15 \times 9 = 1.35 \text{ watts}$$

The resistor must dissipate 1.35 watts, but for practical purposes, you would use a two-watt resistor. [Figure 3-31]

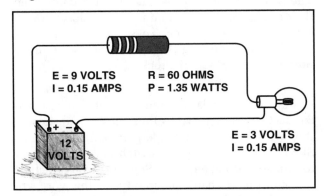

Figure 3-31. Determining the characteristics of a resistor needed to drop voltage in a circuit.

CIRCUIT ELEMENTS

All complete electrical circuits consist of at least three elements. They are a source of electrical energy, a load device to use the electrical energy produced by the source, and conductors to connect the source to the load. However, these circuit elements do not comprise a practical electrical circuit. For example, in order to control the flow of electrons, a control device, such as a switch, is placed in most circuits. Fuses or circuit breakers also are provided to protect the circuit wiring in the event of an overload or malfunction. [Figure 3-32]

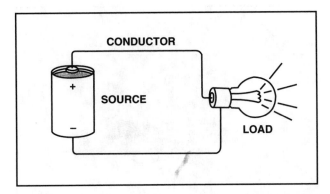

Figure 3-32. All complete circuits must have a source of electrical energy, a load to use the energy, and conductors to join the source and the load.

CONDUCTORS

The purpose of a conductor is to provide a path for electrons to flow from a source, through the load, and back to the source with minimum resistance. However, other factors such as load carrying ability and durability must also be considered. Therefore, the choice of a conductor is often a compromise. Most aircraft electrical systems are of the single-wire type, meaning the aircraft structure provides the path through which the current returns to the source. Although this type of system saves a great deal of weight, it is extremely important that a good connection exist between the aircraft structure and the battery, generator, and all devices using current.

The resistance of a conductor is affected by three things, its physical characteristics, its dimensions, and temperature.

PHYSICAL CHARACTERISTICS

As mentioned earlier, you want a conductor to carry an electrical load and provide minimum resistance. **Resistivity** is the resistance of a standard length and cross-sectional area of a conductor. Most practical aircraft circuits use either copper or aluminum conductors. Copper wire has about two-thirds the resistance of an equivalent gauge of aluminum wire, and, therefore, is most generally used. However, for applications requiring a great deal of current, aluminum wire is often used. Although the resistivity of aluminum is higher than copper and a larger conductor is needed to carry the same current, aluminum weighs much less than copper. Therefore, a great deal of weight is saved by its use. [Figure 3-34]

DIMENSIONS

For most conductors, the amount of resistance varies directly with the conductor's length. That is, as length increases for a given conductor, its resistance increases.

On the other hand, the resistance of a conductor varies inversely with its cross-sectional area. In other words, as a conductor's cross-sectional area increases, resistance decreases. Aircraft wire is measured by the **American Wire Gage (AWG)** system, with the larger numbers representing the smaller wires. The smallest size wire normally used in aircraft is 22-gauge wire, which has a diameter of about

0.025 inch. However, conductors carrying large amounts of current are typically of the 0000, or four aught size, and have a diameter of about 0.52 inch.

A **circular mil** is the standard measurement of a round conductor's cross-sectional area. One mil is equivalent to .001 inches. Thus, a wire that has a diameter of .125 is expressed as 125 mils. To find the cross-sectional area of a conductor in circular mils, square the conductor's diameter. For example, if a round wire has a diameter of 3/8 inch, or 375 mils, its circular area is 140,625 circular mils (375 × 375 = 140,625).

The **square mil** is the unit of measure for square or rectangular conductors such as bus bars. To determine the cross-sectional area of a conductor in square mils, multiply the conductor's length by its width. For example, the cross-sectional area of a strip of copper that is 400 mils thick and 500 mils wide is 200,000 square mils.

It should be noted that one circular mil is .7854 of one square mil. Therefore, to convert a circular mil area to a square mil area, multiply the area in circular mils by .7854 mil. Conversely, to convert a square mil area to a circular mil area, divide the area in square mils by .7854. [Figure 3-33]

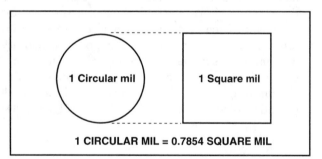

Figure 3-33. Relationship between circular mils and square mils.

TEMPERATURE

Metals have what is known as a **positive temperature coefficient of resistance**. This means that the resistance of the material increases as its temperature increases. This characteristic is used in some temperature measuring instruments where the resistance change in a piece of wire is used to measure temperature. For practical purposes, however, both copper and aluminum exhibit small changes in resistance with the temperatures encountered in flight, and therefore it is normally not considered.

SWITCHES

As mentioned earlier, most practical electrical circuits utilize some sort of switch to safely control the flow of electrons. The following information looks at some of the common switches used in the aviation industry.

TOGGLE OR ROCKER SWITCHES

The two most common switches used to control the flow of electrons in aircraft are the enclosed toggle switch and rocker switch. These switches are actuated by either moving the bat-shaped toggle or by pressing on one side of a rocker.

If a switch controls only one circuit and has two connections and two positions, open and closed, the switch is called a single-pole, single-throw, or SPST switch. This type of switch is generally used to turn something on or off. Some switches are used to control more than one circuit and, therefore, are referred to as double-pole switches. A double-pole switch can have either a single- or double-throw. The double-pole, single-throw, DPST switch is generally used to control both the battery and generator circuit so they both turn ON and OFF at the same time. On the other hand, a double-pole, double-throw, DPDT switch controls two circuits and has either two or three positions. Some toggle and rocker switches have one or both of their positions spring-loaded so they return to a desired position when your finger is removed. [Figure 3-35]

WAFER SWITCHES

When a switch is used to select one of a number of conditions, a wafer switch is often used. These switches have several wafers stacked on a common

Figure 3-35. Most switches found in modern aircraft are either the toggle or the rocker type.

COPPER ELECTRIC WIRE CURRENT-CARRYING CAPACITY

Wire size—Specification MIL-W-5086	Single wire in free air-maximum amperes	Wire in conduit or bundled-maximum amperes	Maximum resistance-ohms/1.000 feet (20 °C)	Nominal conductor area-circular mills
AN-20	11	7.5	10.25	1,119
AN-18	16	10	6.44	1,779
AN-16	22	13	4.76	2,409
AN-14	32	17	2.99	3,830
AN-12	41	23	1.88	6,088
AN-10	55	33	1.10	10,443
AN-8	73	46	.70	16,864
AN-6	101	60	.436	26,813
AN-4	135	80	.274	42,613
AN-2	181	100	.179	66,832
AN-1	211	125	.146	81,807
AN-0	245	150	.114	104,118
AN-00	283	175	.090	133,665
AN-000	328	200	.072	167,332
AN-0000	380	225	.057	211,954

ALUMINUM ELECTRIC WIRE CURRENT-CARRYING CAPACITY

Wire size—Specification MIL-W-7072	Single wire in free air-maximum amperes	Wire in conduit or bundled-maximum amperes	Maximum resistance-ohms/1.000 feet (20 °C)	Nominal conductor area-circular mills
AL-6	83	50	0.641	28,280
AL-4	108	66	.427	42,420
AL-2	152	90	.268	67,872
AL-0	202	123	.169	107,464
AL-00	235	145	.133	138,168
AL-000	266	162	.109	168,872
AL-0000	303	190	.085	214,928

Figure 3-34. Characteristics of aircraft copper and aluminum wire.

shaft, and each wafer can have as many as twenty positions. Wafer switches are seldom used for carrying large amounts of current and are most generally open with wires soldered to the terminals on the wafers. [Figure 3-36]

PRECISION SWITCHES

Some electrical circuits require a switch to be actuated by the movement of some mechanism. In these applications it is usually important that the switch actuate when the mechanism reaches a very definite

Figure 3-36. Wafer switches are used when it is necessary to select any of a large number of circuit conditions.

Figure 3-37 Precision switches snap open and close with an extremely small amount of movement of the operating control.

and specific location. Snap-acting switches have a wide use in these applications. These types of switches typically have a plunger that requires an extremely small movement to trip and drive the contacts together. When the plunger is released, a spring snaps the contacts apart. [Figure 3-37]

RELAYS

It is often necessary to open or close a circuit carrying a large amount of current from a remote location. An example is the starter circuit for an aircraft engine. A starter motor requires a great deal of current and, therefore, a large conductor is required. To prevent having to run a large conductor to the instrument panel where the battery switch is located and back down to the starter, a relay is used. A **relay** is simply an electrical switch that is operated from a remote location. With a relay, a small amount of current energizes an electromagnet which, in turn, closes a set of contacts to complete a second circuit. [Figure 3-38]

Figure 3-38. Starter solenoid switches control large amounts of current, but they are operated by a very small current.

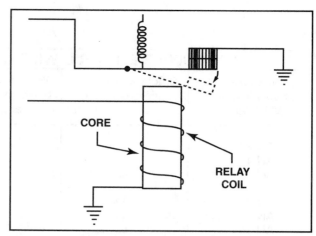

Figure 3-39. With a fixed-core electromagnetic switch, once sufficient current flows through the coil, the resulting magnetic field opens the contacts.

There are two general types of relays used in the aviation industry; those having fixed cores, and those with movable cores. Both use a coil of wire, or **solenoid**, that surrounds a soft iron core. In a **fixed-core electromagnet** the core remains stationary at all times. However, once current flows through the coil, the magnetic force produced opens or closes a set of contacts to complete another circuit. [Figure 3-39]

On relays that utilize a **movable-core electromagnet** the core of the electromagnet is typically held out from the center of the coil by a spring. However, once current flows through the coil, a strong magnetic field is produced that overcomes the spring and pulls the core into the center of the coil. When this occurs, a set of contacts attached to the core are pulled down to complete another circuit. [Figure 3-40]

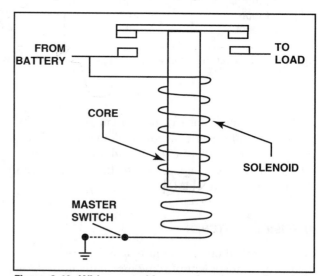

Figure 3-40. With a movable-core electromagnetic switch, once sufficient current flows through the solenoid, the resulting magnetic field closes the contacts.

PROTECTIVE DEVICES

Protective devices are installed in electrical circuits to prevent damage caused by **overloading** a circuit or a **short** in a circuit. Overloading a circuit results from connecting loads that are too large for the wiring. A short, on the other hand, occurs when part of a circuit in which full system voltage is present comes in direct contact with the return side of the circuit. When a short occurs, a path for current flow with little or no resistance is established. This results in large amounts of current flow and conductor heating.

FUSES

One of the simplest devices used to protect a circuit is the fuse. A fuse is made of a low-melting-point alloy enclosed in a glass tube. The fuse is installed in a circuit and, when current flow becomes excessive, the metal alloy melts and opens the circuit. Some fuses are designed to withstand a momentary surge of current, but create an open if the current is sustained. These slow-blow fuses have a small spring attached to a link so when the sustained current softens the link, the spring pulls the link apart and opens the circuit. [Figure 3-41]

CIRCUIT BREAKERS

Because it is often inconvenient to replace a fuse in flight, most aircraft circuits are protected by circuit breakers. Like fuses, circuit breakers automatically open a circuit if current flow becomes excessive. However, once the circuit cools, the breaker is easily reset by moving the operating control. If a breaker trips because of a surge of voltage or some isolated and nonrecurring problem, the circuit breaker remains in and the circuit operates normally. However, if an actual fault such as a short cir-

cuit exists, the breaker trips again. If a breaker trips shortly after it is reset, it should be left open until the problem is isolated.

Aircraft circuit breakers are of the trip-free type which means that once the breaker opens, the circuit remains open until the circuit cools regardless of the position of the operating control. With this type of breaker, it is impossible to hold the circuit closed if an actual fault exists.

Circuit breakers operate on either thermal or magnetic principles. **Thermal breakers** open a circuit when excess current heats an element in the breaker causing the contacts to open. On the other hand, **magnetic breakers** utilize the magnetic field caused by the current in the circuit to open the contacts. In addition to the classification by operating principle just described, three basic types of circuit breakers are used in aircraft. They are the **push/pull**, **push-to-reset**, and **toggle** types. [Figure 3-42]

Another type of circuit breaker is the **automatic reset circuit breaker**. Like other circuit breakers, the automatic reset type opens when excess current flows through the breaker. However, once the circuit cools, the breaker automatically closes.

Figure 3-42. Most circuit protection for modern aircraft is provided by circuit breakers that can be reset in flight. The push/pull type is shown in the upper panel with the push-to-reset type in the lower panel.

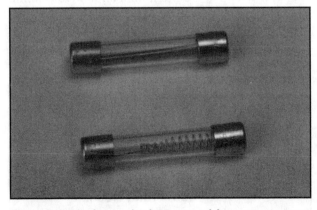

Figure 3-41. In fuses, the heat caused by excess current melts a fusible link and opens a circuit.

RESISTORS

Resistors are used in electrical circuits to control the amount of current flow. They do this by converting some of the electrical energy that flows through the circuit into heat. The resistors used in aircraft are generally classified as fixed or variable.

FIXED RESISTORS

The three most common fixed resistors used in aviation include the composition resistor, the film resistor, and the wire-wound resistor. The **composition resistors** used to control small amounts of current are made of a mixture of carbon and an insulating material. The relative percentage of the two materials in the mix determines the amount of resistance a resistor has. Composition resistors are normally available in sizes from 1/8 watt up to 2 watts. The larger the physical size of the resistor, the more power it can dissipate. Most modern resistors have **axial-leads**, which means one lead comes out of each end of the resistor. Resistance is measured in ohms and the resistance provided by a specific resistor is indicated by a set of three or four colored bands on the resistor. Each color represents a value from one to nine. The band nearest the end of the resistor is the first significant figure of the resistance, the second band represents the second figure, and the third band tells the number of zeros to add to the two numbers. For example, if the first band is green (5), the second is brown (1), and the third yellow (0000), the resistor has a resistance of 510,000 ohms. [Figure 3-43]

Some resistors have a fourth color band that is used as a multiplier for tolerance. For example, a fourth band that is gold indicates a tolerance of plus or minus 5 percent, whereas silver indicates plus or minus 10 percent. If there is no fourth band, the tolerance is plus or minus 20 percent.

Composition resistors with an ohmic value less than 10 have silver or gold as the third band. If the third band is silver, multiply the first two significant figures by 0.01, and if the third band is gold, multiply them by 0.1. For example, a resistor with a yellow (4) first band, a violet (7) second band, and a gold (.1) third band has a resistance of 4.7 ohms. If the third band were silver, the resistance would be 0.47 ohms.

Some composition resistors have leads coming off their body radially instead of parallel to the resistor axis. These **radial-lead** resistors are color-coded with the same colors. However, the color of the body is the first significant figure of the resistance, the color of the end is the second significant figure, and a dot or band of paint around the middle of the resistor indicates the number of zeros to be added. For example, a radial-lead resistor with a red (2) body, green (5) end, and yellow (4) dot has a resistance of 250,000 ohms. [Figure 3-44]

Some radial-lead resistors have a paint mark opposite the colored end which indicates tolerance. Like composition resistors, silver indicates a tolerance of plus or minus 10 percent, gold is plus or minus 5 percent, and no mark is plus or minus 20 percent of the indicated resistance.

Carbon resistors have been replaced in some modern electronics equipment by **film resistors**. Film resistors consist of a thin layer, or film, of resistive material wrapped around a nonconductive ceramic core material. The resistor leads, usually of the axial type, are inserted into a cap and placed onto the ends of the ceramic core. Film resistors are gener-

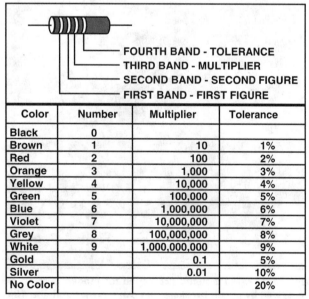

Color	Number	Multiplier	Tolerance
Black	0		
Brown	1	10	1%
Red	2	100	2%
Orange	3	1,000	3%
Yellow	4	10,000	4%
Green	5	100,000	5%
Blue	6	1,000,000	6%
Violet	7	10,000,000	7%
Grey	8	100,000,000	8%
White	9	1,000,000,000	9%
Gold		0.1	5%
Silver		0.01	10%
No Color			20%

Figure 3-43. Color code marking for axial-lead resistors and resistor color-code values.

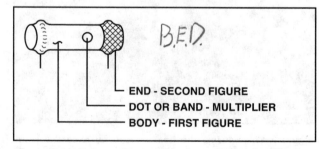

Figure 3-44. Color code markings for radial-lead resistors.

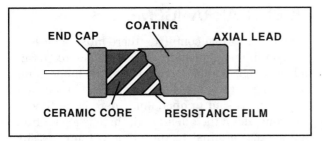

Figure 3-45. The core of a film resistor is typically constructed out of a nonconductive ceramic material that is wrapped by a resistor film.

Figure 3-47. Variable resistors allow the amount of resistance in a circuit to be changed by rotating the shaft.

ally available in the same range of resistance values as carbon resistors. [Figure 3-45]

When a great deal of power needs to be dissipated, special resistors made of highly resistive wire wound over hollow ceramic tubes are used. Some **wire-wound** resistors are tapped along their length to provide different values of resistance. Others have a portion of the wire left bare, so a metal band can be slid over the resistor. This allows the resistor to be set to any desired value. [Figure 3-46]

VARIABLE RESISTORS

When it is necessary to change the amount of resistance in a circuit, variable resistors are used. Variable resistors are either the composition or the wire-wound type. In a variable composition resistor, the mixture of carbon and insulating material is bonded to an insulating disk, and a wiper, or sliding contact, is rotated by the shaft to vary the amount of material between the two terminals. The farther the sliding contact is from the fixed contact, the greater is the resistance. [Figure 3-47]

Rheostats are variable resistors that have only two terminals, one at the end of the resistance material

and the other at a sliding contact. Most rheostats are wire wound and, therefore, can dissipate a great deal of power. Rheostats vary the amount of current flow in circuits and are commonly used in aircraft to control cockpit lighting. [Figure 3-48]

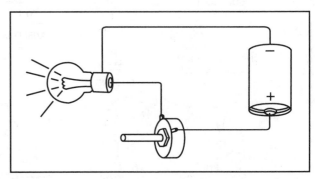

Figure 3-48. Rheostats are used to vary the amount of resistance in a circuit.

If a resistor has three terminals, one on each end of the resistance material, and one on the slider, it is called a **potentiometer**. Potentiometers change the amount of voltage in a circuit, and are often used as voltage dividers. [Figure 3-49]

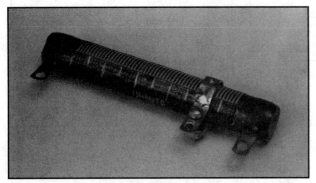

Figure 3-46. Wire-wound resistors are used when there is a great deal of power that must be dissipated. Some wire-wound resistors have a portion of the wire exposed to incorporate a movable tap.

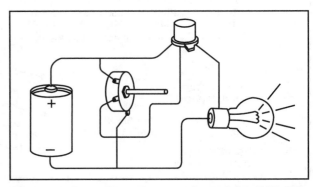

Figure 3-49. Potentiometers are used as voltage dividers in a circuit.

COMPONENT SYMBOLOGY

All components used in electrical circuits are represented by symbols in drawings, blueprints, and illustrations in schematic form. Because of this, you must become familiar with the components commonly used in basic circuits, together with their schematic symbols. [Figure 3-50]

CIRCUIT ARRANGEMENTS

For a circuit to be complete, there must be at least one continuous path from one of the source terminals, through the load and back to the other terminal. If there is any interruption or break in the path, the circuit is said to **open**, and there is no flow of electrons. If, on the other hand, there is a path from one source terminal to the other without passing

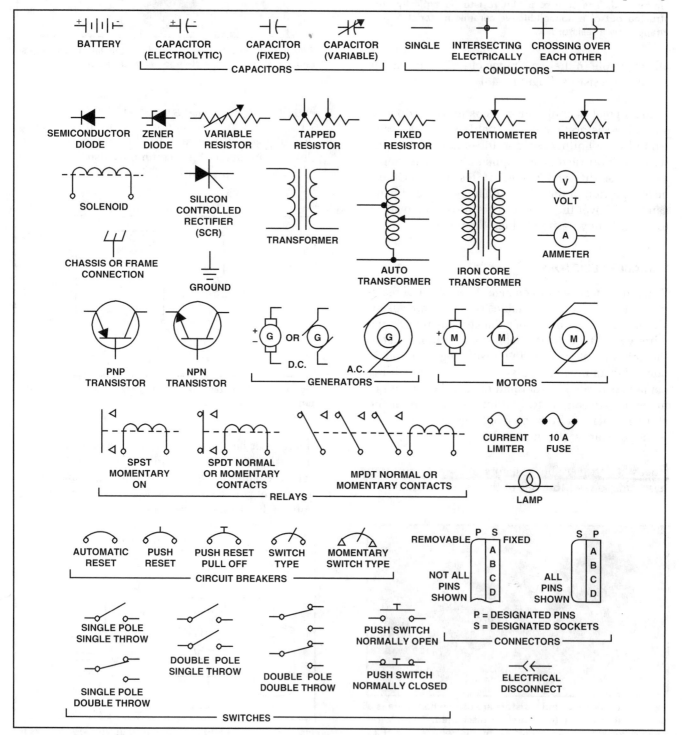

Figure 3-50. Electrical symbols.

through the load the circuit is **shorted**. Not only is there no work being done, but the absence of resistance in the circuit allows excessive current to flow. In this case, unless a fuse or circuit breaker opens the circuit, the wiring and the source can be damaged. [Figure 3-51]

METER USAGE IN CIRCUIT ANALYSIS

Voltmeters, ammeters, and ohmmeters are used to analyze values in electrical circuits. Some basic rules should be followed in their use to prevent injury to the technician or damage to the meters and circuit components.

1. Voltage is always measured across a component. In other words, the probes of the voltmeter go across, or parallel to, the component being measured.
2. Amperage is measured by placing the meter in series with a component. The technician must be sure the ammeter is able to handle the current in the circuit being measured.
3. Resistance is measured with power off the circuit. Individual components must be isolated from the circuit if their resistances are to be checked.
4. Observe proper polarities when connecting a

meter to an operating circuit or to a power supply. Attempting to measure voltage with the meter's probes reversed drives the indicator in the wrong direction and damages the meter.

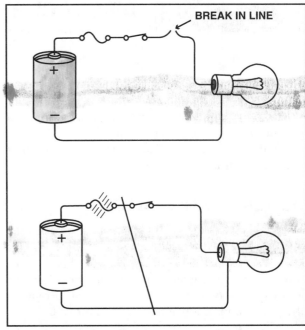

Figure 3-51. (A) — An open circuit allows no current flow and the circuit cannot function. (B) — A short circuit causes excessive current flow, and can damage the circuit components or wiring unless a protective device such as a fuse opens the circuit.

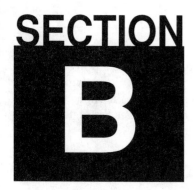
DIRECT CURRENT

Direct current (DC) electricity is nonvarying in nature, such as that obtained from a battery or filtered power supply. In other words, the amplitude of voltage and current remain steady. This is often referred to as "pure DC," meaning that no alternating current or noise is present.

DC TERMS AND VALUES

DC electricity in which either the current or voltage vary from a zero reference level to a maximum or peak value is termed **pulsating DC.** This type of direct current is generally produced by rectifiers in a power supply and is typically filtered to remove the pulses and produce pure DC. [Figure 3-52]

The **average value** of DC is the average of the current or voltage excursion made by a pulsating DC waveform as it moves from zero to its maximum value. The average value is computed by multiplying the maximum value of the pulsating waveform by 0.637. For example, assume the maximum value for a DC waveform is 20 volts. By multiplying 20 volts by 0.637, the average value is calculated to be 12.74 volts. [Figure 3-53]

The **polarity** of DC is expressed as being either positive or negative. It is determined by establishing a reference point (usually ground) and measuring a voltage from that point. For example, battery voltage is measured from the battery's negative terminal, which is connected to ground, to the battery's positive terminal. A typical battery has a voltage of 12 volts and positive polarity with respect to ground.

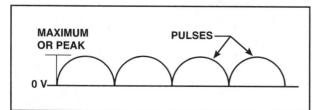

Figure 3-52. As seen on an oscilloscope, pulsating DC begins at a zero reference level and then peaks at a maximum value before it returns to the initial reference level. Pulsating DC never drops below the zero reference level.

SERIES DC CIRCUITS

A **series circuit** is a circuit that has only one path for electrons to flow. Consider a typical circuit in which a battery is the source of power and a lamp is the load which is in series with a rheostat, a switch, and a fuse. Since there is only one path for electron flow, if either the switch is open or the fuse is blown, the lamp cannot illuminate. By the same token, if two lamps are connected in series and one burns out, the circuit opens and the second lamp cannot illuminate. [Figure 3-54]

When a rheostat is in a series circuit, it acts as an electron control device. When it is set for minimum resistance, the maximum amount of current flows through the lamp, and the lamp burns with full brilliance. However, when the rheostat's resistance is increased, part of the power from the battery is dissipated in the resistor in the form of heat. This leaves less power available for the lamp, and therefore, the lamp burns with less than full brilliance.

VOLTS

The German physicist Gustav Robert Kirchhoff helped explain the behavior of voltage and current in electrical circuits. **Kirchhoff's voltage law** states that "the algebraic sum of the applied voltage and the voltage drop around any closed circuit is equal to zero." In other words, the voltage across each load must be exactly the same as the voltage supplied by the source. For example, a six-volt battery is connected in series with a switch, three lamps, and a milliammeter. Each lamp is rated at two volts

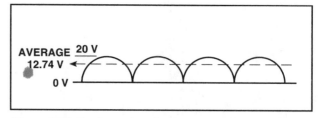

Figure 3-53. To determine the average value of pulsating DC, multiply the peak volts or amperes by .637.

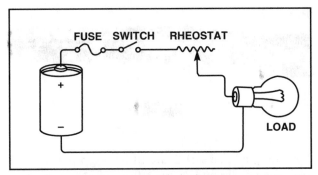

Figure 3-54. In a series circuit there is only one path for the electrons to flow. In this circuit, the flow of electrons begins at the battery's negative terminal and continues through the lamp to the rheostat, switch, fuse, and back to the battery. If the switch is open or the fuse is blown, electrons cannot flow.

and requires 300 milliamps of current to burn at full brilliance. An analysis of this circuit with a voltmeter proves Kirchhoff's voltage law. When a voltmeter is placed across the switch, the full source voltage of six volts is measured when the switch is open. However, when the switch is closed, all of the source voltage flows in the circuit and the meter reads zero volts. Furthermore, with the switch closed and the lamps burning, the measured voltage drop across each lamp is two volts. This indicates that the resistance of each lamp dissipates enough power to drop two volts. To verify this, the total voltage drop across all three lamps is measured and determined to be six volts. [Figure 3-55]

AMPERES

Kirchhoff's current law states that "The algebraic sum of the currents at any junction of conductors in

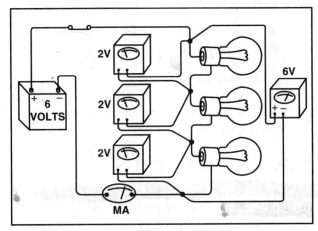

Figure 3-55. In a series circuit, the voltage drop across all resistances equals the system voltage. In the above illustration, the voltage drop across each lamp is two volts, whereas the voltage drop across all the lamps is six volts.

a circuit is zero." This means that the amount of current flowing away from a point in a circuit is equal to the amount flowing to that point. Since there is only one path in a series circuit, the current remains constant throughout the circuit regardless of the number of components. While it is true that an increase in the number of circuit components increases the resistance to current flow, the amount of current flow remains the same value at all points in the circuit.

RESISTANCE

In order to determine the current flow in a circuit, you must know how much resistance a circuit contains. Total resistance in a series circuit equals the sum of the individual resistances in the circuit. This is illustrated in the formula:

$$R_T = R_1 + R_2 + R_3 ...$$

Therefore, a circuit with three resistances of 10 ohms, 20 ohms, and 30 ohms respectively has a total resistance of 60 ohms (10 + 20 + 30 = 60).

Once you know any two of the values for volts, amperes, or resistance, you can use Ohm's law to calculate the third. For example, if a circuit has a voltage of 24 volts and a resistance of 12 ohms, current can be determined.

Example:

Given:
E = 24 volts
R = 12 ohms

$$E = IR$$
$$I = E/R$$
$$I = 24 \text{ volts}/12 \text{ ohms}$$
$$I = 2 \text{ amps}$$

The current flow is 2 amps and is constant throughout the circuit. However, if the resistance is doubled and the voltage remains constant, the value for amps decreases.

Example:

Given:
E = 24 volts
R = 24 ohms
$$I = E/R$$
$$I = 24 \text{ volts}/24 \text{ ohms}$$
$$I = 1 \text{ amp}$$

Therefore, whenever total circuit resistance is doubled, current reduces to half its value. On the other hand, if resistance is reduced to half its former value, the current doubles.

By the same token, if a circuit's resistance is held constant and voltage is doubled, the current flow also doubles. For example, if 48 volts are applied to the earlier example, the current increases to 4 amps.

Example:

Given:
E = 48 volts
R = 12 ohms

$$I = E/R$$
$$I = 48 \text{ volts}/12 \text{ ohms}$$
$$I = 4 \text{ amps}$$

Thus, if resistance remains constant, and voltage increases, current must also increase. On the other hand, if voltage decreases, current decreases.

Ohm's law also allows you to determine the voltage drop across each resistor in a series circuit. This is done by applying Ohm's law to each of the resistances. [Figure 3-56]

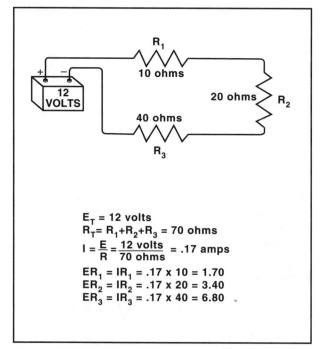

$E_T = 12 \text{ volts}$
$R_T = R_1 + R_2 + R_3 = 70 \text{ ohms}$
$I = \dfrac{E}{R} = \dfrac{12 \text{ volts}}{70 \text{ ohms}} = .17 \text{ amps}$
$ER_1 = IR_1 = .17 \times 10 = 1.70$
$ER_2 = IR_2 = .17 \times 20 = 3.40$
$ER_3 = IR_3 = .17 \times 40 = 6.80$

Figure 3-56. Once total resistance is calculated, current flow can be determined. To determine the voltage drop across each resistor, apply Ohm's law to each resistor.

POWER

Once volts and current are known, the amount of power available in the circuit is calculated using the formula:

$$P = IE$$

For example, in a series circuit having a power supply of 12 volts and current flow of .17 amps, the power available is 2.04 watts. This same formula is used to calculate the power dissipated by each resistor. However, the power dissipated is calculated by multiplying the total current by the voltage drop across each resistor. For example, in our series circuit with three resistors of 10 ohms, 20 ohms, and 40 ohms, the power dissipated is .29 watts, .58 watts, and 1.16 watts respectively.

Example:

Given:
E = 12 volts
I = .17 amps

$P = IE = 12 \text{ volts} \times .17 \text{ amps} = 2.04 \text{ watts}$
$P_{R1} = IE_{R1} = .17 \text{ amps} \times 1.70 \text{ volts} = .29 \text{ watts}$
$P_{R2} = IE_{R2} = .17 \text{ amps} \times 3.40 \text{ volts} = .58 \text{ watts}$
$P_{R3} = IE_{R3} = .17 \text{ amps} \times 6.80 \text{ volts} = 1.16 \text{ watts}$

To briefly summarize the characteristics of a series circuit:

1. There is only one path for the electrons to follow from the source, through the load back to the source.
2. The current is the same wherever it is measured in a series circuit.
3. The sum of all the voltage drops equals the source voltage.
4. The total resistance of the circuit is the sum of the individual load resistances.
5. The total power dissipated in the circuit is the sum of the power dissipated in each of the individual load resistances.

PARALLEL DC CIRCUITS

The most widely used circuit arrangement is the parallel circuit. All of the load components in a parallel circuit are directly across the source, and if one component fails, it has no effect on the others. Therefore, in a parallel circuit, if a lamp burns out, it has no effect on the others.

VOLTS

In a parallel circuit, there are separate paths in which electrons can flow from the source through a load and back to the source. Each path must obey Kirchhoff's voltage law. That is, since each path has only one load device, the voltage must equal the source voltage. Thus, in a circuit powered by a two-volt source, each load has the full two volts across it. Therefore, in a parallel circuit the voltage is the same throughout the circuit. [Figure 3-57]

AMPERES

The behavior of amperes in a parallel circuit is explained in part by **Kirchhoff's current law**. For example, in figure 3-57, all of the current flows through the fuse and the switch. It then splits up, with some passing through each of the lamps. The amount that passes through each lamp is determined by the lamp's resistance. Since all of the lamps have the same resistance in this example, the current flow through each is identical. However, in a parallel circuit with different resistances, the branch containing the small resistance will have a greater current flow than a branch containing a high resistance.

RESISTANCE

Unlike a series circuit, the more resistance added to a parallel circuit, the lower the total resistance. If the resistances are equal in a parallel circuit, total resistance is found by dividing the value of a single resistor by the number of resistors. The formula is:

$$R_T = r/n$$

Where:
r = resistance of one resistor
n = number of resistances

If there are two unlike resistors in a parallel circuit, find the total resistance by dividing the product of the individual resistances by their sum:

$$R_T = \frac{R_1 \times R_2}{R_1 + R_2}$$

For example, if a 100-ohm resistor is connected in parallel with a 200-ohm resistor, their total resistance is 66.7 ohms. [Figure 3-58]

To find the total resistance of two or more unlike resistors in parallel, take the reciprocal of the sum of the reciprocals of the individual resistances. This requires the use of the formula:

$$R_T = \frac{1}{\dfrac{1}{R_1} + \dfrac{1}{R_2} + \dfrac{1}{R_3} + \dots}$$

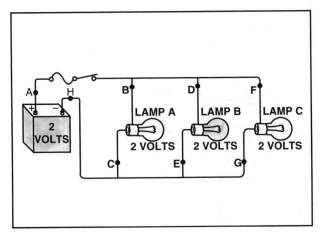

Figure 3-57. In this parallel circuit, there are three separate paths the electrons can follow. If you measure the voltage between A and H, B and C, D and E, and F and G, all read two volts. This demonstrates that in a parallel circuit, the voltage across each path is the same as the source voltage.

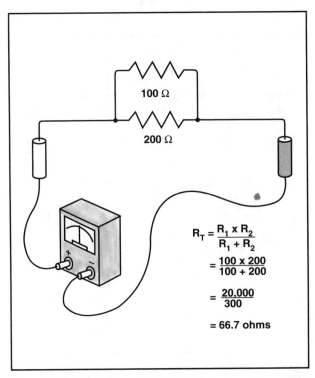

$$R_T = \frac{R_1 \times R_2}{R_1 + R_2}$$
$$= \frac{100 \times 200}{100 + 200}$$
$$= \frac{20{,}000}{300}$$
$$= 66.7 \text{ ohms}$$

Figure 3-58. Finding the equivalent resistance of two unlike resistors in parallel.

Assume a 10-ohm, 20-ohm, and 25-ohm resistor are in parallel. Using these values in the formula given, the total resistance is 5.26 ohms. [Figure 3-59]

An analysis of these equations shows that total resistance is always less than the smallest resistance in a parallel circuit. Furthermore, if a resistor is added to a parallel circuit, total resistance decreases. If a resistor is removed, total resistance increases.

Like a series circuit, once you know two of the three values of volts, amperes, or resistance, the third value can be determined using Ohm's law. Furthermore, Ohm's law is used to determine the current flowing through each branch of a parallel circuit. For example, given a parallel circuit with a 24 volt power supply and three resistors of 10 ohms, 20 ohms, and 50 ohms, determine the total current and the current flowing through each branch. [Figure 3-60]

As a check on the accuracy of your work, add the currents flowing through each branch. The sum of the branch currents should equal the total circuit current.

POWER

Like series circuits, once volts and current are known, the amount of power generated by the circuit is calculated using the formula:

$$P = IE$$

This same formula is also used to calculate the power dissipated by each resistor. However, the current flow in each branch is used instead of the total current. For example, in our series circuit with three resistors of 10 ohms, 20 ohms, and 50 ohms, the power dissipated is 57.6 watts, 28.8 watts, and 11.52 watts respectively.

Example:

$$P_{R1} = I_{R1}E = 2.4 \text{ amps} \times 24 \text{ volts} = 57.6 \text{ watts}$$

$$P_{R2} = I_{R2}E = 1.2 \text{ amps} \times 24 \text{ volts} = 28.8 \text{ watts}$$

$$P_{R3} = I_{R3}E = .48 \text{ amps} \times 24 \text{ volts} = 11.52 \text{ watts}$$

To summarize the characteristics of a parallel circuit:

1. There is more than one path for the electrons to follow from the source, through part of the load, back to the source.
2. The voltage is the same across any of the paths.
3. The current through each path is inversely proportional to the resistance of the path.
4. The total current is the sum of the current flowing through each of the individual paths.

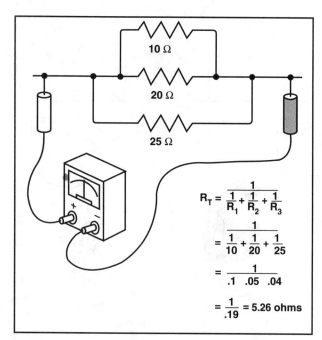

Figure 3-59. The total resistance of two or more unlike resistors is found by taking the reciprocal of the sum of the reciprocals of the resistances.

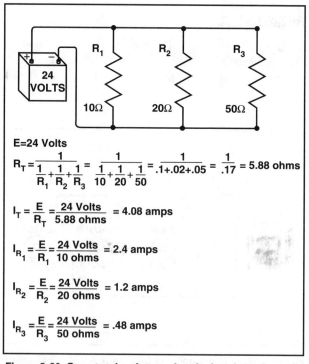

Figure 3-60. Once total resistance is calculated, total current is determined. To determine the amount of current flowing through each branch, apply Ohm's law to each individual resistor.

5. The total resistance of the circuit is less than the resistance of any of the paths.

6. The total power dissipated in the circuit is the sum of the power dissipated in each of the individual load resistances.

COMPLEX DC CIRCUITS

As you will see when you begin working on aircraft, most electrical circuits are not strictly series and parallel. Instead, they are usually complex circuits consisting of both series and parallel circuits. In other words, a complex circuit consists of parallel resistors connected in series with other resistors. [Figure 3-61]

While complex circuits may appear confusing, the same rules used to determine volts, amps, and resistance for series and parallel circuits are applicable. For example, to calculate total resistance in a complex circuit, you can break the circuit down into equivalent series and parallel circuits. Once this is done, you can find the equivalent resistance of each parallel circuit. When doing this, it is typically easiest to start at the parallel branch farthest from the power source.

For example, R_5, R_6, and R_7 comprise a parallel branch with resistances of 40 ohms, 20 ohms, and 40 ohms. Therefore the combined resistance is 10 ohms.

$$R_{5\text{-}6\text{-}7} = \frac{1}{\frac{1}{R_5} + \frac{1}{R_6} + \frac{1}{R_7}}$$

$$= \frac{1}{\frac{1}{40} + \frac{1}{20} + \frac{1}{40}}$$

$$= \frac{1}{.025 + .05 + .025}$$

$$= \frac{1}{.1}$$

$$= 10 \text{ ohms}$$

Now, find the equivalent resistance of the parallel branch containing R_2 and R_3 with resistances of 60 ohms and 30 ohms respectively.

$$R_{2\text{-}3} = \frac{1}{\frac{1}{60} + \frac{1}{30}}$$

$$= \frac{1}{.017 + .033}$$

$$= \frac{1}{.05}$$

$$= 20 \text{ ohms}$$

To make the problem easy to follow, re-draw the circuit using the two equivalent resistances instead of the original combinations. Once this is done, you will see that $R_{5\text{-}6\text{-}7}$ and R_4 are in series with each other and should be combined. This results in an equivalent resistance of 20 ohms. [Figure 3-62]

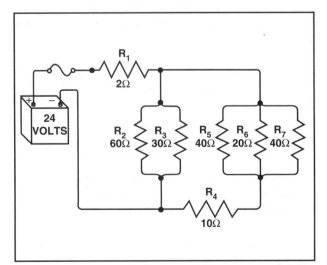

Figure 3-61. This is an example of a typical complex circuit. Notice that resistors R_1 and R_4 are connected in series while resistors R_2, R_3, R_5, R_6, and R_7 are connected in parallel.

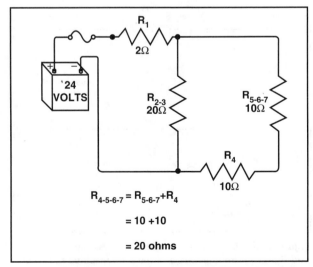

Figure 3-62. When you redraw the circuit using the equivalent resistances, you can combine the resistors in series.

You now have a parallel combination of $R_{4-5-6-7}$ which is 20 ohms and R_{2-3} of 20 ohms. The equivalent resistance of these two combinations is 10 ohms.

$$R_{2-3-4-5-6-7} = \cfrac{1}{\cfrac{1}{R_{4-5-6-7}} + \cfrac{1}{R_{2-3}}}$$

$$= \cfrac{1}{\cfrac{1}{20} + \cfrac{1}{20}}$$

$$= \frac{1}{.05 + .05}$$

$$= \frac{1}{.1}$$

$$= 10 \text{ ohms}$$

Now, redraw the circuit a final time with the equivalent resistances. All that remains is to find the equivalent resistance of the series combination of R_1, which is 2 ohms, and $R_{2-3-4-5-6-7}$, which is 10 ohms. The total circuit resistance is 12 ohms. [Figure 3-63]

You now know the voltage and total resistance of the circuit; therefore, Ohm's law can be used to determine the circuit's total current of 2 amps.

$$E = 24 \text{ volts}$$

$$R = 12 \text{ ohms}$$

$$I = \frac{E}{R} = \frac{24 \text{ volts}}{12 \text{ ohms}} = 2 \text{ amps}$$

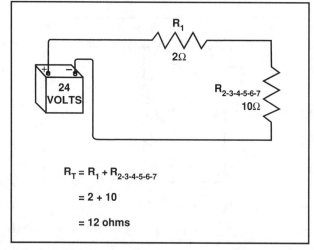

$$R_T = R_1 + R_{2-3-4-5-6-7}$$

$$= 2 + 10$$

$$= 12 \text{ ohms}$$

Figure 3-63. With the circuit redrawn, you can see that R_1 and $R_{2-3-4-5-6-7}$ are in series and have a combined resistance of 12 ohms. This is the equivalent resistance of the entire circuit.

To determine the voltage drop and current across each resistor, you must apply Kirchhoff's voltage and current law to each of the series and parallel circuits. For example, since R_1 is connected in series, all of the current flows through it. However, if you remember, the voltage across a resistor in a series circuit varies. Since R_1 has a resistance of 2 ohms, the voltage drop across it is 4 volts.

$$E_{R_1} = IR_1$$

$$= 2 \text{ amps} \times 2 \text{ ohms}$$

$$= 4 \text{ volts}$$

Since R_1 dissipates 4 volts, only 20 volts are applied to both the combinations R_{2-3} and $R_{4-5-6-7}$. Since R_1 is in series, the current flowing out of R_1 remains at 2 amps. However, R_{2-3} and $R_{4-5-6-7}$ are parallel and, therefore, a portion of the current flows through each branch. Since both branches have the same resistance, the current through each is 1 amp.

$$E = 20 \text{ volts}$$
$$I = 2 \text{ amps}$$
$$R_{2-3} = 20 \text{ ohms}$$
$$R_{4-5-6-7} = 20 \text{ ohms}$$

$$I_{R_{2-3}} = \frac{E}{R_{2-3}} = \frac{20 \text{ volts}}{20 \text{ ohms}} = 1.0 \text{ amp}$$

$$I_{R_{4-5-6-7}} = \frac{E}{R_{4-5-6-7}} = \frac{20 \text{ volts}}{20 \text{ ohms}} = 1.0 \text{ amp}$$

Since R_4 is in series with R_{5-6-7}, the entire 1 amp flows through R_4. However, the resistor is connected in series and, therefore, voltage must drop across it. The voltage drop is determined by multiplying the current entering the resistor (1 amp) by the resistor's resistance.

$$ER_4 = I_{5-6-7} R_4$$

$$= 1 \times 10$$

$$= 10 \text{ volts}$$

This same 1 amp flows through the combination R_{5-6-7}, and since its equivalent resistance is 10 ohms, 10 volts are dropped across it.

$$E_{5-6-7} = I_{5-6-7} \times R_{5-6-7}$$

$$= 1 \text{ amp} \times 10 \text{ ohms}$$

$$= 10 \text{ volts}$$

You can now find the current through resistors R_{5-6-7}.

$$I_{R_5} = \frac{E_{5-6-7}}{R_5} = \frac{10}{40} = .25 \text{ amp}$$

$$I_{R_6} = \frac{E_{5-6-7}}{R_6} = \frac{10}{20} = .5 \text{ amp}$$

$$I_{R_7} = \frac{E_{5-6-7}}{R_7} = \frac{10}{40} = .25 \text{ amp}$$

The total current through these three resistors is the same 1 amp that flowed through R_4. To find the current through R_2 and R_3, use the formulas:

$$I_{R2} = E/R_2 = 20/60 = 0.33 \text{ amp}$$

$$I_{R}3 = E/R_3 = 20/30 = 0.67 \text{ amp}$$

POWER

To find the power dissipated in each resistor, simply multiply the current through each of the resistors by the voltage dropped across it. [Figure 3-64]

You can check your analysis of a complex circuit by determining that the following statements are true:

1. The total power is equal to the sum of the power dissipated in each of the resistors.

    48 watts = 48 watts

2. The voltage drops across R_1, R_4, and either R_5, R_6, or R_7 must equal the source voltage.

    24 volts = 24 volts

3. The voltage drops across R_1, and either R_2 or R_3 must equal the source voltage.

    24 volts = 24 volts

4. The current through R_4 must be the same as the sum of the current through R_2, R_3 or R_5, R_6, and R_7.

    1 amp = 1 amp.

$E_1 = 4$	$E_5 = 10$	$I_1 = 2$	$I_5 = .25$
$E_2 = 20$	$E_6 = 10$	$I_2 = .33$	$I_6 = .5$
$E_3 = 20$	$E_7 = 10$	$I_3 = .67$	$I_7 = .25$
$E_4 = 10$		$I_4 = 1$	

$$P_1 = E_1 \times I_1 = 4 \times 2 = 8 \text{ WATTS}$$
$$P_2 = E_2 \times I_2 = 20 \times .33 = 6.6 \text{ WATTS}$$
$$P_3 = E_3 \times I_3 = 20 \times .67 = 13.4 \text{ WATTS}$$
$$P_4 = E_4 \times I_4 = 10 \times 1 = 10 \text{ WATTS}$$
$$P_5 = E_5 \times I_5 = 10 \times .25 = 2.5 \text{ WATTS}$$
$$P_6 = E_6 \times I_6 = 10 \times .5 = 5 \text{ WATTS}$$
$$P_7 = E_7 \times I_7 = 10 \times .25 = 2.5 \text{ WATTS}$$
$$P_T = E_T \times I_T = 24 \times 2 = 48.0 \text{ WATTS}$$

Figure 3-64. The power dissipated in each resistor is calculated by multiplying the voltage applied to the resistor times the resistor's resistance.

5. The current through R_1 is equal to the sum of the current through R_2, R_3, R_5, R_6, and R_7.

    2 amps = 2 amps

Item	Voltage Volts	Resistance Ohms	Current Amps	Power Watts
Source	24V	12Ω	2A	48W
R_1	4	2	2.00	8
R_2	20	60	.33	6.6
R_3	20	30	.67	13.4
R_4	10	10	1.00	10.0
R_5	10	40	.25	2.5
R_6	10	20	.5	5.0
R_7	10	40	.25	2.5

Figure 3-65. Circuit relationships in a complex circuit.

VOLTAGE DIVIDERS

It is often necessary to have a series of different voltages in a given circuit. This is accomplished with a series of resistors across the power source called a voltage divider. If a voltage divider consists of three 1,000-ohm resistors across a 24-volt battery, a current of .008 amps flows through the divider. This current produces an eight-volt drop across each of the resistors. [Figure 3-66]

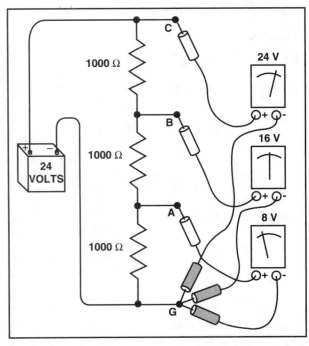

Figure 3-66. An unloaded voltage divider with a 24-volt source and three 1,000 ohm resistors has a current flow of .008 amps and produces a voltage drop of 8 volts across each resistor.

When a load is placed across any terminal of the divider, it acts in parallel with that portion of the resistance and lowers the total resistance. This increases the current through the circuit. If, in figure 3-67, a load made up of a 1,000-ohm resistor is placed between the ground terminal and terminal A, the resistance between A and G drops to 500 ohms and the current increases from eight milliamps to 9.6 ma. The voltage across R_3 with the load is 4.8 volts instead of the eight volts that was between these same terminals without the load. [Figure 3-67]

It is sometimes necessary in electronic circuits to have voltages that operate on either side of a reference value. For example, if you want voltages of -8 to +16 volts without a load, you can use a voltage divider made up of three 1,000-ohm resistors across the 24-volt source. Rather than using the negative terminal of the battery as ground, you can use the junction between R_2 and R_3.

The lower end of R_3 is eight volts negative with respect to the ground. The junction between R_2 and R_1 is eight volts positive with respect to ground, and the

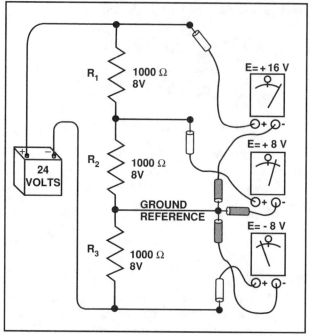

Figure 3-68. A voltage divider that provides voltages on either side of ground, or the reference voltage.

top of R_1 is 16 volts positive with respect to ground. These values change when a load is applied, but the polarity with respect to ground remains constant. [Figure 3-68]

CHANGING DC TO AC

It is often necessary to change direct current into alternating current. For example, many aircraft require alternating current to power equipment such as flight instruments and navigation receivers. During an emergency, when normal aircraft power is not available, power is taken from the battery to operate all electrical loads. Since batteries are capable of storing only direct current, a means must be provided to change DC to AC.

The device used to change DC to AC is called an **inverter**. There are two types of inverters: the rotary inverter and the static inverter. **Rotary inverters** are essentially DC motors with an AC generator built in. They are powered by a DC source, and have AC as an output.

Static inverters are electronic devices containing a specialized circuit known as an oscillator. An **oscillator** is capable of changing DC to AC through electronics, which is discussed in greater detail in a later section of this manual. Oscillators are used in conjunction with amplifiers to produce the correct value of AC from the DC input provided to it. The static inverter has replaced the rotary inverter in most applications, as it is much quieter and more efficient.

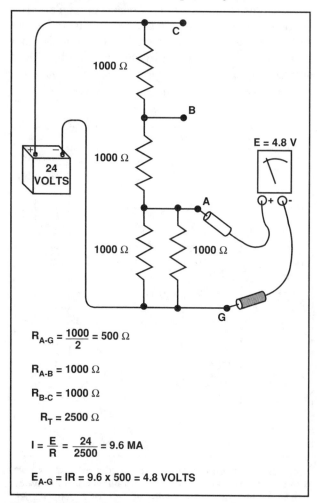

$$R_{A-G} = \frac{1000}{2} = 500 \ \Omega$$

$$R_{A-B} = 1000 \ \Omega$$

$$R_{B-C} = 1000 \ \Omega$$

$$R_T = 2500 \ \Omega$$

$$I = \frac{E}{R} = \frac{24}{2500} = 9.6 \ MA$$

$$E_{A-G} = IR = 9.6 \times 500 = 4.8 \ VOLTS$$

Figure 3-67. A loaded voltage divider.

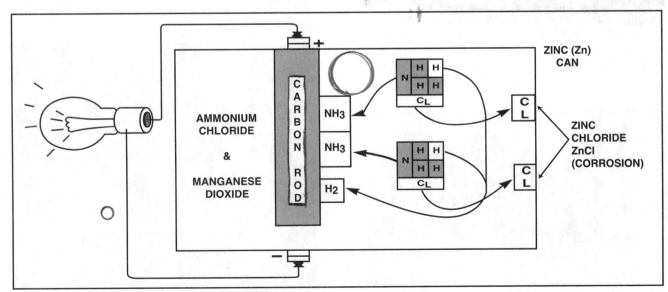

BATTERIES

A **battery** is a device composed of two or more **cells** that convert chemical energy into electrical energy. The chemical nature of battery components provides an excess of electrons at one terminal and a deficiency at the other. Therefore, when the two terminals are joined by a conductor, electrons flow. However, as the electrons flow, the chemical composition of the active material changes and, over time, the active elements become exhausted. In a **primary cell**, the active material cannot be restored. However, in a **secondary cell**, electricity from an external source can restore the active material to its original, or charged, condition.

PRIMARY CELLS

The most common battery in use today is the primary cell or dry cell. Its compact size and low weight make it ideal for use in several electrical devices that require a low power output to operate.

CARBON-ZINC CELLS

The carbon-zinc dry cell is the most commonly used primary cell. A carbon rod is supported in a zinc container by a moist paste containing ammonium chloride, manganese dioxide, and granulated carbon. When a conductor joins the zinc case and the carbon rod, electrons leave the zinc and flow to the carbon. This leaves the zinc with positive ions which attract negative chlorine ions from the ammonium chloride electrolyte. The zinc and chlorine combine to form zinc chloride, which is a form of corrosion that eats away the zinc container. Once the chlorine ions leave the electrolyte, positive ammonium ions remain. These positive ions move toward the carbon rod where they accept the arriving electrons. As the ammonium ions are neutralized, they break down into ammonia and hydrogen gases and are absorbed into the moist manganese dioxide.

Carbon-zinc cells produce one and a half volts regardless of their size. However, the size of the cell does determine the amount of current it supplies. [Figure 3-69]

At one time, leakage was a problem with carbon-zinc batteries. However, this problem has been minimized through the use of effective seals and by enclosing the zinc container inside a steel jacket.

Figure 3-69. In a carbon-zinc cell, as electrons leave the zinc can, negative chlorine ions (Cl) break away from the ammonium chloride and attach to the zinc. The positive ammonium ions (NH$_3$) then attach to the carbon rod to accept the electrons that flow to the positive terminal. Once the ammonium ions are neutralized, they break down into ammonia and hydrogen gases.

ALKALINE CELLS

Alkaline cells provide longer life than the less expensive carbon-zinc cell. The modern alkaline cell uses a zinc rod as the center electrode supported in a manganese dioxide container and immersed in a potassium hydroxide electrolyte solution. The assembly is housed inside a steel case with an insulating disc isolating the center electrode from the case. [Figure 3-70]

Potassium hydroxide has a lower resistance than ammonium chloride. As a result, alkaline cells produce more load current than carbon-zinc cells. This makes them excellent for use in tape recorders and other motor-driven devices.

Since alkaline cells use zinc as a center electrode and manganese dioxide-lined steel as a container, their polarity is reversed from that of a carbon-zinc cell. As a result, alkaline cells must be built differently to be interchangeable with carbon-zinc cells. This is done by mounting the cell in an insulated steel outer case. The negative center terminal of the cell bears against the outer case through a spring, and the case of the cell contacts only the center conductor of the outer shell. As a result, both types of cells have negative outer cases and positive center terminals.

MERCURY CELLS

Another type of alkaline cell is the mercury cell. Mercury cells are used in hearing aids, cameras, and other applications where the need for high capacity and small size outweighs their higher cost. A pellet of mercuric oxide is placed inside a steel container which serves as the positive terminal. A porous insulator, or separator, is then placed over the pel-

let. A roll of extremely thin corrugated zinc is placed over the insulator and is saturated with an electrolyte solution of potassium hydroxide. An insulated steel cap encloses the cell and contacts the zinc, to serve as the negative terminal. For applications that require a positive cap, polarity is reversed in the same manner as in potassium hydroxide cells. [Figure 3-71]

SECONDARY CELLS

Primary cells consume cell materials as they produce electricity. The material cannot be restored nor the cell recharged. However, in secondary cells, the chemical action that releases electron flow is reversible. It is important to note that secondary cells do not produce electrical energy, but merely store it in chemical form. This is why batteries consisting of secondary cells are called storage batteries.

LEAD-ACID BATTERIES

The most commonly used storage battery in aircraft is the lead-acid battery. A typical lead-acid battery consists of 6 or 12 cells, each of which produces approximately 2.1 volts. Each cell consists of a series of positive and negative plates. The positive plates are made up of a grid of lead and antimony filled with lead peroxide. The negative plate uses a similar grid, but its open spaces are filled with spongy lead. An expander material keeps the spongy lead from compacting and losing surface area.

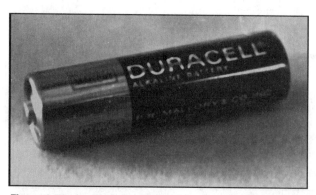

Figure 3-70. An alkaline cell consists of a manganese dioxide container, a zinc center electrode, and a potassium hydroxide electrolyte.

Figure 3-71. A mercury cell is used when a high capacity is needed in a very small cell.

The positive plates are joined together and interlaced between a stack of negative plates. Porous separators keep the plates apart and hold a supply of electrolyte consisting of sulfuric acid and water in contact with the active material. This construction permits electrolyte to circulate freely and provide a path for sediment to settle to the bottom of the cell.

Formerly made of hard rubber, battery cases are now constructed of a high-impact plastic with individual compartments for the cells. Connector straps join the cells and provide the battery's external terminals. A cover seals the cells in the case, and holes in the cover provide access to the cells for servicing and inspection. The cell openings on aircraft batteries are closed with vented screw-in type caps. To prevent electrolyte spillage in unusual flight attitudes, the caps have lead weights inside them that close the vent when the battery is tipped. The complete battery assembly is enclosed in a metal battery box which provides electrical shielding and mechanical protection.

CHEMICAL CHANGES DURING DISCHARGE

When a conductor connects the positive and negative terminals of the battery, electrons flow from the negative lead plates to the lead peroxide in the positive plates. As the electrons leave the negative plates, positive ions form and attract negative sulfate ions from the sulfuric acid in the electrolyte. The combination of these two elements forms lead sulfate on the negative plates. The electrons arriving at the positive plate drive the negative oxygen radicals from the lead peroxide into the electrolyte. This oxygen combines with hydrogen that has lost its sulfate radical and becomes water (H_2O). The positive lead ions that are left on the positive plates also attract and combine with sulfate radicals from the electrolyte and become lead sulfate. Once lead sulfate collects on both the positive and negative plates, and the electrolyte becomes diluted by the water that has formed in it, the battery is considered discharged. When this happens, the water-diluted electrolyte becomes more susceptible to freezing. [Figure 3-72]

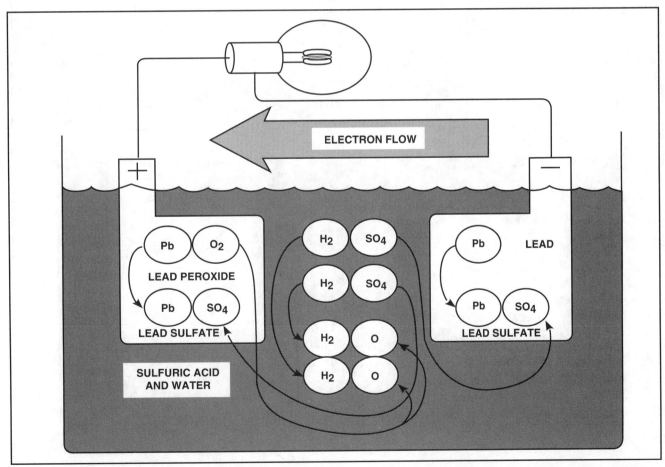

Figure 3-72. When electrons leave the negative lead (Pb) plate, positive ions form and attract sulfate ions (SO_4). These combine to form lead sulfate ($PbSO_4$). As the electrons flow to the positive plate, negative oxygen radicals are forced into the electrolyte where they combine with hydrogen (H) to form water. The positive lead ions left on the positive plate attract sulfate radicals and form lead sulfate.

CHEMICAL CHANGES DURING CHARGE

A discharged battery is recharged using a direct current of the proper voltage. When the positive plates of the battery are connected to the positive terminal of the source, electrons are drawn from the positive plates and forced onto the negative plates. Electrons arriving at the negative plates drive the negative sulfate ions out of the lead sulfate back into the electrolyte. The sulfate ions then join with hydrogen to form sulfuric acid, H_2SO_4.

When the electrons flow from the positive plates, they leave behind positively charged lead atoms. These atoms attract oxygen from the water in the electrolyte and combine to form lead peroxide, PbO_2. When the battery is fully charged, the positive plate again becomes lead peroxide and the negative plate becomes lead. The electrolyte becomes a high concentration of sulfuric acid and, therefore, freezes at a much lower temperature. During the charging process, hydrogen gas is released from the electrolyte and bubbles to the surface. As a battery nears a full charge, the amount of hydrogen released increases, resulting in more bubbling. [Figure 3-73]

DETERMINING THE CONDITION OF CHARGE

The open-circuit voltage of a lead-acid battery remains relatively constant at about 2.1 volts per cell and, consequently, does not reflect a battery's state of charge. However, since the concentration of acid in the electrolyte changes as the battery is used, the electrolyte's specific gravity gives a good indication of the state of charge. If you remember from Chapter 2, specific gravity is the ratio of the weight

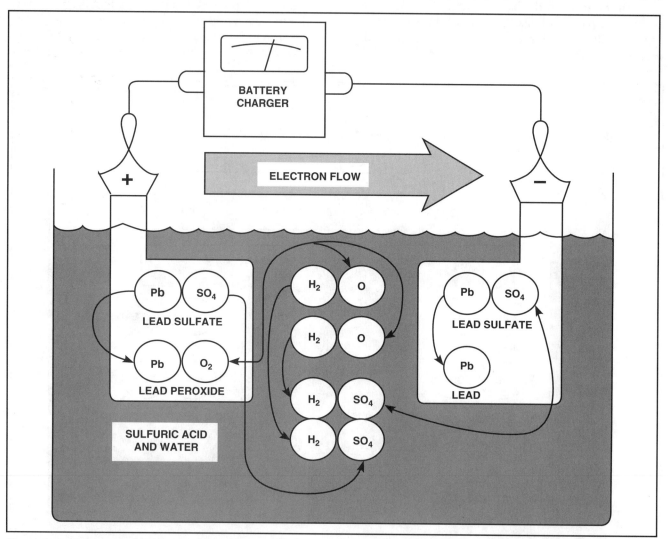

Figure 3-73. When a lead-acid battery is charged, electrons are drawn from the positive plate leaving positively charged lead (Pb) atoms which attract oxygen (O) from the water (H_2O) in the electrolyte. When joined, the lead and oxygen become lead peroxide (PbO_2). As the electrons are forced into the negative plate, they drive the negative sulfate radicals (SO_4) out of the lead sulfate ($PbSO_4$) back into the electrolyte.

of a given volume of a material to the same volume of pure water. In a fully charged, new battery, the electrolyte is approximately 30 percent acid and 70 percent water (by volume). This results in a specific gravity of between 1.275 and 1.300 with an electrolyte temperature of 80 degrees Fahrenheit. As a battery is used, the electrolyte becomes diluted by water and the specific gravity reading decreases. When an electrolyte's specific gravity drops below 1.150, the battery is considered to be discharged. At this level, there is not enough chemical strength in the electrolyte to convert the active materials into lead sulfate.

Battery Testing

Specific gravity is checked with a **hydrometer** which measures the depth a calibrated float sinks in a sample of electrolyte. For example, once a sample of electrolyte is drawn from a cell into a hydrometer, the float and electrolyte are visible in the hydrometer. The graduation on the float's stem that is even with the liquid shows the specific gravity of the electrolyte. The more dense the liquid, the higher the specific gravity reading. In other words, the more bouyant the hydrometer bulb, the more dense the liquid. [Figure 3-74]

As discussed in Chapter 2, the temperature of the electrolyte affects its specific gravity. Therefore, a standard of 80° Fahrenheit is used as the reference. If an electrolyte's temperature is something other than 80° F, a correction must be applied to the hydrometer reading. The electrolyte is less dense at higher temperatures and more dense at lower temperatures. [Figure 3-75]

BATTERY RATINGS

The open-circuit voltage of a lead-acid battery is 2.10 volts per cell when the electrolyte has a specific gravity of 1.265. The physical size of the cell or the number of plates has no effect on this voltage.

As discussed earlier, when a load is placed on a battery, the active material begins to convert into lead sulfate. As the lead sulfate forms, it increases the battery's **internal resistance** and causes the closed-circuit terminal voltage to drop. You can calculate a battery's internal resistance using Ohm's law. For example, a lead-acid battery with 12 cells has a no-load voltage of 2.1 volts per cell. If the battery delivers 5 amps to a load having a resistance of 3 ohms, what is the internal resistance?

Figure 3-74. The specific gravity of the electrolyte is measured with a hydrometer.

Electrolyte Temperature		Correction Points Add or Subtract
°C	°F	
60	140	+24
55	130	+20
49	120	+16
43	110	+12
38	100	+8
33	90	+4
27	80	0
23	70	−4
15	60	−8
10	50	−12
5	40	−16
−2	30	−20
−7	20	−24
−13	10	−28
−18	0	−32
−23	−10	−36
−28	−20	−40
−35	−30	−44

Figure 3-75. This is an example of a correction chart used when determining an electrolyte's specific gravity. For example, suppose the specific gravity read 1240 at a temperature of 60°F. According to the chart, a correction of –8 points should be made. The corrected specific gravity is 1232 which is equivalent to 1.232.

From the information given, the total no-load voltage is determined to be 25.2 volts (12 cells × 2.1 volts/cell). By applying Ohm's law (E=IR), the voltage under load is the product of amps times resistance. Therefore, the load voltage is 15 volts (5 amps × 3 ohms = 15 volts). The internal resistance in the battery thus causes a voltage drop of 10.2 volts (25.2 volts − 15 volts = 10.2 volts). To determine the battery's internal resistance, apply Ohm's law to the voltage drop (10.2 volts) and the amperes the battery delivers (5 amps). The internal resistance is 2.04 ohms.

Example:

$$E = IR$$

$$10.2 \text{ volts} = 5 \text{ amps} \times R$$

$$R = 2.04 \text{ ohms}$$

CAPACITY

The capacity of a battery is its ability to produce a given amount of current for a specified time. Capacity is measured in ampere-hours with one **ampere-hour** equaling the amount of electricity that is put into or taken from a battery when a current of one ampere flows for one hour. Any combination of flow and time that moves this same amount of electricity is referred to as one ampere-hour. For example, a flow of one-half amp for two hours or two amps for one-half hour is one ampere-hour. In theory, a 100 ampere-hour battery can produce 100 amps for one hour, 50 amps for two hours, or 20 amps for five hours.

The capacity of a battery is affected by four things, the amount of active material, the plate area, the quantity of electrolyte, and the temperature. An increase in the amount of active material, the plate area, or the quantity of electrolyte results in an increase in capacity. On the other hand, using a battery in cold temperatures effectively decreases its capacity. For example, at 50°F, a fully charged battery may be able to provide power for 5 hours. However, at 0°F, the same battery may only supply power for 1 hour. The reason for this is that as the temperature drops, the chemical reactions within a battery slow.

Five-Hour Discharge

The standard rating used to specify the capacity of a battery is the **five-hour discharge rating**. This rating represents the number of ampere-hours of capacity when there is sufficient current flow to drop the voltage of a fully charged battery to a completely discharged condition over the course of five hours. For example, a battery that supplies 5 amps for 5 hours has a capacity of 25 ampere-hours.

A battery's capacity decreases when it is discharged at a higher rate. For example, if the same 25 ampere-hour battery supplies 48 amps for 20 minutes, its capacity drops to 16 ampere-hours. If the battery is discharged in 5 minutes, the capacity drops to 11.7 ampere-hours. This is due to heat, sulfation of the plates, and a tendency of the electrolyte to become diluted immediately around the plates. [Figure 3-76]

CELL TEST

If a battery's construction is such that the voltage of an individual cell can be measured, you can get a good indication of the cell behavior under load. To do this, first verify the electrolyte is at the proper level. Then, apply a heavy load to the battery for about three seconds by cranking the engine with the ignition switch off, or the mixture in the idle cutoff position. Now, turn on the landing lights and taxi lights to draw about ten amps. While the load current is flowing, measure the voltage of each cell. A fully charged cell in good condition should have a voltage of 1.95 volts and all cells should be within 0.05 volt of each other. If some of the cells are below 1.95 volts, but all are within 0.05 volt of each other, the cells are in good condition but the battery is somewhat discharged. If any of the cells read higher

BATTERY VOLTAGE	PLATES PER CELL	DISCHARGE RATE					
		5-HOUR		20-MINUTE		5-MINUTE	
		A.H.	AMPS	A.H.	AMPS	A.H.	AMPS
12	9	25	5	16	48	11.7	140
24	9	17	3.4	10.3	31	6.7	80
* Battery is considered discharged when closed-circuit voltage drops to 1.2 volts per cell.							

Figure 3-76. Relationship between ampere-hour capacity and discharge rate.

than 1.95 volts and there is more than a 0.05-volt difference between any of them, there is a defective cell in the battery.

SERVICING AND CHARGING

One of the most important aspects of battery servicing is keeping the battery clean and all of the terminals tight and free of corrosion. If any corrosion exists on the battery terminals or within the battery box it should be removed. To do this, scrub the battery box and the top of the battery with a soft bristle brush and a solution of sodium bicarbonate (baking soda) and water. When washing the top of the battery, avoid getting any baking soda in the cells since it neutralizes the electrolyte. After the battery and box are clean, rinse them with clean water and dry thoroughly. Coat the battery terminals with petroleum jelly or general purpose grease, and touch up any paint damage to the battery box or adjacent area with an acid-resistant paint.

The electrolyte in each cell should just cover the plates. Most batteries have an indicator to show the correct level. If the electrolyte level is low, add distilled or demineralized water. Never add acid to the battery unless it has been spilled, and then, follow the recommendations of the battery manufacturer in detail. The normal loss of liquid in a battery is the result of water decomposing during charging. [Figure 3-77]

Figure 3-77. The electrolyte level should come up to the indicator inside the cell.

Most new batteries are received in a dry-charged state with the cells sealed. When putting a new battery into service, remove the cell seals and pour in the electrolyte that is shipped with the battery. In order to ensure a fully charged battery, the battery must be given a slow freshening, or boost, charge. Once this is done, allow the battery to sit for an hour or so and then adjust the electrolyte level.

It is normally not necessary to mix electrolyte. However, if it should ever become necessary to dilute acid, it is extremely important that the acid be added to the water, and never the other way around. If water is added to the acid, the water, being less dense, floats on top of the acid, and a chemical action takes place along the surface where they meet. This action can generate enough heat to boil the water and splash acid out of the container causing serious injury if it gets on your skin or in your eyes. If acid should get into your eyes, flush them with generous amounts of clean water and get medical attention as soon as possible.

When acid is added to water, the acid mixes with the water and distributes the heat generated by the chemical action throughout the battery. This action still causes the water temperature to rise, but not enough to cause boiling or a violent reaction.

Automotive and aircraft electrolytes are different and should not be mixed. Automotive electrolyte has a lower specific gravity when charged and, therefore, an aircraft battery may never obtain a full charge with automotive electrolyte.

BATTERY CHARGERS

A storage battery is charged by passing direct current through the battery in a direction opposite to that of the discharge current. Because of the battery's internal resistance, the voltage of the external charging source must be greater than the open-circuit voltage. For example, the open-circuit voltage of a fully charged 12-cell, lead-acid battery is approximately 25.2 volts (12 × 2.1 volts). However, the battery's internal resistance causes a voltage drop of 2.8 volts. Therefore, approximately 28 volts are required to charge the battery. Batteries are charged by either the constant-voltage or constant-current method.

Constant-Current Charging

The most effective way to charge a battery is by inducing current back into it at a constant rate. The amount of current induced is typically specified by the manufacturer. However, in the absence of manufacturer information, you should use a current value of no more than seven percent of the battery's ampere-hour rating. For example, if you are charging a 40-ampere-hour battery and do not have specific information from the battery manufacturer, you should charge it at a rate not exceeding 2.8 amperes (40 ampere-hour × .07 = 2.8 amps).

As a battery begins to charge, the no-load voltage increases. Therefore, the voltage on a constant current charger must be varied in order to maintain a constant current throughout the charge. Because of this, a constant current charger usually requires more time to complete and additional attention.

When charging more than one battery with a constant-current charger, connect the batteries in series. One way to remember this is to recall that current remains constant in a series circuit and, therefore, a constant current charge requires multiple batteries to be connected in series. The batteries being charged can be of different voltages, but they should all require the same charging rate. When charging multiple batteries, begin the charge cycle with the maximum recommended current for the battery with the lowest capacity. Then, when the cells begin gassing freely, decrease the current and continue the charge until the proper number of ampere hours of charge is reached.

Constant-Voltage Charging

The generating system in an aircraft charges a battery by the constant voltage method. This method utilizes a fixed voltage that is slightly higher than the battery voltage. The amount of current that flows into a battery being charged is determined by a battery's state-of-charge. For example, the low voltage of a discharged battery allows a large amount of current to flow when the charge first begins. Then, as the charge continues and the battery voltage rises, the current decreases. The voltage produced by a typical aircraft generating system is usually high enough to produce about one ampere of current flow even when a battery is fully charged.

Constant-voltage chargers are often used as shop chargers. However, care must be exercised when using them since the high charging rate produced when the charger is first connected to a discharged battery can overheat a battery. Another thing to keep in mind is that the boost charge provided by a constant-voltage charger does not fully charge a battery. Instead, it usually supplies enough charge to start the engine and allow the aircraft generating system to complete the charge.

Like constant-current chargers, you can also charge several batteries simultaneously with a constant-voltage charger. However, since the voltage supplied to each battery must remain constant, the batteries must have the same voltage rating and be connected in parallel.

CHARGING PRECAUTIONS

Whenever you are working around lead-acid batteries, there are several precautions that must be observed, especially when charging. As mentioned earlier, when a battery is charging, gaseous hydrogen and oxygen are released by the battery cells. Since these gases are explosive, it is essential that you always charge a battery in a well-ventilated place isolated from sparks and open flames. To prevent sparking from the battery, always turn off the battery charger before you connect or disconnect the charging leads. Furthermore, when removing a battery from an aircraft, always disconnect the negative lead first. When installing a battery, connect the negative lead last.

Since lead-acid battery electrolyte is extremely corrosive and will burn skin, you should always wear eye and hand protection whenever working with batteries. If electrolyte is spilled from a battery, it should be neutralized with sodium bicarbonate (baking soda) and rinsed with water.

BATTERY INSTALLATION

Before installing any battery in an aircraft, be sure the battery is correct for the aircraft. The voltage and ampere-hour ratings must meet the manufacturer's specifications. Some aircraft use two batteries connected in parallel to provide a reserve of current for starting and for extra-heavy electrical loads. When installing batteries in this type of system, be sure they are the type specified in the aircraft service manual.

Most aircraft use a single-wire electrical system with the negative terminal of the battery connected to the aircraft structure. As mentioned earlier, when installing a battery, always connect the positive lead first. This way, if you make contact between the battery and the aircraft with a wrench, you will not cause a spark. When removing a battery, always disconnect the negative lead first.

Be sure that the battery box is properly vented and that the battery box drain extends through the aircraft skin. Some batteries are of the manifold type which do not require a separate battery box. With this type of battery, a cover is placed over the cells, and the area above the cells is vented outside the aircraft structure.

The fumes emitted from storage batteries are highly corrosive and toxic. Therefore, they typically must be neutralized before they are released into the atmosphere. To do this, many battery installations vent the fumes inside the battery box through sump jars containing absorbent pads moistened with a solution of sodium bicarbonate and water.

After installing a battery, make certain that it supplies enough current to crank the engine. Also, be sure that the aircraft generating system keeps the battery charged. If an aircraft ammeter shows a full charge rate, but the battery discharges rapidly, it is most likely that the battery is shorted internally.

NICKEL-CADMIUM BATTERIES

Turbine engines require extremely high current for starting. However, high-rate discharges cause the plates of lead-acid batteries to build up sulfate deposits, thereby increasing internal resistance and causing a subsequent voltage drop. This drawback spurred the development of an alkaline battery for aircraft use. The nickel-cadmium or **ni-cad** battery has a very distinct advantage in that its internal resistance is very low, so its voltage remains constant until it is almost totally discharged. This low resistance is also an advantage in recharging, as it allows high charging rates without damage.

While high discharge and charging rates are favorable, there are dangers involved. These dangers begin with the high temperatures associated with nickel-cadmium batteries. For example, the discharge or charging cycle of a nickel-cadmium battery produces high temperatures that break down the cellophane-like material that separates the plates within the cell. The breakdown of the cell separator creates a short circuit allowing current flow to increase. The increased current flow creates more heat, causing further breakdown of the separator material. This condition is aggravated by the fact that the internal resistance of a ni-cad battery drops as the temperature rises. These factors all contribute to the process known as vicious-cycling, or **thermal runaway**. Some nickel-cadmium battery installations are required to have temperature monitoring equipment that enables the flight crew to recognize an overheat condition that can lead to thermal runaway. New cell separator materials and advanced on-board charging equipment have reduced the likelihood of thermal runaway.

CONSTRUCTION

Most nickel-cadmium batteries are made up of individual removable cells. The positive plates are made of powdered nickel (plaque) fused, or sintered, to a porous nickel mesh. This porous mesh is impregnated with nickel hydroxide. The negative plates are made of the same type of porous plaque as on the positive plates, but are impregnated with cadmium hydroxide. Separators of nylon and cellophane keep the plates from touching each other. The cluster of plates and separators is assembled into a polystyrene or nylon cell case and the case is sealed.

A thirty-percent-by-weight solution of potassium hydroxide and distilled water serves as the electrolyte. The specific gravity of this liquid is between 1.24 and 1.30 at room temperature. Since the electrolyte acts only as a conductor during charging and discharging, its specific gravity is no indication of a battery's state of charge.

An individual cell produces an open circuit voltage of between 1.55 and 1.80 volts, depending on the manufacturer. Batteries used in 12-volt aircraft systems use either 9 or 10 cells, while batteries used in 24-volt aircraft systems are made up of 19 or 20 individual cells. [Figure 3-78]

Figure 3-78. Nickel-cadmium batteries are made up of individual cells in transparent cases.

CHEMICAL CHANGES DURING DISCHARGE

As a nickel-cadmium battery discharges, metallic cadmium on the negative plates combines with hydroxide ions in the electrolyte. This releases electrons which flow to the negative terminal. During this process, the cadmium is converted to cadmium hydroxide. At the same time, hydroxide ions leave the positive plates and go into the electrolyte solution. This allows the electrolyte solution to remain about the same. Therefore, specific gravity readings of electrolyte in nickel-cadmium batteries do not indicate the battery's state of charge. However, as a nickel-cadmium battery discharges, the plates absorb some of the electrolyte, therefore the electrolyte level is lowest when a nickel-cadmium battery is completely discharged.

CHEMICAL CHANGES DURING CHARGE

When charging current flows into a nickel-cadmium battery, oxygen is driven from the cadmium oxide on the negative plate leaving metallic cadmium. The nickel hydroxide on the positive plate accepts some of the released oxygen and becomes more highly oxidized. This process continues until all of the oxygen is removed from the negative plate. If charging is continued, gassing occurs in the cell as the water in the electrolyte is decomposed by electrolysis.

During the charging cycle, electrolyte is driven from the positive and negative plates. Therefore, the electrolyte is at its highest level immediately after a charging cycle. Because of this, the electrolyte level is checked and water added only when a nickel-cadmium battery is fully charged.

CELL IMBALANCE

One characteristic of a nickel-cadmium cell being charged is that the negative plate controls the cell's voltage characteristics. This, coupled with a slightly lower charge efficiency in the positive plates, results in an imbalance between the negative and positive plates in each cell. Constant-voltage charging is unable to recognize this condition. Voltages appear normal and the battery appears to be fully charged.

As long as the battery stays on a constant-voltage charge, the imbalance condition becomes worse each time the battery is cycled. Eventually the imbalance reduces the battery's available capacity to the point where there is not enough power to crank the engines or supply emergency power.

Cell imbalance problems have been greatly reduced by more sophisticated charging techniques. For example, in a pulse charging system, battery voltage is monitored and charging current regulated accordingly.

Another way to reduce the chance of cell imbalance is to terminate a constant-voltage charge prior to the battery obtaining a full charge. Then, complete the charge at a constant current rate equivalent to approximately ten percent of the battery's ampere-hour capacity. This technique drives the negative plates into a controlled overcharge and allows the positive plates to be brought to full charge without generating excessive gas and damaging the battery.

SERVICING NICKEL-CADMIUM BATTERIES

The electrolytes used by nickel-cadmium and lead-acid batteries are chemically opposite, and either type of battery can be contaminated by fumes from the other. For this reason, it is extremely important that separate facilities be used for servicing nickel-cadmium batteries and lead-acid batteries.

The alkaline electrolyte used in nickel-cadmium batteries is corrosive. It can burn your skin or cause severe injury if it gets into your eyes. Be careful when handling this liquid. If any electrolyte is spilled, neutralize it with vinegar or boric acid, and flush the area with clean water.

Nickel-cadmium battery manufacturers supply detailed service information for each of their products, and these directions must be followed closely. Every nickel-cadmium battery should have a service record that follows the battery to the service facility each time it is removed for service or testing. It is very important to perform service in accordance with the manufacturer's instructions, and to record all work on the battery service record.

It is normal for most nickel-cadmium batteries to develop an accumulation of potassium carbonate on top of the cells. This white powder forms when electrolyte spewed from the battery combines with carbon dioxide. The amount of this deposit is increased by charging a battery too fast, or by the electrolyte level being too high. If there is an excessive amount of potassium carbonate, check the voltage regulator and the level of electrolyte in the cells. Scrub all of the deposits off the top of the cells with a nylon or other type of nonmetallic bristle brush. Dry the battery thoroughly with a soft flow of compressed air.

Internal short circuits can occur between the cells of a ni-cad battery and are indicated when the battery won't hold a charge. Check for electrical leakage between the cells and the steel case by using a milliammeter between the positive terminal of the battery and the case. If there is more than about 100 milliamps of leakage, the battery should be disassembled and thoroughly cleaned. [Figure 3-79]

Check the condition of all the cell connector hardware and verify there is no trace of corrosion. Dirty contacts or improperly torqued nuts can cause overheating and burned hardware. Heat or burn marks on nuts and contacts indicates the hardware was torqued improperly.

The only way to determine the actual condition of a nickel-cadmium battery is to fully charge it, and then discharge it at a specified rate to measure its ampere-hour capacity. When charging, use the five-hour rate and charge the battery until the cell voltage is that specified by the manufacturer. When the battery is fully charged, and immediately after it is taken off the charger, measure the level of the electrolyte. Ni-cad cell plates absorb electrolyte as a battery discharges or when it sits for long periods. However, the plates release electrolyte as the cells charge. If the level is not checked immediately after the charge is completed, the level drops and the correct level is difficult, if not impossible, to ascertain. Spewing of water and electrolyte during charge is a good sign that water was added while the battery was partially discharged. When water is added, the amount and cell location must be recorded on the battery service record.

When the battery is fully charged and the electrolyte adjusted, it must be discharged at a specified rate and its ampere-hour capacity measured. If the capacity is less than it should be, it is an indication that an imbalance exists. In this situation, the cells must be equalized through a process known as **deep-cycling**. To deep-cycle a battery, continue to discharge it at a rate somewhat lower than that used for the capacity test. When the cell voltage decreases to approximately 0.2 volts per cell, short across each cell with shorting straps. Leave the straps across the cells for three to eight hours to completely discharge them. This process is known as equalization. [Figure 3-80]

After equalization, the battery is ready to charge. Nickel-cadmium batteries may be charged using either the constant-voltage or constant-current methods. The constant-voltage method results in a faster charge; however, the constant-current is most widely used. For either system, the battery manufacturer's service instructions must be followed.

Monitor the battery during charge, and measure individual cell voltages. The manufacturer specifies a maximum differential between cells during the charging process. If a cell exceeds the specification, it must be replaced. Battery manufacturers specify the maximum number of cells that can be replaced before the battery must be retired. Cell replacement should be entered in the battery service record.

As a battery nears the completion of a charge, the cells release gases. This is normal, and must occur before the cell is fully charged. It is normal to overcharge a nickel-cadmium battery to 140 percent of its amp-hour capacity. If the battery has been prop-

Figure 3-79. Cell-to-case leakage should be measured with a milliammeter. If there is more than 100 milliamps of leakage, the battery should be disassembled and cleaned.

Figure 3-80. When the cell voltage falls to approximately 0.2 volts per cell, shorting straps are used to short the cells and ensure that they are equally discharged.

erly serviced and is in good condition, each cell should have a voltage of between 1.55 and 1.80 at a temperature between 70 and 80 degrees Fahrenheit. However, the actual voltage of a charged cell does vary with temperature and the method used for charging. [Figure 3-81]

When working with nickel-cadmium batteries, it is sometimes helpful to have a troubleshooting chart. These charts allow you to associate a probable cause and corrective action to an observed condition. [Figure 3-82]

Figure 3-81. Complete the deep-cycle operation by charging the battery to 140 percent of its ampere-hour capacity.

OBSERVATION	PROBABLE CAUSE	CORRECTIVE ACTION
High-trickle charge – When charging at constant voltage of 28.5 volts (±0.1) volts, current does not drop below 1 amp after a 30-minute charge.	Defective cells.	While still charging, check individual cells. Those below .5 volts are defective and should be replaced. Those between .5 and 1.5 volts may be defective or may be imbalanced, those above 1.5 volts are alright.
High-trickle charge after replacing defective cells, or battery fails to meet amp-hour capacity check.	Cell imbalance.	Discharge battery and short out individual cells for 8 hours. Charge battery using constant-current method. Check capacity and if OK, recharge using constant-current method.
Battery fails to deliver rated capacity.	Cell imbalance or faulty cells.	Repeat capacity check, discharge and constant-current charge a maximum of three times. If capacity does not develop, replace faulty cells.
No potential available.	Complete battery failure.	Check terminals and all electrical connections. Check for dry cell. Check for high-trickle charge.
Excessive white crystal deposits on cells. (There will always be some potassium carbonate present due to normal gassing.)	Excessive spewage.	Battery subject to high charge current, high temperature, or high liquid level. Clean battery constant-current charge and check liquid level. Check charger operation.
Distortion of cell case.	Overcharge or high heat.	Replace cell.
Foreign material in cells – black or gray particles.	Impure water, high heat, high concentration of KOH, or improper water level.	Adjust specific gravity and electrolyte level. Check battery for cell imbalance or replace defective cell.
Excessive corrosion of hardware.	Defective or damaged plating.	Replace parts.
Heat or blue marks on hardware.	Loose connections causing overheating of intercell connector or hardware.	Clean hardware and properly torque connectors.
Excessive water consumption. Dry cell.	Cell imbalance.	Proceed as above for cell imbalance.

Figure 3-82. Nickel-cadmium troubleshooting chart.

ALTERNATING CURRENT

Alternating current is current flow which continually changes its value and periodically reverses direction. It has many advantages over direct current. For example, AC is much easier to generate in the large quantities needed for homes and industries, and for large transport aircraft. More important though, is the ease with which AC current and voltage can be changed to get the most effective use of electrical energy. For example, since the current flowing in a conductor determines the amount of heat generated, current dictates the size of conductor needed to transport it. Delivering the same amount of power with less current would allow the use of smaller conductors, saving money and weight. Power companies use this principle for cross-country transmission of electrical power. The voltage of the electricity carried in transmission lines is boosted up to several thousand volts. At 15,000 volts, 0.067 amp delivers one kilowatt of electrical power. Before the electricity is brought into homes or shops, it is transformed down to a safer and more convenient 115 volts. Voltage and current change many times between the generation of alternating current and its final use. The transformers that accomplish this are quite efficient and lose very little energy. [Figure 3-83]

GENERATION OF AC ELECTRICITY

If you recall from your study of electrical theory, there is a close relationship between magnetism and electricity. Any time electrons flow in a conductor, a magnetic field surrounds the conductor. The amount of electron flow determines the strength of the magnetic field. You also learned that when a magnetic field is moved across a conductor, electrons are forced to flow within the conductor. The rate at which the lines of magnetic flux are cut by the conductor determines the amount of flow. Therefore, electron flow is increased by increasing the strength of the magnetic field, or by increasing the speed of movement of the conductor through the lines of flux.

Try this simple experiment to illustrate alternating current. Wind a conductor into a coil and attach it to a voltmeter. When a magnet is moved back and forth through the coil, the meter deflects from side to side. This shows that electrons flow in one direction when the magnet is moved into the coil, then reverse and flow in the opposite direction when the magnet is withdrawn. [Figure 3-84]

Common household electricity is produced by a rotary generator in which a coiled conductor rotates inside a magnetic field. The changing values of voltage produced as the coil rotates can be observed on an oscilloscope. The values start at zero, rise to a

Figure 3-83. Transformers are used to change the values of alternating current and voltage.

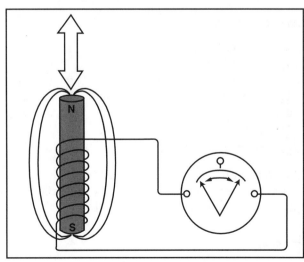

Figure 3-84. Current flows in a conductor when lines of magnetic flux from a permanent magnet cut across it.

peak, and then drop back off to zero. As the coil continues to rotate, the voltage builds up in the opposite direction to a peak and then back to zero. One complete cycle of voltage change is produced with each complete revolution of the coil. [Figure 3-85]

The AC wave form produced by a rotary generator is called a **sine wave**. From the wave illustrated in figure 3-85, you can see that one cycle begins at 0 degrees and ends at 360 degrees. The values of alternating current follow the sine wave. This can be seen through the use of a generator consisting of a single loop of wire that is rotated in a magnetic field. When the loop is parallel with the lines of flux within a magnetic field, it does not pass through any lines of flux, and no voltage is generated. This is the starting point, or the **zero-degree angle**. As the loop rotates to 45 degrees, it cuts across some of the lines of flux. The voltage generated at this point is 0.707 times the peak amount. As the loop continues to rotate to the 90 degree point, it cuts across the maximum number of flux lines for each degree it rotates. It is here that the peak voltage is produced. Further rotation decreases the number of flux lines cut for each degree of rotation. Once the loop reaches 180 degrees, it cuts no flux lines and the output is again zero. Rotation beyond this point brings the opposite side of the loop down through the flux lines near the south pole of the magnet. The voltage builds in the opposite direction and changes in a continuous and smooth manner. [Figure 3-86]

A sine wave's **instantaneous value** is found by multiplying its peak value by the sine of the angle through which the loop rotates beyond its zero-voltage position. For example, if the peak value of a sine wave is 115 volts, the value at 45 degrees equals the sine of 45 degrees times the peak value. [Figure 3-87]

CYCLE

As mentioned earlier, a cycle is one complete sequence of voltage or current change from zero, through a positive peak, back to zero, through a negative peak, and back to zero again. The sequence then repeats.

ALTERNATION

An **alternation** is one-half of an AC cycle in which the voltage or current rises or falls from zero to a peak and back to zero. [Figure 3-88]

PERIOD

The time required for one cycle of events to occur is called the **period** of the alternating current or voltage.

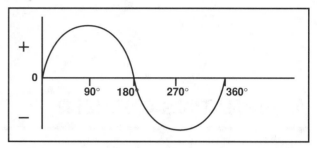

Figure 3-85. As seen on an oscilloscope, alternating current begins at a zero reference line, peaks at a positive value, and returns to zero. It then builds to a negative value exactly opposite the positive value and returns to the reference line.

FREQUENCY

The frequency of AC is the number of cycles completed in one second. Frequency is expressed in **hertz** (Hz) with one hertz equal to one cycle per second. The frequency of alternating current produced by a generator is determined by the number of pairs of magnetic poles in the generator and the number of revolutions completed per minute by the rotating coils. Frequency is found by the formula:

$$\text{Frequency (Hz)} = \frac{\text{Poles}}{2} \times \frac{\text{rpm}}{60}$$

The frequency of commercial alternating current in the United States is 60 hertz, while in some foreign countries it is 50 hertz. The frequency of AC power used in most aircraft is 400 hertz.

SINE WAVE VALUES

As discussed, the **peak value** of a sine wave is the maximum value of voltage or current in either the positive or the negative direction. The difference between the positive and the negative peak values is called the **peak-to-peak value** and is equivalent to twice the peak value.

If all of the instantaneous values of current or voltage in one alternation of a sine wave are averaged together, they have a value of 0.637 times the peak value. This is referred to as an **average value** and has very little practical use for making computations.

The **effective value** of AC is the value that produces the same amount of heat as a corresponding amount of DC. To determine the effective value, square all of the instantaneous values in one alternation, find the average of these squared values, and calculate the square root of this average. The effective value is sometimes referred to as the **root mean square** or **RMS** value and is 0.707 of the peak value. Therefore, an effective value is always less than the peak value.

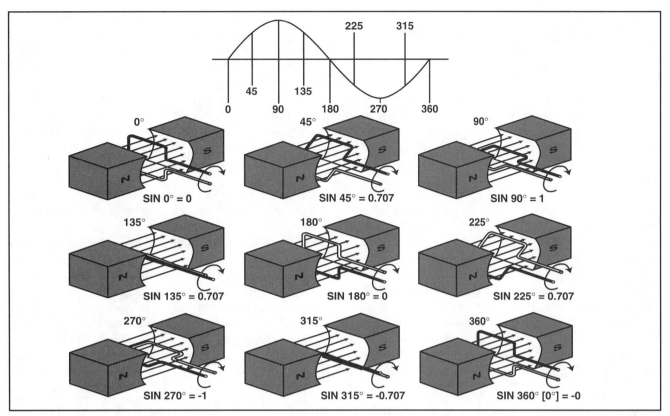

Figure 3-86. At zero degrees, a single loop of wire within a magnetic field cuts no flux and, therefore, no voltage is produced. However, as the loop rotates, it begins cutting lines of flux. At 45 degrees, .707 of the peak voltage is obtained. At 90 degrees, the loop cuts the maximum number of flux lines and produces the maximum voltage. As the loop continues to rotate, fewer lines of flux are cut and the output decreases. At 180 degrees, the output is again zero. Once past the 180 degree point, the loop begins cutting flux lines again and voltage is produced.

Furthermore, unless otherwise specified, all values given for current or voltage are assumed to be effective values. Based on this definition, an AC sine wave having a peak value of 100 volts produces the same amount of heat as 70.7 volts of DC.

When an oscilloscope is used to measure voltage, it displays peak-to-peak voltage. However, an AC volt-

meter measuring the same voltage indicates the effective, or RMS, voltage. Therefore, a peak-to-peak value of 200 volts on an oscilloscope is equivalent

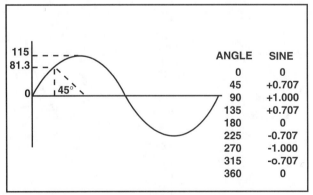

Figure 3-87. The sine of the angle through which a loop rotates is the ratio of the length of the side opposite the angle to the length of the hypotenuse. The peak value of the sine wave is 115 volts and the sine of 45 degrees is .707. Therefore, the instantaneous value at 45 degrees is 81.3 volts.

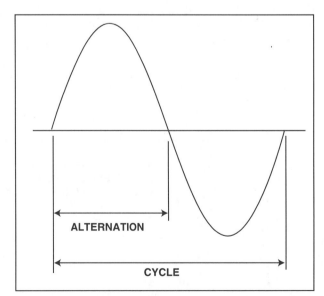

Figure 3-88. One complete cycle of AC begins at zero, rises to a positive peak and returns to zero, then passes through a negative peak as it approaches zero to complete the cycle. An alternation is equivalent to one-half a cycle.

to a peak value of 100 volts, which, as you just saw, is the same as 70.7 volts on a voltmeter. [Figure 3-89]

PHASE

An oscilloscope traces sine wave patterns of AC voltage and current on its screen. When the sine waves cross the zero line at the same time, the voltage and current are said to be in phase. In other words, both voltage and current follow the identical sine wave.

In alternating current where the values are constantly changing, certain circuit components cause a phase shift between the voltage and the current. The amount of shift is referred to as the **phase angle**. For example, some electrical components cause the current to reach its maximum value 90 degrees before the voltage. In this situation, there is a 90 degree phase angle between the current and voltage and the current **leads** the voltage. Other components cause the voltage to change before the current, and the current is said to **lag** the voltage. [Figure 3-90]

POWER

In the study of direct current, electrical power is the product of voltage and current and is measured in watts. However, with alternating current, the values for both voltage and current are given in effective values. The product of these effective values is called the **apparent power** and is expressed in **volt-amps** rather than in watts.

In a circuit that contains only resistance, the current is in phase with voltage and the power developed at any instant is the product of the voltage and the current. As long as the voltage and the current are in phase, the power is positive. [Figure 3-91]

True power is the actual AC power in current when phase is taken into account. For example, if the current and the voltage are not in phase, that is, if the current either leads or lags the voltage, there is at least part of a cycle in which the voltage or current is positive and the other is negative. Since the product of unlike signed numbers is always negative, the power produced during this portion of a cycle is negative power. This means that the load forces power back into the source. True power is expressed in watts and is the product of voltage and that portion of the current that is in phase with the voltage. In a reactive or inductive circuit, true power is always less than apparent power.

The ratio of true power to apparent power is called the **power factor** and, when multiplied by the current, indicates the amount of current that is in phase

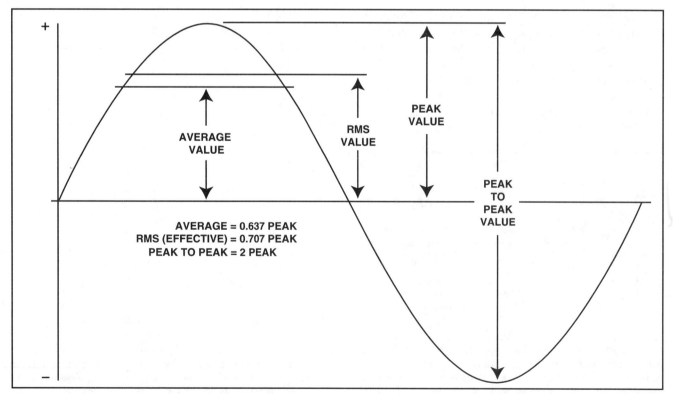

Figure 3-89. Values of an alternating current sine wave.

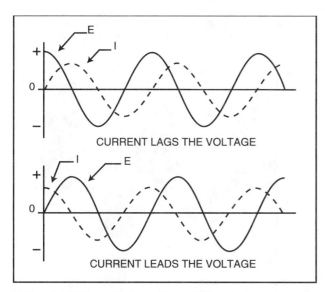

Figure 3-90. There can be a phase difference between voltage and current, two alternating currents, or two voltages.

with the voltage. For example, if the power factor is 0.5, only 50 percent of the current is in phase with the voltage. If all of the current is in phase with the voltage, as it is in a circuit having no opposition other than resistance, the power factor is 1.0.

With this in mind, it can be reasoned that the true power in a circuit is calculated by multiplying the product of the voltage and current by the power factor.

$$\text{True power} = E \times I \times \text{power factor}$$

If neither the true power nor the apparent power of a circuit is known, but the amount of phase shift between the voltage and the current is known, the power factor can be determined by taking the cosine of the phase angle. For example, if the phase angle

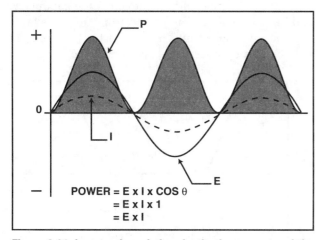

Figure 3-91. In a purely resistive circuit, the current and the voltage are in phase (the phase angle is zero). The apparent power is the product of the effective current and the effective voltage.

between voltage and current is 30 degrees, then the power factor is equivalent to the cosine of 30 degrees. [Figures 3-92]

DEG.	SIN.	COS.	TAN.	COT.	
0	.0	1.000	.0	$+\infty$	90
1	.0175	.9999	.0175	57.29	89
2	.0349	.9994	.0349	28.64	88
3	.0523	.9986	.0524	19.08	87
4	.0698	.9976	.0699	14.30	86
5	.0872	.9962	.0875	11.43	85
6	.1045	.9945	.1051	9.514	84
7	.1219	.9926	.1228	8.144	83
8	.1392	.9903	.1405	7.115	82
9	.1564	.9877	.1584	6.314	81
10	.1737	.9848	.1763	5.671	80
11	.1908	.9816	.1944	5.145	79
12	.2079	.9782	.2126	4.705	78
13	.2250	.9744	.2309	4.331	77
14	.2419	.9703	.2493	4.011	76
15	.2588	.9659	.2680	3.732	75
16	.2756	.9613	.2868	2868	74
17	.2924	.9563	.3057	3.271	73
18	.3090	.9511	.3249	3.078	72
19	.3256	.9455	.3433	2.904	71
20	.3420	.9397	.3640	2.747	70
21	.3584	.9336	.3839	2.605	69
22	.3746	.9272	.4040	2.475	68
23	.3907	.9205	.4245	2.356	67
24	.4067	.9236	.4452	2.246	66
25	.4226	.9063	.4663	2.145	65
26	.4384	.8988	.4877	2.050	64
27	.4540	.8910	.5095	1.963	63
28	.4695	.8830	.5317	1.881	62
29	.4848	.8746	.5543	1.804	61
30	.5000	.8660	.5744	1.732	60
31	.5150	.8572	.6009	1.664	59
32	.5299	.8481	.6249	1.600	58
33	.5446	.8387	.6494	1.540	57
34	.5592	.8290	.6745	1.483	56
35	.5736	.8192	.7002	1.428	55
36	.5878	.8090	.7265	1.376	54
37	.6018	.7986	.7536	1.327	53
38	.6157	.7880	.7813	1.280	52
39	.6293	.7772	.8098	1.235	51
40	.6429	.7660	.8391	1.192	50
41	.6561	.7547	.8693	1.150	49
42	.6691	.7431	.9004	1.111	48
43	.6820	.7314	.9325	1.072	47
44	.6947	.7193	.9657	1.036	46
45	.7071	.7071	1.0000	1.000	45
	COS.	SIN.	COT.	TAN.	DEG.

Figure 3-92. Using this trigonometric function table, you can determine that the cosine of 30 degrees is .866. This is the power factor in an AC circuit with a 30 degree phase shift.

As mentioned earlier, when the current and voltage are in phase, the phase angle is zero and the power factor is 1.0. Under these circumstances, the true power is equal to the apparent power. However, with a 45 degree phase angle, the true power is only 0.707 of the apparent power. [Figure 3-93]

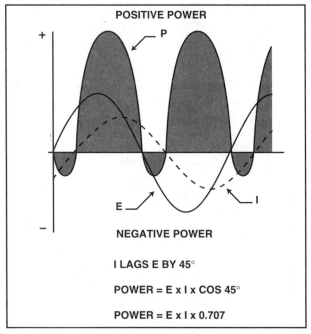

Figure 3-93. When the voltage and current are 45 degrees out of phase, the true power is 0.707 times the apparent power.

The cosine of 90 degrees is zero. Therefore, when the current and voltage are 90 degrees out of phase, the power factor is also zero. In this situation, there is no real power produced in a circuit, even though voltage is present and current is flowing. [Figure 3-94]

RESISTANCE

Circuit components such as light bulbs, heaters, and composition resistors provide resistance to an AC circuit. Circuits containing only these types of devices are called **resistive circuits**. In a resistive circuit, the current and voltage are in phase. In other words, they both pass through zero in the sine wave at the same time and go in the same direction. The power factor in a purely resistive circuit is one, so the apparent power and true power are the same. To calculate the power in watts in a resistive circuit, multiply the effective value of voltage by the effective value of current. [Figure 3-95]

INDUCTANCE

As discussed earlier, any time current flows in a conductor, it produces a magnetic field that surrounds the conductor. The strength of this field is

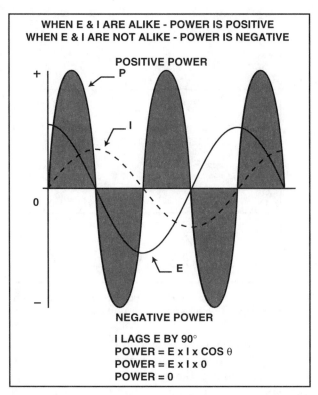

Figure 3-94. The negative power equals the positive power when the current and the voltage are 90 degrees out of phase.

determined by the amount of current flow. To determine the direction the lines of flux encircle the conductor, use the **left-hand rule**. This rule states that if you hold a conductor in your left hand so your thumb points in the direction of electron flow (toward the positive terminal), your fingers encircle the conductor in the direction of the lines of flux. [Figure 3-96]

When the current flow through a conductor changes, the magnetic field expands or contracts as appropriate. As it does, the lines of flux cut across the conductor and induce a voltage into it. According to **Lenz's law**, the voltage induced into the conductor is of such a polarity that it opposes the change that caused it. For

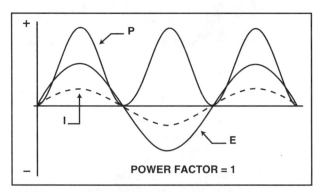

Figure 3-95. In a purely resistive circuit, the current and voltage are in phase, and the power factor equals one.

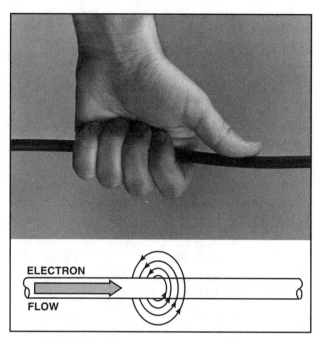

Figure 3-96. When you hold a conductor in your left hand with your thumb pointing in the direction of electron flow, from negative to positive, your fingers encircle the conductor in the same direction as the lines of magnetic flux.

example, as the voltage begins to rise and the current increases, the expanding lines of flux cut across the conductor and induce a voltage that opposes, or slows down, the rising voltage. This induced voltage is sometimes referred to as a **counter-electromotive force** since it opposes the applied voltage. When the current flow in a conductor is steady, the lines of flux do not cut across the conductor and induce a voltage. However, when the current decreases, the lines of flux cut across the conductor as they collapse and induce a voltage that opposes the decrease. [Figure 3-97]

When a conductor carries alternating current, both the amount and the direction of the current continually change. Therefore, an opposing voltage is constantly induced into the conductor. This induced voltage acts as an opposition to the flow of current, and is discussed in detail under its proper name, **inductive reactance**.

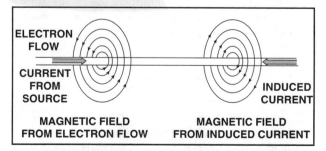

Figure 3-97. The magnetic field induced by current encircles a conductor in the direction opposite that of the magnetic field produced by the source current.

FACTORS AFFECTING INDUCTANCE

The symbol for inductance is **L** and is measured in **henrys**. One henry of inductance generates one volt of induced voltage when the current changes at the rate of one ampere per second. All conductors have the characteristic of inductance since they all generate a back voltage any time the current flow changes. Anything that concentrates the lines of flux or causes more flux lines to cut across the conductor increases the amount of inductance. For example, if a conductor is formed into a coil, the lines of flux surrounding any one of the turns cut not only across the conductor itself, but across each of the turns as well. Therefore, a much greater induced current is generated to oppose the source current. If a soft iron core is inserted into a coil, it further concentrates the lines of flux. This causes an even higher induced current and allows even less source current to flow.

A coil's inductance is determined by the number of turns in the coil, the spacing between the turns, the number of layers of winding, and the wire size. The ratio of the diameter of the coil to its length and the type of material used in a coil's core also affect inductance. Since all of these factors are variable, there is no simple formula available for determining a coil's inductance. [Figure 3-98]

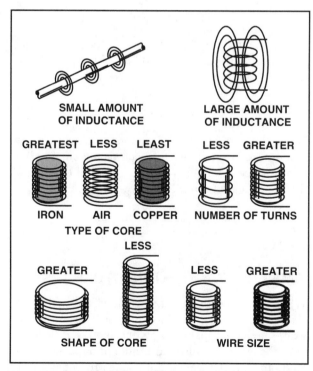

Figure 3-98. A conductor's inductance is increased by forming it into a coil. The inductance of the coil is determined by the type of core, the number of turns, the shape of the coil, and the wire size.

SERIES AND PARALLEL INDUCTORS

Inductors may be connected in a circuit in the same manner as resistors. If inductors are connected in series in such a way that the changing magnetic field of one does not affect the others, the total inductance is equal to the sum of the individual inductances. [Figure 3-99]

$$L_T = L_1 + L_2 + L_3 \ldots$$

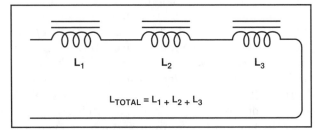

Figure 3-99. Inductances in series.

When inductors are connected in parallel, the total inductance is less than that of any of the individual inductors. The formulas used for finding total inductance in a parallel circuit are the same as those used for finding total resistance in a parallel circuit. For example, if the inductors are all the same, the total inductance is found by dividing the inductance of one inductor by the number of inductors:

$$L_T = \frac{L}{n}$$

If there are only two inductors, the total inductance is found by dividing the product of the two inductors by their sum:

$$L_T = \frac{L_1 \times L_2}{L_1 + L_2}$$

When more than two inductors having different amounts of inductance are connected in parallel, total inductance equals the reciprocal of the sum of the reciprocal of the inductances. [Figure 3-100]

$$L_T = \frac{1}{\dfrac{1}{L_1} + \dfrac{1}{L_2} + \dfrac{1}{L_3 + \ldots}}$$

TIME CONSTANT OF INDUCTORS

When a source of direct current is applied to an inductor, the current does not rise instantly. For example, at the instant a switch is closed in a circuit, the current finds a minimum of opposition and starts to flow. However, the change in current flow rate from zero to maximum induces a maximum back voltage that opposes the current flow. Therefore, current does not begin to flow at its max-

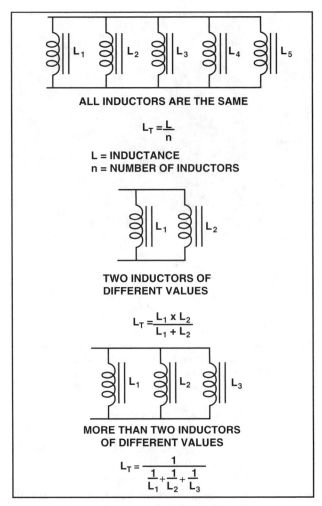

ALL INDUCTORS ARE THE SAME

$$L_T = \frac{L}{n}$$

L = INDUCTANCE
n = NUMBER OF INDUCTORS

TWO INDUCTORS OF DIFFERENT VALUES

$$L_T = \frac{L_1 \times L_2}{L_1 + L_2}$$

MORE THAN TWO INDUCTORS OF DIFFERENT VALUES

$$L_T = \frac{1}{\dfrac{1}{L_1} + \dfrac{1}{L_2} + \dfrac{1}{L_3}}$$

Figure 3-100. The formulas used for determining total inductance in a parallel circuit are the same as those used for finding total resistance in a parallel circuit.

imum rate instantly. The time required for the current to rise to 63.2 percent of its peak value is known as the **time constant** of the circuit and is determined by the value of inductance and resistance in the circuit.

$$\text{Time Constant} = \frac{L \text{ (henrys)}}{R \text{ (ohms)}}$$

For example, a circuit containing 2 henrys of inductance and 50 ohms of resistance has a time constant of 0.04 second, or 40 milliseconds.

$$\text{Time Constant} = \frac{2 \text{ henrys}}{50 \text{ ohms}} = .04 \text{ seconds}$$

Therefore, the current rises to 63.2 percent of its peak value in a period of time equal to one time constant (0.04 second). In two time constants, or 0.08 second, the current rises to 86.5 percent of its peak value. In three time constants (0.12 seconds) it rises to 95 percent, and in four time constants (0.16 seconds) to 98 percent. It takes five time constants, or 0.2 seconds, for the current to approach the peak value of the source.

By the same token, when a switch is opened in a circuit, the changing current induces a voltage that opposes the change, and current drop is slowed. [Figure 3-101]

INDUCTIVE REACTANCE

As mentioned earlier, **inductive reactance** is the opposition to the flow of alternating current caused by the generation of a back voltage. Inductive reactance is represented by the symbol X_L and is measured in ohms. The formula for determining inductive reactance is:

$$X_L = 2\pi f L$$

Where:

$\pi = 3.1416$
f = frequency in cycles per second
L = inductance in henrys

As you can see, inductive reactance is proportional to the inductance within a circuit and the frequency of the alternating current. Simple multiplication also tells you that if the frequency is zero, as it is in

direct current, there is no inductive reactance. This makes sense since there is no changing magnetic field in direct current. However, as the frequency increases above zero, the amount of change in the magnetic field increases. This, in turn, causes the inductive reactance to increase. An infinite frequency would produce an infinite amount of inductive reactance. [Figure 3-102]

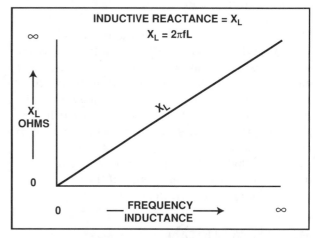

Figure 3-102. Inductive reactance is a function of the amount of inductance in a circuit and the frequency of the alternating current. X_L increases with an increase in both frequency and inductance.

Remember that the opposition produced by induction is caused by the generation of a counter, or back, electromotive force. Unlike the opposition caused by resistance, no heat is generated in a circuit with inductance, and, therefore, no power is dissipated.

If a circuit is purely inductive — that is, there is no resistance present — current does not begin to flow until the voltage rises to its peak value. Then, as the voltage begins to drop off, the current rises until the voltage passes through zero. On a sine wave, this condition is represented by a 90 degree shift in phase. In other words, in a purely inductive circuit, the change in current lags the change in voltage by 90 degrees. [Figure 3-103]

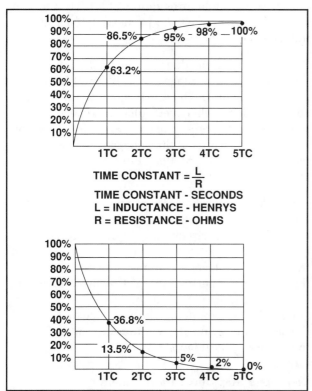

Figure 3-101. These two graphs illustrate the number of time constants required for current to reach its maximum and minimum values once power is supplied or removed from a circuit containing both resistance and inductance.

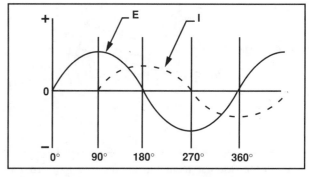

Figure 3-103. In an inductive circuit, the current lags behind the voltage by 90 degrees.

With a 90 degree phase angle the power factor is zero and there can be no true power. This is because the negative power produced cancels the positive power. In other words, the load returns as much power as it receives. If an inductor of the proper size is placed in series with a light bulb, the inductive reactance causes most of the source voltage to be dropped across the inductor. In this case, the bulb burns very dimly, if at all.

Most aircraft use 400-hertz alternating current. This is because the inductive reactance at this frequency is high enough to allow smaller transformers and motors to be used. However, if a transformer designed for 400-hertz AC is used in a 60-hertz circuit, the lower inductive reactance caused by the lower frequency allows enough current flow to burn out the transformer. But if a 60-hertz transformer is used in a 400-hertz circuit, there is so much inductive reactance that the efficiency of the transformer becomes too low for practical use.

MUTUAL INDUCTANCE

When alternating current flows in a conductor, the changing lines of flux radiate out and cut across other nearby conductors. Any time flux lines cut across a conductor, they generate a voltage even though there is no electrical connection between the two. This process is known as mutual inductance, and is the basis for transformer operation. [Figure 3-104]

PRACTICAL INDUCTORS

Inductors are often used to filter out unwanted audio and radio frequency energy in AC circuits. However, they are most commonly used to change the values of alternating current and voltage.

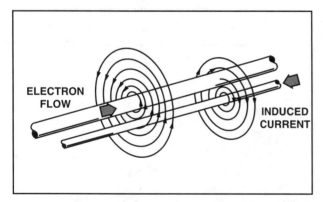

Figure 3-104. Mutual induction causes a voltage to be induced into a conductor that is not electrically connected to the conductor through which the source current flows.

CHOKES

Inductors that filter out certain frequencies are called chokes. When alternating current with a frequency of 50 or 60 hertz is converted to direct current, the output is in the form of pulsating direct current. However, if an inductor is placed in series with the load, the pulsating current induces a back voltage that tends to smooth out the pulsations. Chokes of this type have laminated iron cores and often have an inductance of more than one henry.

TRANSFORMERS

Transformers allow the values of AC voltage and current to be changed through the use of mutual inductance. A typical transformer consists of two coils of wire wound around a common core, but not connected electrically. The coil in which alternating current flows into a transformer is called the **primary coil**, whereas the coil in which the alternating current flows out is the **secondary coil**.

When an alternating current flows in the primary coil, a voltage is induced into the secondary coil. The amount of voltage generated in the secondary coil is equal to the voltage in the primary times the **turns ratio** between the two coils. For example, 100 turns in a primary coil and 1,000 turns in a secondary equates to a turns ratio of 1:10. Therefore, if 115 volts flows across the primary, 1,150 volts are induced across the secondary.

Since a transformer does not generate any power, the product of the voltage and the current in the secondary coil must be the same as that in the primary coil. Therefore, whenever volts are increased in a transformer, amperes must decrease by the same ratio. In other words, if the voltage is increased by a ratio of 1:10, the current must decrease by a ratio of 10:1. [Figure 3-105]

A transformer can have its primary coil connected directly across an AC power line and, as long as there is an open circuit in the secondary coil, the back voltage produced in the primary coil blocks the source voltage so almost no current flows through the primary winding. However, when the circuit is complete in the secondary coil, secondary current flows, producing lines of flux that oppose the back voltage and allow source current to flow in the primary coil. [Figure 3-106]

Step-Up or Step-Down Transformers

In a step-up transformer there are more turns in the secondary coil than in the primary coil. This results in an increase in voltage and a decrease in current.

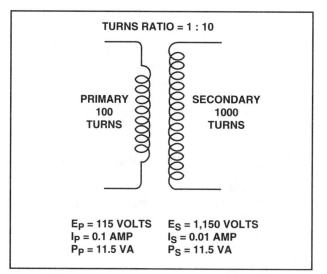

TURNS RATIO = 1 : 10

PRIMARY
100
TURNS

SECONDARY
1000
TURNS

E_P = 115 VOLTS E_S = 1,150 VOLTS
I_P = 0.1 AMP I_S = 0.01 AMP
P_P = 11.5 VA P_S = 11.5 VA

Figure 3-105. Voltage increases and decreases directly with the turn ratio between the primary and secondary coils. However, since a transformer cannot generate power, the current must decrease or increase by the same ratio.

However, in a step-down transformer, the secondary coil has fewer turns than the primary coil. In this case the voltage decreases and current rises. Step-down transformers are often used to get the high current necessary for operating some motors.

Autotransformers

An autotransformer is a form of variable transformer. In an autotransformer there is only one coil that acts as both a primary and secondary coil. One lead of the coil is in common with both the primary and secondary, while the other secondary lead connects to a movable brush that makes contact with a bare spot in the primary coil. The position of this brush determines the amount of secondary voltage. The greater the number of turns between the two leads, the greater the secondary voltage. [Figure 3-107]

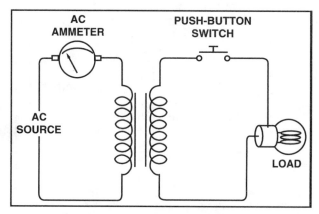

AC AMMETER

PUSH-BUTTON SWITCH

AC SOURCE

LOAD

Figure 3-106. Primary current flows only when the secondary circuit is completed. When the switch in the secondary circuit is pressed, current flow is indicated on the AC ammeter in the primary circuit.

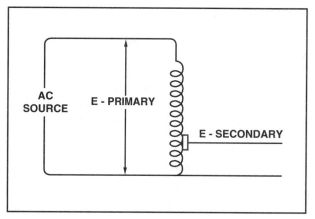

AC SOURCE

E - PRIMARY

E - SECONDARY

Figure 3-107. An autotransformer uses one winding for both the primary and secondary.

Current Transformers

Clip-on type ammeters use current transformers to measure the amount of AC current flowing in a line. The probe consists of a coil that opens so it can be clamped over a wire carrying the current to be measured. The flow of alternating current induces a small current into the coil that is proportional to the amount of load current flow. A calibrated meter displays the value of the load current. [Figure 3-108]

Figure 3-108. A clip-on ammeter is a form of current transformer.

CAPACITANCE

As you know, electrical energy can be stored in batteries through an electrochemical process. However, this is not the only means of storing electricity. For example, it can also be stored in **electrostatic fields** caused by an accumulation of electrical charges. The strength of an electrostatic field is determined by the amount of voltage contained by the static charges. A **capacitor**, sometimes called a **condenser**, is a device that stores electrical energy in the electrostatic fields that exist between two conductors that are separated by an insulator, or **dielectric**.

The principles of a capacitor are simple. Two flat metal plates face each other and are separated by an insulator. One of the plates is attached to the positive terminal of the power source and the other to the negative terminal. In this configuration, electrons are drawn from the plate attached to the positive terminal and flow to the plate attached to the negative terminal. Although there is no flow across the insulator, the plates become charged. In fact, if a voltmeter reading were to be taken across the plates, it would be exactly the same as one taken across the battery. Current flows while the plates are being charged, but stops when they become fully charged. [Figure 3-109]

If a power source connects to a capacitor through a resistor, the resistor limits the amount of current that initially flows to the capacitor. However, the resistor does not prevent the voltage across the capacitor from rising to that of the source.

If a switch is then added to the same circuit, as illustrated in figure 3-110, the capacitor voltage rises to the source voltage with the switch in position A.

Once the voltages match, the current flow stops. If the switch is then moved to position B, the capacitor immediately discharges through the light and causes it to flash. If the switch is placed in its neutral position when the capacitor is charged, it remains charged until the electrons eventually leak off through the dielectric. [Figure 3-110]

FACTORS AFFECTING CAPACITANCE

A capacitor's ability to store an electrical charge is measured in **farads**. One farad is the capacity required to hold one **coulomb** of electricity (6.28×10^{18} electrons) under a force of one volt.

$$C = \frac{Q}{E}$$

Where:
C = capacity in farads
Q = charge in coulombs
E = voltage in volts

The farad typically stores too many electrons for use in practical circuits. Therefore, most capacitors are measured in **microfarads** which are millionths of a farad, or in **picofarads** which are millionths of millionths of a farad. Picofarads were formerly called micro-micro farads and are still referred to in this way in some texts. The Greek letter mu (μ) is used to represent the prefix micro.

1 microfarad (μF) = 1×10^{-6} farad

1 picofarad (pF or $\mu\mu$f) = 1×10^{-12} farad

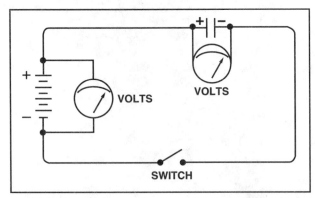

Figure 3-109. When two plates of a capacitor are initially attached to a battery, electrons are drawn from the plate attached to the positive terminal and flow to the plate attached to the negative terminal. This process continues until the plates become fully charged. Once charged, the voltage across the capacitor will match the voltage across the source.

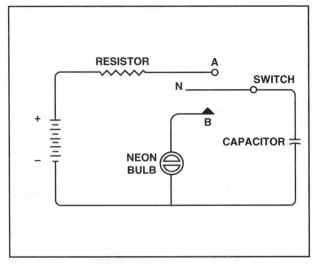

Figure 3-110. The resistor limits the rate at which the capacitor charges when the switch is in position A. When the switch is in position B, the capacitor discharges through the neon bulb, causing it to flash.

Capacity is affected by three variables: the area of the storage plates, the separation between the plates, and the composition of the dielectric. The larger the plates, the more electrons can be stored. One very common type of capacitor has plates made of two long strips of metal foil separated by waxed paper and rolled into a tight cylinder. This construction provides the maximum plate area for its small physical size. [Figure 3-111]

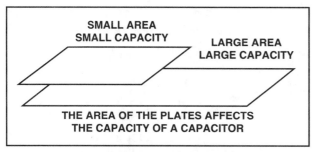

Figure 3-111. **The greater the area of the plates, the greater the capacity of the capacitor.**

The distance between the plates determines the strength of the electrostatic field between them which, in turn, affects capacity. For example, if the plates are far apart, a weak electrostatic field is produced and fewer electrons are pulled onto the negative plate. If, on the other hand, the plates are close together, the attraction caused by the unlike charges between the plates produces a strong electrostatic field in the dielectric. This allows more electrons to be held on the negative plate. The strength of the electrostatic field increases inversely with the separation between the plates. In other words, when the space between the plates is cut in half, the strength of the electrostatic field doubles. However, if the space between the plates doubles, the electrostatic field strength decreases to half its original value. [Figure 3-112]

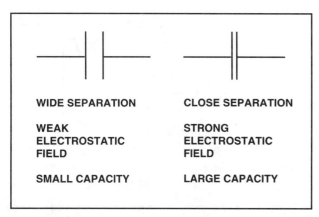

Figure 3-112. **The closer the plates in a capacitor, the greater its capacity.**

There is a limit to how close the plates in a capacitor can be. For example, if the plates get too close, the electrostatic field may become so strong that electrons cross the insulator and actually flow to the positive plate. When this happens, the dielectric typically becomes damaged and a conductive path is set up that shorts the capacitor and makes it useless. For this reason, all capacitors are rated with regard to their working voltage, which must be at least 50 percent above the highest voltage applied in the circuit. The rating is a DC measurement that indicates the strength of the dielectric.

The third factor affecting the capacity of a capacitor is the composition of the dielectric. Capacitors store energy in two ways. One way is through the electrostatic attraction across the dielectric. The second is through the distortion of the electron orbits of the atoms within the dielectric material. For example, as a capacitor charges, the electrons within the dielectric are attracted to the positive plate and the protons are attracted to the negative plate. This distortion, sometimes called **dielectric stress**, stores electrostatic charges similar to the way the plates do. [Figure 3-113]

The higher the dielectric stress, the greater the insulator's capacity. The numbers used to express dielectric stress are referred to as "K" values. Air is used as the reference for measuring the dielectric constant and is given a value of one. If glass, which has a dielectric constant of eight, is substituted for

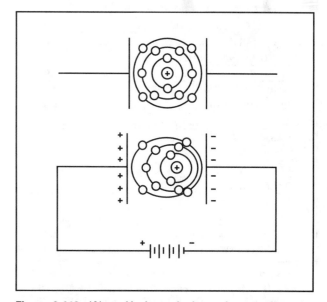

Figure 3-113. **(A) — Uncharged plates do not affect the orbital pattern of the material in the dielectric. (B) — However, when a capacitor is charged, the orbits of the electrons in the dielectric are distorted and energy is stored in the distorted field.**

air as an insulator, capacity increases eight times due to its ability to store energy in the distortion of the electron orbits. [Figure 3-114]

CAPACITORS IN SERIES

It is often necessary to connect multiple capacitors into a circuit. When this is done, the effect is comparable to increasing the separation between the capacitor plates. In other words, a circuit's total capacity decreases when capacitors are connected in series. Furthermore, the total capacity will be less than that of any of the series capacitors. The formulas used for finding total capacitance in a series circuit are the same as those used for finding the total resistance in a parallel circuit. For example, when multiple capacitors of equal value are connected in series, the total capacitance is found by dividing the value of one capacitor by the number of capacitors in series:

$$C_T = \frac{C}{n}$$

If there are two unlike capacitors, the total is found by dividing the product of the two by their sum.

$$C_T = \frac{C_1 \times C_2}{C_1 + C_2}$$

KIND OF DIELECTRIC	APPROXIMATE* K VALUE
AIR (AT ATMOSPHERIC PRESSURE)	1.0
BAKELITE	5.0
BEESWAX	3.0
CAMBRIC (VARNISHED)	4.0
FIBER (RED)	5.0
GLASS (WINDOW OR FLINT)	8.0
GUTTA PERCHA	4.0
MICA	6.0
PARAFFIN (SOLID)	2.5
PARAFFIN COATED PAPER	3.5
PORCELAIN	6.0
PYREX	4.5
QUARTZ	5.0
RUBBER	3.0
SLATE	7.0
WOOD (VERY DRY)	5.0

*THESE VALUES ARE APPROXIMATE, SINCE TRUE VALUES DEPEND UPON QUALITY OR GRADE OF MATERIAL USED, AS WELL AS MOISTURE CONTENT, TEMPERATURE, AND FREQUENCY CHARACTERISTICS OF EACH.

Figure 3-114. Table of dielectric constants.

When there are more than two unlike capacitors, their total capacity is the reciprocal of the sum of the reciprocals of the individual capacitors. [Figure 3-115]

$$C_T = \frac{1}{\frac{1}{C_1} + \frac{1}{C_2} + \frac{1}{C_3} \cdots}$$

$$C_T = \frac{C}{n} = \frac{0.05}{5} = 0.01 \ \mu F$$

$$C_T = \frac{C_1 \times C_2}{C_1 + C_2} = \frac{0.10 \times 0.25}{0.10 + 0.25}$$

$$= \frac{0.025}{0.35} = 0.0714 \mu F$$

$$C_T = \frac{1}{\frac{1}{C_1} + \frac{1}{C_2} + \frac{1}{C_3}} = \frac{1}{\frac{1}{0.05} + \frac{1}{0.10} + \frac{1}{0.25}}$$

$$= \frac{1}{34} = 0.029 \mu F$$

Figure 3-115. Capacitors connected in series.

CAPACITORS IN PARALLEL

Connecting capacitors in parallel has the same effect as adding the areas of their plates. Therefore, the total capacity is equivalent to the sum of the individual capacitors. [Figure 3-116]

$$C_T = C_1 + C_2 + C_3 \cdots$$

$$T = C_1 + C_2 + C_3$$
$$= 0.005 + 0.005 + 0.005$$
$$= 0.015 \mu F$$

Figure 3-116. When capacitors are connected in parallel, total capacitance is calculated by adding the capacities.

TIME CONSTANT OF CAPACITORS

Recall from the study of inductors that when a voltage source is placed across an inductor, the inductance slows the rise of current in the circuit. As a result, changes in current lag behind changes in voltage. In a capacitive circuit the results are opposite. In other words, changes in current lead changes in voltage. For example, when the power is added to a capacitive circuit, current immediately begins to flow as electrons move from the positive plate to the negative plate. However, the voltage across the plate does not immediately increase. Instead it rises as the plates become charged.

The **time constant** of a capacitive circuit is the time, in seconds, required for the voltage across the capacitor to reach 63.2 percent of the source voltage. It is determined by multiplying a circuit's capacitance by its resistance. [Figure 3-117]

$$TC = R \times C$$

Where:

TC = Time constant in seconds
R = Resistance in ohms
C = Capacitance in farads

Timing circuits are often made using a capacitor and a resistor in series. For example, when a 100,000-ohm resistor is connected in series with a 100-microfarad capacitor across a 100-volt power source, the current begins to flow when the circuit is closed. However, the amount of current flow is limited by the opposition caused by the resistor.

The time constant in this circuit is 10 seconds (100,000 ohms × .00001 farad = 10 seconds), therefore, in 10 seconds, the voltage rises to 63.2 volts. In 20 seconds, it rises to 86.5 volts; in 30 seconds, to 95 volts; in 40 seconds to 98 volts; and in 50 seconds the voltage equals the source voltage and current flow ceases. This same time constant applies when discharging the capacitor. In other words, when the discharge cycle begins, it takes 10 seconds to drain the capacitor to 36.8 volts. At this point, the voltage drop starts to slow. In 20 seconds, the capacitor discharges down to 13.5 volts; in 30 seconds, to 5 volts; and in 40 seconds to 2 volts. At 50 seconds, or 5 time constants, all the current flows through the resistor leaving zero volts in the capacitor. [Figure 3-118]

CAPACITIVE REACTANCE

Capacitive reactance is the opposition to the flow of alternating current caused by the capacitance in a circuit. The symbol for capacitive reactance is X_C and it is measured in ohms.

A circuit's capacitive reactance is inversely proportional to its capacitance and frequency. The reason for this is that as a circuit's capacitance increases, more current must flow to charge the capacitor. At the same time, if a circuit's frequency increases, a capacitor charges and discharges more often, resulting in more current flow. The greater the current flow, the less the capacitive reactance. The formula for capacitive reactance shows this relationship:

$$X_c = \frac{1}{2 \pi f C}$$

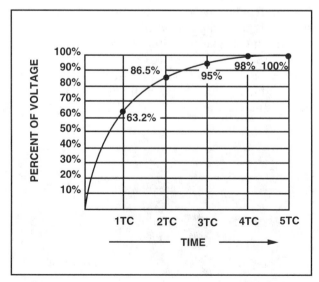

Figure 3-117. Time constant curve of a capacitor during charge.

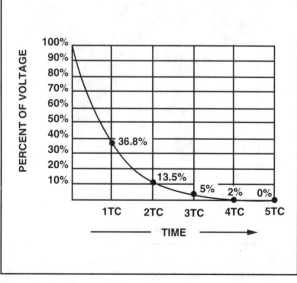

Figure 3-118. Time constant of a capacitor being discharged.

PHASE SHIFT

In a purely capacitive circuit, the current leads the voltage by 90 degrees. In other words, the current must flow into the capacitor before the voltage across it can rise. When the capacitor is fully charged, the voltage is at maximum and the current is at zero. Then, as the capacitor begins to discharge, the current begins to flow and the voltage starts to drop. The current flow is greatest as the voltage passes through zero. As the voltage begins to build in the opposite direction, the current flow starts to drop off until the capacitor is fully charged. [Figure 3-119]

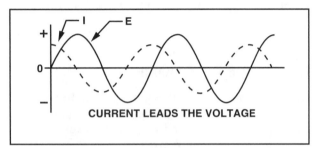

Figure 3-119. The current leads the voltage by 90 degrees in a capacitive circuit.

Since the current leads the voltage by 90 degrees in a purely capacitive circuit, the power factor is zero and no real power is produced. This is because the negative power equals the positive power. [Figure 3-120]

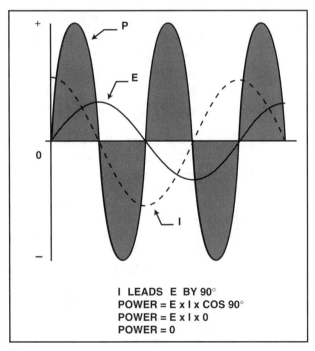

Figure 3-120. In a purely capacitive circuit, the negative power equals the positive power, and no true power is produced.

PRACTICAL CAPACITORS

Capacitors are divided into two types, fixed and variable. The fixed capacitors are further divided into electrolytic and nonelectrolytic types.

NONELECTROLYTIC CAPACITORS

Nonelectrolytic capacitors are used when relatively low values of capacitance are needed. One of the most common types of nonelectrolytic capacitors is the **paper capacitor**. The plates in a paper capacitor are made of two strips of very thin metal foil separated by a strip of waxed paper. These three strips are coiled into a tight roll, and wire leads are attached to the plates. The assembly may be encapsulated in plastic, or, as in the case of an aircraft magneto capacitor, sealed in a metal can. [Figure 3-121]

Figure 3-121. A magneto capacitor is a paper capacitor sealed in a metal container.

Capacitors requiring a smaller capacity but a higher working voltage are made using stacks of thin metal foil sandwiched between thin sheets of mica. This stack is then encapsulated in plastic to form a rectangular block-like capacitor. [Figure 3-122]

For high-voltage applications, paper capacitors are enclosed in a metal container filled with an insulating oil. If a voltage surge breaks through the insula-

Figure 3-122. Block-type mica capacitor.

tor, the oil flows in and restores its insulating characteristics. These are sometimes referred to as **self-healing capacitors**. [Figure 3-123]

Figure 3-123. High voltage, oil-filled paper capacitor.

High-voltage, low-capacitance capacitors are made of either a disc or a tube of ceramic material plated with silver on each end to form the plates. The leads are attached to the silver, and the entire unit is covered with a protective insulation. [Figure 3-124]

ELECTROLYTIC CAPACITORS

Electrolytic capacitors are used when it is necessary to have a large amount of capacity with a relatively low working voltage. These capacitors are polarized, meaning they act as capacitors only when they

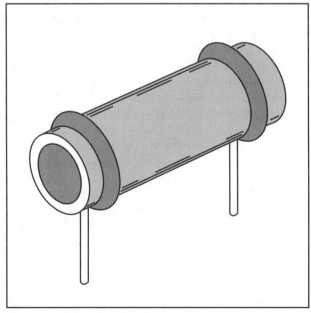

Figure 3-124. Tubular ceramic capacitor.

are properly connected into a circuit. Therefore, electrolytic capacitors can be used only in DC circuits. If an electrolytic capacitor is installed with the wrong polarity, current flows through the capacitor causing it to overheat and explode. [Figure 3-125]

Figure 3-125. Electrolytic capacitors are polarized and, therefore, can only be used in DC circuits.

The reason electrolytic capacitors have such a high capacity for their small physical size is their extremely thin dielectric. The positive plate is made of aluminum foil with an extremely thin oxide film deposited on it to serve as the dielectric. A liquid or paste electrolyte contacts both the positive plate and the negative container the capacitor is sealed in. This allows the electrolyte to form the second plate of the capacitor. This combination of metal plate, oxide film, and conductive liquid, or paste, results in a capacitor that has a great deal of capacity for its size.

VARIABLE CAPACITORS

As discussed earlier, a capacitor's capacity is determined by three things: the area of the plates, the distance between of the plates, and the type of dielectric. If you are able to change any of these factors, you can change the capacity.

Most radios utilize a tuner that varies capacitance by changing the plate area. One set of plates, called **rotors**, are made of thin sheets of aluminum that are meshed together with another group of fixed plates called **stators**. The rotors are mounted on a rotatable shaft and the air between the plates serves as the dielectric. When the plates are fully meshed, the capacity is at its maximum. However, as the shaft is rotated, the meshed plate area decreases and the capacity drops. [Figure 3-126]

Figure 3-126. A variable-plate-area tuning capacitor, using air as the dielectric.

While variable-area capacitors are typically used as the main tuning capacitor for a radio, small **trimmer** and **padder** capacitors accomplish fine tuning. These small capacitors are made up of a stack of foil plates separated by thin sheets of mica. A screw adjustment allows the plates to be squeezed together to increase capacitance, or relaxed to decrease capacitance.

The last way to change capacity is by changing the dielectric constant. The most common fuel quantity measuring system uses a capacitor that allows just that. The measuring units are capacitors in the form of probes in the fuel tanks. Each probe is made up of two concentric tubes which fit across the tank from top to bottom. Each tube acts as one plate of the capacitor, and both the area and the separation between the plates are fixed. When the tank is empty, the dielectric is air, which has a dielectric constant of one. When the tank is full, the dielectric is the fuel, which has a dielectric constant of approximately two. The fuel indicator in the cockpit measures the capacitance of the probes and converts it into a number that reflects the amount of fuel in the tanks. [Figure 3-127]

IMPEDANCE

You now know that the flow of current in an AC circuit is opposed by resistance (R), capacitance (C), and inductance (L). A circuit containing all three of these is often referred to as an **RCL circuit**. On the

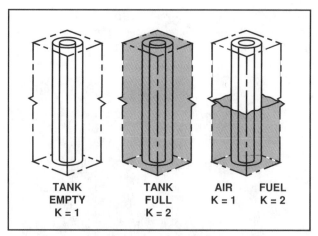

Figure 3-127. The capacitance-type fuel gauging system uses capacitors as probes in the fuel tank. When the tank is empty, the dielectric is air. When the tank is full, fuel is the dielectric.

other hand, circuits containing only two of the three oppositions are referred to as **resistive capacitive (RC)** and **resistive inductive (RL)** circuits.

In section A of this chapter you learned that, through Ohm's law, the current in a circuit is equal to the voltage divided by the resistance. However, in an AC circuit you must also consider the effects of both capacitive and inductive reactance. The combined effect of resistance, capacitance reactance, and inductance reactance is called **impedance** and is represented by the letter **Z**. Like resistance, the unit of measure for impedance is the ohm.

At first it may appear that you can just add the sum of the individual oppositions. However, this is not true since inductive reactance and capacitive reactance have opposite effects on a circuit. Because of this, you must determine the net effect of the two reactances. For example, if you consider inductive reactance (X_L) positive because it causes the voltage to lead the current, and capacitive reactance (X_C) as negative because it causes the current to lead the voltage, you can add the two quantities using the formula:

$$\text{Total Reactance} = X_L + (-X_C)$$

Since inductive and capacitive reactance cause 90 degree phase shifts, they cannot be algebraically added to resistance to calculate impedance. However, they can be considered as two forces acting at right angles to each other. This is best illustrated through vectors. If you remember from Chapter 2, a **vector** is a quantity that has both magnitude and direction and is usually represented as an arrow. The length of the arrow reflects the strength or size of the quantity, and the direction of the arrow represents the direction of the quantity. A circuit's resistance is plotted on a horizontal line extending to the right of the zero point. Since reactance acts 90 degrees to resistance, it is plotted on a vertical line extending up from the zero point. Using vector addition, you can now combine resistance and reactance into a resultant force which represents impedance. [Figure 3-128]

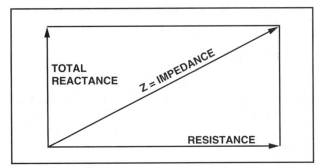

Figure 3-128. The total opposition, or the impedance of a circuit is the vector sum of the total reactance and the resistance.

Using the pythagorean theorem, $A^2 + B^2 = C^2$, you can determine the length of the resultant or impedance vector. This is illustrated in the formula:

$$R^2 + (X_L - X_C)^2 = Z^2$$

Simplified, the formula for impedance in a series circuit reads:

$$Z = \sqrt{R^2 + (X_L - X_C)^2}$$

It should be noted that this formula can only be used for a series circuit. With this in mind, consider a 400-hertz, 115 volt AC circuit with a total resistance of 100 ohms, inductance of 20 millihenries, and capacitance of 5 microfarads. [Figure 3-129]

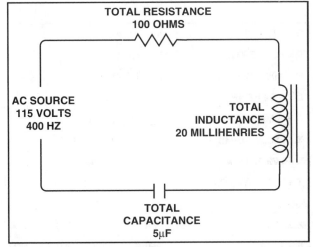

Figure 3-129. A series circuit with 100 ohms of resistance, 20 millihenries of inductance, and 5 microfarads of capacitance.

To begin, you must find the inductive and capacitive reactance. The inductive reactance is 50.25 ohms and the capacitive reactance is 79.62 ohms.

Example:

Inductive reactance

$$X_L = 2\pi f L$$

$$= 6.28 \times 400 \times .020$$

$$= 50.25 \text{ ohms}$$

Capacitive reactance

$$X_C = \frac{1}{2\pi f C}$$

$$= \frac{1}{6.28 \times 400 \times .000005}$$

$$= 79.62 \text{ ohms}$$

Since the capacitive reactance is larger than the inductive reactance, the circuit is said to be capacitive and the amps lead the volts.

Now that you know the values for resistance, inductive reactance, and capacitive reactance, you can calculate impedance using the formula:

$$Z = \sqrt{R^2 + (X_L - X_C)^2}$$

$$= \sqrt{100^2 + (50.25 - 79.62)^2}$$

$$= \sqrt{10,000 + 862.60}$$

$$= \sqrt{10,862.60}$$

$$= 104.22 \text{ ohms}$$

CURRENT

Once impedance is found, you can use Ohm's law to determine circuit current. However, since the symbol "Z" is used to represent total opposition in an AC circuit, it takes the place of "R" in Ohm's law. The total circuit current is 1.10 amps.

$$I = E/Z$$

$$= 115 \text{ volts}/104.22 \text{ ohms}$$

$$= 1.10 \text{ amps}$$

VOLTAGE DROP

To calculate the voltage drop across each component, use Ohm's law and multiply the total circuit amperage by each of the values of resistance, inductive reactance, and capacitive reactance.

Since the inductor and capacitor cause a phase shift, the voltage dropped across the inductor and capacitor are not in phase with the current through each. However, the voltage and current are in phase through the resistor. Because of this, the sum of the voltage across the three components is greater than the source voltage.

$$E_R = I_T \times R = 1.10 \times 100 = 103.0 \text{ volts}$$

$$E_L = I_T \times X_L = 1.10 \times 50.25 = 55.28 \text{ volts}$$

$$E_C = I_T \times X_C = 1.10 \times 79.62 = 87.58 \text{ volts}$$

APPARENT POWER

Apparent power is the product of the source voltage times the total current. In this circuit, the apparent power is 126. 5 voltamps.

$$P_{apparent} = EI$$

$$= 115 \text{ volts} \times 1.1 \text{ amps}$$

$$= 126.5 \text{ voltamps}$$

POWER FACTOR

Not all of the current and voltage in this circuit are in phase, and, by definition, power is produced only by those portions of voltage and current that are in phase. You must therefore find the percentage of voltage and current that are in phase.

You saw earlier that a circuit's power factor is the ratio of the true power to the apparent power. However, since the voltage varies in a series circuit, it is also the ratio of the voltage drop across the resistor to the total voltage. In this problem, the voltage drop across the resistor is 110 volts and the total circuit voltage is 115 volts. This results in a power factor of .96, which means that 96 percent of the current is in phase with the voltage. [Figure 3-130]

$$\text{POWER FACTOR} = \frac{\text{TRUE POWER}}{\text{APPARENT POWER}} = \frac{E_R}{E_T}$$

$$= \frac{110}{115}$$

$$= 0.96$$

Figure 3-130. The power factor is equivalent to the ratio between the voltage drop across the resistor to the circuit's total voltage.

PHASE ANGLE

The phase angle is the angle whose cosine is the power factor. The trigonometric function chart in figure 3-92 reveals that 0.96 is the cosine of 16 degrees. Therefore, in this circuit, the current lags behind the voltage by 16 degrees.

TRUE POWER

The true power developed in this circuit is found by the formula P = E × I × power factor, and is equal to 121.44 watts. [Figure 3-131]

$$P_{true} = E \times I \times \text{Power Factor}$$

$$= 115 \times 1.1 \times .96$$

$$= 121.44 \text{ watts}$$

PARALLEL AC CIRCUITS

Almost all of the alternating current circuits in shops and homes have their components connected in parallel rather than in series. Parallel AC circuits are handled in much the same way as parallel DC circuits. The main exception is that you must take

E = 115 VOLTS	I_R = 1.1 AMP
F = 400 HERTZ	I_L = 1.1 AMP
R = 100 OHMS	I_C = 1.1 AMP
L = 0.02 HENRY	E_R = 110 VOLTS
C = 5 MICROFARAD	E_C = 87.58 VOLTS
X_L = 50.25 OHMS	E_L = 55.28 VOLTS
X_C = 79.62 OHMS	P_F = 0.96 (96%)
X = 29.33 OHMS-CAPACITIVE	$P_{apparent}$ = 126.5 VOLT-AMPS
Z = 104.21 OHMS	P_{true} = 121.4 WATTS
h = 1.1 AMPS	PHASE ANGLE = 16° LEADING

Figure 3-131. Values for the series R-L-C circuit.

into consideration the phase shifts that occur among the current flow in each of the three components. [Figure 3-132]

The circuit has a 115-volt, 400-hertz power source and consists of a 50-ohm resistor, a 40-millihenry inductor, and a 10-microfarad capacitor, all in parallel.

The reactance of the inductor and capacitor is calculated using the same formulas as before. The inductive reactance is 100.48 ohms and the capacitive reactance is 39.81ohms.

Example:

Inductive reactance

$$X_L = 2\pi f L$$

$$= 6.28 \times 400 \times .040$$

$$= 100.48 \text{ ohms}$$

Capacitive reactance

$$X_C = \frac{1}{2\pi f C}$$

$$= \frac{1}{6.28 \times 400 \times .000010}$$

$$= 39.81 \text{ ohms}$$

As mentioned earlier, impedance is calculated with a different formula in a parallel circuit than in a series circuit. The difference in formulas is similar to the difference between the formulas for calculating total resistance in a series and parallel circuit. In other words, impedance in a parallel circuit equals the reciprocal of the sum of the reciprocals of the individual components. This requires the use of the formula:

$$Z = \frac{1}{\sqrt{(\frac{1}{R})^2 + (\frac{1}{X_L} - \frac{1}{X_C})^2}}$$

Using this formula, you can calculate the impedance within the parallel circuit.

Example:

$$Z = \frac{1}{\sqrt{(\frac{1}{50})^2 + (\frac{1}{100.48} - \frac{1}{39.81})^2}}$$

$$= \frac{1}{\sqrt{(.02)^2 + (.01 - .025)^2}}$$

$$= \frac{1}{\sqrt{.0004 + .0002}}$$

$$= \frac{1}{.0245}$$

$$= 40.82 \text{ ohms}$$

CURRENT

Once impedance is found, you can use Ohm's law to determine circuit current. The total current flowing through the circuit is 2.82 amps.

$$I = E/Z$$

$$= 115 \text{ volts}/40.82 \text{ ohms}$$

$$= 2.82 \text{ amps}$$

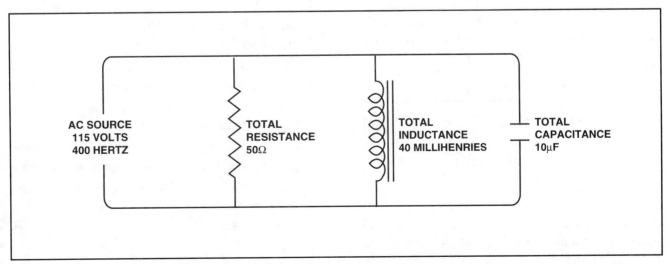

Figure 3-132. This parallel R-L-C circuit consists of a resistor, an inductor, and a capacitor placed in parallel with each other.

CURRENT DROP

To calculate the current drop across each component, use Ohm's law and divide the total circuit voltage by each of the values for total resistance, inductive reactance, and capacitive reactance.

$$I_R = E/R = 115/50 = 2.3 \text{ amps}$$

$$I_L = E/X_L = 115/100.48 = 1.14 \text{ amps}$$

$$I_C = E/X_C = 115/39.81 = 2.89 \text{ amps}$$

APPARENT POWER

Apparent power is the product of the source voltage times the total current. In this circuit, the apparent power is 324.3 voltamps.

$$P_{apparent} = EI$$

$$= 115 \text{ volts} \times 2.82 \text{ amps}$$

$$= 324.3 \text{ volt-amps}$$

POWER FACTOR

As with the earlier series circuit, not all of the current and voltage are in phase. Therefore, power is produced only by those portions of voltage and current that are in phase. You must therefore find this percentage, which is the power factor.

Since the current varies in a parallel circuit, the power factor is equivalent to the ratio of the current drop across the resistor to the total current. In this problem, the current drop across the resistor is 2.3 amps and the total circuit current is 2.82 amps. This results in a power factor of .82, which means that 82 percent of the current is in phase with the voltage. [Figure 3-133]

$$\text{POWER FACTOR} = \frac{I_R}{I_T}$$

$$= \frac{2.30}{2.82}$$

$$= 0.82$$

Figure 3-133. The power factor is equivalent to the ratio between the current drop across the resistor to the circuit's total current.

PHASE ANGLE

The phase angle is the angle whose cosine is the power factor. The trigonometric function chart of figure 3-92 reveals that .82 is the cosine of 35 degrees. In this circuit, the current lags behind voltage by 35 degrees.

TRUE POWER

The true power developed in this circuit is found by the formula P = E × I × power factor, and is equal to 265.9 watts. [Figure 3-134]

$$P_{true} = E \times I \times \text{Power Factor}$$

$$= 115 \text{ volts} \times 2.82 \text{ amps} \times .82$$

$$= 265.9 \text{ watts}$$

RESONANCE IN AN AC CIRCUIT

Inductive reactance in a coil is zero when the frequency is zero. However, as the frequency increases, the inductive reactance increases. Therefore, the higher the frequency, the more back voltage the inductor generates, and less current flows. This continues until the back voltage equals the source voltage, and no current flows.

The reactance in a capacitor varies in the opposite way. For example, at a frequency of zero, no current flows through a capacitor, and therefore reactance is

E = 115 VOLTS	I_L = 1.14 AMPS
f = 400 HERTZ	I_C = 2.89 AMPS
R = 50 OHMS	I_X = -1.75 AMPS
L = 0.04 HENRY	I_T = 2.82 AMPS
C = 10μF	Z = 40.82 OHMS
X_L = 100.48 OHMS	P_F = .82 (82%)
X_C = 39.81 OHMS	PHASE ANGLE = 35° LAGGING
E_R = 115 VOLTS	$P_{APPARENT}$ = 324.5 VOLT-AMPS
E_L = 115 VOLTS	P_{TRUE} = 265.9 WATTS
E_C = 115 VOLTS	
I_R = 2.3 AMPS	

Figure 3-134. Values for a parallel circuit.

infinite. But, as the frequency increases, the capacitive reactance decreases until there is no capacitive reactance. Both of these relationships can be plotted on a graph. [Figure 3-135]

The lines representing the two reactances cross at the **resonant frequency**. In other words, a circuit's resonant frequency is that frequency where inductive and capacitive reactance are the same. The resonant frequency is expressed in hertz and is found by dividing 1 by the constant 2 π times the square root of the product of the inductance in henrys and the capacitance in farads. This is expressed in the formula:

$$F_R = \frac{1}{2 \pi \sqrt{LC}}$$

SERIES RESONANT CIRCUIT

In a series R-L-C circuit at its resonant frequency, the current flowing in the inductor and the capacitor are equal. However, they are 180 degrees out of phase with each other. The inductive and capacitive reactances are also exactly the same, but because of the phase difference they cancel each other, leaving a total reactance of zero. In this case, the total opposition offered to the flow of AC is that of the resistance. Therefore, a circuit's impedance is minimum when at its resonant frequency and is equal to the circuit resistance.

The voltage drop across the resistor equals the source voltage and the current flow remains constant. The voltage across either of the reactances can be higher than the source voltage. However, since they are 180 degrees out of phase, their polarities are opposite and they cancel each other. Therefore, the sum of the individual voltages does not equal the source voltage as it does in a DC circuit.

A series resonant circuit acts as a pure resistance circuit. The source voltage and current are in phase, so the power factor of the circuit is one.

PARALLEL RESONANT CIRCUIT

A large amount of current flows between the capacitor and the inductor in a parallel R-L-C circuit at its resonant frequency. This allows energy to first be stored in the electrostatic field of the capacitor and then in the electromagnetic field around the inductor. If there were no resistance in the circuit, once the exchange of energy between the two types of fields started, the circulating current would continue to flow back and forth indefinitely. But in practice, all circuits have some resistance which causes this current to die down, unless extra energy is added from the source.

At the resonant frequency, the circulating current in the inductor and capacitor is high. There is almost no current supplied from the source, though, so the source sees the parallel circuit as having a high impedance. The reactances cancel each other, and so the opposition is purely resistive. The power factor of the circuit is one.

CONVERTING AC TO DC

It is often necessary to convert alternating current into direct current to power various circuits in the aircraft or within electronic equipment. The conversion of alternating to direct current is accomplished by a circuit referred to as a **rectifier**. Rectifier circuits employ vacuum tube or solid-state diodes that allow current flow in only one direction. Rectifier circuits are discussed in detail in Section E of this chapter.

THREE-PHASE AC

When it is necessary to get the maximum amount of power from alternating current, it is typically converted into three-phase AC. Generators that produce three-phase AC have three sets of output windings excited by a single rotating field. The voltage in each winding is 120 degrees out of phase with that in the other windings. Three-phase AC offers several advantages. For example, when it is rectified

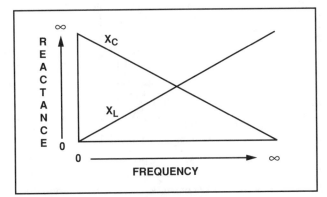

Figure 3-135. This graph illustrates how inductive reactance increases as the frequency increases and how capacitive resistance decreases as the frequency increases.

into direct current, there are three times as many pulses of rectified current as there are in single-phase AC. The pulses overlap so the current never drops to an instantaneous value of zero. Furthermore, the higher the pulse frequency, the easier it is to filter the AC and make it smooth. [Figure 3-136]

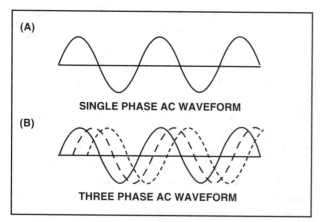

Figure 3-136 (A) — With single-phase AC, the current drops below zero making it impossible to smooth out. (B) — With three-phase AC, current flows in the circuit at all times resulting in minimal drop from peak current per cycle.

There are two ways of connecting the three output windings of an alternator. The first is a **Y** connected hookup. With this type of connection, one end of each winding is connected at a common point. The other ends of the three windings are brought out as the output leads. Each output lead crosses two of the windings in series. A fourth neutral lead coming off the common point is also established. Since the voltage across each winding is 120 degrees out of phase with that in the other windings, the output voltage is never twice that of one of the phase windings. Instead, it is 1.73 times that of a single-phase winding. Therefore, if 120 volts is produced across each phase, the voltage between any two of the leads is 208 volts. Since the windings are in series between the output leads, the output current is the same as the phase current. [Figure 3-137]

The second method for connecting the phase windings is through a **delta connection**. With this method, both ends of each phase winding are connected to the ends of the other windings to form a loop. An output lead is brought from each junction so that the output voltage is always the same as the phase voltage. However, there are two windings in series across, or in parallel, with the third winding. Since the current in each of these windings is 120

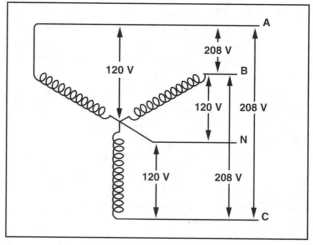

Figure 3-137. In a Y-connected, three-phase alternator, 208 volts is obtained between each phase winding and 120 volts is obtained between the neutral and one phase winding.

degrees out of phase with that in the other windings, the output current is 1.73 times that of the current in the phase winding. [Figure 3-138]

THREE-PHASE TRANSFORMERS

Three-phase transformers can have their primary and secondary windings connected in either a Y or a delta to provide the needed output.

In a delta-to-delta connection, both the primary and secondary windings have their leads connected in the delta form. If there are the same number of turns in the secondary as there are in the primary, both

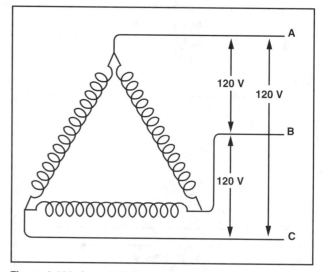

Figure 3-138. In a delta-connected, three-phase alternator, the output voltage is the same as the phase voltage. However, since two windings parallel the third, the current is 1.73 times the phase winding.

the secondary voltage and current are the same as the primary voltage and current. [Figure 3-139]

If both the primary and secondary windings are Y-connected, the secondary voltage and current are the same as the voltage and current in the primary. [Figure 3-140]

Connecting the primary as a delta and the secondary as a Y produces a secondary voltage that is 1.73

times as high as the primary voltage. But remember, a transformer is not capable of producing power. Therefore, when the secondary voltage is higher than the primary, the secondary current must be lower. With this type of connection, the current in the secondary is only 0.578 times the primary current.

If the primary winding is connected as a Y and the secondary as a delta, the secondary voltage is again only 0.578 of the primary voltage. The secondary current becomes 1.73 times the primary current.

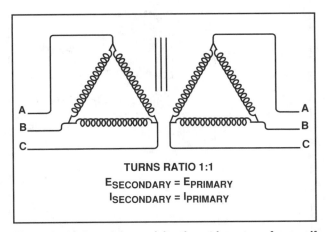

Figure 3-139. In a delta-to-delta three-phase transformer, if the number of windings in the secondary are the same as the primary, there is no change in the output.

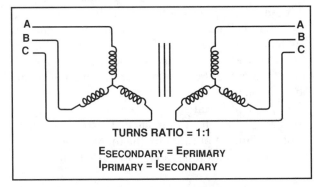

Figure 3-140. Y-to-Y three-phase transformer.

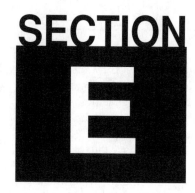

SECTION E

ELECTRON CONTROL DEVICES

VACUUM TUBES = *control Grid*

Dr. Lee DeForest ushered in the age of electronics in 1907 with the discovery of the **vacuum tube**, or **audion**. This device controlled the flow of electrons by electrical charges rather than by mechanical switches and power dissipating resistors. Vacuum tubes served well and played a major role in the development of radio communication, television, and the electronic computer. However, because of their large physical size, the amount of power they require, and the heat they generate, they have been replaced almost entirely by the more modern solid-state devices and integrated circuit (IC) chips.

DIODES

The simplest vacuum tube is the diode, or two-element tube. The vacuum tube diode consists of a glass or metal container with two active elements, a cathode and a plate, and a small electric heater. When power is applied to a vacuum tube diode, the cathode is heated. When this happens, the electrons accelerate until they leave the cathode. To enhance this process, the cathode is typically coated with a material that weakens the bonds of electrons. When a DC source is connected across the cathode and the plate with the positive side of the voltage connected to the plate, the electrons emitted by the cathode are attracted to the plate. Therefore, current flows through the tube when the plate is positive with respect to the cathode. This characteristic allows the diode to rectify AC into DC. Diode rectifiers may be used as either half-wave or full-wave rectifiers. In a half-wave rectifier, current flows only during that half-wave when the plate is positive and the cathode is negative. However, during the other half of the cycle, the negative plate repels the electrons back to the cathode and there is no flow. This results in an output of pulsating direct current. [Figure 3-141]

In a vacuum tube connected as a full-wave rectifier, current flows to the load on both half cycles of the alternating voltage. For example, current

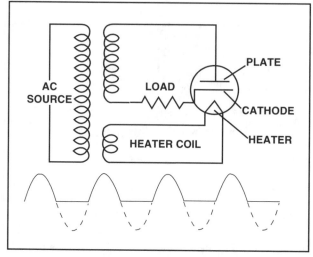

Figure 3-141. In a half-wave vacuum tube diode, the cathode releases electrons on the first alternation, allowing current to flow through the load and back into the diode. This produces pulsating DC current.

flows from the top plate through the load on one alternation, and on the next alternation, current flows to the lower plate and through the load in the same direction.

Due to the limitations discussed earlier, vacuum tube rectifiers have almost entirely been replaced in aircraft by semiconductor diodes. These will be discussed later in this section.

THE DRY-DISC RECTIFIER

Another type of rectifier you may see is the dry-disc rectifier. Three types of dry-disc rectifiers may be found on aircraft: copper-oxide, selenium, or magnesium. The copper-oxide rectifier consists of small copper discs with an oxide film deposited on one side, and a lead disc pressed against the oxide to form a conductor. When an AC source is supplied, electrons flow during the half-cycle when the copper is negative, and cease during the half-cycle when the copper is positive. This results in a pulsating DC output. Dry-disc rectifiers are not very efficient because they provide a great deal of resistance and, therefore, they generate a substantial amount of heat. For this reason, cooling plates are required to dissipate the excess heat so the rectifier won't become damaged. Another problem is the relatively low voltage the rectifier can withstand without breaking down. Because of this characteristic, a number of disks must be stacked in series so the voltage drop across each does not exceed each disk's relatively low breakdown voltage. [Figure 3-142]

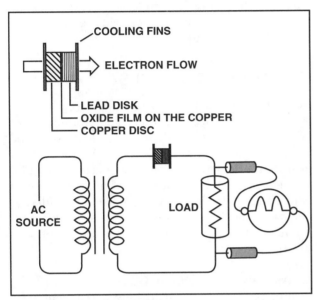

Figure 3-142. A dry-disc rectifier.

TRIODES

The triode vacuum tube contains three elements: a cathode, a plate, and a **control grid** located between the cathode and the plate. This allows the triode to act more like a control valve rather than a check valve. For example, if a negative voltage is placed on the grid, the electrons are repelled back to the cathode so they cannot be influenced by the positive plate. In fact, a high enough negative potential on the grid can completely stop the flow of electrons. However, if a positive potential is placed on the grid, electrons are accelerated on their way to the plate, thereby increasing the current flow. The grid is so small relative to the plate, and the positive potential on the grid is so much lower than that on the plate, that only a few electrons are attracted to the grid. The grid current is negligible compared to the flow between the cathode and the plate. [Figure 3-143]

If a low-voltage AC signal is placed across the grid, the amount of current through the tube varies in the same way as the voltage on the grid. [Figure 3-144]

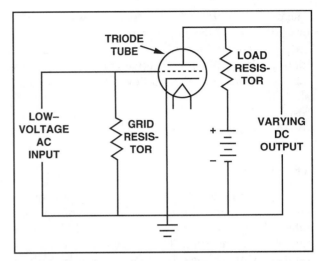

Figure 3-143. The plate in the triode tube is connected to the positive terminal of a high-voltage DC power supply through the load resistor. The cathode is connected to the negative terminal, or the ground, of the power source. When the heater warms up, the cathode emits electrons which are attracted to the positively charged plate. There is then electron flow within the tube.

The most common use of a triode is as an amplifier tube. When a resistance is connected in series in the plate circuit, the voltage drop across the plate can be changed by varying the grid voltage. A small change in grid voltage causes a large change in the voltage drop across the plate resistance. Therefore, the voltage applied to the grid is amplified in the plate circuit.

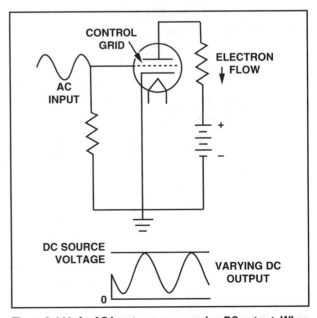

Figure 3-144. An AC input causes a varying DC output. When there is no signal voltage on the grid, the output is a steady flow of direct current. However, when the grid has a positive voltage on it, the electron flow through the tube increases, and the voltage drop across the load resistor increases.

The polarity of the voltage across the load resistor is opposite that of the power source. As a result, the output becomes less positive as the grid becomes more positive. When a negative voltage on the grid makes it so negative that all flow through the tube is shut off, there is no voltage drop across the load resistor, and the output voltage rises to the value of the DC source.

TETRODES

At high frequencies the capacitive reactance between the control grid and the plate of a triode tube is low enough that voltage can feed back from the plate to the grid. This feedback generates an unwanted AC voltage that causes the tube to oscillate. Therefore, to prevent oscillation, an extra grid is built into the tube between the plate and the control grid. This second grid is called a **screen grid**, and because the tube now has four active electrodes, it is called a tetrode.

The screen grid is connected to the plate through a resistor and to ground through a capacitor. Since the screen grid is between the control grid and the plate, the grid intercepts any AC signal fed back from the plate and bypasses it to ground through the capacitor. This way, feedback never reaches the control grid to cause oscillation. [Figure 3-145]

PENTODES

When the power handled by a vacuum tube is increased, electrons are drawn to the plate with such velocity that some of them bounce off and are attracted to the positively charged screen grid. This secondary emission can cause an unwanted screen current. To control this, a **suppressor grid** is added between the plate and the screen grid. This grid is connected either directly to the cathode inside the tube, or, in some cases, to the ground outside the tube. When electrons bounce off the plate, they are forced back to the plate by the negative charge on the suppressor grid before they have the chance to attach to the positively charged screen grid. Vacuum tubes containing a suppressor grid are called pentodes. [Figure 3-146]

SOLID-STATE DEVICES

Although vacuum tubes ushered in the age of electronics, they did have their shortcomings. For example, because of their very large size and power requirements, they were not practical for use in the somewhat compact electronic equipment used in modern aircraft. Therefore, almost all vacuum tubes have been replaced by solid-state devices. The term solid-state refers to any electronic mechanism that utilizes a solid material to control the flow of electrons.

SEMICONDUCTOR DIODES

Before you can understand how a semiconductor diode rectifies AC to DC, you must gain an understanding of what takes place within the solid materials that make up a diode. To begin, a semiconductor material is one that has four electrons in the outer, or valence shell of each atom. Two types of material that exist in this configuration are **silicon**

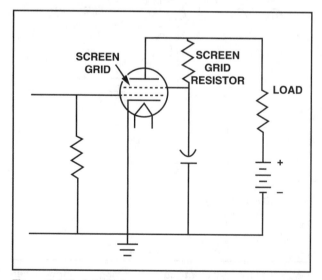

Figure 3-145. A tetrode vacuum tube uses a screen grid to intercept AC feedback and send it through a capacitor to the ground. When this is done, interelectrode capacitance between the plate and the control grid is minimized.

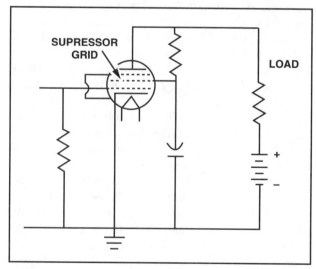

Figure 3-146. The suppressor grid is connected either to the cathode or to the ground, giving the grid a negative charge that suppresses secondary emissions back into the plate.

(Si) and **germanium (Ge)**. Because of the number of electrons in the valence shell and the strong covalent bonds formed when the valence electrons in one atom combine with those in another, neither silicon nor germanium conduct electricity. [Figure 3-147]

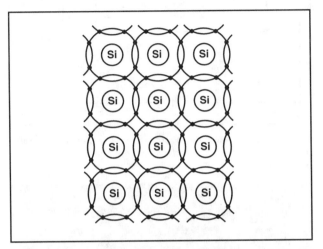

Figure 3-147. Here you can see the shared valence electrons of a semiconductor material. Since there is only room for eight electrons in a valence shell, the material will not accept electrons from an outside source.

The only way silicon or germanium can carry a current is if you combine or **dope** them with another material containing atoms with five electrons in their valence shells. Common elements that are doped with silicon and germanium include arsenic, bismuth, and antimony. When this is done, spare, or free electrons exist after the covalent bonds are formed. These electrons are free to move, so the material is called a **donor**, or an **N-type material**. [Figure 3-148]

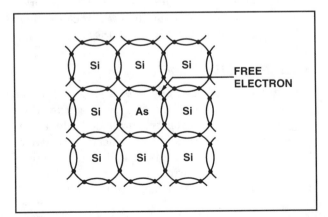

igure 3-148. When an insulator is doped with an element having five valence electrons, it produces a material with an excess of electrons. This type of material is referred to as an N-type material.

Elements such as boron, indium, and gallium have only three valence electrons. When any of these are doped with silicon or germanium, there are areas where covalent bonds can not form due to the shortage of electrons. These areas are called **holes** and accept electrons from an outside source. Materials doped in this way are called **acceptors**, or **P-type material**. [Figure 3-149]

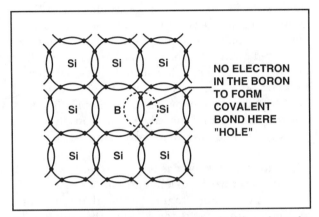

Figure 3-149. When an insulator is doped with an impurity having only three valence electrons, it produces a material that accepts electrons. This type of material is referred to as a P-type material.

The N- and P-type materials can be joined either by a junction or a point contact to form a **semiconductor diode**. However, when this is done, the holes, or positive charges, in the P-type material attract the electrons, or negative charges, in the N-type material. The unlike charges combine at the junction, leaving a **depletion area** where there are no more free electrons or holes. [Figure 3-150]

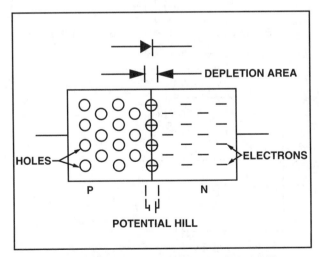

Figure 3-150. When a piece of P-type and N-type silicon are joined, the electrons and holes combine along the junction and form a depletion area that can act as either a conductor or insulator.

As the charges move toward the junction, a large number of positive ions are left in the N-type material farthest from the junction and a large number of negative ions are left in the P-type material farthest from the junction. These stationary ions provide charges that stop the movement of electrons and holes across the junction.

When a voltage source is attached to a semiconductor diode with its positive terminal connected to the P material and its negative terminal to the N material, electrons combine with the positive ions to neutralize their hold on the electrons in the depletion area. This allows the electrons to flow across the depletion area to occupy the holes in the P material and flow to the battery's positive terminal. In this example, the diode is said to be **forward biased**, and electrons flow from the N material to the P material. [Figure 3-151]

When the power source is turned around so that the positive terminal is attached to the N material and the negative terminal to the P material, the electrons and holes are attracted away from the junction. In this situation, the diode is said to be **reversed biased**, and no electrons or holes can combine. Therefore, electron flow in the external circuit stops. [Figure 3-152]

DIODE TESTING

Since a diode only allows electrons to flow in one direction, it can be tested by applying a current across it. The easiest way to do this is with an ohmmeter. To test a diode, set the ohmmeter to a low resistance scale. Next, place the positive lead on one

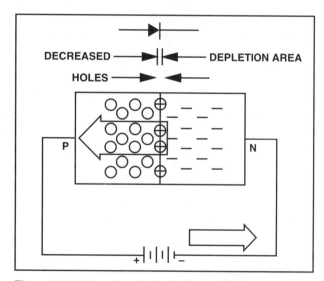

Figure 3-151. A semiconductor is forward biased when the positive terminal of the power source is attached to the P-type material and the negative terminal is attached to the N-type material.

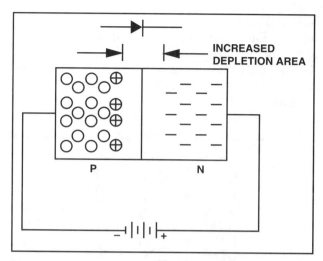

Figure 3-152. When a conductor is reverse biased, the holes and electrons are attracted away from the junction and no current can flow.

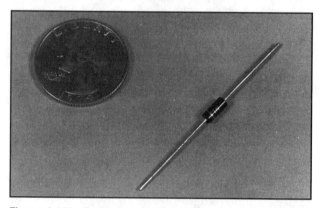

Figure 3-153. A semiconductor diode acts as an electron check valve, allowing current flow in one direction. Its small size makes it more practical for use in modern electronics.

end of the diode and the negative lead on the other and note the indication. Now, reverse the leads and note the indication again. With a good diode, the ohmmeter should indicate a low resistance when forward-biased and a high indication when reverse-biased. In a defective diode, the resistance readings will be nearly the same. [Figure 3-154]

HALF-WAVE RECTIFIER

A half-wave rectifier circuit uses a single diode in series with an AC source and a load. Electrons flow only during the half-cycle when the cathode, represented by the bar across the arrowhead, is negative. The output waveform of this type of rectifier is one-half the alternating-current wave, making it inefficient for many applications. [Figure 3-155]

FULL-WAVE RECTIFIER

A full-wave rectifier contains two diodes that change both halves of an AC cycle into DC.

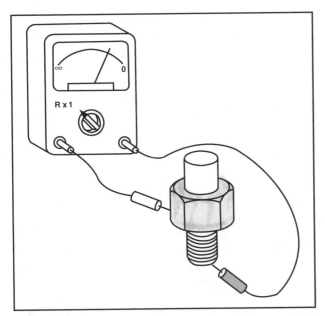

Figure 3-154. To test a diode with an ohmmeter, place the test leads on opposite terminals and note the reading. Then, reverse the leads and note the second indication. A good diode has low resistance when forward biased and high resistance when reverse biased.

Therefore, full-wave rectification is more efficient and produces a much smoother output. The diodes are connected to the secondary coil of a transformer that is tapped at its center. [Figure 3-156]

When analyzing a schematic, it is usually easier if you assume a current flows from positive to negative, following the arrowheads in the diode symbols. For example, during the first half-cycle, when the top of the secondary coil is positive, current flows through diode D_1, and passes through the load from top to bottom. This causes the top of the load to be positive. After leaving the load, the current flows into the secondary coil at the negative center tap.

During the next half-cycle, the bottom of the secondary winding becomes positive and the center tap

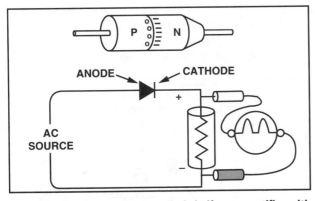

Figure 3-155. A semiconductor diode half-wave rectifier with its output waveform.

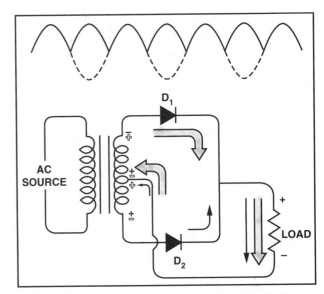

Figure 3-156. In a basic full-wave rectifier, two diodes are connected to each end of a transformer's secondary coil. After the current passes through the load, it flows back to the coil through a center tap.

negative with respect to the bottom. Current flows through diode D_2 and through the load resistor in the same direction it passed during the first half-cycle. The output waveform is pulsating direct current with a frequency twice that of the pulsating DC produced by a half-wave rectifier.

BRIDGE-TYPE RECTIFIER

The two-diode full-wave rectifier requires a transformer to produce the desired output voltage across one-half of a secondary coil. To overcome this inefficiency, four diodes must be arranged into a bridge-type rectifier circuit. [Figure 3-157]

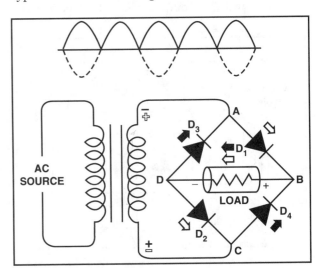

Figure 3-157. A bridge-type rectifier utilizes four diodes and an AC power source connected at points A and C. The load is then connected at points B and D.

During the half-cycle when the top of the secondary coil is positive, current flows through diode D_1 and through the load resistor from right to left, and then down through diode D_2 and back to the negative end of the secondary coil. The polarity of the secondary coil reverses during the next half-cycle, and current flows through diode D_4 to the positive side of the load, and up through diode D_3 to the opposite side of the coil. The output waveform is similar to that produced by the two-diode full-wave rectifier, but the voltage is much higher because the entire secondary coil is used.

THREE-PHASE RECTIFIER

Almost all aircraft alternators produce three-phase AC, therefore, a three-phase rectifier must be used to produce direct current in modern aircraft. A typical three-phase rectifier in an aircraft alternator uses a three-phase stator and six silicon diodes. [Figure 3-158]

Examine the current flow through the load resistor for one complete cycle of all three phases. Remember that you are tracing conventional current so you can follow the direction indicated by the arrowheads in the diode symbols. In that portion of the cycle when the output end of phase A is positive, current leaves coil A and flows through diode D_1 to the load. After leaving the load, the current flows through diode D_2 and coil C, whose output lead is negative. As the alternator field rotates, it causes the output end of coil B to become positive and the output of coil A to be negative. Current flows out of coil B and passes through diode D_3, the load, diode D_4, and back through coil A. Continued rotation causes the output of coil C to be positive

and coil B to be negative causing current to leave coil C and pass through diode D_5, the load, diode D_6 and back into coil B.

The output waveform of a three-phase rectifier gives a very steady direct current as the current from the three phases overlap. There is never a time when the current drops to zero.

ZENER DIODES

You have just seen that a semiconductor diode conducts when it is forward biased and does not conduct when it is reverse biased. However, there is an exception to this. When a specific voltage, or **zener voltage** is placed across a diode in its reverse bias direction, the covalent bonds between the atoms break down and the diode allows current to flow.

Figure 3-159 shows a 15-volt zener diode in series with a bleeder resistor across a 24-volt direct current power supply. The anode of the zener diode is connected to the negative terminal. This is the reverse bias direction. The zener diode breaks down at 15 volts and allows enough current to flow through the bleeder resistor to maintain a nine volt drop across it. The load is connected across the zener diode, and since it maintains a 15-volt drop across it, there is always exactly 15 volts across the load. When the load current increases, the current through the zener diode decreases enough to maintain the nine volt drop across the bleeder resistor. When the load current decreases, rather than allowing the voltage to rise, the current through the zener diode increases. The current through the zener diode varies so that the total current through the load and the zener diode is the correct amount to maintain the nine volt drop across the bleeder resistor.

TRANSISTORS

The transistor is an electronic device that is capable of performing most of the functions of vacuum tubes. However, transistors are very small, lightweight, and do not require a heating element. Furthermore, transistors are mechanically rugged and do not pick up stray signals. [Figure 3-160]

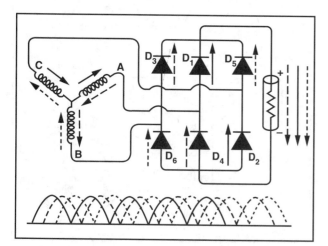

Figure 3-158. Most aircraft direct current is produced by a three-phase alternator. To rectify three-phase current, a three-phase rectifier with six-diodes is used.

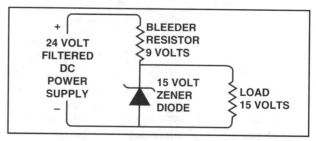

Figure 3-159. A zener diode voltage regulator circuit.

Figure 3-160. The transistor serves as an electron control valve in much the same way as a vacuum tube, but it is much smaller and does not require as much power to operate.

A transistor is essentially a sandwich of N-type silicon or germanium between two pieces of P-type material, or a piece of P-type material between two pieces of N-type material. Transistors having a P-type material between two N-type materials are called **NPN** transistors, whereas a **PNP** transistor consists of an N-type material between two P-type materials. The center piece of material is called the **base**, and is the control element of a transistor. One end piece is the **emitter** and the other is the **collector**. In circuit diagrams, NPN and PNP transistors are represented by different symbology. [Figure 3-161]

The operation of NPN and PNP transistors is essentially the same, except the polarity of the power sources required to provide the bias for their operation. For example, for a transistor to conduct current

between its emitter and its collector, the emitter-base junction must be forward biased and the collector-base junction reverse biased. On an NPN transistor, the emitter-base is forward biased when the base is positive with respect to the emitter and the collector-base junction is reverse biased when the collector is positive with respect to the base. On a PNP transistor, the emitter-base is forward biased when the base is negative with respect to the emitter and the collector-base is reverse biased when the collector is negative with respect to the base.

In most applications, the output circuit path you wish to control is from the emitter to the collector. The base is the control element that turns the main stream current on and off. This is done by injecting electrons into the base or pulling them out, depending on whether the transistor is PNP or NPN, respectively. For example, when the emitter-base of a PNP transistor is forward biased and a small amount of current flows, the depletion area at the emitter-base junction becomes extremely narrow. This allows the relatively large reverse bias collector-base current to force electrons to the emitter-base junction where they are attracted to the positive voltage source attached to the emitter. Electrons leaving the collector return to the positive terminal of the emitter source, rather than returning to the positive terminal of the collector source. [Figure 3-162]

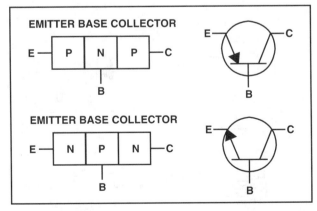

Figure 3-161. You can differentiate the symbols for PNP and NPN transistors by referring to the arrowhead on the emitter. For a PNP transistor, the arrowhead points to the base, whereas the emitter arrowhead on an NPN transistor points away from the base.

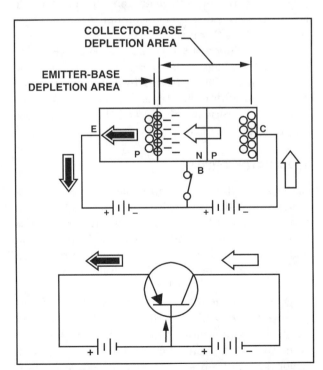

Figure 3-162. When the emitter-base junction of a transistor is forward biased, current flows. The large emitter-collector current is controlled by varying the much smaller emitter-base current.

If the base circuit is opened, as it is in figure 3-163, there is no longer a force to keep the emitter-base depletion area reduced. Therefore, there is no attraction for the electrons from the negative terminal of the collector source across both depletion areas to the positive terminal of the emitter source. Therefore, when there is no base current, there can be no flow between the emitter and the collector. [Figure 3-163]

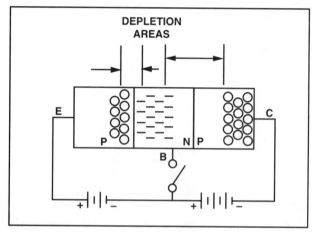

Figure 3-163. When the base circuit is open, no base current flows, and there is no emitter-collector current.

NPN transistors are used in circuit applications that require a positive voltage to cause a transistor to conduct. NPN transistors are similar in almost every way to the PNP transistor, except for the arrangement of the doped areas. When biasing an NPN transistor, the voltage polarities are exactly opposite those of the PNP transistor. However, for maximum conduction with either type, the emitter-base junction must be forward biased and the collector-base junction reverse biased.

TRANSISTOR TESTING

Like diodes, transistors can be tested with an ohmmeter. To do this, begin by removing the transistor from the circuit and measuring the resistance between the emitter and collector. Since no current is flowing to the base, the resistance should be high across the emitter-collector in both directions.

Now, check the emitter-base junction with both forward and reverse biasing. Since current flows easily when the emitter-base is forward biased, the resistance indicated should be low. On the other hand, when the emitter-base is reverse biased, no current flows and a high resistance is indicated. If the resistance readings in both directions are equal, the transistor is defective.

The last junction to check is the collector-base. Like the emitter-base junction, the resistance should be high in one direction and low in the opposite direction. If this is not the case, the transistor is defective.

OTHER SOLID-STATE DEVICES

In addition to diodes and transistors, there are several other solid-state devices that control the flow of electrons. As an aircraft technician, you should have a working understanding of each of these.

SILICON CONTROLLED RECTIFIERS

At times, it is necessary to limit the amount of current that flows to some electrical devices such as lights and some types of motors. One way to do this is by dropping the voltage supplied to these devices through a relatively large resistor which, in turn, generates a great deal of heat. Another way to accomplish the same task is to use a silicon controlled rectifier, or SCR. Instead of dropping the voltage, an SCR decreases the amount of current supplied to a device by controlling the time cycle of AC.

A silicon controlled rectifier is similar to a silicon diode in its outward appearance, except for its extra terminal, or gate. Another difference is that the case of a stud-type SCR is its anode, while on a regular diode, the case is the cathode. [Figure 3-164]

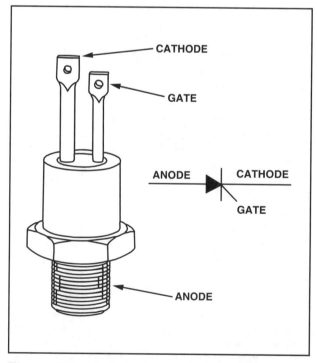

Figure 3-164. A typical silicon controlled rectifier and its symbol.

An SCR is a special type of diode having three junctions. Two of these junctions are forward biased, and one is reverse biased so no electrons can flow through the load. If the gate is momentarily connected to a positive voltage at the anode, the reverse biased junction becomes forward biased, and electrons flow through the SCR. Once this flow starts, it maintains the forward bias and current flow continues until the voltage across the SCR is removed. This feature makes SCRs useful as warning circuits on aircraft. For example, if an overtemp condition momentarily occurs in a turbine engine, an engine, or overheat warning light will illuminate in the cockpit. The light will remain illuminated until the pilot opens the circuit carrying voltage to the light. [Figure 3-165]

Since the SCR is a rectifier, it can also be used to convert AC to DC. In fact, an SCR not only produces DC, but it is selective in the amount that it produces. The diode D_1, figure 3-166, and the voltage divider R_1, R_2, and R_3 provide an adjustable direct current to trigger the SCR. Observe the waveform of the output controlled by the SCR, and note that there is no current flow in the half cycle when the anode is negative. There is also no current flow during the half-cycle when the anode is positive until the voltage across the divider rises sufficiently to charge capacitor C_1 enough to trigger the gate of the SCR. When the voltage rises high enough, the gate is triggered and the charged capacitor provides the needed pulse to start the SCR conducting. Once it starts conducting, it continues until the supply voltage reverses, as it does in the next half-cycle. [Figure 3-166]

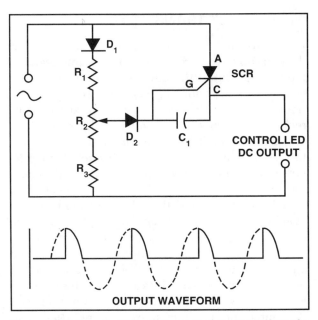

Figure 3-166. The position of the control R_2 determines the time in the cycle the silicon controlled rectifier conducts. Current flows only during the positive half-cycle after the SCR has been triggered.

TRIACS

One of the limitations of a silicon-controlled rectifier is that it controls only one-half of the cycle of AC. A triac, on the other hand, overcomes this by acting as two SCRs connected side by side, in opposite directions. As explained earlier, an SCR requires a positive pulse to trigger its gate. However, a triac can be triggered by a pulse of either polarity. Therefore, a triac controls a full cycle of AC. [Figure 3-167]

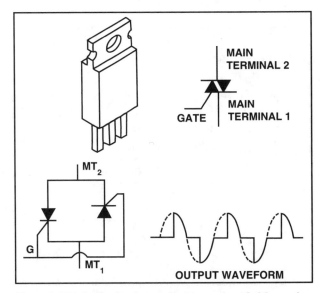

Figure 3-167. A triac is triggered with a pulse of either polarity, and it conducts during the remainder of the cycle after it has been triggered. To develop full power, the triac must be triggered at the beginning of the cycle. However, if it is triggered later in the cycle as seen here, only about one-half of the current flows.

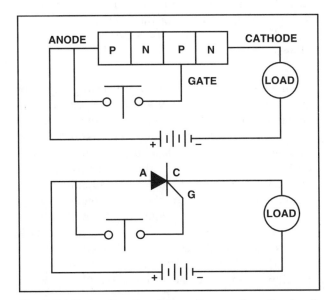

Figure 3-165. A pulse of positive voltage on the gate of a silicon controlled rectifier causes it to conduct until the voltage across the anode and the cathode is removed.

FIELD-EFFECT TRANSISTORS

A transistor is a low-impedance device that depends on the control of current flow into or out of the base to control the flow of current between the emitter and the collector. However, some applications require a solid-state device that controls current flow by controlling voltage. The field effect transistor, or FET, has been developed to do just this. An FET is constructed of a channel of either N-type or P-type silicon with a gate sitting in the channel and acting as a valve. One end of the channel is called the **source** and the other end the **drain**. A channel constructed of N-type material has a P-type gate. Therefore, when a positive voltage is applied to the gate, the FET is forward-biased, resulting in a greater flow of electrons between the source and the drain. If a negative voltage is applied to the gate, the FET becomes reverse-biased, and the flow between the source and the drain slows. [Figure 3-168]

UNIJUNCTION TRANSISTORS

A unijunction transistor, or UJT, is sometimes called a double-base diode. It is made up of a single crystal of uniformly doped N-type silicon, and has contacts at each end. A small emitter made of P-type material is located near the middle. A UJT acts as an insulator until the voltage at the emitter becomes high enough to allow the transistor to conduct with a minimum of resistance. UJTs are used in circuits where it is necessary to provide short, high intensity current pulses when the control voltage rises to a

given value. They are also used to provide the gate pulses for silicon controlled rectifiers and triac circuits. [Figure 3-169]

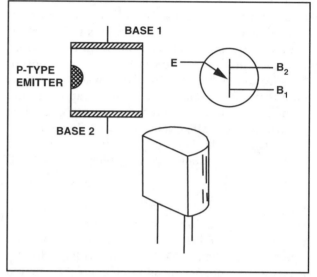

Figure 3-169. A unijunction transistor produces a pulse output when the voltage between its emitter and base rises to a predetermined value.

MAGNETIC AMPLIFIERS

Vacuum tubes and most semiconductor devices are used to control relatively small amounts of current. To control large amounts of current, the magnetic amplifier, or saturable reactor is often employed. A magnetic amplifier, or magamp, is a special form of transformer in which one coil is supplied with a flow of direct current. The DC input can be varied to change the permeability of the core which, in turn, controls the amount of alternating current that is allowed to flow.

To understand the operation of a magamp, consider an AC circuit with a coil of wire in series with an electric light bulb. If the coil's core has a low permeability, it has little inductance and, therefore, produces a small amount of back voltage. In this type of circuit, most of the source voltage is dropped across the light bulb as it burns brightly. However, if a highly permeable core is inserted into the coil to concentrate the lines of flux, a significant amount of back voltage is generated within the coil. Therefore, less voltage flows to the light. The amount of inductance in series with an electrical load determines the amount of current flow through the load. [Figure 3-170]

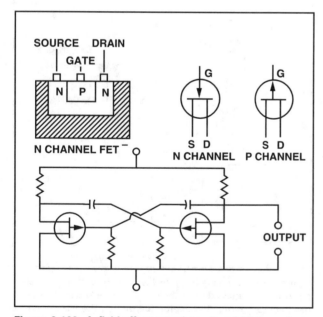

Figure 3-168. A field-effect transistor is a high-impedance control device that uses a change in voltage on its gate to control the current between the source and the drain.

Now, if you add a second coil to the core that is attached to a DC power source, you can control the

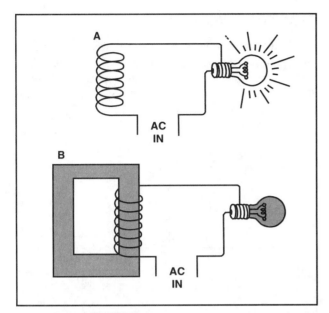

Figure 3-170. (A) — The amount of inductance that occurs in a coil with a low permeable core may not generate enough back voltage to oppose the flow of alternating current. In this situation, almost all of the voltage flows to the load. (B) — However, with a highly permeable core, the lines of flux become concentrated and produce enough back voltage to prevent the flow of alternating current.

core's permeability and, therefore, control the amount of alternating current that flows to the load. For example, when the DC supply is off, the iron core accepts lines of flux more readily from the AC and concentrates them to produce a back voltage that opposes the source voltage. However, when the DC current is increased, its magnetic field saturates the core so it no longer accepts the flux lines from the AC coil. The core now appears to have a low permeability, so a minimal back voltage is generated and the bulb illuminates. Regulating a small amount of DC allows a large amount of AC to be controlled. It is for this reason that a saturable reactor is called a magnetic amplifier. [Figure 3-171]

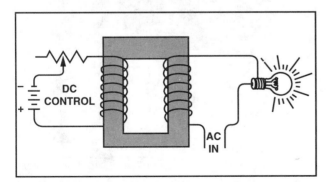

Figure 3-171. Direct current flowing in the control coil can magnetically saturate a core and decrease its permeability. Therefore, if you can control the amount of DC current, you can control an AC coil's output.

Magnetic amplifiers are often wound on a ring-shaped soft iron core to form a **toroidal coil**. This type of core concentrates lines of magnetic flux and provides maximum inductance by reducing the number of flux lines that are lost. Toroidal coils can be mounted close together without there being any magnetic reaction or coupling between them. [Figure 3-172]

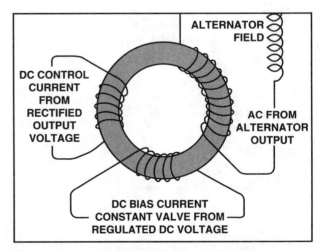

Figure 3-172. A toroidal coil concentrates the lines of magnetic flux and prevents any stray fields from interfering with other coils.

A practical example of a magnetic amplifier circuit in an aircraft is the voltage sensing circuit in some AC voltage regulators. A toroidal coil is wound with three windings, one of which is a DC bias-voltage coil supplied with a constant amount of current from a regulated DC source. A second control coil is supplied with direct current that varies with the alternator output voltage. The polarity of these two coils is such that they tend to cancel each other. The third coil has the excitation current for the alternator field. [Figure 3-173]

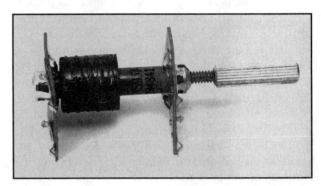

Figure 3-173. A magnetic amplifier-type voltage regulator consists of three coils on a single core. One coil contains a constant DC bias-voltage supplied by a DC source. The second coil is a control coil that is supplied with a variable DC current. The third coil contains the AC output from the alternator.

When the alternator produces the proper voltage, the magnetizing force from the opposing DC bias current and the DC control current maintain the permeability of the core so the correct amount of field current passes through the AC output coil. However, if the load on the alternator decreases and the voltage rises, the magnetizing force from the control coil increases and opposes the magnetizing effect of the bias coil. This decreases the magnetization in the core and, in effect, increases its permeability so more back voltage is produced. The increased back voltage then decreases the alternator field current which lowers the alternator output. On the other hand, when the alternator load increases, the output voltage drops and the magnetizing effect of the control coil decreases. This allows the bias coil to nearly saturate the core which decreases the permeability, so less back voltage is generated. This increases the alternator field current and brings the output voltage back to the desired value.

PHOTODIODES

Light is increasingly used as both a power source and a control in sophisticated electronic circuits. Light energy is electromagnetic in nature and has the ability to increase the reverse, or leakage, current in a semiconductor device. A photodiode is a special diode that is triggered by light. Light shining through an aperture in its case releases free electrons into the diode's depletion area and causes it to conduct. When no light shines, no current can flow. [Figure 3-174]

PHOTOTRANSISTOR

A phototransistor is a transistor that incorporates a photocell at its emitter-base junction. When light strikes the photocell, energy is released to forward-bias the emitter-base junction and allow current to flow through the emitter-collector circuit. [Figure 3-175]

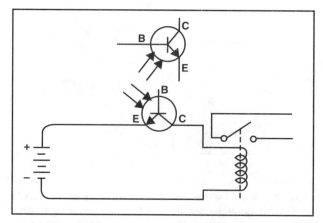

Figure 3-175. A phototransistor conducts when sufficient light strikes its emitter-base junction to provide a forward bias.

LIGHT-EMITTING DIODES

One of the most familiar optoelectronic devices is the light-emitting diode, or **LED**. An LED consists of tiny bars or dots of photo-conductive material that light up when forward-biased current flows through them. However, when an LED is reverse-biased, no current can flow and the diode is dark. LEDs are commonly used in aircraft instruments. [Figure 3-176]

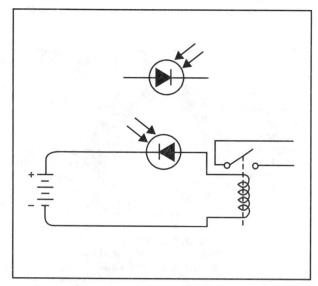

Figure 3-174. When the photodiode is dark, no current can flow and the relay is open. However, when light shines through the aperture, the diode breaks down and allows current to flow from the battery through the relay coil and close the contacts.

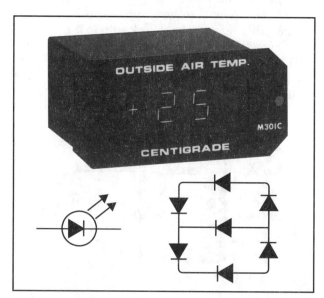

Figure 3-176. Light emitting diodes (LEDs) emit light when they are forward-biased and current flows through them. The most common layout for LEDs will display both numbers and letters.

FILTERS

As mentioned, the output waveform of a single-phase rectifier is pulsating direct current which drops to zero, then rises to a peak. In a half-wave rectifier, the voltage drops to zero and remains there during one-half of the cycle before it rises to a peak, and back to zero. In order to make this type of voltage output useful, some form of filter must be used to smooth out the DC so it remains at a fairly constant value.

CAPACITOR-INPUT FILTERS

The capacitor-input filter smooths out pulsating DC by using a large capacity electrolytic capacitor across the load. The capacitor charges as the voltage rises, and then, as the voltage drops, it discharges to keep the voltage from dropping to zero. The rate at which the capacitor discharges is based on the amount of current the load requires. For example, since a capacitor can only store a given amount of current, a large flow of load current causes a capacitor to discharge more rapidly than a small current. [Figure 3-177]

When a capacitor is installed in a full-wave rectifier, the action of the capacitor-input filter is the same as that in a half-wave circuit, except that the frequency of the output ripple is twice as high. [Figure 3-178]

CHOKE-INPUT FILTER

As discussed in Section D, a choke is a type of inductor that when placed in series with the load,

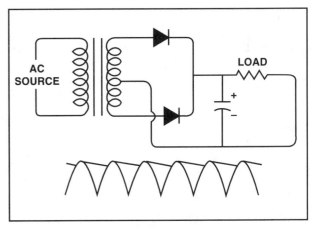

Figure 3-178. When a capacitor-input filter is installed on a full-wave rectifier, the action of the capacitor is the same as that in a half-wave circuit.

opposes any change in the load current. During the part of the cycle in which the current increases, the magnetic field in the inductor induces a current that opposes the rise. By the same token, when the source current starts to drop off, the inductor produces a current that opposes the drop-off. This action results in a waveform of low alternating voltage whose frequency is twice that of the source. [Figure 3-179]

PI FILTER

An even more effective way to smooth out, or filter DC current is to combine a capacitor and an inductor. There are a number of arrangements of capacitors and inductors that can be used. One of the most common is the capacitor-input filter whose circuit

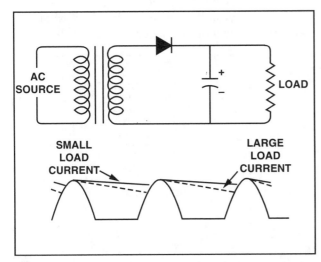

Figure 3-177. One way to smooth out the flow of pulsating DC is to install a capacitor across the load. When this is done, the capacitor discharges when the voltage drops and keeps the load voltage from dropping to zero.

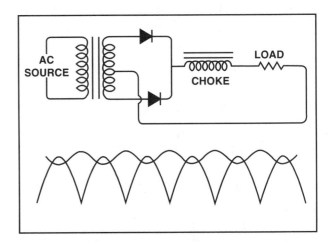

Figure 3-179. When a choke-input filter is installed on a full-wave rectifier, the magnetic field in the choke induces a current that opposes the rise and drop-off of current. This results in a low alternating voltage with a frequency twice that of the source.

diagram resembles the Greek letter pi (π). In this type of circuit, two high-capacity capacitors are connected in parallel and one high impedance inductor is connected in series with the load. The capacitors offer a minimum reactance to the ripple frequency of the rectified AC load, but do not allow any flow of DC to ground. The inductor, on the other hand, has a high opposition to the AC caused by the ripple, but offers very little opposition to the flow of the DC. With this arrangement, the ripple frequency AC is passed to ground and leaves an almost pure DC output. [Figure 3-180]

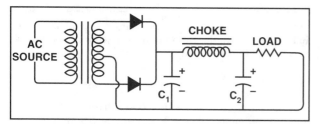

Figure 3-180. Pi-filter on a full-wave rectifier.

AMPLIFIERS

An amplifier uses a very small change in either input voltage or current to produce a large change in the output voltage or current. A transistor is a low-impedance device, which means that it uses a small change in current in the emitter-base circuit to cause a large change in the current in the emitter-collector circuit.

In a common-emitter amplifier such as that in figure 3-181, the transistor is installed in such a way that the forward-bias for the emitter-base junction is provided by the voltage drop across resistor R_B, and the reverse bias for the collector-base junction is provided by the voltage drop across resistors R_A and R_C. When there is no signal voltage across R_B, there is

a steady current flowing through the transistor and resistor R_C, whose voltage drop is opposite in polarity to that of the battery. This results in a steady negative output current that is not as negative as the battery.

When a negative signal is put on the input between the base and the emitter across R_B, the forward bias across the emitter-base increases. This results in more current flow through the transistor and R_C. The increased flow increases the voltage drop across R_C which causes the output to become less negative. During the half-cycle when the input signal is positive, the forward bias of the emitter-base decreases and the transistor conducts less. This decreases the voltage drop across resistor R_C, causing the output to become more negative.

In a common-emitter transistor amplifier, the emitter-collector current is much greater than the emitter-base current. The output current is similar to, but opposite in phase to the input current. [Figure 3-182]

In studying this basic transistor amplifier, you should note a very important principle. When the base of a PNP transistor is negative with respect to its emitter, the transistor conducts. When it is not negative, it does not conduct. An NPN transistor works in the same way, except its base must be positive for it to conduct.

OSCILLATORS

In order to better understand electronic oscillation, let's briefly review resonance in a parallel L-C circuit, or **tank circuit**. You saw earlier that when the value of a capacitor and an inductor are chosen so

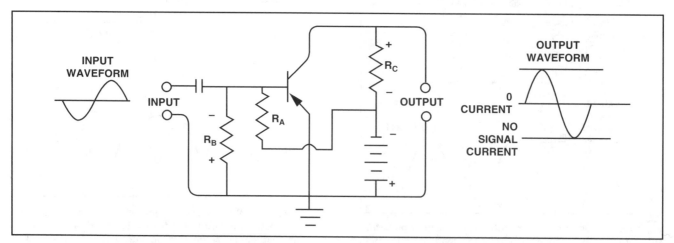

Figure 3-181. A small voltage applied to the input of a transistor amplifier causes a large change in the output current. However, the output voltage is 180 degrees out of phase.

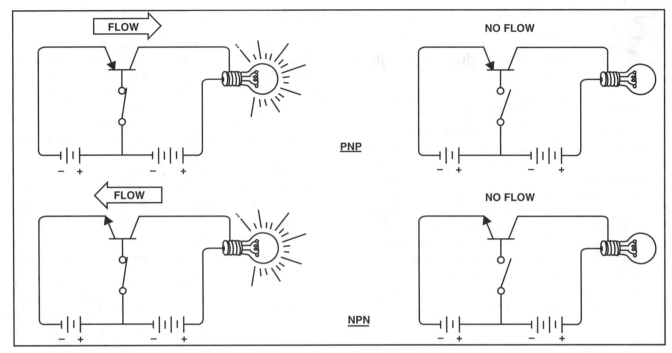

Figure 3-182. Current flows in the emitter-collector circuit of a transistor amplifier when current flows in the emitter-base circuit.

that their reactances at a particular frequency are the same, the energy stored in the electromagnetic field around the inductor is exactly the same as the energy stored in the electrostatic field across the plates of the capacitor. When this occurs, a large amount of current circulates back and forth between the inductor and the capacitor, while little source current flows through the circuit. Also recall that the current would continue to flow back and forth, or oscillate, indefinitely if there were no circuit resistance. However, as you know, all practical circuits have some resistance. Therefore, in order for an electronic oscillator to function, the feedback from the output back into the input must be amplified with the proper phase to replace the energy that is lost in the resistance.

When switch S in figure 3-183 is closed, current begins to flow in the circuit. The transistor is biased through the voltage divider, R_B and R_A, and current flows through the lower half of the tapped coil at L_1. As the current increases in the coil, it induces a voltage into L_2, the upper half of the coil. Voltage flows through L_2 and charges capacitor C_2 which increases the forward bias on the transistor and further increases the current flow. By the time maximum current flows in the circuit, there is no more voltage induced into L_2 from L_1. At this point, all of the energy stored in the tank circuit is in the electrostatic fields in the capacitor C_1. Since there is no excess force left to push electrons into capacitor C_1, it discharges, and the energy lost is made up by that

stored in capacitor C_2 as it discharges. Furthermore, as C_2 discharges, the forward bias on the transistor's emitter-base junction decreases and the transistor conducts less, thereby decreasing the current in L_1. This decrease in current induces a voltage into L_2 which charges capacitor C_2 in such a direction that it decreases the transistor's forward bias and eventually reverse-biases the emitter-base junction causing current to stop flowing. When this happens, C_1 and C_2 are both fully charged and the cycle repeats itself. [Figure 3-183]

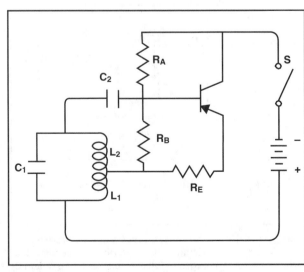

Figure 3-183. A Hartley oscillator using a transistor.

The frequency of the oscillation is determined by the value of the capacitor and the inductor. By varying either one, the frequency of the alternating current produced by the oscillator can be changed.

FULL-WAVE VOLTAGE DOUBLER

A circuit consisting of three capacitors and two diodes can be used to provide direct current with a voltage almost twice the peak input AC voltage. An illustration of the circuit arrangement needed to do this is in figure 3-184. During the half cycle when the top of the power source is positive, current flows through diode D_1 to fully charge capacitor C_1. During the next half-cycle, when the bottom terminal of the power source is positive, current flows through D_2 and fully charges capacitor C_2. Capacitors C_1 and C_2 are in series across capacitor C_3, which charges to twice the voltage of the input.

The DC output voltage depends to a great extent on the amount of load current that flows. When a low-output current is produced, the DC voltage is approximately twice the peak value of the source AC.

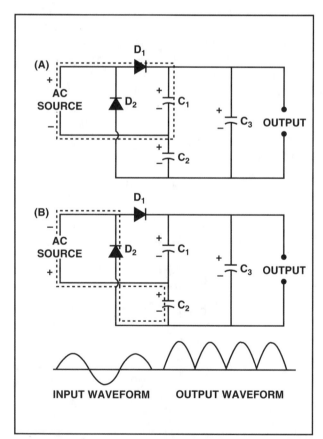

Figure 3-184. (A) — During the first half-cycle, capacitor C_1 charges to full line voltage. (B) — During the second half-cycle, capacitor C_2 charges to full line voltage. Since C_1 and C_2 are in series across C_3, it charges to double the line voltage.

LOGIC GATES

Logic gates represent digitally controlled binary circuits used in computers and microprocessors. To understand their function and purpose, recall the discussion on the binary number system from Chapter 1. If you recall, the only real option in an electrical circuit is ON or OFF. Therefore, a number system based on only two digits is used to create electronic calculations. For example, when a circuit is ON, a 1 is represented, and when a circuit is OFF, a 0 is indicated. By converting these ON or OFF messages to represent numbers found in the decimal system, a computer can perform complex tasks. [Figure 3-185]

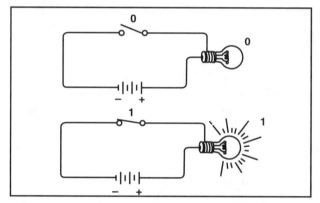

Figure 3-185. In a simple circuit with a switch representing the input and a light representing the output, when the input is 0, the output will be 0. If the input is 1 then the output will be 1.

A **truth table** depicts the combinations of switch positions and their resulting outputs. At times, a truth table can be a helpful means of understanding how a specific logic gate operates.

The most common logic gates used are the AND gate, OR gate, NAND gate, NOR gate, and the exclusive OR gate. The name given to each gate represents the task it performs. For example, in an **AND gate**, every input must be 1 (on) in order for the output to be 1 (on). This is similar to having two switches connected in series with a light. In this type of circuit, switch number 1 AND switch number 2 must be on (1) before the light can turn on (1). If either switch is off (0), the light remains off (0). [Figure 3-186]

The **OR gate** requires that one OR more of the inputs be 1 (on) to produce a 1 (on) output. In other words, if input number 1 OR input number 2 is 1 (on), the output will be 1 (on). Another way of expressing the OR function is if all inputs are 0 (off), the output

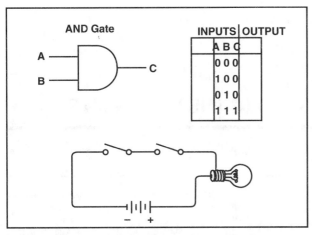

Figure 3-186. The symbol for a two-input AND gate is illustrated along with its corresponding truth table. The AND gate is similar to a simple series circuit consisting of two switches and a light. Both inputs must be 1 (on) to get an output of 1 (on).

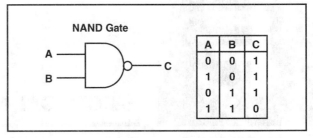

Figure 3-188. The symbol for a two-input NAND gate is illustrated along with its truth table. A NAND gate produces a 1 (on) output when either input is 0 (off).

the output side. The circle indicates the negation or inversion of the symbol. [Figure 3-188]

The operation of a **NOR gate**, or **NOT OR gate**, is the inverse of the operation of an OR gate. A NOR gate gives a 0 (off) output anytime there is a 1 (on) input. [Figure 3-189]

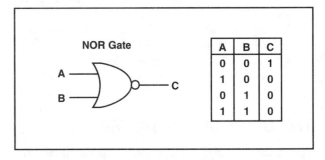

Figure 3-189. The symbol for a two-input NOR gate is illustrated along with its truth table. For a NOR gate, if either input is 1 (on), the output is 0 (off).

will be 0 (off). An electrical circuit that is similar to an OR gate is represented by two switches connected in parallel with a light. In this situation, if either switch number 1 OR switch number 2 is on (1), the light will turn on (1). [Figure 3-187]

The function of a **NAND gate**, or **NOT AND gate**, is opposite that of an AND gate. For example, the output of a NAND gate is 1 (on) when input number 1 or input number 2 are NOT 1 (off). In other words, at least one of the inputs must be 0 (off) to produce an output of 1 (on). The symbol for a NAND gate is the same as the AND symbol with a small circle on

The **EXCLUSIVE OR gate** produces an output of 1 (on) when one and only one of its inputs is 1 (on). In other words, if both inputs are 1 (on) or 0 (off), the output is 0 (off). [Figure 3-190]

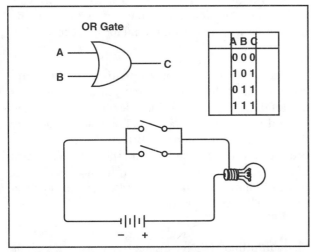

Figure 3-187. The symbol for a two-input OR gate is illustrated along with its corresponding truth table. The OR gate is similar to a circuit with two parallel switches connected in parallel with a light. If either input is 1 (on), the output will be 1 (on).

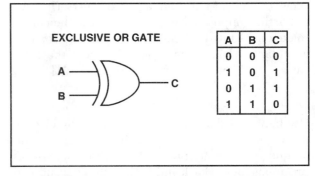

Figure 3-190. The symbol for a two-input EXCLUSIVE OR gate is illustrated along with its truth table. An EXCLUSIVE OR gate produces a 1 (on) output when one and only one of its inputs is 1 (on).

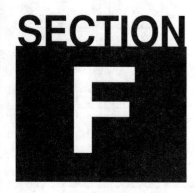
ELECTRICAL MEASURING INSTRUMENTS

Measurement of electrical quantities is essential to the maintenance of any modern device. As a technician you must be able to measure each of the four electrical variables: current, voltage, resistance, and power. There have been a number of principles used for these measurements, but by far the most common is electromagnetism. It is based on two fundamental assumptions:

1. The strength of an electromagnetic field is proportional to the amount of current that flows in the coil.
2. Voltage, resistance, and power all relate to a flow of current, and if the amount of current is known, the other values may be found.

THE D'ARSONVAL METER

The most widely used meter movement is the D'Arsonval movement, whose pointer deflects an amount proportional to the current flowing through its moving coil. A reference magnetic field is created by a horseshoe-shaped permanent magnet, and its field is concentrated by a cylindrical keeper in the center of the open end.

The current being measured flows through the coil and creates a magnetic field whose polarity is the same as that of the permanent magnet. The two fields thus oppose each other and cause the coil to rotate on its low-friction bearings until the force of a calibrated hairspring exactly balances the force caused by the magnetic fields. [Figure 3-191]

Oscillation of the pointer is minimized by electromagnetic damping. The moving coil is wound around a thin aluminum bobbin, or frame, and as this frame moves back and forth in the concentrated magnetic field, eddy currents are generated within the bobbin that produce their own fields which oppose the movement.

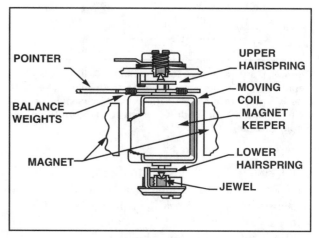

Figure 3-191. As current flows through the coil of a D'Arsonval-type meter, the coil obtains a magnetic field with the same polarity as the permanent magnet. The like magnetic fields oppose each other causing the meter indicator to rotate.

METER RATINGS AND TERMS

Before using a basic meter you need to understand some of the terms associated with it. For example, **full-scale current** is the amount of current that must flow through the meter coil to cause a full scale deflection. The amount of current required to do this varies with the scale the meter is set to.

The measurement of **meter sensitivity** is the reciprocal value of the full-scale current and represents the total amount of resistance for each volt needed to produce a full-scale current. For example, a meter that requires 1 milliamp (.001 amp) of current to produce a full-scale deflection requires 1,000 ohms of resistance to limit the current through the meter to 1 milliamp. This meter is said to have a sensitivity of 1,000 ohms per volt.

Many multimeters have a sensitivity of 20,000 ohms per volt. These meters require 50 microamps of current to move the pointer full scale. Highly sensitive meters are used for applications that require precise measurement of very small electrical quantities.

The total resistance of a meter, also known as **meter resistance**, must be considered when making computations regarding the current through the meter. Both the moving coil and the hairsprings have resistance, and in some meters there is a temperature compensating resistor in series with the coil. This resistor is made of a material whose resistance decreases with an increase in temperature, which is opposite to the change in resistance in a coil. As a result, the meter resistance remains constant as the temperature changes.

AMMETERS, MILLIAMMETERS, AND MICROAMMETERS

If the range of current to be measured is greater than the full-scale current of a particular meter, a **shunt** must be installed in parallel with the meter. A shunt is a type of resistor that is connected in parallel with a meter that increases the amount of current it can measure. The load current flowing through a shunt produces a voltage drop that is proportional to the current. The meter displays this voltage drop in terms of amps, milliamps, or microamps. The standard aircraft shunt produces a voltage drop of 50 millivolts when its rated current flows through it. [Figure 3-192]

It is sometimes necessary to extend the range of an ammeter by using a precision resistor as a shunt. For example, assume that you want a meter to deflect full scale when 10 milliamps flows through the meter and shunt combination. If the meter requires 1 milliamp for full-scale deflection and

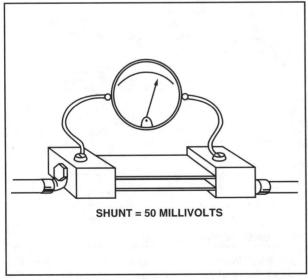

Figure 3-192. The voltage drop across an ammeter shunt is proportional to the amount of current flowing through it.

SHUNT = 50 MILLIVOLTS

has an internal resistance of 50 ohms, you can use Ohm's law to determine that a full scale deflection occurs when the meter is connected across a voltage of 50 millivolts.

$$E = IR$$

$$= .001 \times 50$$

$$= .050 \text{ volts}$$

Therefore, if you want a full-scale deflection to occur when 10 milliamps flows through the meter, the shunt must produce a voltage drop of 50 millivolts when 9 milliamps flows through it. To determine the resistance needed to do this, use Ohm's law again.

$$R_{shunt} = E \div I$$

$$= 0.050 \div 0.009$$

$$= 5.55 \text{ ohms}$$

In this example, when the meter is connected in parallel with a 5.55-ohm shunt, full scale deflection occurs when 1 milliamp flows through the meter and 9 milliamps through the shunt.

VOLTMETERS

A D'Arsonval meter can be used to measure voltage by connecting resistance in series with the meter movement. This limits the current flow to a value which results in full scale deflection. For example, if a 1 milliampere meter with a resistance of 1,000 ohms is used to measure the voltage across a 1.5 volt battery, how much additional resistance must be connected in series with the meter to limit the current to 1 milliampere? To determine this you must first calculate the total resistance required using Ohm's law.

Given:
E = 1.5 volts
I = .001 amps

$$E = IR$$

$$R = 1.5 \div .001$$

$$R = 1,500 \text{ ohms}$$

A total of 1,500 ohms of resistance is required to limit the current to 1 milliampere. However, since the meter already has a resistance of 1,000 ohms only 500 ohms of additional resistance is required. [Figure 3-193]

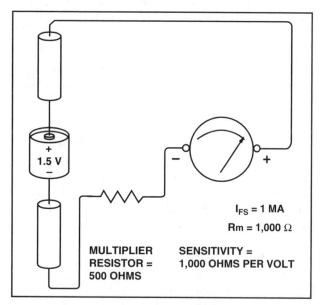

Figure 3-193. If 1 milliamp produces a full-scale deflection in a meter with a sensitivity of 1,000 ohms per volt and a resistance of 1,000 ohms, then an additional 500 ohms of resistance is required to measure the voltage across a 1.5 volt battery.

A resistor that is placed in series with the meter movement is called a **multiplier resistor**, or **multiplier**, because it multiplies a meter's basic range. Multi-range voltmeters use one meter movement with several different multipliers. These multipliers are usually arranged so the current for each succeedingly higher range flows through the multipliers for all of the lower ranges. [Figure 3-194]

Instead of using separate terminals to measure different voltages, most multi-range meters use a selector switch. When using this type of meter, set the switch to the voltage range that is higher than that anticipated. After the meter is connected and the needle is deflected, select the range which results in a needle deflection in the center third of the scale.

OHMMETER

Resistance is most easily determined by measuring the current through an unknown resistor when a known voltage is placed across it. The **series ohmmeter** uses small flashlight or penlight batteries connected in series with a fixed resistor, an adjustable resistor, and a meter. If a meter uses a 3.0 volt battery and has a sensitivity of 1,000 ohms per volt, the total resistance required to produce a full scale deflection equals 3,000 ohms. [Figure 3-195]

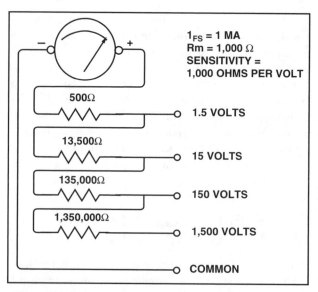

Figure 3-194. When the meter is used to measure 1.5 volts, the test leads are connected to the common terminal and the 1.5-volt terminal. Current then flows through the meter and the 500-ohm resistor. To measure 15 volts, the current flows through the meter, the 500-ohm, and the 13,500-ohm multiplier resistors. The total resistance in the circuit is 15,000 ohms. To measure voltages as high as 1,500 volts, the current must flow through all of the resistors in the voltmeter circuit.

Because the battery voltage changes with use, the variable resistor is used to "zero," or standardize, the meter before each use. To set the meter up for use, hold the test leads together without touching the metal leads and turn the zero adjusting knob until the meter indicates an exact full-scale deflection. When the leads are separated, the needle

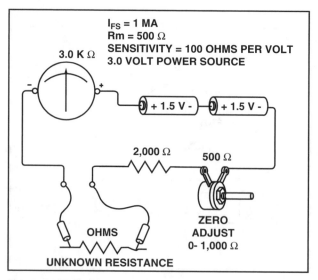

Figure 3-195. In this circuit, if a resistance of 3,000 ohms, the same as the total circuit resistance, is placed between the test leads, the current drops to one-half of its original value. At this point, the meter indicates 3.0 kilo-ohms.

drops back to the opposite side of the scale which is sometimes marked with the symbol ∞. This is the symbol for infinity, and indicates that there is an infinite resistance between the test leads, and no current is flowing.

Since this meter has 500 ohms of meter resistance and uses a 2,000 ohm fixed resistor, the variable resistor must provide 500 ohms resistance with a fully-charged battery to zero the meter. When a resistor is placed between the test leads, the meter circuit is completed and the meter indicator deflects an amount proportional to the voltage dropped by the unknown resistance. In this example, the indicator moved to the middle of the scale, meaning that the unknown resistance dropped exactly the same voltage as the internal resistance of the meter. Therefore, the value of the unknown resistance is 3,000 ohms.

The scale on a series ohmmeter is nonlinear, meaning that there is no uniform distance between the graduations. The numbers are widely separated at the low-range end and are close at the high end. To obtain the most accurate measurement of resistance, you should use a scale that results in a pointer deflection in the center third of the dial. Different resistance ranges are selected by using different values of battery voltage and fixed resistance.

POTENTIOMETER-TYPE OHMMETER

The series-type ohmmeter has a shortcoming in that resistances on the high end of the scale are crowded together on the meter face. This problem is solved to some extent by the potentiometer-type ohmmeter. Although the scale on a potentiometer-type ohmmeter is still nonlinear, the meter face is not crowded nearly as bad. To accomplish this, a low resistance resistor is connected in series with the battery and the resistance to be measured. This sets up a voltage divider circuit. [Figure 3-196]

Rather than being calibrated in volts or millivolts, the meter dial on an ohmmeter is calibrated in ohms. When the test leads are shorted together, all of the battery voltage is dropped across the standard resistor, and the meter is adjusted with the zero-set variable resistor. At this point, the indicator reads zero since there is no resistance between the leads. When the leads are separated, the battery circuit is open and no current can flow. The point at which the needle rests is marked ∞, indicating that there is an infinite resistance between the leads.

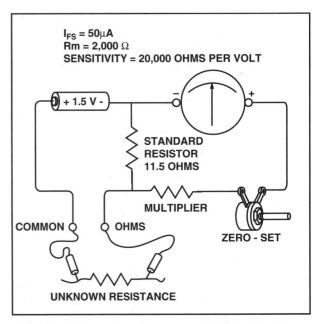

Figure 3-196. The voltage across the standard resistor is proportional to the current through the unknown resistance. A sensitive meter having a high internal resistance is used to measure the voltage drop across the standard resistor; hence the name potentiometer-type ohmmeter.

SHUNT-TYPE OHMMETER

It is sometimes necessary to measure very low resistances, such as that of the primary winding of a magneto coil. To do this, a shunt-type ohmmeter is used. The shunt-type ohmmeter uses a meter with a very low internal resistance connected in series with a switch, a fixed resistor, a variable resistor, and a power source. [Figure 3-197]

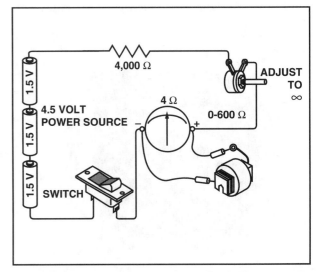

Figure 3-197. This shunt-type ohmmeter consists of a meter with an internal resistance of 4 ohms connected in series with a switch, a 4,000-ohm fixed resistor, a 600-ohm variable resistor, and a 4.5 volt power source.

The unknown resistance is placed between the terminals in parallel with the meter movement. The smaller the resistance value being measured, the less current flows through the meter movement.

The value of the fixed resistor is usually large compared to the resistance of the meter movement. This keeps the current drawn from the battery practically constant. Thus, the value of the unknown resistor determines how much constant current flows through the meter and how much through the unknown resistor.

MEGOHMMETER

It is sometimes necessary to measure very high resistance values that require a voltage in excess of that provided by a standard ohmmeter. For this application the megohmmeter, or **megger**, is used. A hand-cranked generator with a slip clutch allows the operator to produce a voltage of several hundred volts. When the leads are separated and the crank is turned, the pointer deflects fully to the left, indicating that there is an infinite resistance between the leads. When a high resistance is placed between the leads, a second coil within the meter pulls the needle to the proper resistance measurement. Meggers are often used for measuring insulation resistance in ignition systems and other high-voltage circuits.

It is important that you exercise caution when testing resistance with a megger. The high voltage generated by the megger can arc to ground through defective insulation in a wire being tested and, if conditions are right, cause damage to equipment or injury to personnel. Some maintenance organizations or companies limit or prohibit the use of meggers. [Figure 3-198]

MULTIMETERS

The most versatile electrical measuring instrument used by the aircraft technician is the multimeter. This handy tool has a single meter movement and a selector switch which is used to select what you are measuring as well as the appropriate range. [Figure 3-199]

ANALOG MULTIMETERS

Analog multimeters typically have voltage ranges from 0 to 2.5, 10, 50, 250, 1,000, and 5,000 volts for both AC and DC. They can measure amperes in ranges of 100 microamperes; 10, 100, and 500 milliamps; and 10 amps. Ranges for resistance typically include from 0-2,000 ohms, 0-200,000 ohms, and 0-20 megohms. Most analog meters usually have a sensitivity of 20,000 ohms per volt for measuring DC. However, because of the rectifier circuit, the sensitivity for AC is typically 1,000 ohms per volt. All analog multimeters have a "zero adjust" knob to reset the scale as the internal battery discharges with time and usage.

DIGITAL MULTIMETERS

In addition to the analog multimeters, there are several digital multimeters, or **DMMs** that are used within the industry. In addition to the tasks an analog meter can perform, several upper-end digital

Figure 3-198. A megohmmeter measures high resistances by applying a high voltage across the resistance being measured.

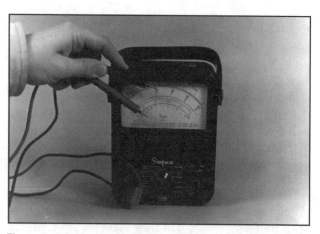

Figure 3-199. The multimeter is one of the most useful electrical measuring instruments used by the A&P technician.

multimeters can measure frequency and test diodes. However, the primary difference between analog and digital multimeters is the way the information is displayed. [Figure 3-200]

RESOLUTION

Resolution refers to how small a measurement a meter can make. The terms **bits** and **counts** are used to describe a meter's resolution. Digital multimeters are grouped by the number of counts or digits they display. A 3-1/2 digit meter, for example, can display three full digits ranging from 0 to 9, and one "half" digit which displays a 1 or is left blank. Therefore, a 3-1/2 digit meter displays up to 1,999 counts of resolution. It is more precise to describe a meter by counts of resolution rather than by 3-1/2 or 4-1/2 digits. Today's 3-1/2 digit meters may have enhanced resolution of up to 3,200 or 4,000 counts.

ACCURACY

A meter's accuracy is the largest allowable error that occurs under specific operating conditions. In other words, it is an indication of how close a meter's displayed measurement is to the actual value of the signal being measured. Accuracy for a meter is usually expressed as a **percent of reading**. An accuracy of ±1 percent means that for a displayed reading of 100.0 volts, the actual value of the voltage could be anywhere between 99.0 volts to 101.0 volts. Specifications may also include a range of digits added to the basic accuracy specification. This indicates how many counts the digit to the extreme right of the display may vary. Therefore, an accuracy of ±(1 percent + 2) indicates that for a display reading of 100.0 volts the actual voltage is between 98.8 volts and 101.2 volts.

For high accuracy and resolution, the digital display excels, showing three or more digits for each mea-

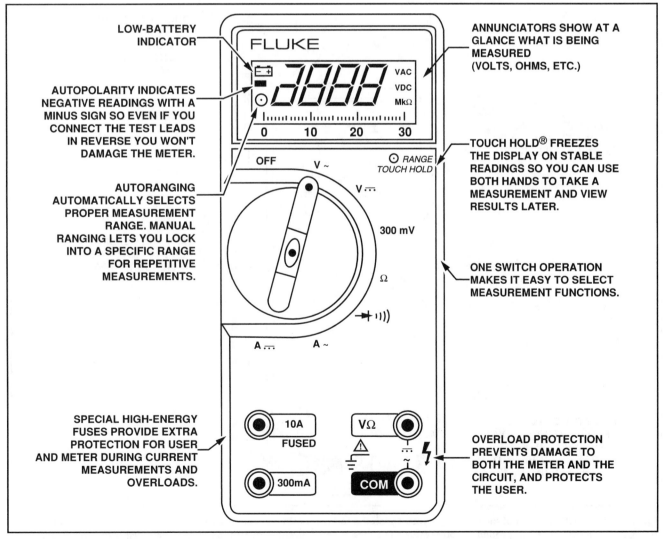

Figure 3-200. A digital multimeter can have a great many special features that differ from an analog multimeter.

surement. The analog needle display is less accurate and has lower effective resolution since you must estimate values between the lines.

Some DMMs have a bar graph display. A bar graph shows changes and trends in a signal just like an analog needle, but is more durable and less prone to damage.

MEASURING VOLTAGE

One of the most basic tasks a multimeter performs is measuring voltage. Testing for proper supply voltage is usually the first thing measured when troubleshooting a circuit. If there is no voltage present,

or if it is too high or too low, the voltage problem should be corrected before investigating further. [Figure 3-201]

The waveforms associated with AC voltages are either sinusoidal (sine waves) or non sinusoidal (sawtooth, square, ripple, etc.). DMMs display the root-mean-square, or RMS value of these voltage waveforms. The RMS value is the effective or equivalent DC value of the AC voltage. Most meters, called "average responding," give accurate RMS readings if the AC voltage signal is a pure sine wave. Averaging meters are not capable of measuring non-sinusoidal signals accurately. Special DMMs, called "true-RMS" DMMs, accurately measure the correct RMS value, regardless of the waveform, and should be used for non sinusoidal signals.

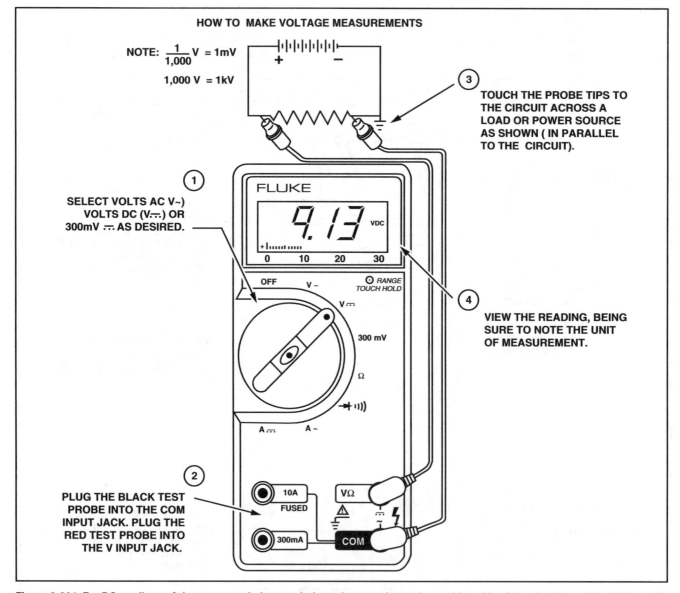

HOW TO MAKE VOLTAGE MEASUREMENTS

NOTE: $\frac{1}{1,000}$ V = 1mV

1,000 V = 1kV

③ TOUCH THE PROBE TIPS TO THE CIRCUIT ACROSS A LOAD OR POWER SOURCE AS SHOWN (IN PARALLEL TO THE CIRCUIT).

① SELECT VOLTS AC V~) VOLTS DC (V⎓) OR 300mV ⎓ AS DESIRED.

④ VIEW THE READING, BEING SURE TO NOTE THE UNIT OF MEASUREMENT.

② PLUG THE BLACK TEST PROBE INTO THE COM INPUT JACK. PLUG THE RED TEST PROBE INTO THE V INPUT JACK.

Figure 3-201. For DC readings of the correct polarity, touch the red test probe to the positive side of the circuit, and the black probe to the negative side or circuit ground. If you reverse the connections, a DMM with auto polarity merely displays a minus sign indicating negative polarity. However, with an analog meter you risk damaging the meter.

A DMM's ability to measure AC voltage can be limited by the frequency of the signal. Most DMMs can accurately measure AC voltages with frequencies from 50 Hz to 500 Hz, while others can measure AC voltages with frequencies from 20 Hz to 100 kHz. DMM accuracy specifications for AC voltage and AC current should state the frequency range of a signal the meter can accurately measure.

RESISTANCE

Resistance measurements allow a technician to determine the resistance of a conductor, the value of a resistor, or check the operation of a variable resistor.

As you know, resistance is measured in ohms (Ω). Resistance values may vary from a few milliohms (mΩ) for contact resistance to billions of ohms for insulators. Most DMMs can measure a resistance as small as 0.1 ohm, and some measure as high as 300 megaohms (300,000,000 ohms). Infinite resistance is read as "OL" on some displays and means the resistance is greater than the meter can measure. Open circuits also read OL on some displays.

Resistance measurements must be made with the circuit power off. If power is left on, you may damage the meter and the circuit. Some DMMs provide protection in the ohms mode in case of accidental contact with voltages. The level of protection varies greatly between different models. [Figure 3-202]

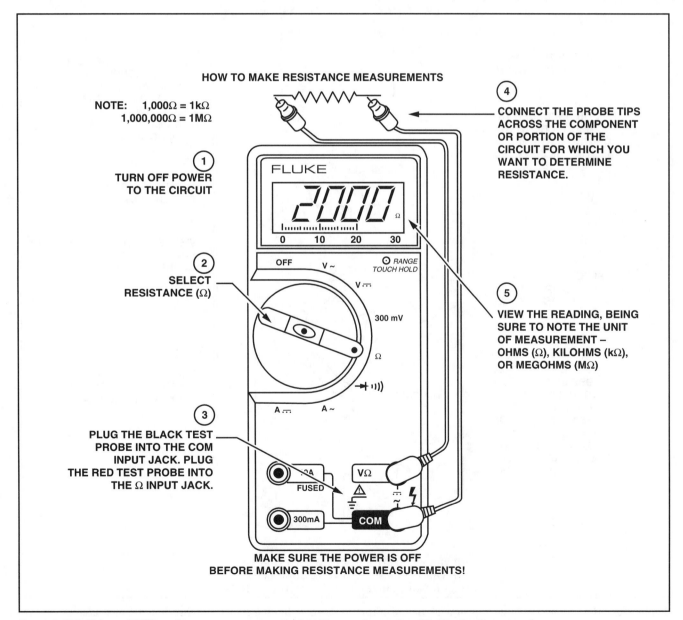

Figure 3-202. Using a DMM, resistance measurements should never be made with the circuit powered up.

For accurate low resistance measurements, the resistance in the test leads must be subtracted from the total resistance measured. Typical test lead resistance is between 0.2 and 0.5 ohms. If the resistance is greater than 1 ohm, they should be replaced. If a multimeter supplies less than 0.3 volts DC test voltage for measuring resistance, it can measure the values of resistors that are isolated in a circuit by diodes from semiconductor junctions. This often allows you to test resistors on a circuit board without unsoldering them.

CONTINUITY

Continuity tests distinguish good fuses from bad ones, open or shorted conductors, the operation of switches, and facilitate the tracing of circuit paths. A DMM with a continuity beeper allows you to conduct many continuity tests easily and quickly. The meter beeps when it detects a closed circuit, so you don't have to look at the meter as you test. The level of resistance required to trigger the beeper varies from model to model of DMM.

MEASURING CURRENT

Current tests help determine circuit overloads, circuit operating currents, or current in different branches of a circuit. Current measurements are different from other measurements made with a multimeter in that current is measured in series, unlike

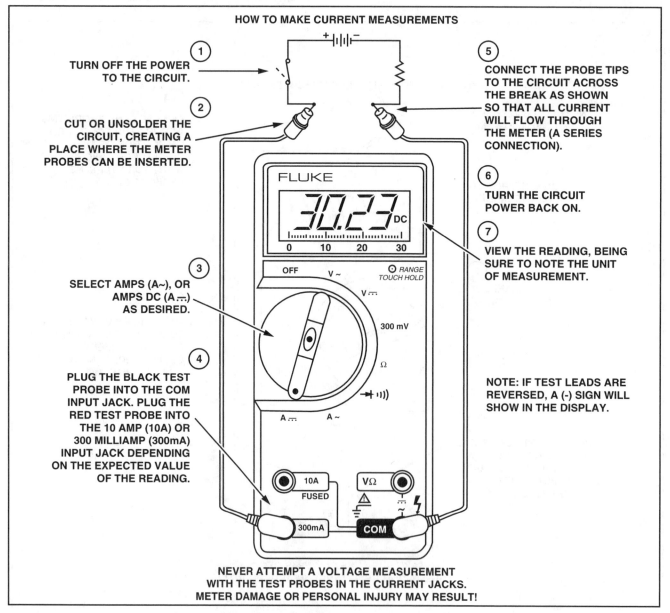

HOW TO MAKE CURRENT MEASUREMENTS

① TURN OFF THE POWER TO THE CIRCUIT.

② CUT OR UNSOLDER THE CIRCUIT, CREATING A PLACE WHERE THE METER PROBES CAN BE INSERTED.

③ SELECT AMPS (A~), OR AMPS DC (A⎓) AS DESIRED.

④ PLUG THE BLACK TEST PROBE INTO THE COM INPUT JACK. PLUG THE RED TEST PROBE INTO THE 10 AMP (10A) OR 300 MILLIAMP (300mA) INPUT JACK DEPENDING ON THE EXPECTED VALUE OF THE READING.

⑤ CONNECT THE PROBE TIPS TO THE CIRCUIT ACROSS THE BREAK AS SHOWN SO THAT ALL CURRENT WILL FLOW THROUGH THE METER (A SERIES CONNECTION).

⑥ TURN THE CIRCUIT POWER BACK ON.

⑦ VIEW THE READING, BEING SURE TO NOTE THE UNIT OF MEASUREMENT.

NOTE: IF TEST LEADS ARE REVERSED, A (-) SIGN WILL SHOW IN THE DISPLAY.

NEVER ATTEMPT A VOLTAGE MEASUREMENT WITH THE TEST PROBES IN THE CURRENT JACKS. METER DAMAGE OR PERSONAL INJURY MAY RESULT!

Figure 3-203. Always make sure the power is off before cutting or unsoldering the circuit and inserting a multimeter for current measurements. Even small amounts of current can be dangerous.

voltage or resistance measurements, which are made in parallel. This allows the entire current being measured to flow through the meter. On most multimeters, the test leads must be plugged into a different set of input jacks to measure current. [Figure 3-203]

SAFETY CHECKLIST

Meters must frequently be used in operating electric circuits. As a result, the risk of electric shock is often present. To avoid injury to personnel and damage to equipment, follow these basic safety rules when using measuring instruments.

1. Use a meter that meets accepted safety standards.
2. Use a meter with fused current inputs and be sure to check the fuses before making current measurements.
3. Inspect test leads for physical damage before making a measurement.
4. Use the meter to check continuity of the test leads.
5. Only use test leads that have shrouded connectors and finger guards.
6. Only use meters with recessed input jacks.
7. Select the proper function and range for your measurement.
8. Follow all equipment safety procedures.
9. Always disconnect the "hot" (red) test lead first.
10. Don't work alone.
11. Use a meter which has overload protection on the ohms function.
12. When measuring current without a current clamp, turn the power off before connecting into the circuit.
13. Be aware of high current and high voltage situations and use the appropriate equipment, such as high voltage probes and high current clamps.

ELECTRODYNAMOMETER WATTMETER

An electrodynamometer operates in a manner similar to a D'Arsonval meter, except that an electromagnet is used instead of a permanent magnet to produce the fixed field. The electromagnet consists of a large coil of heavy wire connected in series with the load. Since the electromagnet is connected to the load, the strength of the magnetic field is proportional to the amount of current flowing through the load. The movable voltage coil, on the other

hand, is connected across the load, and its magnetic strength is proportional to the amount of voltage dropped across the load. The magnetic fields caused by the current and the voltage react with each other to move the pointer an amount that is proportional to the power dissipated by the load. [Figure 3-204]

Electrodynamometer wattmeters may be used in either DC or AC circuits. In an AC circuit they measure true power, because even if the current and voltage are out of phase, they will also be out of phase within each of the coils of the instrument. The resultant field causes the pointer to deflect an amount proportional to the true power rather than the apparent power. The apparent power in an AC circuit is found by measuring the current with an AC ammeter and the voltage with an AC voltmeter. The product of these two values is the apparent power.

$$\text{Apparent power (volt-amps)} = \text{volts} \times \text{amps}$$

The power factor of a circuit can be found as the quotient of the true power divided by the apparent power.

$$\text{Power factor} = \frac{\text{True power (watts)}}{\text{Apparent power (volt-amps)}}$$

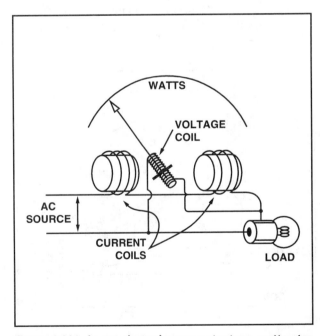

Figure 3-204. In an electrodynamometer-type wattmeter, coils of wire replace the permanent magnet. Since these coils are connected in series with the load, the force of the electromagnetic field varies with the current through the load. The movable voltage coil is connected across the load and is therefore controlled by voltage.

ELECTRODYNAMOMETER VOLTMETERS AND AMMETERS

Electrodynamometers are used as voltmeters and ammeters to measure both DC and AC values. An electrodynamometer can measure AC since the polarity of both the fixed and movable fields reverse at the same time. The sensitivity of this type of meter is considerably lower than that of the D'Arsonval-type meter. [Figure 3-205]

REPULSION-TYPE MOVING-VANE METERS

The repulsion-type moving-vane meter, like the electrodynamometer, can be used to measure either AC or DC voltage or current. If the meter is used as an ammeter, its coil has relatively few turns of heavy wire. However, if it is designed as a voltmeter, the coil has several turns of fine wire. Inside the coil there are two vanes, one fixed and the other movable. The pointer staff is attached to the movable vane by a calibrated hairspring. When current flows in the coil, both the fixed and moving vanes are magnetized with the same polarity, therefore, they repel each other. This action drives the pointer. The greater the current, the farther the pointer deflects. It makes no difference in which direction the current flows, the pointer always deflects upscale. It is for this reason that moving-vane meters can be used on AC circuits without a rectifier. [Figure 3-206]

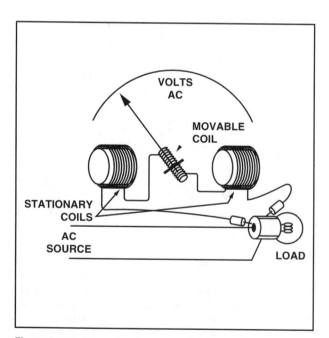

Figure 3-205. An electrodynamometer-type voltmeter can be used in either AC or DC circuits.

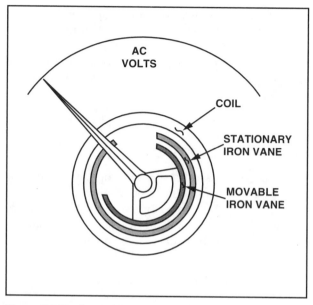

Figure 3-206. A repulsion-type moving-vane meter can be used in either AC or DC circuits. This principle is used for ammeters and voltmeters.

D'ARSONVAL METERS WITH RECTIFIERS

D'Arsonval meter movements can be adapted for use in AC circuits by using a rectifier to change AC into DC before it flows through the meter coil. Once this is done, meter indications are identical to those obtained when measuring direct current. Most D'Arsonval meters used in AC circuits employ a four-diode full-wave bridge-type rectifier. [Figure 3-207]

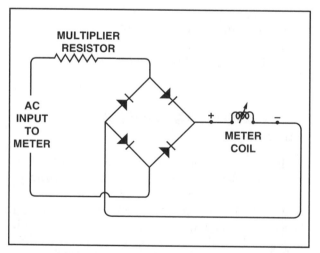

Figure 3-207. D'Arsonval meters must use a rectifier such as this four-diode, bridge-type, full-wave rectifier when they are used in an AC circuit.

THERMOCOUPLE-TYPE AMMETERS

Low-frequency alternating current can be measured with an electrodynamometer or by a repulsion-type moving-vane meter. However, these meters do not work when the frequency is in the kilohertz or megahertz range. For these applications, the **thermocouple type indicator** is used. In a thermocouple-type indicator, the alternating current being measured passes through a small piece of resistance wire inside the meter case. The greater the amount of current, the more the wire is heated. A thermocouple made of two dissimilar metals welded together is attached to the resistance wire. The other ends of the thermocouple are connected to the moving coil of a D'Arsonval-type meter movement. A voltage is generated in the thermocouple that is proportional to the difference in the temperature between the two junctions. Since the junction at the meter movement has a relatively constant temperature, the amount of voltage, and therefore the current, through the meter is proportional to the temperature of the resistance wire. The meter scale is calibrated in amperes, and since the amount of heat produced in the wire is a function of the square of the current ($P = I^2 \times R$), the scale is not uniform. Therefore, the numbers are close together on the low end of the scale and spread out as the current increases. As a result, when the current doubles, there is four times as much deflection. [Figure 3-208]

VIBRATING-REED FREQUENCY METERS

For precise frequency measurement, integrated circuit chips having clock circuits are used to actually count the cycles in a given time period. However, a much simpler type of frequency meter is used for determining the frequency of the AC produced by aircraft alternators. These frequency meters use a series of metal reeds of different lengths. The center reed has a resonant frequency of exactly 400 hertz while the reeds on one side have a higher resonant frequency, and those on the opposite side have a lower resonant frequency.

Alternating current flows through a coil that is wound around the fixture holding the reeds. The magnetic fields generated by the AC cause the fixture to vibrate at the applied AC frequency. The reed whose natural resonant frequency is that of the AC vibrates with a large amplitude and shows up as a blur. The other reeds remain stationary or move with far less amplitude. [Figure 3-209]

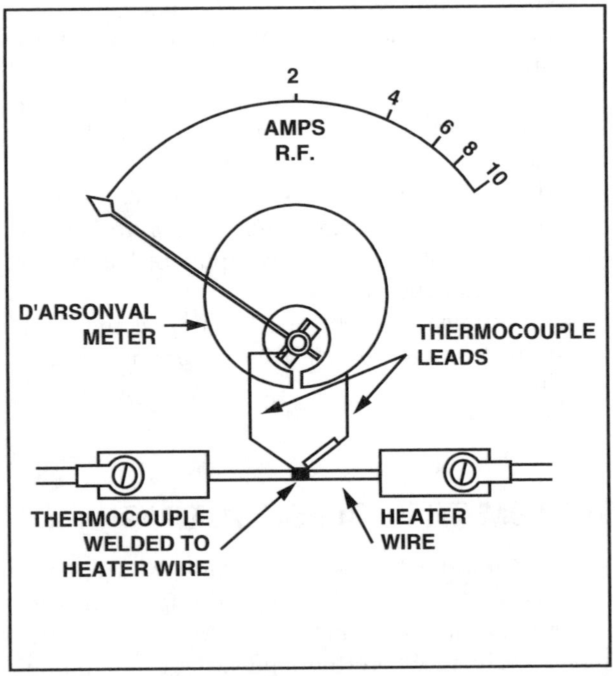

Figure 3-208. Thermocouple-type ammeters are useful for measuring high-frequency alternating current. Their operation is based on the amount of heat a given value of AC produces in a resistance wire. The heat transfers to a thermocouple which produces a small voltage that causes the needle to deflect.

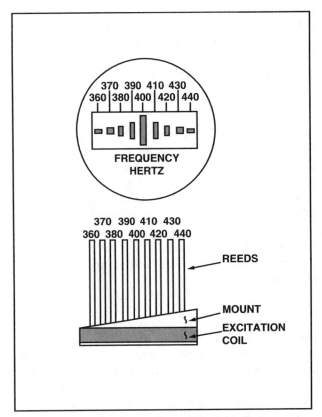

Figure 3-209. A vibrating-reed type frequency meter.

SECTION G

CIRCUIT ANALYSIS AND TROUBLESHOOTING

One of an aviation maintenance technician's most important jobs is the analysis of faults, or as it is more commonly called, troubleshooting. By definition, **troubleshooting** is the process of locating the cause or causes of a malfunction. Systematic troubleshooting allows you to find the cause of a problem and intelligently choose the correct solution.

PRINCIPLES OF TROUBLESHOOTING

Electrical troubleshooting is very similar to that of any other system. When there is a malfunction or a problem, you must first find out exactly how the system should operate. Only when you clearly understand what should be happening are you able to find the problem. The difference between what is actually happening and what should happen is a clear statement of the problem. More often than not, when a problem is clearly stated, its cause and solution are evident.

Before you can begin troubleshooting, there are two basic terms you must be familiar with. The first is the short circuit. A **short circuit** is simply a low resistance path. It can be across a power source or between two points in a circuit. When a short circuit is present, circuit resistance decreases and current flow increases. The increased flow can result in circuit over-heating and even burning of the conductor and circuit components.

An **open circuit** is a circuit that is not continuous or complete. An open circuit is typically the result of a broken conductor or a damaged component. For example, a burned out light bulb, a loose connection, or burned out fuse all create an open within a circuit.

USING METERS

When it comes to troubleshooting electrical circuits, there are two basic instruments used. They are the voltmeter and the ohmmeter. It is extremely important for you to understand that the voltmeter is used to find circuit problems with power on the circuit, while the ohmmeter is used only when the power is removed from the circuit. If an ohmmeter is used on a circuit that is energized, the meter will probably be damaged. When using a meter for troubleshooting there are a few rules to remember.

1. Voltage measurements are taken by placing the meter across (parallel to) the component.
2. The battery negative terminal, or ground, is considered to be the zero reference for voltage.
3. When a voltmeter is placed across an open component in a series circuit, it reads the battery, or applied voltage.
4. When a voltmeter is placed across a functioning component in an open circuit, the voltmeter reads zero.
5. When an ohmmeter is properly connected across a circuit component and a resistance reading is obtained, the component has continuity and is not open.
6. Current measurements are taken by connecting the meter in series with the component being tested.

SYSTEM TROUBLESHOOTING

When troubleshooting a system fault, you should never assume you know what is wrong before you have all pertinent information. As an example, consider a problem where the navigation lights on an airplane do not illuminate. When confronted with a problem such as this, the first thing to do is review the nav light electrical system. One way to do this is to list the conditions that describe the situation as it should be:

1. The battery should be in the airplane and connected.
2. The master switch should be on.
3. The navigation light circuit breaker must be in.
4. The navigation light switch should be on.
5. All three navigation lights should illuminate.

The next step is to go to the airplane and see what is actually happening:

1. Turn the master switch on. If the battery is in the airplane and is properly connected, you should hear the master contactor close. Some of the instruments and cockpit lights will give an indication that power is actually being supplied to the aircraft power bus. If you do not get any indication of power, the problem is between the battery and the main bus. In this example, assume there is power to the bus. This takes care of items one and two on your list.
2. Check the navigation light circuit breaker to be sure that it is not tripped. If the breaker is in, power can flow. However, if the breaker is out, the power stops flowing at the breaker. If a circuit breaker is out, you may reset the breaker one time. However, if it pops a second time, there may be a short and further investigation is necessary. In this example, assume the circuit breaker is in. This clears item three.
3. Be sure the navigation light switch is on, and then go out and actually look at the three navigation lights. If you go to the left wing tip and see that the red light is not burning, but the green light on the right wing and the white light on the tail are burning, you know that the problem is isolated to the left wing.

The most logical condition, if only one light does not illuminate, is a burned-out light bulb. At times it can be difficult to visually check if a bulb's filament is damaged. Therefore, to check a bulb you should remove it and check the continuity of the filament with an ohmmeter. With the meter set on the R × 1 scale, touch one lead to the base of the bulb and the other lead to the bulb's end. If some value of resistance is indicated there is continuity. If there were no continuity, the ohmmeter would read an infinite resistance. [Figure 3-210]

If the bulb has continuity, the trouble is in the aircraft wiring. To see if current is getting to the light socket, connect a voltmeter across the socket. To do this, set the multimeter on a DC voltage scale that is higher than the aircraft system voltage, and place the negative test lead against the light socket and the positive lead against the center conductor. Assume that there is no voltage here. Now, have someone turn on the master and the nav light switch. If there is no voltage reading, current is being stopped somewhere between the switch and the light socket. Figure 3-211 illustrates a typical nav light circuit. You know there is voltage at point A because the green and white lights are burning. However, there is no voltage between points B and C. In this situation, there are two possibilities. The wire connecting the lamp socket to the airframe may not be making a good ground connection, or there is no connection between point A and B. The easiest place to check for a good ground is between point C, the case of the lamp socket, and the airframe. Set the multimeter to the R × 1 ohmmeter scale and hold one test lead on the aircraft structure while you touch the lamp socket with the other lead. If the ground connection is good, the ohmmeter should give an indication of zero resistance. If you find that the lamp fixture is properly grounded, the trouble is most

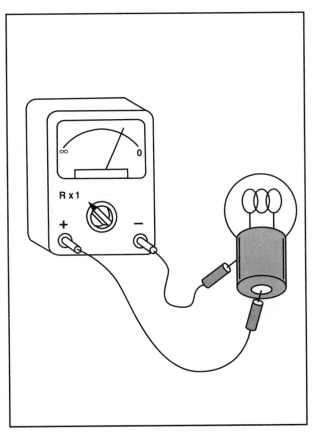

Figure 3-210. To measure the continuity of a light bulb, touch one lead to the base of the bulb and the other lead to the bulb's end. If some value of resistance is indicated, the bulb is good.

likely in the quick-disconnect fitting in the wing root. This fitting allows the electrical circuits in the wing to be disconnected when the wing is removed. [Figure 3-211]

As an aviation maintenance technician you will frequently encounter more complex circuit analysis problems than the one just discussed. For example, consider the circuit diagram in figure 3-212. The diagram shows the electrical diagram for the landing gear system of a tricycle gear aircraft. Suppose the flight crew reported that when they put the gear switch in the down position, the gear came down and locked, but the green gear down light did not illuminate. This condition indicates that there is a problem within the indicator portion of the circuit. Like the navigation light circuit discussed earlier, the most logical thing to check first is if the indicator light is burned out. To do this, begin by tracing

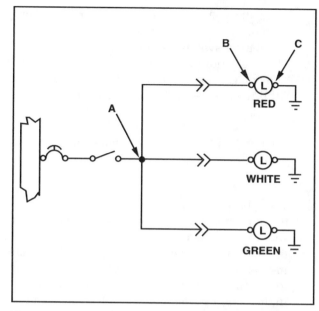

Figure 3-211. A typical aircraft navigation light circuit.

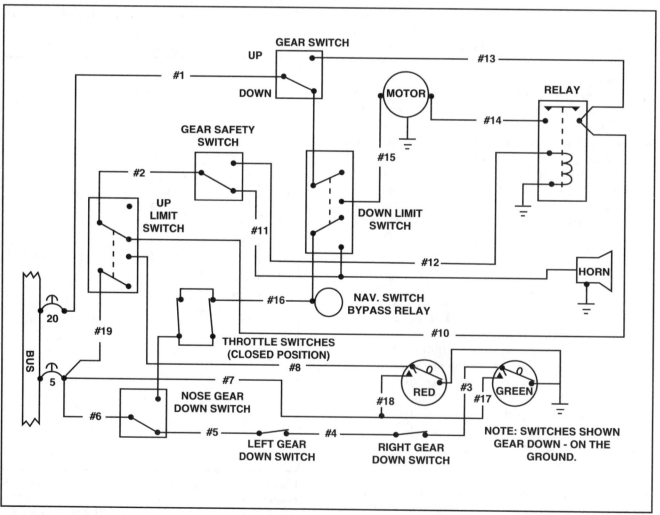

Figure 3-212. A complex landing gear circuit.

the press-to-test circuit. Power comes off the bus through a 5 amp circuit breaker and goes through wires #7 and #17 to power the press-to-test function. If the bulb illuminates when you press it, you know that the circuit breaker and lamp are good. Further analysis of the circuit reveals that the only path current can flow to the green light passes through wire #6, the nose gear down switch, wire #5, the left gear down switch, wire #4, the right gear down switch, and finally wire #3 to the lamp. Since this is the only path for current, there must be either an open in one of the wires or a defective gear down switch. To determine where the problem is, begin checking for power at each of the switches beginning at the circuit breaker.

To become familiar with circuit troubleshooting procedures, try analyzing the affect a given fault would have on a system. For example, what would be the result of a break in wire #12?

Wire #12 powers the relay that supplies one side of the gear actuator motor. Trace current flow from the bus, through the 20 amp circuit breaker, to the gear switch. When the aircraft is in the air and the gear up position is selected, current continues through wire #13, wire #10, the up-limit switch, the gear safety switch which changes position as soon as weight is off the landing gear, and finally through wire #12 to the relay coil. With the coil energized the relay closes and current flows through wire #14 to the motor and ground allowing the gear to be retracted. Based on this, if there were a break in wire #12, the relay would not close and power would not engage the motor to retract the gear.

SOLENOIDS AND RELAYS

In most aircraft, the main battery solenoid is connected in series between the battery and bus. To energize the solenoid, the ground circuit for the coil is completed through the aircraft master switch. Therefore, when the master switch is closed, current flows through the solenoid's coil, causing the contacts to close and complete the circuit from the battery to the bus. [Figure 3-213]

If, when the master switch is closed, there is no voltage on the main bus, you can easily check the solenoid with a voltmeter. Since the solenoid is connected directly to the battery, there should always be battery voltage at terminal B.

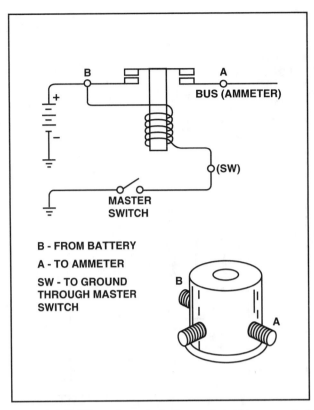

Figure 3-213. When the main battery solenoid is connected in an aircraft electrical system as shown, a relatively small amount of current is needed to close the contacts and supply power to the bus.

When the master switch is OFF, there should be battery voltage at terminal SW. However, there should be no voltage between terminal A and the ground. If there is no voltage at the SW terminal when the master switch is off, there is a possibility of an open coil. To find out if this is the problem, disconnect the wire from the SW terminal and, with the battery disconnected, measure the continuity between terminals B and SW with an ohmmeter. If the coil is intact, a low resistance will be indicated. However, if there is an infinite resistance reading, the coil is open, and the solenoid is defective.

When the master switch is ON, there should be no voltage at terminal SW since the master switch connects the SW end of the coil to ground. If there is voltage at SW with the master switch ON, there is a problem in the master switch circuit. With the coil energized, the solenoid closes and supplies battery voltage between A and ground. If no voltage exists at terminal A, there is a problem with the solenoid.

Relays pose special challenges when troubleshooting complex circuits because they typically open or close several switches at one time. As a result, actu-

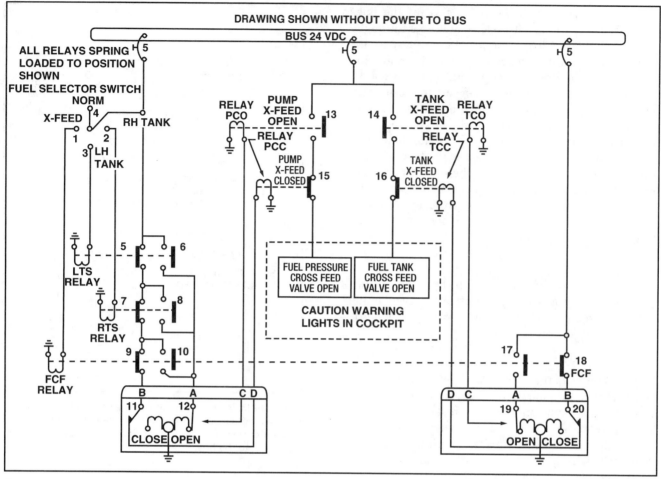

Figure 3-214. A complex fuel crossfeed system schematic with multiple relays.

ating one relay often changes an entire circuit's operating characteristics. For example, figure 3-214 shows a fuel system circuit containing several relays. The notes in the upper right and top of the schematic state that the circuit is shown with no power to the bus, and that all relays are spring loaded to the positions shown.

When power is supplied to the bus, current flows through the 5 amp circuit breaker on the right, through the contacts at 18 FCF, through switch 20, and actuates relay TCC to open the tank cross-feed switch at 16. At the same time, current flows through the 5 amp circuit breaker on the left side of the bus and travels through points 5, 7, and 9, through switch 11 in the fuel pressure cross-feed valve, and powers relay PCC to the open position. In this configuration, current does not flow to the cross-feed valve indicator lights in the cockpit.

When the fuel selector switch is placed in the cross-feed position, current from the bus flows through

point 1 and powers the FCF relay, which opens switches 9 and 18. This cuts power to relays PCC and TCC, which close switches 15 and 16. At the same time, switches 10 and 17 close, powering the

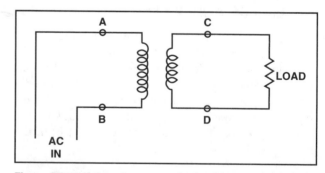

Figure 3-215. If there is an open in the primary winding, full source voltage exists between points A and B, but if the source is disconnected at either A or B, there is an infinite resistance between them. In an open secondary winding no voltage will exist across points C and D. However, if the load is disconnected at either point and some resistance is indicated, the transformer is good.

motors in the fuel pressure and fuel tank crossfeed valves to the open position. When the valves fully open, switches 12 and 19 close, and current flows to relays PCO and TCO which open the pump and tank cross-feeds and close switches 13 and 14. The fuel system is now cross-feeding fuel, and the cockpit warning lights illuminate.

TRANSFORMERS

There are two problems that could cause a transformer to function improperly. Either one of the windings could be open, or one of them could be shorted. If either winding of a transformer is open, no voltage flows out of the transformer to the load. For example, with an open in the primary winding, no voltage is induced into the secondary winding. By the same token, an open in the secondary winding allows no power to flow to the load. [Figure 3-215]

In a step-up transformer, the secondary winding usually has a higher resistance than the primary. However, in a step-down transformer, the secondary winding usually has a very low resistance. When measuring the resistance of either winding, it is important that the transformer be removed from the circuit. If this is not done, the shunting effect from the rest of the circuit affects the resistance reading. For example, if you attempt to measure the resistance of the secondary winding without disconnecting the load, the ohmmeter reads the resistance of the load as well as that of the winding.

It is often difficult to find a shorted transformer winding by measuring its resistance unless you have an extremely accurate ohmmeter and know exactly what the resistance should be. Therefore, it is easier to determine if a coil is shorted by feeling the transformer after it has been operating for a while. If a winding is shorted, the transformer heats up.

Another way to check for a shorted winding is to measure the output voltage. If the input voltage is the proper value and the output voltage is low, it is a good indication that a winding is short-circuited.

CAPACITORS

It is difficult to check low-capacity capacitors for open circuits without a special tester. This is because even good capacitors indicate some value

of resistance with an ohmmeter. As a general rule, if a capacitor does not indicate infinite resistance, it is probably short-circuited.

An electrolytic capacitor, however, can be tested by using an ohmmeter on its R × 1 scale. When measuring its resistance, the ohmmeter needle should deflect slightly up-scale as the capacitor charges, and then come to rest at a high value of resistance. Now, reverse the leads. The needle should read far up-scale (low resistance) for a short time and then come back to a high-resistance reading. This large and temporary deflection of the pointer is caused by the capacitor discharging before it recharges in the opposite direction. [Figure 3-216]

Another thing you must be aware of is that anytime you handle a high-voltage capacitor, you should short across its terminals before removing it from a circuit. The reason for this is that some capacitors can store enough of a charge that you can be injured if it discharges through your body.

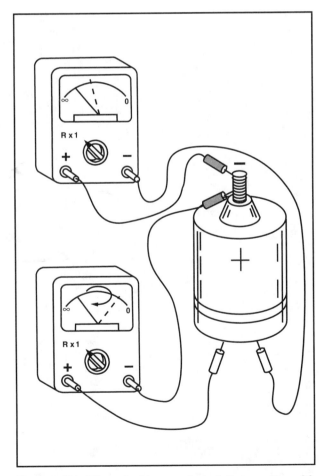

Figure 3-216. When checking an electrolytic capacitor with an ohmmeter, the indicator will deflect momentarily toward a low resistance value and then come to rest at a high resistance value.

ELECTRICAL INSTRUMENTS

A word of warning is needed here about checking circuits with an electrical measuring instrument, such as a voltmeter or ammeter. Many of these meters use external multiplier resistors or shunts, and if an attempt is made to check continuity through the meter itself, it is possible for the ohmmeter to put out enough current when set on its R × 1 range to damage the meter being examined. If it is necessary to measure across a meter, be sure that the ohmmeter is on its highest range. In this range, the minimum current flows through the test leads, and most meters can be checked for an open circuit without damaging them.

CHAPTER 4

ELECTRICAL GENERATORS AND MOTORS

INTRODUCTION

As an aviation maintenance technician, you must be familiar with modern aircraft electrical systems, and their means of generating and sustaining power. Through the study of electrical generators and motors this can be accomplished and, even though most of the work done on these systems by the aviation maintenance technician is limited, a thorough understanding of these concepts is required.

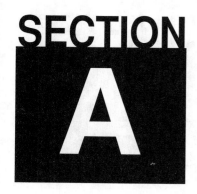

SECTION A

GENERATORS

Energy for the operation of most electrical equipment on large aircraft and some small aircraft is supplied by a generator. A **generator** is any piece of equipment which converts mechanical energy into electrical energy by electromagnetic induction. Generators designed to produce direct current are called **DC generators** whereas generators that produce alternating current are called **AC generators**.

On many older aircraft, the DC generator is the source of electrical energy. With this type of system, one or more DC generators are driven by the engine(s) to supply power for all electrical equipment as well as for charging the battery. In most cases only one generator is driven by each engine; however, some large aircraft have two generators that are driven by a single engine.

THEORY OF OPERATION

After it was discovered that electric current flowing through a conductor creates a magnetic field around the conductor, there was considerable scientific speculation about whether a magnetic field could create current flow. The English scientist, Michael Faraday, demonstrated in 1831 that this, in fact, could be accomplished. This discovery is the basis for the operation of the generator.

To show how an electric current is created by a magnetic field, several turns of wire are wrapped around a cardboard tube, and the ends of the conductor are connected to a galvanometer. A bar magnet is then moved through the tube. As the magnet's lines of flux are cut by the turns of wire, the galvanometer deflects from its zero position. However, when the magnet is at rest inside the tube, the galvanometer shows a reading of zero, indicating no current flow. When the magnet is moved through the tube in the opposite direction, the galvanometer indicates a deflection in the opposite direction. [Figure 4-1]

The same results are obtained by holding the magnet stationary and moving the coil of wire. This indicates that current flows as long as there is relative motion between the wire coil and the magnetic field. The strength of the induced current depends

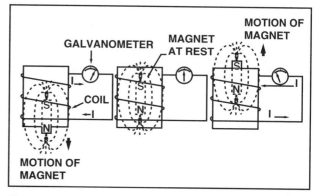

Figure 4-1. Current flow induced into a coil by magnetic flux lines is seen on a galvanometer. The direction of current flow is dependent upon how the magnetic fields cross the conductor.

on both the strength of the magnetic field and the speed at which the lines of flux are cut.

When a conductor is moved through a magnetic field, an electromotive force (EMF) is induced into the conductor. The direction, or polarity, of the induced EMF is determined by the direction the conductor is moved in relation to the magnetic flux lines.

The left-hand rule for generators is one way to determine the direction of the induced EMF. For example, if you point your left index finger in the direction of the magnetic lines of flux (north to south), and your thumb in the direction the conductor is moved through the magnetic field, your second finger indicates the direction of the induced EMF when extended perpendicular to your index finger. [Figure 4-2]

When a conductor in the shape of a single loop is rotated in a magnetic field, a voltage is induced in each side of the loop. Although the two sides of the loop cut the magnetic field in opposite directions, the induced current flows in one continuous direction within the loop. This increases the value of the induced EMF. [Figure 4-3]

When the loop is rotated half a turn so that the sides of the conductor have exchanged positions, the induced EMF in each wire reverses its direction. This is because the wire formerly cutting the lines of

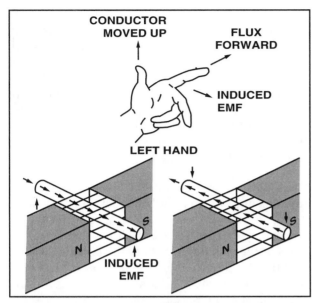

Figure 4-2. When applying the generator left-hand rule, your index finger points in the direction the lines of magnetic flux travel, your thumb indicates the conductor's direction of movement, and your second finger indicates the direction of induced EMF.

flux in an upward direction is now moving downward and the wire formerly cutting the lines of flux in a downward direction is now moving upward. In other words, the sides of the loop cut the magnetic field in opposite directions.

In a simple generator, two sides of a wire loop are arranged to rotate in a magnetic field. When the sides of the loop are parallel to the magnetic lines of flux, the induced voltage causes current to flow in one direction. Maximum voltage is induced at this position because the wires are cutting the lines of flux at right angles. This means that more lines of flux per second are cut than in any other position relative to the magnetic field.

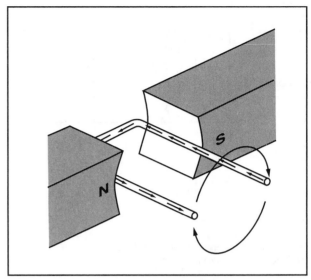

Figure 4-3. When a loop of wire is rotated in a magnetic field, the current induced in the wire flows in one continuous direction.

As the loop approaches the vertical position, the induced voltage decreases. This is because both sides of the loop become perpendicular to the lines of flux; therefore, fewer flux lines are cut. When the loop is vertical, the wires momentarily travel perpendicular to the magnetic lines of flux and there is no induced voltage.

As the loop continues to rotate, the number of flux lines being cut increases until the 90 degree point is reached. At this point, the number of flux lines cut is maximum again. However, each side of the loop is cutting the lines of flux in the opposite direction. Therefore, the direction, or polarity, of the induced voltage is reversed. Rotation beyond the 90 degree point again decreases the number of flux lines being cut until the induced voltage becomes zero at the vertical position. [Figure 4-4]

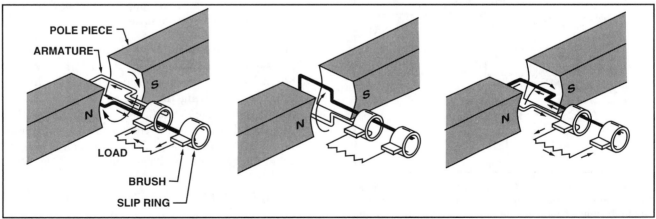

Figure 4-4. In a simple generator, the magnets are called pole pieces and the loop of wire is called the armature. Attached to each end of the loop is a slip ring on which a set of brushes ride to complete a circuit through a load. Maximum voltage is induced into the armature when it is parallel with the flux lines. Once the armature is perpendicular with the flux lines, no lines of flux are cut and no voltage is induced. As the armature rotates to the 90 degree point, the maximum number of flux lines are being cut, but in the opposite direction.

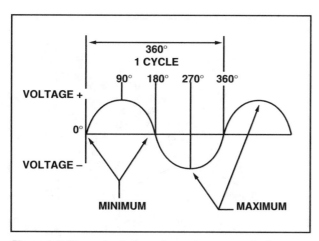

Figure 4-5. The output of an elementary generator is represented by the sine curve as a loop rotates 360 degrees through the lines of magnetic flux.

When the voltage induced throughout the entire 360 degrees of rotation is shown on an oscilloscope, the curve increases from a zero value at zero degrees, to a maximum positive voltage at 90 degrees. Once beyond 90 degrees the curve decreases until it reaches 180 degrees where the output is again zero. As the curve continues beyond 180 degrees the amount of negative voltage produced increases up to the 270 degree point where it is maximum. The amount of negative voltage then decreases to the zero point at 360 degrees. [Figure 4-5]

As you can see in the illustration, the output produced by a single loop rotating in a magnetic field is alternating current. By replacing the slip rings of the basic AC generator with two half-cylinders, commonly referred to as a single **commutator**, a basic DC generator is obtained. [Figure 4-6]

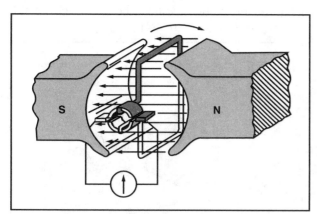

Figure 4-6. In this illustration the black side of the coil is connected to the black segment and the white side of the coil to the white segment. The segments are insulated from each other and two stationary brushes are placed on opposite sides of the commutator. The brushes are mounted so that each brush contacts each segment of the commutator as it revolves with the loop.

To explain how DC current is obtained, start with the armature at zero degrees where no lines of flux are cut and therefore, no output voltage is obtained. Once the armature begins rotating, the black brush comes in contact with the black segment of the commutator and the white brush comes in contact with the white segment of the commutator. Furthermore, the lines of flux are cut at an increasing rate until the armature is parallel to the lines of flux and the induced EMF is maximum.

Once the armature completes 180 degrees of rotation, no lines of flux are being cut and the output voltage is again zero. At this point, both brushes are contacting both the black and white segments on the sides of the commutator. After the armature rotates past the 180 degree point, the brushes contact only one side of the commutator and the lines of flux are again cut at an increasing rate.

The switching of the commutator allows one brush to always be in contact with that portion of the loop that travels downward through the lines of flux and the other brush to always be in contact with the half of the loop that travels upward. Although the current reverses its direction in the loop of a DC generator, the commutator action causes current to flow in the same direction. [Figure 4-7]

The variation in DC voltage is called **ripple**, and is reduced by adding more loops. As the number of loops increases, the variation between the maximum and minimum values of voltage is reduced. In fact, the more loops that are used, the closer the output voltage resembles pure DC. [Figure 4-8]

As the number of armature loops increases, the number of commutator segments must also increase. For example, one loop requires two commutator segments, two loops requires four segments, and four loops requires eight segments.

The voltage induced in a single-turn loop is small. Increasing the number of loops does not increase the maximum value of the generated voltage. However, increasing the number of turns in each loop does increase the voltage value. This is because voltage is obtained as an average only from the peak values. The closer the peaks are to each other, the higher the generated voltage value.

DC GENERATOR CONSTRUCTION

Generators used on aircraft differ somewhat in design because they are made by various manufacturers. However, all are of the same general construction and operate similarly. The major parts, or

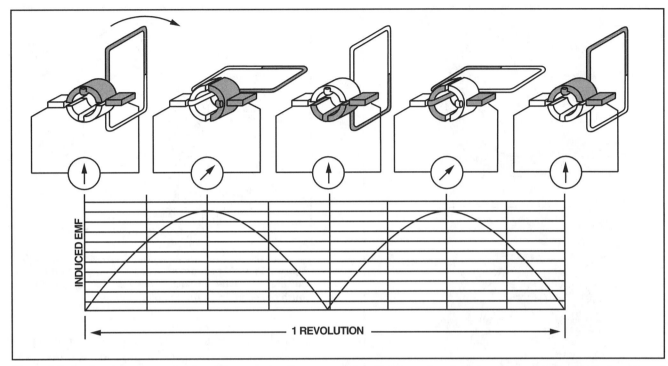

Figure 4-7. As the armature rotates in a DC generator, the commutator allows one brush to remain in contact with that portion of the loop that moves downward through the flux lines and the other brush to remain in contact with the portion of the loop that moves upward. This commutator action produces pulsating DC voltage that varies from zero to a maximum twice in one revolution.

assemblies, of a DC generator include the field frame, rotating armature, and brush assembly. [Figure 4-9]

FIELD FRAME

The field frame, or yoke, constitutes the foundation for the generator. The frame has two primary functions; 1) it completes the magnetic circuit between the poles and 2) it acts as a mechanical support for the other parts. In small generators, the frame is made of one piece of iron; however, in larger generators, it is usually made up of two parts bolted together. The frame is highly permeable and, together with the pole pieces, forms the majority of the magnetic circuit.

The magnetizing force inside a generator is produced by an electromagnet consisting of a wire coil

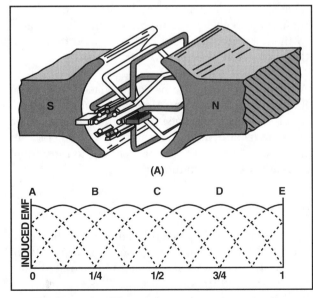

Figure 4-8. Increasing the number of loops in an armature reduces the ripple in DC voltage.

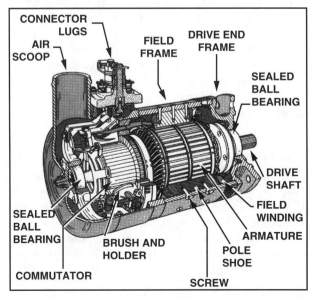

Figure 4-9. Cross section of a typical 24-volt generator with major parts labeled.

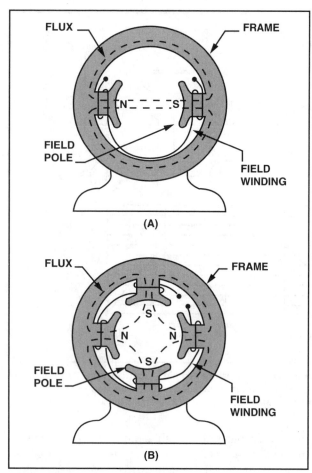

Figure 4-10. Generator field frames typically house either two or four pole shoes. These shoes are not permanent magnets, and rely on the field windings to produce a magnetic field. To try and produce the magnetic field necessary with only permanent magnets would greatly increase the physical size of the generator.

called a **field coil** and a core called a **field pole**, or **shoe**. The pole shoes are bolted to the inside of the frame and are usually laminated to reduce eddy current losses and concentrate the lines of force produced by the field coils. The frame and pole shoes are made from high quality magnetic iron or sheet steel. There is always one north pole for each south pole, so there is always an even number of poles in a generator. [Figure 4-10]

Note that the pole shoes project from the frame. The reason for this is that since air offers a great deal of resistance to a magnetic field, most generator designs reduce the length of air gap between the poles and the rotating armature. This increases the efficiency of the generator. When the pole pieces are made to project inward from the frame, they are called **salient poles**.

The field coils are made up of many turns of insulated wire. The coils are wound on a form that is

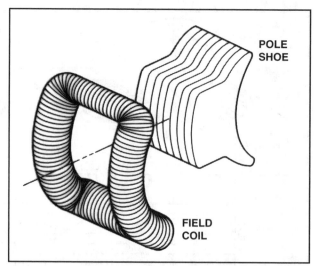

Figure 4-11. Field coils are form fitted around the pole shoes and are connected in such a manner that the north and south poles are in alternate polarity order.

securely fastened over the iron core of the pole shoes. The current used to produce the magnetic field around the shoes is obtained from an external source or from the current generated by the unit itself. Remember that the magnetic field created flows through the field coils and there is no electrical connection between the windings of the field coils and the pole shoes. [Figure 4-11]

ARMATURE

As mentioned earlier, the armature assembly consists of the armature coils, the commutator, and other associated mechanical parts. The armature is mounted on a shaft which rotates in bearings located in the generator's end frames. The core of the armature acts as a conductor when it is rotated in the magnetic field and it is laminated to prevent the circulation of eddy currents. [Figure 4-12]

A **drum-type armature** has coils placed in slots in the core of the armature. However, there is no electrical connection between the coils and core. The coils are usually held in the slots by wooden or fiber wedges. The coil ends are brought out to individual segments of the commutator.

COMMUTATORS

The commutator is located at one end of the armature and consists of wedge-shaped segments of hard-drawn copper. Each segment is insulated from the other by a thin sheet of mica. The segments are held in place by steel V-rings or clamping flanges fitted with bolts. Rings of mica also insulate the seg-

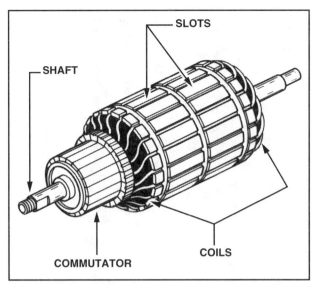

Figure 4-12. The armature rotates within the frame assembly and current is induced into it by the electromagnetic field created by the field coils and pole shoes.

ments from the flanges. The raised portion of each segment is called the **riser**, and the leads from the armature coils are soldered to each riser. In some generators, the segments have no risers. In this situation the leads are soldered to short slits in the ends of the segments. [Figure 4-13]

One end of a single armature coil attaches to one commutator segment while the other end is soldered to the adjacent segment. In this configuration,

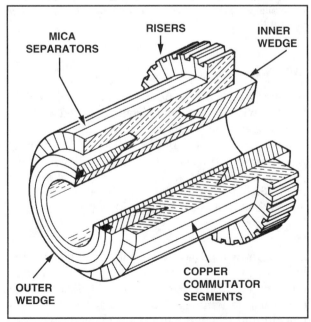

Figure 4-13. Each segment of a commutator is mounted to an inner wedge and separated by thin pieces of insulating mica. The risers on each segment hold the leads coming from the armature coils.

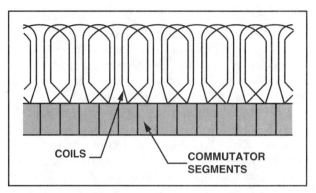

Figure 4-14. Lap winding connects one end of two coils to each commutator segment and the ends of each coil to adjacent segments.

each coil laps over the preceding one. This is known as **lap winding**. When an armature rotates at operational speed, the magnetic fields that it produces lags behind the speed of rotation. Lap winding is a method for stabilizing these armature magnetic fields. [Figure 4-14]

BRUSHES

Brushes ride on the surface of the commutator and act as the electrical contact between armature coils and an external circuit. A flexible braided-copper conductor, called a **pig-tail**, connects each brush to the external circuit. The brushes are made of high-grade carbon and held in place by spring-loaded brush holders that are insulated from the frame. The brushes are free to slide up and down in their holders so they can follow any irregularities in the commutator's surface and allow for wear. A brushes' position is typically adjustable so that the pressure on the commutator can be varied, and so the brush position with respect to the risers can be changed as necessary. [Figure 4-15]

Figure 4-15. Carbon brushes connect to an external circuit through pig tails. The brushes are typically adjustable to allow varied pressure on the commutator and position on the segments.

The constant making and breaking of connections to the coils of the armature necessitates the use of a brush material that has a definite contact resistance. This material must also have low friction to prevent excessive wear. The high-grade carbon used to make brushes must be soft enough to prevent undue wear to the commutator and yet hard enough to provide reasonable brush life. The contact resistance of carbon is fairly high due to its molecular structure and the commutator surface is highly polished to reduce friction as much as possible. Oil or grease must never be used on a commutator and extreme care must be used when cleaning a commutator to avoid marring or scratching its surface.

TYPES OF DC GENERATORS

There are three types of DC generators. They are the series-wound, shunt-wound, and shunt-series or compound-wound. The difference between each depends on how the field winding is connected to the external circuit.

SERIES-WOUND

The field winding of a series-wound generator is connected in series with the external load circuit. In this type of generator, the field coils are composed of a few turns of large wire because magnetic field strength depends more on current flow than the number of turns in the coil.

Because of the way series-wound generators are constructed, they possess poor voltage regulation capabilities. For example, as the load voltage increases, the current through the field coils also increases. This induces a greater EMF which, in turn, increases the generator's output voltage. Therefore, when the load increases, voltage increases; likewise, when the load decreases, voltage decreases.

One way to control the output voltage of a series-wound generator is to install a rheostat in parallel with the field windings. This limits the amount of current that flows through the field coils thereby limiting the voltage output. [Figure 4-16]

Since series-wound generators have such poor voltage regulation capabilities, they are not suitable for use in aircraft. However, they are suitable for situations where a constant RPM and constant load are applied to the generator.

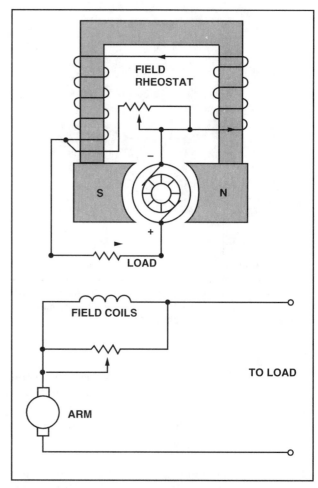

Figure 4-16. The diagram and schematic of a series-wound generator show that the field windings are connected in series with the external load. A field rheostat is connected in parallel with the field windings to control the amount of current flowing in the field coils.

SHUNT-WOUND

A generator having a field winding connected in parallel with the external circuit is called a shunt-wound generator. Unlike the field coils in a series-wound generator, the field coils in a shunt-wound generator contain many turns of small wire. This permits the field coil to derive its magnetic strength from the large number of turns rather than the amount of current flowing through the coils. [Figure 4-17]

In a shunt-wound generator the armature and the load are connected in series; therefore, all the current flowing in the external circuit passes through the armature winding. However, due to the resistance in the armature winding some voltage is lost. The formula used to calculate this voltage drop is:

IR drop = current $\times$ armature resistance

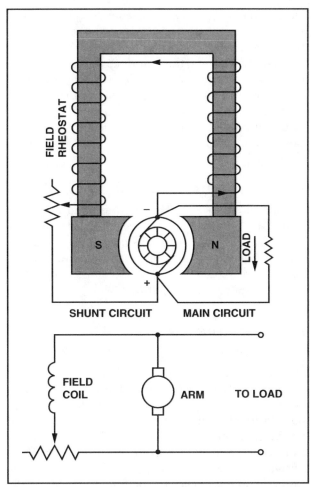

Figure 4-17. In a shunt-wound generator the field windings are connected in parallel with the external load.

From this formula you can see that as the load, or current, increases the IR drop in the armature also increases. Since the output voltage is the difference between induced voltage and voltage drop, there is a decrease in output voltage with an increased load. This decrease in output voltage causes a corresponding decrease in field strength because the current in the field coils decreases with a decrease in output voltage. By the same token, when the load decreases, the output voltage increases accordingly, and a larger current flows in the windings. This action is cumulative and, if allowed, the output voltage would rise to a point called **field saturation**. At this point there is no further increase in output voltage. Because of this, a shunt-wound generator is not desired for rapidly fluctuating loads.

To control the output voltage of a shunt generator, a rheostat is inserted in series with the field winding. In this configuration, as armature resistance increases, the rheostat reduces the field current which decreases the output voltage. For a given setting on the field rheostat, the terminal voltage at the

armature brushes is approximately equal to the generated voltage minus the IR drop produced by the armature resistance. However, this also means that the output voltage at the terminals drops when a larger load is applied. Certain voltage-sensitive devices are available which automatically adjust the field rheostat to compensate for variations in load. When these devices are used, the terminal voltage remains essentially constant.

The output and voltage-regulation capabilities of shunt-type generators make them suitable for light to medium duty use on aircraft. However, most of these units have generally been replaced by DC alternators.

COMPOUND-WOUND

A compound-wound generator combines a series winding and a shunt winding so that the characteristics of each are used. The series field coils consist of a relatively small number of turns made of large copper conductor, either circular or rectangular in cross section. As discussed earlier, series field coils are connected in series with the armature circuit. These coils are mounted on the same poles as the shunt field coils and, therefore, contribute to the magnetizing force, or **magnetomotive force**, which influences the generator's main field flux. [Figure 4-18]

If the ampere-turns of the series field act in the same

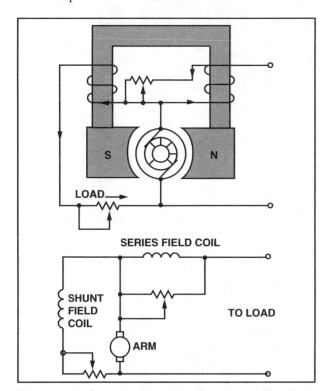

Figure 4-18. Compound-wound generators utilize both series and shunt windings. In this type of generator, voltage regulation is controlled by a diverter.

direction as those of the shunt field, the combined magnetomotive force is equal to the sum of the series and shunt field components. Load is added to a compound-wound generator in the same manner as a shunt-wound generator; by increasing the number of parallel paths across the generator. When this is done, the total load resistance decreases causing an increase in armature-circuit and series-field circuit current. Therefore, by adding a series field, the field flux increases with an increased load. Thus, the output voltage of the generator increases or decreases with load, depending on the influence of the series field coils. This influence is referred to as the **degree of compounding**.

The amount of output voltage produced by a compound-wound generator depends on the degree of compounding. For example, a **flat-compound** generator is one in which the no-load and full-load voltages have the same value. However, an **under-compound** generator has a full-load voltage less than the no-load voltage, and an **over-compound** generator has a full-load voltage that is higher than the no-load voltage.

Generators are typically designed to be over-compounded. This feature permits varied degrees of compounding by connecting a variable shunt across the series field. Such a shunt is sometimes called a **diverter**. Compound generators are used where voltage regulation is of prime importance.

If, in a compound-wound generator, the series field aids the shunt field, the generator is said to be **cumulative-compounded**. However, if the series field opposes the shunt field, the generator is said to be **differentially compounded**. [Figure 4-19]

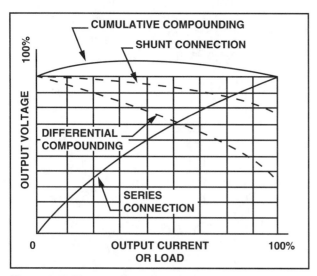

Figure 4-19. The generator characteristics chart above illustrates a summary of characteristics involving various types of generators.

Differentially compounded generators have somewhat the same characteristics as series generators in that they are essentially **constant-current generators**. In other words, they produce the same amount of current regardless of the load size. However, they do generate voltage when no load is applied and the voltage drops as the load current increases. Constant-current generators are ideally suited as power sources for electric arc welders and are used extensively for this task.

If the shunt field of a compound-wound generator is connected across both the armature and the series field, it is known as a **long-shunt connection**. However, if the shunt field is connected across the armature alone, it is called a **short-shunt connection**. These connections produce essentially the same generator characteristics.

STARTER GENERATORS

Many small turbine engines are equipped with starter generators rather than separate starters and generators. This saves appreciably in weight, as both starters and generators are very heavy. A typical starter generator consists of at least two sets of windings and one armature winding. When acting as a starter, a high current flows through both sets of field windings and the armature to produce the torque required to start the engine. However, in the generator mode, only the high resistance shunt-winding receives current while the series-winding receives no current. The current flowing through the shunt-winding is necessary to produce the magnetic field that induces voltage into the armature. Once power is produced, it flows to the primary bus.

ARMATURE REACTION

As you know, anytime current flows through a conductor, a magnetic field is produced. Therefore, it stands to reason that when current flows through an armature, electromagnetic fields are produced in the windings. These fields tend to distort or bend the lines of magnetic flux between the poles of the generator. This distortion is called **armature reaction**. Since the current flowing through the armature increases as the load increases, the distortion becomes greater with larger loads. [Figure 4-20]

Armature windings of a generator are spaced so that during rotation there are certain positions when the brushes contact two adjacent segments on the commutator, thereby shorting the armature windings. When the magnetic field is not distorted, there is no voltage induced in the shorted windings and no harmful results occur. However, when the field is distorted by armature reaction, a voltage is induced in the shorted windings and sparking takes place between the brushes and the commutator segments. Consequently, the commutator becomes pitted, the wear on the brushes becomes excessive, and the output of the generator is reduced.

To correct this condition, the brushes are set so that the plane of the coils being shorted is perpendicular to the distorted magnetic field. This is accomplished by moving the brushes forward in the direction of rotation. This operation is called **shifting the brushes** to the neutral plane, or **plane of commutation**. The neutral plane is the position where the plane of the two opposite coils is perpendicular to the magnetic field in the generator. On a few generators, the brushes are shifted manually ahead of the normal neutral plane to the neutral plane caused by field distortion. On nonadjustable brush generators, the manufacturer sets the brushes for minimum sparking.

In some generators, special field poles called **interpoles** are used to counteract some of the effects of field distortion when the speed and load of the generator are changing constantly. An interpole is another field pole that is placed between the main poles. [Figure 4-21]

An interpole has the same polarity as the next main pole in the direction of rotation. The magnetic flux produced by an interpole causes the current in the armature to change direction as the armature winding rotates under the interpole's field. This cancels

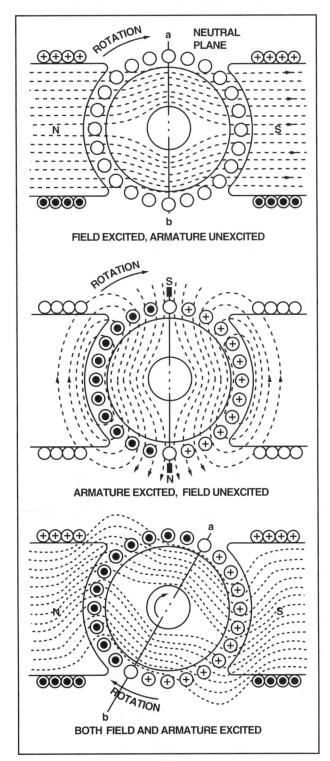

FIELD EXCITED, ARMATURE UNEXCITED

ARMATURE EXCITED, FIELD UNEXCITED

BOTH FIELD AND ARMATURE EXCITED

Figure 4-20. The lines of flux in the field coil flow in a horizontal path from north to south and induce voltage into the armature. However, as this is done, magnetic fields are produced in the armature that tend to distort or bend the lines of flux produced by the field coil.

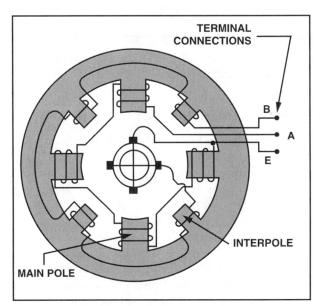

Figure 4-21. This generator has four poles and four interpoles. The interpoles are used to counter the effect of armature reaction.

the electromagnetic fields produced by the armature windings. The interpoles are connected in series with the load and, therefore, the magnetic strength of the interpoles varies with the generator load. Since the field distortion also varies with the load, the magnetic field of the interpoles counteract the effects of the field around the armature windings and minimize distortion. In other words, the interpoles keep the neutral plane in the same position for all loads.

GENERATOR RATINGS

A generator is rated according to its power output. Since a generator is designed to operate at a specified voltage, the rating is usually given as the number of amperes the generator can safely supply at its rated voltage. For example, a typical generator rating is 300 amps at 28.5 volts. A generator's rating and performance data are stamped on the name plate attached to the generator. When replacing a generator make sure it is the proper rating.

The rotation of generators is termed as either clockwise or counterclockwise, as viewed from the driven end. If no direction is stamped on the data plate, the rotation is marked by an arrow on the cover plate of the brush housing. To maintain the correct polarity, it is important to use a generator with the correct direction of rotation.

The speed of an aircraft engine varies from idle rpm to takeoff rpm; however, the majority of flight, is conducted at a constant cruising speed. The generator drive is usually geared between 1-1/8 and 1-1/2 times the engine crankshaft speed. Most aircraft generators have a speed at which they begin to produce their normal voltage. This is termed the **coming-in speed**, and is typically around 1,500 rpm.

GENERATOR TERMINALS

On large 24-volt generators, electrical connections are made to terminals marked B, A, and E. The positive armature lead connects to the B terminal, the negative armature lead connects to the E terminal, and the positive end of the shunt field winding connects to terminal A. The negative end of the shunt field winding is connected to the negative terminal brush. Terminal A receives current from the negative generator brush through the shunt field winding. This current passes through the voltage regulator and back to the armature through the positive brush. Load current, which leaves the armature through the negative brush, comes out of the E lead

and passes through the load before returning through the positive brush.

GENERATOR VOLTAGE REGULATION

Efficient operation of electrical equipment in an aircraft depends on a voltage supply that varies with a system's load requirements. Among the factors which determine the voltage output of a generator, the strength of the field current is the only one that is conveniently controlled.

One way to control the field current is to install a rheostat in the field coil circuit. When the rheostat is set to increase the resistance in the field circuit, less current flows through the field coils and the strength of the magnetic field decreases. Consequently, less voltage is induced into the armature and generator output decreases. When the resistance in the field circuit is decreased with the rheostat, more current flows through the field coils, and the magnetic field becomes stronger. This allows more voltage to be induced into the armature which produces a greater output voltage. [Figure 4-22]

One thing to keep in mind is that, the weaker the magnetic field is, the easier it is to turn the armature. On the other hand, if the strength of the magnetic field is increased, more force is required to turn the armature. This means that, when the load on a generator increases, additional field current must be supplied to increase the voltage output as well as overcome the additional force required to turn the armature.

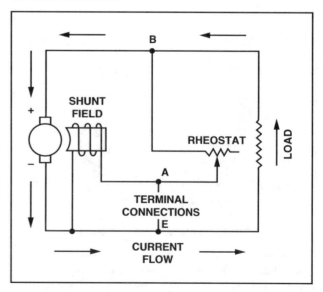

Figure 4-22. **When generator voltage is regulated by field rheostat, more resistance results in less output voltage while less resistance results in more output voltage.**

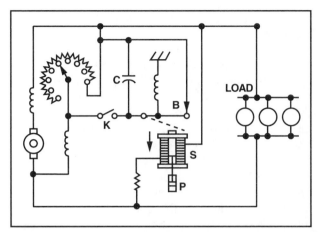

Figure 4-23. With the generator running at normal speed and switch K open, the field rheostat is adjusted so that the output voltage is about 60 percent of normal. At this level, solenoid S is weak and contact B is held closed by the spring. However, when K is closed, a short circuit is placed across the field rheostat. This action causes the field current to increase and the output voltage to rise.

This principle is further developed by the addition of a solenoid which electrically connects or removes the field rheostat from the circuit as the voltage varies. This type of setup is found in a **vibrating-type voltage regulator**. [Figure 4-23]

When the output voltage rises above a specified critical value, the downward pull of the solenoid's coil exceeds the spring tension and contact B opens. This reinserts the field rheostat in the field circuit. The additional resistance reduces the field current and lowers output voltage. When the output voltage falls below a certain value, contact B closes, shorting the field rheostat and the terminal voltage starts to rise. Thus, an average voltage is maintained with or without load changes. The dashpot P provides smoother operation by acting as a dampener to prevent hunting, and capacitor C across contact B helps eliminate sparking.

With a vibrating-type voltage regulator, contact B opens and closes several times per second to maintain the correct generator output. Based on this, if the solenoid should malfunction or the contacts stick closed, excess current would flow to the field and generator output would increase.

Certain light aircraft employ a **three-unit regulator** for their generator systems. This type of regulator includes a current limiter, a reverse current cutout, and a voltage regulator. [Figure 4-24]

The action of the voltage regulator unit is similar to the vibrating-type regulator described earlier. The **current limiter** is the second of three units, and it limits the generator's output current. The third unit is a **reverse-current cutout** which disconnects the

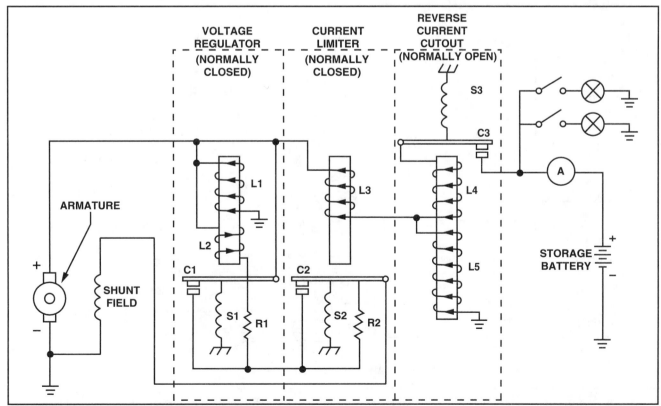

Figure 4-24. A voltage regulator contains three coils: a voltage regulator coil, a current limiter coil, and a reverse current cutout coil.

battery from the generator when the generator output is lower than the battery output. If the battery were not disconnected, it would discharge through the generator armature when the generator voltage falls below that of the battery. When this occurs the battery attempts to drive the generator as a motor. This action is called **motoring** the generator and, unless prevented, the battery discharges in a short time.

Since contacts have a tendency to pit or burn when large amounts of current flow through them, vibrating-type regulators and three-unit regulators cannot be used with generators that require a high field current. Therefore, heavy-duty generator systems require a different type of regulator, such as the **carbon-pile** voltage regulator. The carbon-pile voltage regulator relies on the resistance of carbon disks arranged in a pile or stack. The resistance of the carbon stack varies inversely with the pressure applied. For example, when the stack is compressed, less air exists between the carbon disks and the resistance decreases. However, when the pressure is reduced, more air is allowed between the disks causing the resistance to increase.

Pressure on the carbon pile is created by two opposing forces: a spring and an electromagnet. The spring compresses the carbon pile, and the electro-

magnet exerts a pull on the spring which decreases the pressure. [Figure 4-25].

Whenever the generator voltage varies, the pull of the electromagnet varies. If the generator voltage rises above a specific amount, the pull of the electromagnet increases, thereby decreasing the pressure exerted on the carbon pile and increasing its resistance. Since this resistance is in series with the field, less current flows through the field winding and there is a corresponding decrease in field strength. This results in a drop in generator output. On the other hand, if the generator output drops below a specified value, the pull of the electromagnet decreases and the carbon pile places less resistance in the field winding circuit. This results in an increase in field strength and a corresponding increase in generator output. A small rheostat provides a means of adjusting the current flow through the electromagnet coil.

DC GENERATOR SERVICE AND MAINTENANCE

Because of their relative simplicity and durable construction, generators operate many hours without trouble. The routine inspection and service done at each 100-hour or annual inspection interval is generally all that is required to keep a generator in good

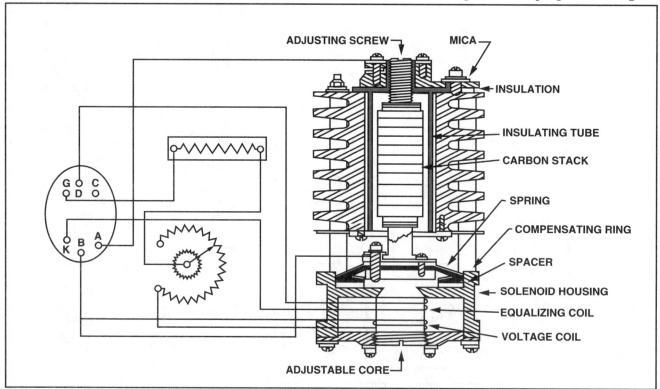

Figure 4-25. A carbon pile voltage regulator relies on the amount of air space within a stack of carbon disks to control generator voltage. Pressure is maintained on the disks by a spring while an electromagnet controls spring tension

working order. Generator overhaul is often accomplished at the same time as engine overhaul. This minimizes aircraft down time and increases the likelihood of trouble free operation when the aircraft is placed back in service.

ROUTINE INSPECTION AND SERVICING

The 100-hour and annual inspection of a generator should includethe following items:

1. Inspect generator for security of mounting, check the mounting flange for cracks and loose mounting bolts.
2. Inspect mounting flange area for oil leaks.
3. Inspect generator electrical connections for cleanliness and security of attachment.
4. Remove band covering the brushes and commutator. Use compressed air to blow out accumulated dust. Inspect brushes for wear, and freedom of movement. Check tension of the brush springs, using a spring scale.
5. Inspect commutator for cleanliness, wear, and pitting.
6. Inspect area around the commutator and brush assemblies for any solder particles. The presence of solder indicates that the generator has overheated and melted the solder attaching the armature coils to the risers. When this happens, an open is created in the armature.

If a DC generator is unable to keep an aircraft battery charged, and if the ammeter does not show the proper rate of charge, you should first check the aircraft electrical system associated with the battery and generator. Physically check every connection in the generator and battery circuit and electrically check the condition of all fuses and circuit breakers. Check the condition of all ground connections for the battery, battery contacts, and the generator control units. When you have determined that there is no obvious external problem and the generator armature turns when the engine is cranked, check the generator and the voltage regulator.

One of the easiest ways to determine which unit is not operating is to connect a voltmeter between the G terminal of the voltage regulator and ground. This checks the generator's output voltage. However, because this check requires the generator to be turning it must be accomplished with the engine running, or on an appropriate test stand. In either case, observe proper safety precautions. Even when the field winding is open, or the voltage regulator is malfunctioning, the generator should produce residual voltage. In other words, the voltage produced by

the armature cutting across the residual magnetic field in the generator frame produces voltage. This should be around one or two volts.

If there is no residual voltage, it is possible that the generator only needs the residual magnetism restored. Residual magnetism is restored by an operation known as **flashing the field**. This is accomplished by momentarily passing current through the field coils in the same way that it normally flows. The methods vary with the internal connections of the generator, and with the type of voltage regulator used. For example, in an "A" circuit generator the field is grounded externally, and you must touch the positive terminal of the battery to the armature. You also may have to insulate one of the brushes by inserting a piece of insulating material between the brush and commutator. To flash the field in an internally grounded "B" circuit generator, the positive battery terminal is touched to the field. Whatever the case, be certain to follow the specific manufacturer instructions. Failure to do so could result in damage to the generator and/or voltage regulator.

If the generator produces residual voltage and no output voltage, the trouble could be with the generator or the regulator. To determine which, operate the engine at a speed high enough for the generator to produce an output, and bypass the voltage regulator with a jumper wire. This method varies with the type of generator and regulator being used, and should be performed in accordance with the manufacturer's recommendations. If the generator produces voltage with the regulator shorted, the problem is with the voltage regulator. If this is the case, be sure that the regulator is properly grounded, because a faulty ground connection prevents a regulator from functioning properly. It is possible to service and adjust some vibrator-type generator controls. However, due to expense, time involved, and test equipment needed to do the job properly, most servicing is done by replacing a faulty unit with a new one.

If the generator does not produce an output voltage when the regulator is bypassed, remove the generator from the engine and overhaul or replace it with an overhauled unit.

GENERATOR OVERHAUL

Generator overhaul is accomplished any time a generator is determined to be inoperative, or at the same time the aircraft engine is overhauled. Although an overhaul can be done in some aircraft repair facilities, it is more often the job of an FAA Certified Repair Station licensed for that operation.

The steps involved in the overhaul of a generator are the same for the overhaul of any unit: (1) disassembly, (2) cleaning, (3) inspection and repair, (4) reassembly, and (5) testing.

DISASSEMBLY

Disassembly instructions for specific units are covered in the manufacturer's overhaul manual and must be followed exactly. Specialized tools are sometimes required for removing pole shoes since the screws holding these in place are usually staked to prevent them from accidentally backing out. Special instructions must also be followed when removing bearings. If the incorrect procedures or tools are used, damage to the bearings or their seating area could result.

CLEANING

Care must be taken when cleaning electrical parts. The proper solvents must be used, and generally parts are not submerged in solvent tanks. Using the wrong solvent could remove the lacquer-type insulation used on field coils and armatures resulting in short circuits after the generator is reassembled.

INSPECTION AND REPAIR

Inspect components for physical damage, corrosion, or wear, and repair or replace as required. Testing for proper operation of electrical components is accomplished using a growler and an electrical multimeter. A **growler** is a specially designed test unit for DC generators and motors and a variety of tests on the armature and field coils are performed using this equipment. Growlers consist of a laminated core wound with many turns of wire that are connected to 110 volts AC. The top of the core forms a vee into which the armature of a DC generator fits. The coil and laminated core of the growler form the primary of a transformer, while the generator armature becomes the secondary. Also included on most growlers is a 110 volt test lamp. This is a simple series circuit with a light bulb that illuminates when the circuit is complete. [Figure 4-26]

To test an armature for an open, place it on an energized growler. Using the probes attached to the test lamp, test each armature coil by placing the probes on adjacent segments. The lamp should light with

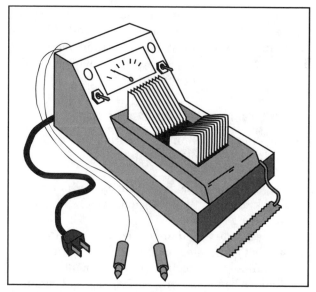

Figure 4-26. A growler is used to test the armature of a DC generator for open circuits.

each set of commutator bars. Failure of the test lamp to illuminate indicates an open in that coil, and replacement of the armature is called for.

Armatures are also tested for shorts by placing them on a growler, energizing the unit and holding a thin steel strip, typically a hacksaw blade, slightly above the armature. Slowly rotate the armature on the growler, if there are any shorts in the armature windings, the blade vibrates vigorously. [Figure 4-27]

A third test for armature shorts is accomplished using a 110 volt test lamp to check for grounds. To use a test lamp, one lead is touched to the armature shaft while the second lead is touched to each commutator segment. If a ground exists between any of the windings and the core of the armature, the test lamp illuminates. [Figure 4-28]

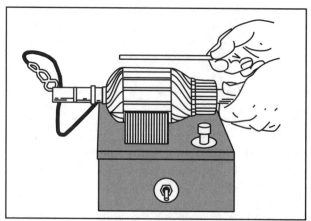

Figure 4-27. When a short exists in the armature windings the hacksaw blade vibrates vigorously.

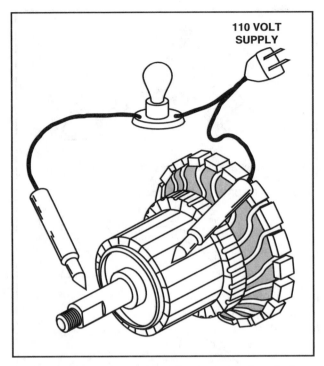

Figure 4-28. When a ground exists between the windings and the armature core, the test light illuminates. This test is also accomplished using an ohmmeter.

You can also test a generator's field coil for shorts by using a test lamp. To do this, one probe is placed at the field winding and the other probe at the generator frame. If the light illuminates, a short exists. [Figure 4-29]

You can also test the field coil for continuity using an ohmmeter set to the low-ohms scale. A shunt

field coil should indicate between 2 and 30 ohms, depending on the specific coil. A series type field coil shows almost no resistance. In some cases a current draw test is specified by the manufacturer. This test is accomplished by connecting a battery of proper voltage across the field coils and measuring the current flow. This value must be within the limits specified in the manufacturer's test specifications. [Figure 4-30]

To resurface or remove irregularities and pitting from a commutator, an armature is turned in a special armature lathe, or an engine lathe equipped with a special holding fixture. When doing this, remove only enough metal to smooth the commutator's surface. If too much material is removed, the security of the coil ends are jeopardized. If the commutator is only slightly roughened, it is smoothed using No. 000 sandpaper. Never use emery cloth or other conductive material since shorting between commutator segments could result.

With some commutators, the mica insulation between the commutator bars may need to be **undercut**. However, when doing this, you must follow the most recent instructions provided by the manufacturer. This operation is accomplished using a special attachment for the armature lathe, or a hack-saw blade. When specified, the mica is undercut about the same depth as the mica's thickness, or approximately 0.020 inch.

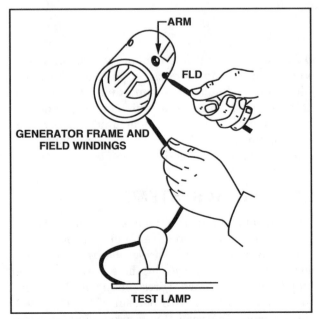

Figure 4-29. In order for the frame and field windings to be considered good, the light should not illuminate.

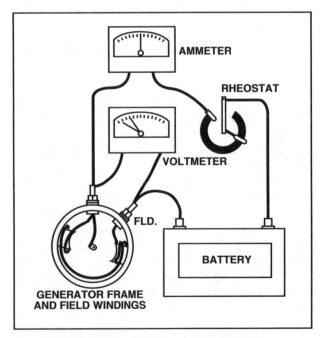

Figure 4-30. To test for shorted turns in a field coil, measure the current drawn by the field at a specific voltage. The current drawn by the field must be within the manufacturer's prescribed range in order for the field windings to be good.

REASSEMBLY

Prior to reassembly, the painted finish on the exterior of the frame is restored. In certain cases the special insulated coatings on the interior surfaces are renewed. Furthermore, all defective parts are replaced according to the reassembly order specified by the manufacturer.

When reassembling a generator, make certain that all internal electrical connections are properly made and secured, and that the brushes are free to move in their holders. Check the pigtails on the brushes for freedom, and make sure they do not alter or restrict the brushes' free motion. The purpose of the pigtail is to conduct current and help eliminate any current in the brush springs that could alter its spring action. The pigtails also eliminate possible sparking caused by movement of the brush within the holder, thus minimizing brush side wear.

Generator brushes are normally replaced at overhaul, or when half worn. When new brushes are installed they must be seated, or contoured, to maximize the contact area between the face of the brush and the commutator. Seating is accomplished by lifting the brush slightly to permit the insertion of No. 000, or finer sandpaper, rough side out. With the sandpaper in place, pull the sandpaper in the direction of armature rotation, being careful to keep the ends of the sandpaper as close to the commutator as possible to avoid rounding the edges of the brush. When pulling the sandpaper back to the starting point, the brush is raised so it does not ride on the sandpaper. [Figure 4-31]

Brush spring tension is checked using a spring scale. A carbon, graphite, or light metal brush should only exert a pressure of 1-1/2 to 2-1/2 psi on the commutator. If the spring tension is not within the limits set by the manufacturer, the springs must be replaced. When a spring scale is used, the pressure measurement exerted by a brush is read directly on the spring scale. The scale is attached at the point of contact between the spring arm and the top of the brush, with the brush installed in the guide. The scale is then pulled up until the arm just lifts the brush off the commutator. At that instant, the force on the scale is read.

After the generator has run for a short period, the brushes should be reinspected to ensure that no pieces of sand are embedded in the brush. Under no circumstance should emery cloth or similar abra-

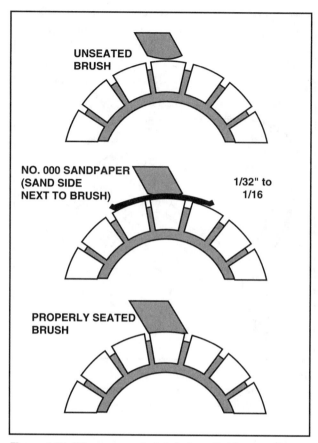

Figure 4-31. When a new brush is installed in a generator, the brush must be contoured to fit the commutator. To do this, insert a piece of No. 000 sandpaper between the brush and the commutator and sand the brush in the direction of rotation.

sive be used for seating brushes or smoothing commutators, since they contain conductive materials.

TESTING

Operational testing of generators is accomplished on test benches built for that purpose. Bench testing allows the technician the opportunity to flash the field, and ensure proper operation of the unit before installation. Generator manufacturers supply test specifications in their overhaul instructions that should be followed exactly.

GENERATOR SYSTEMS

When installed on most aircraft, the output of a generator typically flows to the aircraft's bus bar where it is distributed to the various electrical components. In this type of system, the allowable voltage drop in the main power wires coming off the generator to the bus bar is two percent of the regulated voltage when the generator is producing its rated current. As added insurance to make sure that a given electrical load does not exceed a generator's output capability, the total continuous electrical load permitted in a given system is limited to 80 percent of the total rated generator output. For example, if an aircraft has a 60 amp generator installed, the maximum continuous load that can be placed on the electrical system is 48 amps.

In the event a generator quits producing current or produces too much current, most aircraft systems have a generator master switch that allows you to disconnect the generator from the electrical system. This feature helps prevent damage to the generator or to the rest of the electrical system. On aircraft that utilize more than one generator connected to a common electrical system, the Federal Aviation Regulations require individual generator switches that can be operated from the cockpit. This allows you to isolate a malfunctioning generator and protect the remaining electrical system.

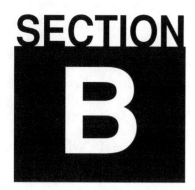
ALTERNATORS

There are two types of alternators used in today's aircraft. They are the DC alternator and the AC alternator. DC alternators produce relatively small amounts of current and, therefore, are typically found on light aircraft. AC alternators, on the other hand, are capable of producing a great deal of power and, therefore, are typically found on larger aircraft and military aircraft. Furthermore, since AC electricity can be carried through smaller conductors, AC alternators allow an appreciable weight savings.

DC ALTERNATORS

DC alternators do the same thing as DC generators. They produce AC that is then converted to DC before it enters an aircraft's electrical system. The difference, however, is that in an alternator the magnetic poles rotate and induce voltage into a fixed, or stationary winding. Furthermore, the AC current produced is rectified by six solid-state diodes instead of a commutator. [Figure 4-32]

All alternators are constructed in basically the same way. The primary components of an alternator include the rotor, the stator, the rectifier, and the brush assembly.

ROTORS

An alternator **rotor** consists of a wire coil wound on an iron spool between two heavy iron segments with interlacing fingers. Some rotors have four fingers while others have as many as seven. Each finger forms one pole of the rotating magnetic field. [Figure 4-33]

The two coil leads pass through one segment and each lead attaches to an insulated **slip ring**. The slip rings, segments, and coil spool are all pressed onto a hardened steel rotor shaft which is either splined or has a key slot. In an assembled alternator, this shaft is driven by an engine accessory pad, or fitted with a pulley and driven by an accessory belt. The slip-ring end of the shaft is supported in the housing with a needle bearing and the drive end with a ball bearing. Two carbon brushes ride on the smooth slip rings to bring a varying direct current into the field and carry it out to the regulator.

STATOR

As the rotor turns, the load current is induced into stationary stator coils. The coils making up the sta-

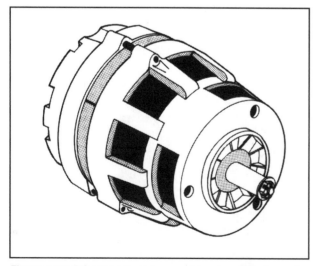

Figure 4-32. DC alternators are used in aircraft that require a low or medium amount of electrical power.

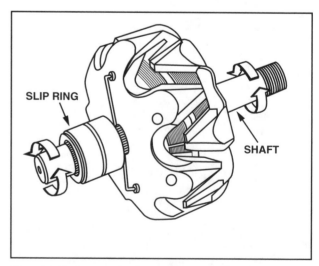

Figure 4-33. Rotors consist of interlacing fingers that form the poles of the rotating magnetic field.

tor are wound in slots around the inside periphery of the stator frame, which is made of thin laminations of soft iron. Most alternators are three-phase alternators. This means that the stator has three separate coils that are 120 degrees apart. To do this, one end of each coil is brought together to form a common junction of a Y-connection. [Figure 4-34]

With the stator wound in a three-phase configuration, the output current peaks in each set of windings every 120 degrees of rotation. However, after the output is rectified, the DC output becomes much smoother. [Figure 4-35]

Because an alternator has several field poles and the large number of stator windings, most alternators produce their rated output at a relatively low rpm. This differs from a generator which must rotate at a fairly high speed to produce its rated output.

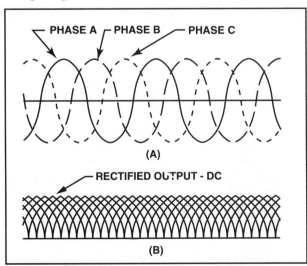

Figure 4-34. Three sets of coils in a stator are typically brought together to form a Y-connection.

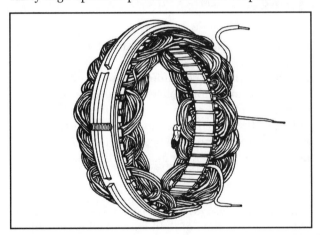

Figure 4-35. (A)—The waveform produced by a three-phase stator winding results in a peak every 120 degrees of rotation. (B)—Once rectified, the output produces a relatively smooth direct current, with a low-amplitude, high-frequency ripple.

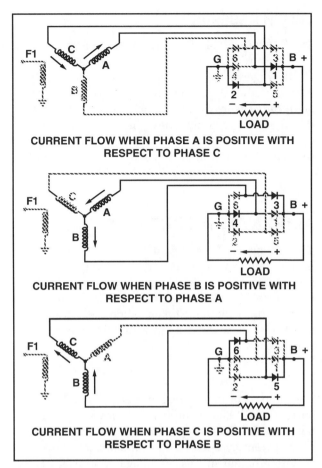

CURRENT FLOW WHEN PHASE A IS POSITIVE WITH RESPECT TO PHASE C

CURRENT FLOW WHEN PHASE B IS POSITIVE WITH RESPECT TO PHASE A

CURRENT FLOW WHEN PHASE C IS POSITIVE WITH RESPECT TO PHASE B

Figure 4-36. This circuit diagram illustrates how the AC produced in each winding is rectified into DC.

RECTIFIERS

The three-phase, full-wave rectifier in an alternator is made up of six heavy-duty silicon diodes. Three of the diodes are pressed into the slip-ring end frame, and the other three are pressed into a heat sink that is electrically insulated from the end frame. [Figure 4-36]

By referring to figure 4-36, you can see that at the instant the output terminal of winding "A" is positive with respect to the output end of winding "C," current flows through diode 1 to the load, and back through diode 2 which is pressed into the alternator end frame. From this diode, it flows back through winding "C."

As the rotor continues to turn, winding "B" becomes positive with respect to winding "A" and the current flows through diode 3 to the load, and then back through diode 4 and winding "A."

When the rotor completes 240 degrees of rotation, "C" becomes positive with respect to "B." When this occurs, current flows through diode 5 to the load, and back through diode 6. After 360 degrees of rotation, the process begins again.

BRUSH ASSEMBLY

The brush assembly in an alternator consists of two brushes, two brush springs, and brush holders. Unlike a generator which uses brushes to supply a path for current to flow from the armature to the load, the brushes in an alternator supply current to the field coils. Since these brushes ride on the smooth surface of the slip rings, the efficiency and service life of alternator brushes is typically better.

ALTERNATOR CONTROLS

The voltage produced by an alternator is controlled in the same way as in a generator, by varying the DC field current. Therefore, when the output voltage rises above the desired value, the field current is decreased. By the same token, when the output voltage drops below the desired value, the field current is increased.

The process of increlasing and decreasing the field current could be accomplished in low-output alternators with vibrator-type controls that interrupt the field current by opening a set of contacts. However, a more efficient means of voltage control has been devised that uses a transistor to control the flow of field current.

The transistorized voltage regulator utilizes both vibrating points and transistors for voltage control. The vibrating points operate the same as they do in vibrator-type voltage regulators. However, instead of the field current flowing through the contacts, the transistor base current flows through them. Since this current is small compared to the field current that flows through the emitter-collector there is no arcing at the contacts. [Figure 4-37]

In a completely solid-state voltage regulator, semiconductor devices replace all of the moving parts. These units are very efficient, reliable, and generally have no serviceable components. Therefore, if a completely solid-state unit becomes defective, it is typically removed from service and replaced.

Alternator control requirements are different from those of a generator for several reasons. For exam-

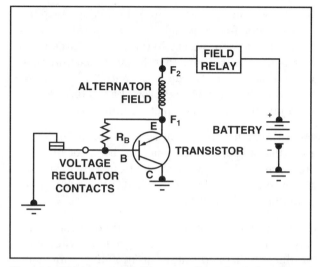

Figure 4-37. Alternator field current flows from the emitter to the collector only when the voltage regulator contacts are closed and current flows to the base.

ple, since an alternator uses solid-state diodes for rectification, current cannot flow from the battery into the alternator. Therefore, there is no need for a reverse-current cutout relay. Furthermore, since the alternator field is excited by the system bus whose voltage is limited, there is no way an alternator can put out enough current to burn itself out. Because of this, there is no need for a current limiter.

With an alternator there must be some means of shutting off the flow of field current when the alternator is not producing power. To do this, most systems utilize either a field switch or a field relay that is controlled by the master switch. Either setup allows you to isolate the field current and shut it off if necessary.

Another control that most aircraft alternator circuits employ is some form of overvoltage protection. This allows the alternator to be removed from the bus should a malfunction occur that increases the output voltage to a dangerous level. This function is often handled by an **alternator control unit** or **ACU**. Basically, the ACU drops the alternator from the circuit when an overvoltage condition exists.

DC ALTERNATOR SERVICE AND MAINTENANCE

When an alternator fails to keep the battery charged, you should first determine that the alternator and

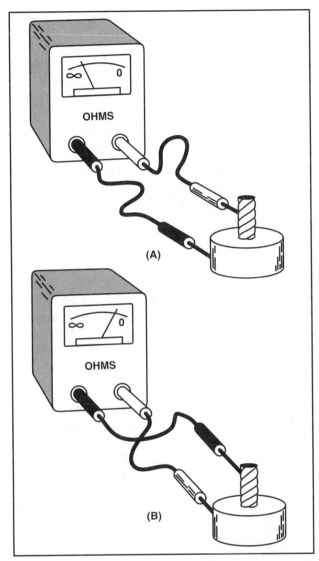

Figure 4-38. (A) — A good diode produces a high resistance reading when reverse biased. (B) — On the other hand, resistance should be low when forward biased. A shorted diode typically causes a low resistance reading when checked in the forward and reverse biased direction while an open diode produces high resistance readings in either direction.

ground. To accomplish this, set the ohmmeter on the R × 1 scale and measure resistance. Then, reverse the ohmmeter leads and measure the resistance again. If the diodes are good, you get a relatively low resistance reading when the diodes are forward biased, and an infinite or very high reading when the diodes are reverse biased. If an infinite or very high reading is not obtained, one or more of the diodes are shorted. [Figure 4-38]

Since the diodes in an alternator are connected in parallel, an open diode cannot be detected with an ohmmeter. However, if the diodes are checked individually, an ohmmeter will identify an open diode.

Solid-state diodes are quite rugged and have a long life when properly used. However, they can be damaged by excessive voltage or reverse current flow. For this reason you should never operate an alternator that is not connected to an electrical load.

Since alternators receive their field current from the aircraft bus and do not rely on residual magnetism to be started, you must never flash the field or polarize an alternator.

To aid in systematic alternator troubleshooting, some manufacturers have specialized test equipment. This test equipment is usually plugged into the aircraft electrical system between the voltage regulator and the aircraft bus. Through the use of indicator lights, the test equipment tells you whether a problem exists in the voltage regulator, the overvoltage sensing circuit, or the alternator field/output circuit. By using this type of test equipment, you can save time and avoid the unnecessary replacement of good components.

To avoid burning out the rectifying diodes during installation, it is extremely important that the battery be connected with the proper polarity. In addition, anytime an external power source is connected to the aircraft, ensure correct polarity is applied.

AC ALTERNATORS

Direct current is used as the main electrical power for small aircraft because it is storable and aircraft engines are started through the use of battery power. However, large aircraft require elaborate ground service facilities and external power sources for starting. Therefore, they can take advantage of the appreciable weight savings provided by using alternating current as their primary power source.

In addition to saving weight, alternating current has the advantage over direct current in that its voltage

battery circuits are properly connected. This includes checking for open fuses or circuit breakers. If everything is connected properly, check for battery voltage at the alternator's battery, or "B" terminal and at the "Batt" or "+" terminal of the voltage regulator.

There are basically two problems that prevent an alternator from producing electrical power. The most likely is a shorted or open diode in the rectifying circuit. The other problem is the possibility of an open circuit in the field.

To check for a shorted circuit, measure the resistance between the alternator's "B" terminal and

is easily stepped up or down. Therefore, when needed, AC carries current a long distance by passing it through a step-up transformer. This promotes additional weight savings since high voltage AC is conducted through a relatively small conductor. Once the voltage arrives at its destination it passes through a step-down transformer where voltage is lowered and current is stepped up to the value needed.

In some situations, such as charging batteries or operating variable speed motors, direct current is required. However, by passing AC through a series of semiconductor diodes it is easily changed into DC with relatively little loss. This is another advantage of AC as compared to DC.

TYPES OF AC ALTERNATORS

AC alternators are classified in order to distinguish differences. One means of classification is by the output voltage phase numbers. Alternating current alternators can be single-phase, two-phase, three-phase, and sometimes even six-phase or more. However, almost all aircraft electrical systems use a three-phase alternator.

In a **single-phase alternator**, the stator is made up of several windings connected in series to form a single circuit. The windings are also connected so the

AC voltages induced into each winding are in phase. This means that, to determine a single-phase alternator's total output, the voltage induced into each winding must be added. Therefore, the total voltage produced by a stator with four windings is four times the single voltage in any one winding. However, since the power delivered by a single-phase circuit is pulsating, this type of circuit is impractical for many applications. [Figure 4-39]

Two-phase alternators have two or more single-phase windings spaced symmetrically around the stator so that the AC voltage induced in one is 90 degrees out of phase with the voltage induced in the other. These windings are electrically separate from each other so that when one winding is cutting the maximum number of flux-lines, the other is cutting no flux lines.

A **three-phase** or **polyphase** circuit is used in most aircraft alternators. The three-phase alternator has three single-phase windings spaced so that the voltage induced in each winding is 120 degrees out of phase with the voltage in the other two windings. [Figure 4-40]

The three individual phase voltages produced by a three-phase alternator are similar to those generated by three single-phase alternators, whose voltages are out of phase by 120 degrees. In a three-phase alternator, one lead from each winding is connected to form a common junction. When this is done, the stator is Y- or star-connected. A three-phase stator can also be connected so that the phases are end-to-end. This arrangement is called a delta connection.

Still another means of classifying alternators is to distinguish between the type of stator and rotor used. When done this way, there are two types of alternators, the revolving-armature type and the revolving-field type.

The **revolving-armature type** alternator is similar in construction to the DC generator, in that the armature rotates within a stationary magnetic field. This

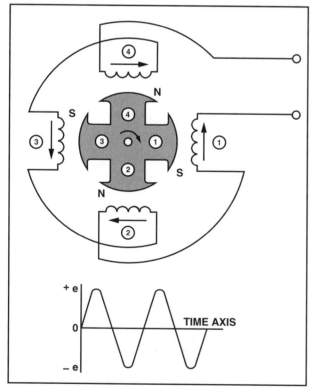

Figure 4-39. In this single-phase alternator the rotating field induces voltage into the stationary stator. The number of stator windings determines the output voltage.

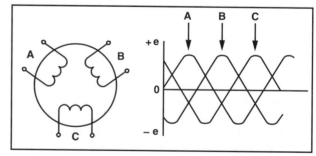

Figure 4-40. Sinewave "A" is 120 degrees out of phase with sinewave "B," and sinewave "B" is 120 degrees out of phase with sinewave "C."

type of setup is typically found only in alternators with a low power rating and generally is not used.

The **revolving-field type** alternator has a stationary armature winding (stator) and a rotating-field winding (rotor). The advantage of this configuration is that the armature is connected directly to the load without sliding contacts in the load circuit. Direct connection to the armature circuit makes it possible to use large cross-section conductors that are adequately insulated for high voltage. [Figure 4-41]

BRUSHLESS ALTERNATORS

The AC alternators used in large jet-powered aircraft are of the brushless type and are usually air cooled. Since the brushless alternators have no current flow between brushes or slip rings they are very efficient at high altitudes where brush arcing is often a problem.

As discussed earlier, alternator brushes are used to carry current to the rotating electromagnet. However, in a brushless alternator, current is induced into the field coil through an exciter. A brushless alternator consists of three separate fields, a permanent magnetic field, an exciter field, and a main output field. The permanent magnets furnish the magnetic flux to start the generator producing an output before field current flows. The magnetism produced by these magnets induces voltage into an armature that carries the current to a generator control unit, or GCU. Here, the AC is rectified and sent to the exciter field winding. The exciter field then induces voltage into the exciter output winding.

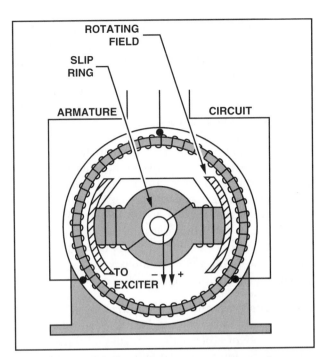

Figure 4-41. In a revolving-field type alternator, the armature is directly connected to the load without the use of sliding contacts. The rotating-field alternator is used almost universally in aircraft systems.

The output from the exciter is rectified by six silicon diodes, and the resulting DC flows through the output field winding. From here, voltage is induced into the main output coils. The permanent magnet, exciter output winding, six diodes, and output field winding are all mounted on the generator shaft and rotate as a unit. The three-phase output stator windings are wound in slots that are in the laminated frame of the alternator housing. [Figure 4-42]

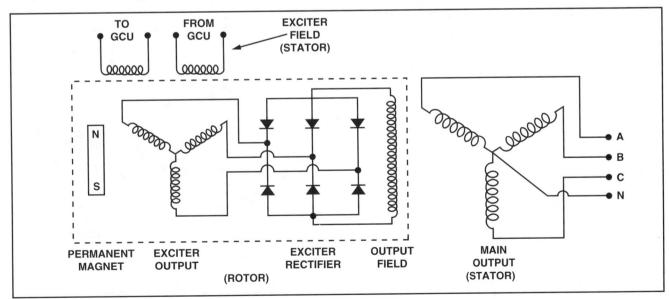

Figure 4-42. In a brushless alternator there are actually three generators. The permanent magnet generator, the exciter generator, and the main generator. The permanent magnet generator induces voltage into the exciter generator which in turn supplies the field current for the main generator.

The main output stator winding ends of a brushless alternator are connected in the form of a Y, and in the case of the previous figure, the neutral winding is brought to the outside of the housing along with the three-phase windings. These alternators are usually designed to produce 120 volts across a single phase and 208 volts across two phases.

The GCU actually monitors and regulates the main generator's output by controlling the amount of current that flows into the exciter field. For example, if additional output is needed, the GCU increases the amount of current flowing to the exciter field winding which, in turn, increases the exciter output. A higher exciter output increases the current flowing through the main generator field winding thereby increasing alternator output.

Since brushless alternators utilize a permanent magnet, there is no need to flash the field. In addition, the use of a permanent magnet eliminates the need to carry current to a rotating assembly through brushes.

ALTERNATOR RATINGS

AC alternators are rated in volt-amps which is a measure of the apparent power being produced by the generator. Because most AC alternators produce a great deal of power, their ratings are generally expressed in kilo-volt amperes or KVA. A typical Boeing 727 AC alternator is rated at 45 KVA.

FREQUENCY

The AC frequency produced by an AC generator is determined by the number of poles and the speed of the rotor. The faster the rotor, the higher the frequency. By the same token the more poles on a rotor, the higher the frequency for any given speed. The frequency of AC generated by an alternator is determined using the equation:

$$F = \frac{P}{2} \times \frac{N}{60} = \frac{PN}{120}$$

Where:
F = Frequency
P = the number of poles
N = the speed in rpm

With this formula you can determine that a two-pole, 3,600 rpm alternator has a frequency of 60 hertz.

$$F = \frac{2 \times 3,600}{120} = 60 \text{ Hz}$$

Since AC constantly changes rate and direction, it always produces a back voltage that opposes current flow. If you recall from your general studies, this opposition is called inductive reactance. The higher the AC frequency produced by an alternator, the more times the current switches direction and the greater the back voltage, or inductive reactance. Because aircraft systems use 400-hertz AC, inductive reactance is high and current is low. However, because of this higher frequency, motors wound with smaller wire produce a high torque value. Furthermore, transformers are made much smaller and lighter.

To provide a constant frequency as engine speed varies and maintain a uniform frequency between multiple generators, most AC generators are connected to a **constant speed drive** unit, or **CSD**. Although CSDs come in a variety of shapes and sizes, their principle of operation is essentially the same. The drive units consist of an engine-driven axial-piston variable-displacement hydraulic pump that supplies fluid to an axial-piston hydraulic motor. The motor then drives the generator. The displacement of the pump is controlled by a governor which senses the rotational speed of the AC generator. The governor action holds the output speed of the generator constant and maintains an AC frequency at 400-hertz, plus or minus established tolerances. [Figure 4-43]

Some modern jet aircraft produce AC with a generator called an **Integrated Drive Generator** or **IDG**.

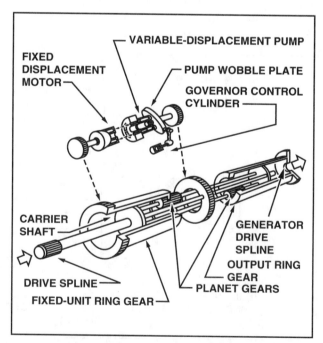

Figure 4-43. A constant-speed drive axial-gear differential, like the one above, is used in the Sunstrand Integrated Drive Generator.

An IDG differs from a CSD in that both the constant speed drive unit and the generator are sealed in the same housing. [Figure 4-44]

AC ALTERNATOR MAINTENANCE

Maintenance and inspection of alternator systems is similar to that of DC systems. The proper maintenance of an alternator requires the unit to be kept clean and all electrical connections tight and in good repair. Alternators and their drive systems differ in design and maintenance requirements; therefore, specific information is found in the manufacturer's service publications and in the maintenance program approved for a particular aircraft.

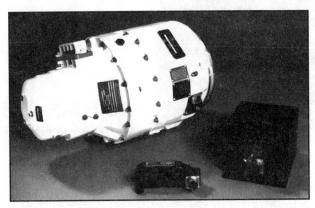

Figure 4-44. The constant speed drive unit is included in the housing with the Sunstrand Integrated Drive Generator above. Also shown is the generator's control unit and a current transformer assembly.

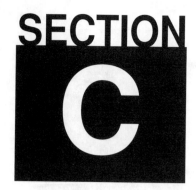
Many aircraft functions require an application of force greater than a pilot can perform manually. For example, raising and lowering the landing gear by hand or extending and retracting the flaps would take a great deal of time and effort on larger, high performance aircraft. Electric motors can perform this and many other operations quickly and easily.

DC MOTORS

Many devices in an airplane, from the starter to the auto-pilot, depend upon the mechanical energy furnished by direct-current motors. A DC motor is a rotating machine that transforms direct-current electrical energy into mechanical energy.

MOTOR THEORY

As you know, the lines of flux between two magnets flow from the north pole to the south pole. At the same time, when current flows through a wire, lines of flux set up around the wire. The direction these flux lines encircle the wire depends on the direction of current flow. When the wire's flux lines and the magnet's flux lines are placed together, a reaction occurs. For example, when the flux lines between two magnetic poles are flowing from left to right and the lines of flux encircling a wire between the mag-

netic poles flow in a counterclockwise direction, the flux lines reinforce each other at the bottom of the wire. This happens because the lines of flux produced by the magnet and the flow of flux lines at the bottom of the wire are traveling in the same direction. However, at the top of the wire the flux lines oppose, or neutralize, each other. The resulting magnetic field under the wire is strong and the magnetic field above the wire is weak. Consequently, the wire is pushed away from the side where the field is the strongest. [Figure 4-45]

Using this same principle, if the current flow through the wire were reversed, the flux lines encircling the wire would flow in the opposite direction. The resulting combination of the magnetic flux lines and wire flux lines would create a strong magnetic field at the top of the wire and a weak magnetic field at the bottom. Consequently, the wire is pushed downward away from the stronger field.

PARALLEL CONDUCTORS

When two current-carrying wires are in the vicinity of one another they exert a force on each other. This force is the result of the magnetic fields set up around each wire. When the current flows in the same direction, the resulting magnetic fields

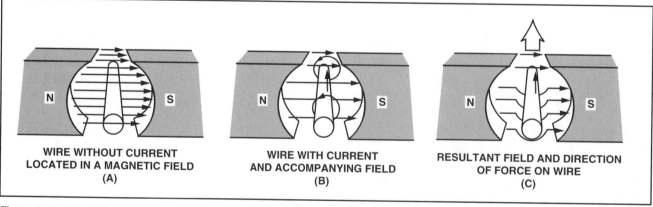

WIRE WITHOUT CURRENT
LOCATED IN A MAGNETIC FIELD
(A)

WIRE WITH CURRENT
AND ACCOMPANYING FIELD
(B)

RESULTANT FIELD AND DIRECTION
OF FORCE ON WIRE
(C)

Figure 4-45. (A) — When no current flows through a wire that is between two magnets, the lines of flux flow from north to south without being disturbed. (B) — When current flows through the wire, magnetic flux lines encircle it. (C) — The flux lines from the magnet and the flux lines encircling the wire react with one another to produce a strong magnetic field under the wire and a weak magnetic field above it.

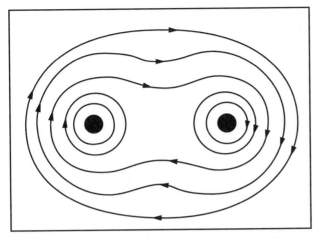

Figure 4-46. When two parallel wires have current flowing through them in the same direction they are forced in the direction of the weaker field, which is toward each other.

encompass both wires in a clockwise direction. These fields oppose each other and, therefore, cancel each other out. [Figure 4-46]

When the electron flow in the wires is opposite, the magnetic field around one wire radiates outward in a clockwise direction while the magnetic field around the second wire rotates counterclockwise. These fields combine, or reinforce each other between the wires. [Figure 4-47]

DEVELOPING TORQUE

When a current-carrying coil is placed in a magnetic field, the magnetic fields produced cause the coil to rotate. The force that produces rotation is called **torque**. [Figure 4-48]

The amount of torque developed in a coil depends on several factors including, the strength of the mag-

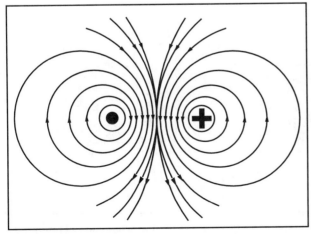

Figure 4-47. When the flow of electrons in two wires is opposite, the resulting magnetic fields force the wires apart.

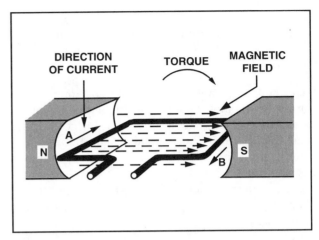

Figure 4-48. The coil above has current flowing inward on side A and outward on side B. Therefore, the magnetic field encircling wire A is counterclockwise while the magnetic field encircling wire B is clockwise. The resultant force pushes wire B downward, and wire A upward. This causes the coil to rotate until the wires are perpendicular to the magnetic flux lines. In other words, torque is created by the reacting magnetic fields around the coil.

netic field, the number of turns in the coil, and the position of the coil in the field.

The **right-hand motor rule** is used to determine the direction a current-carrying wire moves in a magnetic field. If you point your right index finger in the direction of the magnetic field and your second finger in the direction of current flow, your thumb indicates the direction the wire moves. [Figure 4-49]

BASIC DC MOTOR

Torque is the technical basis governing the construction of DC motors. Recall that a coil only rotates when it is at a 90 degree angle to the mag-

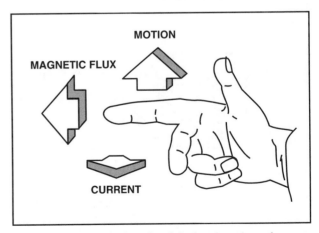

Figure 4-49. When using the right-hand motor rule, your right index finger indicates the direction the magnetic flux flows, while your second finger indicates the direction of current flow and your thumb the direction the coil rotates.

netic field produced by two magnets. Therefore, when a coil lines up with the magnetic field it does not rotate. This is because the torque at that point is zero.

The coil of a motor must rotate continuously in order for the motor to operate efficiently. Therefore, it is necessary for a device to reverse the coil current just as the coil becomes parallel with the magnet's flux lines. When the current is reversed, torque is again produced and the coil rotates. When a current-reversing device is set up to reverse the current each time the coil is about to stop, the coil rotates continuously.

One way to reverse the current in a coil is to attach a commutator similar to what is used on a generator. When this is done, the current flowing through the coil changes direction continuously as the coil rotates, thus preserving torque. [Figure 4-50]

A more effective method of ensuring continuous coil torque is to have a large number of coils wound on an armature. When this is done, the coils are spaced so that, for any position of the armature, a coil is near the magnet's poles. This makes torque

both continuous and strong. However, it also means that the commutator must contain several segments.

To further increase the amount of torque generated, the armature is placed between the poles of an electromagnet instead of a permanent magnet. This provides a much stronger magnetic field. Furthermore, the core of an armature is usually made of soft iron that is strongly magnetized through induction.

DC MOTOR CONSTRUCTION

The major parts in a practical motor are the armature assembly, the field assembly, the brush assembly, and the end frames. This arrangement is very similar to a DC generator.

ARMATURE ASSEMBLY

The armature assembly contains a soft iron core, coils, and commutator mounted on a rotatable steel shaft. The core consists of laminated stacks of soft iron that are insulated from each other. Solid iron is not used because it generates excessive heat that uses energy needlessly. The armature windings are

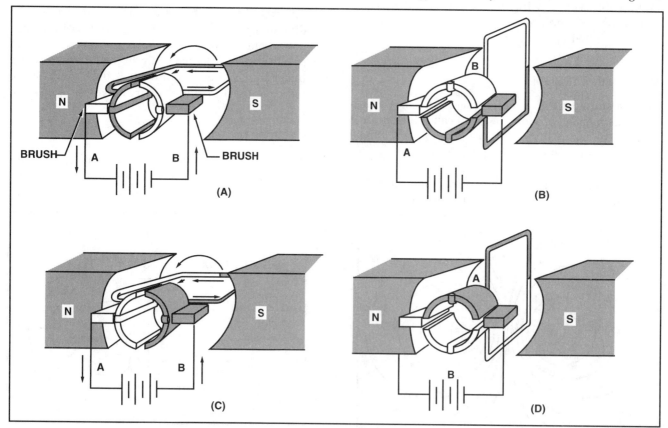

Figure 4-50. (A) — As current flows through the brushes to the commutator and coil, torque is produced and the coil rotates. **(B)** — As the coil becomes parallel with the magnetic lines of flux, each brush slides off one terminal and connects the opposite terminal to reverse the polarity. **(C)** — Once the current reverses, torque is again produced and the coil rotates. **(D)** — As the coil again becomes parallel with the flux lines, current is again reversed by the commutator and torque continues to rotate the coil.

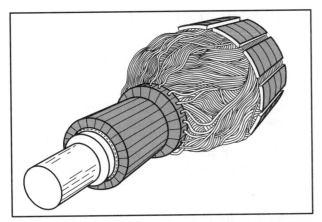

Figure 4-51. As you can see, the armature of a typical DC motor is similar to that of a DC generator.

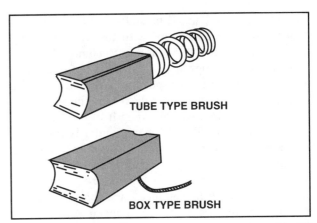

Figure 4-52. Two of the most common brushes used in DC motors are the tube type and box type.

made of insulated copper wire that is inserted into slots and protected by a fiber paper that is sometimes referred to as fish paper. The ends of the windings are physically connected to the commutator segments with wedges or steel bands. The commutator consists of several copper segments insulated from each other and the armature shaft by pieces of mica. Insulated wedge rings hold the segments in place. [Figure 4-51]

FIELD ASSEMBLY

The field assembly consists of the field frame, a set of pole pieces, and field coils. The field frame is located along the inner wall of the motor housing and contains the laminated steel pole pieces on which the field coils are wound. The field coils consist of several turns of insulated wire that fit over each pole piece. Some motors have as few as two poles, while others have as many as eight.

BRUSH ASSEMBLY

The brush assembly consists of brushes and their holders. The brushes are usually made of small blocks of graphitic carbon because of its long service life. The brush holders permit the brushes to move somewhat and utilize a spring to hold them against the commutator. [Figure 4-52]

END FRAME

The end frame is the part of the motor that the armature assembly rotates in. The armature shaft, which rides on bearings, extends through one end frame and is connected to the load. Sometimes the drive end frame is part of the unit driven by the motor.

MOTOR SPEED AND DIRECTION

Certain applications call for motors whose speed or direction are changeable. For example, a landing gear motor must be able to both retract and extend

the gear while a windshield wiper motor must have variable speeds to suit changing weather conditions. Certain internal or external changes need to be made in the motor design to allow these operations.

CHANGING MOTOR SPEED

A motor in which the speed is controlled is called a **variable speed motor** and is either a shunt or series motor. Motor speed is controlled by varying the current in the field windings. For example, when the amount of current flowing through the field windings is increased, the field strength increases, causing the armature windings to produce a larger counter EMF which slows the motor. Conversely, when the field current is decreased, the field strength decreases, and the motor speeds up because the counter EMF is reduced. [Figure 4-53]

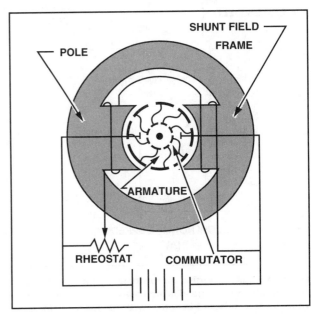

Figure 4-53. A shunt motor with variable speed control uses a rheostat to control the motor's speed.

In a shunt motor, speed is controlled by a rheostat that is connected in series with the field winding. Therefore, the speed depends on the amount of current flowing through the rheostat to the field windings. To increase motor speed, the resistance in the rheostat is increased. This decreases the field current which decreases the strength of the magnetic field and counter EMF. This momentarily increases the armature current and torque which, in turn, causes the motor to speed up until the counter EMF increases and causes the armature current to decrease to its former value. Once this occurs, the motor operates at a higher fixed speed.

To decrease motor speed, resistance in the rheostat is decreased. This action increases the current flow through the field windings and increases the field strength. The higher field strength causes a momentary increase in the counter EMF which decreases the armature current. As a result, torque decreases and the motor slows until the counter EMF decreases to its former value. Once the counter EMF and armature current are balanced, the motor operates at a lower fixed speed than before.

In a series motor, the rheostat speed control is connected in one of three ways. The rheostat is either connected in parallel or in series with the motor field, or in parallel with the motor armature. Each method of connection allows for operation in a specified speed range. [Figure 4-54]

REVERSING MOTOR DIRECTION

The direction of a DC motor's rotation is reversed by reversing the direction of current flow in either the armature or the field windings. In both cases, this reverses the magnetism of either the armature or the magnetic field the armature rotates in. If the wires connecting the motor to an external source are interchanged, the direction of rotation is not reversed since these wires reverse the magnetism of both the field and armature. This leaves the torque in the same direction.

One method for reversing the direction of rotation employs two field windings wound in opposite directions on the same pole. This type of motor is called a **split field motor**. A single-pole, double-throw switch makes it possible to direct current to either of the two windings. [Figure 4-55]

Some split field motors are built with two separate field windings wound on alternate poles. An example of this is the armature in a four-pole reversible motor. In this configuration, the armature rotates in one direction when current flows through one set of windings and in the opposite direction when current flows through the other set of windings.

Another method of reversal is called the **switch method**. This type of motor reversal employs a double-pole, double-throw switch that changes the

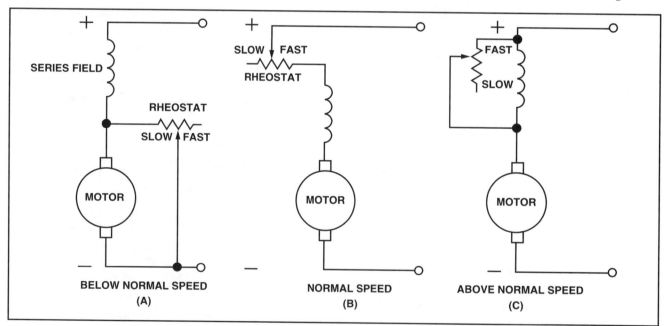

Figure 4-54. (A) — With a motor that is to be operated below normal speed, the rheostat is connected in parallel with the armature and the motor speed is increased by decreasing the current. **(B)** — When a motor is operated in the normal speed range, the rheostat is connected in series with the motor field. In this configuration, motor speed is increased by increasing the voltage across the motor. **(C)** — For above normal speed operation, the rheostat is connected in parallel with the series field. In this configuration, part of the voltage bypasses the series field causing the motor to speed up.

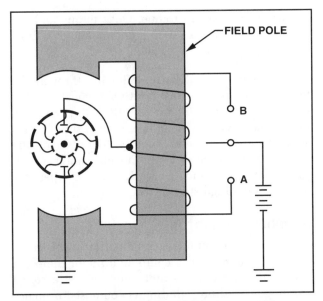

Figure 4-55. When the switch is in the lower position, current flows through the lower field winding creating a north pole at the lower field winding and at the lower pole piece. However, when the switch is placed in the up position, current flows through the upper field winding. This reverses the field magnetism and causes the armature to rotate in the opposite direction.

direction of current flow in either the armature or the field. [Figure 4-56]

TYPES OF DC MOTORS

DC motors are classified by the type of field-armature connection used and by the type of duty they are designed for. For example, there are three basic types of DC field-armature connections. They are the series, shunt, and compound.

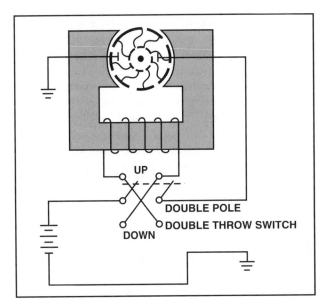

Figure 4-56. In the illustration above, a double-pole, double-throw switch is used to reverse the current through the field. When the switch is in the "up" position, current flows through the field windings. This establishes a north pole on the right side of the motor. When the switch is moved to the "down" position, polarity is reversed and the armature rotates in the opposite direction.

SERIES DC MOTOR

In a series motor, the field windings consist of heavy wire with relatively few turns that are connected in series with the armature winding. This means the same amount of current flows through the field windings and the armature windings. In this configuration, an increase in current causes a corresponding increase in the magnetism of both the field and armature. [Figure 4-57]

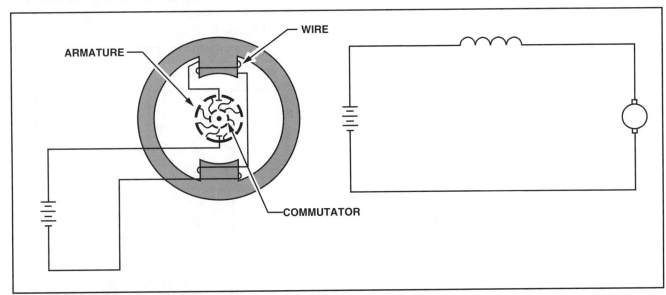

Figure 4-57. Since the field windings and armature in a series motor are connected in series, an increase of current through the field windings results in an increase of current in the armature.

The series motor is able to draw a large starting current because of the winding's low resistance. This starting current passes through both the field and armature windings and, therefore, produces a high starting torque. For this reason, series motors are often used in aircraft as starters and for raising and lowering landing gear, cowl flaps, and wing flaps. However, as the speed of a series motor increases, the counter EMF builds and opposes the applied EMF. This, in turn, reduces the current flow through the armature which reduces the current draw.

The speed of a series motor depends on the load applied. Therefore, any change in load is accompanied by a substantial change in speed. In fact, if the load is removed entirely, a series motor will operate at an excessively high speed and the armature could fly apart. In other words, a series motor needs resistance to stay within a safe operating range.

SHUNT DC MOTOR

In a shunt motor, the field winding is connected in parallel with the armature winding. To limit the amount of current that passes through the field, the resistance is high in the field winding. In addition, because the field is connected directly across the power supply, the amount of current that passes through the field is constant. In this configuration, when a shunt motor begins to rotate, most of the current flows through the armature, while relatively little current flows to the field. Because of this, shunt motors develop little torque when they are first started. [Figure 4-58]

As a shunt-wound motor picks up speed, the counter EMF in the armature increases causing a decrease in the amount of current draw in the armature. At the same time, the field current increases slightly which causes an increase in torque. Once torque and the resulting EMF balance each other, the motor will be operating at its normal, or rated, speed.

Since the amount of current flowing through the field windings remains relatively constant, the speed of a shunt motor varies little with changes in load. In fact, when no load is present, a shunt motor assumes a speed only slightly higher than the loaded speed. Because of this, a shunt motor is well suited for operations where a constant speed is desired and a high starting torque is not.

COMPOUND DC MOTOR

The compound motor is a combination of the series and shunt motors. In a compound motor there are two field windings: a shunt winding and a series winding. The shunt winding is composed of many turns of fine wire and is connected in parallel with the armature winding. On the other hand, the series winding consists of a few turns of large wire and is connected in series with the armature winding. The starting torque is higher in a compound motor than in a shunt motor and lower than in the series motor. Furthermore, variation of speed with load is less than in a series-wound motor but greater than in a shunt motor. The compound motor is used whenever the combined characteristics of the series and shunt motors are desired. [Figure 4-59]

TYPE OF DUTY

Electric motors must operate under various conditions. For example, some motors are used for intermittent operations while others operate continuously. In most cases, motors built for intermittent duty may only be operated for short periods of time before they must be allowed to cool. On the other

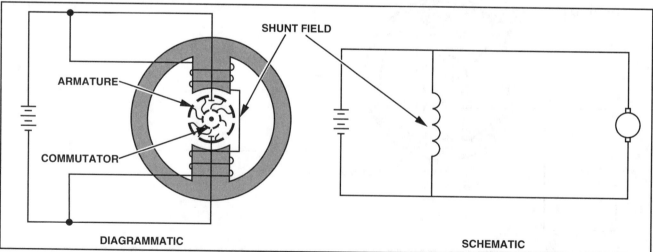

Figure 4-58. In a shunt-wound motor, the field winding is connected in parallel with the armature winding. Because of this, the amount of current that flows to the field when the motor is started is limited, and the resulting torque is low.

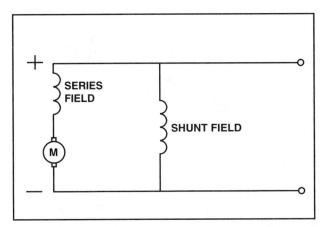

Figure 4-59. In a compound motor, one field winding is connected to the armature winding in series while the other is connected in parallel.

hand, motors built for continuous duty are operable at their rated power for long periods.

ENERGY LOSSES IN MOTORS

When electrical energy is converted to mechanical energy in a motor, some losses do occur. By the same token, losses also occur when mechanical energy is converted to electrical energy. Therefore, in order for machines to be efficient, both electrical and mechanical losses must be kept to a minimum. **Electrical losses** are classified as either copper losses or iron losses, while **mechanical losses** originate from the friction of various moving parts.

Copper losses occur when electrons are forced through the copper armature and field windings. These losses occur because some power is dissipated in the form of heat due to the inherent resistance possessed by copper windings. The amount of loss is proportional to the square of the current and is calculated with the formula:

$$\text{Copper Loss} = I^2R$$

Iron losses are divided into hysteresis and eddy current losses. **Hysteresis losses** result from the armature revolving in an alternating magnetic field and becoming magnetized in two directions. Since some residual magnetism remains in the armature after its direction is changed, some energy loss does occur. However, since the field magnets are always magnetized in one direction by DC current, they produce no hysteresis losses. **Eddy current losses** occur because the armature's iron core acts as a conductor revolving in a magnetic field. This sets up an EMF across portions of the core causing currents to flow within the core. These currents heat the core and, when excessive, can damage the windings. To keep eddy current losses to a

minimum, a laminated core made of thin, insulated sheets of iron is used. The thinner the laminations, the greater the reduction in eddy current losses.

INSPECTION AND MAINTENANCE OF DC MOTORS

The inspection and maintenance of DC motors should be in accordance with the guidelines established by the manufacturer. The following is indicative of the types of maintenance checks typically called for:

1. Check the unit driven by the motor in accordance with the specific installation instructions.
2. Check all wiring, connections, terminals, fuses, and switches for general condition and security.
3. Keep motors clean and mounting bolts tight.
4. Check the brushes for condition, length, and spring tension. Procedures for replacing brushes, along with their minimum lengths, and correct spring tensions are given in the applicable manufacturer's instructions. If the spring tension is too weak, the brush could begin to bounce and arc causing commutator burning.
5. Inspect the commutator for cleanliness, pitting, scoring, roughness, corrosion, or burning. Check the mica between each of the commutator segments. If a copper segment wears down below the mica, the mica will insulate the brushes from the commutator. Clean dirty commutators with the recommended cleaning solvent and a cloth. Polish rough or corroded commutators with fine sandpaper (000 or finer) and blow out with compressed air. Never use emery paper because it contains metal particles which can cause shorts. Replace the motor if the commutator is burned, badly pitted, grooved, or worn to the extent that the mica insulation is flush with the commutator surface.
6. Inspect all exposed wiring for evidence of overheating. Replace the motor if the insulation on the leads or windings is burned, cracked, or brittle.
7. Lubricate only if called for by the manufacturer's instructions. Most motors used today do not require lubrication between overhauls.
8. Adjust and lubricate the gearbox or drive unit in accordance with the applicable manufacturer's instructions.

Troubleshoot any problems and replace the motor only when the trouble is due to a defect in the motor itself. In most cases, motor failure is caused by a defect in the external electrical circuit or by mechanical failure in the mechanism driven by the motor.

AC MOTORS

AC motors have several advantages over DC motors. For example, in many instances AC motors do not use brushes or commutators and, therefore, cannot spark like a DC motor. Furthermore, AC motors are well suited for constant-speed applications although some are manufactured with variable speed characteristics. Other advantages some AC motors have include their ability to operate on single or multiple phase lines as well as at several voltages. In addition, AC motors are generally less expensive than comparable DC motors. Because of these advantages, many aircraft are designed to use AC motors.

Because the subject of AC motors is very extensive, this text does not attempt to cover everything. However, those types of AC motors common to aircraft systems are covered in detail.

Because aircraft electrical systems typically operate at 400 hertz AC, an electric AC motor operates at about seven times the speed of a 60-hertz commercial motor with the same number of poles. In fact, a 400-hertz induction type motor typically operates at speeds ranging from 6,000 rpm to 24,000 rpm. This high rotation speed makes AC motors suitable for operating small high-speed rotors. Furthermore, through the use of reduction gears, AC motors are made to lift and move heavy loads such as wing flaps and retractable landing gear, as well as produce enough torque to start an engine.

TYPES OF AC MOTORS

There are three basic types of AC motors. They are the universal motor, the induction motor, and the synchronous motor. Each type represents a variation on basic motor operating principles.

UNIVERSAL MOTORS

Fractional horsepower AC series motors are called universal motors. A unique characteristic of universal motors is that they can operate on either alternating or direct current. In fact, universal motors resemble DC motors in that they have brushes and a commutator. Universal motors are used extensively to operate fans and portable tools like drills, grinders, and saws. [Figure 4-60]

INDUCTION MOTORS

The most popular type of AC motor is the induction motor. In an induction motor there is no need for an electrical connection between the motor housing and the rotating elements. Therefore, there are no

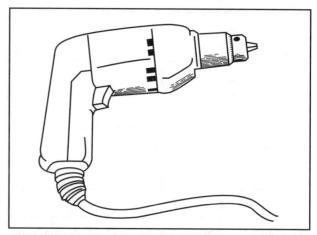

Figure 4-60. An electric drill uses a universal motor which is similar in construction to a series-wound DC motor.

brushes, commutators, or slip rings to contend with. Induction motors operate at a fixed rpm that is determined by their design and the frequency of AC applied. In addition, an induction motor can be operated on either single-phase or three-phase alternating current.

A single-phase induction motor is used to operate devices like surface locks, intercooler shutters, oil shutoff valves, and places where the power requirements are low. Single-phase induction motors require some form of starting circuit that automatically disconnects after the motor is running. Single-phase induction motors operate well in either rotational direction, with the direction determined by the starting circuit.

Unlike single-phase induction motors, three-phase induction motors are self-starting and are commonly used when high power is needed. Common applications for three-phase induction motors include engine starting, operating flaps and landing gear, and powering hydraulic pumps.

Construction

The two primary parts of an induction motor are the stator and the rotor. The stator is unique in the fact that instead of having field poles that extend outward, windings are placed in slots around the stator's periphery. These windings comprise a series of electromagnets that produce a magnetic field.

The rotor of an induction motor consists of an iron core made of thin circular laminations of soft steel that are keyed in to a shaft. Longitudinal slots are cut into the rotor's circumference and heavy copper or aluminum bars are embedded in them. These bars are welded to a heavy ring of high conductivity on either end. [Figure 4-61]

When AC is applied to the stator, the strength and polarity of the electromagnets changes with the excitation current. Furthermore, to give the effect of a rotating magnetic field, each group of poles is attached to a separate phase of voltage.

When the rotor of an induction motor is subjected to the revolving magnetic field produced by the stator windings, a voltage is induced in the longitudinal bars. This induced voltage causes current to flow through the bars and produce its own magnetic field which combines with the stator's revolving field. As a result, the rotor revolves at nearly a synchronous speed with the stator field. The only difference in the rotational speed between the stator field and the rotor is that necessary to induce the proper current into the rotor to overcome mechanical and electrical losses.

If a rotor were to turn at the same speed as the rotating field, a resonance would set up. When this happens, the rotor conductors are not cut by any magnetic lines of flux and no EMF is induced into them. Thus, no current flows in the rotor, resulting in no torque and little rotor rotation. For this reason, there must always be a difference in speed between the rotor and the stator's rotating field.

The difference in rotational speed is called **motor slip** and is expressed as a percentage of the synchronous speed. For example, if the rotor turns at 1,750 rpm and the synchronous speed is 1,800 rpm, the difference in speed is 50 rpm. The slip is therefore equal to 50/1,800 or 2.78 percent.

Single-Phase Induction Motor

A single-phase motor differs from a multi-phase motor in that the single-phase motor has only one stator winding. In this configuration, the stator winding generates a field that pulsates. This generates an expanding and collapsing stator field that induces currents into the rotor. These currents generate a rotor field opposite in polarity to that of the stator. The opposition of the field exerts a turning force on the upper and lower parts of the rotor which try to turn it 180 degrees from its position. Since these forces are exerted in the center of the rotor, the turning force is equal in each direction. As a result, the rotor will not begin turning from a standing stop. However, if the rotor starts turning, it continues to rotate. Furthermore, since the turning force is aided by the rotor momentum, there is no opposing force.

Shaded-Pole Induction Motor

The first effort in the development of a self-starting, single-phase motor was the shaded-pole induction motor. Like the generator, the shaded-pole motor has field poles that extend inward from the motor housing. In addition, a portion of each pole is encircled with a heavy copper ring. [Figure 4-62]

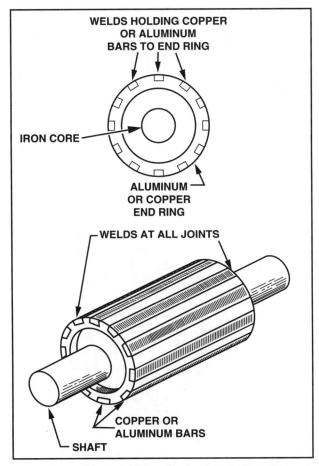

Figure 4-61. The complete rotor in an induction motor is sometimes called a squirrel cage and, therefore, motors containing this type of rotor are often called squirrel cage induction motors.

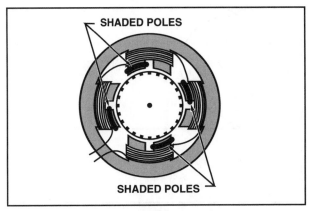

Figure 4-62. In a shaded-pole induction motor, a portion of each salient pole is encircled by a copper ring.

The presence of the copper ring causes the magnetic field through the ringed portion of the pole face to lag appreciably behind that of the other half of the pole. This results in a slight component of rotation in the field that is strong enough to cause rotation. Although the torque created by this field is small, it is enough to accelerate the rotor to its rated speed. [Figure 4-63]

Split-Phase Motor

There are various types of self-starting motors, one type is known as split-phase motor. Split-phase motors have a winding that is dedicated to starting the rotor. This "start" winding is displaced 90 degrees from the main, or run, winding and has a fairly high resistance that causes the current to be out of phase with the current in the run winding. The out of phase condition produces a rotating field that makes the rotor revolve. Once the rotor attains approximately 25 percent of its rated speed, a centrifugal switch disconnects the start winding automatically.

Capacitor-Start Motor

With the development of high-capacity electrolytic capacitors, a variation of the split-phase motor was made. Motors that use high-capacity electrolytic capacitors are known as capacitor-start motors. Nearly all fractional horsepower motors in use today on refrigerators, oil burners, and other similar appliances are of this type.

In a capacitor-start motor, the start and run windings are the same size and have identical resistance

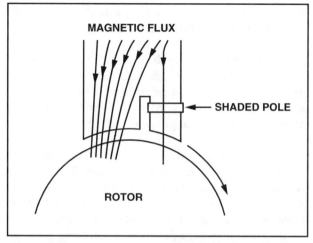

Figure 4-63. The portion of magnetic flux lines that pass through the shaded pole lag behind the opposite pole, thereby creating a slight component of rotation.

values. The phase shift between the two windings is obtained by using capacitors connected in series with the start winding. [Figure 4-64]

Capacitor-start motors have a starting torque comparable to their rated speed torque and are used in applications where the initial load is heavy. Again, a centrifugal switch is required for disconnecting the start winding when the rotor speed is approximately 25 percent of the rated speed.

Although some single-phase induction motors are rated as high as 2 horsepower, most produce 1 horsepower or less. A voltage rating of 115 volts for the smaller sized motors and 110 to 220 volts for one-fourth horsepower and up is normally sufficient. Poly-phase motors are used for higher power

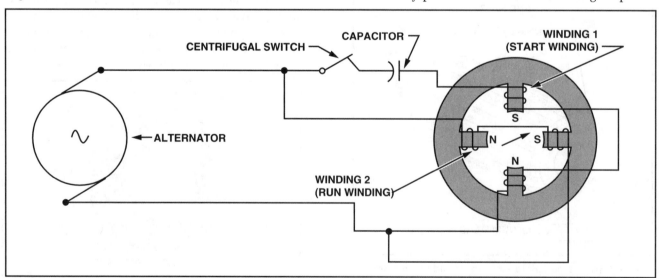

Figure 4-64. A single-phase motor with capacitor start windings has a capacitor connected in series with an alternator and the start winding.

ratings since they have much better starting torque characteristics.

Direction of Rotation

The direction of rotation for a three-phase induction motor is changed by reversing two of the motor leads. The same effect is obtained in a two-phase motor by reversing the connections on one phase. In a single-phase motor, reversing the connections to the start winding reverses the direction of rotation. Most single-phase motors designed for general application are built so you can readily reverse the connections to the start winding. On the other hand, a shaded-pole motor cannot be reversed because its rotational direction is determined by the physical location of the copper shaded ring.

After starting, if one phase of a three-phase motor is rendered useless, the motor will continue to run. However, it delivers only one-third of its rated power. On the other hand, a two-phase motor that loses one phase delivers only one-half its rated power. One thing to keep in mind is that neither motor will start if a wire connection to a phase is broken.

SYNCHRONOUS MOTORS

Like the induction motor, a synchronous motor uses a rotating magnetic field. However, the torque developed by a synchronous motor does not depend on the induction of currents in the rotor. Instead, the principle of operation of the synchronous motor begins with a multi-phase source of AC applied to a series of stator windings. When this is done, a rotating magnetic field is produced. At the same time, DC current is applied to the rotor winding producing a second magnetic field. A synchronous motor is designed so that the rotor is pulled by the stator's rotating magnetic field. The rotor turns at approximately the same speed as the stator's magnetic field. In other words, they are synchronized.

To understand the operation of a synchronous motor, use the following example. Assume that poles A and B in figure 4-65 are physically rotated clockwise in order to produce a rotating magnetic field. These poles induce the opposite polarity in the soft-iron rotor between them thereby creating an attraction between the rotating poles and the rotor. This attraction allows the rotating poles to drag the rotor at the same speed. [Figure 4-65]

When a load is applied to the rotor shaft, its axis momentarily falls behind that of the rotating field.

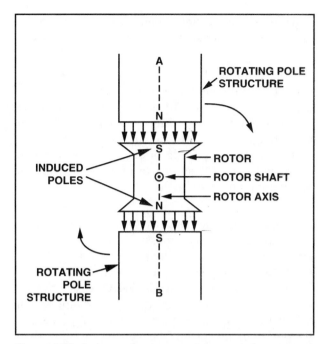

Figure 4-65. In a synchronous motor, a rotating magnet induces opposite magnetic fields in a soft iron rotor. The rotor then turns at the same speed as the magnet.

However, the rotor catches up and again rotates with the field at the same speed, as long as the load remains constant. If the load is too large, the rotor pulls out of sync with the rotating poles and is unable to rotate at the same speed. In this situation the motor is said to be overloaded.

The idea of using a mechanical means to rotate the poles is impractical because another motor would be required to do this. Therefore, a rotating magnetic field is produced electrically by using phased AC voltages. In this respect, a synchronous motor is similar to the induction motor.

The synchronous motor consists of a stator field winding that produces a rotating magnetic field. The rotor is either a permanent magnet or an electromagnet. If permanent magnets are used, the rotor's magnetism is stored within the magnet. On the other hand, if electromagnets are used, the magnets receive power from a DC power source through slip rings.

Since a synchronous motor has little starting torque, it requires assistance to bring it up to synchronous speed. The most common method of doing this is to start the motor with no load, allow it to reach full speed, and then energize the magnetic field. The magnetic field of the rotor then locks with the magnetic field of the stator and the motor operates at synchronous speed. [Figure 4-66]

The magnitude of the induced rotor poles is so small that sufficient torque cannot be developed for most practical loads. To avoid the limitation on motor operation, a winding is placed on the rotor and energized with direct current. To adjust the motor for varying loads, a rheostat is placed in series with the DC source to provide a means of varying the pole's strength.

A synchronous motor is not self-starting. Since rotors are heavy, it is impossible to bring a stationary rotor into magnetic lock with a rotating magnetic field. As a result, all synchronous motors have some kind of starting device. One type of simple starter used is another AC or DC motor which brings the rotor up to approximately 90 percent of its synchronous speed. The starting motor is then disconnected and the rotor locks in with the rotating field. Another method utilizes a second squirrel-cage type winding on the rotor. This induction winding brings the rotor almost to synchronous speed before the direct current is disconnected from the rotor windings and the rotor is pulled into sync with the field.

MAINTENANCE OF AC MOTORS

The inspection and maintenance of AC motors is very simple. For example, since most bearings are sealed and lubricated at the factory, they generally require no further attention. Be sure coils are kept dry and free from oil or other abuse.

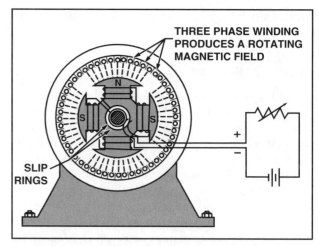

Figure 4-66. The weak field induced in the rotor poles limits the poles ability to produce torque. This problem is overcome by applying DC current through a rheostat to vary the pole's field strength.

The temperature of a motor is its only operating factor. A good rule of thumb to use is "if a motor is too hot to touch, it's too hot for safety."

Next to temperature, the sound of a motor or generator is the best trouble indicator. When operating properly, a motor should hum evenly. On the other hand, if a motor is overloaded it "grunts." A three-phase motor with one lead disconnected refuses to turn and "growls." If a knocking sound is present it generally indicates a loose armature coil, shaft out of alignment, or an armature dragging because of worn bearings.

The inspection and maintenance of all AC motors should be performed in accordance with the applicable manufacturer's insturctions.

CHAPTER 5

AIRCRAFT DRAWINGS

INTRODUCTION

When an aircraft is conceived in the minds of engineers and designers, it is nothing but an idea until put on paper. After a drawing is made others can add their expertise and ideas. These thoughts can then be translated into an aluminum or steel part. When an aircraft is designed, detailed drawings of every single part are made on translucent vellum or Mylar. Once the drawings are made, they are sent to the reproduction department. Here, copies are printed by a process called blueprinting, where the black lines from the translucent vellum or Mylar are printed as blue lines on paper. Every aircraft drawing is numbered so that a careful record is maintained. This way, if any changes are required, the information is passed on to affected personnel who incorporate the changes into the parts as they are built.

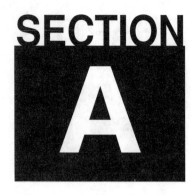

SECTION A

DRAWINGS

Typically, all aircraft factories have a drawing room manual that details all lines, symbols, and conventions used by that company. However, in an effort to establish an industry standard, applications for aircraft drawings have been standardized. It is these applications and standards that are covered by this text.

TYPES OF DRAWINGS

As an aircraft technician there are several types of drawings and graphic representations you must become familiar with. Each type of drawing is designed to transmit a certain piece of information. The most common type of drawing you will use is the **working drawing**. There are three classes of

working drawings, the detail drawing, the assembly drawing, and the installation drawing. Other types of drawings include sectional drawings, exploded-view drawings, block diagrams, logic flowcharts, electrical wiring diagrams, pictorial diagrams, and schematic diagrams. Each type of drawing is designed to transmit a certain type of information.

DETAIL DRAWINGS

When an aircraft is designed, a detail drawing is made for every part. A detail drawing supplies all the information required to construct a part, including all dimensions, materials, and type of finish. When needed, an enlarged section or a drawing of another view is added to make the drawing easier to understand. [Figure 5-1]

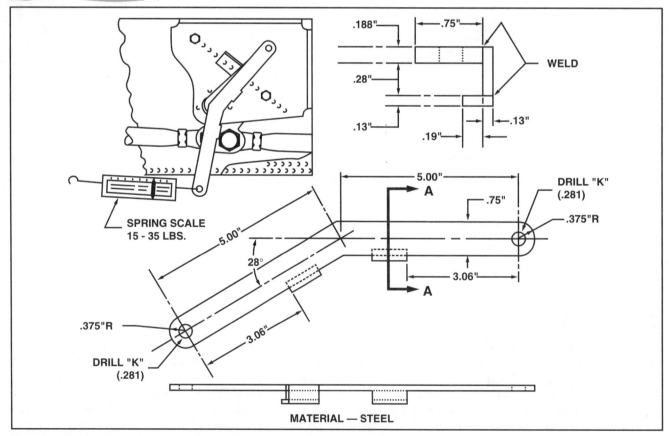

Figure 5-1. A detail drawing includes enough information to fabricate an individual part. If necessary, another view is added to enhance understanding.

When a detail drawing is made, it is carefully and accurately drawn to scale and dimensioned. However, when a print is made, the paper the copy is made on tends to shrink or stretch. Therefore, a measurement should never be scaled from a print. Instead, all measurements should be derived from the dimensions given.

ASSEMBLY DRAWING

After individual parts are fabricated, they are assembled into various subassemblies with the aid of an assembly drawing. An assembly drawing depicts the relationship between two or more parts. These drawings reference individual parts by their part number and specify the type and number of fasteners needed to join them. Because there are detail drawings for each component, no materials are specified and only those dimensions needed to assemble the parts are included. [Figure 5-2]

INSTALLATION DRAWINGS

All subassemblies are brought together in an installation drawing. This type of drawing shows the general arrangement or position of parts with respect to an aircraft and provides the information needed to install them. Like the assembly drawing, the bill of material on an installation drawing lists the fasteners needed, as well as any instructions required for the installation. Dimensions are given only for those adjustments necessary for the part to function. Often, portions of an aircraft that are not involved in the installation are shown using phantom lines. This helps you locate where a part is installed. Parts that are used only as a reference are

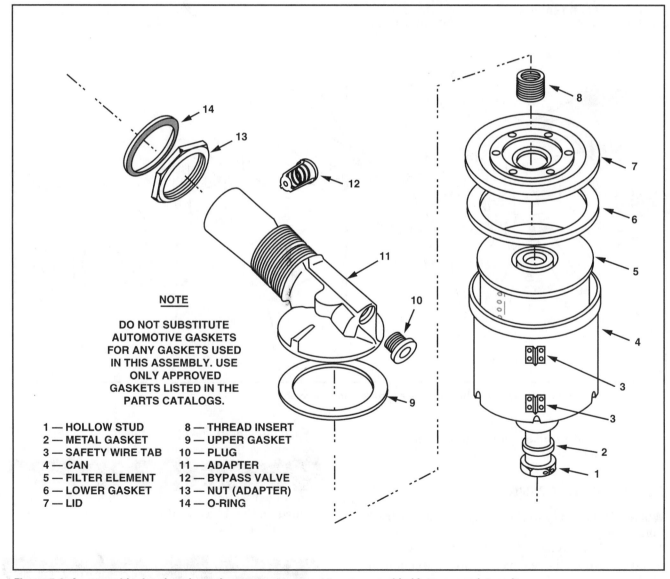

NOTE

DO NOT SUBSTITUTE AUTOMOTIVE GASKETS FOR ANY GASKETS USED IN THIS ASSEMBLY. USE ONLY APPROVED GASKETS LISTED IN THE PARTS CATALOGS.

1 — HOLLOW STUD
2 — METAL GASKET
3 — SAFETY WIRE TAB
4 — CAN
5 — FILTER ELEMENT
6 — LOWER GASKET
7 — LID
8 — THREAD INSERT
9 — UPPER GASKET
10 — PLUG
11 — ADAPTER
12 — BYPASS VALVE
13 — NUT (ADAPTER)
14 — O-RING

Figure 5-2. An assembly drawing shows how two or more parts are assembled into a complete unit.

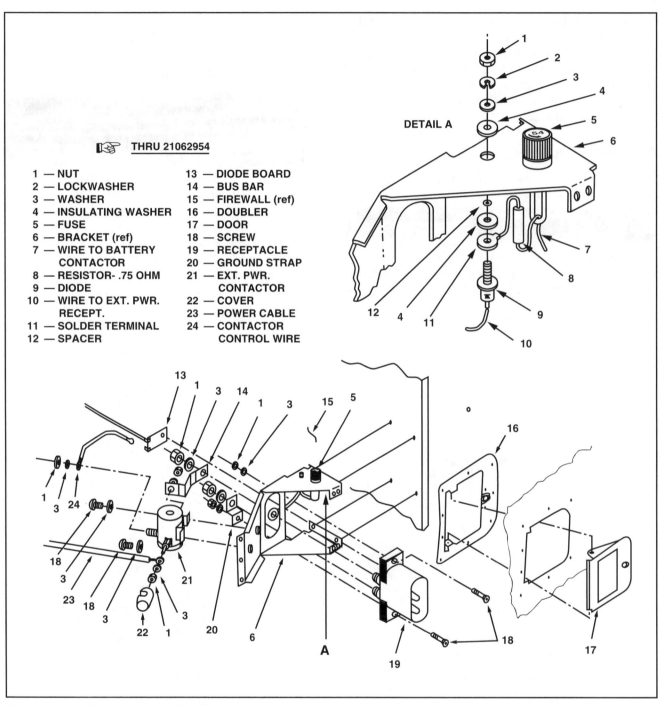

Figure 5-3. Installation drawings show how parts are installed on an aircraft.

often identified by their part name and the word "Ref" is noted beside it. [Figure 5-3]

SECTIONAL DRAWINGS

When it is necessary to show the internal construction or shape of a part a sectional drawing is used. There are four types of sectional drawings, the revolved section, the removed section, the complete section, and the half section.

When only the shape of a part needs to be shown, it is shown with either a revolved or removed section. The **revolved section** drawing is often used to illustrate simple items with no interior parts. Basically, a revolved section drawing shows how a part is sectioned and revolved to illustrate it from a different view. [Figure 5-4]

Like the revolved section drawing, the **removed section** drawing is also used to illustrate simple

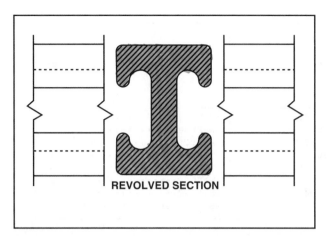

Figure 5-4. In a revolved section, a portion of an object is turned or revolved to show a different view. Here, the I-beam has been broken in two places with long break lines and the cross section is shown between the breaks.

objects. However, to do this, the object is cut by a cutting plane line and a section is removed to illustrate another angle. [Figure 5-5]

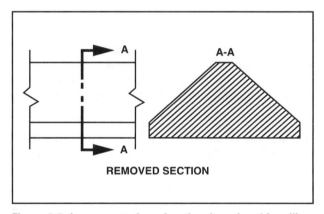

Figure 5-5. In a removed section drawing, the object illustrated is cut and a section is removed to illustrate another angle.

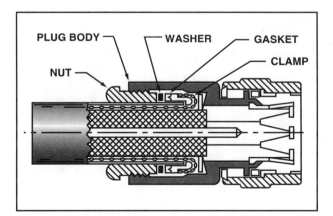

Figure 5-6. The use of a sectional view to illustrate a cable connector makes it easy to identify the unit's separate parts.

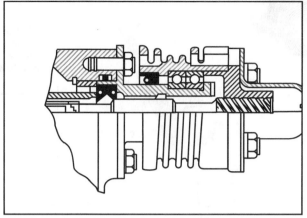

Figure 5-7. The half-sectional view allows the inside and outside of a part to be seen at the same time.

Complex assemblies like cable connectors are typically shown in **complete sectional views**. With this type of view, it is easy to identify individual parts within a complex assembly. This feature is further enhanced through the proper use of section lines. [Figure 5-6]

When it is helpful to see the outside of a part as well as the inside, a **half-sectional view** is made. With this type of drawing, typically the upper half of a drawing shows the internal construction of the assembly, while the lower half shows the entire assembly as it appears from the outside. [Figure 5-7]

EXPLODED-VIEW DRAWING

Illustrated parts drawings often make use of exploded view drawings to show every part in an assembly. In this type of drawing, all parts are typically in their relative positions and expanded outward. Each part is identified by both its physical appearance and its reference number which is used on the parts list. [Figure 5-8]

BLOCK DIAGRAMS

With electrical systems and electronic components becoming more complex, procedures and graphical aids have been developed to aid you in locating problems. One such aid is the block diagram. A block diagram consists of individual blocks that represent several components such as a printed circuit board or some other type of replaceable module. Since most of the maintenance needed on complex systems consists of identifying a malfunctioning subassembly and replacing it, block diagrams greatly enhance this process. When using a block

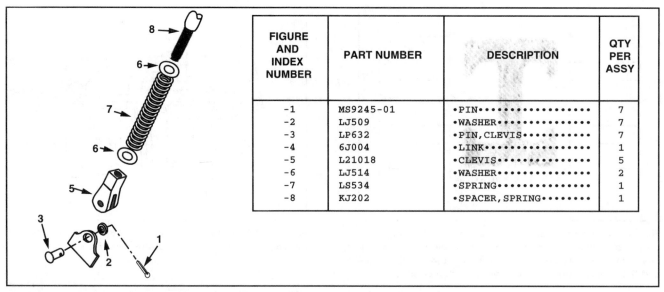

FIGURE AND INDEX NUMBER	PART NUMBER	DESCRIPTION	QTY PER ASSY
-1	MS9245-01	•PIN••••••••••••••••	7
-2	LJ509	•WASHER••••••••••••	7
-3	LP632	•PIN,CLEVIS••••••••••	7
-4	6J004	•LINK•••••••••••••••	1
-5	L21018	•CLEVIS•••••••••••••	5
-6	LJ514	•WASHER••••••••••••	2
-7	LS534	•SPRING••••••••••••	1
-8	KJ202	•SPACER,SPRING••••••••	1

Figure 5-8. Exploded-view drawings are typically found in illustrated parts catalogs. They show a part's relative position within a unit.

diagram you must trace the problem to the module that receives the correct input, but does not produce the required output. Once this is done, the module is removed as a whole and replaced. [Figure 5-9]

LOGIC FLOWCHARTS

Logic flowcharts are another aid used in troubleshooting. A logic flowchart represents the mechanical, electrical, or electronic action of a system without expressing construction or engineering information. When using a logic flowchart, go to the oblong START symbol and follow the arrows through the logical testing sequence.

On most flow charts rectangular boxes explain a procedure, while diamonds identify questions that require a specific answer. In other words, after using a rectangular box to test something, you must match the existing condition before proceeding to the next course of action. Each diamond has one input and at least two outputs. In order to assure that all discrepancies are addressed, you must follow a flow chart to the oblong END OF TEST symbol.

In addition to identifying the probable cause of a problem, many flow charts specify a fix for each circumstance. By using this information, troubleshooting time is reduced to a minimum. [Figure 5-10]

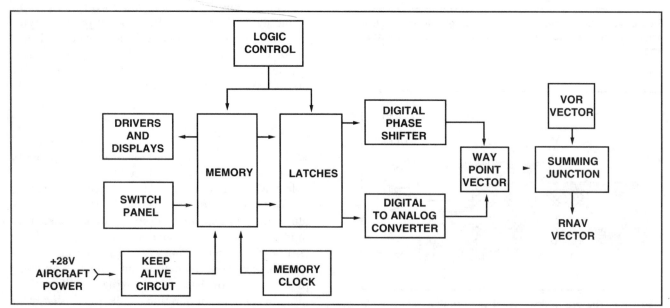

Figure 5-9. The module that receives the correct input but does not produce the required output is the unit to remove and replace.

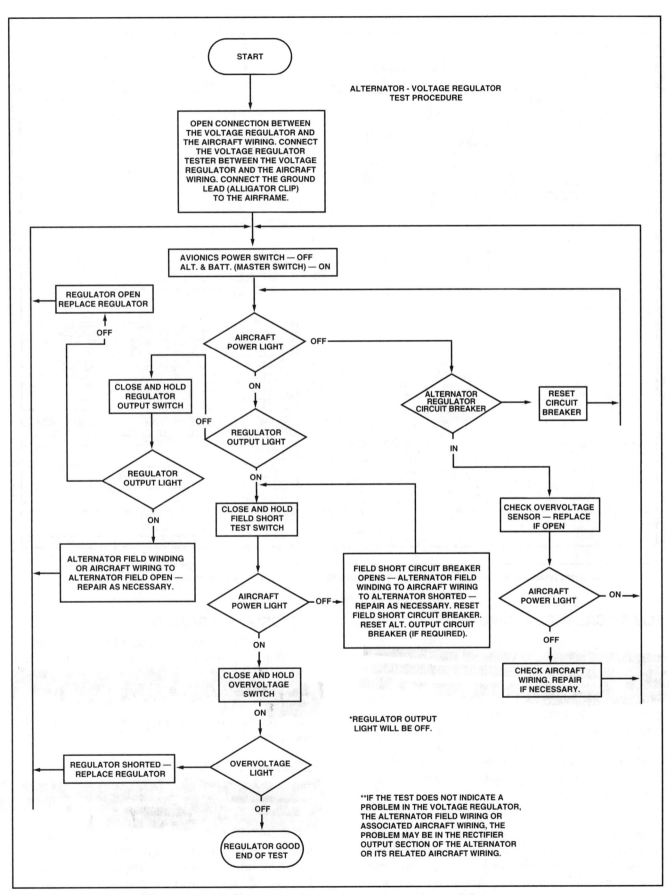

Figure 5-10. To aid in troubleshooting, logic flowcharts give step-by-step instructions to follow.

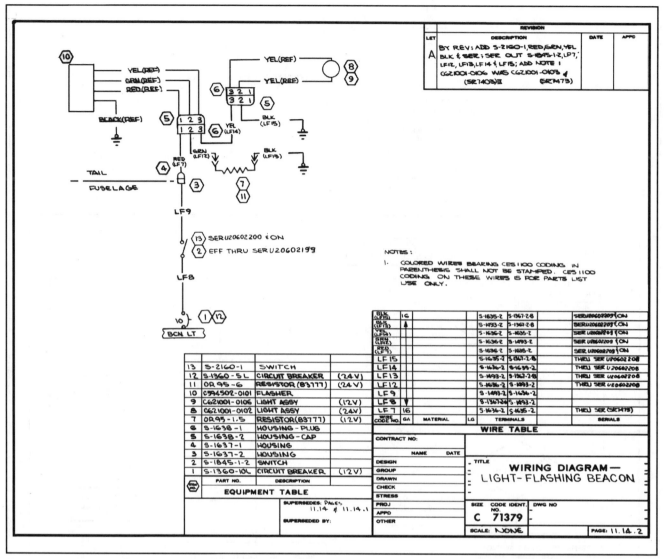

Figure 5-11. To identify and troubleshoot electrical parts and circuits an electrical wiring diagram is typically used.

ELECTRICAL WIRING DIAGRAMS

Electrical wiring diagrams are included in most aircraft service manuals and specify things like the size of wire and type of terminals to be used for a particular application. Furthermore, wiring diagrams typically identify each component within a system by its part number and its serial number, including any changes that were made during the production run of an aircraft. Because of this, wiring diagrams are extremely valuable for troubleshooting. [Figure 5-11]

There are several types of electrical wiring diagrams. Some diagrams show only one circuit while others show several circuits within a system. More detailed diagrams show the connection of wires at splices or the arrangement of parts.

PICTORIAL DIAGRAMS

Pilot's handbooks and some training manuals often use pictorial diagrams of electrical and hydraulic systems. In a pictorial diagram pictures of components are used instead of the conventional electrical symbols found in schematic diagrams. In most cases, pictorial diagrams help a person visualize the operation of a specific system. [Figure 5-12]

SCHEMATIC DIAGRAMS

A schematic diagram is used to illustrate a principle of operation and, therefore, does not show parts as they actually appear or function. However, schematic diagrams do indicate the location of components with respect to each other and in the case of a hydraulic system, the direction of fluid flow. Because of this, schematic diagrams are best utilized for troubleshooting. [Figure 5-13]

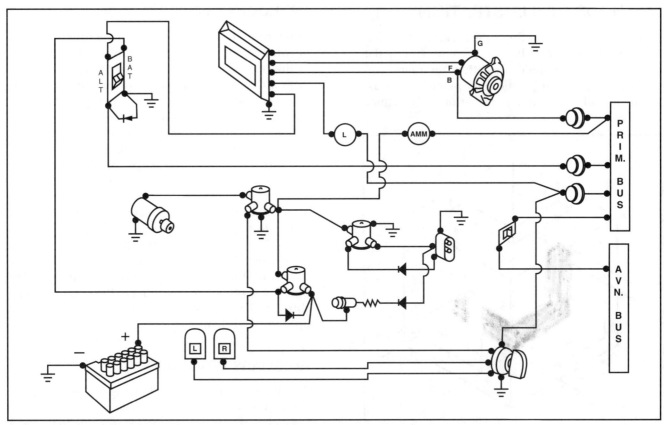

Figure 5-12. This pictorial diagram of an electrical system allows you to visualize the components involved in the system.

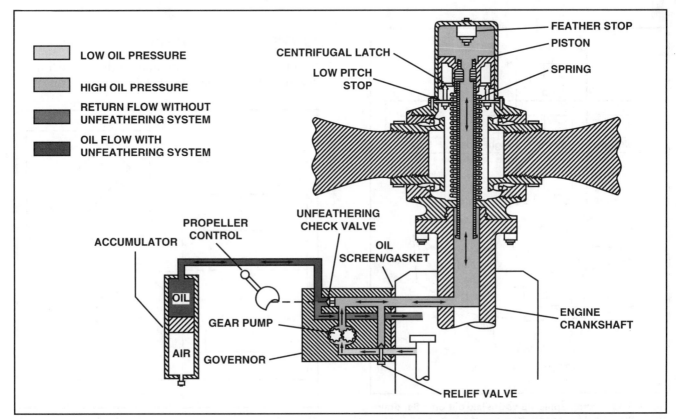

Figure 5-13. This schematic diagram of a constant-speed propeller illustrates all components with respect to each other to show how a constant speed propeller operates.

METHODS OF ILLUSTRATION

The methods of illustrating a part refer to its orientation with respect to how it is viewed on a flat plane. The orthographic projection, auxiliary, isometric, oblique, and perspective methods of illustration are all common to the aviation maintenance industry.

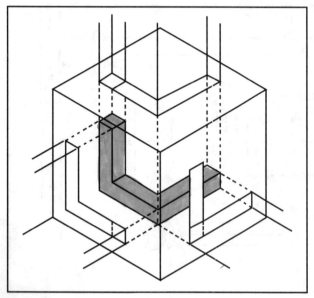

Figure 5-14. With full orthographic drawings, you see all sides of an object.

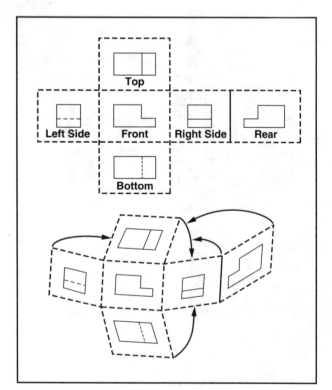

Figure 5-15. Orthographic views are laid out on a flat drawing as they would appear if the sides of a box enclosing the object were opened out.

ORTHOGRAPHIC PROJECTION

Most drawings used in the construction of a detailed part are drawn using the orthographic projection method of illustration. In orthographic projection there are six possible views from which an object can be drawn: the front, rear, top, bottom, left side, and right side. Each view is drawn as if you put an object in a transparent box and viewed it from one of the box faces. All rays extending from the part are parallel and perpendicular to the side they are viewed from. [Figure 5-14]

More often than not, six views are not needed to illustrate a part. In fact, one-view, two-view, and three-view drawings are the most common. In a three-view drawing the front, right side, and top views are illustrated. When drawn, these views are positioned on paper according to the same relationship they have if the sides of the transparent box are opened out. [Figure 5-15]

AUXILIARY DRAWINGS

Although an orthographic drawing can represent up to six individual views, it is sometimes necessary to see a view that is not at a 90 degree angle to the face of an object. In this situation, an auxiliary drawing is used. [Figure 5-16]

ISOMETRIC DRAWINGS

The form of pictorial presentation that is most used for aircraft drawing and sketching is the isometric

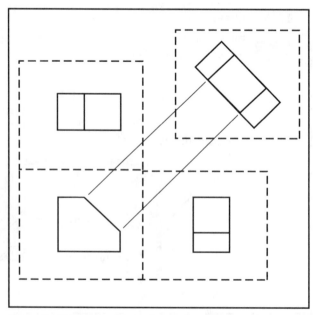

Figure 5-16. In an auxiliary view, a drawing is made at some angle other than 90 degrees from an object's face.

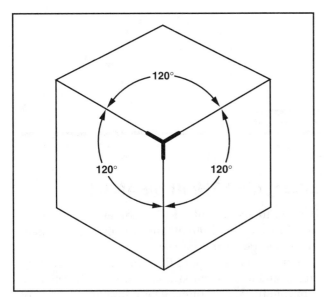

Figure 5-17. The most common type of isometric drawing is the cube. When drawing an isometric drawing, the angles formed by the three sides of an object are equal.

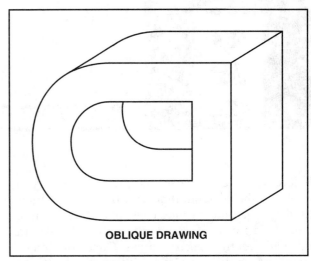

Figure 5-18 An oblique drawing is similar to an isometric drawing in that three sides of the object are visible. However, one of the object faces is parallel to the drawing plane.

drawing. An isometric drawing is a projection of a three-dimensional object on a flat plane. With this type of drawing an object is rotated so three sides are visible. In other words, to make an isometric drawing, an object is rotated so that three views are visible and touching the drawing plane. When doing this, you must ensure that the edges all form the same angle to the drawing plane. [Figure 5-17]

In an isometric drawing all distances are the same length as the actual sides. This makes an isometric drawing fairly easy since no changes are made to any dimensions. Since an isometric drawing allows you to visualize a part, most pictorial drawings are illustrated in this way.

OBLIQUE DRAWINGS

An oblique drawing is an isometric drawing with one object face parallel to the drawing plane. In other words, two axes are perpendicular to each other, with the front of the object identical to the front view of an orthographic drawing. The depth axis of the oblique drawing is typically any convenient angle and most often about 30 degrees. [Figure 5-18]

There are two special types of oblique drawings. They are the cabinet drawing, and the cavalier drawing. A **cabinet drawing** gets its name from drawings used for cabinet work. In these drawings, the oblique side is at a 45 degree angle to the front side and is 1/2 the scale. This allows for an accurate

and undistorted front view. The remainder of the drawing is present only to illustrate depth.

Cavalier drawings use the same scale for the front view as the oblique side lines. However, the oblique sides are still set at a 45 degree angle to the front view. This creates a distorted picture of an object's true proportions. These drawings are primarily used when detailing is required on the oblique side.

PERSPECTIVE DRAWINGS

A perspective drawing is used when you need to see an object similar to the way the human eye sees it. The basic difference between a perspective drawing and an oblique or isometric drawing is that on a perspective drawing the lines, or rays of an object meet at a distant point on the horizon. This point is referred to as the **vanishing point**. Perspective drawings are not generally used in aircraft drawings. [Figure 5-19]

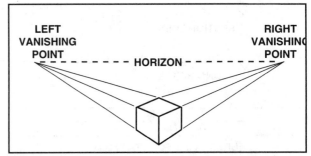

Figure 5-19. In perspective drawings the rays that project from the drawing intersect at a vanishing point on the horizon.

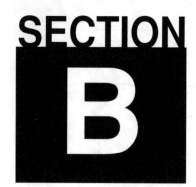

SECTION B

DRAWING PRACTICES

Before you can properly interpret drawings, you must first become familiar with the types of lines used to illustrate various concepts. Different line widths, arrowheads, and alternating breaks in lines all identify specific things. Once you understand the various aspects of aircraft drawings you will be ready to begin sketching parts and repairs on your own.

LINES AND THEIR MEANING

In order to display information contained in a drawing, lines with different appearances are needed. Lines can be in the form of a solid line, a dashed line, or a combination of the two. Furthermore, several drawings use three line widths or intensities, thin, medium, and thick. The following is a list of line types used on aircraft drawings. [Figure 5-20]

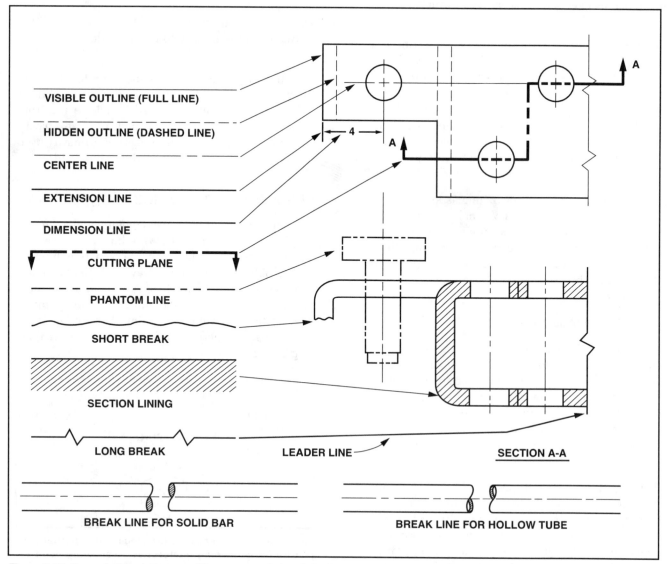

Figure 5-20. Several different types of lines are needed to fully illustrate a part or unit.

Visible lines on outlines are used to illustrate a visible part. A visible line consists of a medium-weight solid line and is the most common type of line used on most drawings.

Hidden lines indicate invisible edges or contours. Hidden lines consists of a dashed line of medium-weight.

Center lines are made up of alternating long and short dashes and are used to show the middle of a symmetrical part. In the case of a hole, the exact center is marked by the intersection of two short dashes.

Extension lines are light lines that extend from the point where a measurement is made. These lines do not actually touch the visible lines of an object, but are approximately 1/16 inch from a part's edge.

Dimension lines are light lines that are broken in the center so a dimension can be inserted. Typically, dimension lines have an arrowhead placed at each end and touch an extension line. This shows the exact location from which the dimension is made. All dimensions are placed so that they read from left to right.

The dimension of an angle is indicated by placing the degree of the angle in its arc. Circular part dimensions are always given in terms of the circle diameter and are usually marked with the letter "D" or the abbreviation "DIA." The dimension of an arc is given in terms of its radius and is marked with the letter R following the dimension. Parallel dimensions are placed so that the longest dimension is farthest from the outline and the shortest dimension is closest to the object outline. On a drawing showing several views, dimensions are placed on each view to show all details.

Cutting-plane lines consist of medium or heavy alternating long dashes and two short dashes with an arrowhead at each end. A cutting-plane line is used to indicate the plane in which a sectional view of an object is taken. The arrowheads show the direction in which the view is seen and have letters to identify the section shown.

Phantom lines are light lines made of alternating long dashes and two short dashes. These lines indicate the presence of another part and are included for reference or to indicate a part's alternate position. For example, a movable part is illustrated by solid lines in one position, and by phantom lines for its alternate position.

Short **break lines** are used across small dimensions to show that a part continues. Break lines are medium weight lines that are often drawn freehand. Long break lines are used across a large part and consist of a light line with a series of irregular breaks or zigzags. Long break lines usually extend beyond the solid lines indicating the edges of the part.

Leader lines are light lines with arrowheads that extend from a note, number, or information box to a part. To minimize confusion, leader lines should never cross a dimension line, an extension line, or another leader line.

Section lines are used to show differences in types of materials or exposed surfaces. Although various materials can be illustrated by different section lines, if the materials used are listed in the bill of materials, the symbol for cast iron is frequently used to represent all metals. [Figure 5-21]

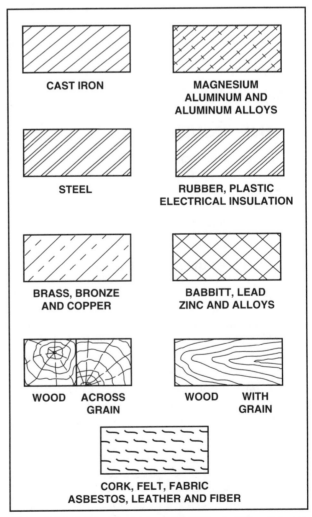

Figure 5-21. Section lines are used to show different types of materials and exposed surfaces.

ABCDEFGHIJKLMNOPQR
STUVWXYZ 1234567890

ABCDEFGHIJKLMNOPQR
STUVWXYZ 1234567890

Figure 5-22. For ease of reading, single-stroke Gothic letters are used on most aircraft drawings.

The most important consideration for an aircraft drawing is that it accurately portrays information. Therefore, lettering is often used to help identify some items. For legibility and speed, all lettering is done freehand, using single-stroke Gothic upper-case letters. [Figure 5-22]

When it comes to placing letters on a diagram it is common practice to draw very light guidelines and to space letters so there is approximately the same distance between them for uniformity. Appearance is what makes the lettering attractive and easy to read. Words should be separated by the amount of space required for the letter "I" with space on each side of it. Fractions are always made with a horizontal division line and numbers should be two-thirds as high as whole numbers.

DIMENSIONING

In order for a drawing to be meaningful it must show the shape of a part as well as accurately give all needed dimensions. Dimensions that appear on a drawing represent the perfect size, and are called basic, or **nominal dimensions**.

ALLOWANCE AND TOLERANCE

Parts that have a maximum and minimum allowable size are still considered acceptable if their size falls within the range given. The difference between the nominal dimension and the upper or lower limit is called the allowance. It represents the tightest permissible fit for proper construction and operation of mating parts. For example, if a dimension is depicted as .3125 ±.0005, the allowable dimensions are between .3120 and .3130 inches. [Figure 5-23]

Tolerance is the difference between the extreme permissible dimensions. For example, given the dimension .281 ±.0005, the extreme permissible dimensions are .2805 and .2815. The difference between these two extremes is .0010 inches. Therefore, the tolerance is .0010 inches.

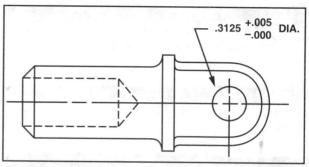

Figure 5-23. The allowance of a part is indicated with typical dimensioning practices.

Another way to calculate the tolerance is to subtract the allowance figures. Using the previous example with an allowance of +.0005 and −.0005, the difference between these two figures is .0010 inches.

In aircraft drawings, any dimension given as a common fraction normally assumes a dimensional tolerance of plus or minus 1/64 inch. If a dimension is given with one decimal, such as 2.5 inches, the tolerance is plus or minus 0.1 inch. If the dimension appears as 2.50 inches, the tolerance is plus or minus .04 inch. More precise measurements are made using three decimal places, such as 2.500, which denotes a tolerance of plus or minus 0.010 inch. If still more exact measurement is needed, the dimensions may specify the limits as:

$$2.500 + .0005 - .0000$$

This dimension requires that the part be no larger than a half-thousandth inch over the base dimension and that it be no smaller than the given dimension.

PLACEMENT OF DIMENSIONS

Most aircraft drawings are dimensioned using a reference edge from which all dimensions are made. Holes are typically located with reference to one corner of a part. This makes finding the center of each hole easier to locate, with no cumulative errors. Cumulative errors exist if one hole is measured from the center of the adjacent hole. [Figure 5-24]

There are two ways of placing dimensions on a drawing. One way is to write all dimensions perpendicular to the dimension lines. When this is done, the numbers are parallel to the right edge of the drawing for vertical dimension lines and at various angles across the drawing for parts that have angled surfaces. The second and more conventional

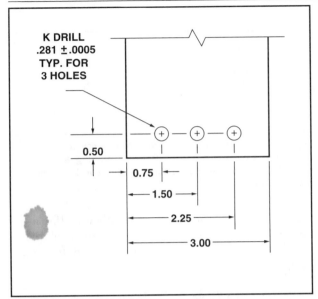

Figure 5-24. Notice that the holes are drilled with reference to the end of the part and not the center of an adjacent hole. This helps to minimize cumulative errors.

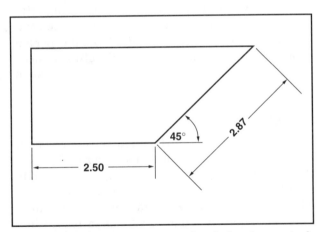

Figure 5-25. All dimensions on an aircraft drawing are typically made parallel to the bottom of the drawing.

method is to write all dimensions parallel to the bottom of the drawing. [Figure 5-25]

AIRCRAFT PRODUCTION DRAWINGS

As stated earlier, every aircraft factory has its own drawing room. Likewise, most drawing rooms have a manual that lists the standards used by the company when making drawings. The information given in this text is typical and, while it may differ from some company manuals, it applies to most of them.

TITLE BLOCK

A title block is generally printed in the lower right-hand corner of every aircraft drawing. It contains the information necessary to manufacture the illustrated part. If special or additional information is needed, it is typically listed to the left of the title block. [Figure 5-26]

TITLE BOX

The part name is put in the box labled "title". In most cases the subject is listed first. For example, a drawing of an autopilot servo bracket, would be titled "Bracket, Autopilot Servo."

SIZE

Letters are typically used to specify the size of a drawing. An A-size drawing is 8 1/2 × 11 inches, a B-size is 11 × 17 inches, a C-size is 17 × 22 inches, and a D-size drawing is 22 × 34 inches. Larger drawings are made on paper 36 to 42 inches wide and are specified as R-size drawings.

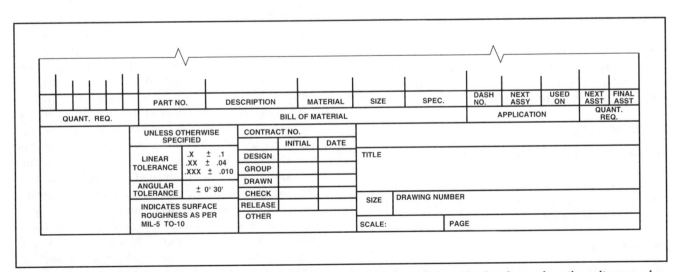

Figure 5-26. The title block of an aircraft drawing contains important information about the drawing such as the unit name, size, drawing number, scale, and page.

DRAWING NUMBER

A drawing number or part number is assigned to each drawing. The part carries this number throughout its entire life. If an aircraft has a left-hand and a right-hand part, they typically have the same part number. However, the left-hand part number is followed by a -1, while the right-hand part number is followed by a -2.

SCALE

The scale indicates the size of a drawing and is noted with comments such as FULL, HALF, or 1 INCH = 1 FOOT. If the drawing is essentially full-scale, typically there are some details that are drawn to another scale. The notation indicating this is "FULL EXCEPT AS NOTED," with a note placed next to each detail that is not drawn to full-scale. If the drawing does not include any parts that are drawn to scale, the word "NONE" is used in this blank.

PAGE

When drawings are assembled into a book, as with electrical wiring diagrams, this space on the title block is used for the page number of the book.

RESPONSIBILITY

The columns to the left of the title are available for the initials of the people responsible for all aspects of the part. This includes the person who designed the part, the group that approved the design, the person who made the drawing, the person who checked the drawing, and the person who released the drawing.

STANDARDS

In the column entitled "UNLESS OTHERWISE SPECIFIED," are the standards of manufacturing tolerances used by the design company. Here, the manufacturer specifies the linear and angular dimensional tolerances along with a note specifying the standard for surface roughness.

BILL OF MATERIAL

Just above the title block and extending upward as needed, is a list of every material needed to manufacture the part. The quantity required, part number, description, size, and specification number for these materials are all indicated.

APPLICATION

The block marked "APPLICATION" indicates where the part is used. It shows the model of aircraft the

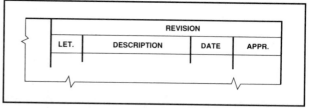

Figure 5-27. Most aircraft drawings have a revision block in the upper right-hand corner. Prior to making a part you should check the revision block to verify you have the latest revision.

part is used on and the part number of the next assembly. Also indicated is the quantity of parts needed for both the next assembly and the final assembly.

REVISION BLOCK

In the upper right-hand corner of a drawing is a block where revisions are recorded. For example, if revision A is issued by the engineering department the letter A is indicated in this space along with the description of the revision, the date the revision was released, and the initials of the person approving the revision. You should always check to make sure you have the most current revision. [Figure 5-27]

NOTES

Notes are added to a drawing to identify a deviation from the norm, give additional information, list alternatives, call attention to an item, or specify modifications to an original design. Notes are related to the drawing or other related drawings. Notes are typically found in a block or placed in a location that does not interfere with the drawing.

ZONES

Aircraft drawings are usually 36 to 42 inches wide and are several feet long. Since these drawings are so large, it is often difficult to find detailed views without a system of location. Therefore, a system similar to the grid on a map is used. The grid is made by marking the edges of the drawing every 12 inches both vertically and horizontally. The vertical marks are identified by letters, with A being the bottom 12 inches, B the next 12 inches, and so on. The horizontal marks are identified by numbers beginning with 1 on the extreme right side. The combination of a vertical and horizontal coordinate identifies a **zone**. For example, if a revision is noted as a dimension change in zone C-7, you should know that this change is made somewhere in the section three feet up from the bottom and seven feet from the right edge.[Figure 5-28]

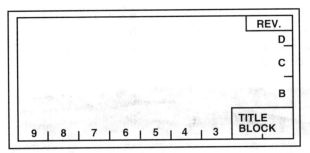

Figure 5-28. Details on a large drawing are easy to locate through the use of zone numbers and letters.

LOCATION IDENTIFICATION ON AIRCRAFT

The location of a part on an aircraft is typically specified by fuselage station numbers, water lines, and buttock lines.

Fuselage station numbers identify locations fore and aft along the fuselage. All station numbers are measured from a reference called station zero. This reference, often called the **datum,** is typically on the fuselage or ahead of it. For example, if the datum is six inches ahead of the fuselage nose and the center line of the main spar is 137 inches from the datum, its fuselage station number is 137. All fuselage frames and bulkheads are identified by fuselage station numbers.

Vertical locations are identified with **water lines**. Like station numbers, water lines are measured from a zero reference. However, in this case the zero reference is called water line zero. For example, if the floor of the main cabin must be installed at water line -16, the floor is 16 inches below water line zero.

Distances to the right or left of the fuselage centerline are measured by **buttock lines** and are referenced from an aircraft's longitudinal centerline. For example, if the tip of a horizontal surface is located at buttock line 108.88, it means that it is 108.88 inches from the fuselage centerline.

Wing stations are measured from the centerline of the fuselage, or buttock line zero. They indicate the distance in inches along the wing toward the wing tip. For example, if the right edge of the aileron is at wing station 123, the right edge of the aileron is located 123 inches from the aircraft's longitudinal centerline. [Figure 5-29]

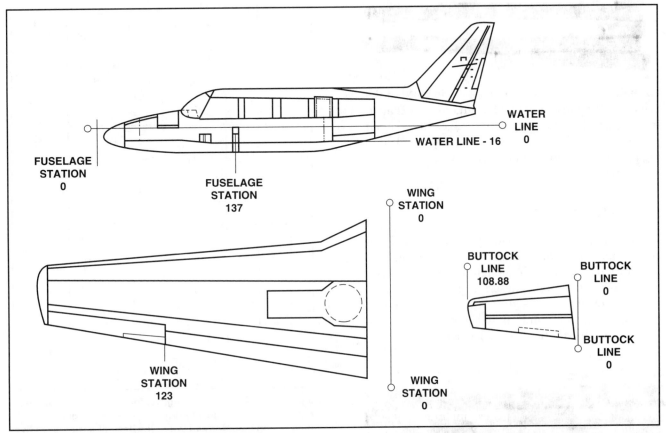

Figure 5-29. Fuselage station numbers are used to locate items along the fuselage. Water lines locate positions vertically on the fuselage. Buttock lines locate points to the right or left of the fuselage centerline. Wing stations are measurements along the span of the wing, with wing station zero the centerline of the fuselage.

BASIC SKETCHING

As an aircraft maintenance technician you are not required to have the skills of a professional draftsman. However, you must be able to graphically express yourself well enough to describe a repair and make a simple sketch of a part. When making a sketch, a simple one will do as long as it contains the information needed to make the part or repair.

The most common means of illustrating something is with a sketch. Most sketches made for aircraft maintenance are either orthographic or isometric. When sketching an orthographic three-view drawing, look at the part and make a sketch showing the shape of the object from the front, side, and top. Once this is done, place extension lines and dimensions where needed.

The competency level required to make a sketch is easy to attain if you follow four basic steps.

1. Determine what views are necessary to portray the object then block in the views using light construction lines.
2. Complete the details, and darken the object outline.
3. Sketch extension and dimension lines, and add detail.
4. Complete the drawing by adding dimensions, notes, a title, and a date.

By their very nature, sketches are made without the use of drafting instruments. For this reason, sketching on graph paper typically makes the layout process easier. Another way to simplify the sketching process is to break an object down into individual shapes.

REPAIR SKETCHES

When a major repair is made to a certificated aircraft, an FAA Form 337 must be completed that describes the repair with enough detail for the FAA to approve it. Part of this description includes drawing a sketch that shows the location of the repair, the materials used, and enough detail so that the repair could be duplicated if necessary.

BASIC SHAPES

Almost all objects are comprised of one or a combination of six basic shapes. These include the triangle, circle, cube, cylinder, cone, and sphere. If you master these shapes you should be able to acceptably sketch most parts.

Angles are easily estimated on square-ruled paper by making **triangles**. For example, a 45 degree angle is formed by a triangle having two sides of equal length. [Figure 5-30]

A **circle** is easily sketched by marking a center point and then making four marks on the horizontal and vertical lines equal distances from the center.

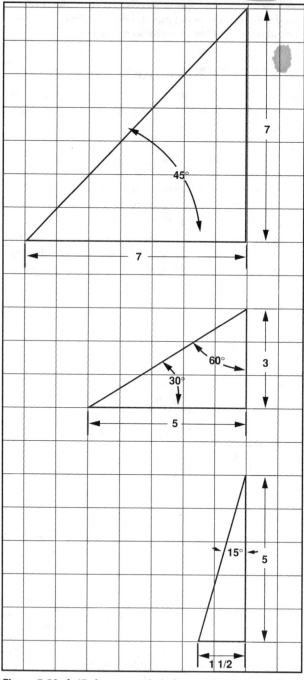

Figure 5-30. A 45 degree angle is formed by a triangle having two sides of equal length. A 30 degree angle and a 60 degree angle are formed when the base of a triangle is five units long and the height is three units high. A 15 degree angle is approximated by making the base one and a half units and the height five units.

Between each of these marks, make four more marks that are the same distance from the center. Now, lightly sketch a curved line through all eight points.

A **cube** is drawn by first laying out a vertical center line. This is followed by drawing the edges of the two top sides so they extend out at equal angles and distances from the top of the vertical line. Complete the top of the cube with lines parallel to the edges just drawn. Next, form the outer edges by drawing vertical lines that are parallel with the center line, and draw in the bottom edges parallel to the top edges. To make a cube look more like a solid object, shade the side so it appears as though light were striking it. [Figure 5-31]

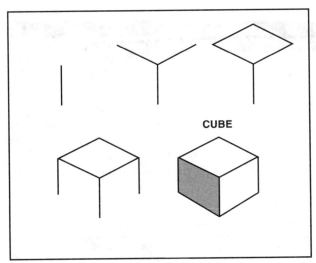

Figure 5-31. The steps for drawing a cube are simple when broken down into a step-by-step process.

To draw a **cylinder** begin by sketching the top of a cube and then draw an ellipse that touches the center of each side. Next, draw parallel sides that go straight down. Finish the cylinder by drawing half an ellipse to form the cylinder bottom. Finally, shade the sides of the cylinder with a series of arcs so it looks as though it were three-dimensional rather than flat.

A **cone** is made in much the same way as the cylinder with the sides tapering up to a point. Again, shading gives it the appearance of being three-dimensional.

A **sphere** is simply a circle with shading in the lower quarter. [Figure 5-32]

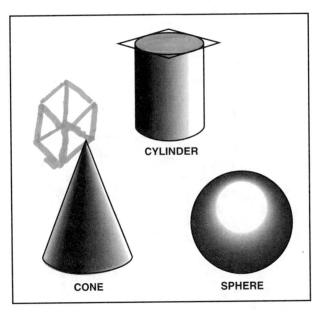

Figure 5-32. A cylinder, cone, and sphere all take a three-dimensional look when shading is applied correctly.

APPLIED GEOMETRY

In addition to knowing how to sketch individual shapes, it is also helpful if you know how to apply simple geometry in the drawing process. For example, to find the center of a line, use a compass that is open to a distance longer than halfway across the line and draw two arcs that intersect the line at a 90 degree angle. Next, connect the intersection of the two arcs with a straight line. This line, known as the **perpendicular bisector**, divides the first line into two equal parts. [Figure 5-33]

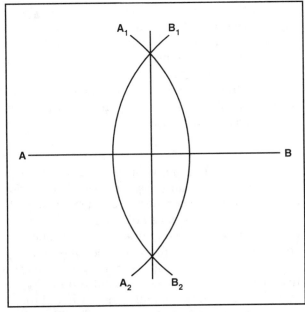

Figure 5-33. By drawing arcs A_1-A_2 and B_1-B_2 and connecting their intersections, the center of line A-B is located.

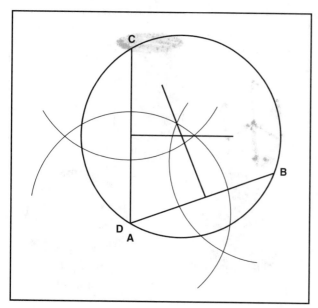

Figure 5-34. To find the center of a circle, draw two chord lines, and their corresponding perpendicular bisectors.

To find the center of a circle begin by drawing two chord lines across the circle. Next, draw perpendicular bisectors for each of the chord lines. If necessary, extend each of the perpendicular bisectors until they intersect. The point at which the perpendicular bisectors cross is the center of the circle. [Figure 5-34]

Another thing geometry allows you to do is draw a line perpendicular to another line through a specific point. To do this, refer to figure 5-35. Begin by opening a compass to a distance that is greater than the space from line A-B to point C. Using C as the center, draw arcs 1 and 2 on line A-B. Using the same compass setting, put the point of the compass where line 1 intersects line A-B and make an arc 1_a-1_b on the opposite side of point C. Now, draw a second arc using the same setting on the compass and the intersection of line A-B and arc 2. The line that joins the intersection of these arcs with point C is perpendicular to line A-B and passes through point C. [Figure 5-35]

Often, you are given a line of odd length that needs to be divided into an equal number of parts. For example, if you have a line of odd length, such as that in figure 5-36, you can divide it into five equal parts by drawing a second line that is five inches or greater in length, and diagonal from and touching the end of the first line. Now, draw a line from the five-inch mark on line A-C to point B on line A-B. Lines are then drawn parallel to this line through each of the one-inch marks on line A-C. Where these lines cross line A-B, they form division marks. Each

mark is one-fifth the distance between A and B. [Figure 5-36]

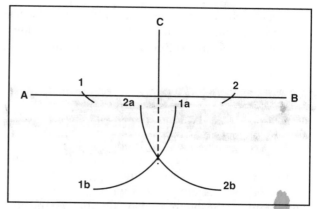

Figure 5-35. Drawing a line through a point that is perpendicular to a baseline requires you to draw 2 sets of arcs.

In addition to allowing you to divide lines, geometry is used to divide or **bisect** angles. To do this, refer to figure 5-37. Begin by using A as the center and draw a set of arcs at a convenient compass distance. Make arc 1 in line A-C and arc 2 on line A-D. Now using the same setting on the compass and point 1 as the center, make an arc that is between line A-C and A-B. Continue by drawing another arc with point 2 as the center that crosses the first arc. Finally, draw a line from the intersection of the two arcs to point A. This line bisects angle A-B-C. [Figure 5-37]

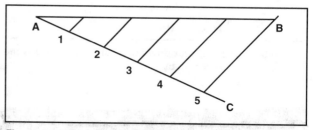

Figure 5-36. Dividing a line into an equal number of parts requires a line of even length that is drawn diagonally from the original line.

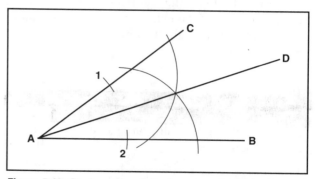

Figure 5-37. To draw line A-D, draw arcs 1 and 2, then arc 1a-b and 2a-b. The line drawn from the intersection of arcs 3 and 4 to point A is the bisecting line.

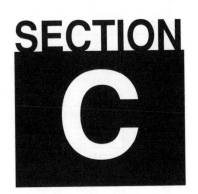
CHARTS AND GRAPHS

Graphs and charts are pictorial representations of data. They enable you to quickly visualize certain relationships, complete complex calculations, and predict trends. Furthermore, charts allow you to see the rate and magnitude of changes.

Information is presented graphically in many different forms. Graphs are often found in the form of **bar-graphs**, **pictographs**, **broken-line graphs**, continuous-curve graphs, and the circular graph or **pie chart**. [Figure 5-38]

NOMOGRAMS

The need to show how two or more variables affect a value is common in the maintenance industry. Nomograms are a special type of graph that enable you to solve complex problems involving more than one variable.

Most nomogram charts contain a great deal of information and require the use of scales on three sides of the chart, as well as diagonal lines. In fact, some charts contain so much information that it is very

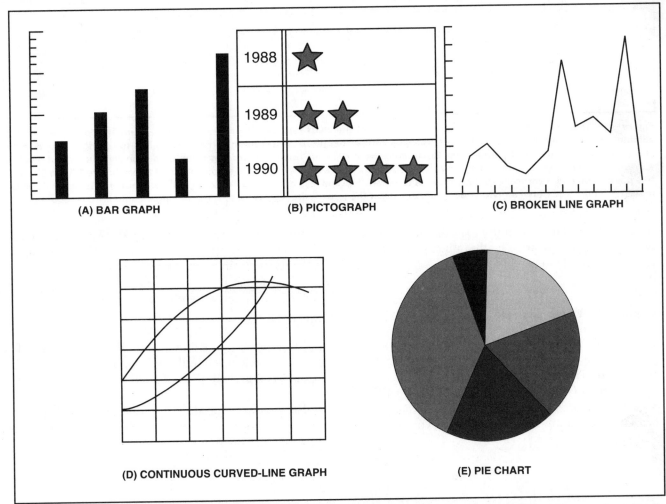

Figure 5-38. Bar graphs, pictographs, broken-line graphs, continuous curved-line graphs, and pie charts are all ways of graphically representing numerous calculations.

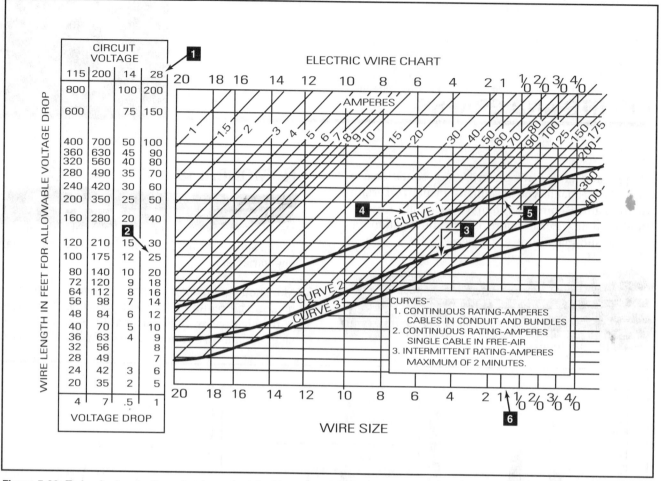

Figure 5-39. To begin, locate the column on the left side of the chart representing a 28 volt system (item 1). Move down in this column until you find the horizontal line representing a wire length of 25 feet (item 2). Follow this line to the right until it intersects the diagonal line for 125 amps (item 3). Because the wire is in a bundle and carries a continuous current, you must be at or above curve 1 on the chart (item 4). Follow along the diagonal line representing 125 amps until it intersects curve 1 (item 5). From this point, drop down vertically to the bottom of the chart. The line falls between wire sizes 1 and 1/0 (item 6). Whenever the chart indicates a wire size between two sizes, you must select the larger wire. In this case, a 1/0, or single aught wire is required.

important for you to carefully read the instructions before using the chart. On the other hand, some charts are simple to use.

ELECTRIC WIRE CHART

An example of a nomogram chart that is used extensively in the maintenance industry is the electric wire chart. This chart is made up of vertical lines that represent the American Wire Gauge (AWG) wire sizes. Horizontal lines represent the length of wire in feet that produces an allowable voltage drop

for each electrical system listed. Drawn diagonally across the chart is a series of parallel lines representing current flow. A common use for this chart is to find the wire size required to carry a given amount of current without exceeding the allowable voltage drop.

For example, determine the minimum size wire of a single cable in a bundle carrying 125 amps 25 feet in a 28-volt system. [Figure 5-39]

Notice that the three curves extend diagonally across the chart from the lower left corner to the right side of the chart. These curves represent the ability of a wire to carry the current without overheating. Curve one represents the continuous rating of a wire when routed in bundles or conduit. If the intersection of the current and wire length lines are above this curve, the wire can carry the current without generating excessive heat.

If the intersection of the current and wire length lines falls between curve one and two, the wire can only be used to carry current continuously in free air. If the intersection falls between curves two and three, the wire can only be used to carry current intermittently.

BRAKE-HORSEPOWER CHARTS

Another common type of graph you will encounter as a technician is the performance chart. One common performance chart is the brake-horsepower chart. These charts represent many hours of calculation by engineers but are presented so that you can quickly determine if the performance being observed is acceptable. For this sample chart, assume you have an engine that has a 2,000 cubic-inch displacement and develops 1,500 brake-horsepower at 2,400 rpm. [Figure 5-40]

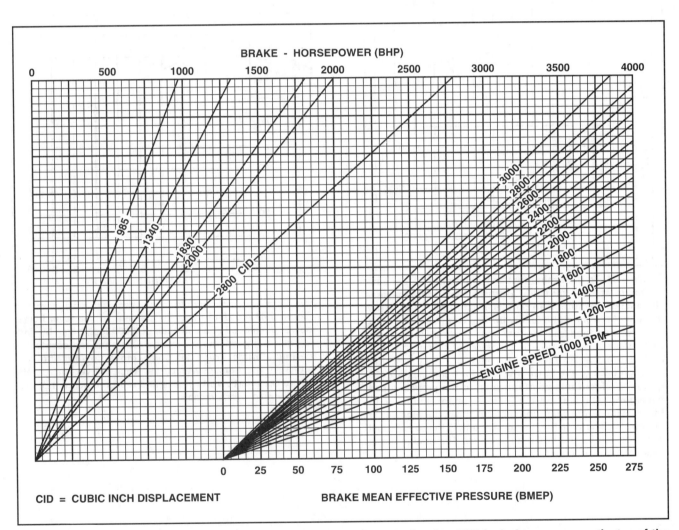

Figure 5-40. To calculate the brake mean effective pressure, BMEP, begin by locating 1,500 brake-horsepower on the top of the chart. From this value, drop down vertically until you reach the line representing 2,000 cubic inches of displacement. From this intersection, extend a line horizontally to the right until you intercept the line representing 2,400 rpm. Now, drop down vertically to read the brake mean effective pressure on the bottom line of the chart. The brake mean effective pressure is approximately 248.

FUEL CONSUMPTION CHARTS

The fuel consumption chart is another type of performance chart that you must be familiar with. For this sample chart, assume that you are trying to determine how much fuel an engine consumes when it is operating at a cruise of 2,400 rpm. [Figure 5-41]

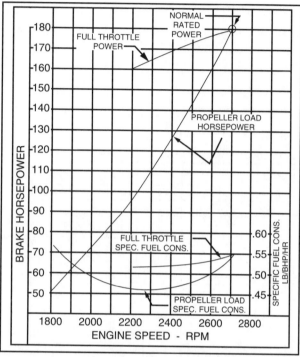

Figure 5-41. To determine fuel consumption for an engine operating at 2,400 rpm, you must first determine the specific fuel consumption. To do this, locate 2,400 rpm on the bottom of the chart and follow the line up until it intersects the propeller load specific fuel consumption curve. From this intersection, extend a line to the right side of the chart and read a specific fuel consumption of .47 LB/BHP/HR. Now, go back to the bottom of the chart and locate 2,400 rpm again. From this point move up to the propeller load horsepower curve. From this intersection, extend a line to the left side of the chart and read the brake horsepower of 127 hp. To determine the fuel burn, multiply the specific fuel consumption by the brake horsepower. The engine burns 59.69 pounds per hour (.47 × 127 = 59.69).

ENGINE HORSEPOWER/ALTITUDE

This chart represents the relationship between engine horsepower and altitude. For this sample chart, assume you are doing an engine run-up at an altitude of 7,000 feet. [Figure 5-42]

There are many other ways of presenting information with graphs. Pie or circular charts can show the percentage of an item to the whole. Graphs show the relationship of two or more variables.

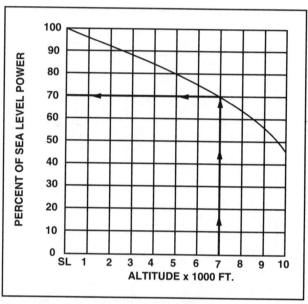

Figure 5-42. To determine the percent of sea level horsepower developed at an altitude of 7,000 feet, begin by finding the point on the horizontal axis that represents the desired altitude. From this point, move upward until you intersect the horsepower curve. Then move horizontally left to the chart's vertical axis and read the percent of sea-level horsepower available.

WEIGHT AND BALANCE

INTRODUCTION

As an aircraft maintenance technician you will often perform repairs or alterations that change an aircraft's weight and balance. This in turn changes the way an aircraft's weight must be effectively distributed, as well as its flight characteristics. Therefore, it is the responsibility of the aircraft maintenance technician who makes a repair or alteration to change the aircraft weight and balance paperwork. This paperwork must reflect the new computations for weight and balance, and indicate the aircraft is safe to fly. This information then allows the pilot to make an informed decision as to the airworthiness of the aircraft.

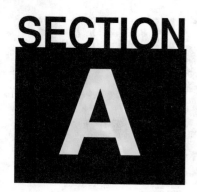

SECTION A

WEIGHING PROCEDURES

When calculating a weight and balance report, certain procedures must be followed in order to make the report accurate. These procedures include more than just weighing the aircraft, they include understanding weight and balance terms, knowing how to prepare the aircraft for weighing, and how to set up and use the necessary equipment. There are also different FAA regulations that apply to general aviation as compared to commercial aviation that must be adhered to. These items along with some safety precautions are included in the following text.

AIRCRAFT WEIGHT

The starting point for weight computation is the weight of the aircraft before passengers, cargo, and fuel are added. The term **basic empty weight** includes the weight of the standard aircraft, any optional or special equipment, fixed ballast, unusable fuel, and full operating fluids including oil, hydraulic fluid, and other fluids required for normal operation of aircraft systems except potable water, lavatory precharge water, and water intended for injection in the engines.

Aircraft certified prior to March 1, 1978, might use the term **licensed empty weight**, which is similar to basic empty weight except that it does not include full engine oil. For these aircraft, the licensed empty weight includes only the weight of undrainable oil. When in question as to what operating fluids are included in the empty weight, check the aircraft Type Certificate Data Sheets. In fact, in order to obtain all information regarding the weight and balance data for a particular aircraft you should check all the following sources.

Aircraft specifications
Aircraft operating limitations
Aircraft flight manual
Aircraft weight and balance report
Aircraft Type Certificate Data Sheets

Another weight term often used is **standard empty weight**. This is simply the weight of an aircraft without optional equipment, and is obtained from aircraft manufacturers.

Standard weights are used for computing the weight of fuel, oil, crew, water, and baggage. For general weight and balance purposes, the following weights are considered standard:

ITEM	QTY
Avgas	6 pounds per gallon
Turbine fuel	6.7 pounds per gallon
Lubricating oil	7.5 pounds per gallon
Water	8.35 pounds per gallon
Crew & Passengers	170 pounds per person
	190 pounds for utility/aerobatic aircraft

FAR 135 covers air taxi operators and commercial operators of small aircraft. This regulation has added the following standard weights:

ITEM	QTY
Adults (summer)	170 pounds per person
Adults (winter)	175 pounds per person
Flight crew (male)	170 pounds per person
Flight crew (female)	150 pounds per person
Female flight attendants	130 pounds per person
Male flight attendants	150 pounds per person
Check-in baggage	23.5 pounds per item
Carry-on baggage	10 pounds per item

Another weight term, **payload**, refers to the weight of the flight crew, passengers, and any cargo or baggage; **useful load** is the difference between maximum takeoff weight and basic empty weight. Useful load includes payload, usable fuel, and full operating fluids. **Usable fuel** is the fuel available for the flight. It does not include **unusable fuel**, which is the quantity of fuel that cannot be safely used during flight. **Zero fuel weight**, on the other hand, is the operational weight of the aircraft including the payload but excluding the fuel load. The following summary should help you understand how various weights are calculated:

Basic Empty Weight
+ Payload
_____
= Zero Fuel Weight
+ Usable Fuel
_____
= Ramp Weight
– Fuel Used for Start, Taxi, and Engine Runup
_____
= Takeoff Weight
– Fuel Used During Flight
_____
= Landing Weight

WEIGHT LIMITATIONS

Some weight and balance terms may be preceded by the word **maximum**. When this word is used, it indicates a limitation which must not be exceeded. For example, the **maximum ramp weight** is the maximum weight approved for ground operations, while **maximum takeoff weight** is the maximum weight approved for the start of the takeoff roll. As another example, **maximum landing weight** is the maximum allowable weight at which an aircraft can be landed. The maximum landing weight is usually less than maximum takeoff weight because the stresses during landing are greater than those during takeoff.

Weight limitations are necessary to guarantee the structural integrity of an airplane, as well as enable the pilot to predict aircraft performance accurately. These structural limitations are based on an aircraft being operated at or below its **maximum weight**. Maximum weight is the authorized weight of the aircraft and its contents as indicated in the aircraft specifications. There are variations to the maximum allowable weight depending on the aircraft purpose and flight conditions. As an example, a certain aircraft may have a maximum gross weight of 2,750 pounds when flown in the normal category, but when flown in the utility category, the same aircraft's maximum allowable gross weight decreases to 2,175 pounds.

Operating an aircraft above its maximum weight could result in structural deformation or failure during flight in adverse conditions such as, strong wind gusts or turbulence.

As an aviation technician, you will probably weigh an aircraft sometime during your career, and there are some additional weights you must be familiar with. These include scale weight, tare weight, and net weight. As the name implies, the **scale weight** represents the reading taken from the scales. It includes the weight of all items on the scales including the aircraft, chocks, and jacks. The **tare weight**, on the other hand, includes the weight of all items on the scales that are not part of the aircraft such as jacks, blocks, and chocks. The weight of these items must be deducted from the scale weight to obtain the actual or **net weight**.

NEED FOR REWEIGHING

The empty weight and corresponding CG of all civil aircraft is determined at the time of certification. Furthermore, an accurate record of changes must be maintained throughout the life of the aircraft. A manufacturer is required to weigh one aircraft out of each 10 produced. The remaining nine aircraft are issued a computed weight and balance report based on the averaged figures of aircraft that are actually weighed. The condition of the aircraft at the time of determining the empty weight must be one that is well defined so that loading requirements can be easily computed.

Once an aircraft is placed in service, most equipment changes and modifications do not require aircraft reweighing. However, they do require a change to the aircraft's weight and balance information. These changes are often calculated by aircraft maintenance technicians and entered in the aircraft's permanent weight and balance records. Since these records stay with the aircraft forever, they must reflect current aircraft status.

Privately owned and operated aircraft are not required by regulation to be weighed periodically because they are usually weighed when originally certificated. In fact, about the only time a general aviation aircraft must be weighed and a new set of records computed is when the weight and balance records are lost and cannot be duplicated from any source. However, after making major alterations that affect the weight and balance, weighing should be accomplished to ensure that the maximum weight and CG limits are not exceeded during operation. Furthermore, since aircraft have a tendency to gain weight due to the accumulation of dirt, greases, and other trapped debris, periodic aircraft weighing is desired for light aircraft.

Unlike privately owned aircraft, air carrier and air taxi aircraft are required by the Federal Aviation Regulations (FARs) to be weighed periodically. The exact interval varies from operator to operator, but is typically done on an annual basis. Furthermore, air carrier and air taxi aircraft (scheduled and non-scheduled) that carry passengers or cargo are required to show that the aircraft is loaded properly and will not exceed the authorized weight and balance limitations during operation.

CENTER OF GRAVITY

The point along an aircraft's longitudinal axis at which an aircraft's weight is considered to be concentrated is called the aircraft's center of gravity (CG). For light aircraft, the CG is typically expressed in terms of inches from a predetermined reference point.

Aircraft engineers not only compute the CG, they also compute the most forward and rearward CG limits. The distance between these limits is called the **CG range**. An aircraft is considered to be in balance when the average moment for the loaded aircraft falls within this range. Operation of an aircraft that has a CG outside of its CG range is prohibited and, as an aircraft technician, you are responsible for assuring that a repair or alteration does not shift an aircraft's CG beyond its limits.

CG DESIGN LIMITS

When an aircraft is designed, limits are put on its maximum weight, and restrictions are set up regarding the range within which the CG may fall. A part of the certification procedure for an airplane is to determine that its weight and balance are within the allowable limits, and this information is furnished with the aircraft as part of its operations manual.

It is the responsibility of the pilot to know before each flight that the aircraft is properly loaded, that it does not exceed the allowable gross weight, and that the CG is within the allowable range.

AIRCRAFT LOADING ASPECTS

The weight of an aircraft changes during its operational life as equipment is added or removed and as repairs are made. It is the responsibility of the aviation maintenance technician to see that the permanent aircraft records reflect these changes.

The fuel tanks in most small training aircraft are located in the wing, while the seats and baggage compartments are located directly below the wing. In this configuration, it is not likely that an aircraft will be loaded in such a way that the CG falls outside its allowable range. Most larger aircraft, however, have several rows of seats and baggage compartments, some of which are ahead of the forward CG limit and some behind the most rearward limit. This wide range of loading possibilities makes the use of charts or other aids to loading a necessity for the pilot to be sure that the CG is within the allowable range.

The problems occurring in typical weight and balance scenarios fall into one of three categories. The aircraft is either loaded over its maximum weight, the CG is too far forward, or the CG is too far aft. The following conditions are associated with each.

OVERLOADED AIRCRAFT

An overloaded aircraft must accelerate to a higher-than-normal speed to generate sufficient lift for flight. This means the aircraft requires more runway for takeoff, since it must attain a higher speed. The additional weight also reduces acceleration during the takeoff roll, and this adds to the total takeoff distance.

Problems do not end once the aircraft is airborne. An overloaded aircraft suffers a reduction in climb performance, and its service ceiling is decreased. This is because excess power and thrust are limited. In addition, exceeding the maximum allowable weight may cause the aircraft to become unstable and difficult to fly.

CG TOO FAR FORWARD

If an aircraft is loaded so the CG is forward of the forward CG limit, it will be too nose heavy. Although this tends to make an aircraft seem stable, adverse side effects include longer takeoff distance and higher stalling speeds. The condition gets progressively worse as the CG moves to an extreme forward position. Eventually, elevator (stabilator) effectiveness will be insufficient to lift the nose.

CG TOO FAR AFT

A CG located aft of the approved CG range is even more dangerous than a CG that is too far forward. With an aft CG, the aircraft becomes tail heavy and very unstable in pitch, regardless of speed. Furthermore, as the CG moves aft, stabilator effectiveness decreases. When the CG is at the aft limit, stabilator effectiveness is adequate; but, when the CG is beyond the aft limit, the stabilator may be ineffective for stall or spin recovery.

THE DATUM

The datum is an imaginary vertical plane from which all horizontal measurements are taken with the aircraft in a level flight attitude. The reference datum sits at a right angle to the aircraft's longitudinal axis. For each aircraft make and model, the location of all items including equipment, tanks, baggage compartments, seats, engines, and propellers are listed in the Aircraft Specifications or Type Certificate Data Sheets as being so many inches from the datum.

There is no fixed rule for the location of a datum. It may be located on the nose of the aircraft, the firewall, the leading edge of the wing, or even at a point in space ahead of the aircraft. The manufacturer chooses a location for the datum where it is most convenient for measurement, the location of equipment, and for weight-and-balance computation.

The datum location is indicated on most aircraft specifications. However, on some older aircraft, the datum may not be listed. In this situation, any convenient datum may be selected. However, once the datum is selected, it must be properly identified so that anyone who reads the figures will have no doubt about its exact location. [Figure 6-1]

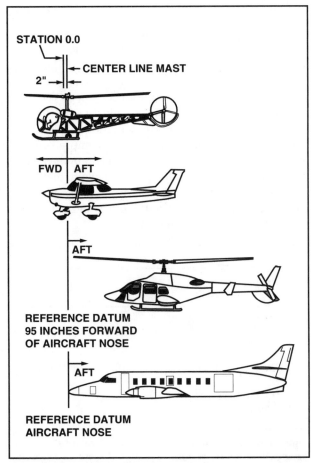

Figure 6-1. Once the reference datum is set, all measurements to the right are positive, while all measurements to the left are negative.

WEIGHT AND BALANCE PRINCIPLES

One method used in understanding weight and balance is to visualize a playground teeter-totter. This child's toy represents a practical example of weight and balance fundamentals. For example, when a large child and a small child get on a teeter-totter, the large child must sit closer to the support, or **fulcrum**, in order to balance the small child sitting farther away from the fulcrum. The distance each child is from the fulcrum is called the **arm**, lever arm, or station, and is typically measured in inches or meters. In any case, arm measurements carry their corresponding algebraic sign to indicate which side of the datum each child is sitting on. For example, on an aircraft all items aft of the datum carry a positive measurement and all items ahead of the datum carry a negative measurement. [Figure 6-2]

When an object's weight is multiplied by its arm, the result is known as its **moment** and is expressed in pound-feet, pound-inches, or in gram- or kilogram-meters. In other words, moment is a force which results from an object's weight acting at a distance. Moment is also referred to as the tendency of an object to rotate or pivot about a point. The farther an object is from a pivotal point, the greater its force. The point about which the algebraic sum of all moments is equal to zero is called the object's balance point.

Just like the children on the teeter-totter, each component of an aircraft, as well as each object it carries, exerts a moment proportional to its weight and distance from the designated reference datum. By totaling the weight and moments of all components and objects carried, you can determine the point where a loaded aircraft would balance on an imaginary fulcrum. This is the aircraft's CG.

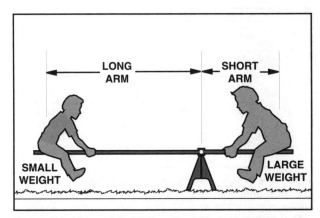

Figure 6-2. The small child on the large arm balances the large child on the short arm.

CALCULATING WEIGHT AND BALANCE

To better understand the principles of weight and balance, consider a teeter-totter with weight on each end. To illustrate this, imagine a board has a weight of 25 pounds on the left, and a weight of 50 pounds on the right. Between the center of the two weights there is 12 feet. In order to find out where the fulcrum must be placed to balance the two weights, choose an arbitrary location for the datum and construct a chart that lists the two weights, their arms, and their moments. For this example, the datum is located at the center of the 25 pound weight. [Figure 6-3]

Since weight A is directly over the datum, its arm and moment are zero. However, the arm of weight B is 12 feet and, therefore, has a moment of 600 pound-feet (12 ft. × 50 lbs. = 600 pound-feet.

To find the balance point, divide the total moment by the total weight. The total moment is 600 pound-feet, and the total weight is 75 pounds. This places the balance point eight feet to the right of the datum.

To check your calculations and prove that the board balances at the 8-foot point, make a chart similar to

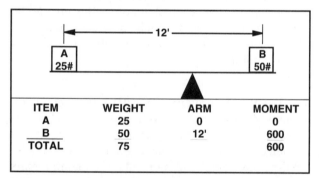

ITEM	WEIGHT	ARM	MOMENT
A	25	0	0
B	50	12'	600
TOTAL	75		600

Figure 6-3. A lever problem is simplified if you draw a picture of the problem and construct a chart. Notice that weight A is used as a datum.

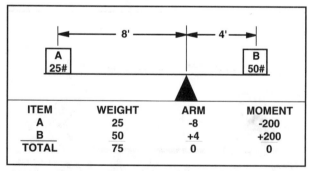

ITEM	WEIGHT	ARM	MOMENT
A	25	-8	-200
B	50	+4	+200
TOTAL	75	0	0

Figure 6-4. Weight A has an arm of negative eight feet, and a corresponding moment of negative 200 pound-feet. The arm of weight B is a positive four feet, and has a moment of 200 pound-feet. The sum of the moments is zero and, therefore, the board balances.

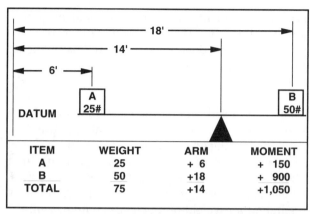

ITEM	WEIGHT	ARM	MOMENT
A	25	+ 6	+ 150
B	50	+18	+ 900
TOTAL	75	+14	+1,050

Figure 6-5. The moment of A is +150 pound-feet, whereas, the moment of weight B is +900 pound-feet. This equates to a total moment of +1,050 pound-feet. When this is divided by the total weight, the balance point is found to be 14 feet to the right of the datum. This is the same location found in the previous example, eight feet to the right of weight A.

the previous. However, this time use the 8-foot point as the datum. When this is done, all distances to the right are considered positive, and all distances to the left are negative. [Figure 6-4]

When aircraft manufacturers place the datum a given distance ahead of the aircraft to make all moments positive, the balance point is still calculated the same way. For example, assume the datum is located six feet to the left of weight A. [Figure 6-5]

Up to this point, the discussion has involved only two weights. However, this is almost never the case when computing aircraft weight and balance changes. Therefore, in this next example, assume you have a 50-pound weight that is 25 inches to the left of the fulcrum, a 40-pound weight that is 45 inches to the right of the fulcrum, and a third, 50-pound weight that you want to place on the board to make it balance. [Figure 6-6]

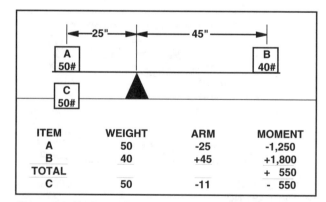

ITEM	WEIGHT	ARM	MOMENT
A	50	-25	-1,250
B	40	+45	+1,800
TOTAL			+ 550
C	50	-11	- 550

Figure 6-6. To determine where to place weight C, add the moments of both weight A and B together. Since the moment of weight B is greater than that of A, there is a net force, or moment of +550 pound-inches to the right of the fulcrum.

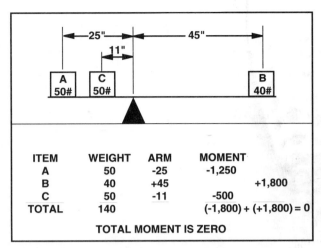

Figure 6-7. The sum of the moments left of the fulcrum is −1,800 pound-inches, and the moment right of the fulcrum is +1,800 pound-inches, therefore, the board balances.

With the moment calculated, the next step is to determine where to place weight C. In order to make the board balance, a force of −550 pound-inches must be exerted left of the fulcrum. To determine where weight C must be placed, divide the force (moment) needed by 50 pounds. The center of weight C must be 11 inches to the left of the fulcrum. To prove this, calculate the total moment on each side of the fulcrum. [Figure 6-7]

WEIGHING PROCEDURES

Before weighing an aircraft, you must become familiar with the weight and balance data for that particular aircraft. This information is found in FAA documentation and in the manufacturer's manuals such as:

1. Aircraft Specifications.
2. Type Certificate Data Sheets.
3. Manufacturer's Maintenance Manual.
4. Approved Airplane Flight Manual

The Type Certificate Data Sheets and the Aircraft Specifications contain basically the same information. However, sometimes the Aircraft Specifications have more detail concerning optional equipment. This information is sometimes furnished by the manufacturer on aircraft that have a Type Certificate Data Sheet but, in either case, both documents should be researched to obtain the following pertinent information:

1. CG range.
2. Empty weight CG range.
3. Leveling means.
4. Maximum weight.
5. Seats and location.
6. Baggage capacity.
7. Fuel capacity.
8. Datum location.

Some aircraft do not have a specified empty weight CG range. This is usually the case with aircraft that have a loading graph that is available to the pilot. However, when an empty weight CG range is indicated, it is valid only when the aircraft is loaded according to standard specifications. The installation of items not listed in the specifications do not permit the use of this range. This is important because, if the empty weight CG falls within its range, it is impossible to exceed the empty weight CG limits using specified loading arrangements.

Much of the information necessary to weigh and perform the necessary computations is self-explanatory. However, some information may need some clarification. For example, when looking at the top block of a Type Certificate Data Sheet, several aircraft are often listed in the same data sheet. This is typically done when multiple variations of the same aircraft are manufactured under the same Aircraft Specification or data sheet. When several models are specified on one data sheet, the pertinent information for a particular model is listed in one area.

PREPARATION

Aircraft accumulate enough dirt over time to give an inaccurate weight. Therefore, the first step when preparing an aircraft to be weighed is to clean the inside and outside. At the same time, a check of the equipment list must be made to ensure all required equipment is installed, and that there is no additional equipment installed that is not on the current equipment list.

As stated earlier, all fluid reservoirs and tanks must be filled to the level specified in the Type Certificate Date Sheets. For example, hydraulic reservoirs must be filled and if the aircraft is equipped with an anti-icing system, it also should be full. Since only unusable fuel is included in an aircraft's empty weight, the fuel tanks should be drained. However, since draining fuel tanks is often impractical, it is permissible to fill fuel tanks completely and then subtract out the weight of the usable fuel as specified in the Type Certificate Data Sheets. Furthermore, since the weight of fuel varies with temperature, the fuel temperature should be taken when the aircraft is weighed and a correction applied as necessary.

The oil tanks or sumps for aircraft certificated under FAR Part 23 are required to be filled. However, since only undrainable oil is included in the empty weight of aircraft that are not certificated under this part, their sumps are drained and the drain plugs left out or the drain valves left open. Again, if this is

	A12CE
	Revision 11
	BEECH
	60
	A60
	May 7, 1973

DEPARTMENT OF TRANSPORTATION
FEDERAL AVIATION ADMINISTRATION

TYPE CERTIFICATE DATA SHEET NO. A12CE

This data sheet which is part of type certificate No. A12CE prescribes conditions and limitations
under which the product for which the type certificate was issued meets the airworthiness requirements
of the Federal Aviation Regulations.

Type Certificate Holder   Beech Aircraft Corporation
                                 Wichita, Kansas 67201

I – Model 60, 4 or 6 PCLM (Normal Category), Approved February 1, 1968
    Model A60, 4 or 6 PCLM (Normal Category), Approved January 30, 1970

Engines	Lycoming T10-541-E1A4 or T10-541-E1C4 (2 of either or 1 of each)
Fuel	100/130 minimum grade aviation gasoline
Engine limits	For all operations, 2900 r.p.m. (380 b.hp.)
Propeller and	(a) Two (in any combination) Hartzell three-blade propellers
propeller limits	Diameter: 74 in., (Normal) Minimum allowable for repair 73 1/2 in.
	(No further reduction permitted)
	Pitch settings at 30 in. sta.:
	low 14°, high 81.7°
	HC-F 3 YR-2/C7479-2R
	or HC-F-3 YR-2/C7497B-2R
	or HC-F 3 YR-2F/FC 7479B-2R
	or HC-F 3 YR-2F/FC 7479B-2R

Airspeed limits		
Never exceed	235 knots
Maximum structural cruising	208 knots
Maneuvering	161 knots
Maximum flap extension speed	
Approach position 15°	175 knots
Full down position 30°	135 knots
Landing gear extended	174 knots
Landing gear operating	174 knots

C.G. range (landing
gear extended)

(+134.2) to (+139.2) at 6725 lb.
(+128.0) to (+139.2) at 5150 lb. or less
Straight line variation between points given
Moment change due to retracting landing gear (+857 in.-lb.)

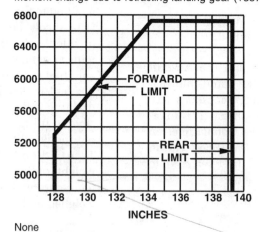

Empty wt. C.G. range	None
Maximum weight	Takeoff and landing 6725 lb.
Ramp weight	6819 lb.

Figure 6-8. An aircraft's Type Certificate Data Sheets contain information necessary to perform a weight and balance.

- 2 -   A12CE

No. of seats	4 (2 at +141, 2 at +173) (add 2 at +205)
Maximum baggage (structural limit)	500 lb. at +75 (nose compartment
	655 lb. at +212 (aft area of cabin)
Fuel capacity	142 gal. (+138) comprising two interconnected cells in each wing
	or
	204 gal. (+139) comprising three cells in each wing and one cell in each nacelle (four cells interconnected)
	See NOTE 1 for data on system fuel
Oil capacity (wet sump)	26 qt. (+88)
Max. operating limit	30,000 ft. pressure altitude

Control surface
movements

Wing flaps		Maximum 30°		
Aileron	Up	25°	Down	15°
Aileron tab (L.H. only)	Up	10°	Down	10°
Aileron tab anti-servo	Up	12°	Down	7°
Elevator	Up	17°	Down	15°
Elevator tab (L.H. only)	Up	10°	Down	30°
Elevator tab servo	Up	6°	Down	7°
Rudder	Right	33°	Left	28°
Rudder tab	Right	26°	Left	26°

Serial Nos. eligible	Model 60: P-3 thru P-125 (except P-123)
	Model A60: P-123, P-127 and up (see NOTE 3)
Datum	Located 100 in. forward of front pressure bulkhead
Leveling means	Drop plumb line between leveling screws in cabin door frame rear edge
Certification basis	Part 23 of the Federal Aviation Regulations effective February 1, 1965 as amended by 1, 2, 3, and 12; and Special Conditions dated May 16, 1967, forwarded with FAA letter dated June 1, 1967; approved for flight into known icing conditions when equipped as specified in the approved airplane flight manual.
	Application for Type Certificate dated December 22, 1965.
	Type Certificate No. A12CE issued February 1, 1968, obtained by the manufacturer under delegation option procedures.
Production basis	Production Certificate No. 8 issued and Delegation Option Manufacturer No. CE-2 authorized to issue airworthiness certificates under delegation option provisions of Part 21 of the Federal Aviation Regulations.
Equipment	The basic required equipment as prescribed in applicable airworthiness regulations (see Certification basis) must be installed in the aircraft for certification. This equipment must include, for all operations, Airplane Flight Manual P/N 60-590000-5D dated January 15, 1971, amended July 1, 1971, or later issue.

In addition:

1. For flights into known icing conditions, these flight manual
supplements and the equipment noted therein:

      60-590001-17   Flight Into Known Icing Conditions.

      60-590001-11   Continuous Pressure Operated Surface Deice System.

      60-59-001-13   Goodrich Electrothermal Propeller Deice System.

2. For all other operations:

      Pre-stall warning indicator P/N 151-6, 151-7, or 190-2 (Safe
Flight Corporation).

Figure 6-8. Continued.

- 3 -   A12CE

NOTE 1.   Current weight and balance data including list of equipment included in certificated empty weight and loading instructions when necessary must be provided for each aircraft at the time of original certification.

            The certificated empty weight and corresponding center of gravity locations must include unusable fuel of 24 lb. at (+135).

NOTE 2.   The following placard must be displayed in front of an in clear view of the pilot:

            "This airplane must be operated in the normal category in compliance with the operation limitations stated in the form of placards, markings and manuals."

NOTE 3.   Fuselage pressure vessel structural life limit — refer to the latest revision of the Airplane Flight Manual for mandatory retirement time.

NOTE 4.   Model 60 (S/N P-3 thru P-126 except P-123) when modified to Beech dwg. 60-5008 and Model A60 (S/N P-123, P-127 and up) eligible for a maximum weight of 6775 lb.

NOTE 5.   A landing weight of 6435 lb. must be observed if 10 PR tires are installed on aircraft not equipped with 60-810012-15 (LH) or 60-810012-16 (RH) shock struts.

. . . END . . .

Figure 6-8. Continued.

not practical, the sumps should be filled and the weight deducted from the scale weight. Since the amount of oil is quite small compared with the amount of fuel carried, it is not necessary to account for weight differences caused by variations in temperature. However, be aware that the amount of oil required is typically given in quarts while the standard weight for oil is given in gallons. Therefore, a conversion from quarts to gallons is required.

On some aircraft the position of the flight controls and seats are essential prior to weighing. These positions, when necessary, are noted in the Aircraft Specifications or the manufacturer's maintenance manual.

Prior to weighing an aircraft, you must determine where the weighing points are located. **Weighing points** are specific points on an aircraft where the scales must be placed for weighing. These weighing points are indicated in the aircraft's Type Certificate Data Sheets and should be clearly indicated on the aircraft weighing form.

EQUIPMENT

The two types of scales most commonly used to weigh aircraft are platform scales and electronic load cells. **Platform scales** are normally used with smaller aircraft. With this type of scale, aircraft are typically lifted off the floor with jacks and then lowered onto platform scales that are placed under each wheel. [Figure 6-9]

Once on the scales, chocks are used on the scale platforms to prevent the aircraft from rolling. The brakes must be released to prevent an uneven application of force to the platforms. The weight of the chocks is part of the tare weight that must be subtracted from the scale reading to get the aircraft's actual, or net weight.

Larger aircraft are weighed by placing **load cells** between the jack and the jack-pad of the aircraft. These load cells are strain-gauge capsules whose resistance changes proportionally to the amount of load imposed on them. An electronic bridge circuit converts this change in resistance into the aircraft weight readout.

Prior to recording the scale weights, an aircraft must be leveled according to instructions in the Type Certificate Data Sheets. The procedure for leveling some aircraft is accomplished by placing a level across two screws on the side of the fuselage. Other aircraft are leveled by dropping a plumb bob from a

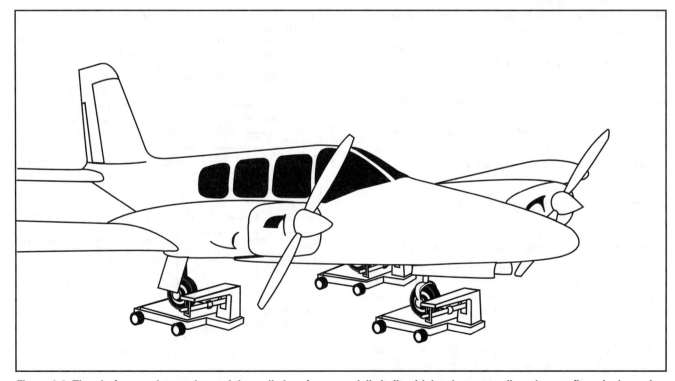

Figure 6-9. The platform scales used to weigh small aircraft are specially built with low beams to allow them to fit under low-wing aircraft.

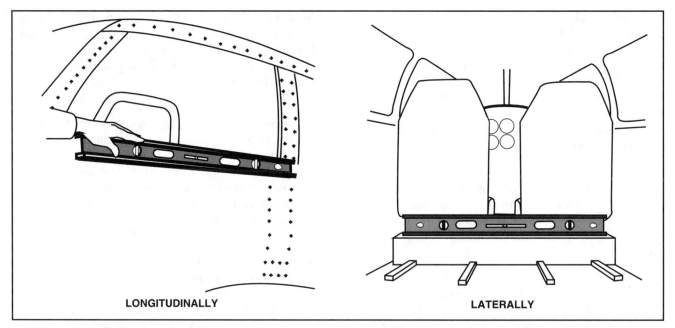

Figure 6-10. The leveling means for an aircraft are specified in its Type Certificate Data Sheets.

specified point within a cabin door frame or landing gear well. The longitudinal level is the most important condition when weighing. However, the aircraft should be laterally level as well. [Figure 6-10]

If an aircraft is weighed on jacks, it is typically leveled by adjusting the jack extension. However, if the aircraft is resting on its landing gear while on the weighing scales, adjustments are made by changing the amount of tire or strut inflation. In either case, it is extremely important when jacking or weighing an aircraft that the recommendations of the aircraft manufacturer be followed in detail.

When an aircraft is on the scales and level, plumb bobs are dropped from the datum and from the weighing points. This allows you to draw chalk lines on the floor to determine the distance the weighing points are from the datum.

RECORDING THE DATA

A weighing form is useful to systematically record all information for weight and balance. Most forms include space for the aircraft make, model, serial number, and N number of the aircraft being weighed as well as the datum location which is indicated on the Type Certificate Data Sheet. [Figure 6-11]

Also on the weighing form is space for the weighing conditions. This includes items such as the location of the weighing points in inches forward or aft of the datum, the scale reading, the tare weight, the

arm, and the moment. This format allows you to record the scale reading for the left, right, and nose or tail weighing points. Once this is done, the weight of any tare used to hold the aircraft on the scales is subtracted from the scale reading to get the net weight at each weighing point.

WEIGHING FORM

MAKE_____MODEL_____SERIAL_____N_____
DATUM LOCATION: _____
WEIGHING CONDITIONS:
   MAIN WEIGHING POINT IS LOCATED:
   (- FORWARD_____") (+ AFT_____") OF DATUM.
   TAIL OR NOSE WEIGHING POINT IS LOCATED:
   (-_____" FORWARD) (+_____" AFT) OF DATUM.

WEIGHING POINT	SCALE READING	- TARE	= NET WEIGHT	x ARM	= MOMENT
LEFT MAIN					
RIGHT MAIN					
NOSE OR TAIL					
TOTAL AS WEIGHED					

SPACE FOR LISTING OF ITEMS WHEN AIRCRAFT IS NOT WEIGHED EMPTY.

ITEM	NET WEIGHT	ARM	MOMENT
AIRCRAFT AS WEIGHED			
FUEL GAL. TEMP.°F #/GAL.			
AIRCRAFT EMPTY WEIGHT & C.G.			

MAXIMUM ALLOWABLE
GROSS WEIGHT_____COMPUTED BY:_____
                       A & P NUMBER:_____
USEFUL LOAD_____   DATE:_____

Figure 6-11. A typical weighing form is useful to systematically record all weighing information.

Once the net weight is found, the distance as measured along the floor between the datum and the weighing points is recorded in the arm column. If the weighing point is ahead of the datum, the arm is negative, but if it is behind the datum, the arm is positive. The moments are then calculated by multiplying the arm by the net weight. Next, the algebraic sum of the net weights and moments is calculated and recorded. With the total net weight and moment known, the next step is to find the CG by dividing the total moment by the total net weight.

If the aircraft was weighed empty, the result represents the aircraft's empty weight CG (EWCG). However, if the aircraft was not weighed empty, the weight and moments of all items not included in the aircraft empty weight must be subtracted out. The lower half of the weighing form is used to accomplish this. In this example, assume the aircraft was weighed with full fuel. Therefore, usable fuel must be subtracted. To accomplish this, the aircraft's total weight and moment are recorded in the lower chart. Then the weight, arm, and moment of the usable fuel is recorded and subtracted out.

If usable oil must be subtracted out, remember that the arm for oil on a small aircraft is typically negative. This means that when the oil is subtracted, the negative weight and negative arm produce a positive moment.

When the algebraic sum of the moments in the lower chart is divided by the total net weight, the result is the empty weight CG. The weighing form is completed by noting the maximum allowable gross weight as found in the Type Certificate Data Sheet, and the useful load, which is the difference between the maximum allowable gross weight and the aircraft empty weight.

LOCATING THE CG

It is sometimes more convenient to locate the datum and CG using the centerline of the main wheels as a reference. When this is done, the empty weight CG is determined through the use of formulas established for tailwheel- and nosewheel-type aircraft. The formulas are based on whether the CG is located forward or aft of the main wheels. [Figure 6-12]

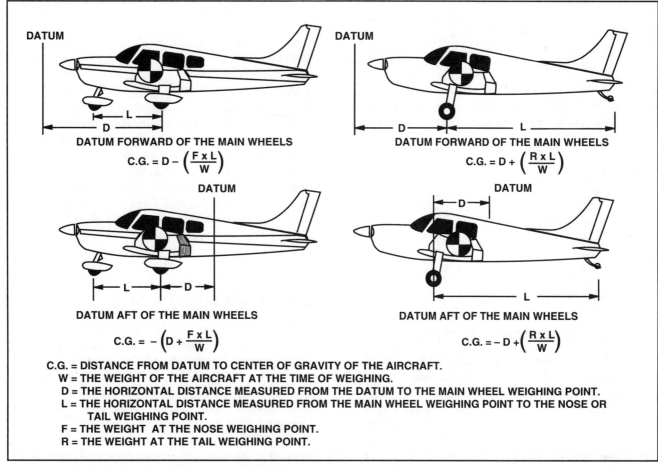

DATUM FORWARD OF THE MAIN WHEELS

$$C.G. = D - \left(\frac{F \times L}{W} \right)$$

DATUM AFT OF THE MAIN WHEELS

$$C.G. = - \left(D + \frac{F \times L}{W} \right)$$

DATUM FORWARD OF THE MAIN WHEELS

$$C.G. = D + \left(\frac{R \times L}{W} \right)$$

DATUM AFT OF THE MAIN WHEELS

$$C.G. = - D + \left(\frac{R \times L}{W} \right)$$

C.G. = DISTANCE FROM DATUM TO CENTER OF GRAVITY OF THE AIRCRAFT.
W = THE WEIGHT OF THE AIRCRAFT AT THE TIME OF WEIGHING.
D = THE HORIZONTAL DISTANCE MEASURED FROM THE DATUM TO THE MAIN WHEEL WEIGHING POINT.
L = THE HORIZONTAL DISTANCE MEASURED FROM THE MAIN WHEEL WEIGHING POINT TO THE NOSE OR TAIL WEIGHING POINT.
F = THE WEIGHT AT THE NOSE WEIGHING POINT.
R = THE WEIGHT AT THE TAIL WEIGHING POINT.

Figure 6-12. The formulas for locating the centerline of an aircraft are based on the type of wheel configuration the aircraft has.

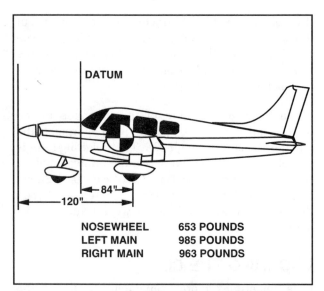

Figure 6-13. It is often easier to compute weight and balance information when a picture is drawn that displays the given data.

For example, assume you are weighing a nosewheel-type aircraft with the datum located ahead of the main wheels. After dropping a plumb bob, the datum is determined to be 84 inches ahead of the main gear weighing points and the distance between the nose gear and main gear is 120 inches. In addition, assume the scale reading is 653 pounds for the nose wheel, 985 pounds for the left main gear, and 963 pounds for the right main gear. [Figure 6-13]

Based on this information, the aircraft's CG is calculated to be 53.87 inches to the right or aft of the datum.

$$CG = D - \frac{F \times L}{W}$$

$$= 84 - \frac{653 \times 120}{2,601}$$

$$= 84 - 30.13$$

$$= 53.87 \text{ Inches}$$

MEAN AERODYNAMIC CHORD (MAC)

The CG location in a transport aircraft is given in terms of percent of the mean aerodynamic chord (%MAC) of the wing. If you remember, a wing's chord is the distance from the leading edge to the trailing edge.

The **mean aerodynamic chord** (MAC) is the chord drawn through the center of the wing plan area.

Since the wing on most aircraft is not a rectangle, the mean aerodynamic chord is determined for weight and balance and aerodynamic purposes. The center-of-gravity range on most large aircraft is expressed with respect to the CG location on the mean aerodynamic chord. For example, if the center of gravity is 15 percent aft of the leading edge of a mean aerodynamic chord which is 100 inches long, the center of gravity would be 15 inches aft of the leading edge of the mean aerodynamic chord.

The leading edge of the MAC is referred to as LEMAC and the trailing edge of the MAC is designated as TEMAC. The location of LEMAC is usually expressed as a body station number to aid in determining the body station number of the center of gravity. [Figure 6-14]

The length of the MAC is established by the manufacturer and is found in the aircraft's Type Certificate Data Sheets and maintenance manuals. When the length of the MAC and the CG position are known, it is an easy task to determine the CG in percent MAC.

At the maximum landing weight of 137,500 pounds for this aircraft, the center of gravity range with the gear and flaps in the landing position is from 14% MAC to 36.5% MAC. By computing the percentage figure in inches, it can be determined that the forward CG limit is 25.3 inches aft of the LEMAC and the aft CG limit is 66 inches aft of LEMAC. By adding these dimensions to the station number of LEMAC (860.2 inches), it is determined that the forward center of gravity limit at the maximum landing weight with the gear and flaps in the landing position is at station number 885.5 and the aft CG limit is at station number 926.2.

EMPTY WEIGHT CG RANGE

Light aircraft with fuel tanks located in the wing, and seats that are side-by-side with a small baggage compartment immediately behind the seats, have a relatively limited CG range. For this reason, the manufacturer includes an empty weight CG range in the Aircraft Specification Sheets. If the empty weight CG of an aircraft falls within the empty weight CG range, the aircraft cannot be legally loaded so that its CG in flight falls outside of the loaded CG range.

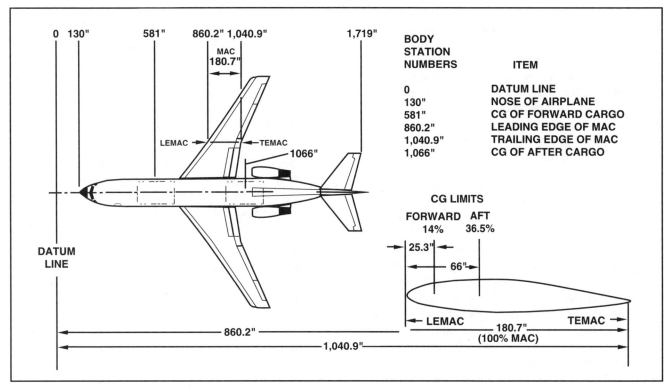

Figure 6-14. The LEMAC on this aircraft is at station 860.2 and the TEMAC is at body station number 1,040.9. The mean aerodynamic chord of this wing is 180.7 inches and represents the chord through the center of the wing.

LOADED CG RANGE

Larger aircraft with several rows of seats and both forward and aft baggage compartments typically have a loaded CG graph contained in the Type Certificate Data Sheets. Above the graph are the figures the graph is based on. [Figure 6-15]

The information from this graph applies to the aircraft with its landing gear extended. When the gear is retracted there is a moment change of −857 pound-inches. This results in the CG moving aft when the gear is retracted.

The aircraft has a ramp weight of 6,819 pounds and a gross allowable takeoff-and-landing weight of 6,725 pounds. Once the aircraft reaches this weight, it can only be loaded with additional fuel to reach the maximum ramp weight. This means the aircraft may be loaded with 15.7 gallons of fuel above the allowable gross weight. However, all of this fuel must be consumed by the time the aircraft taxies into the takeoff position.

WEIGHT DISTRIBUTION

It is the responsibility of the aviation maintenance technician to ensure that weight and balance information for an aircraft is correct and updated.

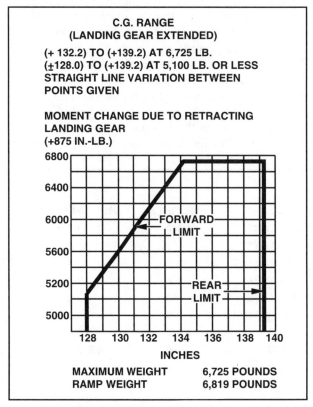

Figure 6-15. The graph indicates that up to a gross weight of 5,150 pounds, the forward CG limit is +128 inches, and the aft limit is +139.2 inches. However, at the gross weight of 6,725 pounds, the forward limit shifts aft to +134.2 inches while the aft limit remains at +139.2 inches.

Changes to the aircraft's weight and balance resulting from equipment being added or removed, or a repair or alteration, must be recorded in the weight and balance documentation. Furthermore, all old computations should be marked to ensure that correct information is used by the pilot. The flight manual for the aircraft normally has a weight and balance loading graph and a moment envelope. Before each flight where weight and balance is critical, the pilot should complete a weight and balance form.

To make computations easier, a **reduction factor** is applied to all moments. This results in what is called a **moment index**. For example, a moment of 12,500 pound-inches that has a reduction factor of 1,000 applied to it results in a moment index of 12.5 pound-inches. A typical loading problem determines whether an aircraft weighing 1,340 pounds, having a moment index of +51.6, and a maximum takeoff weight of 2,400 pounds is within its legally loaded envelope when it carries four standard-weight occupants (170 pounds each), full fuel, full oil and 80 pounds of baggage. To begin, record the empty weight of the airplane and its moment index in a weight and balance chart. [Figure 6-16]

Next, find the moment index of the eight quarts, or two gallons, of oil. To do this, multiply two gallons by 7.5 pounds per gallon to obtain 15 pounds, and then locate 15 pounds on the chart's vertical scale. From here, move right horizontally until you intersect the oil line. Now, drop down vertically and read the moment index of −1.0 pound-inches.

Two occupants with a combined weight of 340 pounds are in the front seats and they have a moment index of +11.8 pound-inches. The rear seat also carries 340 pounds, but has a moment index of +24.3 pound-inches. The 80 pounds of baggage has a moment index of +7.3 pound-inches, while 40 gallons of fuel weighs 240 pounds and has a moment index of +11.5 pound-inches. The total weight is calculated at 2,355 pounds, and the total moment index is +105.5 pound-inches.

To determine whether or not the loaded aircraft falls within its allowable CG range, transfer the total weight and total moment index to a loaded moment envelope graph. [Figure 6-17]

This loaded aircraft moment chart is approved for both normal and utility categories. In the normal category the allowable gross weight is 2,400 pounds with a moment index of 110 pound-inches. However, in the utility category, the maximum gross weight decreases to 2,050 pounds and the moment index of approximately 84 pound-inches results.

AIR TAXI LOADING

FAR Part 135 governs Air Taxi and Commercial Operators. It dictates that passenger and cargo manifests be completed before each flight. These manifests allow the pilot to know the exact takeoff weight of an aircraft, and if the aircraft is loaded within allowable CG limits. Air taxi operators are allowed to use either actual or standard weights when calculating the weight of passengers. The moment indexes of actual weights are calculated by using a moment index chart. [Figure 6-18]

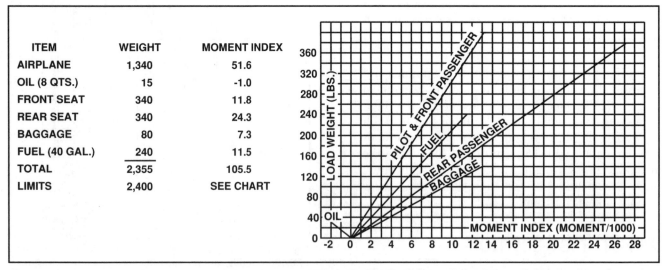

ITEM	WEIGHT	MOMENT INDEX
AIRPLANE	1,340	51.6
OIL (8 QTS.)	15	-1.0
FRONT SEAT	340	11.8
REAR SEAT	340	24.3
BAGGAGE	80	7.3
FUEL (40 GAL.)	240	11.5
TOTAL	2,355	105.5
LIMITS	2,400	SEE CHART

Figure 6-16. (A)— A typical weight and balance chart begins with the aircraft empty weight and moment index. (B)— To use the moment index graph start on the vertical axis with an object's weight. Then move right horizontally until you intersect the proper vertical line. From this intersection, drop down to the bottom of the chart to read the item's moment index.

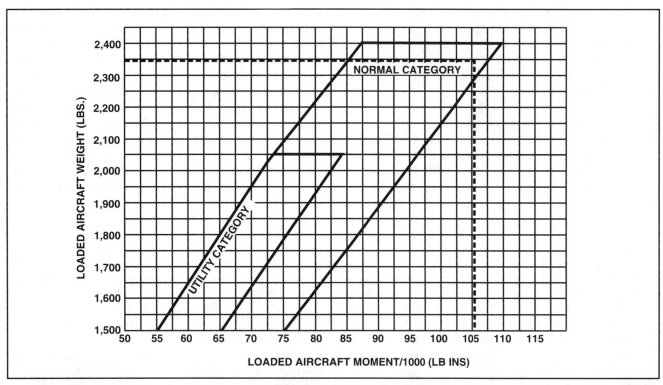

Figure 6-17. In this case, draw a horizontal line to the right through the 2,355-pound weight and a vertical line up through the 105.5 moment index line. These two lines must intersect within the envelope in order for the aircraft to be loaded properly. In this case, the intersection is within the envelope, near the rearward CG limit.

PASSENGER-CARGO LIST	WEIGHT	SEAT COMPT.	INDEX
	180	1	126.0
	170	2	119.0
	140	3	126.0
Passenger Names	150	4	135.0
Entered Here	200	5	220.0
	160	6	176.0
	210	7	273.0
	130	8	169.0
	120	9	180.0
	140	10	210.0
	130	11	221.0

				WEIGHT	SEAT COMPT.	INDEX
PASSENGER- CARGO-MAIN CABIN TOTAL						
			BAGGAGE COMPT	100	NOSE	20.0
WEIGHT		INDEX	BAGGAGE COMPT	200	REAR	380.0
300		270.0	PILOT-COPILOT	330		165.0
300		270.0	OIL	85		7.7
120		144.0	FUEL	840	TOTAL	828.0
120		144.0	EMPTY WEIGHT	6,150		5,842.5
			TAKEOFF WEIGHT	9,435		9,198.2
840		828.0	TAKEOFF LIMITS	10,000		8,467.9 9,694.5

Figure 6-18. In a typical manifest, the weight and associated moment indexes are entered into the appropriate spaces. They are then totaled and compared to the aircraft's limits.

After the fuel weight and moment indexes are found they are entered into the appropriate blanks of the cargo manifest as well as the weight and moment indexes of the crew, the oil and the empty airplane. The total weight and moment index is the sum of these individual values.

After computing the takeoff or loaded weight, find the allowable forward and aft CG limits. In this example, the loaded weight is computed at 9,435 pounds with a moment index of 9,198.2 pound-inches. To determine if the aircraft is loaded properly, apply these figures to a weight/moment index chart. [Figure 6-19]

The notes at the bottom of the chart indicate that the difference in the minimum index for each 100 pounds is 89.8 index numbers. Since the aircraft's weight of 9,435 pounds is 35 pounds above 9,400 pounds, find that 35% of 89.8 is 31.4 and add it to 8,436.5 pound-inches to determine the minimum index of 8,467.9 pound-inches. The maximum index interpolation requires 102.8 index numbers for each 100 pounds. Therefore, since 35% of 102.8 is 36.0, the maximum index at 9,435 pounds is 9,694.5 pound-inches. Both the loaded weight and loaded moment indexes of the problem are within the allowable limits.

WEIGHT	MIN. INDEX	MAX. INDEX	WEIGHT	MIN. INDEX	MAX. INDEX	WEIGHT	MIN. INDEX	MAX. INDEX
7000	6282.5	7192.5	8000	7180.0	8220.0	9000	8077.5	9247.5
7050	6327.4	7243.9	8050	7224.9	8271.4	9050	8122.4	9298.9
7100	6372.4	7295.3	8100	7296.8	8322.8	9100	8167.3	9350.3
7150	6417.1	7346.6	8150	7314.6	8374.1	9150	8212.1	9401.6
7200	6462.0	7398.0	8200	7359.5	8425.5	9200	8257.0	9453.0
7250	6506.9	7449.4	8250	7404.4	8476.9	9250	8301.9	9504.4
7300	6551.8	7500.8	8300	7449.3	8528.3	9300	8346.8	9555.8
7350	6596.6	7552.1	8350	7494.1	8579.6	9350	8391.6	9607.1
7400	6641.5	7603.5	8400	7539.0	8631.0	9400	8436.5	9658.5
7450	6686.4	7654.9	8450	7583.9	8682.4	9450	8481.4	9709.9
7500	6731.3	7706.3	8500	7628.8	8733.8	9500	8526.3	9761.3
7550	6776.1	7757.6	8550	7673.6	8785.1	9550	8571.1	9812.6
7600	6821.0	7809.0	8600	7718.5	8836.5	9600	8616.0	9864.0
7650	6865.9	7860.4	8650	7763.4	8887.9	9650	8660.9	9915.4
7700	6910.8	7911.8	8700	7808.3	8939.3	9700	8705.8	9966.8
7750	6955.6	7963.1	8750	7853.1	8990.6	9750	8750.6	10018.1
7800	7000.5	8014.5	8800	7898.0	9042.0	9800	8795.5	10069.5
7850	7045.4	8065.9	8850	7942.9	9093.4	9850	8840.4	10120.9
7900	7090.3	8117.3	8900	7987.8	9144.8	9900	8885.3	10172.3
7950	7135.1	8168.6	8950	8032.6	9196.1	9950	8930.1	10223.6
						10000	8975.0	10275.0

THE DIFFERENCE IN MIN. INDEX FOR EACH 100 POUNDS IS 89.8 INDEX NUMBERS
THE DIFFERENCE IN MAX. INDEX FOR EACH 100 POUNDS IS 102.8 INDEX NUMBERS

TO FIND LIMITS FOR A SPECIFIC WEIGHT: FOR EXAMPLE 9435 FIND LIMITS FOR 9400
8436.5 - 9658.5
FOR 9435 POUNDS, ADD 35% OF 89.8 OR 31.4 TO 8436.5 = 8467.9
ADD 35% OF 102.8 OR 36.0 TO 9658.3 = 9694.5

Figure 6-19. To determine if the aircraft is loaded properly, find the weight that matches the aircraft's loaded weight. Since 9,435 pounds is between 9,400 and 9,450 pounds, you must interpolate the minimum and maximum moment index numbers.

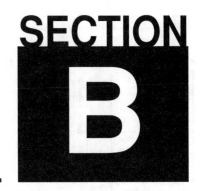
SHIFTING THE CG

Whenever something is added to an aircraft, such as a beacon or strobe light, the center of gravity changes. Because of this, a weight and balance report should be calculated before a repair or installation is accomplished to ensure that the CG will fall within its specified range. Completing these calculations early also allows you the option of trying different installation locations before any work is accomplished.

BALLAST

Often it is necessary to install ballast, or weight in an aircraft to bring the CG into its desired range. **Permanent ballast**, like its name implies, is permanently installed in an aircraft and must not be removed. **Temporary ballast**, on the other hand, is typically installed to bring the CG within its range for a specific flight condition. For example, temporary ballast may need to be installed when an aircraft has only one person in it. However, if two or more people are on board the ballast is removed.

In order to determine the amount of ballast needed for a particular scenario, it is usually helpful to construct a drawing to illustrate the problem. For example, assume you have an aircraft that weighs 1,490 pounds and balances at a point 37.5 inches aft of the datum and you want it to balance at a point 42.5 inches aft. Anytime you need to shift the CG rearward, ballast must be installed aft of the datum. In this example, there is a convenient location for ballast at 170 inches aft of the datum. The goal then, is to find the amount of weight needed 170 inches from the datum that shifts the CG 5 inches aft.

The formula used to accomplish this task is:

$$\text{Ballast} = \frac{\text{Aircraft weight} \times \text{Desired CG change}}{\text{arm of ballast} - \text{arm of desired CG}}$$

When the figures indicated are plugged into this equation, a ballast weight of 58.43 pounds at 170 inches is required to move the balance point 5 inches aft. [Figure 6-20]

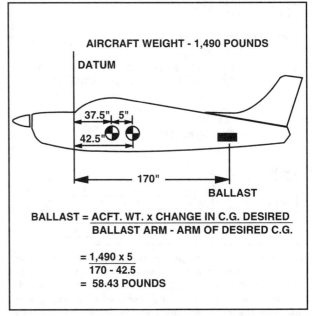

AIRCRAFT WEIGHT - 1,490 POUNDS

$$\text{BALLAST} = \frac{\text{ACFT. WT.} \times \text{CHANGE IN C.G. DESIRED}}{\text{BALLAST ARM} - \text{ARM OF DESIRED C.G.}}$$

$$= \frac{1,490 \times 5}{170 - 42.5}$$

$$= 58.43 \text{ POUNDS}$$

Figure 6-20. To determine how much ballast must be placed at station 170 to move the CG 5 inches aft, use the ballast formula.

INSTALLING BALLAST

The aircraft structure where ballast is attached must be strong enough to support the ballast weight under all flight conditions. Ballast is normally painted red and marked "Permanent Ballast—Do Not Remove." Temporary ballast must also be clearly marked, stating that it is to be carried in the aircraft only during the flight condition where it is required. When ballast is needed, it should be installed as far aft as possible. This allows the weight to be kept to a minimum.

The formula for determining the amount of temporary ballast required is the same as the one for permanent ballast. In order for some tandem-seat aircraft to be flown solo from the front seat, temporary ballast must be carried in the baggage compartment behind the rear seat. For example, assume that an aircraft with pilot and full fuel weighs 1,045 pounds and its CG is located at +10 inches. The CG needs to move 2 inches in order to be within its range, and

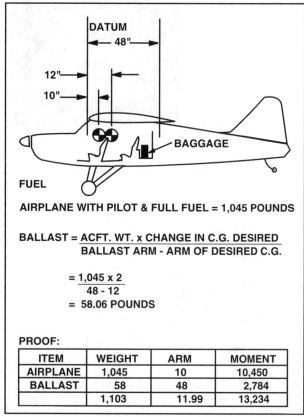

AIRPLANE WITH PILOT & FULL FUEL = 1,045 POUNDS

$$\text{BALLAST} = \frac{\text{ACFT. WT. x CHANGE IN C.G. DESIRED}}{\text{BALLAST ARM - ARM OF DESIRED C.G.}}$$

$$= \frac{1,045 \times 2}{48 - 12}$$

$$= 58.06 \text{ POUNDS}$$

PROOF:

ITEM	WEIGHT	ARM	MOMENT
AIRPLANE	1,045	10	10,450
BALLAST	58	48	2,784
	1,103	11.99	13,234

Figure 6-21. It is often helpful to use graphic illustration aids in determining the temporary ballast needed to correct an unusual flight condition. In this example, 58.06 pounds must be carried in the baggage compartment during solo flight from the front seat. Also, always proof the formula by using a typical weight and balance chart.

the temporary ballast is carried in the baggage compartment at 48 inches. [Figure 6-21]

SHIFTING WEIGHT

Large aircraft have several rows of seats and more than one baggage compartment that weight can be moved between in order to maintain a desired CG. For example, assume a large aircraft has a baggage compartment at station 26 and one at station 246. The aircraft's CG is 1.5 inches ahead of the forward limit and, therefore, baggage must be moved from the forward compartment to the aft compartment. [Figure 6-22]

The amount of weight that needs to shift is based on the fact that the ratio of the amount of weight shifted to the total aircraft weight is proportional to the ratio of the change in CG required to the distance the weight is shifted. This is represented by the formula:

$$\frac{\text{Weight shifted}}{\text{Total weight}} = \frac{\text{CG change required}}{\text{Distance weight is shifted}}$$

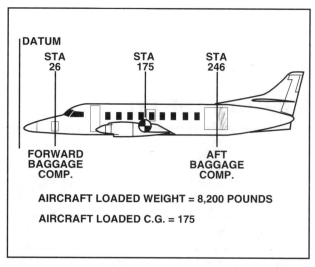

AIRCRAFT LOADED WEIGHT = 8,200 POUNDS

AIRCRAFT LOADED C.G. = 175

Figure 6-22. In order to bring the CG of a large aircraft within the CG range, weight is shifted from one location to another. In this case, the weight is moved from the forward baggage compartment located at station 26 to the aft baggage compartment at station 246.

When algebraically manipulated the formula for determining the weight shifted is:

$$\text{Weight shifted} = \frac{\text{Total weight} \times \text{CG change required}}{\text{Distance weight is shifted}}$$

Based on this, 55.9 pounds must be shifted to the aft baggage compartment.

$$= \frac{8,200 \times 1.5}{220}$$

$$= 55.9 \text{ pounds}$$

WEIGHT AND BALANCE CHANGES AFTER AN ALTERATION

One of the most important weight and balance problems an aircraft maintenance technician performs is finding the new empty weight and empty weight CG after altering an aircraft. Normally, if the weights and location of all items installed or removed from the aircraft are known or the precise location of an alteration is known, a new empty weight and empty weight CG is computed without re-weighing the aircraft. For example, a specific alteration may require the removal of a radio from the instrument panel, and its power supply from the baggage compartment. In its place, a smaller radio that does not require a separate power sup-

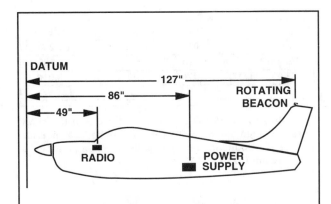

ITEM	WEIGHT	ARM	MOMENT
AIRPLANE	2,024	+ 52.5	+ 106,260
REMOVE RADIO	(−) 16	+ 49	(−) 784
REMOVE POWER SUPPLY	(−) 12	+ 86	(−) 1,032
ROTATING BEACON	(+) 4	+127	+ 508
RADIO	+ 9	+ 49	+ 441
	2,009	52.46	+ 105,393

Figure 6-23. The empty weight of the airplane is 2,024 pounds and its empty weight CG is +52.5 inches. The radio removed from the instrument panel has a weight of 16 pounds and an arm of +49 inches. The power supply weight is 12 pounds and has an arm of +86. The new radio is installed where the old radio was, but has a weight of +9 pounds. The anticollision beacon weighs +4 pounds and is installed at station +127.

ply is installed. At the same time, an anticollision beacon is installed on the tail. [Figure 6-23]

Once the weight and location of the alterations are known, the new empty weight and moment are calaculated. The sum of the new moments is +105,393 pound-inches while the new empty weight is 2,009 pounds. When the total moment is divided by the total weight, the new empty weight CG is found to be 52.46 inches aft of the datum.

ADVERSE-LOADED CG

As discussed earlier, when you change the weight of an aircraft, a new CG must be determined. If, after computation, this new CG does not fall within the empty weight CG range, it is then necessary to determine if an adverse-loaded condition exists. This is accomplished by computing a **forward adverse-loading check**, a **rearward adverse-loading check**, and a **maximum gross weight check**. If you can compute a weight and balance load that falls out-

side the loaded CG range, the aircraft is placarded as appropriate to prevent a pilot from loading the aircraft improperly. In extreme cases, fixed ballast is installed in the aircraft to prevent it from falling outside of the CG range.

MINIMUM FUEL

When doing an adverse CG check the term "minimum fuel" is used extensively. Mimimum fuel, as it applies to weight and balance, is the amount of fuel that must be shown on the weight and balance report when the aircraft is loaded for an extreme condition check. The minimum fuel load for a small aircraft with a reciprocating engine is based on engine horsepower. It is calculated in terms of METO (maximum except take-off) horsepower and is the figure used when the fuel load must be reduced to obtain the most critical loading on the CG limit being investigated. An aircraft's METO rating is found in the Type Certificate Data Sheets, on the engine data tag, or in the approved flight manual.

There are two ways to calculate minimum fuel. One is to divide the METO horsepower by 12 to get the number of gallons required. This is then multiplied by 6 to convert the gallons into pounds. This is represented by the formula:

$$\text{Minimum fuel} = \frac{\text{METO}}{12} \times 6$$

The second method allows you to convert directly to pounds by dividing the METO horsepower by 2. This is illustrated in the formula:

$$\text{Minimum fuel} = \frac{\text{METO}}{2}$$

For turbine aircraft, the minimum fuel required for an adverse CG check is specified by the aircraft manufacturer.

To conduct an adverse loading check we will use a typical four-place airplane powered by an engine having 230 METO horsepower, carrying 50 gallons of fuel, four occupants, and a rear baggage compartment that is limited to 120 pounds of baggage. The forward CG limit is at +52 inches, and the aft limit is at +58 inches. The airplane has an empty weight of 2,024 pounds and an empty weight CG of +52.5.

The maximum allowable gross weight for this airplane is 3,150 pounds. [Figure 6-24]

FORWARD ADVERSE-LOADING CHECK

To determine if it is possible to legally load this aircraft in such a way that its CG falls ahead of the forward limit, you must load everything ahead of the forward limit to its maximum weight, and everything aft of the forward CG to a minimum weight. One way to compile the information for an adverse check is with a weight and balance chart. For example, for the forward adverse-loading check enter the airplane with its weight, arm, and moment. Fuel is the next item on the list. Since the fuel is located aft of the forward limit the minimum fuel figures are used for this computation. In this example, the engine has 230 METO horsepower, so the minimum fuel weighs 115 pounds.

The two front seats are also located behind the forward CG, however, since a pilot is always needed to fly the aircraft, one occupant is included in the forward adverse-loading check. Since the exact weight of the pilot is unknown, the standard weight of 170 pounds is used.

The rear seats and baggage compartment are both behind the forward CG limit, therefore, they should be loaded with a minimum of weight. Since nothing

ITEM	WEIGHT	ARM	MOMENT
AIRPLANE	2,024	52.5	106,260
FUEL (MIN.)	115	57	6,555
F. SEAT (PILOT)	170	53	9,010
	2,309	52.76	121,825

Figure 6-25. Once all the items are entered into the chart the total weight and moment are found to be 2,309 pounds and 121,825 pound-inches respectively. This equates to a most forward CG of +52.76 inches.

required to fly the aircraft is located in either of those areas they are left empty. [Figure 6-25]

Since the most forward CG check produced a CG that is behind the forward limit, it is not possible to legally load the aircraft and exceed the forward CG limit.

REARWARD ADVERSE-LOADING CHECK

The rearward adverse-loading check is accomplished in the same manner as the forward adverse-loading check. However, this time everything aft of the rearward CG limit is loaded to a maximum while everything forward of the aft limit is loaded to a minimum. For example, since the CG of the fuel tank is ahead of the rear limit, minimum fuel is used. The front seat is ahead of the rear limit, so the minimum or only one occupant (the pilot) is used. Both the rear seat and the baggage compartment are behind the rear CG limit. Therefore, two occupants are calculated to be in the rear seat and the baggage area is loaded to 120 pounds. [Figure 6-26]

MAXIMUM GROSS WEIGHT CHECK

The reason for doing a maximum gross weight check is to determine if it is possible to load the aircraft in such a way that the aircraft's gross weight

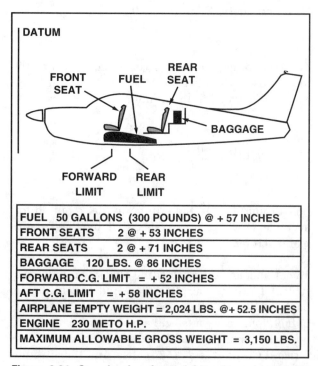

FUEL   50 GALLONS (300 POUNDS) @ + 57 INCHES
FRONT SEATS   2 @ + 53 INCHES
REAR SEATS   2 @ + 71 INCHES
BAGGAGE   120 LBS. @ 86 INCHES
FORWARD C.G. LIMIT   = + 52 INCHES
AFT C.G. LIMIT   = + 58 INCHES
AIRPLANE EMPTY WEIGHT = 2,024 LBS. @ + 52.5 INCHES
ENGINE   230 METO H.P.
MAXIMUM ALLOWABLE GROSS WEIGHT = 3,150 LBS.

Figure 6-24. Sample aircraft used for adverse-loaded CG checks.

ITEM	WEIGHT	ARM	MOMENT
AIRPLANE	2,024	52.5	106,260
FUEL (MIN.)	115	57	6,555
F. SEAT (PILOT)	170	53	9,010
R. SEAT (MAX.)	340	71	24,140
BAGGAGE (MAX.)	120	86	10,320
	2,769	56.44	156,285

Figure 6-26. The total weight and moment are found to be 2,769 pounds and 156,285 pound-inches respectively. This equates to a most rearward CG of +56.44 inches. This is well within the rearward CG limit.

can be exceeded. To do this, the total weight and moment of the airplane, full fuel, maximum baggage, and four occupants are compiled into a weight and balance chart. [Figure 6-27]

The maximum weight is calculated at 3,124 pounds which is less than the maximum allowable gross weight. The fully loaded CG is +56.29 inches and well within the allowable CG range. If the conditions of any adverse-loading check is not within the CG range the aircraft must be placarded to indicate the condition.

ITEM	WEIGHT	ARM	MOMENT
AIRPLANE	2,024	52.5	106,260
FUEL (MAX.)	300	57	17,100
F. SEAT (PILOT) +1	340	53	18,020
R. SEAT (MAX.)	340	71	24,140
BAGGAGE (MAX.)	120	86	10,320
	3,124	56.29	175,840

Figure 6-27. The maximum gross weight check is accomplished by loading everything with the maximum possible.

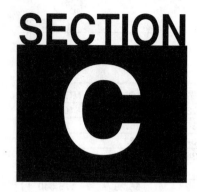
HELICOPTER WEIGHT AND BALANCE

The weight and balance principles and procedures used for aircraft generally apply to helicopters as well. However, there are some differences. For example, although every model helicopter is certificated for a specific maximum gross weight, helicopters cannot be operated at this maximum weight under all conditions. Combinations of high altitude, high temperature, and high humidity increase density altitude for a particular location thus decreasing performance. This, in turn, critically affects the hovering, takeoff, climb, autorotation, and landing performance of a helicopter. Furthermore, a heavily loaded helicopter has less ability to withstand shocks and additional loads caused by turbulent air. The heavier the load, the less the margin of safety for supporting structures like the main rotor, fuselage, and landing gear. Another difference exists in the fact that in addition to computing the weight and CG for the longitudinal axis, you must compute it for the lateral axis as well.

CG

Most helicopters have a much smaller CG range than airplanes. In some cases, this range is less than 3 inches. The exact location and length of the CG range is specified for each helicopter and usually extends a short distance fore and aft of the main rotor mast or the center of a dual rotor system.

Ideally, a helicopter should have such perfect balance that the fuselage remains horizontal while in hover. In this situation, only cyclic adjustments would be required to compensate for the wind.

The fuselage acts as a pendulum suspended from the rotor. Any change in the CG changes the angle at which it hangs from its pivot point or support. For this reason, modern helicopters have loading compartments and fuel tanks located at or near the balance point. If a helicopter is loaded improperly, the fuselage does not hang horizontally in a hover. For example, if the CG is too far aft, the nose tilts up, and excessive forward cyclic control is required to maintain a stationary hover. Conversely, if the CG is too far forward, the nose tilts down and excessive aft cyclic control is required. In extreme out-of-balance conditions, full fore or aft cyclic control is insufficient to maintain control. Similar lateral balance problems are encountered if external loads are carried. For this reason, a lateral CG check is sometimes necessary.

CALCULATING THE CG

To determine a helicopter's longitudinal and lateral CG, a sample problem is used for a typical corporate helicopter. In preparing to find the empty weight and CG, the following information is taken from the helicopter's Type Certification Data Sheet. [Figure 6-28]

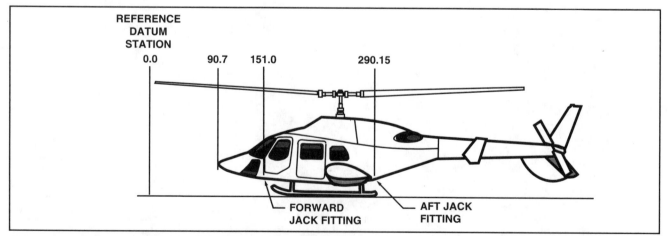

Figure 6-28. On this sample helicopter the datum is located 90.7 inches in front of the nose while one jacking point is located at +151 inches and two others at +290.15 inches.

Datum: Location is +90.7 inches forward of the radome nose.

Leveling means: Plumb line from right inside top of baggage compartment.

Maximum weight: 8,250 pounds.

Minimum crew: 1 pilot weighing 165 pounds.

Maximum baggage: 500 pounds located between station +268 and station +324.

Fuel capacity: 247.1 gallons of usable fuel at station +263.3.

Oil capacity: 3.7 gallons at +263.3.

Jack points: The forward jack point is located at station +151.0 and butt line zero. The aft jack points are located at station +290.15 and butt line +23.0 and −23.0

Like any aircraft, the helicopter is prepared for weighing by the following steps:

Clean and remove all loose articles.

Drain fuel from tanks and fill the hydraulic fluid and transmission oil.

Place the aircraft in a closed hangar and remove the rotor tiedowns.

Zero the scales and place one load cell between each jack pad and jack.

Level the helicopter longitudinally and laterally. [Figure 6-29]

Once the entire weight of the helicopter is resting on the load cells and the aircraft is leveled, record the scale readings. Since electronic scales are used, there is no tare weight. In this example, the scale weights are:

Front jack point	2,405 lb. at station +151
Left aft jack point	2,718 lb. at station +290.15
Right aft jack point	2,855 lb. at station +290.15
Total weight	7,978 lb.

To determine the empty weight CG use the formula:

$$CG = \frac{(\text{Fwd. scale wt} \times \text{arm}) + (\text{Aft scale wt} \times \text{arm})}{\text{Total weight}}$$

$$CG = \frac{(2,405 \times 151) + (5,573 \times 290.15)}{7,978}$$

$$= \frac{363,155 + 1,617,005.95}{7,978}$$

$$= 248.20$$

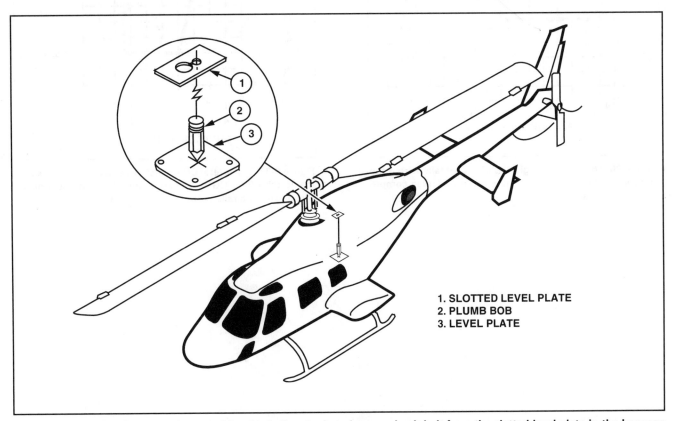

Figure 6-29. The leveling procedure used for this helicopter is to hang a plumb bob from the slotted level plate in the baggage compartment. The helicopter is level when the plumb bob is directly over the intersection of the insribed lines.

1. SLOTTED LEVEL PLATE
2. PLUMB BOB
3. LEVEL PLATE

The CG of 248.20 falls within the empty weight CG range. [Figure 6-30]

Calculation of the lateral CG is similar to the longitudinal CG except that Butt Line "0" is used as the datum. Butt lines to the right are designated (+) while butt lines to the left are designated (–). Since the forward jack point is located on butt line zero it has no bearing on the lateral CG. The formula used to calculate a lateral CG is:

$$\text{Lateral CG} = \frac{(\text{BL} \times \text{C}) + (\text{BR} \times \text{D})}{\text{W}}$$

Where:

BL = Butt measurement left.
C = Left scale weight.
BR = Butt measurement right.
D = Right scale weight.
W = Weight of aircraft.

$$\text{CG} = \frac{(-23 \times 2{,}718) + (23 \times 2{,}855)}{7{,}978}$$

$$\text{CG} = \frac{-62{,}514 + 65{,}665}{7{,}978}$$

$$\text{CG} = \frac{3151}{7{,}978} = .4$$

The lateral CG of +0.4 along with the longitudinal CG are checked against a graph to verify the new CG is within limits. [Figure 6-31]

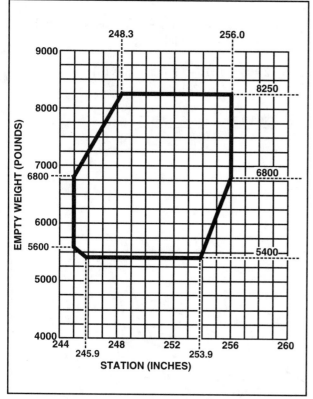

Figure 6-30. The empty weight CG of 248.20 falls within the empty weight CG envelope. Therefore no adverse computations are required.

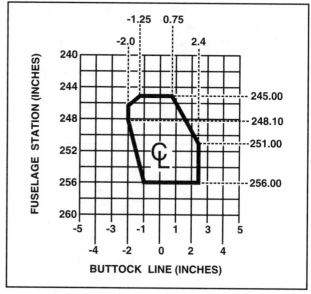

Figure 6-31. The lateral CG is calculated to be 0.4 inches and the longitudinal CG is 248.20 inches. The intersection of these two CGs falls within the acceptable range.

AIRCRAFT STRUCTURAL MATERIALS

CHAPTER 7

INTRODUCTION

The techniques and materials used in the early years of aviation were quite primitive by modern standards. The Wright brothers' "Flyer," for example, was made from steel, wire, cable, silk, and wood. However, as aircraft development advanced, a breakthrough occurred in the aircraft aluminum industry. Metallurgists found that mixing, or alloying aluminum with other metals resulted in a much stronger material. In fact, alloying increased the tensile strength of pure aluminum from about 13,000 psi to a tensile strength of 65,000 psi or greater, which is equivalent to structural steel. As the need for aluminum alloys grew, manufacturers continued to refine them to produce materials with better corrosion resistance and greater strength. Today, military aircraft are constructed of about 65 percent aluminum and 35 percent of other alloys, including titanium, inconel, silver, and nickel. Civilian aircraft are approximately 80 percent aluminum alloy and 20 percent other alloys. In addition to aluminum alloys, composite materials are being used for more applications in aircraft structures. For example, a typical Boeing 737-300 aircraft utilizes graphite, Kevlar, and fiberglass composites in flight controls and fairings. The use of these materials saves more than 600 pounds in weight over traditional sheet metal parts. However, while aircraft manufacturers continue to find more uses for composite materials, aluminum and steel alloys remain the most popular structural materials.

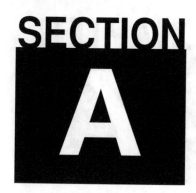

SECTION A

METALS

Today, most aircraft are constructed of various types of metals and metal alloys. Because of this, you, as an aviation maintenance technician, must possess a thorough knowledge of metals. However, before you can develop a complete understanding of metals used in the aviation industry, you must first become familiar with some of the properties metals have.

PROPERTIES OF METALS

A given metal can possess several properties. Among these are strength, hardness, malleability, ductility, brittleness, conductivity, expansion, elasticity, toughness, fusibility, and density.

STRENGTH

One way to classify metals is according to the amount of strength they possess. A metal's strength is determined by the percentage of parent metal and other elements used to make an alloy.

TENSILE STRENGTH

When a piece of sheet metal is pulled from each end, the resultant force is called **tension**. The ability to withstand tension is called tensile strength, and is measured in pounds per square inch. Since the strength values for several metals are rated to several thousand psi, the letter "K" is often used to represent the last three zeros of the psi rating. Therefore, a tensile strength of 70,000 psi is written as 70 KSI.

YIELD STRENGTH

The ability of a metal to resist deformation is called its yield strength. For example, when a tensile load

is applied to a material, the material resists any deformation until its yield point is reached. However, once the yield point is reached the metal stretches without an increase in the applied stress. Furthermore, as the metal stretches, the molecular structure changes enough to increase the metal's strength and, therefore, resist further deformation. This continues until the ultimate load is reached, at which time the material breaks.

SHEAR STRENGTH

Shear strength describes a metal's ability to resist opposing forces. A rivet holding two or more sheets of metal together resisting the force of the sheets trying to slide apart is an example of a shear load. When the rivets installed in a joint have more strength than the metal in the surrounding joint, the joint is said to be loaded in shear.

BEARING STRENGTH

Bearing strength is the ability of a joint to withstand any form of crushing or excessive compressive distortion. Material under a compression load usually fails by buckling or bending. The force at which something buckles while being compressed varies with an object's length, cross-sectional area, and shape.

HARDNESS

A metal's hardness refers to its ability to resist cutting, penetration, or abrasion. The tensile strength of steel relates directly to its hardness, but for most metals this relationship is not absolute. Some metals are hardened through heat-treating or work-hardening, while others are softened by a process called **annealing**.

MALLEABILITY

A material's ability to be bent, formed, or shaped without cracking or breaking is called malleability. Hardness and malleability are generally considered opposite characteristics. To help increase malleability, several metals are annealed, or softened. In this condition complex shapes can be formed. After forming is complete, the metal is then heat treated to increase its strength. A metal may be fully annealed when the forming is started, but hammering and shaping can harden it to such an extent that it must be re-annealed before forming is completed.

DUCTILITY

The ability of metal to be drawn into wire stock, extrusions, or rods is called ductility. Ductile metals are preferred for aircraft use because of their ease of forming and resistance to failure under shock loads. For this reason, aluminum alloys are often used for cowlings, fuselage and wing skins, and formed or extruded parts such as ribs, spars, and bulkheads.

BRITTLENESS

Brittleness describes a material's tendency to break or shatter when exposed to stress, and is the opposite of ductility and malleability. A brittle metal is more apt to break or crack before it changes shape. Because structural metals are often subjected to shock loads, brittleness is not a desirable property. Cast iron, cast aluminum, and very hard steel are examples of brittle metals.

CONDUCTIVITY

Conductivity is the property which enables a metal to carry heat or electricity. If a metal is able to transmit heat it is said to be **thermally conductive**. However, before a metal can carry heat away from its source, it must first absorb it. This ability to conduct heat away is called **heat exchange**. The fins on the cylinder heads of an air cooled piston engine remove heat in this fashion.

Metals that can carry heat also carry electrons, making them good electrical conductors. **Electrical conductivity** is the measure of a material's ability to allow electron flow. A metal conductor can be a wire, an aircraft frame, or an engine. If you recall from your study of electricity, electrons flow much easier in some metals than they do in others. Because of their molecular structures, the best electrical conductors are gold, silver, copper, and aluminum.

THERMAL EXPANSION

The property of a metal to expand when heated and shrink when cooled is called thermal expansion. The amount of expansion or contraction is predictable at specific temperatures and is called its **coefficient of expansion**. All aircraft experience thermal expansion and contraction as the ambient temperature changes.

ELASTICITY

Elasticity describes a metal's tendency to return to its original shape after normal stretching and bending. The flexibility of spring steel used for the construction of landing gear is a good example of elasticity. Another form of elasticity is demonstrated when aircraft skins expand and contract when an aircraft is pressurized.

A metal's **elastic limit** is the point beyond which the metal does not return to its original shape after a deforming force is removed. Soft materials such as lead, copper, and pure aluminum have very low elastic limits, while the elastic limit of hard spring steel is very high.

TOUGHNESS

Toughness is a material's ability to resist tearing or breaking when it is bent or stretched. Hammer faces and wrenches are examples of metal that must be tough as well as hard to be useful.

FUSIBILITY

The ability of metal to be joined by heating and melting is defined as fusibility. To fuse metal means to melt two or more compatible pieces of metal into one continuous part. The correct term is called fusion joining or welding.

DENSITY

Density is a material's mass per unit volume, and throughout this section the term is used to compare the weights of various metals. The standard from which a metal's density is determined is water. For example, one cubic centimeter of pure water weighs one gram and, therefore, has a density of one. Aluminum has a density of 2.7; therefore, a cubic centimeter weighs 2.7 grams. In the English system, a cubic inch of pure water weighs 0.03611 pound, and aluminum with its density of 2.7 weighs 2.7 times this amount, or 0.0975 pound. [Figure 7-1]

METALWORKING PROCESSES

After metal alloys are produced, they must be formed into useful shapes. **Wrought** objects are those formed by physically working the metal into shape, whereas **cast** items are formed by pouring molten metal into molds. When it comes to mechanically working metal into a desired shape, there are three methods commonly used. They are hot-working, cold-working, and extruding.

Hot-working is the process of forming metal at an elevated temperature when it is in its annealed, or soft condition. Almost all steel is hot-worked from the ingot into a form which is either hot- or cold-worked to a finished shape. When an ingot is stripped from its mold, its surface is solid, but its interior is still molten. The ingot is then placed in a soaking pit to slow the cooling process until the molten interior gradually solidifies. After soaking, the temperature is equalized throughout the ingot, then it is worked into its desired shape through

METAL	DENSITY	WEIGHT OF ONE CUBIC INCH (POUNDS)
WATER (REFERENCE)	1.0	0.03611
ALUMINUM	2.7	0.0975
TITANIUM	4.4	0.159
STEEL	7.9	0.285
COPPER	8.9	0.321
LEAD	11.4	0.412
GOLD	19.3	0.697

Figure 7-1. A material's mass per unit volume is called its density. Water, which is used as a standard, weighs one gram per cubic centimeter and has a density of 1.

rolling and forging. As the name implies, **rolling** consists of forming hot metal ingots with rollers to form sheets, bars, and beams. **Forging**, on the other hand, is a process wherein a piece of metal is worked at temperatures above its critical range. Forging is typically used to form intricate shapes and is accomplished through either pressing or hammering.

Pressing is used to form large and heavy parts. Since a press is slow acting, its force is uniformly transmitted to the center of the material being pressed. This affects the interior grain structure resulting in the best possible structure throughout.

Drop forging is a hammering process whereby a hot ingot is placed between a pair of formed dies in a machine called a drop hammer and a weight of several tons is dropped on the upper die. This results in the hot metal being forced to take the form of the dies. Because the process is very rapid, the grain structure of the metal is altered, resulting in significant increases in the strength of the finished part.

Hammering is a type of forging that is usually used on small parts because it requires a metalworker to physically hammer a piece of metal into its finished shape. The advantage of hammering is that the operator has control over both the amount of pressure applied and the finishing temperature. This type of forging is usually referred to as **smith forging** and is used extensively where only a small number of parts are needed. In addition to the forming operation, hammering hardens the metal.

Cold-working is performed well below a metal's critical temperature and ranges from the manual bending of sheet metal for skin repairs to drawing seamless tubing and wire. Cold-working strain hardens the worked metal, increasing its strength and hardness but decreasing ductility. For this reason, cold worked material usually must be heat treated before use. While there are several cold-working processes, the two that are most common are cold-rolling and cold-drawing. **Cold-rolling** usually refers to the rolling of metal at room temperature to its approximate size. Once this is done, the metal is pickled to remove any scale, and then passed through chilled finishing rolls. This results in a smooth surface and extremely accurate dimensions.

Cold-drawing is used in making seamless tubing, wire, streamlined tie rods, and other forms of stock. Wire is made from hot-rolled rods of various diameters. The size of rod used for drawing depends on the diameter wanted in the finished wire. To reduce a rod to a specific size, it is drawn cold through a die. To do this, one end of the rod is filed or hammered to a point and slipped through the die opening. Here it is gripped by a set of jaws and pulled through the die. In order to reduce the rod gradually to the desired size, it is necessary to draw the wire through successively smaller dies. Because each draw reduces the wire's ductility, it must be annealed occasionally.

In making seamless aircraft tubing, the tubing is cold-drawn through a ring-shaped die with a mandrel or metal bar inside the tubing to support it while it is drawn. This forces the metal to flow between the die and the mandrel and affords a means of controlling the wall thickness as well as the inside and outside diameters.

Extrusion is the process of forcing metal through a die which imparts a required cross-section to the metal. Some metals such as lead, tin, and aluminum may be extruded cold, however, most metals are heated. The principal advantage of the extrusion process is its flexibility. For example, because of its workability, aluminum can be economically extruded to more intricate shapes and larger sizes than is practicable with other metals. Many structural parts, such as channels, angles, T-sections, and Z-sections are formed by the extrusion process.

NONFERROUS METALS

Much of the metal used on today's aircraft contains no iron. The term that describes metals which have elements other than iron as their base is nonferrous. Aluminum, copper, titanium, and magnesium are some of the more common nonferrous metals used in aircraft construction and repair.

ALUMINUM AND ITS ALLOYS

Pure aluminum lacks sufficient strength to be used for aircraft construction. However, its strength increases considerably when it is **alloyed**, or mixed, with other compatible metals. For example, when aluminum is mixed with copper or zinc, the resultant alloy is as strong as steel with only one third the weight. Furthermore, the corrosion resistance possessed by the aluminum carries over to the newly formed alloy.

ALLOYING AGENTS

Aluminum alloys are classified by their major alloying ingredient. The elements most commonly used for aluminum alloying are copper, magnesium, manganese, and zinc. Wrought aluminum and wrought aluminum alloys are identified by a four-digit index system. The first digit of a designation identifies the major alloying element used in the formation of the alloy. The most common alloying elements used are as follows:

> 1xxx — aluminum
>
> 2xxx — copper
>
> 3xxx — manganese
>
> 4xxx — silicon
>
> 5xxx — magnesium
>
> 6xxx — magnesium and silicon
>
> 7xxx — zinc
>
> 8xxx — other elements

The second number represents a specific alloy modification. For example, if this digit is zero, it indicates there were no special controls over individual impurities. However, a digit of 1 through 9 indicates the number of controls over impurities in the metal.

The last two numbers of the 1xxx group of alloys are used to indicate the hundredths of 1 percent above the original 99 percent pure aluminum. For example, if the last two digits are 75, the alloy contains 99.75 percent pure aluminum. However, in the 2xxx through 8xxx groups the last two digits identify the different alloys in the group. [Figure 7-2]

The **1xxx series** of aluminum alloys represents commercially pure aluminum, of 99 percent or higher purity. Pure aluminum offers high corrosion resistance, excellent thermal and electrical properties, and is easily worked. However, pure aluminum is very low in strength.

Alloys within the **2xxx series** utilize copper as the principle alloying agent. When aluminum is mixed with copper, certain metallic changes take place in the resultant alloy's grain structure. For the most part, these changes are beneficial and produce greater strength. However, a major drawback to aluminum-copper alloys is their susceptibility to intergranular corrosion when improperly heat-treated. Most aluminum alloy used in aircraft structures is an aluminum-copper alloy. Two of the most commonly used in the construction of skins and rivets are 2017 and 2024.

The **3xxx series** alloys have manganese as the principle alloying element, and are generally considered nonheat treatable. The most common variation is 3003, which offers moderate strength and has good working characteristics.

The **4xxx series** aluminum is alloyed with silicon, which lowers a metal's melting temperature. This results in an alloy that works well for welding and brazing.

Magnesium is used to produce the **5xxx** series alloys. These alloys possess good welding and corrosion-resistance characteristics. However, if the metal is exposed to high temperatures or excessive cold working, its susceptibility to corrosion increases.

If silicon and magnesium are added to aluminum, the resultant alloy carries a **6xxx series** designation. In these alloys, the silicon and magnesium form magnesium silicide which makes the alloy heat-treatable. Furthermore, the 6xxx series has medium strength with good forming and corrosion-resistance properties.

When parts require more strength and little forming, harder aluminum alloys are employed. The **7xxx series** aluminum alloys are made harder and stronger by the addition of zinc. Some widely used forms of zinc-aluminum alloys are 7075 and 7178. The aluminum-zinc alloy 7075 has a tensile strength of 77 KSI and a bearing strength of 139 KSI. However, the alloy is very hard and is difficult to bend. An even stronger zinc alloy is 7178 which has a tensile strength of 84 KSI and a bearing strength of 151 KSI.

CLAD ALUMINUM ALLOY

Most external aircraft surfaces are made of **Alclad** aluminum. Alclad is a pure aluminum coating that is rolled on to the surface of heat-treated aluminum alloy. The thickness of this coating is approximately

NON-HEAT TREATABLE ALLOY	PERCENT OF ALLOYING ELEMENTS					
	COPPER	**SILICON**	**MANGANESE**	**MAGNESIUM**	**ZINC**	**CHROMIUM**
1100						
3003			1.2			
5052				2.5		0.25
HEAT TREATABLE ALLOY						
2017	4.0		0.5	0.5		
2117	2.5			0.3		
2024	4.5		0.6	1.5		
6061	0.25	0.6		1.0		0.25
7075	1.6			2.5	5.6	0.3

Figure 7-2. A variety of elements are used to produce aluminum alloys.

5 percent of the alloy's thickness on each side. For example, if an alclad sheet of aluminum is .040 inches thick, 5 percent, or .002 inches of pure aluminum is applied to each side. This results in an alloy thickness of .036 inches.

This clad surface greatly increases the corrosion resistance of an aluminum alloy. However, if it is penetrated, corrosive agents can attack the alloy within. For this reason, sheet metal should be protected from scratches and abrasions. In addition to providing a starting point for corrosion, abrasions create potential stress points.

HEAT TREATMENT

Heat treatment is a series of operations involving the heating and cooling of metals in their solid state. Its purpose is to make the metal more useful, serviceable, and safe for a definite purpose. By heat treating, a metal can be made harder, stronger, and more resistant to impact. Heat treating can also make a metal softer and more ductile. However, one heat-treating operation cannot produce all these characteristics. In fact, some properties are often improved at the expense of others. In being hardened, for example, a metal may become brittle.

All heat-treating processes are similar in that they involve the heating and cooling of metals. They differ, however, in the temperatures to which the metal is heated and the rate at which it is cooled.

There are two types of heat treatments used on aluminum alloys. One is called solution heat treatment, and the other is known as precipitation heat treatment. Some alloys, such as 2017 and 2024, develop their full properties as a result of solution heat treatment followed by about 4 days of cooling, or **aging**, at room temperature. However, other alloys, such as 2014 and 7075, require both heat treatments.

SOLUTION HEAT-TREATMENT

When aluminum is alloyed with materials such as copper, magnesium, or zinc, the resultant alloys are much stronger than aluminum. To understand why this happens, it is necessary to examine the microscopic structure of aluminum. Pure aluminum has a molecular structure that is composed of weakly bonded aluminum atoms and, therefore, is extremely soft. Aluminum alloys, on the other hand, consist of a base metal of aluminum and an alloying element that is dispersed throughout the

structure. In this configuration, when the aluminum alloy is subjected to stress, these alloying particles adhere to the aluminum molecules and resist deformation. However, special processes must be used to allow the base metal and alloy to mix properly. For example, when aluminum is alloyed with copper through conventional processes, approximately .5 percent of the copper dissolves, or mixes with the aluminum. The remaining copper takes the form of the compound $CuAl_2$. However, when the aluminum alloy is heated sufficiently, the remaining copper enters the base metal and hardens the alloy.

The process of heating certain aluminum alloys to allow the alloying element to mix with the base metal is called solution heat treating. In this procedure, metal is heated in either a molten sodium or potassium nitrate bath or in a hot-air furnace to a temperature just below its melting point. The temperature is then held to within about plus or minus 10 degrees Fahrenheit of this temperature and the base metal is **soaked** until the alloying element is uniform throughout. Once the metal has sufficiently soaked, it is removed from the furnace and cooled or **quenched**. It is extremely important that no more than about ten seconds elapse between removal of an alloy from the furnace and the quench. The reason for this is that when metal leaves the furnace and starts to cool, its alloying metals begin to precipitate out of the base metal. If this process is not stopped, large grains of alloy become suspended in the aluminum and weaken the alloy. Excessive precipitation also increases the likelihood of intergranular corrosion.

To help minimize the amount of alloying element that precipitates out of a base metal, a quenching medium is selected to ensure the proper cooling rate. For example, a water spray or bath provides the appropriate cooling rate for aluminum alloys. However, large forgings are typically quenched in hot water to minimize thermal shock that could cause cracking. Thin sheet metal normally warps and distorts when it is quenched, so it must be straightened immediately after it is removed from the quench. After the quench, all metals must be rinsed thoroughly since the salt residue from the sodium or potassium nitrate bath can lead to corrosion if left on the alloy.

PRECIPITATION HEAT-TREATMENT

Heat-treatable aluminum alloys are comparatively soft when first removed from a quench. With time, however, the metal becomes hard and gains strength. When an alloy is allowed to cool at room temperature, it is referred to as **natural aging** and can take several hours or several weeks. For example, aluminum alloyed with copper gains about 90 percent of its strength in the first half-hour after it is removed from the quench, and becomes fully hard in about four or five days.

An alloy's aging process time can be lengthened or shortened. For example, the aging process can be slowed by storing a metal at a sub-freezing temperature immediately after it is removed from the quench. On the other hand, the aging process can be accelerated by reheating a metal and allowing it to soak for a specified period of time. This type of aging is identified by several terms such as **artificial age-hardening, percipitation-hardening** or **precipitation heat treatment**. This process develops hardness, strength, and corrosion resistance by locking a metal's grain structure together.

Naturally aged alloys, such as the copper-zinc-magnesium alloys, derive their full strength at room temperature in a relatively short period and require no further heat treatment. However, other alloys, particularly those with a high zinc content, need thermal treatment to develop full strength. These alloys are called artificially aged alloys.

ANNEALING

Annealing is a process that softens a metal and decreases internal stress. In general, annealing is the opposite of hardening. To anneal an aluminum alloy, the metal's temperature is raised to an annealing temperature and held there until the metal becomes thoroughly heat soaked. It is then cooled to 500°F at a rate of about 50° per hour. Below 500°F, the rate of cooling is not important.

When annealing clad aluminum metals, they should be heated as quickly and as carefully as possible. The reason for this is that if clad aluminum is exposed to excessive heat, some of the core material tends to mix with the cladding. This reduces the metal's corrosion resistance. [Figure 7-3]

NON-HEAT-TREATABLE ALLOYS

Commercially pure aluminum does not benefit from heat treatment since there are no alloying materials in its structure. By the same token, 3003 is an almost identical metal and, except for a small amount of manganese, does not benefit from being heat treated. Both of these metals are lightweight

and somewhat corrosion resistant. However, neither has a great deal of strength and, therefore, their use in aircraft is limited to nonstructural components such as fairings and streamlined enclosures that carry little or no load.

Alloy 5052 is perhaps the most important of the nonheat-treatable aluminum alloys. It contains about 2.5 percent magnesium and a small amount of chromium. It is used for welded parts such as gasoline or oil tanks, and for rigid fluid lines. Its strength is increased by cold working.

HEAT-TREATMENT IDENTIFICATION

Heat-treatable alloys have their hardness condition designated by the letter **-T** followed by one or more numbers. A list of these designations includes:

-T   Solution heat treated

-T2   Annealed (cast products only)

-T3   Solution heat-treated, followed by strain hardening. Different amounts of strain hardening of the heat-treated alloy are indicated by a second digit. For example, -T36 indicates that the material has been solution heat-treated and has had its thickness reduced 6 percent by cold rolling.

-T4   Solution heat-treated, followed by natural aging at room temperature to a stable condition.

-T5   Artificially aged after being rapidly cooled during a fabrication process such as extrusion or casting.

THERMAL TREATMENT				
ALLOY	**ANNEALING**	**SOLUTION HEAT TREAT**	**PRECIPITATION HEAT TREAT**	
	TEMP - °F	TEMP - °F	TEMP - °F	TIME - HOURS
1100	650°			
3003	775°			
5052	650°			
2017	775°	940°		
2117	775°	940°		
2024	775°	920°	375°	7 - 9
6061	775°	970°	350°	6 - 10
7075	775°	870°	250°	24 - 28

Figure 7-3. Aluminum alloys are heat treated to increase their strength and improve their working characteristics. Heat treatment temperatures and soak times are critical to disperse the alloying elements.

-T6 Solution heat-treated, followed by artificial aging (precipitation heat-treated).

-T7 Solution heat-treated and then stabilized to control its growth and distortion.

-T8 Solution heat-treated, strain hardened, and then artificially aged.

-T9 Solution heat-treated, artificially aged, and then strain-hardened.

-T10 Artificially aged and then cold worked.

REHEAT TREATMENT

Material which has been previously heat-treated can generally be reheat treated any number of times. As an example, rivets made of 2017 or 2024 are extremely hard and typically receive several reheat treatments to make them soft enough to drive.

As discussed earlier, the number of solution heat-treatments allowed for clad materials is limited due to the increased diffusion of core material into the cladding. This diffusion results in decreased corrosion resistance. As a result, clad material is generally limited to no more than three reheat treatments.

STRAIN HARDENING

Both heat-treatable and nonheat-treatable aluminum alloys can be strengthened and hardened through strain hardening, also referred to as **cold working** or **work hardening**. This process requires mechanically working a metal at a temperature below its critical range. Strain hardening alters the grain structure and hardens the metal. The mechanical working can consist of rolling, drawing, or pressing.

Heat-treatable alloys have their strength increased by rolling after they have been solution heat-treated. On the other hand, nonheat-treatable alloys are hardened in the manufacturing process when they are rolled to their desired dimensions. However, at times these alloys are hardened too much and must be partially annealed.

HARDNESS DESIGNATIONS

Where appropriate, a metal's hardness, or **temper**, is indicated by a letter designation that is separated from the alloy designation by a dash. When the

basic temper designation must be more specifically defined, one or more numbers follow the letter designation. These designations are as follows:

-F As fabricated.

-O Annealed, recrystallized (wrought materials only).

-H Strain hardened.

-H1 Strain hardened only.

-H2 Strain hardened and partially annealed.

-H3 Strain hardened and stabilized.

The digit following the designations H1, H2, and H3 indicate the degree of strain hardening. For example, the number 8 represents the maximum tensile strength while O indicates an annealed state. The most common designations include:

-Hx2 Quarter-hard

-Hx4 Half-hard

-Hx6 Three-quarter hard

-Hx8 Full-hard

-Hx9 Extra-hard

MAGNESIUM AND ITS ALLOYS

Magnesium alloys are used for castings and in its wrought form is available in sheets, bars, tubing, and extrusions. Magnesium is one of the lightest metals having sufficient strength and suitable working characteristics for use in aircraft structures. It has a density of 1.74, compared with 2.69 for aluminum. In other words, it weighs only about 2/3 as much as aluminum.

Magnesium is obtained primarily from electrolysis of sea water or brine from deep wells, and lacks sufficient strength in its pure state for use as a structural metal. However, when it is alloyed with zinc, aluminum, thorium, zirconium, or manganese, it develops strength characteristics that make it quite useful.

The American Society for Testing Materials (ASTM) has developed a classification system for magnesium alloys that consists of a series of letters and numbers to indicate alloying agents and temper condition. [Figure 7-4]

Magnesium has some rather serious drawbacks that had to be overcome before it could be used successfully. For example, magnesium is highly susceptible to corrosion, and tends to crack. The cracking contributes to its difficulty in forming and limits its use for thin sheet metal parts. However, this tendency is overcome to a great extent by forming parts while the metal is hot. The corrosion problem is minimized by treating the surface with chemicals that form an oxide film to prevent oxygen from reaching the metal. When oxygen is excluded from the surface, no corrosion can form. Another important step in minimizing corrosion is to always use hardware such as rivets, nuts, bolts, and screws that are made of a compatible material.

In addition to cracking and corroding easily, magnesium burns readily in a dust or small particle form. For this reason, caution must be exercised when grinding and machining magnesium. If a fire should occur, extinguish it by smothering it with dry sand or some other dry material that excludes air from the metal and cools its surface. If water is used, it will only intensify the fire.

Solution heat-treatment of magnesium alloys increases tensile strength, ductility, and resistance to shock. After a piece of magnesium alloy has been solution heat-treated, it can be precipitation heat treated by heating it to a temperature lower than that used for solution heat treatment, and holding it at this temperature for a period of several hours. This increases the metal's hardness and yield strength.

ALLOYING ELEMENTS	TEMPER CONDITION
A – ALUMINUM	F – AS FABRICATED
E – RARE EARTH	O – ANNEALED
H – THORIUM	H24 – STRAIN HARDENED AND PARTIALLY ANNEALED
K – ZIRCONIUM	T4 – SOLUTION HEAT-TREATED
M – MANGANESE	T5 – ARTIFICIALLY AGED ONLY
Z – ZINC	T6 – SOLUTION HEAT-TREATED AND ARTIFICIALLY AGED

Figure 7-4. Magnesium alloys use a different designation system than aluminum. For example, the designation AZ31A-T4 identifies an alloy containing 3 percent aluminum and 1 percent zinc that has been solution heat-treated.

TITANIUM AND ITS ALLOYS

Titanium and its alloys are light weight metals with very high strength. Pure titanium weighs .163 pounds per cubic inch, which is about 50 percent lighter than stainless steel, yet it is approximately equal in strength to iron. Furthermore, pure titanium is soft and ductile with a density between that of aluminum and iron.

Titanium is a metallic element which, when first discovered, was classified as a rare metal. However, in 1947 its status was changed due to its importance as a structural metal. In the area of structural metallurgy, it is said that no other structural metal has been studied so extensively or has advanced aircraft structures so rapidly.

In addition to its light weight and high strength, titanium and its alloys have excellent corrosion resistance characteristics, particularly to the corrosive effects of salt water. However, since the metal is sensitive to both nitrogen and oxygen, it must be converted to titanium dioxide with chlorine gas and a reducing agent before it can be used.

Titanium is classified as alpha, alpha beta, and beta alloys. These classifications are based on specific chemical bonding within the alloy itself. The specifics of the chemical composition are not critical to working with the alloy, but certain details should be known about each classification.

Alpha alloys have medium strengths of 120 KSI to 150 KSI and good elevated-temperature strength. Because of this, alpha alloys can be welded and used in forgings. The standard identification number for alpha titanium is 8A1-1Mo-1V-Ti, which is also referred to as Ti-8-1-1. This series of numbers indicates that the alloying elements and their percentages are 8 percent aluminum, 1 percent molybdenum, and 1 percent vanadium.

Alpha-beta alloys are the most versatile of the titanium alloys. They have medium strength in the annealed condition and much higher strength when heat treated. While this form of titanium is generally not weldable, it has good forming characteristics.

Beta alloys have medium strength, excellent forming characteristics, and contain large quantities of high-density alloying elements. Because of this, beta titanium can be heat-treated to a very high strength.

The grain size of titanium is refined when aluminum is added to the alloy mixture. However, when copper is added to titanium, a precipitation-hardening alloy is produced. Titanium added to high temperature nickel-cobalt-chromium alloy produces a precipitation-hardening reaction which provides strength at temperatures up to 1,500°F. [Figure 7-5]

Because of its high strength-to-weight ratio, titanium is now used extensively in the civilian aerospace industry. Although once rare on commercial aircraft, modern jet transports now utilize alloys containing 10 to 15 percent titanium in structural areas.

NICKEL AND ITS ALLOYS

As an aircraft technician, you need to be familiar with two nickel alloys. They are monel and inconel. **Monel** contains about 68 percent nickel and 29 percent copper, along with small amounts of iron and manganese. It can be welded and has very good machining characteristics. Certain types of monel, especially those containing small percentages of aluminum, are heat-treatable to tensile strengths equivalent to steel. Monel works well in gears and parts that require high strength and toughness, as well as for parts in exhaust systems that require high strength and corrosion resistance at elevated temperatures.

The International Nickel Company, Inc., produces a series of high strength, high temperature alloys containing approximately 80 percent nickel, 14 percent chromium, and small amounts of iron and other elements. The alloys, commonly referred to as **inconel**, find frequent use in turbine engines because of their ability to maintain their strength and corrosion resistance under extremely high temperature conditions.

Inconel and stainless steel are similar in appearance and are frequently used in the same areas. Therefore, it is often necessary to use a test to differentiate between unknown metal samples. A common test involves applying one drop of cupric chloride and hydrochloric acid solution to the unknown metal and allowing it to remain for two minutes. At the end of the dwell period, a shiny spot indicates that the material is inconel, whereas a copper-colored spot identifies stainless steel.

COPPER AND ITS ALLOYS

Neither copper nor its alloys find much use as structural materials in aircraft construction. However, due to its excellent electrical and thermal conductivity, copper is the primary metal used for electrical wiring.

Of the several alloys that use copper as a base, brass, bronze, and beryllium are the primary alloys used on aircraft. **Brass** is a copper alloy containing zinc and small amounts of aluminum, iron, lead, manganese, magnesium, nickel, phosphorous, and tin. Brass with a zinc content of 30 to 35 percent is very ductile, while brass containing 45 percent zinc has relatively high strength.

Bronze is a copper alloy that contains tin. A true bronze consists of up to 25 percent tin and, along with brass, is used in bushings, bearings, fuel-metering valves, and valve seats. Bronzes with less than 11 percent tin are used in items such as tube fittings.

ALLOY	COMPOSITION	TENSILE STRENGTH	ELONGATION
ALPHA	5% AL - 2.5% SN	130 KSI	15%
ALPHA-BETA	6% AL - 4% V	140 KSI	15%
ALPHA-BETA HEAT-TREATED	6% AL - 4% V	180 KSI	7%
BETA	13% V - 11% CR - 3% AL	150 KSI	15%
BETA HEAT-TREATED	13% V - 11% CR - 3% AL	200 KSI	6%
DEFINITIONS OF LETTERS AND PERCENTAGES ARE: AL - ALUMINUM CR - CHROMIUM V - VANADIUM SN - TIN			

Figure 7-5. This table illustrates the composition, tensile strength, and elongation of titanium alloys. The degree of strength is denoted by the smaller hole elongation percentage shown in the last column. The titanium alloy most commonly used by the aerospace industry is an alpha-beta heat-treated alloy called 6Al-4V. This alloy has a tensile strength of 180 KSI, or 180,000 pounds per square inch. It is frequently used for special fasteners.

Beryllium copper is probably one of the most used copper alloys. It consists of approximately 97 percent copper, 2 percent beryllium, and sufficient nickel to increase its strength. Once heat treated, beryllium copper achieves a tensile strength of 200,000 psi and 70,000 psi in its annealed state. This makes beryllium extremely useful for diaphragms, precision bearings and bushings, ball cages, and spring washers.

Copper tubing was once used extensively in aircraft fluid lines, but because of its weight and tendency to become brittle when subjected to vibration, it has been almost entirely replaced by aluminum alloy. Brass fittings for fluid lines have also been replaced with either aluminum alloy or steel fittings.

FERROUS METAL

Any alloy containing iron as its chief constituent is called a **ferrous** metal. The most common ferrous metal in aircraft structures is steel, an alloy of iron with a controlled amount of carbon added.

IRON

Iron is a chemical element which is fairly soft, malleable, and ductile in its pure form. It is silvery white in color and is quite heavy, having a density of 7.9 grams per cubic centimeter. Iron combines readily with oxygen to form iron oxide, which is more commonly known as rust. This is one reason why iron is usually mixed with various forms of carbon and other alloying agents or impurities. Iron poured from a furnace into molds is known as cast iron and normally contains more than two percent carbon and some silicon. Cast iron has few aircraft applications because of its low strength-to-weight ratio. However, it is used in engines for items such as valve guides where its porosity and wear characteristics allow it to hold a lubricant film. It is also used in piston rings.

Iron is produced by mixing iron ore with coke and limestone and submitting it to hot air. The coke burns and forms superheated carbon monoxide which absorbs oxygen from the ore causing the molten iron to sink to the bottom of the furnace. The limestone reacts with impurities in the iron and coke to form a slag which floats on top of the molten iron. The slag is removed and the refined metal is then poured from the furnace. The resulting metal is known as pig iron and is typically remelted and cast into cast-iron components, or converted into steel.

STEEL

To make steel, pig iron is re-melted in a special furnace. Pure oxygen is then forced through the molten metal where it combines with carbon and burns. A controlled amount of carbon is then put back into the molten metal along with other elements to produce the desired characteristics. The molten steel is then poured into molds where it solidifies into ingots. The ingots are placed in a soaking pit where they are heated to a uniform temperature of about 2,200°F. They are then taken from the soaking pit and passed through steel rollers to form plate or sheet steel.

Much of the steel used in aircraft construction is made in electric furnaces, which allow better control of alloying agents than gas-fired furnaces. An electric furnace is loaded with scrap steel, limestone, and flux. Carbon electrodes are lowered into the steel, producing electric arcs between the steel and the carbon. The intense heat from the arcs melts the steel and the impurities mix with the flux. Once the impurities are removed, controlled quantities of alloying agents are added, and the liquid metal is poured into molds.

SAE CLASSIFICATION OF STEELS

The **Society of Automotive Engineers (SAE)** has classified steel alloys with a four-digit numerical index system. For example, one common steel alloy is identified by the designation SAE 1030. The first digit identifies the principal alloying element in the steel, the second digit denotes the percent of this alloying element, and the last two digits give the percentage in hundredths of a percent of carbon in the steel. [Figure 7-6]

ALLOYING AGENTS IN STEEL

As discussed earlier, iron has few practical uses in its pure state. However, adding small amounts of other materials to molten iron dramatically changes its properties. Some of the more common alloying agents include carbon, sulfur, silicon, phosphorous, nickel, and chromium.

CARBON

Carbon is the most common alloying element found in steel. When mixed with iron, compounds of iron carbides called **cementite** form. It is the carbon in steel that allows the steel to be heat-treated to obtain

TYPES OF STEEL	NUMERALS AND DIGITS
CARBON STEELS	1xxx
PLAIN CARBON STEELS	10xx
FREE CUTTING STEELS	11xx
MANGANESE STEELS (MANGANESE 1.60 TO 1.90%)	13xx
NICKEL STEELS	2xxx
3.50% NICKEL	23xx
5.00% NICKEL	25xx
NICKEL CHROMIUM STEELS	3xxx
9.70% NICKEL, 0.07% CHROMIUM	30xx
1.25% NICKEL, 0.60% CHROMIUM	31xx
1.75% NICKEL, 1.00% CHROMIUM	32xx
3.50% NICKEL, 1.50% CHROMIUM	33xx
CORROSION AND HEAT RESISTING	30xxx
MOLYBDENUM STEELS	40xx
CHROMIUM MOLYBDENUM STEELS	41xx
NICKEL CHROMIUM MOLYBDENUM STEELS	43xx
NICKEL MOLYBDENUM STEELS	
1.75% NICKEL, 0.25% MOLYBDENUM	46xx
3.50% NICKEL, 1.50% CHROMIUM	48xx
CHROMIUM STEELS	5xxx
LOW CHROMIUM	51xx
MEDIUM CHROMIUM	52xxx
CORROSION AND HEAT RESISTING	51xxx
CHROMIUM VANADIUM STEELS	6xxx
1.00% CHROMIUM	61xxx
NATIONAL EMERGENCY STEELS	8xxx
SILICON MANGANESE STEELS	9xxx
2.00% SILICON	92xx

Figure 7-6. The Society of Automotive Engineers has established a classification system for steel alloys. For example, SAE 1030 identifies plain carbon steel containing .30 percent carbon.

varying degrees of hardness, strength, and toughness. The greater the carbon content, the more receptive steel is to heat treatment and, therefore, the higher its tensile strength and hardness. However, higher carbon content decreases the malleability and weldability of steel.

Low-carbon steels contain between 0.10 and 0.30 percent carbon and are classified as SAE 1010 to SAE 1030 steel. These steels are primarily used in safety wire, cable bushings, and threaded rod ends. In sheet form, these steels are used for secondary structures where loads are not high. Low-carbon steel is easily welded and machines readily, but does not accept heat treatment well.

Medium-carbon steels contain between 0.30 and 0.50 percent carbon. The increased carbon helps these steels accept heat treatment, while still retaining a reasonable degree of ductility. This steel is especially adaptable for machining or forging and where surface hardness is desirable.

High-carbon steels contain between 0.50 and 1.05 percent carbon, and are very hard. These steels are primarily used in springs, files, and some cutting tools.

SULFUR

Sulfur causes steel to be brittle when rolled or forged and, therefore, it must be removed in the refining process. If all the sulfur cannot be removed its effects can be countered by adding manganese. The manganese combines with the sulfur to form manganese sulfide, which does not harm the finished steel. In addition to eliminating sulfur and other oxides from steel, manganese improves a metal's forging characteristics by making it less brittle at rolling and forging temperatures.

SILICON

When silicon is alloyed with steel it acts as a hardener. When used in small quantities, it also improves ductility.

PHOSPHOROUS

Phosphorous raises the yield strength of steel and improves low carbon steel's resistance to atmospheric corrosion. However, no more than 0.05 percent phosphorous is normally used in steel, since higher amounts cause the alloy to become brittle when cold.

NICKEL

Nickel adds strength and hardness to steel and increases its yield strength. It also slows the rate of hardening when steel is heat-treated, which increases the depth of hardening and produces a finer grain structure. The finer grain structure reduces steel's tendency to warp and scale when heat-treated. SAE 2330 steel contains 3 percent nickel and 0.30 percent carbon, and is used in producing aircraft hardware such as bolts, nuts, rod ends, and pins.

CHROMIUM

Chromium is alloyed with steel to increase strength and hardness as well as improve its wear and corrosion resistance. Because of its characteristics, chromium steel is used in balls and rollers of antifriction bearings.

In addition to its use as an alloying element in steel, chromium is electrolytically deposited on cylinder walls and bearing journals to provide a hard, wear-resistant surface.

NICKEL-CHROMIUM STEEL

As mentioned earlier, nickel toughens steel, and chromium hardens it. Therefore, when both elements are alloyed they give steel desirable characteristics for use in high-strength structural applications. Nickel-chrome steels such as SAE 3130, 3250, and 3435 are used for forged and machined parts requiring high strength, ductility, shock resistance, and toughness.

STAINLESS STEEL

Stainless steel is a classification of corrosion-resistant steels that contain large amounts of chromium and nickel. Their strength and resistance to corrosion make them well suited for high-temperature applications such as firewalls and exhaust system components. These steels can be divided into three general groups based on their chemical structure: austenitic, ferritic, and martensitic.

Austenitic steels, also referred to as 200 and 300 series stainless steels, contain a large percentage of chromium and nickel, and in the case of the 200 series, some manganese. When these steels are heated to a temperature above their critical range and held there, a structure known as austenite forms. Austenite is a solid solution of **pearlite**, an alloy of iron and carbon, and **gamma iron**, which is a nonmagnetic form of iron. Austenitic stainless steels can be hardened only by coldworking while heat treatment serves only to anneal them.

Ferritic steels are primarily alloyed with chromium but many also contain small amounts of aluminum. However, they contain no carbon and, therefore, do not respond to heat treatment.

The 400 series of stainless steel is a **martensitic steel**. These steels are alloyed with chromium only and therefore are magnetic. Martensitic steels become extremely hard if allowed to cool rapidly by quenching from an elevated temperature.

The corrosion-resistant steel most often used in aircraft construction is known as 18-8 steel because it contains 18 percent chromium and 8 percent nickel. One of the distinctive features of 18-8 steel is that its strength may be increased by cold-working.

MOLYBDENUM

One of the most widely used alloying elements for aircraft structural steel is molybdenum. It reduces the grain size of steel and increases both its impact strength and elastic limit. Molybdenum steels are extremely wear resistant and possess a great deal of fatigue strength. This accounts for its use in high-strength structural members and engine cylinder barrels.

Chrome-molybdenum (chrome-moly) steel is the most commonly used alloy in aircraft. Its SAE designation of 4130 denotes an alloy of approximately 1 percent molybdenum and 0.30 percent carbon. It machines readily, is easily welded by either gas or electric arc, and responds well to heat treatment. Heat-treated SAE 4130 steel has an ultimate tensile strength about four times that of SAE 1025 steel, making it an ideal choice for landing gear structures and engine mounts. Furthermore, chrome-moly's toughness and wear resistance make it a good material for engine cylinders and other highly stressed engine parts.

VANADIUM

When combined with chromium, vanadium produces a strong, tough, ductile steel alloy. Amounts up to 0.20 percent improve grain structure and increase both ultimate tensile strength and toughness. Most wrenches and ball bearings are made of chrome-vanadium steel.

TUNGSTEN

Tungsten has an extremely high melting point and adds this characteristic to steel it is alloyed with. Because tungsten steels retain their hardness at elevated operating temperatures, they are typically used for breaker contacts in magnetos and for high-speed cutting tools.

HEAT TREATMENT OF STEEL

Iron is an **allotropic** metal, meaning it can exist in more than one type of lattice structure, depending on temperature. Pure molten iron begins to solidify at 2,800° F. Its structure at this point is known as the Δ (Delta) form. However, if cooled to 2,554°F, the atoms rearrange themselves into a Γ (Gamma) form. Strangely enough, iron in this form is nonmagnetic. When nonmagnetic gamma iron in this form is cooled to 1,666°F, another change occurs and the iron is transformed into a nonmagnetic form of the A (Alpha) structure. As cooling continues to 1,414°F, the material becomes magnetic with no further changes in its lattice structure.

There are two basic forms of steel that you should be aware of when it comes to heat treatment. They are ferrite and austenite. **Ferrite** is an alpha solid solution of iron containing some carbon and exists at temperatures below the lower critical temperature. Above this lower critical temperature, the steel begins to turn into **austenite**, which consists of gamma iron containing carbon. As the temperature increases, the transformation of ferrite into austenite continues until the upper critical temperature is reached. Above the upper critical temperature, the entire structure consists of austenite.

Below the alloy's lower critical temperature, the carbon, which exists in the steel in the form of iron carbides (Fe_3C), is scattered throughout the iron matrix as a physical mixture. When the steel is heated to its upper critical temperature, this carbon dissolves into the matrix and becomes a solid solution, rather than a physical mixture. The steel has now become austenite, and the iron is in its gamma form.

When the temperature of the steel drops below this critical value, the carbide particles precipitate out of the solution. If steel is cooled slowly, the particles are quite large and the steel is soft. On the other hand, if it is cooled rapidly through quenching, the carbon particles remain extremely fine and effectively bind the molecular structure of the steel together, making it hard and strong.

The critical temperature and ultimate strength a steel develops varies with different alloying agents, but the most important factor is the amount of carbon. Low-carbon steel does not heat-treat satisfactorily because of the small amount of carbon. But as carbon content increases, the steel gains the ability to be hardened and strengthened by heat treatment. This occurs up to about 0.80 percent carbon. While hardness does not increase beyond this concentration, wear resistance improves with an increase in the amount of carbon.

ANNEALING

Annealing softens steel and relieves internal stress. To anneal steel, it is heated to about 50°F above its critical temperature, soaked for a specified time, then cooled. The soaking time is typically around one hour per inch of material thickness.

As the steel cools, the carbon that precipitates out forms large particles that do not bind tightly with the iron. The resulting steel is soft, ductile, and can be easily formed. The steel can be cooled by leaving it in the furnace and allowing both the furnace and steel to cool together or by packing the steel in hot sand or ash so the heat is conducted away slowly.

NORMALIZING

The processes of forging, welding, or machining usually leave stresses within steel that could lead to failure. These stresses are relieved in ferrous metals by a process known as normalizing. To normalize steel, it is heated to about 100°F above its upper critical temperature and held there until the metal is uniformly heat soaked. The steel is then removed from the furnace and allowed to cool in still air. Although this process does allow particles of carbon to precipitate out, the particles are not as large as those formed when steel is annealed.

One of the most important uses of normalizing in aircraft work is on welded parts. When a part is welded, internal stresses and strains set up in the adjacent material. In addition, the weld itself is a cast structure whereas the surrounding material is wrought. These two types of structures have different grain sizes and, therefore, are not very compatible. To refine the grain structure as well as relieve the internal stresses, all welded parts should be normalized after fabrication.

HARDENING

Pure iron, wrought iron, and extremely low-carbon steels cannot be hardened by heat treatment since they contain no hardening element. Cast iron, on the other hand, can be hardened, but the amount and type of heat treatment used is limited. For example, when cast iron is cooled rapidly, it forms white iron, which is hard and brittle. However, when cooled slowly, gray iron forms, which is soft but brittle under impact.

Carbon steel can be hardened readily. The maximum hardness obtained by carbon steel depends almost entirely on the amount of carbon content. For example, as the carbon content increases, the ability of a steel to be hardened increases. However, this increase continues only to a certain point. In practice, that point is 0.85 percent carbon content.

To harden steel, it is heated above its critical temperature so carbon can disperse uniformly in the iron matrix. Once this occurs, the alloy is cooled rapidly by quenching it in water, oil, or brine. The speed of the quench is determined by the quenching medium. Oil provides the slowest quench, and brine the most rapid.

If the quench is too quick, insufficient time is allowed for the carbon to precipitate out leaving it trapped in the alloy. The resultant structure is known as **martensite**, a supersaturated solid solution of carbon in an iron matrix. Although martensite is the hardest possible alloy, it is far too hard and brittle for most applications.

TEMPERING

Tempering reduces the undesirable qualities of martensitic steel. To temper an alloy, it is heated to a level considerably below its critical temperature and held there until it becomes heat soaked. It is then allowed to cool to room temperature in still air. Tempering not only reduces hardness and brittleness, but also relieves stress and improves a steel's ductility and toughness.

DETERMINING STEEL TEMPERATURE

When steel must be heat treated without the aid of a pyrometer, its temperature can be estimated fairly accurately through the use of commercial crayons, pellets, or paints that melt at specific temperatures. The least accurate method of estimating temperature is by observing the color of the material being heated. Some of the reasons why color observation is inaccurate include the fact that the observed color is affected by the amount of artificial and natural light and the ambient air temperature. However, as a last resort, when annealing or hardening nonstructural components, color observations can be used. [Figure 7-7]

TEMPERING SMALL TOOLS

As a technician, you frequently need a special tool to suit a unique application. It is often more convenient to manufacture such a tool rather than purchase one. To produce a tool that will not harm the work and is safe to use, a fabricated tool must be tempered.

Tool-steel rod can be used to make screwdriver blades, chisels, and punches. This steel has a high carbon content and is usually obtained in its normalized condition. After shaping the tool, heat it to a cherry red color and quench it in oil. Next, polish a portion and reheat it until the proper oxide color appears. When the tool is quenched again, it should have the proper hardness. [Figure 7-8]

COLOR OF STEEL	TEMPERATURE OF STEEL	
	(°F)	(°C)
FAINT RED	900	482
BLOOD RED	1,050	566
DARK CHERRY	1,075	579
MEDIUM CHERRY	1,250	677
CHERRY OR FULL RED	1,375	746
BRIGHT RED	1,550	843
SALMON	1,650	899
ORANGE	1,725	941
LEMON	1,825	996
LIGHT YELLOW	1,975	1,079
WHITE	2,200	1,204
DAZZLING WHITE	2,350	1,288

Figure 7-7. To anneal a piece of chrome-molybdenum steel (SAE 4130), it must be heated to between 1,525°F and 1,575° F. This temperature is achieved when the metal is heated to a bright red.

CASE HARDENING

Certain components in aircraft engines and landing gear systems require metal with hard, durable bearing surfaces and core material that remains tough. This is accomplished through a process called case hardening. The steels best suited for case-hardening are the low-carbon and low-alloy steels. If high-carbon steel is case-hardened, the hardness penetrates the core and causes brittleness. The two methods presently used to case harden steel are carburizing and nitriding.

CARBURIZING

Carburizing forms a thin layer of high-carbon steel on the exterior of low-carbon steel. This form of case-hardening is accomplished through one of three methods: pack carburizing, gas carburizing, and liquid carburizing. **Pack carburizing** is done by enclosing the metal in a fire-clay container and packing it with a carbon-rich material such as charcoal. The container is then sealed, placed in a furnace, and heated to 1,700°F. As the charcoal heats up, carbon monoxide gas forms and combines with the gamma iron in the metal's surface. The depth to which the carbon penetrates depends upon the soaking time.

TOOL	OXIDE COLOR
SCRIBE OR HAMMER FACE	PALE YELLOW
CENTER PUNCH	GOLDEN YELLOW
COLD CHISEL OR DRIFT	BROWN
SCREWDRIVER	PURPLE

Figure 7-8. Common tools can be tempered by heating the metal until it turns the proper color, and then quenching.

Gas carburizing is similar to pack carburizing, except the carbon monoxide is produced by a gas rather than a solid material. In this process, carbon from carbon monoxide gas combines with gamma iron and forms a high-carbon surface.

Liquid carburizing produces a high-carbon surface when a part is heated in a molten bath of sodium cyanide or barium cyanide. Either of the components supplies the carbon needed to harden the metal's surface.

At times, only a portion of a part must be carburized. Since carbon does not infuse into a copper-plated surface, any portion of a part that you do not want case-hardened should be copper-plated.

NITRIDING

Nitriding differs from carburizing in that a part is first hardened, tempered, and then ground to its finished dimensions before it is case hardened. Once these steps are accomplished, the part is placed in a special furnace, heated to a temperature of approximately 1,000°F, and then surrounded by ammonia gas (NH_3). The high temperature breaks the ammonia down into nitrogen and hydrogen. The nitrogen is absorbed into the steel as iron nitride. Most steels can be nitrided, however, special alloys are required for best results. These special alloys contain aluminum as one of the alloying elements and are called "nitralloys." The depth of a nitrided surface depends on the length of time it is exposed to the ammonia gas. Surface hardness gradually decreases with depth until it is the same as the core.

Aircraft engine crankshafts and cylinder walls are commonly nitrided for increased wear resistance. However, since nitrided surfaces are highly susceptible to pitting corrosion they must be protected from the air with a coating of oil.

RELATIONSHIP BETWEEN HARDNESS AND STRENGTH

As noted in the section on the heat treatment of steel, the smaller the carbon particles introduced into steel, the harder and stronger the steel becomes. Because of this, there is a definite relationship between the strength and hardness of steel. This relationship does not necessarily exist, however, with metals other than steel.

Two systems of measuring the hardness of steel are commonly used in shops where steel is heat-treated. They are the Brinell and Rockwell systems.

THE BRINELL HARDNESS SYSTEM

The Brinell hardness tester uses a hydraulic force to impress a spherical penetrator into the surface of a sample. The amount of force used is approximately 3,000 kilograms for steel, and 500 kilograms for nonferrous metals. This force is hydraulically applied by a hand pump and read on a pressure gauge. When the sample is removed from the tester, the diameter of the impression is measured with a special calibrated microscope. This diameter is converted into a Brinell number by using a chart furnished with the tester. [Figure 7-9]

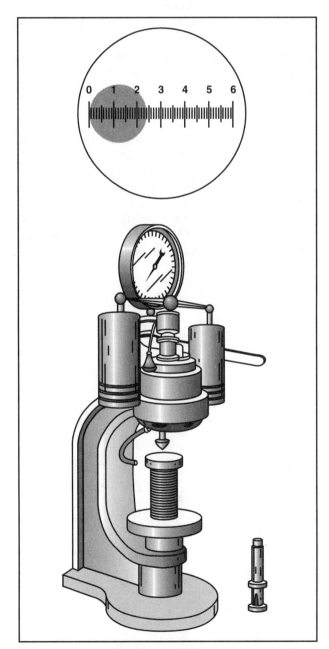

Figure 7-9. A Brinell hardness tester determines hardness by forcing a penetrator into a sample and measuring the diameter of the resulting impression.

THE ROCKWELL HARDNESS SYSTEM

The Rockwell hardness tester gives the same information the Brinell tester gives, except that it measures the depth to which the penetrator sinks into the material rather than the diameter of the impression. [Figure 7-10]

To use a Rockwell hardness tester, a sample is thoroughly cleaned, the two opposite surfaces are ground flat and parallel, and all scratches are polished out. The sample is then placed on the anvil of the tester and raised up against the penetrator. A 10-kilogram load, called the **minor** load, is applied and the machine is zeroed. A **major** load is then applied and the dial on the tester indicates the depth the penetrator sinks into the metal. Instead of indicating the depth of penetration in thousandths of an inch, it indicates in Rockwell numbers on either a red or a black scale.

Rockwell testers use three types of penetrators: a conical diamond, a 1/16 inch ball, and a 1/8 inch ball. There are also three major loads: 60 kilograms, 100 kilograms, and 150 kilograms. The two most commonly used Rockwell scales are the B-scale for soft metals, which uses a 1/16 ball penetrator and a 100 kg major load, and the C-scale for hard metals, which uses the conical diamond penetrator and a 150 kg major load.

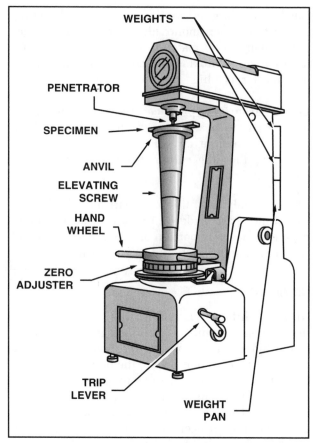

Figure 7-10. A Rockwell hardness tester produces the same information as a Brinell tester by measuring the depth of the impression in a sample.

NONMETALLIC STRUCTURAL MATERIALS

Nonmetallic structural materials played a very important role in the early days of aviation. However, after the introduction of aluminum, nonmetallic materials saw considerably less use. Today, aluminum is still the most widely used material in the construction and repair of aircraft. However, since its introduction, several new materials have come into use, many of which are spin-offs from the space program. These materials once again have the aviation industry taking a close look at nonmetallic materials for use on aircraft.

WOOD

As you know, far fewer aircraft today are made of wood. One reason why this is true is because of the greater economy of building aircraft of sheet metal. Wooden structures require a great deal of handwork and, therefore, are extremely expensive. It is possible, however, that as an aviation maintenance technician, you could be involved in the restoration of a classic wooden aircraft, or be asked to make repairs to a wooden homebuilt aircraft. Therefore, it is important to be aware of the materials and techniques used on wood aircraft structures.

TYPES OF WOOD

There are three basic types of wood used for aircraft structure. They are solid wood, laminated wood, and plywood. **Solid wood** is used for some aircraft wing spars and, as the name implies, is made of one solid piece cut from a log. Most solid wood used on aircraft is cut from a tree by quarter sawing. When planks are cut from a log in this fashion, the annual rings cross the plank at an angle greater than 45 degrees. This reduces the chance of warpage as compared to planks that are cut straight across, or plain sawed. [Figure 7-11]

Wing spars require long pieces of extremely high-quality wood that is free from imperfections. Since wood of this size and quality is very expensive, many manufacturers use **laminated wood.** Laminated wood is made of two or three pieces of

thin wood glued together with the grain running in the same direction. Each of the laminations can be spliced, so large components, such as spars, can be made from pieces of wood that are not large enough for a solid spar. The FAA allows solid and laminated spars to be used interchangeably, as long as they are both of the same quality and neither has a grain deviation that exceeds the allowable limits.

The third type of wood used for aircraft construction is **plywood**. Plywood consists of three or more layers of thin veneer glued together so the grain of each successive layer crosses the others at an angle of 45 degrees or 90 degrees.

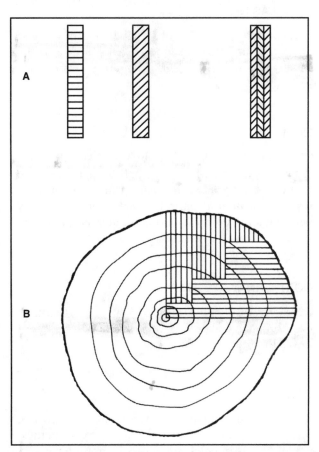

Figure 7-11. (A) — Quarter-sawing produces a plank more resistant to warpage. (B) — The annual rings of a quarter-sawed plank cross at an angle more than 45 degrees.

SPECIES OF WOOD

Two basic species of wood are used for aircraft construction: **hardwood** and **softwood**. Hardwoods come from deciduous trees having broad leaves, while softwoods come from coniferous trees with needle-like or scale-like leaves. These terms are quite deceptive, as some of the so-called hardwoods are softer than some of the woods that are classified as softwoods. For example, balsa wood is extremely light and soft, but, because of its type of leaf, is classified as a hardwood.

SPRUCE

Sitka spruce is the most common wood used in aircraft structures. Some of the reasons why sitka spruce is so widely used is because it is relatively free from defects, has a high strength-to-weight ratio, and is available in large sizes. In fact, since sitka spruce is such a high quality wood, the FAA has chosen it as the reference wood for aircraft construction.

DOUGLAS FIR

The strength properties of Douglas fir exceed those of spruce; however, it is much heavier. Furthermore, it is more difficult to work than spruce, and has a tendency to split.

NOBLE FIR

This type of fir is slightly lighter than spruce and is equal or superior to spruce in all properties except hardness and shock resistance. Noble fir is often used for structural parts that are subject to heavy bending and compression loads such as spars, spar flanges, and cap strips.

NORTHERN WHITE PINE

Northern white pine possesses excellent working and gluing characteristics, but its strength is slightly lower than that of spruce.

BALSA

Balsa is an extremely light wood that is grown in South America. Because balsa lacks structural strength, it is often sliced across its grain for use as a core material for sandwich-type panels that require light weight and rigidity.

MAHOGANY

This hardwood is heavier and stronger than spruce. Mahogany's primary use in aircraft construction is for face sheets of plywood used as aircraft skin.

BIRCH

Birch is a heavy hardwood with very good shock resistant characteristics. It is recommended for the face plies of plywood used as reinforcement plates on wing spars and in the construction of wooden propellers. [Figure 7-12]

QUALITY OF WOOD

Any wood approved for aircraft construction or structural repair must meet specific requirements regarding quality and allowable defects. Some of the categories a wood's quality is based on include how straight the grain is, the number of knots, pitch pockets, splits, and the presence of decay.

GRAIN DEVIATION

Regardless of the species of wood used for aircraft structure, it must have a straight grain. This means all of the wood fibers must be oriented parallel to the material's longitudinal axis. A maximum deviation of 1:15 is allowed. In other words, the grain must not slope more than one inch in 15 inches. In some woods it is difficult to determine exactly

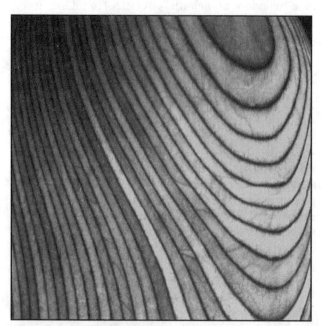

Figure 7-12. This laminated wood propeller is made of thin strips of birch wood glued together, with the grain of all strips running in the same direction.

which way the grain runs, but you can see the grain direction by putting a bit of free-flowing ink on the wood. The ink is absorbed into the fibers along the grain making it easier to identify.

KNOTS

A knot in a piece of wood identifies a point where a branch grew from the tree trunk. Knots may be round, oval, or spiked in shape, depending on the way the lumber was cut. For example, spike knots are knots that have been cut parallel to a branch. Spike knots weaken a board and are not acceptable in wood for aircraft construction or repair.

Hard knots up to 3/8 inch at their maximum dimension are acceptable in aircraft wood with the following restrictions:

1. These knots must not be in the projecting portion of I-beams, along the edges of rectangular or beveled unrouted beams, or along the edges of the flanges of box beams, except in low-stress areas.
2. They must not cause a grain divergence at the edge of the board or in the flanges of a beam more than 1:15.
3. They must be in the center third of the beam and no closer than 20 inches to another knot or other defect.

Small pin-knot clusters are acceptable if they do not cause excessive grain deviation. Although knots of up to 3/8 inch are allowed in some places, any wood having knots larger than 1/4 inch in any direction should be used only with extreme caution.

PITCH POCKETS

Small openings within the annual rings of a tree can fill with resin and form pitch pockets. These pockets weaken a piece of wood slightly and, therefore, are acceptable only in the center portion of a beam, provided they are at least 14 inches apart and measure no more than 1 1/2 inches by 1/8 inch by 1/8 inch deep.

COMPRESSION WOOD

If a tree leans in its growth, its annual rings are not concentric. Furthermore, the summer wood, which is represented by the heavier and darker rings, will be much larger on the lower side of the tree. Wood having extra wide summer-growth rings indicates the wood was subjected to compressive stresses during its growth, and is denser and weaker than wood that grew normally. Because of this weakness, compression wood cannot be used in aircraft construction or repair.

COMPRESSION FAILURE

Compression failure occurs when a tree falls and must not be confused with compression wood. If a tree falls across a log, some of the wood fibers on the tree's lower side may buckle. This buckling severely weakens the wood and is identified by irregular, thread-like lines across the grain. Any wood showing this type of failure has been weakened to an extent that it is unacceptable for use in aircraft construction or repair.

CHECKS, SHAKES, AND SPLITS

A check is a crack that runs across the annual rings of a board and occurs during the seasoning process. A shake, on the other hand, is a crack or separation that occurs when two annual rings separate along their boundary. A split is a lengthwise separation of the wood caused by the wood fibers tearing apart. Wood having any of these defects is not allowed in aircraft structure.

STAINS AND DECAY

Stains caused by decay usually appear as streaks in the grain. These streaks may extend several feet from the decayed area. As a rule, stains that uniformly discolor the annual rings are evidence of decay. Rotted wood characteristically ranges from red to white in color. For example, a stain caused by red rot in its early stages is reddish-purple. However, in its developed stage the heart-wood becomes honeycombed with small, soft, white pockets. Even in its early stages, decay seriously reduces the toughness of wood, and in its well-developed state leaves wood with little or no strength.

Decay is caused by fungi that grow in damp wood, and is prevented by proper seasoning and dry storage. A simple way of identifying decayed wood is to pick at a suspected area with the point of a knife. Sound wood will splinter, while a knife point will bring up a chunk of decayed wood.

DRY ROT

Dry rot is a misnomer, because dry wood, particularly kiln dried wood, does not rot. For fungi to attack wood and cause decay, moisture must be present long enough to thoroughly saturate the wood. Dry rot can be detected by inserting a knife point into the area and prying a small piece out. If the wood is good, it splinters, while decayed wood crumbles.

Prevention of dry rot is simply a matter of using kiln dried wood with a moisture content of 20 percent or less, and protecting it from the elements. Application of a good finish such as a varnish or urethane and adequate placement and maintenance of drain holes are also a necessity in preventing dry rot. The only repair for dry rot is replacement of the affected material.

PLASTICS

Just as plastic has become a major part of our everyday life, it also has become an important element in the construction of aircraft. Plastics, or **resins**, fall into two major classifications: thermoplastic and thermosetting.

THERMOPLASTIC RESINS

Thermoplastic materials in their normal state are hard, but become soft and pliable when heated. When softened, thermoplastic materials can be molded and shaped, and retain their shape when cooled. Unless their heat limit is exceeded, this process can be repeated many times without damaging the material.

Two types of transparent thermoplastic materials are used for aircraft windshields and side windows. They are cellulose acetate and acrylic. Early aircraft used **cellulose acetate** plastic because of its transparency and light weight. However, it has a tendency to shrink and turn yellow and, therefore, has almost completely been replaced. Cellulose acetate can be identified by its slightly yellow tint and the fact that, if a scrap of it is burned, it produces a sputtering flame and dark smoke. Another way to identify cellulose acetate is with an acetone test. When acetone is applied to cellulose acetate it softens.

Acrylic plastics are identified by such trade names as Lucite or Plexiglas, or in Britain by the name Perspex. Acrylic is stiffer than cellulose acetate, more transparent and, for all practical purposes, is colorless. It burns with a clear flame and produces a fairly pleasant odor. Furthermore, if acetone is applied to acrylic it leaves a white residue but remains hard. [Figure 7-13]

THERMOSETTING RESINS

Thermosetting materials usually have little strength in themselves and are generally used to impregnate linen, paper, or glass cloth. For example, both fiberglass cloth and mat have a great deal of strength for their weight, but lack rigidity. To convert fiberglass into a useful structural material, it is impregnated with **polyester resin** and molded into a desired form. Polyesters cure by chemical action and, therefore, differ from materials that cure by evaporation of an oil or solvent. Polyester resin has a complex molecular structure that, in its pure form, is thick and unmanageable. For this reason, a **styrene monomer** is added to polyester resin to thin it and make it more workable.

If left alone, a mixture of polyester and styrene eventually hardens into a solid mass. To prevent this, inhibitors are added to resins to extend their working time. However, when the resin is to be used, a **catalyst** must be introduced to suppress the inhibitors and initiate the curing process. Furthermore, the resin's curing time can be appreciably shortened by the addition of a measured amount of an **accelerator**. The amount of accelerator needed depends on the ambient temperature and the thickness of the resin layer.

Figure 7-13. Modern aircraft construction makes extensive use of plastic materials for engine cowlings, wheel fairings, windshields, and side windows.

The actual cure of polyester resin occurs when a chemical reaction between the catalyst and accelerator generates heat within the resin. The less surface area there is for heat to escape, the more heat remains in the resin and the faster it cures. Therefore, when submitted to identical conditions, a thick layer of resin cures more rapidly than a thin layer.

EPOXY RESIN

Another type of resin that can be used in place of polyester in laminated structures is epoxy resin. Epoxy has a low percentage of shrinkage, high strength for its weight, and the ability to adhere to a wide variety of materials. Unlike polyester resins which require a catalyst, epoxy resins require a hardener, or curing agent.

Another difference between epoxy and polyester is the ratio in which each resin is mixed with its catalyst or hardener. For polyesters, about one ounce of catalyst is used with 64 ounces of resin. Epoxies, on the other hand, use a ratio of about one quart of curing agent with four quarts of resin. [Figure 7-14]

THIXOTROPIC AGENTS

Some plastic resins are extremely sensitive to temperature changes. For example, at temperatures around 60°F some resins become extremely thick, while at temperatures above 90°F they become very thin. Since heat is used to effect the cure, many resins are likely to run off any inclined or vertical surface before they have a chance to set. To eliminate this problem, a **thixotropic** agent is added to

the resin to increase its viscosity. Among the more familiar thixotropic agents used in aviation maintenance are **microballoons**. These hollow glass or phenolic balls range in diameter from about 10 to 300 microns, or about the size of fine sand. However, microscopic inspection reveals that microballoons are perfect spheres.

After an epoxy or polyester resin is properly mixed, a thixotropic agent is carefully added to make a paste. This paste has enough body to remain in place when it is troweled onto a surface. After the resin cures, it can be sanded to the required contour.

REINFORCING MATERIALS

For many years aircraft control cable pulleys have been made from thermosetting resin reinforced with layers of linen **cloth**. The resin-impregnated layers of cloth are cured in a mold under high temperature and pressure to form pulleys that have high strength, but do not wear control cables. [Figure 7-15]

When layers of paper are impregnated with a thermosetting resin such as **phenol-formaldehyde** or **urea-formaldehyde**, they can be molded into either flat sheets or complex shapes. Once cured, this material is stiff and acts as an exceptionally good electrical insulator. It is, however, quite heavy. Electrical terminal strips have traditionally been made of paper-reinforced resin.

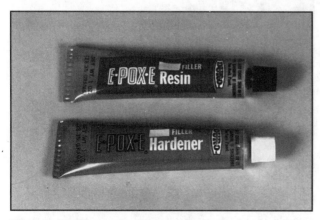

Figure 7-14. When mixed in proper portions, epoxy resins form a high-strength adhesive and provide a good bond between many types of materials.

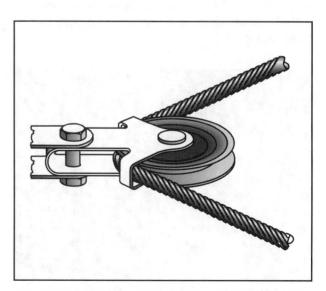

Figure 7-15. Phenol-formaldehyde resin is used to impregnate linen cloth to make control cable pulleys. These pulleys have high strength but do not wear the control cable.

Glass fiber greatly enhances the strength and durability of thermosetting resin. For high strength requirements, the glass fibers are woven into a cloth. On the other hand, where cost is of greater importance than strength, the fibers are gathered into a loose mat which is saturated with resin and molded into a desired shape. [Figure 7-16]

Ceramic fiber is a form of glass fiber designed for use in high-temperature applications. It can withstand temperatures approaching 3,000°F, making it an effective laminate for use around engines and exhaust systems. However, its disadvantages include its weight and expense.

Kevlar™ fiber is one of the most commonly used cloth reinforcing fabrics. In its cloth form, Kevlar is a soft yellow organic fiber that is extremely light, strong, and tough. Its great impact resistance makes it useful in areas where damage from sand or other debris can occur. These areas include around landing gear and behind propellers. Kevlar is rather difficult to work with, however, and does not perform well under compressive loads.

Graphite fibers are manufactured by heating and stretching Rayon™ fibers. This produces a change in the fiber's molecular structure that makes it extremely lightweight, strong, and tough.

LAMINATED STRUCTURAL MATERIALS

To increase the strength and rigidity of many nonmetallic structures, one of several structural materials are laminated between two layers of cloth and resin. Some of the more common **core** materials include honeycomb, aluminum, wood, and metal-faced honeycomb.

The cellular core for laminated honeycomb material may be made of resin-impregnated kraft paper or glass cloth. When this material is cut to the proper shape and bonded between two face sheets of resin-impregnated fiberglass cloth, it makes an electrically transparent material that has a high strength-to-weight ratio and is extremely rigid. [Figure 7-17]

Most **metal cores** are made of a cellular material consisting of extremely thin aluminum. Like the honeycomb, the metal core also is bonded between face sheets of fiberglass-reinforced resin.

A third type of core is the **wood core**. Most wood cores consist of balsa wood that has been cut across its grain and the end-grain slabs bonded between two face sheets of fiberglass or metal. This provides an exceptionally strong and lightweight structural material that is used for floors, wall panels, and aircraft skins.

The honeycomb materials used in fiberglass laminations can also be sandwiched between thin sheets of aluminum alloy. This type of core is referred to as **metal-faced honeycomb** and is used where abrasion resistance is important.

NONMETALLIC COMPONENTS

In addition to the nonmetallic materials used in aircraft structures, modern aircraft utilize a great deal of nonmetallic materials on a variety of important components. Therefore, you, as a technician, should be familiar with these materials as well as their proper use and handling.

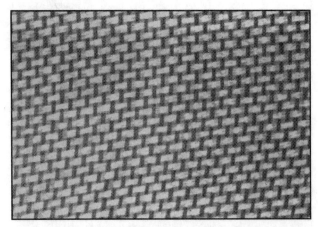

Figure 7-16. Thin filaments, or fibers, of glass are woven into a cloth to reinforce thermosetting resins to form various types of aircraft structures.

Figure 7-17. A high-strength structural material used to provide an electrically transparent, streamlined housing for radar antenna is made of fiberglass-reinforced honeycomb.

SEALS

Seals or packings serve both to retain fluids in their operating systems and to exclude air and contaminants. However, with the increased use of high-pressure fluid systems, packings and gaskets must also be able to perform under a wide range of operating pressures and temperatures. As a result, packings are made in the form of O-rings, V-rings, and U-rings, each designed for a specific purpose. As an example, synthetic or natural rubber packings are generally used as "running seals" in units that contain moving parts, such as actuating cylinders, pumps, and selector valves. [Figure 7-18]

O-RING PACKINGS

O-ring packings effectively seal in both directions and, therefore, are used to prevent both internal and external leakage. O-rings are the most commonly used seals in aviation. In installations subject to pressures above 1,500 psi, **backup** rings are used with O-rings to prevent the O-ring from being forced out, or **extruded**.

Most O-rings are similar in appearance and texture. However, the material an O-ring is made of is typi-

cally designed for use under various operating conditions and temperatures. For example, an O-ring intended specifically as a stationary seal generally will not seal a moving part such as a hydraulic piston. In other words, an O-ring is useless if it is not compatible with the system it is used in.

Advances in aircraft design necessitate new O-ring compositions to meet changing operating conditions. Hydraulic O-rings were originally established under AN specification numbers 6227, 6230, and 6290 for use with MIL-H-5606 hydraulic fluid, at temperatures ranging from −65°F to +160°F. However, new component designs raised operating temperatures to as high as 275° F, requiring the development of new compounds that can withstand these harsher conditions.

Manufacturer's color coding on some O-rings is not a reliable or complete means of identification. Color codes identify only system fluid or vapor compatibility and, in some cases, the manufacturer. In addition, the coding on some O-rings is not permanent. On some O-rings the color may be omitted due to manufacturing difficulties, or interference with operation. Furthermore, the color-coding system provides no means to establish an O-ring's age or temperature limitations.

To avoid these problems, O-rings are available in individual hermetically sealed envelopes labeled with all pertinent data. The part number on a sealed envelope provides the most reliable compound identification.

Although an O-ring may appear perfect at first glance, slight surface flaws are often present that can lead to leakage under various operating pressures. Therefore, before installing an O-ring inspect it for cuts, abrasions, or surface imperfections that could affect its performance. Rolling an O-ring on an inspection cone or dowel reveals its inner diameter surface. Inspect this area for small cracks, particles of foreign material, or other irregularities that can cause leakage or shortened life.

BACKUP RINGS

For applications such as actuators that subject a seal to pressure from two sides, two backup rings must be used. When an O-ring is subject to pressure on only one side, a single backup ring is generally adequate. In this case, the backup ring is placed on the side of the O-ring away from the pressure.

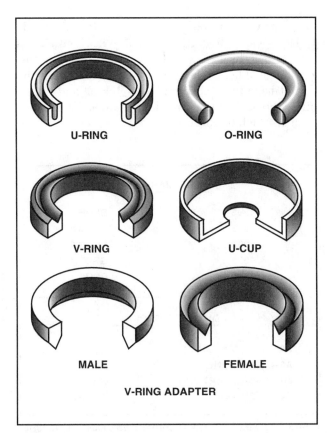

Figure 7-18. Packing rings are manufactured with different profiles to suit varying operating requirements.

Backup rings are commonly made of Teflon™ and, therefore, do not deteriorate with age, are unaffected by any system fluid or vapor, and tolerate temperature extremes well in excess of those encountered in high-pressure hydraulic systems. Their dash numbers indicate both their size and the dash number of the O-ring for which they are dimensionally suited. Any Teflon backup ring can be used to replace any other Teflon backup ring if it is of proper overall dimension required to support the applicable O-ring. Backup rings are not color coded or marked and must be identified from package labels.

When inspecting backup rings, ensure that the surfaces are free from irregularities, that the edges are clean and sharp, and that scarf cuts are parallel. When checking Teflon spiral backup rings, make sure that the coils do not separate more than 1/4 inch when unrestrained.

V-RING PACKINGS (CHEVRON)

V-ring packings (AN6225) are one-way seals that are installed with the open end of the "V" facing the pressure. Each V-ring packing consists of a male and female adapter that hold the packings in the proper position after installation. In order for a seal to perform properly, it is also necessary to torque the seal retainer to the value specified by the manufacturer of the component being serviced. [Figure 7-19]

U-RING PACKINGS

U-ring packings (AN6226) and **U-cup** packings are used in brake assemblies and brake master cylinders. Since U-ring and U-cup packings seal in only one direction, their concave surfaces must face toward the pressure. U-ring packings are primarily low-pressure packings that are used with pressures less than 1,000 psi.

SHOCK ABSORBER CORD

To absorb the shock loads associated with landing, some older aircraft utilize shock absorber cord on their landing gear. The cord is installed between the two main gears and stretches under heavy loads allowing the main gear to spread out and absorb sudden shocks.

Shock absorber cord is made from natural rubber strands encased in a braided cover of woven cotton cords treated to resist oxidation and wear. To achieve a high degree of tension and elongation, rubber strands are stretched to three times their original length before the woven jacket is applied.

Three colored threads are braided into a cord's outer cover. Two of these threads are the same color and represent the year the cord was manufactured. The third thread, which is a different color, represents the quarter of the year the cord was made. The colors cover a 5-year period and then repeat. [Figure 7-20]

There are two types of elastic shock-absorbing cord. **Type I** is a straight cord, and **Type II** is a continuous ring, known as a "bungee." Type II cords have the advantage of being easily and quickly replaced, and do not have to be secured by stretching and whipping. Shock cord is available in standard diameters from 1/4 inch to 13/16 inch.

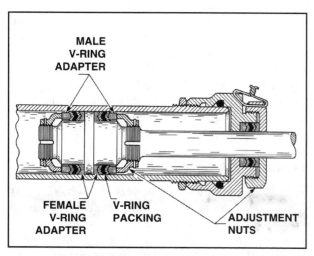

Figure 7-19. A chevron seal must be installed with the female adapter facing the pressure.

YEAR MARKING	
1995	BLUE
1996	YELLOW
1997	BLACK
1998	GREEN
1999	RED
QUARTER MARKING	
JAN, FEB, MARCH	RED
APR, MAY, JUNE	BLUE
JULY, AUG, SEPT	GREEN
OCT, NOV, DEC	YELLOW

Figure 7-20. Since a shock absorber cord's elasticity deteriorates with age, a color code indicates its date of manufacture.

GASKETS

Gaskets are used in fluid systems around the end caps of actuating cylinders, valves, and other units. A gasket is shaped specifically for its intended use and sometimes resembles the shape of an O-ring. Gasket materials used in stationary applications between two flat surfaces include copper, cork, and rubber.

SOLID COPPER WASHER

A solid copper washer is commonly used for spark plug gaskets and some hydraulic fittings that require a noncompressible yet semi-soft gasket. When compressed between two solid surfaces, the copper deforms to provide a tight seal.

CORK GASKETS

When mating surfaces are uneven or rough, a gasket must be placed between the two surfaces. Furthermore, the gasket material must be able to expand or compress to provide an adequate seal. Cork works well under these circumstances and is often used in areas such as between an engine crankcase and accessories, or anywhere a gasket is required that can occupy an uneven or varying space.

RUBBER SHEETING

Rubber sheeting is often used anywhere a compressible gasket is required. Its flexibility allows it to compress easily to provide a tight seal. However, if rubber seals are exposed to gasoline, oil, or some types of hydraulic fluid, they will deteriorate.

WIPERS

Wipers clean and lubricate the exposed portions of piston shafts. By literally wiping a surface, they prevent dirt from entering the system and help protect the piston shaft from scoring. While wipers can be either metallic or felt, the two are sometimes used together, with a felt wiper installed behind a metallic wiper.

SEALING COMPOUNDS

Certain areas of all aircraft are sealed to withstand pressurization, prevent fuel leakage or passage of fumes, or to prevent corrosion by sealing against the weather. Most sealants consist of two or more ingredients that are compounded to produce a desired combination of strength, flexibility, and adherence.

Some materials are ready for use as packaged, but others require mixing before application.

ONE-PART SEALANT

One-part sealants are prepared by the manufacturer and are ready for application as packaged. However, the consistency of some of these compounds can be altered to satisfy a particular application method. For example, if thinning is required, a thinner recommended by the sealant manufacturer is mixed.

TWO-PART SEALANT

Two-part sealants are compounds requiring separate packaging to prevent curing prior to application. The two parts are identified as the **base** sealing compound and the **accelerator**. Two-part sealants are generally mixed by combining equal portions (by weight) of the base and accelerator compounds and any deviation from the prescribed ratios can reduce the material's quality.

To ensure the proper ratio is used, all sealant material should be carefully weighed in accordance with the manufacturer's recommendations. Sealant material is usually weighed with a balance scale equipped with weights specially prepared for various quantities of sealant and accelerator. To ensure a proper mix ratio, the base and accelerator should be thoroughly stirred before they are weighed. Accelerator which is dried out, lumpy, or flaky should be discarded. Some manufacturers produce pre-weighed sealant kits that require no weighing. In this case the entire quantity provided is mixed.

After the proper amount of base compound and accelerator have been determined, the accelerator is added to the base compound. Immediately after adding the accelerator, thoroughly mix the two parts by stirring or folding, depending on the material's consistency. Mix the material carefully to prevent air entrapment in the mixture. Avoid rapid or prolonged stirring, since it shortens the sealant's available working time. To ensure a compound is well-mixed, test a small portion by smearing it on a clean, flat metal or glass surface. If flecks or lumps are found, continue mixing. If the flecks or lumps cannot be eliminated, the batch should be rejected.

The working life of mixed sealant typically ranges from a half hour to four hours, depending on the sealant class. Therefore, mixed sealant should be applied as soon as possible or placed in refrigerated storage.

SEALANT CURING

The curing rate of mixed sealants varies with temperature and humidity. For example, at temperatures below 60°F curing is extremely slow. However, a temperature above 70°F typically results in a faster curing time. A temperature of 77°F with 50 percent relative humidity is the ideal condition for curing most sealants.

If you must accelerate the curing time of a sealant, you may increase the temperature by applying heat. However, the temperature should not be allowed to exceed 120°F at any time in the curing cycle. Heat can be applied by using infrared lamps or heated air.

If heated air is used, it must be properly filtered to remove moisture and dirt.

Heat should not be applied to any faying surface sealant installation until all work is completed. All faying surface applications must have all attachments, permanent or temporary, completed within the application limitations of the sealant.

Sealant must be cured to a tack-free condition before applying brush top coatings. Tack-free consistency is the point at which a sheet of cellophane pressed onto the sealant no longer adheres.

CHAPTER 8

AIRCRAFT HARDWARE

INTRODUCTION

The term aircraft hardware describes the various types of fasteners and miscellaneous small items used in the manufacture and repair of aircraft. The importance of aircraft hardware is often overlooked because of its small size; however, the safe and efficient operation of all aircraft depends on the correct selection and use of aircraft hardware. Today there are more than 30,000 different fasteners available for aerospace applications. For ease of understanding, this chapter is broken out into two sections. The first section discusses the various types of rivets used in the construction and repair of aircraft, while Section B looks at the various other fasteners available in the aviation industry.

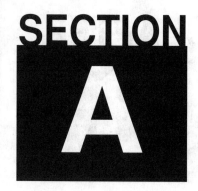
AIRCRAFT RIVETS

While a number of methods are available for joining metal parts, few are ideal for joining aluminum sheets. The most common technique of joining sheets of aluminum is riveting. A rivet is a metal pin with a formed head on one end. A rivet is inserted into a drilled hole, and its shank is then deformed by a hand or pneumatic tool. Rivets create a union at least as strong as the material being joined.

SPECIFICATIONS AND STANDARDS

Before the Federal Aviation Administration issues a Type Certificate for an aircraft, the manufacturer must demonstrate that the aircraft conforms to all airworthiness requirements. These requirements pertain not only to performance, but to structural strength and integrity as well. To meet these requirements, each individual aircraft produced from a given design must meet the same standards. To accomplish this, all materials and hardware must be manufactured to a standard of quality. Specifications and standards for aircraft hardware are generally identified by the organization originating them. Some of the most common are:

AMS   Aeronautical Material Specifications
AN   Air Force-Navy
AND   Air Force-Navy Design
AS   Aeronautical Standard
ASA   American Standards Association
ASTM   American Society for Testing and Materials
MS   Military Standard
NAF   Naval Aircraft Factory
NAS   National Aerospace Standard
SAE   Society of Automotive Engineers

When an MS20470-AD4-4 rivet is required, specifications have already been written for it and are available to both the aircraft manufacturer and rivet producer. These specifications stipulate the material to be used as well as the rivet dimensions. By using these specifications and calling for standard hardware, aircraft manufacturers are able to build reproducible aircraft at an economical cost.

SOLID SHANK RIVETS

The solid shank rivet has been used since sheet metal was first utilized in aircraft, and remains the single most commonly used aircraft fastener today. Unlike other types of fasteners, rivets change in dimension to fit the size of a hole. [Figure 8-1]

When a rivet is driven, its cross sectional area increases along with its bearing and shearing strengths. Solid shank rivets are available in a variety of materials, head designs, and sizes to accommodate different applications.

RIVET CODES

Rivets are given part codes that indicate their size, head style, and alloy material. Two systems are in use today, the Air Force - Navy, or AN system, and the Military Standards 20 system, or MS20. While there are minor differences between the two systems, both use the same method for describing rivets. As an example, consider the rivet designation AN470-AD4-5.

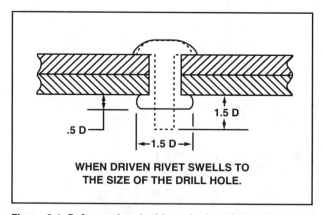

WHEN DRIVEN RIVET SWELLS TO THE SIZE OF THE DRILL HOLE.

Figure 8-1. Before a rivet is driven, it should extend beyond the base material at least one and a half times the rivet's diameter. Once driven, the rivet shank expands to fill the hole and the bucktail expands to one and a half times its original diameter. Once the bucktail expands to the appropriate diameter it should extend beyond the base material by at least one half the original rivet diameter.

The first component of a rivet part number denotes the numbering system used. As discussed, this can either be AN or MS20. The second part of the code is a three-digit number that describes the style of rivet head. The two most common rivet head styles are the universal head, which is represented by the code 470, and the countersunk head, which is represented by the code 426. Following the head designation is a one or two-digit letter code representing the alloy material used in the rivet. These codes will be discussed in detail later.

After the alloy code, the shank diameter is indicated in 1/32 inch increments and the length in increments of 1/16 inch. Therefore, in this example the rivet has a diameter of 4/32 inch and is 5/16 of an inch long. [Figure 8-2]

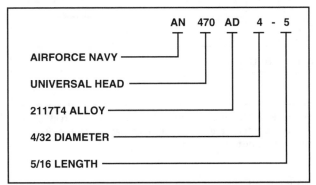

AN 470 AD 4 - 5

AIRFORCE NAVY

UNIVERSAL HEAD

2117T4 ALLOY

4/32 DIAMETER

5/16 LENGTH

Figure 8-2. Rivet identification numbers indicate head style, material, and size.

The length of a universal head (AN470) rivet is measured from the bottom of the manufactured head to the end of the shank. However, the length of a countersunk rivet (AN426) is measured from the top of the manufactured head to the end of the shank. [Figure 8-3]

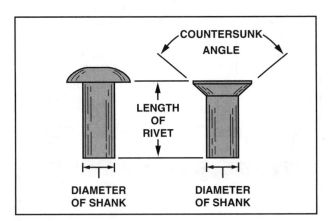

COUNTERSUNK ANGLE

LENGTH OF RIVET

DIAMETER OF SHANK

DIAMETER OF SHANK

Figure 8-3. Universal and countersunk rivet diameters are measured in the same way, but their length measurements correspond to their grip length.

HEAD DESIGN

As mentioned, solid shank rivets are available in two standard head styles, **universal** and **countersunk**, or flush. The AN470 universal head rivet now replaces all previous protruding head styles such as AN430 round, AN442 flat, AN455 brazier, and AN456 modified brazier. [Figure 8-4]

AN426 countersunk rivets were developed to streamline airfoils and permit a smooth flow over an aircraft's wings or control surfaces. However, before a countersunk rivet can be installed, the metal must be countersunk or dimpled. **Countersinking** is a process in which the metal in the top sheet is cut away in the shape of the rivet head. On the other hand, **dimpling** is a process that mechanically "dents" the sheets being joined to accommodate the rivet head. Sheet thickness and rivet size determine which method is best suited for a particular application.

Joints utilizing countersunk rivets generally lack the strength of protruding head rivet joints. One reason is that a portion of the material being riveted is cut away to allow for the countersunk head. Another

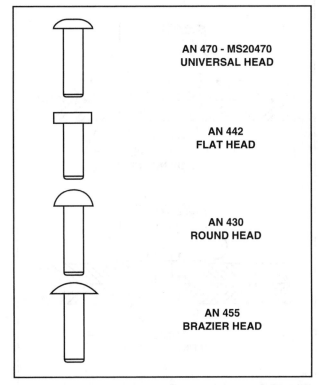

AN 470 - MS20470 UNIVERSAL HEAD

AN 442 FLAT HEAD

AN 430 ROUND HEAD

AN 455 BRAZIER HEAD

Figure 8-4. The AN470 rivet now replaces almost all other protruding head designs. The round head rivet (AN430) was used extensively on aircraft built before 1955, while the flat head rivet (AN442) was widely used on internal structures. Flat head rivets are still used for applications requiring higher head strength.

reason is that, when riveted, the gunset may not make direct contact with the rivet head if the rivet hole was not countersunk or dimpled correctly, resulting in the rivet not expanding to fill the entire hole. To ensure head-to-gunset contact, it is recommended that countersunk heads be installed with the manufactured head protruding above the skin's surface about .005 to .007 of an inch. This ensures that the gunset makes direct contact with the rivet head. To provide a smooth finish after the rivet is driven, the protruding rivet head is removed using a **microshaver**. This rotary cutter shaves the rivet head flush with the skin, leaving an aerodynamically clean surface. [Figure 8-5]

An alternative to leaving the rivet head sticking up slightly is to use the Alcoa **crown flush rivet**. These rivets have a slightly crowned head to allow full contact with the gunset. To drive these rivets, a mushroom-type rivet set is placed directly on the crown flush head. When the rivet is driven, the gun drives the countersunk head into the countersink, while simultaneously completing rivet expansion. This results in a fully coldworked rivet that needs no microshaving. [Figure 8-6]

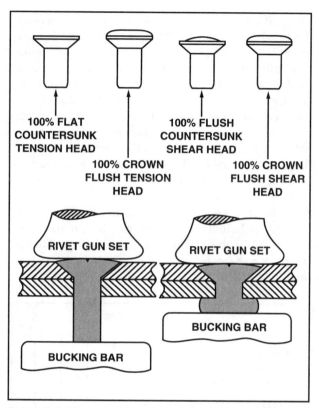

Figure 8-6. The raised head of a crown flush rivet allows greater contact area with a rivet set. This results in a stronger countersunk joint.

RIVET ALLOYS

Most aircraft rivets are made of an aluminum alloy. The type of alloy is identified by a letter in the rivet code and by a mark on the rivet head itself. [Figure 8-7]

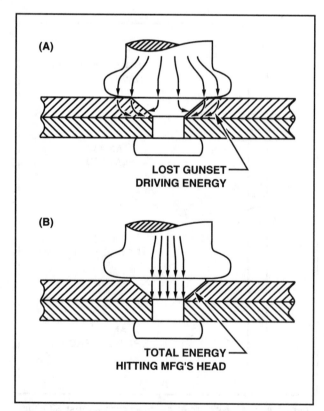

Figure 8-5. (A) — If a countersunk rivet is set with the rivet head flush with the metal's surface, some of the gunset's driving energy is lost. (B) — However, if the rivet head is allowed to protrude above the metal all of the gunset's energy hits the head resulting in a stronger joint.

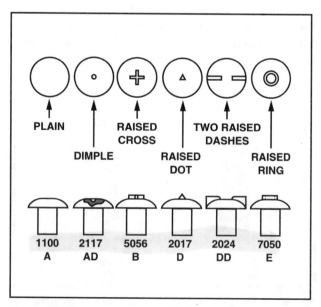

Figure 8-7. Head markings indicate the alloy used in common aircraft rivets.

1100 Aluminum (A)

Rivets made of pure aluminum have no identifying marks on their manufactured head, and are designated by the letter A in the rivet code. Since this type of rivet is made out of commercially pure aluminum, the rivet lacks sufficient strength for structural applications. Instead, 1100 rivets are restricted to nonstructural assemblies such as fairings, engine baffles, and furnishings. The 1100 rivet is driven cold, and therefore, its shear strength increases slightly as a result of cold working.

2117 Aluminum Alloy (AD)

The rivet alloy 2117-T3 is the most widely used for manufacturing and maintenance of modern aircraft. Rivets made of this alloy have a dimple in the center of the head and are represented by the letters AD in rivet part codes. Because AD rivets are so common and require no heat treatment, they are often referred to as "field rivets."

The main advantage for using 2117-T3 for rivets is its high strength and shock resistance characteristics. The alloy 2117-T3 is classified as a heat-treated aluminum alloy, but does not require re-heat-treatment before driving.

5056 Aluminum Alloy (B)

Some aircraft parts are made of magnesium. If aluminum rivets were used on these parts, dissimilar metal corrosion could result. For this reason, magnesium structures are riveted with 5056 rivets which contain about 5 percent magnesium. These rivets are identified by a raised cross on their heads and the letter B in a rivet code. The maximum shear strength of an installed 5056H32 rivet is 28,000 pounds per square inch.

2017 Aluminum Alloy (D)

If you remember from Chapter 7, 2017 aluminum alloy is extremely hard. Rivets made of this alloy are often referred to as D rivets and were widely used for aircraft construction for many years. However, the introduction of jet engines placed greater demands for structural strength on aircraft materials and fasteners. In response to this, the aluminum industry modified 2017 alloy to produce a new version of 2017 aluminum, called the **crack free rivet alloy**. The minimum shear strength of the older 2017T31 rivet alloy is 30 KSI, while that of the new 2017T3 alloy is 34 KSI.

D-rivets are identified by a raised dot in the center of their head and the letter D in rivet codes. Because D-rivets are so hard they must be heat treated before they can be used. [Figure 8-8]

If you recall from your study of heat treatments, when aluminum alloy is quenched after heat treatment it does not harden immediately. Instead, it remains soft for several hours and gradually becomes hard and gains full strength. Rivets made of 2017 can be kept in this annealed condition by removing them from a quench bath and immediately storing them in a freezer. Because of this, D-rivets are often referred to as **icebox rivets**. These rivets become hard when they warm up to room temperature, and may be reheat-treated as many times as necessary without impairing their strength.

ALLOY	LETTER	HEAD MARKING	DRIVEN CONDITION	POUNDS IN KSI
1100	A	PLAIN	1100-F	9.5
2117	AD	DIMPLE	2117T3	30
5056	B	RAISED CROSS	5056H32	28
2017	D	RAISED DOT	2017T31	34
2017	D	RAISED DOT	2017T3	38
2024	DD	TWO RAISED DASHES	2024T31	41
7050	E	RAISED RING	7050T73	43

NOTE: KSI = $\dfrac{psi}{1000}$ (e.g. 30 KSI = 30,000 psi)

Figure 8-8. Different rivet alloys produce different shear strengths in their driven condition.

2024 Aluminum Alloy (DD)

DD-rivets are identified by two raised dashes on their head. Like D-rivets, DD-rivets are also called icebox rivets and must be stored at cool temperatures until they are ready to be driven. The length of time the rivets remain soft enough to drive is determined by the storage temperature. For example, if the storage temperature is -30° F, the rivets will remain soft enough to drive for two weeks. When DD rivets are driven their alloy designation becomes 2024T31.

7050T73 Aluminum Alloy (E)

A new and stronger rivet alloy was developed in 1979 called 7050T73. The letter E is used to designate this alloy, and the rivet head is marked with a raised circle. 7050 alloy contains zinc as the major alloying ingredient and is precipitation heat-treated. This alloy is used by the Boeing Airplane Company as a replacement for 2024T31 rivets in the manufacture of the 767 widebody aircraft.

Corrosion-Resistant Steel (F)

Stainless steel rivets are used for fastening corrosion-resistant steel sheets in applications such as firewalls and exhaust shrouds. They have no marking on their heads.

Monel (M)

Monel rivets are identified with two recessed dimples in their heads. They are used in place of corrosion-resistant steel rivets when their somewhat lower shear strength is not a detriment.

SPECIAL RIVETS

A rivet is any type of fastener that obtains its clamping action by having one of its ends mechanically upset. Conventional solid shank rivets require access to both ends to be driven. However, special rivets, often called **blind rivets** are installed with access to only one end of the rivet. While considerably more expensive than solid shank rivets, blind rivets find many applications in today's aircraft industry.

POP™ RIVETS

Pop rivets have limited use on aircraft and are never used for structural repairs. However, they are useful for temporarily lining up holes. In addition, some "home built" aircraft utilize Pop rivets. They are available in flat head, countersunk head, and modified flush heads with standard diameters of 1/8, 5/32, and 3/16 inch. Pop rivets are made from soft aluminum alloy, steel, copper, and Monel. [Figure 8-9]

FRICTION-LOCK RIVETS

One early form of blind rivet that was the first to be widely used for aircraft construction and repair was the **Cherry friction-lock** rivet. Originally, Cherry friction-locks were available in two styles, hollow shank pull-through and self-plugging types. The pull-through type is no longer common, however, the self-plugging Cherry friction-lock rivet is still used for repairing light aircraft.

Cherry friction-lock rivets are available in two head styles, universal and 100 degree countersunk. Furthermore, they are usually supplied in three standard diameters, 1/8, 5/32, and 3/16 inch. However, larger sizes can be specially ordered in sizes up to 5/16 inch. [Figure 8-10]

A friction-lock rivet cannot replace a solid shank rivet, size for size. When a friction-lock is used to

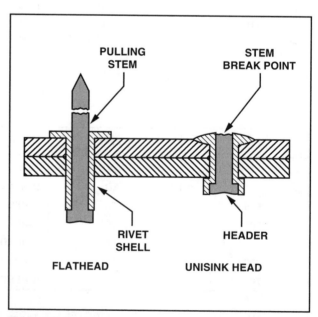

Figure 8-9. Pop rivets are frequently used for assembly and non-structural applications. They must not be used in areas that are subject to moderate or heavy loads.

replace a solid shank rivet, it must be at least one size (1/32 inch) larger in diameter. This is because a friction-lock rivet loses considerable strength if its center stem falls out due to damage or vibration.

MECHANICAL-LOCK RIVETS

Mechanical-lock rivets were designed to prevent the center stem of a rivet from falling out as a result of the vibration encountered during aircraft operation. Unlike the center stem of a friction-lock rivet, a mechanical-lock rivet permanently locks the stem into place and vibration cannot shake it loose.

HUCK-LOKS

Huck-Lok rivets were the first mechanical-lock rivets and are used as structural replacements for solid shank rivets. However, because of the expensive

tooling required for their installation, Huck-Loks are generally limited to aircraft manufacturers and some large repair facilities.

Huck-Loks are available in four standard diameters, 1/8, 5/32, 3/16, and 1/4 inch, and come in three different alloy combinations: a 5056 sleeve with a 2024 pin, an A-286 sleeve with an A-286 pin, and a Monel 400 sleeve with an A-286 pin. [Figure 8-11]

CHERRYLOCKS™

The Cherry mechanical-lock rivet, often called the bulbed CherryLOCK, was developed shortly after the Huck-Lok. Like the Huck-Lok, the CherryLOCK rivet is an improvement over the friction-lock rivet because its center stem is locked into place with a lock ring. This results in shear

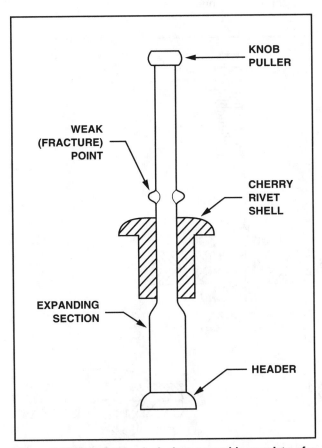

Figure 8-10. The friction-lock rivet assembly consists of a shell and mandrel or pulling stem. The stem is pulled until the header forms a bucktail on the blind side of the shell. At this point, a weak point built into the stem shears and the stem breaks off. After the stem fractures, part of it projects upward. The projecting stem is cut close to the rivet head and the small residual portion is filed smooth.

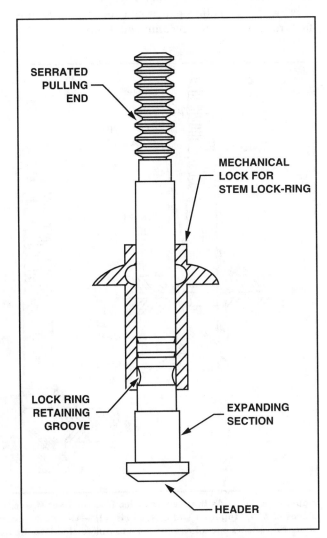

Figure 8-11. Unlike friction lock rivets, Huck-Loks utilize a lock ring that mechanically locks the center stem in place.

and bearing strengths that are high enough to allow CherryLOCKS to be used as replacements for solid shank rivets. [Figure 8-12]

CherryLOCK rivets are available with two head styles, 100 degree countersunk and universal. Like most blind rivets, CherryLOCKs are available with diameters of 1/8, 5/32, and 3/16 inch, with an oversize of 1/64 inch for each standard size. The rivet, or shell, portion of a CherryLOCK may be constructed of 2017 aluminum alloy, 5056 aluminum alloy, Monel, or stainless steel. Installation of CherryLOCK rivets requires a special pulling tool for each different size and head shape. However, the same size tool can be used for an oversize rivet in the same diameter group.

One disadvantage of a CherryLOCK is that if a rivet is too short for an application, the lock ring sets prematurely resulting in a malformed shank header. This fails to compress the joint, leaving it in a weakened condition. To avoid this, always use the proper rivet length selection gauge and follow the manufacturer's installation recommendations.

OLYMPIC-LOKS

Olympic-lok blind fasteners are light weight, mechanically-locking spindle-type blind rivets. Olympic-loks come with a lock ring stowed on the head. As an Olympic-lok is installed, the ring slips down the stem and locks the center stem to the outer shell. These blind fasteners require a specially designed set of installation tools. [Figure 8-13]

Olympic-lok rivets are made with three head styles: universal, 100 degree flush, and 100 degree flush shear. Rivet diameters of 1/8, 5/32, and 3/16 inch are available in eight different alloy combinations of 2017-T4, A-286, 5056, and Monel.

When Olympic-loks were first introduced, they were advertised as an inexpensive blind fastening system. The price of each rivet is less than the other types of mechanical locking blind rivets, and only three installation tools are required. The installation tools fit both countersunk and universal heads in the same size range.

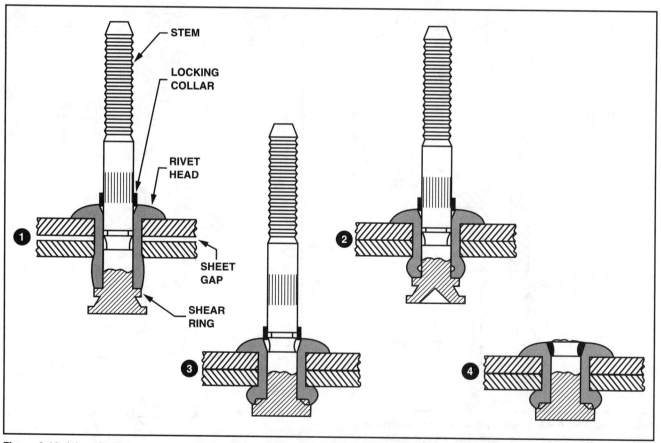

Figure 8-12. (1) — As the stem is pulled into the rivet sleeve, a bulb forms on the rivet's blind side that begins to clamp the two pieces of metal together and fill the hole. (2) — Once the pieces are clamped tightly together, the bulb continues to form until the shear ring shears and allows the stem to pull further into the rivet. (3) — With the shear ring gone, the stem is pulled upward until the pulling head automatically stops at the stem break notch and the locking collar is ready to be inserted. (4) — When completely installed, the locking collar is inserted and the stem is fractured flush with the rivet head.

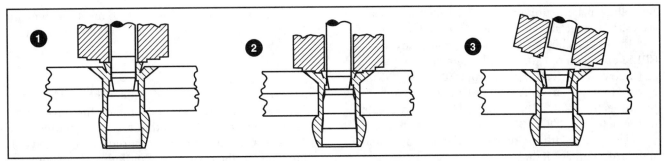

Figure 8-13. (1) — Once an Olympic-Lok rivet is inserted into a prepared hole, the stem is pulled into the sleeve closing any gap between the materials being riveted, filling the hole, and forming a bearing area. (2) — When the stem travel is stopped by the sleeve's internal step, the locking collar shears free and is forced into the locking groove. (3) — Continued pulling breaks the stem flush with the rivet head.

CHERRYMAX™

The CherryMAX rivet is economical to use and strong enough to replace solid shank rivets, size for size. The economic advantage of the CherryMAX system is that one size puller can be used for the installation of all sizes of CherryMAX rivets. A CherryMAX rivet is composed of five main parts: a pulling stem, a driving anvil, a safe-lock locking collar, a rivet sleeve, and a bulbed blind head. [Figure 8-14]

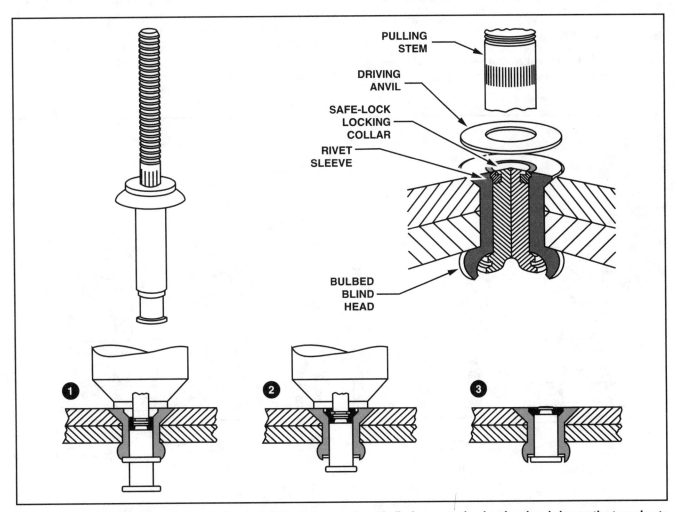

Figure 8-14. (1) — As the stem pulls into the rivet sleeve it forms a large bulb that seats the rivet head and clamps the two sheets tightly together. (2) — As the blind head is completed, the safe-locking collar moves into the rivet sleeve recess. (3) — As the stem continues to be pulled, the safe-lock collar is formed into the head recess by the driving anvil, locking the stem and sleeve securely together. Further pulling fractures the stem, providing a flush, burr-free installation.

Available in both universal and countersunk head styles, the rivet sleeve is made from 5056, monel, and inco 600. The stems are made from alloy steel, CRES, and inco X-750. The ultimate shear strength of CherryMAX rivets ranges from 50KSI to 75KSI. Furthermore, CherryMAX rivets can be used at temperatures from 250° F to 1,400° F. They are available in diameters of 1/8, 5/32, 3/16 and 1/4 inches and are also made with an oversize diameter for each standard diameter listed.

REMOVAL OF MECHANICAL-LOCK RIVETS

To remove mechanical-lock rivets, you must first file a flat spot on the rivet's center stem. Once this is done, a center punch is used to punch out the stem so the lock ring can be drilled out. With the lock ring removed you can tap out the remaining stem, drill to the depth of the manufactured head, and tap out the remaining shank. All brands of mechanical-lock blind rivets are removed using the same basic technique.

HI-SHEAR RIVETS

One of the first special fasteners used by the aerospace industry was the Hi-Shear rivet. Hi-Shear rivets were developed in the 1940s to meet the demand for fasteners which could carry greater shear loads.

The Hi-shear rivet has the same strength characteristics as a standard AN bolt. In fact, the only difference between the two is that a bolt is secured by a nut and a Hi-Shear rivet is secured by a crushed collar. The Hi-Shear rivet is installed with an interference fit, where the side wall clearance is reamed to a tolerance determined by the aircraft builder. When properly installed, a Hi-Shear rivet has to be tapped into its hole before the locking collar is swaged on.

Hi-Shear rivets are made in two head styles, flat and countersunk. As the name implies, the Hi-Shear rivet is designed especially to absorb high shear loads. The Hi-Shear rivet is made from steel alloy having the same tensile strength as an equal size AN bolt. The lower portion of its shank has a specially milled groove with a sharp edge that retains and finishes the collar as it is swaged into the locked position. [Figure 8-15]

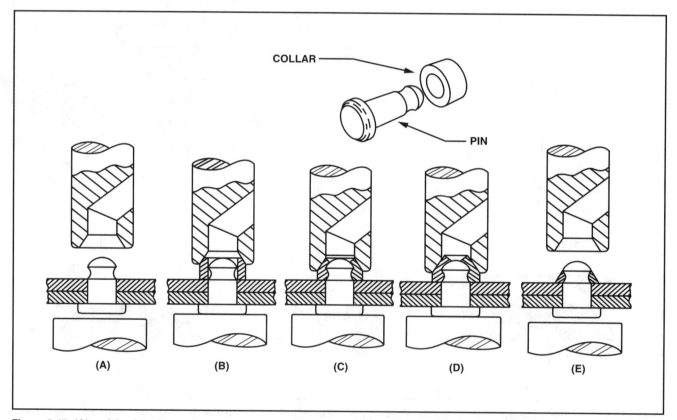

Figure 8-15. (A) — A bucking bar and rivet gun are used to install Hi-Shear rivets. (B) — A collar is placed over the pin's small end. (C) — The rivet gun forces the collar over the pin. (D) — The gunset drives the collar onto the rivet pin and cuts off excess material. (E) — When the collar is fully driven, excess collar material is ejected from the gunset.

SPECIAL FASTENERS

Many special fasteners have the advantage of producing high strength with light weight and can be used in place of conventional AN bolts and nuts. When a standard AN nut and bolt assembly is tightened, the bolt stretches and its shank diameter decreases, causing the bolt to increase its clearance in the hole. Special fasteners eliminate this change in dimension because they are held in place by a collar that is squeezed into position instead of being screwed on like a nut. As a result, these fasteners are not under the same tensile loads imposed on a bolt during installation.

LOCKBOLTS

Lockbolts are manufactured by several companies and conform to Military Standards. These standards describe the size of a lockbolt's head in relation to its shank diameter, as well as the alloy used. Lockbolts are used to assemble two materials permanently. They are lightweight and are as strong as standard bolts.

There are three types of lockbolts used in aviation, they are the pull-type lockbolt, the blind-type lockbolt, and the stump-type lockbolt. The **pull-type lockbolt** has a pulling stem on which a pneumatic installation gun fits. The gun pulls the materials together and then drives a locking collar into the grooves of the lockbolt. Once secure, the gun fractures the pulling pin at its break point. The **blind-type lockbolt** is similar to most other types of blind fasteners. To install a blind lockbolt, it is placed into a blind hole and an installation gun is placed over the pulling stem. As the gun pulls the stem, a blind head forms and pulls the materials together. Once the materials are pulled tightly together, a locking collar locks the bolt in place and the pulling stem is broken off. Unlike other blind fasteners that typically break off flush with the surface, blind lockbolts protrude above the surface. The third type of lockbolt is the **stump-type lockbolt** and is installed in places where there is not enough room to use the standard pulling tool. Instead, the stump-type lockbolt is installed using an installation tool similar to that used to install Hi-Shear rivets. [Figure 8-16]

Lockbolts are available for both shear and tension applications. With shear lockbolts, the head is kept thin and there are only two grooves provided

for the locking collar. However, with tension lockbolts, the head is thicker and four or five grooves are provided to allow for higher tension values.

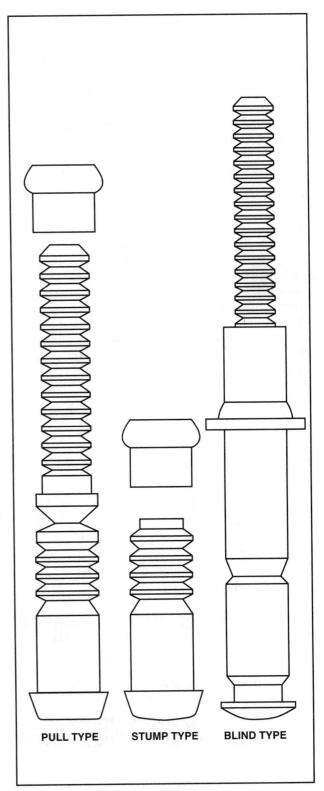

PULL TYPE   **STUMP TYPE**   **BLIND TYPE**

Figure 8-16. Lockbolts are classified as pull type, stump type, and blind type. The bolt used for a particular application depends primarily on access to the work area.

The locking collars used on both shear and tension lockbolts are color coded for easy identification. [Figure 8-17]

HI-LOKS

Hi-Lok bolts are manufactured in several different alloys such as titanium, stainless steel, steel, and aluminum. They possess sufficient strength to withstand bearing and shearing loads, and are available with flat and countersunk heads.

A conventional Hi-Lok has a straight shank with standard threads. Although wrenching lock nuts are usually used, the threads are compatible with standard AN bolts and nuts. To install a Hi-Lok, the hole is first drilled with an interference fit. The Hi-Lok is then tapped into the hole and a shear collar is installed. A Hi-Lok retaining collar is installed using either specially prepared tools or a simple Allen and box end wrench. Once the collar is tightened to the appropriate torque value, the wrenching device shears off leaving only the locking collar. [Figure 8-18]

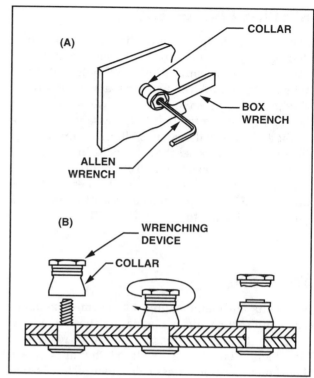

Figure 8-18. (A) — An allen wrench holds a Hi-Lok bolt in place while a wrench is used to tighten the shear nut. (B) — Once a Hi-Lok is installed, the collar is installed and tightened. When the appropriate torque value is obtained, the wrenching device shears off leaving the collar.

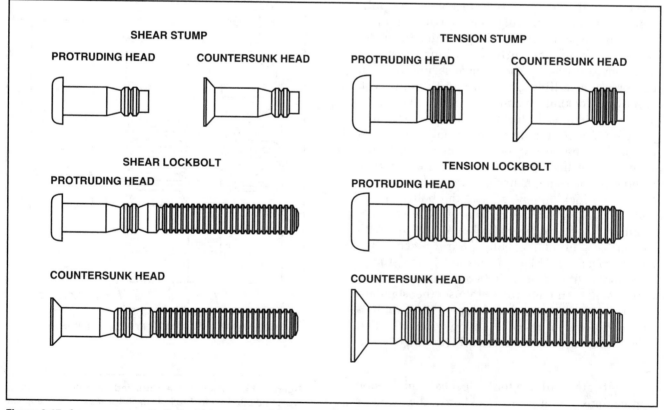

Figure 8-17. As you can see, both the shear and tension type lockbolts come in a variety of sizes with multiple head styles.

HI-LITES

The Hi-Lite fastener is similar to the Hi-Lok except that it is made from lighter materials and has a shorter transition from the threaded section to the shank. Furthermore, the elimination of material between the threads and shank give an additional weight saving with no loss of strength. The Hi-Lite's main advantage is its excellent strength to weight ratio.

Hi-Lites are available in an assortment of diameters ranging from 3/16 to 3/8 inch. They are installed either with a Hi-Lok locking collar or by a swaged collar like the Lockbolt. In either case, the shank diameter is not reduced by stretch torquing.

CHERRYBUCKS

The CherryBUCK is a one-piece special fastener that combines two titanium alloys which are bonded together to form a strong structural fastener. The head and upper part of the shank of a CherryBUCK is composed of 6AL-4V alloy while Ti-Cb alloy is used in the lower shank. When driven, the lower part of the shank forms a bucktail.

An important advantage of the CherryBUCK is the fact that it is a one piece fastener. Since there is only one piece, CherryBUCKs can safely be installed in jet engine intakes with no danger of foreign object damage. This type of damage often occurs when multiple piece fasteners lose their retaining collars and are ingested into a compressor inlet. [Figure 8-19]

TAPER-LOK

Taper-Loks are the strongest special fasteners used in aircraft construction. Because of its tapered shape, the Taper-Lok exerts a force on the conical

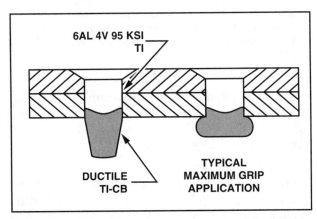

Figure 8-19. CherryBUCK fasteners combine two titanium alloys to produce a one-piece fastener with 95 KSI shear strength.

walls of a hole, much like a cork in a wine bottle. To a certain extent, a Taper-Lok mimics the action of a driven solid shank rivet, in that it completely fills the hole. However, a Taper-Lok does this without the shank swelling.

When a washer nut draws the Taper-Lok into its hole, the fastener pushes outward and creates a tremendous force against the tapered walls of the hole. This creates radial compression around the shank and vertical compression lines as the metals are squeezed together. The combination of these forces generate strength unequaled by any other type of fastener. [Figure 8-20]

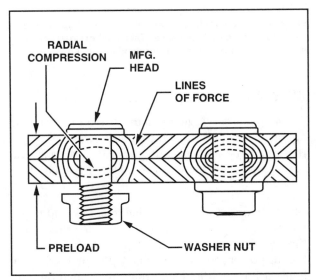

Figure 8-20. The hole for a Taper-Lok is made with a special tapered drill. Once a Taper-Lok is installed and a washer nut is tightened, radial compression forces and vertical compression forces combine to create an extremely strong joint.

HI-TIGUE

The Hi-Tigue fastener has a bead that encircles the bottom of its shank and is a further advancement in special fastener design. This bead preloads the hole it fills, resulting in increased joint strength. During installation, the bead presses against the side wall of the hole, exerting a radial force which strengthens the surrounding area. Since it is preloaded, the joint is not subjected to the constant cyclic action that normally causes a joint to become coldworked and eventually fail.

Hi-Tigue fasteners are produced in aluminum, titanium, and stainless steel alloys. The collars are also composed of compatible metal alloys and are available in two types, sealing and non-sealing. As with Hi-Loks, Hi-Tigues can be installed using an Allen and box end wrench. [Figure 8-21]

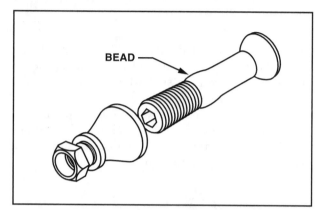

Figure 8-21. This HI-Tigue features a subtly shaped bead at the threaded end of the shank. This bead preloads the hole it is inserted into thereby strengthening the joint.

JO-BOLTS

These patented high-strength structural fasteners are used in close-tolerance holes where strength requirements are high but physical clearance precludes the use of standard AN, MS, or NAS bolts.

The hole for a Jo-Bolt is drilled, reamed, and countersunk before the Jo-Bolt is inserted and held tightly in place by a nose adapter of either a hand tool or power tool. A wrench adapter then grips the bolt's driving flat and screws it up through the nut. As the bolt pulls up, it forces a sleeve up over the tapered outside of the nut and forms a blind head on the inside of the work. When driving is complete, the driving flat of the bolt breaks off. [Figure 8-22]

REMOVAL OF SPECIAL FASTENERS

Special fasteners that are locked into place with a crushable collar are easily removed by splitting the collar with a small cape chisel. After the collar is split, knock away the two halves and tap the fastener from the hole. Fasteners which are not damaged during removal can be reused using new locking collars. The removal techniques of certain special fasteners are basically the same as those used for solid shank rivets. However, in some cases, the manufacturer may recommend that a special tool be used.

Removal of Taper-Loks, Hi-Loks, Hi-Tigues, and Hi-Lites requires the removal of the washernut or lock-

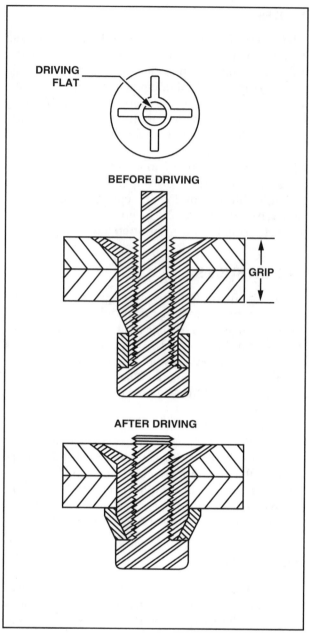

Figure 8-22. Once a Jo-Bolt is inserted into a hole, the bolt is rotated causing the nut to pull up to the metal. As the nut moves upward, a sleeve is forced over the tapered end of the bolt. This creates a blind head that holds the joint together.

ing collar. Both are removed by turning them with the proper size box end wrench or a pair of vise-grips. After removal, a mallet is used to tap the remaining fastener out of its hole.

To remove a Jo-Bolt, begin by drilling through the nut head with a pilot bit followed by a bit of the same size as the bolt shank. Once the nut head is removed, a punch is used to punch out the remaining portion of the nut and bolt. [Figure 8-23]

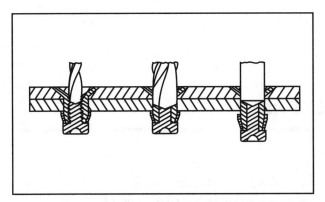

Figure 8-23. Remove flush-type Jo-Bolts by first drilling a pilot hole into the bolt slightly deeper than the inside of the head of the nut. Then, using a drill bit of the same size, drill to the depth of the pilot hole. Drive the shank and blind head from the hole using a pin punch of the proper size.

THREADED RIVETS—RIVNUTS

Goodrich Rivnuts were developed by the B.F. Goodrich Company to attach rubber deicer boots to aircraft wing and tail surfaces. To install a rivnut, a hole is drilled in the skin to accommodate the Rivnut, and a special cutter is used to cut a small notch in the circumference of the hole. This notch locks the Rivnut into the skin to prevent it from turning when it is used as a nut. A Rivnut of the proper grip length is then screwed onto the puller and inserted into the hole with its key aligned with the keyway cut in the hole. When the handle of the puller is squeezed, the hollow shank of the Rivnut upsets and grips the skin. The tool is then unscrewed from the Rivnut, leaving a threaded hole that accepts machine screws for attaching a deicer boot. [Figure 8-24]

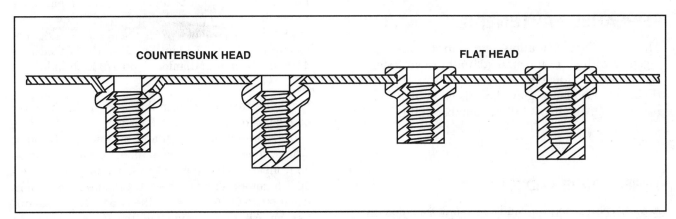

Figure 8-24. Rivnuts are commonly available with flat heads and with 100 degree countersunk heads. Countersunk head Rivnuts are made with both 0.048-inch and 0.063-inch head thicknesses, with the thinner head used when it is necessary to install a Rivnut in a machine countersunk hole in thin material. Closed-end Rivnuts are available for installation in a pressurized structure or sealed compartment.

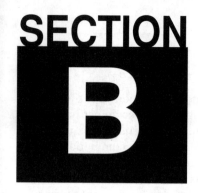
AIRCRAFT FASTENERS

In addition to the numerous permanent fasteners used on aircraft, there is a second type of fastener that, unlike rivets, can be reused. These fasteners include threaded hardware such as bolts and screws and the various types of nuts that secure them.

THREADED FASTENERS

Threaded fasteners allow parts to be fastened together with all of the strength unthreaded fasteners provide. However, unlike rivets and pins, threaded fasteners may be disassembled and reassembled an almost infinite number of times.

THREAD TYPE AND FITS

Aircraft bolts, screws, and nuts are threaded in either the **American National Coarse** (NC), the **American National Fine** (NF), the **American Standard Unified Coarse** (UNC), or the **American Standard Unified Fine** (UNF) series. The difference between the American National series and the American Standard Unified series is the American National series has more threads per inch than the American Standard Unified series. For example, on a one inch diameter bolt, the NF thread specifies 14 threads per inch (1-14NF), while the UNF thread specifies 12 threads per inch (1-12 UNF). Both thread types are designated by the number of times the threads rotate (number of turns) around a 1-inch length of a given diameter bolt or screw.

In addition to being identified as either coarse or fine, threads are also designated by class of fit from one to five. A Class 1 thread is a loose fit, a Class 2 is a free fit, a Class 3 is a medium fit, a Class 4 is a close fit, and a Class 5 fit is a tight fit. A Class 1 fit allows you to turn the nut all the way down using only your fingers. Wing nuts are a good example of a Class 1 fit. A Class 4 and 5 fit requires a wrench to turn a nut down from start to finish. Aircraft bolts

are usually fine threaded with a Class 3 fit, whereas screws are typically a Class 2 or 3 fit.

DESIGNATION CODES

Like rivets, threaded fasteners are given a part code indicating a fastener's diameter in 1/16 inch increments and its length in 1/8 inch increments. For example, an AN4-7 identifies a bolt that measures 4/16 or 1/4 inch in diameter and 7/8 inch in length.

For bolts that are longer than 7/8 inch, the code changes. For example, a 1 inch bolt is identified by a -10 representing 1 inch and no fraction. In other words, there are no -8 or -9 lengths. Dash numbers go from -7 to -10, from -17 to -20, and from -27 to -30. Therefore, a bolt that is 1 1/2 inches long is identified by a -14. A bolt with the code AN5-22 identifies an Air Force-Navy bolt that is 5/16 inch in diameter and 2 1/4 inches long.

Threaded aircraft fasteners 1/4 inch in diameter and smaller are dimensioned in screw sizes rather than 1/8 inch increments. The AN3 bolt is the exception to this rule. These machine screw sizes range from 0 to 12. A number 10 fastener has a diameter of approximately 3/16 inch and a number 5 fastener has a 1/8 inch diameter. [Figure 8-25]

STANDARD AIRCRAFT BOLTS

As you know, a bolt is designed to hold two or more items together. Bolts that are typically used for airframe structural applications have hex heads and range in size from AN3 to AN20.

As discussed earlier, bolts are identified by their diameter and length. A diameter represents the shank diameter while the length represents the dis-

NATIONAL FINE THREAD SERIES MEDIUM FIT. CLASS 3 (NF)			
SIZE AND THREADS	DIAMETER OF BODY FOR THREAD	BODY DRILL	TAP DRILL SIZE
0-80	.060	52	3/64
1-72	.073	47	#53
2-64	.086	42	#50
3-56	.099	37	#46
4-48	.112	31	#42
5-44	.125	25	#38
6-40	.138	27	#33
8-36	.164	18	#29
10-32	.190	10	#21
12-28	.216	2	#15
1/4-28	.250	F	#3
1/16-24	.3125	5/16	I
3/8-24	.375	3/8	Q
7/16-20	.4375	7/16	W
1/2-20	.500	1/2	7/16
9/16-18	.5625	9/16	1/2
5/8-18	.625	5/8	9/16
3/4-16	.750	3/4	11/16
7/8-14	.875	7/8	51/64
1-14	1.000	1.0	49/64

NATIONAL COARSE THREAD SERIES MEDIUM FIT. CLASS 3 (NC)			
SIZE AND THREADS	DIAMETER OF BODY FOR THREAD	BODY DRILL	TAP DRILL SIZE
1-64	.073	47	#53
2-56	.086	42	#51
3-48	.099	37	5/64
4-40	.112	31	#44
5-40	.125	29	#39
6-32	.138	27	#36
8-32	.164	18	#29
10-24	.190	10	#26
12-24	.216	2	#17
1/4-20	.250	1/4	#7
5/16-18	.3125	5/16	#F
3/8-16	.375	3/8	5/16
7/16-14	.4375	7/16	U
1/2-13	.500	1/2	27/64
9/16-12	.5625	9/16	31/64
5/8-11	.625	5/8	17/32
3/4-10	.750	3/4	41/64
7/8-9	.875	7/8	49/64
1-8	1.000	1.0	7/8

Figure 8-25. This table indicates fastener diameters, the number of threads per inch, and the bit size required to drill a hole the fastener will fit into or be tapped. For example, a number 5 fine thread fastener has 44 threads per inch, a body diameter of .125 inch, fits in a hole drilled by a number 25 bit, and is tapped into a hole made by a number 38 bit.

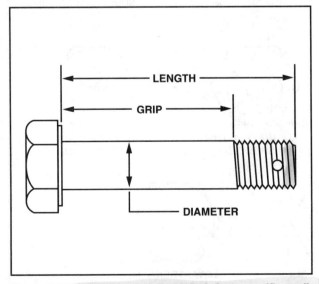

Figure 8-26. When choosing an AN bolt for a specific application, you must know the diameter, length, and grip length required.

tance from the bottom of the head to the end of the bolt. A bolt's **grip length** is the length of the unthreaded portion. [Figure 8-26]

Aircraft bolts are available in cadmium-plated nickel steel, corrosion resistant steel, and in 2024 aluminum alloy. Unless specified, a bolt is made of cadmium-plated nickel steel. A corrosion resistant bolt, on the other hand, is identified by the letter "C" inserted between the diameter and length designations. Aluminum alloy bolts are identified by the letters "DD." For example, a bolt that is 1/4 inch in diameter, 3/4 inch long, and made of cadmium-plated nickel steel is identified by the code AN4-6. However, if the same bolt is made of corrosion resistant steel it carries the code AN4C6, whereas an aluminum alloy bolt would be AN4DD6.

In addition to the designation code, most aircraft bolts have a marking on their head identifying what the bolt is made of and, in many cases, the manufacturer. For example, a standard AN bolt has an asterisk in the center of its manufactured head. [Figure 8-27]

The FAA forbids the use of aluminum alloy bolts and alloy steel bolts smaller than AN3 on structural components. Furthermore, since repeated tightening and loosening of aluminum alloy bolts eventually ruins their threads, they are not used in areas where they must be removed and installed frequently. Aluminum alloy nuts can be used with cadmium-plated steel bolts loaded in shear, but only on land aircraft. However, since exposure to moist air increases the possibility of dissimilar-metal corrosion, they cannot be used on seaplanes.

When hardware was first standardized, almost all nuts were locked onto a bolt with a cotter pin and, therefore, all bolts had holes drilled near the end of their shank to accommodate a cotter pin. However, when self-locking nuts became popular, many standard AN bolts were made without a drilled shank. To help you identify whether or not a bolt has a hole drilled through it, the letter **A** is used in the part code. For example, if an "A" appears immediately after the dash number the bolt does not have a hole. However, the absence of an "A" indicates a hole exists in the shank. As an example, an AN6C-12A bolt is 3/8 inch in diameter, made of corrosion-resistant steel, 1 1/4 inches long, and has an undrilled shank.

Some AN bolts, such as those used to fasten a propeller into a flanged shaft, must be safetied by passing safety wire through holes drilled through the bolt's head. A bolt drilled for this type of safetying has the letter **H** following the number indicating its

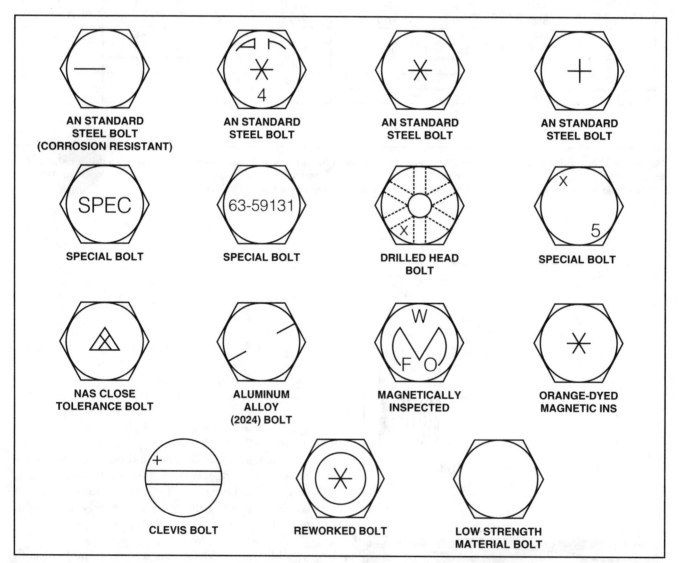

Figure 8-27. To aid in identifying what type of metal a bolt is made of, bolt heads are marked with a symbol.

diameter. For example, the part number AN6H34A identifies a bolt that is 3/8 inch in diameter, made of nickel-steel, has a drilled head, is 3 1/2 inches long, and has an undrilled shank.

DRILLED-HEAD ENGINE BOLTS

AN73 through AN81 bolts are hex-headed nickel-steel bolts that are similar in appearance to the AN3 through AN20 series. However, unlike standard bolts, drilled-head engine bolts have a thicker head that is drilled with a small hole in each of the flats and in the center of the head. As with most bolts, the diameters of drilled-head engine bolts are in $1/16$ increments while bolt lengths are in $1/8$ inch increments. The diameter is indicated by the second number following the "AN" designation while the bolt length is indicated by a dash number. For example, a drilled-head engine bolt designated as AN74-6 has a diameter of $1/4$ inch and a length of $3/4$ inch.

An advantage of drilled-head engine bolts is that they are made with either fine or course threads. A fine threaded bolt is identified by the absence of an "A" preceding the dash number, while a course threaded bolt is identified by the presence of the letter "A" before the dash number. As an example, a drilled-head engine bolt with an AN75A7 designation has a diameter of $5/16$ inch, a length of $7/8$ inch, and course threads. On the other hand, the designation AN75-7 identifies a bolt with the same dimensions and fine threads.

Under MS standards, AN73 through AN81 drilled-head engine bolts have been superseded by MS20073 and MS20074. In this case, MS20073 bolts have fine threads while MS20074 bolts have course threads. The diameter of MS20073 and MS20074 bolts is identified by a dash number. For example, a MS20073-05 identifies a fine threaded bolt with a diameter of $5/16$ inch. Diameters are in $1/16$ inch increments and range from -03 ($3/16$ inch) up to -12, ($3/4$ inch). Bolt length, on the other hand, is indicated by a second dash number whose value is indicated in a chart. This second dash number can range from -02 (.344 inch) to -60 (6.062 inches). [Figure 8-28]

CLOSE TOLERANCE BOLTS

Close tolerance bolts are designated AN173 to AN186 and are ground to a tolerance of +0.000 −0.0005 inch. This is much tighter then standard AN3 through AN14 bolts which are manufactured with a tolerance of +0.000 −0.0025, or AN16 through AN20 bolts which are manufactured with a tolerance of +0.000 −0.0055 inch. Close tolerance bolts must be used in areas that are subject to pounding loads or in a structure that is required to be both riveted and bolted. [Figure 8-29]

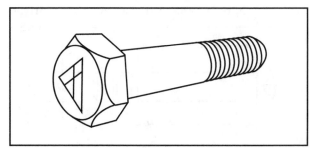

Figure 8-29. Close tolerance bolts carry a triangle mark on their heads and are ground to a much tighter tolerance than standard bolts.

		LENGTH												
		02	03	04	05	06	07	10	11	12	13	14	15	16
DIAMETER	03	.344	.469	.594	.656	.781	.906	1.031	1.156	1.281	1.406	1.531	1.656	1.781
	04		.469	.594	.656	.781	.906	1.031	1.156	1.281	1.406	1.531	1.656	1.781
	05			.609	.672	.797	.922	1.047	1.172	1.297	1.422	1.547	1.672	1.797
	06				.734	.797	.922	1.047	1.172	1.297	1.422	1.547	1.672	1.797
	07					.797	.922	1.047	1.172	1.297	1.422	1.547	1.672	1.797
	08							1.047	1.172	1.297	1.422	1.547	1.672	1.797
	09									1.312	1.438	1.562	1.688	1.812
	10											1.562	1.688	1.812
	12													1.812

Figure 8-28. To determine the length of an MS20073 or MS20074 drilled-head engine bolt, locate the dash number representing the diameter on the left side of the chart and the dash number representing the length across the top. The intersection of the two dash numbers represents the bolt's length. For example, an MS20073-05-12 bolt has fine threads, a diameter of $5/16$ inch, and a length of 1.297 inches.

Since the manufacturing specifications for a close tolerance bolt are so tight, the application of cadmium plating is typically not possible. Therefore, to provide some degree of corrosion protection, a thin layer of grease is usually applied to the shank of a close tolerance bolt prior to installation.

The diameter and length of close-tolerance bolts are measured in the same increments as standard airframe bolts. For example, an AN175-26A designates a 5/16 inch diameter close-tolerance bolt that is 2 3/4 inches long and has an undrilled shank.

CLEVIS BOLTS

All of the bolts discussed so far may be used for either shear or tensile loads. However, some applications require a bolt to be loaded in shear only. For example, a control cable must be attached to a control horn with a bolt that is loose enough to allow the cable to pivot freely as the control surface moves, but not so loose that excess play exists. For these applications a clevis bolt is used. The AN21 through AN36 clevis bolt has a domed head that is typically slotted or recessed to accept a screwdriver.

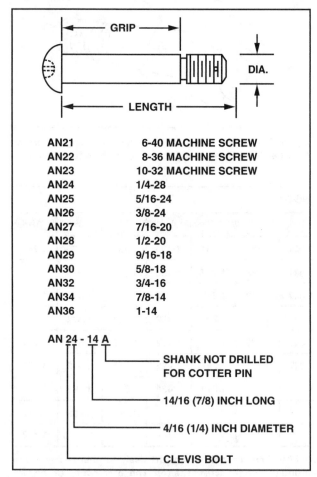

AN21	6-40 MACHINE SCREW
AN22	8-36 MACHINE SCREW
AN23	10-32 MACHINE SCREW
AN24	1/4-28
AN25	5/16-24
AN26	3/8-24
AN27	7/16-20
AN28	1/2-20
AN29	9/16-18
AN30	5/8-18
AN32	3/4-16
AN34	7/8-14
AN36	1-14

AN 24 - 14 A

- SHANK NOT DRILLED FOR COTTER PIN
- 14/16 (7/8) INCH LONG
- 4/16 (1/4) INCH DIAMETER
- CLEVIS BOLT

Figure 8-30. Clevis bolts are loaded in shear only and are frequently used in control cable installations.

A unique feature of a clevis bolt is that only a short portion of the shank is threaded, and there is a small notch between the threads and the shank. This results in a long grip length which increases the bolt's shear strength and allows the bolt to rotate more freely in its hole.

The diameter of a clevis bolt is given in 1/16 inch increments. The length of a clevis bolt is more critical than that of the other types of bolts and, therefore, it is also measured in 1/16 inch increments with a dash number indicating the length. For example, an AN29-20 identifies a 9/16 inch diameter clevis bolt that is 20/16 (1 1/4) inches long. [Figure 8-30]

When installed on forked-end cable terminals, clevis bolts are secured with a shear nut tightened to a snug fit, but with no strain imposed on the fork. The nut on most clevis bolts is safetied on with a cotter pin. Therefore, most clevis bolts have a hole drilled through their end. However, if a self-locking nut is used with a clevis bolt no hole is necessary. If a clevis bolt is not drilled for a cotter pin, the letter A follows the dash number in its part code. For example, an AN22-8A is a clevis bolt that is 2/16 (1/8) inch in diameter, 8/16 (1/2) inch long, and has no hole drilled for a cotter pin.

INTERNAL WRENCHING BOLTS

MS20004 through MS20024 internal wrenching bolts are high-strength steel bolts used primarily in areas that are subjected to high tensile loads. A six-sided hole is machined into the center of their heads to accept an Allen wrench of the proper size. These bolts have a radius between the head and shank and, when installed in steel parts, the hole must be counterbored to accommodate this radius. When an internal wrenching bolt is installed in an aluminum alloy structure, a MS20002C washer must be used under the head to provide the needed bearing area.

The strength of internal wrenching bolts is much higher than that of a standard steel AN bolt and, for this reason, an AN bolt must never be substituted for an internal wrenching type. [Figure 8-31]

BOLT SELECTION AND INSTALLATION

When joining two pieces of material, their combined thickness determines the correct length of bolt to use. If you recall, a bolt's grip length is the length of the shank's unthreaded portion and should be identical to the total thickness of the materials being fastened together. If the grip length is slightly longer than this thickness, washers must be added to ensure that the nut can provide the proper amount

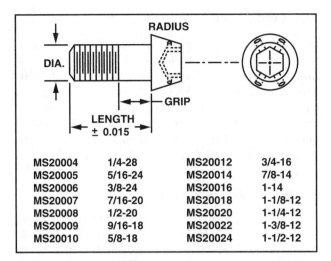

MS20004	1/4-28	MS20012	3/4-16
MS20005	5/16-24	MS20014	7/8-14
MS20006	3/8-24	MS20016	1-14
MS20007	7/16-20	MS20018	1-1/8-12
MS20008	1/2-20	MS20020	1-1/4-12
MS20009	9/16-18	MS20022	1-3/8-12
MS20010	5/8-18	MS20024	1-1/2-12

Figure 8-31. High-strength internal wrenching bolts can bear high tension loads and are frequently used to mount engines.

of pressure when it is tightened. On the other hand, if the grip length is substantially less than the thickness of the materials the bolt's threads will extend into the material, resulting in a weaker joint.

When installing a bolt, washer, and nut combination check the installation drawing to be sure you have the correct hardware. If nothing is specified about the washers being used, it is good practice to use a washer under both the bolt head and nut to protect the material through which the bolt passes. When joining an aluminum alloy or magnesium, the washers should be made of aluminum to minimize the possibility of dissimilar metal corrosion.

Unless otherwise specified in an assembly drawing, bolts should be installed with their head on top or forward. Placing the head in either of these positions makes it less likely that a bolt will fall out of a hole if the nut is lost. An acronym to help remember the proper direction for bolt installation is "IDA," which stands for inboard, down, or aft. Even when bolt direction is not specified, some clearances may be critical, such as where control cables attach to bellcranks. A good general rule to follow is to never assume a control cable is clear of a bolt until the bellcrank is inspected for full travel.

NUTS

All nuts used in aircraft construction must have some sort of locking device to prevent them from loosening and falling off. Many nuts are held on a bolt by passing a cotter pin through a hole in the bolt shank and through slots, or castellations, in the nut. Others have some form of locking insert that grips a bolt's threads or relies on the tension of a spring-type lock-washer to hold the nut tight enough against the threads to keep it from vibrating loose.

There are two basic types of nuts, self-locking and non self-locking. As the name implies, a self-locking nut locks onto a bolt on its own while a nonself-locking nut relies on either a cotter pin, check nut, or lock washer to hold it in place. We'll begin by looking at the most common non self-locking nuts. [Figure 8-32]

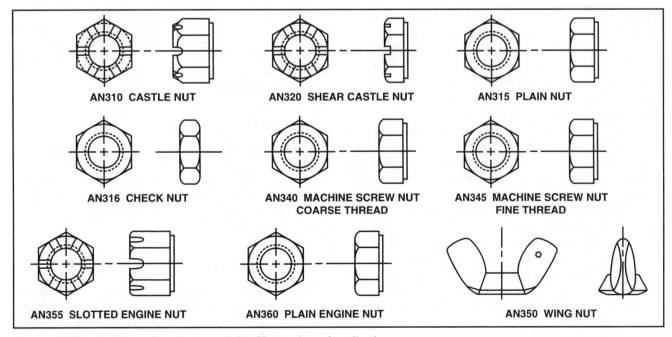

Figure 8-32. Standard aircraft nuts are available for a variety of applications.

AN310 CASTLE NUT

These fine-thread nuts are designed to fit on a standard airframe bolt with a Class 3 fit, and are used when the bolt is subjected to either shear or tensile loads. The size of a nut is indicated in the part code by a dash number which denotes the size of the bolt it fits. For example, an AN310-6 nut fits an AN6 bolt which has a diameter of 3/8 inch. Castle nuts are available in cadmium-plated nickel steel, corrosion-resistant steel, and 2024 aluminum alloy. Unless specified, a castle nut is made of cadmium-plated nickel steel. A corrosion resistant nut, on the other hand, is identified by the letter "C" inserted before the dash number in the part code. Aluminum alloy nuts are identified by the letter "D." For example, the part code AN310D-6 identifies an aluminum alloy nut that has an inside diameter of 6/16 (3/8) inch.

AN320 SHEAR CASTLE NUT

The AN320 shear castle nut is made of the same material and has the same type of thread as a AN310 nut. However, shear castle nuts are much thinner than standard castle nuts and, therefore, are used only for shear loads on clevis bolts. An AN320-6 nut is a shear castle nut that is used on an AN26 clevis bolt. An aluminum alloy (2024) nut is identified as an AN320D6.

AN315 PLAIN NUT

The AN315 plain nut has no castellations and, therefore, cannot be held in place using a cotter pin. Since these fine-thread nuts have no locking provisions, a spring-type lock washer must be used in combination with the nut. The lock washer applies a spring force to prevent the nut from shaking loose. AN315 nuts are used with either tensile or shear loads and are made of either nickel steel, corrosion-resistant steel, and aluminum alloy. The type of material used is indicated in the designation code in the same way it is for bolts. In other words, the absence of an additional letter identifies nickel steel, whereas the letter "C" preceding the dash number identifies corrosion resistant steel, and a "D" identifies 2024 aluminum alloy. Furthermore, plain nuts are made with both right and left-hand threads. For example, an AN315-7R is a nickel steel nut with right-hand threads that fits an AN7 bolt. An AN315C-4L, on the other hand, is a 1/4 inch diameter corrosion-resistant steel plain nut with left-hand threads.

AN316 CHECK NUT

In some instances a plain nut is locked in place using a check nut. A check nut is simply a second nut that is tightened against the primary nut so it cannot turn off. An AN316 check nut is made of cadmium-plated steel and is available in both right- and left-hand threads. An AN316-4R is a right-hand check nut that fits a quarter-inch thread, while an AN316-4L has a left-hand thread.

AN340 MACHINE SCREW NUT

AN340 machine screw nuts are made in machine screw sizes from number 2 up through 1/4 inch and have coarse threads. They are available in carbon steel, corrosion-resistant steel (C), brass (B), and 2024 aluminum alloy (DD). A nut identified as an AN340B-6 is a brass nut that fits a 6-32 machine screw. An AN340DD-416 is an aluminum alloy nut that fits a 1/4-20 thread.

AN345 MACHINE SCREW NUT

These nuts are similar to AN340 nuts except they have national-fine series threads. They are available in cadmium-plated carbon steel, corrosion-resistant steel (C), commercial brass (B), and 2024 aluminum alloy (DD). An AN345B-6 is a brass nut that fits a 6-40 machine screw. An AN345DD-416 is an aluminum alloy nut that fits a 1/4-28 thread per inch machine screw.

AN355 SLOTTED ENGINE NUT

This nut is designed for use on an aircraft engine and is not approved for airframe use. It is made of heat-treated steel and has national fine threads that produce a Class 3 fit. It is available in sizes from AN355-3 (3/16 inch) to AN355-12 (3/4 inch) and has slots cut in it for a cotter pin.

AN360 PLAIN ENGINE NUT

This engine nut is similar to the AN355 in that it is approved for use on engines only. However, an AN360 differs from an AN355 in that it does not have cotter pin slots and has a black rustproof finish. An AN360-7 is a plain engine nut that fits a 7/16 inch bolt.

AN350 WING NUT

Wing nuts are used when it is necessary to remove a part frequently without the use of tools. Aircraft wing nuts are made of either cadmium-plated steel or brass and are available in sizes to fit number six machine screws up to 1/2 inch bolts. All of these nuts have national fine threads that produce a Class 2 fit. Nuts for machine screw sizes are designated by the series number. However, nuts used on bolts have a bolt size given in 1/16 inch increments followed by the number 16. For example, with an AN350-616 wing nut, the -6 indicates that the nut will fit a 3/8 (6/16) inch bolt.

SELF-LOCKING NUTS

Self-locking nuts, or lock nuts, employ a locking device in their design to keep them from coming loose. However, because there are several different types of lock nuts, you must be certain that the proper locknut is used in a given application. Failure to do so could result in failure of the locking provision. The two general types of self-locking nuts used in aviation are the fiber, or nylon type, and the all metal type. [Figure 8-33]

LOW-TEMPERATURE SELF-LOCKING NUTS

AN365 self-locking nuts are used on bolts and machine screws and are held in position by a nylon insert above the threads. This insert has a hole slightly smaller than the thread diameter on which it fits. The nut's Class 3 fit allows it to run down on a bolt's threads easily until the bolt enters the insert. Once this happens, the nylon insert exerts a strong

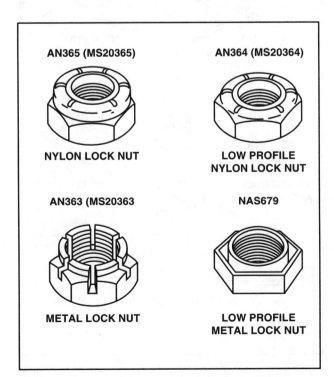

Figure 8-33. As you can see, there are several different types of self-locking nuts available for multiple applications.

downward push on the nut that removes all axial play between the bolt threads and nut. This creates friction between the threads and nut to keep the nut from vibrating loose.

A bolt does not actually cut threads into the insert, but rather forces its way into the resilient material. Since no permanent threads are made in the insert, nuts may be reused as long as there is enough friction between the nut and bolt so that the nut cannot be turned down by hand. A tap must never be run through a self-locking nut to make it easier to screw onto a bolt, since this would destroy its locking ability.

A self-locking nut must be screwed onto a bolt until all of the chamfer on the bolt's end protrudes through the insert. If the bolt is not chamfered, at least one thread but not more than three threads should protrude through the nut. If more than three threads are exposed, you risk the danger of "bottoming out" the nut and undertorqueing the assembly, thus creating a stress point that could fail. If more than three threads are exposed, either replace the bolt with one of the correct length or install a washer.

Some bolts have their shank drilled for a cotter pin. These bolts can be used with self-locking nuts provided the bolt is 5/16 inch or larger in diameter and the edges of the cotter pin hole are chamfered so there are no burrs around the hole.

You may not use self-locking nuts in areas which subject either the bolt or nut to rotation. However, a self-locking nut may be used with antifriction bearings and control pulleys, provided the inner race of the bearing is clamped to the supporting structure by a nut and bolt.

Nylon self-locking nuts should not be used in any location where the temperature could exceed 250°F. However, you may use them on engines in those locations specified by the engine manufacturer.

METAL SELF-LOCKING NUTS

In applications where temperatures exceed 250°F, all-metal lock nuts, such as the AN363, are used. Some of these nuts have a portion of their end slotted and the slots swaged together. This gives the end of the nut a slightly smaller diameter than its body allowing the threads to grip the bolt. Others have the end of the nut squeezed into a slightly oval shape, and as the bolt screws up through the threads it must make the hole round, creating a gripping action. Since both types of self-locking nuts are available in either NF or NC threads, a self-locking nut's dash number specifies both diameter and number of threads per inch. [Figure 8-34]

AN364 SHEAR SELF-LOCKING NUTS

These nuts resemble AN365 self-locking nuts except they have a much lower profile and are approved for shear loads only. Since they are an all-metal self-locking nut they can be used in areas that are subject to high temperatures. However, shear self-locking nuts are typically made to be used on clevis bolts that do not have drilled shanks.

ANCHOR NUTS

Anchor nuts are permanently mounted nut plates that enable inspection plates and access doors to be easily removed and installed. To make the installation of an access door easier where there are a great number of screws, a floating anchor nut is often used. With a floating anchor nut the nut fits loosely into a small bracket which is riveted to the skin. Since the nut is free to move within the bracket it aligns itself with a screw. To speed the production of aircraft, ganged anchor nuts are installed around inspection plate openings. These are floating-type anchor nuts that are installed in a channel that is riveted to the structure. Each nut floats in the channel with enough play so that a screw can move the nut enough to align it. [Figure 8-35]

TINNERMAN NUTS

Tinnerman nuts are cost-economical nuts that are stamped out of sheet metal. Because of their semi-rigid construction, tinnerman nuts can be adapted for use in many situations. For example, tinnerman nuts are commonly used on light aircraft to mount instruments to the instrument panel as well as attach inspection panels and cowlings.

DASH NUMBERS	SIZE AND THREADS	
–440	4-40	MACHINE SCREW (NC)
–448	4-48	MACHINE SCREW (NF)
–632	6-32	MACHINE SCREW (NC)
–640	6-40	MACHINE SCREW (NF)
–832	8-32	MACHINE SCREW (NC)
–840	8-40	MACHINE SCREW (NF)
–1024	10-24	MACHINE SCREW (NC)
–1032	10-32	MACHINE SCREW (NF)
–420	1/4-20	NC
–428	1/4-28	NF
–518	5/16-18	NC
–524	5/16-24	NF
–616	3/8-16	NC
–624	3/8-24	NF
–714	7/16-14	NC
–720	7/16-20	NF
–813	1/2-13	NC
–820	1/2-20	NF
–918	9/16-18	NF
–1018	5/8-18	NF
–1216	3/4-16	NF
–1414	7/8-14	NF
–1614	1-14	NF

Figure 8-34. Part codes for self-locking nuts use dash numbers to indicate size and thread pitch. For example, a -524 represents a self-locking nut that fits a 5/16 inch fine thread bolt with 24 threads per inch.

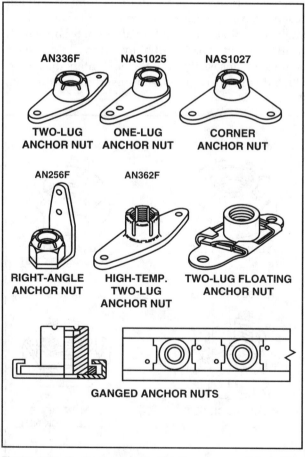

Figure 8-35. Anchor nuts simplify the process of installing and removing inspection plates. Some of the more familiar anchor nuts are shown above.

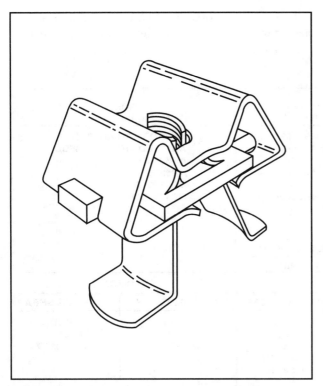

Figure 8-36. To reduce magnetic influences in the cockpit, nonmagnetic mounting nuts secure instruments in a control panel.

Tinnerman nuts used to mount instruments can either be installed in an instrument panel or in the instrument case itself. To reduce the chance of magnetic interference, the nuts are made of brass and the cage that holds the nut is constructed of phosphor bronze. If the instrument is rear mounted, the legs of the nut are long enough to pass through the

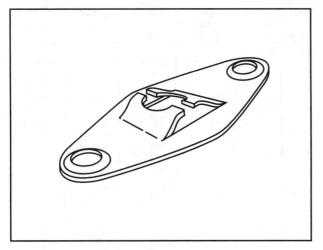

Figure 8-37. Anchor type tinnerman nuts are suitable for nonstructural applications.

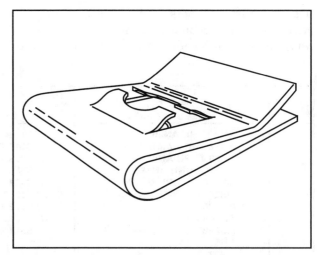

Figure 8-38. U-type Tinnerman nuts provide convenient anchor points for cowlings, fairings, and panels.

instrument case. If the instrument is front mounted, the nut fastens into the screw hole in the instrument panel. [Figure 8-36]

On many light aircraft where cost is a major factor, **tinnerman-type anchor nuts** are riveted to a structure to hold screws used to secure inspection plates. Although these nuts lack the strength of a regular threaded nut plate, they are approved for nonstructural inspection plates where their use protects aircraft skin from damage by repeated insertion and removal of self-tapping screws. [Figure 8-37]

The cowlings on some light aircraft are held on with self-tapping sheet metal screws. To prevent the sheet metal screws from enlarging the holes in the cowling by repeated insertion and extraction, a **U-type Tinnerman nut** is slipped over the edge of the inside cowling so that it straddles the screw hole. When a screw is tightened into the nut, the spring action of the nut holds the screw tight. [Figure 8-38]

TORQUE RECOMMENDATIONS

The strength of a joint held together by a threaded fastener depends upon proper preloading of the fastener's threads. To ensure this preloading is accomplished the nut must be properly torqued onto its bolt. While too much torque can damage threads, too little torque can allow excessive loads to be applied on a bolt resulting in failure. To prevent this, maintenance manuals often specify torque values for most fasteners. Other sources such as AC43.13-1B give recommended minimum and maximum torque values for the most commonly used nut and bolt combinations. Unless otherwise specified, the values are for clean and dry threads. [Figure 8-39]

BOLTS STEEL TENSION		BOLTS STEEL TENSION		BOLTS ALUMINUM	
AN3 THRU AN20 AN42 THRU AN49 AN73 THRU AN81 AN173 THRU AN186 MS20033 THRU MS20046 MS20073 MS20074 AN509 NK9 MS24694 AN525 NK525 MS27039		MS20004 THRU MS20024 NAS144 THRU NAS158 NAS333 THRU NAS340 NAS583 THRU NAS590 NAS624 THRU NAS644 NAS1303 THRU NAS1320 NAS172 NAS174 NAS517 **STEEL SHEAR BOLT** NAS464		AN3DD THRU AN20DD AN173DD THRU AN186DD AN509DD AN525D MS27039D MS4694DD	
NUTS		**NUTS**		**NUTS**	
STEEL TENSION	STEEL SHEAR	STEEL TENSION	STEEL SHEAR	ALUMINUM TENSION	ALUMINUM SHEAR
AN310 AN315 AN363 AN365 NAS1021 MS17825 MS21045 MS20365 MS20050 NAS679	AN320 AN364 NAS1022 MS17826 MS20364	AN310 AN315 AN363 AN365 MS17825 MS20365 MS21045 NAS1021 NAS679 NAS1291	AN320 AN364 NAS1022 MS17826 MS20364	AN365D AN310D NAS1021D	AN320D AN364D NAS1022D
FINE THREAD SERIES												

NUT-BOLT SIZE	TORQUE LIMITS IN-LBS MIN.	MAX.	TORQUE LIMITS IN-LBS MIN.	MAX.	TORQUE LIMITS IN-LBS MIN.	MAX.	TORQUE LIMITS IN-LBS MIN.	MAX.	TORQUE LIMITS IN-LBS MIN.	MAX.	TORQUE LIMITS IN-LBS MIN.	MAX.
8-36	12	15	7	9	—	—	—	—	5	10	3	6
10-32	20	25	12	15	25	30	15	20	10	15	5	10
1/4-28	50	70	30	40	80	100	50	60	30	45	15	30
5/16-24	100	140	60	85	120	145	70	90	40	65	25	40
3/8-24	160	190	95	110	200	250	120	150	75	110	45	70
7/16-20	450	500	270	300	520	630	300	400	180	280	110	170
1/2-20	480	690	290	410	770	950	450	550	280	410	160	260
9/16-18	800	1,000	480	600	1,100	1,300	650	800	380	580	230	360
5/8-18	1,100	1,300	660	780	1,250	1,550	750	950	550	670	270	420
3/4-16	2,300	2,500	1,300	1,500	2,650	3,200	1,600	1,900	950	1,250	560	880
7/8-14	2,500	3,000	1.500	1,800	3,550	4,350	2,100	2,600	1,250	1,900	750	1,200
1-14	3,700	4,500	2,200	3,300	4,500	5,500	2,700	3,300	1,600	2,400	950	1,500
1 1/8-12	5,000	7,000	3,000	4,200	6,000	7,300	3,600	4,400	2,100	3,200	1,250	2,000
1 1/4-12	9,000	11,000	5,400	6,600	11,000	13,400	6,600	8,000	3,900	5,600	2,300	3,650
COARSE THREAD SERIES												
8-32	12	15	7	9	—	—	—	—	—	—	—	—
10-24	20	25	12	15	—	—	—	—	—	—	—	—
1/4-20	40	50	25	30	—	—	—	—	—	—	—	—
5/16-18	80	90	48	55	—	—	—	—	—	—	—	—
3/8-16	160	185	95	110	—	—	—	—	—	—	—	—
7/16-14	235	255	140	155	—	—	—	—	—	—	—	—
1/2-13	400	480	240	290	—	—	—	—	—	—	—	—
9/16-12	500	700	300	420	—	—	—	—	—	—	—	—
5/8-11	700	900	420	540	—	—	—	—	—	—	—	—
3/4-10	1,150	1,600	700	950	—	—	—	—	—	—	—	—
7/8-9	2,200	3,000	1,300	1,800	—	—	—	—	—	—	—	—
1-8	3,700	5,000	2,200	3,000	—	—	—	—	—	—	—	—
1 1/8-8	5,500	6,500	3,300	4,000	—	—	—	—	—	—	—	—
1 1/4-8	6,500	8,000	4,000	5,000	—	—	—	—	—	—	—	—

Figure 8-39. **Manufacturers usually specify torques for threaded fasteners. However, when this information is not available, refer to a standard torque chart.**

When torque is critical for a self-locking nut, measure the amount of torque that is needed to turn the nut onto a bolt before it contacts the surface. Then, add this torque value to the amount of torque that is recommended for the joint.

SCREWS

Screws are probably the most commonly used threaded fastener in aircraft. They differ from bolts in that they are generally made of lower strength materials. Screws are typically installed with a loose-fitting thread, and the head shapes are made to engage a screwdriver or wrench. Some screws have a clearly defined grip length while others are threaded along their entire length.

There are three basic classifications of screws used in aircraft construction: machine screws, which are the most widely used; structural screws, which have the same strength as bolts; and self-tapping screws, which are typically used to join light weight materials.

MACHINE SCREWS

Machine screws are used extensively for attaching fairings, inspection plates, fluid line clamps and other light structural parts. The main difference between aircraft bolts and machine screws is that the threads of a machine screw usually run the full length of the shank, whereas bolts have an unthreaded grip length. Screws normally have a Class 2, or free fit and are available in both national coarse and national fine threads. The most common machine screws used in aviation are the fillister-head screw, the flat-head screw, the round-head screw, and the truss-head screw. [Figure 8-40]

Fillister-Head Machine Screw

Fillister-head screws are slotted and have a hole drilled through their head. They come in both coarse-thread (MS35265) and fine-thread (MS35266), both of which produce a Class 3 fit. Fillister-head screws are available in sizes from 4-40 up to 1/4-20 and have no clearly defined grip.

Flat-Head Machine Screw

These countersunk screws are made of cadmium-plated carbon steel, and are available with a recessed head for cross-point screwdrivers. They are available with either fine or coarse threads and come with either an 82 degree or 100 degree coun-

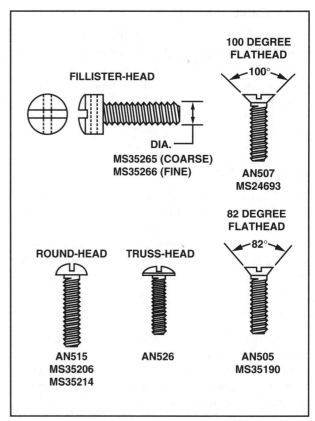

Figure 8-40. Machine screws are the most widely used screw in aircraft applications. A variety of sizes and styles are available.

tersink. The 100 degree flat-head screw carries the part code MS24693. The MS24693S is made of cadmium-plated carbon steel and the MS24693BB is made of black oxide coated brass. These screws are available in sizes from 4-48 through 3/8-16.

Round-Head Machine Screw

The MS35206 round-head machine screw is made of cadmium-plated carbon steel and has either a slotted or recessed head. This screw is also available in brass and is identified by the part designation MS35214. The brass screws are typically coated with a black oxide and are sometimes used to mount instruments. Like most other screws, the round-head machine screw is available with either fine or coarse threads.

Truss-Head Machine Screw

The AN526 truss-head machine screw has a large head that provides good holding ability on thin pieces of metal. This screw is available in cadmium-plated carbon steel in either slotted or cross-recessed heads.

STRUCTURAL SCREWS

Structural screws are made of alloy steel, are heat treated, and can be used as structural bolts. They have a definite grip and the same shear strength as a bolt of the same size. Shank tolerances are similar to AN hex-head bolts, and the threads are National Fine. Structural screws are available with fillister, flat, or washer heads.

Fillister-Head Screw

The fillister-head structural screw is similar in appearance to the fillister-head machine screw except for the cross on its head indicating that it is made of high-strength steel. Structural fillister-head screws carry an AN502 part code if they have coarse threads and AN503 if they have fine threads.

Flat-Head Screw

Structural flat-head screws have an MS24694 part number and are made of heat-treated carbon steel that is cadmium plated. They are distinguished from 100 degree flat-head machine screws by the "X" marked on their head.

Washer-Head Screw

These structural screws have a washer formed onto their head to increase the screw's holding ability. This added bearing area is required when used with some thinner materials. Washer-head screws are made of cadmium-plated high-strength steel. [Figure 8-41]

SELF-TAPPING SCREWS

Self-tapping screws have coarse-threads and are used to hold thin sheets of metal, plastic, or plywood together. The type-A screw has a gimlet

(sharp) point, and the type B has a blunt point with threads that are slightly finer than those of a type-A screw.

There are four types of heads available on self-tapping screws: a round head, a truss head, a countersunk head, which is flat on top, and the countersunk oval screw. The truss-head is rounded, similar to the round head screw, but is considerably thinner. [Figure 8-42]

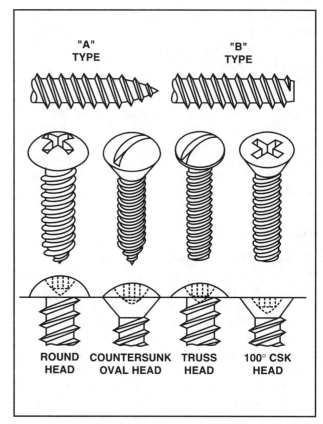

Figure 8-42. Self-tapping sheet metal screws are useful for attaching trim and upholstery.

PINS

The main types of pins used in aircraft structures are the roll pin, clevis pin, cotter pin, and taper pin. Pins are used in shear applications and for safetying.

ROLL PIN

Roll pins are often used to provide a pivot for a joint where the pin is not likely to be removed. A roll pin is made of flat spring steel that is rolled into a cylinder but the two ends are not joined. This allows the pin to compress when it is pressed into a hole and

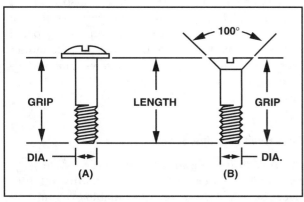

Figure 8-41. Structural screws are available with both protruding and countersunk heads.

create a spring action that holds the pin tight against the edge of the hole. To remove a roll pin, it must be driven from a hole with a proper size pin punch. [Figure 8-43]

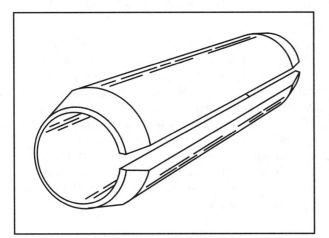

Figure 8-43. MS16562 spring steel rollpins are often used in the movable joints of aircraft seats.

CLEVIS PIN

Clevis, or flat-head, pins are used for hinge pins in some aircraft control systems. They are made of cadmium-plated steel and have grip lengths in 1/16 inch increments. When installing a clevis pin place the head in the up position, place a plain washer over the opposite end, and insert a cotter pin through the hole to lock the pin in place. [Figure 8-44]

COTTER PINS

Castellated nuts are locked onto drilled bolts by passing a cotter pin through the hole and nut castellations and then spreading the ends of the cotter pin. They are made of either cadmium-plated carbon steel or corrosion-resistant steel.

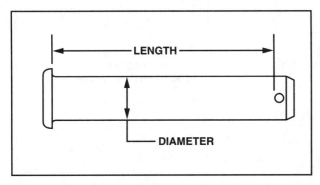

Figure 8-44. AN392 through AN406 (MS20392) clevis pin are often found in control cable systems.

There are two methods of securing cotter pins that are generally acceptable. In the preferred method, one leg of the cotter pin is bent up over the end of the bolt, and the other leg is bent down over one of the flats of the nut. With the second method, the cotter pin is rotated 90 degrees and the legs wrapped around the castellations. It is important to note that nuts should never be overtorqued to make the hole in the bolt align with the castellations. If the castellations in the nut fail to align with the drilled bolt hole, add washers under the nut until a cotter pin can be inserted. [Figure 8-45]

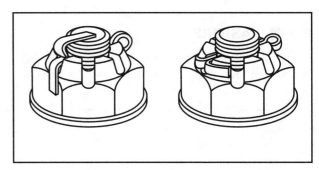

Figure 8-45. You may use either of the two acceptable methods to safety castellated nuts.

TAPER PIN

Both the plain and threaded taper pin are used in aircraft structures to make a joint that is designed to carry shear loads. This type of pin does not allow any loose motion or play. The AN385 plain taper pin is forced into a hole that has been reamed with a Morse standard taper pin reamer and is held in place by friction. It can be safetied by passing safety wire around the shaft and through a hole drilled in its large end. An AN386 taper pin is similar to the AN385 except that its small-end is threaded to accept either a self-locking shear nut (AN364) or a shear castle nut (AN320). [Figure 8-46]

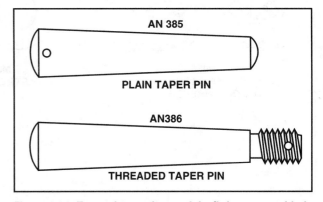

Figure 8-46. Taper pins produce a tight fit in a reamed hole for applications loaded in shear.

WASHERS

Washers provide a bearing surface area for nuts, and act as spacers or shims to obtain the proper grip length for a bolt and nut assembly. They are also used to adjust the position of castellated nuts with respect to drilled cotter pin holes in bolts as well as apply tension between a nut and a material surface to prevent the nut from vibrating loose. The three most common types of washers used in airframe repair are the plain washer, lock washer, and special washer. [Figure 8-47]

PLAIN WASHERS

An AN960 plain washer provides a smooth surface between a nut and the material being clamped.

These washers are made of cadmium-plated steel, commercial brass (B), corrosion-resistant steel (C), and 2024 aluminum alloy (D). They are available in sizes that range from those that fit a number two machine screw to those that fit a one-inch bolt.

If a thin washer is needed, a light series washer that is one-half the thickness of a regular washer is available. An example of where a light series washer should be used is if the castellations of an AN310 nut do not line up with a cotter pin hole when the nut is properly torqued. In this situation a light series washer can be substituted for the regular washer to align the holes. A light series washer is identified by the letter "L" added to the code. For example, the code AN960L identifies a light series washer.

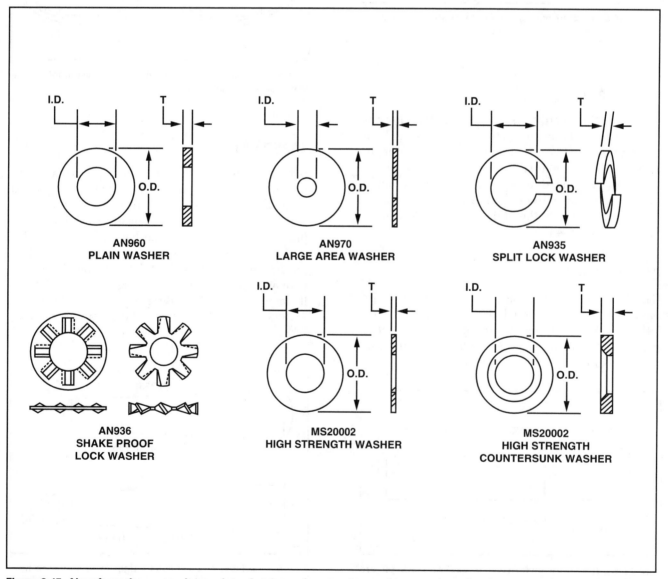

Figure 8-47. Aircraft washers come in a variety of styles and are used to perform a variety of tasks.

When working with wood or composite structures, washers with a large surface area are used to spread the fastener load over a wider area. These large area washers carry the code of AN970 and are all made of cadmium-plated steel with inside diameters from 3/16 to 1/2 inch.

LOCK WASHERS

In some instances it is not convenient to use self-locking nuts or cotter pins on bolts. For these applications, a lock washer is often used between the nut and joint surface if the joint is not structurally critical. Lock washers are made of steel and are twisted so that when a nut is tightened against it, the spring action of the washer creates a strong friction force between the bolt threads and those in the nut.

Two types of lock washers are used in aircraft construction. The most common is the AN935 split lock washer. These washers are available in sizes that fit from a number four machine screw to a 1/2 inch bolt. The second type of lock washer is the thinner AN936 shakeproof lock washer which is available with both internal and external teeth.

SPECIAL WASHERS

Some high-strength internal wrenching bolts have a radius between their shaft and the underside of the bolt head. To provide a tight mating surface, MS20002C countersunk washers are used under the heads of internal wrenching bolts. These washers have a countersunk edge to accommodate the radius on the bolt head. When these bolts are used in aluminum alloy structures, a countersunk washer is used under the bolt head and a plain type washer is used under the nut. Countersunk washers are made of heat-treated steel and are cadmium plated.

Finishing washers are often used in aircraft interiors to secure upholstery and trim. These washers have a countersunk face to accommodate flush screws. Finishing washers bear against a large area to avoid damaging fragile interior components.

Occasionally a nut or bolt cannot be safetied by conventional means. In many instances, **keyed washers** can be used as a safety device. Keyed washers have small keys or protrusions to engage slots cut into bolts or panels. In addition, they have tabs that can be bent

up against a nut or bolt head to keep it from rotating.

HOLE REPAIR HARDWARE

Threaded holes wear out after repeated insertion and extraction of fasteners. In the past, this often meant an expensive part had to be scrapped when a hole was stripped out. However, now, hole repair hardware allows you to make a fast and inexpensive repair to worn or damaged holes.

HELI-COIL™ INSERTS

Many screws and bolts are driven into threads cut into castings made of soft aluminum, magnesium, or plastic. Rather than allowing these soft materials to wear each time the screw is inserted or removed, the holes are often protected with Heli-Coil inserts. These inserts typically consist of a helix of stainless steel having a diamond cross section.

Heli-Coil inserts are typically installed in soft castings when the component is manufactured. However, they may also be installed in the field when threads in a casting are stripped out. If the threads in a hole are stripped, first determine whether increasing the hole diameter for installation of the insert will weaken the casting. If an insert can be used, drill out the stripped threads with the drill furnished in the Heli-Coil installation kit and tap the hole with the special tap. Then, run a gauge through the new threads to be sure that they can accommodate the insert. Next, place the insert on the installation tool and screw it into the tapped hole. The installation tool stretches the insert to reduce its outside diameter enough for it to easily enter the hole. When the insert is below the surface of the casting to the extent specified in the installation instructions, remove the installation tool. When the tool is removed, the natural springiness of the insert holds it tight. Any attempt of the insert to back out causes it to wedge tighter into the hole. After the insert has been driven into a hole to proper depth, the tang on the lower end of the threads which was held by the driving tool is broken off so the bolt can pass completely through the threads.

When Heli-Coils are used in a repair, the repair must be approved by the component manufacturer or by the FAA. Furthermore, the installation instructions furnished by the insert manufacturer must be followed in detail. [Figure 8-48]

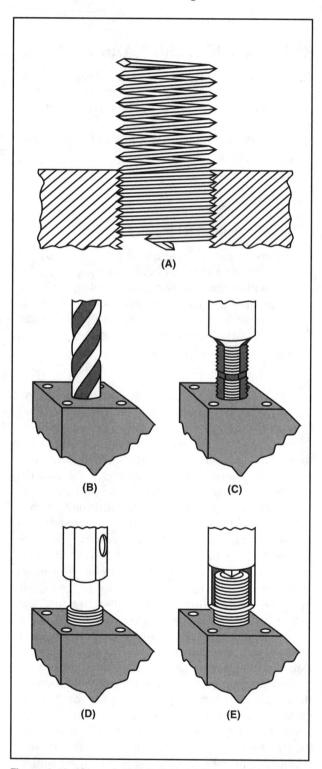

Figure 8-48. (A) — A Heli-Coil insert is a special stainless steel helical insert that screws into a threaded hole to restore damaged threads to their original size. (B) — To insert a Heli-Coil, the damaged threads are first drilled out. (C) — New threads are then cut with a special tap. (D) — The threads are gauged with a thread gauge. (E) — The Heli-Coil insert is screwed into the hole with a special inserting tool.

ACRES SLEEVES

If a hole has worn oversize, or if there is corrosion in the hole that cannot be cleaned out, the part may be made serviceable by installing an Acres sleeve. Acres fastener sleeves are thin walled, tubular elements with a flared end. The sleeves are installed in holes to accept standard bolts and rivet type fasteners. Before installing an Acres sleeve you must first determine that there is enough strength in the material to allow the hole to be enlarged. If there is, drill the hole out approximately 1/64 inch larger than its original diameter. A special aluminum alloy or stainless steel sleeve is then installed in the hole. For certain applications, the sleeve can be bonded into the hole with an adhesive. However, it is typical for the sleeve to be held in the hole by friction. [Figure 8-49]

TURNLOCK FASTENERS

Turnlock fasteners are used to secure inspection plates, doors, cowlings, and other removable panels on aircraft. The most desirable feature of these fasteners is that they permit quick and easy removal of access panels for inspection and servicing purposes. Turnlock fasteners are manufactured and supplied by a number of manufacturers under various trade

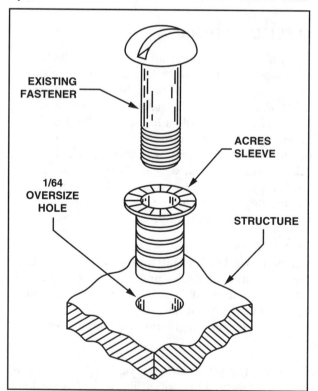

Figure 8-49. Acres sleeves are used to repair holes that have been worn oversize or corroded. The hole is drilled oversize and the sleeve is pressed or bonded in place. The original fastener is then installed.

names. Some of the most commonly used are the Dzus, Airloc, and Camloc.

DZUS FASTENERS

Cowling and other inspection access doors that must be opened frequently can be held with Dzus fasteners that require only a quarter of a turn to lock or unlock. With a Dzus fastener a hard spring-steel wire is riveted across an opening on a fixed part of a fuselage, and a stud is mounted on the removable panel with a metal grommet. When the panel is closed, a slot in the stud straddles the spring. Turning the stud a quarter of a turn pulls the spring up into the slanted slot and locks it as the spring passes over the hump in the slot. [Figure 8-50]

When something is fastened with Dzus fasteners, care must be taken that the stud in every fastener straddles each of the springs rather than passing beside them. In order to be sure that all of the fasteners are properly locked, the slots should all be lined up. Furthermore, when a Dzus fastener is fastened, a distinct click is heard when the spring drops over the hump into the locked position. To aid in assuring that no stud misses the spring, special receptacle-type fasteners are available that guide the stud over the spring. [Figure 8-51]

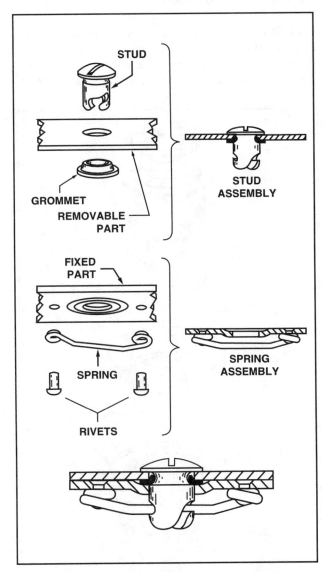

Figure 8-50. With a standard Dzus fastener, a slotted stud engages a spring mounted to the fuselage. As the stud is turned one quarter turn, the spring locks the fastener in place.

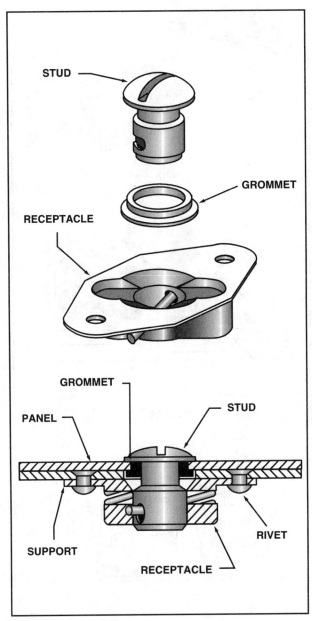

Figure 8-51. The receptacle of a receptacle-type Dzus fastener guides the stud to the exact location it needs to be prior to engaging the spring.

AIRLOC FASTENER

An Airloc fastener consists of a steel stud and crosspin in a removable cowling or door and a sheet spring-steel receptacle in the stationary member. To lock this type of fastener, the stud slips into the receptacle and is rotated a quarter of a turn. The pin drops into an indentation in the receptacle spring and holds the fastener locked. [Figure 8-52]

CAMLOCK FASTENER

The stud assembly of a Camlock fastener consists of a housing containing a spring and a stud with a steel pin. This assembly is held onto the removable portion of the cowling or access door with a metal grommet. The stud fits into a pressed steel receptacle, and a quarter of a turn locks the steel pin in a groove in the bottom of the receptacle. [Figure 8-53]

CONTROL CABLES AND TERMINALS

While a number of different systems are used to actuate flight and engine controls from the cockpit, flexible control cables are by far the most commonly used method. Multiple-strand control cables are simple, strong, and reliable.

Cable has several advantages over other types of linkages. It is strong and light in weight, and its flexibility makes it easy to route. In addition to primary flight controls, cable is used on engine controls,

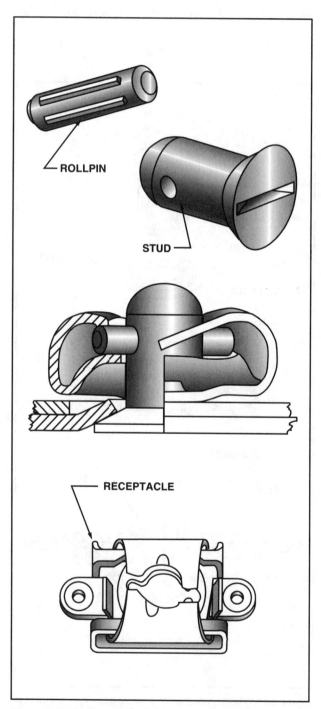

Figure 8-52. Airloc cowling fasteners are similar to Dzus fasteners and are used in many of the same applications.

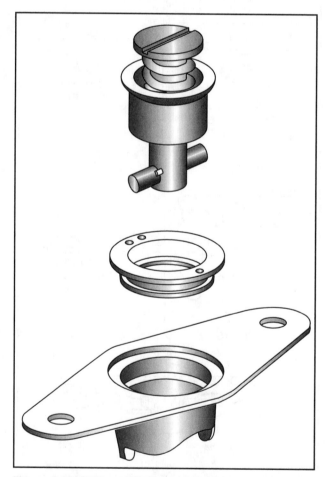

Figure 8-53. With a Camlock cowling fastener the stud assembly can be inserted into the receptacle when the pin is aligned with the slot in the receptacle.

emergency landing gear extension controls, trim tab systems, and various other applications.

One disadvantage of a cable system for control movement relates to thermal contraction. As an aircraft climbs to a high altitude, its temperature drops and its structure contracts. Furthermore, the aluminum structure contracts much more than the small mass of steel in a control cable, and as a result, cables lose their tension. Large aircraft have a rather complex automatic tensioning system to keep control cable tension relatively constant as the aircraft contracts and expands. However, small aircraft must have their cable tension adjusted as a compromise so they are not too tight when the airplane is hot nor too loose when it is cold. [Figure 8-54]

NONFLEXIBLE CABLE

In areas where a linkage does not pass over any pulleys nonflexible cable can be used. It is available in either a 1 × 7 or 1 × 19 configuration. The 1 × 7 cable is made up of one strand comprised of seven individual wires, whereas the 1 × 19 consists of one strand made up of 19 individual wires. Nonflexible cable is available in both galvanized carbon steel and stainless steel.

FLEXIBLE CABLE

Flexible steel cable made up of seven strands of seven wires each is called 7 × 7 or flexible cable,

and is available in 1/16 and 3/32 inch sizes in both galvanized carbon steel and stainless steel. Both types are preformed which means that when the cable is manufactured each strand is formed into a spiral shape. This process keeps strands together when the cable is wound and also helps prevent the cable from spreading out when cut. Furthermore, preforming gives cable greater flexibility and relieves bending stresses when the strands are woven into the cable.

EXTRA-FLEXIBLE CABLE

The most widely used cable, 7 × 19, is available in sizes from 1/8 inch up. It is extra flexible and is made of 133 individual wires wound in seven strands, each strand having 19 wires. These cables are preformed and are available in both galvanized and stainless steel. Galvanized cable is more resistant to fatigue than stainless steel, but in applications where corrosion is a factor, stainless steel is used. [Figure 8-55]

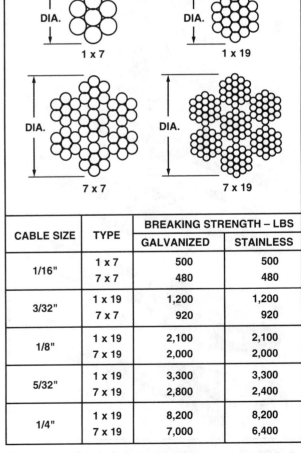

CABLE SIZE	TYPE	BREAKING STRENGTH – LBS	
		GALVANIZED	STAINLESS
1/16"	1 x 7	500	500
	7 x 7	480	480
3/32"	1 x 19	1,200	1,200
	7 x 7	920	920
1/8"	1 x 19	2,100	2,100
	7 x 19	2,000	2,000
5/32"	1 x 19	3,300	3,300
	7 x 19	2,800	2,400
1/4"	1 x 19	8,200	8,200
	7 x 19	7,000	6,400

Figure 8-55. Different steel control cables are available for various applications based on flexibility and breaking strength.

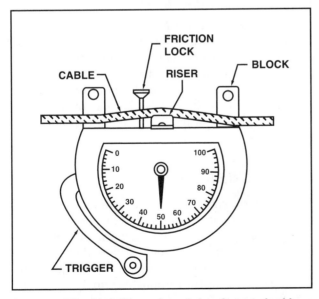

Figure 8-54. To check the tension of aircraft control cables a tensiometer is used. To use a tensiometer a cable is placed between the two blocks on the frame and the riser. The trigger is then pulled to apply pressure to the cable and indicate the cable tension.

ATTACHING CABLES

At one time, most cables were attached to bellcranks, control surfaces, and flight controls with **woven splices**, such as the **Army-Navy five-tuck splice** or the **Roebling roll**. Because both types of woven splices require a great deal of hand work and develop only 75 percent of the cable strength, this method of attaching cables has almost been completely replaced.

SWAGED TERMINALS

The cable fittings used most in large aircraft manufacture are MS-type **swaged cable terminals**. To install these terminals, cut the cable and insert it into the end of a terminal. Then, use either a hand or power swaging tool to force the metal of the terminal down into the cable. This forms a joint that is at least as strong as the cable itself. [Figure 8-56]

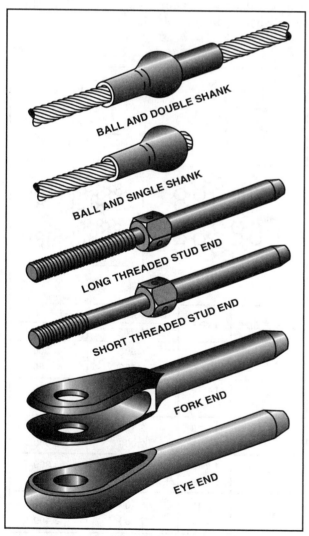

Figure 8-56. Swaged cable terminals have all but replaced woven splices and are as strong or stronger than the cable itself.

To ensure that a terminal is properly swaged, a measurement is made of the swaged terminal with a go/no-go gauge. The swaging process must decrease the terminal's diameter to the extent that the go end of a go/no-go gauge passes over the swaged terminal, but the no-go end does not. As an inspection aid to ensure the cable does not pull out of the terminal, a small mark of paint is placed over the terminal end and onto the cable. A broken paint mark indicates the cable has slipped inside the terminal. [Figure 8-57]

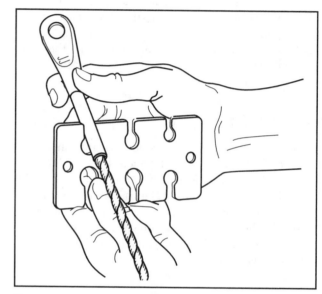

Figure 8-57. A go/no-go gauge ensures that a swaged terminal was installed properly.

NICOPRESS OVAL SLEEVES

Many light aircraft use Nicopress sleeves that are squeezed onto control cables to form terminal ends. A nicopress sleeve is made of copper and has two holes to accommodate a control cable. When a cable is wrapped around an AN100 thimble and properly squeezed with the correct Nicopress squeezer, the terminal develops at least the strength of the cable. [Figure 8-58]

TURNBUCKLES

Turnbuckles are a type of cable fastener that allows cable tension to be adjusted. A complete turnbuckle assembly consists of two ends, one with right-hand threads and the other having left-hand threads, with a brass barrel joining them. Minor cable adjustment is made by rotating the turnbuckle which effectively lengthens or shortens the cable's length.

To ensure that a turnbuckle develops full cable strength, there must be no more than three threads of

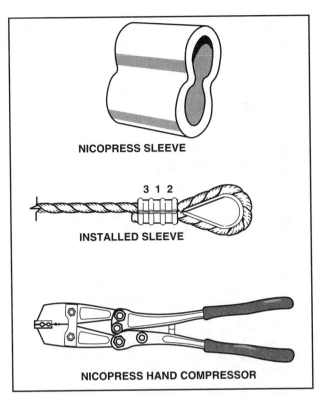

Figure 8-58. To install a nicopress sleeve, slip a sleeve over the cable end and loop the cable back through the sleeve. Next, place a proper size AN100 thimble inside the loop, and pull the cable tight leaving about 1/16 inch of cable protruding from the sleeve. Three squeezes are then applied with the squeezing tool. The first squeeze in the middle, the second on the end nearest the thimble, and the last on the sleeve end nearest the free cable end. After the sleeve is squeezed, check it with a gauge to ensure that it has been forced far enough down into the cable.

either end sticking out of the barrel. After cable tension is adjusted, the turnbuckle barrel is safetied to the two cable ends so that it cannot turn. [Figure 8-59]

CABLE INSPECTION

When inspecting control cables pay particular attention to those sections of cable that pass through

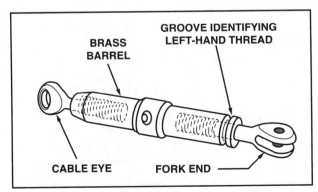

Figure 8-59. Turnbuckles provide a means of adjusting control cables as they age and stretch.

fairleads and around pulleys. To properly inspect each section which passes over a pulley or through a fairlead, remove the cable from the aircraft to the extent necessary to expose that particular section. Examine cables for broken wires by passing a cloth along the length of the cable. This cleans the cable as well as detects broken wires if the cloth snags on the cable. When snags are found, closely examine the cable to determine the full extent of the damage.

Wear normally extends along a cable equal to the distance a cable moves a specific location and may occur on one side of the cable or on its entire circumference. Replace flexible and nonflexible cables when the individual wires in each strand appear to blend together, or when the outer wires are worn 40 to 50 percent.

SAFETYING METHODS

Because aircraft vibrate, there must be some provision for safetying or locking all fasteners to keep them from vibrating loose. Self-locking nuts are used for the vast majority of applications in modern aircraft construction, but there are still places where safety wire or cotter pins are needed. For example, drilled-head bolts are often used in vibration-prone areas and are safety wired together. When installing safety wire, the wire should pull the bolt head in the direction of tightening and should be twisted evenly to the next bolt. After the end of the wire is passed through the head of the second bolt it is again twisted, this time for about three or four turns. Once this is done, the excess is cut off and the ends of the wire are bent back where they cannot cut anyone who passes their hand over the bolts.

In areas where a number of bolts must be safetied, such as a propeller, you may safety wire the bolts in groups of three. If more than three bolts are safetied together it is difficult to get the safety wire tight enough to be effective. [Figure 8-60]

Self-locking nuts should not be used on studs, nor should drilled studs be safetied with cotter pins, since neither of these safetying methods prevents a

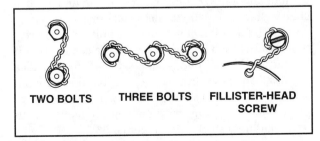

Figure 8-60. Many bolt heads must be safety wired to keep them from vibrating loose. As a general rule, no more than three bolts should be safety wired together.

stud from backing out of its casting. Instead, safety two drilled studs with castellated nuts wired together. This keeps the nut from coming loose, as well as prevents the stud from backing out of the casting. [Figure 8-61]

Figure 8-61. To prevent nuts from becoming loose on studs, use castellated nuts and safety wire them as you would bolt heads.

Electrical connectors can be safetied to drilled-head fillister-head screws. However, when this is done, make sure that the wire pulls the plug in the direction of tightening.

As with any threaded fastener, turnbuckles must be safetied to prevent them from coming loose. The simplest method of safety wiring a turnbuckle is called the **single wrap method** and uses a single piece of wire that passes through the hole in the turnbuckle's center and is wrapped around each end. A similar method, the **single wrap spiral**, also uses a single piece of safety wire, but is spiraled around the turnbuckle barrel and passes through the center hole twice. Two pieces of safety are used in the **double wrap method**, which is basically the same as two single wrap safeties, one in each direction. A **double wrap spiral** is essentially the same as two single wrap spirals, one in each direction.

Before safetying a turnbuckle, the cable must have the correct tension and there must not be more than three threads showing on either side of the turnbuckle barrel. Wrap the wire around the turnbuckle and finish the safety wiring with at least four turns around the shank of the turnbuckle. [Figure 8-62]

Different thicknesses of safety wire are available to safety different sizes of turnbuckles. For example, turnbuckles on 1/8 inch cable are safetied with a single wrap of 0.040 stainless steel or monel safety wire, or 0.057 diameter copper or brass wire. Forty-thousandths-inch copper or brass wire can be used if the turnbuckle is double-wrap safetied. Turnbuckles on 5/32 inch control cable can be double-wrap safetied with 0.040 stainless steel wire or 0.051 copper or brass, or they may be single wrapped with 0.057 stainless steel wire. [Figure 8-63]

Safety clips can be used in place of safety wire to safety turnbuckles if the turnbuckle hardware is

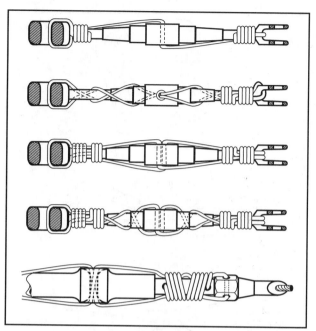

Figure 8-62. Turnbuckles and swaged cable studs must be safetied to prevent them from becoming loose. Of the available methods, double wrapping is preferred.

CABLE SIZE	TYPE OF WRAP	WIRE DIAMETER	MATERIAL
1/16	SINGLE	0.040	BRASS
1/8	SINGLE	0.040	STAINLESS STEEL
1/8	DOUBLE	0.040	BRASS
5/32	SINGLE	0.057 (MIN)	STAINLESS STEEL
5/32	DOUBLE	0.051	BRASS

Figure 8-63. When safety wiring turnbuckles, make sure you use the proper size safety wire. Brass safety wire can be used on turnbuckles, but a larger size or a double wrap is required to provide the same strength as stainless steel wire.

drilled to accommodate this type of clip. The straight part of the clip is inserted in a groove between the barrel and the swaged cable end, and the U-shaped end is pushed into the center hole of the turnbuckle. Locking clips perform the same function as safety wire with much less work. [Figure 8-64]

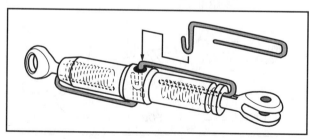

Figure 8-64. Locking clips can be used to safety turnbuckles if the barrel and terminal are notched to accept them.

HAND TOOLS AND MEASURING DEVICES

CHAPTER 9

INTRODUCTION

Hand tools are indispensable for the inspection, maintenance, and repair of aircraft structures and components. Therefore, as a maintenance technician, you must be familiar with the tooling used in the industry and its correct use and care. When working with any tool or measuring device, remember that safety should always be your primary consideration.

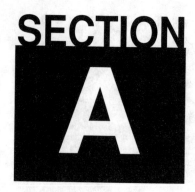

SECTION A

HAND TOOLS

The term "hand tools" encompasses all of the hand-held tools most commonly used in everyday maintenance and repair. Some of these tools are common, while others have a very specialized usage. Regardless of how common the tool, the subtleties for their proper use are not always known. Therefore, the most common uses are listed in the following text.

POUNDING TOOLS

Pounding tools include different types and weights of hammers and mallets, each with a very specific use. Since misuse of pounding tools can result in damage to aircraft components and injury to personnel, it is important that you always use these tools properly.

Before using any hammer or mallet, you should make sure you have the appropriate eye and face protection. Furthermore, you should inspect the tool for any damage that could affect safety. For example, before using a hammer or mallet, you should make sure the handle is secure and in good condition. When striking a blow with a hammer, think of your forearm as an extension of the handle. In other words, swing the hammer by bending your elbow, not your wrist. Always strike the work squarely using the full face of the hammer. To prevent marring the work, keep the face of a hammer or mallet smooth and free of dents.

BALL PEEN HAMMERS

The ball peen hammer ranges in weight from one ounce to two or three pounds. One hammer face is always flat while the other is formed into the shape of a ball. The flat hammer face is used for pounding, but should not be used to drive a nail. The ball end of the hammer is typically used to peen over rivets in commercial sheet metal work. However, this is not the method used for securing rivets in aircraft sheet metal work. [Figure 9-1]

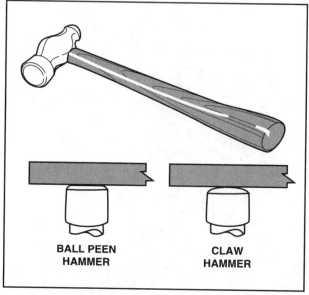

Figure 9-1. A ball peen hammer has a flat face and a ball-shaped end. The flat-faced end should not be used to drive a nail, since the curved face of a claw hammer is better shaped for nail driving control.

CROSS PEEN AND STRAIGHT PEEN HAMMERS

Commercial sheet metal and automotive body work often requires metal to be bent by hammering. This is typically accomplished by using either a cross peen or straight peen hammer. Unlike the ball peen hammer, the cross and straight peen hammers have a wedge-type end that is used to either crease metal to start a bend, or to straighten out a rolled edge. [Figure 9-2]

CLAW HAMMERS

A claw hammer is slightly crowned on the face for nail driving control and has a set of claws opposite the face. The head of a claw hammer is typically hardened, making it more brittle and susceptible to chipping. Therefore, a claw hammer should not be used on hardened steel parts. [Figure 9-3]

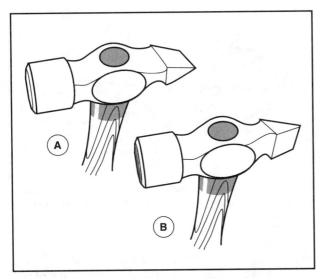

Figure 9-2. (A) — On a cross peen hammer, the wedge is at a right angle to the handle. (B) — On a straight peen hammer, the wedge is parallel to the handle.

SLEDGE HAMMERS

Sledge hammers have two flat faces and are sized according to the weight of the head without the handle. For example, the head of a five-pound sledge hammer weighs five pounds. Sledge hammers are typically used whenever a lot of heavy pounding force is needed, such as when driving stakes. [Figure 9-4]

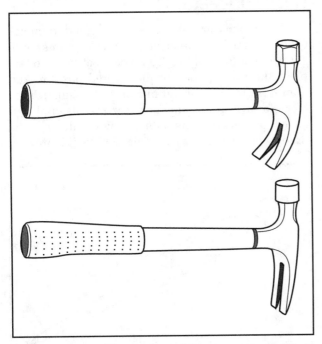

Figure 9-3. There are two types of claw hammers. One type has a curved set of claws that is used for pulling nails, while the other has a straight set of claws that is used to split wood as well as pull nails.

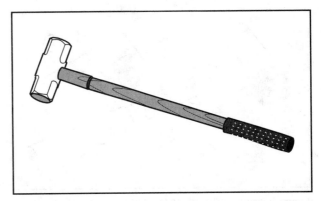

Figure 9-4. The metal head of a sledge hammer is sized according to the weight of the head without the handle, and is rarely used in aviation maintenance.

BODY HAMMERS

Body or **planishing hammers** have large smooth faces and are light weight. These hammers are specifically used to remove small dents and to smooth or stretch sheet metal. Other types of body hammers include riveting, setting, and stretching hammers. [Figure 9-5]

MALLETS

There are two types of mallets, soft-faced and hard plastic tipped. The forming and shaping of a soft aluminum alloy is accomplished with soft-faced mallets which, in early aviation, consisted of a rawhide roll held in a clamp. However, when modern synthetic materials were developed, plastic and rubber mallets virtually replaced those made of rawhide. Hard plastic tip mallets, on the other hand,

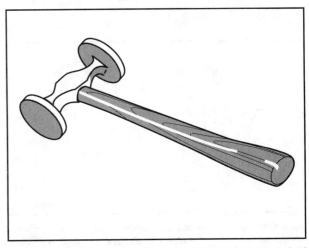

Figure 9-5. Body hammers do not typically have a great deal of weight to them and are used to smooth metal.

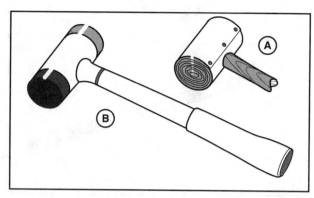

Figure 9-6. (A) — The rawhide mallet today is used more as a specialized hand tool. (B) — Plastic mallets have virtually replaced rawhide mallets for most uses.

have two faces that are often replaceable. One face is made of a resilient rubber-like material, while the other is made of a hard plastic. [Figure 9-6]

PUNCHES

Although punches are not pounding tools, they do allow the force from a hammer blow to be concentrated in the immediate area of a punch tip. This in turn means that the pressure at the end of the punch is increased compared to a hammer blow without a punch. Because of this, eye and face protection should always be worn when using punches or any type of hand tool.

After continuously hammering on a punch shank, the shank end typically deforms to the shape of a mushroom. When this happens, the mushroom shape should be removed and returned to its original crowned shape using a bench grinder. The crowned shape minimizes the chance of the punch splitting or chipping by allowing the hammer to hit it squarely. [Figure 9-7]

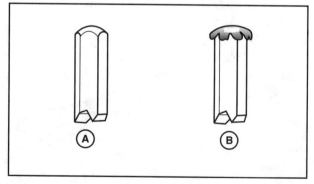

Figure 9-7. (A) — The end of a punch must have a crowned shape in order to minimize the chance of splitting or chipping. (B) — This illustrates what the end of a punch typically looks like after continuous use.

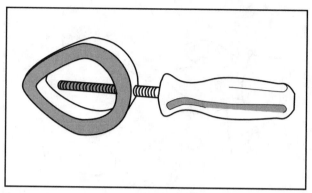

Figure 9-8. A punch holder is used when a hard blow is required.

When you must strike a punch with a hard blow, a **punch holder** is used to minimize the chance of self injury. When using a punch holder, a punch or chisel is clamped in the vise end of the holder allowing a better grip for a harder blow to be applied. [Figure 9-8]

There are several types of punches you must be familiar with. Some are used to mark locations, some to punch bolts, rivets, or pins from their holes, and some are used to indent material so that a drill bit will start in a precise location. Each punch is designed for a specific purpose and each should be used for the purpose it is designed for.

PRICK PUNCHES

Prick punches are small, sharp-pointed punches used to transfer dimensions and locations onto sheet metal for drilling. To mark sheet metal work, place the prick punch in the desired position and then tap the head of the punch with a small hammer. Because prick punches scar the material and have relatively delicate points, they should never be used to drive a pin or rivet from a hole. [Figure 9-9]

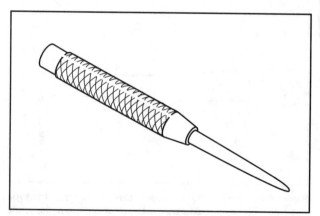

Figure 9-9. A prick punch is used with a light hammer blow to make reference points on metal.

CENTER PUNCHES

A center punch is a relatively sharp-pointed tool used to make indentations in metal. The indentation aids in starting a drill bit when drilling a hole, but should never be used to drive a pin or rivet from a hole. The point of a center punch is ground to an angle of about 60 degrees to provide an indentation that is approximately the cutting angle of a drill. When using a center punch, the punch mark should not distort the surrounding metal but should be deep enough so that the drill starts to cut rather than wander across the material.

There are two general types of center punches: the solid steel punch and the automatic center punch. With a solid steel punch, a hammer is used to strike the punch and make a punch mark. The operation of an automatic punch, on the other hand, is based on a spring loaded trip mechanism in its handle. The point is placed in the exact location a hole is needed and the handle pushed in. When the proper force is reached, a trip mechanism releases and the point is struck with a sharp blow from within the handle. Rotating the punch handle adjusts the spring compression in the handle and controls the amount of impact force applied. This type of punch is relatively low in cost and allows marks to be made accurately and consistently. When a lot of sheet metal work is anticipated, an automatic center punch becomes extremely useful. [Figure 9-10]

PIN PUNCHES

One of the most useful punches for a sheet metal worker is the pin punch. A common use for the pin punch is to remove rivets. To do this, begin by drilling through the rivet head and knocking it off its shank with a small chisel or by snapping the rivet head off by inserting a pin punch and prying

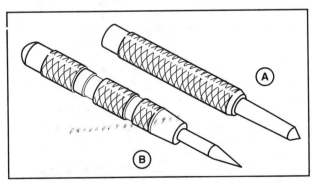

Figure 9-10. (A) — A hammer is used with a solid steel center punch to mark sheet metal. (B) — An automatic center punch uses a spring loaded trip mechanism in its handle to create a punch mark.

sideways. After the rivet head is removed, the rivet shank is punched out of its hole with the proper size pin punch leaving the original size rivet hole. When removing rivets, thin sheet metal must be properly backed to prevent bending when the rivet shank is driven out. [Figure 9-11]

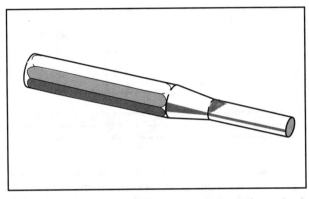

Figure 9-11. A pin punch has a flat tip and a uniform shank that is available in several different sizes.

STARTING PUNCHES

Starting punches are typically used to aid in driving bolts and pins from their holes. To use a starting punch correctly, its tapered shank is driven into a bolt or pin hole until it almost fills it. The starting punch is then removed and the job is finished using a pin punch to gently drive the bolt or pin from the hole. [Figure 9-12]

TRANSFER PUNCHES

A transfer punch is a special form of punch used to mark rivet holes when using an original template to lay out a new skin pattern. The diameter of a transfer punch shank is the same as the original rivet hole with the end ground so that it marks the new skin in the center of the rivet hole. Transfer punches

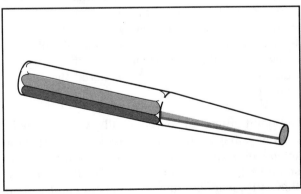

Figure 9-12. As you can see, a starting punch has a flat face and tapers outward to the shank.

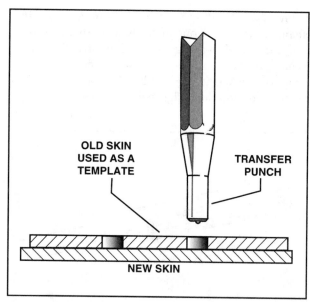

Figure 9-13. A transfer punch has a shank the same size as the rivet hole being transferred. This way, when it is inserted into an old rivet hole, the new skin is marked for the drill center.

are made as both solid steel and automatic punches. [Figure 9-13]

HOLDING TOOLS

Holding tools are very important to an aircraft maintenance technician. These tools come in a variety of shapes and sizes, and are designed for different tasks and needs. The proper use of holding tools helps ensure a professional looking job. However, care must be taken to use the proper tool for the job.

SLIP-JOINT PLIERS

One of the most commonly used holding tools in aviation is the slip-joint pliers. These pliers come in lengths from approximately four inches to more than nine inches. The six-inch size is the most commonly used. Slip-joint pliers gets their name from the double hole used at the pivot. This double hole design allows the jaws to work in either of two positions, increasing the range of material that can be gripped.

Slip-joint pliers are exceptionally versatile tools. However, they should only be used for their specific intended purpose, which is to hold things. You should never use slip-joint pliers to turn a nut as they will invariably round off the nut corners. This is especially true of tubing nuts which are made of a soft aluminum or brass. [Figure 9-14]

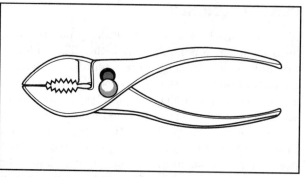

Figure 9-14. The six-inch slip-joint pliers is the size most commonly used in aircraft maintenance. These pliers get their name from the double hole used for their pivot.

INTERLOCKING-JOINT PLIERS

Interlocking-joint pliers are commonly called water pump pliers because they are often used to tighten the packing gland nut around a water pump shaft. These pliers have several curved grooves that make up a series of interlocking joints. Furthermore, the length of the handles allows a great deal of force to be applied to the jaws. Interlocking joint pliers are available in lengths from around five inches up to about 20 inches. [Figure 9-15]

VISE-GRIP™ PLIERS

Vise-Grip is the registered trade name of the Petersen Mfg. Co. for a special compound-action type pliers. The opening of these jaws are adjustable by a knurled screw located in the end of the pliers handles. When these handles are squeezed together, compound leverage multiplies the effort and applies a tremendous force to the jaws. A toggle action clamps the jaws together so they will not open when the handles are released. The jaws are released by a small lever in one of the handles.
Vise-Grip pliers come in a wide variety of lengths and jaw styles. Some are designed to hold pipes, some to cut wire, and others to pinch off hoses. Special forms of Vise-Grip pliers are made with

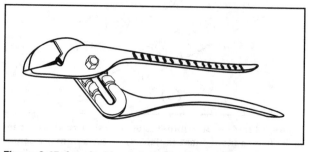

Figure 9-15. Interlocking-joint pliers have various jaw sizes to allow for a variety of grips.

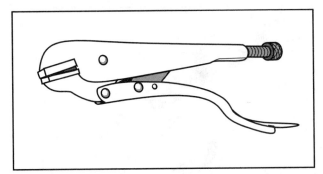

Figure 9-16. One thing you must watch for when using Vise-Grip pliers is that the force applied to the jaws can damage work.

sheet metal bending jaws, while others are made with jaws that serve as welding clamps or C-clamps. [Figure 9-16]

DUCKBILL PLIERS

A special type pliers used in aviation is the duckbill pliers. These long-handled, flat-nose pliers are typically used to twist and help remove safety wire. The jaws of duckbill pliers have serrations to grip safety wire, while the handles are long enough to provide a good tight grip on the wire while it is being twisted. [Figure 9-17]

NEEDLE NOSE PLIERS

Several different designs of needle nose pliers are used in aviation maintenance. Needle nose pliers come in handy for electrical and electronic work because they are typically small enough to grip and hold small components and wires. Some needle nose pliers have long thin jaws that are bent at a right angle to the handle. This allows the pliers to grip a component and hold it without your hand being in the way of the task at hand. [Figure 9-18]

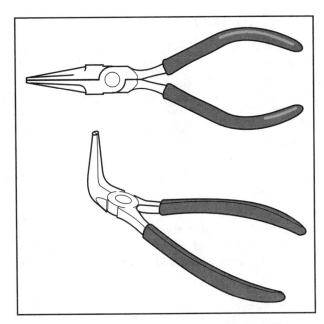

Figure 9-18. Needle nose pliers come in a variety of shapes and sizes and are sometimes spring loaded open or closed to aid in gripping things.

SAFETY WIRE PLIERS

A special tool used by aviation technicians is the safety wire pliers. This tool combines the features of diagonals with those of the duckbills. Safety wire pliers have a special twister built into the handle that quickly and efficiently twists safety wire. To safety wire something, cut the length of wire needed and insert it into whatever is being safetied. The wire is then crossed close to the head of the object being safetied and the ends of wire are gripped with the flat serrated jaws. After locking the handles together, the knob in the pliers handle is pulled and a spiral rod in the handle rotates the pliers, thus twisting the wire. When twisting is completed and the wire is secured, the ends of the wire are cut off with the built-in cutters, and a "pig tail" or curl is put on the end of the wire. [Figure 9-19]

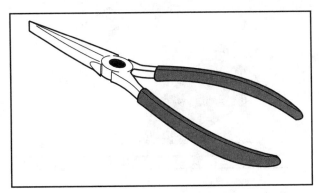

Figure 9-17. Duckbill pliers are an excellent tool for twisting safety wire because they have serrations that grip the wire and their handles are long enough to provide a tight grip.

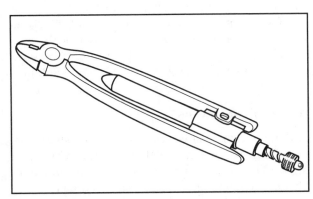

Figure 9-19. Safety wire pliers are used to secure bolt heads or nuts with safety wire.

CUTTING TOOLS

The cutting tools used in aviation maintenance go far beyond the snips and saws that probably come to mind. In fact, cutting tools include any tool that removes or separates material. The specific cutting tools we will discuss in this chapter include chisels, files, shears, saws, drills, hole cutters, reamers, taps, and dies.

CHISELS

The chisel is one of the simplest forms of cutting tool and is made of a high grade tool steel that has been carefully heat-treated and tempered. To utilize the chisel, it is struck with a hammer to shear or cut metal.

FLAT CHISELS

The flat or **cold chisel** is the most common type of chisel used by the aviation technician. Flat chisels are made from square or octagonal stock, ranging from 5/16 inch to 11/16 inch across. The cutting edge of a flat chisel is forged so it is slightly wider than the shank and is ground to an angle of approximately 70 degrees. This angle allows the chisel to cut or shear metal. To use a flat chisel, the cutting edge is held flat against the material, and the depth of the cut is controlled by varying the angle between the chisel and the work. The cutting edge of a flat chisel is ground slightly convex, so that the greatest stress from a hammer blow is directed into the center of the chisel. Thus, the lesser stresses are transmitted to the sides of the cutting edge. [Figure 9-20]

CAPE CHISELS

The cape chisel is forged from the same type of steel as the flat chisel. However, a cape chisel's cutting edge is much narrower and has either a single or double bevel. This bevel makes the cape chisel ideal for cutting keyways and channels. Another common use for cape chisels is to knock the heads off rivets after they have been drilled through. This type chisel is preferred over the wider flat chisel because it is less likely to damage the skin. [Figure 9-21]

DIAMOND-POINT CHISELS

A diamond-point chisel is forged into a sharp-cornered square section and then ground to an acute angle. This forms a cutting edge that is similar in shape to a diamond. These chisels are used to cut V-grooves and sharp corners in square or rectangular grooves. [Figure 9-22]

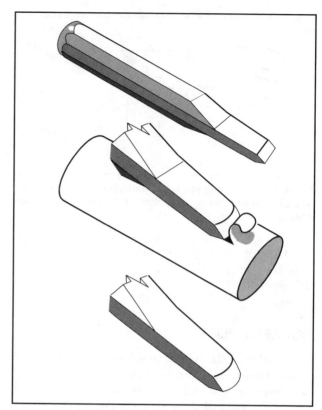

Figure 9-20. A flat or cold chisel has a cutting edge that is ground to about 70 degrees and is slightly convex.

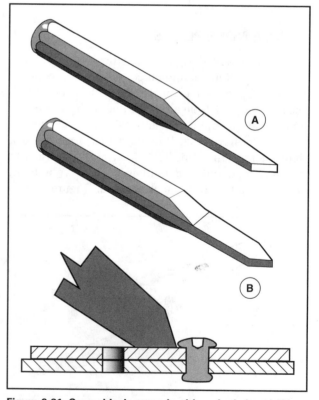

Figure 9-21. Cape chisels come in either single-bevel (A), or double-bevel (B), and are typically used to knock off drilled-through rivet heads.

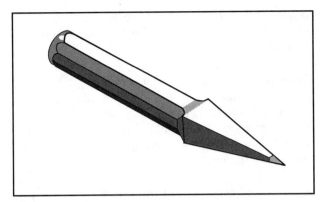

Figure 9-22. The diamond-point chisel gets its name from the shape of its cutting edge.

FILES

Next to the chisel, the most simple cutting tool is the file. A file differs from a chisel in that it has a large number of cutting edges or points rather than a single cutting edge. A file is pushed across a material by hand, and as it moves, the teeth act like small chisels removing small chips of material.

The parts of a file include the tang, heel, face, edge, and point. The **tang** is the end of a file that fits into a handle or is sometimes used as a handle. The **heel**, on the other hand, is that portion of the file above the tang where the main file body begins. The **face** of a file is the long flat surface used as the primary cutting surface. A file's **edge** is the long narrow surface that is perpendicular to the face while the **point** is the end of a file opposite the heel. [Figure 9-23]

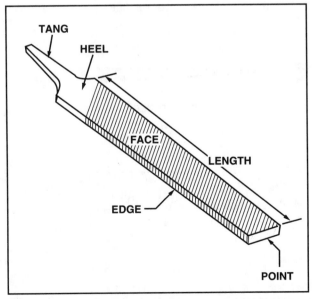

Figure 9-23. Parts of a file include the tang, heel, face, edge, and point. A file's length represents the body length from the heel to the point.

CUT OF A FILE

When referring to the cut of a file, you are actually referring to the number of teeth rows the file has. For example, a **single-cut** file has one row of teeth, while a **double-cut** file has two rows of teeth. The teeth of a single-cut file are cut at an angle between 65 and 85 degrees, and typically produce a smooth material finish. Therefore, most hand files are single cut. With a double-cut file, one set of teeth are set at an angle of about 45 degrees and the other between 70 and 80 degrees. Double-cut files are primarily used to remove relatively large amounts of material. [Figure 9-24]

COARSENESS OF CUT

Generally speaking, double-cut files are used for removing large amounts of material, while single-cut files are used to produce a smooth finish. There are five grades of cut with regard to coarseness. In general, the larger the file, the larger the teeth for any grade of cut. Therefore, a large file is generally more coarse than a small file of the same cut. Cuts of files, when going from the coarsest to the finest are: **coarse cut**, **bastard cut**, **second cut**, **smooth cut**, and **dead smooth cut**.

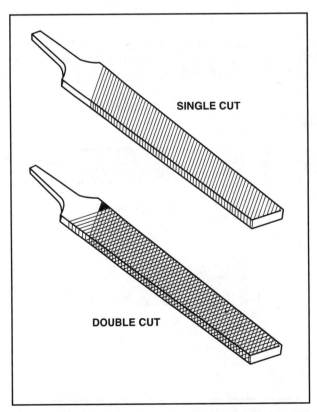

Figure 9-24. This illustration shows typical single-cut and double-cut files.

The **hand file** is uniform in width and tapers in thickness. This type file is used for smoothing flat surfaces and has at least one smooth, or safe edge.

The **half-round file** has one flat and one curved side that tapers in both thickness and width. These files may be single- or double-cut and their shape allows them to be used with several applications.

The **triangular** or three-square file has three tapered cutting sides. This type of file is typically used for filing sharp corners in grooves or keyways and for filing the teeth of handsaws.

The **knife file** is tapered in thickness and width, and has a sharp edge. It is used for filing acute angles.

The **rat-tail** or round file is circular in cross-section and may be tapered or blunt. They come in both single- and double-cut and are typically used for filing small round holes and for cutting a radius in a groove or a keyway.

The **wood rasp** has individual teeth cut into its surfaces rather than rows of teeth. Wood rasps are usually half-round and tapered, and are used to remove wood where a saw or a plane is not practical. The surface they leave is quite rough and needs smoothing with a file or sandpaper.

Vixen files have curved teeth and are used to produce a smooth finish. They slice off very small amounts of material from the surface over which they are moved. Vixen files do not taper in either thickness or width and are often used in a special file holder that slightly arches to allow for more concentrated cutting pressure. Their most common use is in automotive body repair. [Figure 9-25]

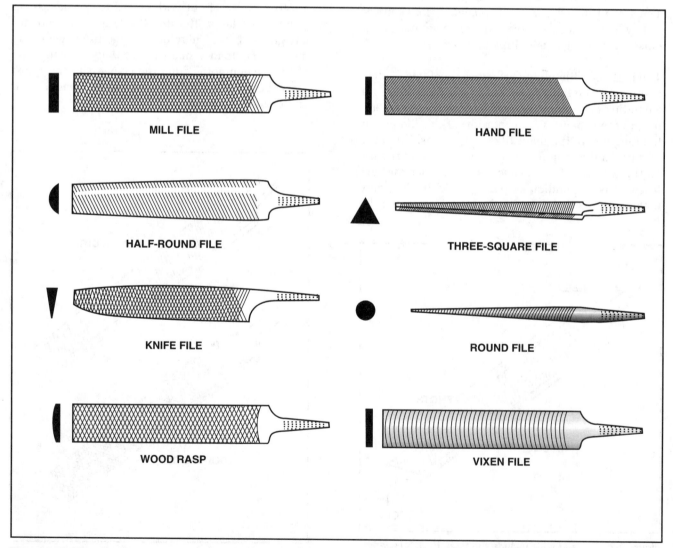

Figure 9-25. Files are classified with regard to their kind, their length, and their cut. The most commonly used files in aviation are the mill file, hand file, half-round file, three-square file, knife file, round file, wood rasp, and the vixen file.

USE OF FILES

A file is a simple cutting tool that is often misused. However, if a few elementary rules are observed in storing and using files, they produce excellent results and have a long service life.

To ensure proper file care, use the correct type file for the job and use a file with a handle. The handle not only protects your hand from being gouged, it also helps you guide the file and control the amount of pressure exerted on the cutting stroke. Secondly, match the length and cut of the file with the type of material being cut and the finish required. This makes completing the job quicker and more efficient.

After a job is completed and a file is ready for storage, it should either be hung on a rack so it does not come in contact with other files or other tools, or it should be wrapped in a paper or plastic envelope to protect its teeth. Furthermore, files should be kept dry and clean to prevent rusting.

To utilize a file properly, the type of material being filed must be taken into consideration. For example, when filing a hard metal, apply pressure only during the forward stroke. Drawing a file back over the material dulls the teeth. However, if filing a very soft metal such as lead or soft aluminum, the file should be drawn back over the material to help remove chips from the teeth.

Generally the weight of a file itself is enough to remove chips from the file teeth. However, if this isn't the case, metal chips can be cleaned out of the teeth using a stiff brush or file card for loose chips and a steel pick for stubborn chips. In either case, keep the file free of chips as they tend to scratch the work surface.

When using a file for cross filing, the handle should be held in the left hand with the thumb on the top of the handle and the fingers below the handle. The right hand should hold the end of the file with the thumb on top of the file and the fingers below it. When the file is held in this manner, it is moved smoothly across the material and pressure is applied only on the forward stroke. [Figure 9-26]

When filing, always use a slow, even stroke, and use the entire length of the file. Also, hold the material being filed firmly in a vise so that the file does not chatter as it moves over the work. If this isn't done, the file tends to dull faster and mar the work. Enough pressure should be held on the file as it moves over the work to prevent slipping. When filing with a

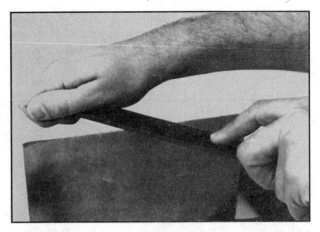

Figure 9-26. When cross filing, the file is moved lengthwise across the work with hands positioned as illustrated.

round file, rotate it with each stroke, and when using a half-round file, use a slicing motion to obtain a smoother cut and to avoid dips and hollows.

An extremely smooth surface is produced by using a smooth or dead-smooth single-cut file and draw filing. In this process, the file is grasped by both ends and is moved sideways across the work. [Figure 9-27]

SHEARS

Shears are another type of cutting tool used on aircraft sheet metal. Long straight cuts across a piece of sheet metal are made on squaring shears which are either foot or power operated. However, the fabrication of a more precision part requires a great deal of hand cutting and fitting. Therefore, the **tinner's snips** or **tin snips** are found in almost every sheet metal shop. These snips range in length from about

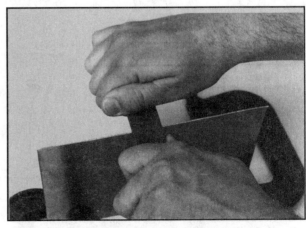

Figure 9-27. In draw filing, the file is moved sideways across the work, with hands positioned as illustrated.

seven inches up to 12 inches and, even though they are basically used for making straight cuts, they are also used to cut curves to either the left or right. [Figure 9-28]

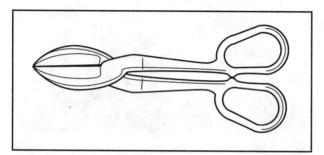

Figure 9-28. Tin snips make straight or curved cuts.

Compound action shears are known throughout the industry as **aviation snips** or **Dutchman snips**. These shears have serrated cutting edges and a leverage multiplication that allows relatively thick sheet metal to be cut without requiring excessive handle force. Aviation snips come in sets of three, one that cuts straight, one that cuts to the right, and one that cuts to the left. The straight snips are color coded with a yellow handle, the right-hand snips have a green handle, and those that cut to the left have a red handle. [Figure 9-29]

DIAGONAL CUTTERS

These long-handled, short-jaw cutters allow you to get into close quarters and clip twisted safety wire.

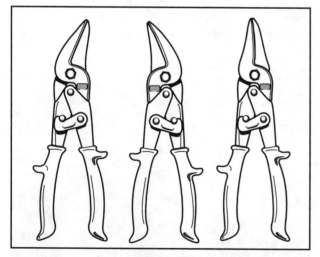

Figure 9-29. In addition to being color-coded, aviation snips can be identified by their shape. For example, a straight snip has a relatively straight nose. However, with right-cut snips, when held in your hand, the lower jaw is on the right whereas the lower jaw is on the left with left-cut snips.

Diagonals, or dikes, are available in extremely small sizes for delicate electronic and instrument work as well as large sizes up to about nine inches in length. [Figure 9-30]

A special form of cutter, quite similar to diagonals is the flush cutter. These cutters leave a perfectly smooth edge on the part they cut, and are often used for cutting the stem off friction-lock rivets or for cutting electrical wire flush with its solder joint.

SAWS

Another type of cutting tool is the saw, and for aviation purposes, saws are divided into two basic types: those designed for cutting wood, and those designed for cutting metal. These saw types are further divided by the number of teeth per inch on their blade, the cut of the blade teeth, and by the type of material used for their blade.

WOOD SAWS

Wood saws are divided into four categories, including crosscut saws, ripsaws, backsaws, and keyhole saws. A handsaw used for cutting across the grain of wood is called a **crosscut saw**. This type of saw typically has from ten to 12 teeth per inch that are filed with alternating knife-like cutting edges. The teeth are set with every other tooth bent outward in the same direction. This offset causes the width of the cut, or the **kerf**, to be wider than the blade to prevent binding. [Figure 9-31]

Ripsaws have fewer teeth than crosscut saws, usually from six to eight and a half teeth per inch. These teeth are cut with an almost straight leading edge while the back tapers. Furthermore, the teeth are filed flat across the front so they act as chisels

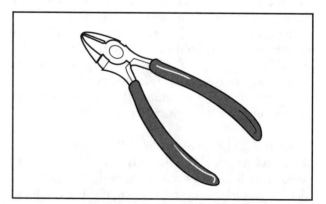

Figure 9-30. Diagonal cutters are commonly used to clip twisted safety wire, and often a version of these cutters are found on needle nose pliers at their pivot point.

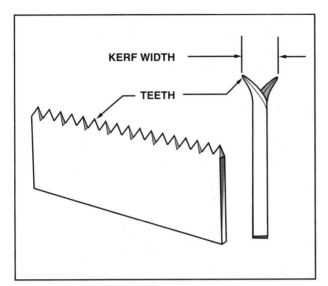

Figure 9-31. The teeth of a crosscut saw act as knives to cut through wood fibers. The kerf of the teeth dictates the width of the cut.

rather than knives and, therefore, require a much smaller amount of offset than that of a crosscut saw. [Figure 9-32]

Backsaws are used for cutting precision parts such as those used in making wooden wing ribs. The blades of a backsaw are stiffened with a steel back so that they cut very straight. Backsaws have from 18 to 32 teeth per inch and very little offset, so that they make a smooth cut. [Figure 9-33]

Keyhole saws are commonly used to start a hole or cut a circular hole. To start a hole, the thin tapered blade is inserted into a drilled hole and the cut is

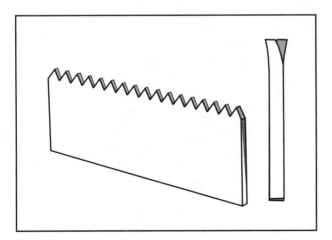

Figure 9-32. The teeth of a ripsaw act as chisels to dig into the wood fibers.

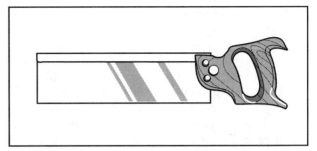

Figure 9-33. Backsaws cut very straight and are therefore used for cutting precision parts.

made as needed. Most keyhole saws have replaceable blades that range from 24 teeth per inch for cutting plastics to eight teeth per inch for cutting wood. [Figure 9-34]

METAL SAWS OR HACKSAWS

The hacksaw is the most widely used metal-cutting hand saw. It consists of an adjustable frame that holds a small flexible steel blade under tension. Hacksaw blades are typically made of either carbon steel or high-speed molybdenum steel. However, when made with high-speed molybdenum the saw teeth must be hardened and the back of the blade annealed. Although the high-speed molybdenum blades are considerably more costly than the carbon steel blades, they typically have a much longer life.

Hacksaw blades are available in both 10- and 12-inch lengths, and have from 18 to 32 teeth per inch. When cutting a piece of soft, narrow metal, it is best to use a blade with as few teeth as possible. This results in a fast cut with plenty of chip clearance. However, for harder materials, a blade with more teeth is needed for a good cut. When cutting thin materials, or odd-shaped materials having thin sections, a blade should be chosen that allows at least two teeth to be on the metal at all times. When cutting thin wall tubing, there should be at

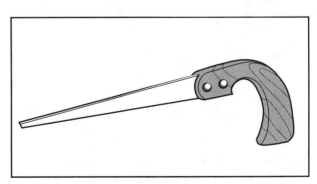

Figure 9-34. A keyhole saw has a thin tapered blade that allows a large hole to be cut when only a small hole exists.

least two teeth on each side of the tube as it is cut. [Figure 9-35]

To properly use a hacksaw, long, slow, steady strokes are made so that as much of the blade is used as possible. As with many cutting tools, all the cutting is done on the forward stroke and the blade is lifted for the return stroke. Short, fast strokes with equal pressure on both the forward and backward stroke rapidly dulls the blade and may cause it to break.

DRILLS

There are generally five types of portable drills used in aviation for holding and turning twist drills. These include the hand drill, the breast drill, the electric drill, the pneumatic drill, and the cordless drill. The popular "egg beater" style hand drill is typically only used to drill holes 1/4 inch and under. The breast drill, on the other hand, is designed to hold larger twist drills than the hand drill. Thus, holes over 1/4 inch are typically drilled with a breast drill. Electric and pneumatic drill motors are available in a variety of types and sizes to suit any job. However, electric drill motors should not be used around flammable materials because they present a fire hazard. Pneumatic drills are better suited for use around flammable materials as they present little chance of spark. Newer battery-powered drill motors, often called cordless drills, offer power and more freedom than electric or pneumatic drills. However, they still should not be used near flammable materials.

WOOD BITS

There are three types of wood bits used in aviation maintenance. They include the auger bit, the Forstner bit, and the flat wood-boring bit. An **auger bit** is a spiral-type bit turned by a bow type brace. These bits are available in 1/16 inch increments, from 1/4 inch to one inch diameter, and range in length from about seven to ten inches. The distinguishing feature of an auger bit is that it utilizes a feed screw as a point that pulls the cutting edges into the wood. The chips that are cut loose when the larger cutting angle comes into contact with the wood, are carried out of the hole by the spiral angled flutes.

The **Forstner bit** has a circular boring end the same size as the hole to be bored. These bits are designed for drilling flat bottomed holes and have one shearing edge to remove chips, with cutting edges on the outside edge of the boring end. [Figure 9-36]

Large holes are typically drilled with an electric drill motor with a **flat wood-boring bit**. These bits resemble a large flat paddle with a tapered pilot in the center of the cutting edge. [Figure 9-37]

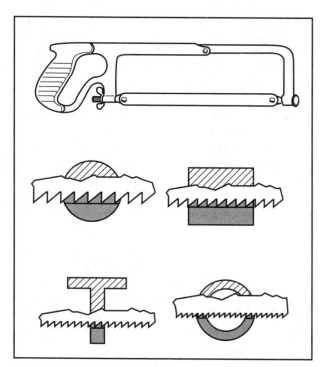

Figure 9-35. When choosing a hacksaw blade you should select one that allows at least two teeth to rest on the work at all times.

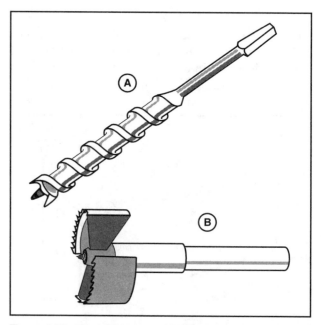

Figure 9-36. (A) — The auger bit is designed to drill holes through a material. (B) — The Forstner bit, on the other hand, is designed to drill flat-bottomed holes.

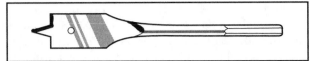

Figure 9-37. A tapered pilot on a flat wooden boring bit aids in keeping the cutting edges centered in the hole.

METAL DRILLS

Metal drills are made of carbon steel, high-speed steel, and steel alloys. Carbon steel drills typically cost less, however, they tend to overheat more readily thereby losing their hardness. Alloy steels such as carbide are used to make drills designed for special applications. For example, solid carbide drills are used for drilling very abrasive materials like printed circuit boards.

A twist drill is made up of three parts: the shank, the body, and the point. The **shank** is the portion held in a chuck. Drills used in aircraft maintenance all have straight shanks, while drills turned by large drill presses typically have tapered shanks. The **body** of a twist drill has spiral **flutes** milled from the point to the shank. These flutes carry chips being cut by the cutting edge out of the hole, as well as carry lubricants to the material and cutting edges. The material between the flutes is called the **land** and the land nearest the cutting edge of the point that is the full drill size is called the **margin**. Immediately behind the margin, the land is ground away slightly to provide body clearance. [Figure 9-38]

Almost all difficulties encountered in drilling holes with twist drills is attributed to a dull or improperly ground point. As an aviation maintenance technician, you should be able to properly grind a new drill point. However, there are machine shops equipped with grinding fixtures that can quickly sharpen a drill point. The best way to ensure a properly sharpened point is to use a grinding fixture that grinds two cutting edges that are exactly the same length, and that meet in the exact center of the drill. [Figure 9-39]

For most materials, cutting edges are ground at an angle of 59 degrees from the central axis of the drill, which gives an included angle of 118 degrees. For drilling soft material such as lead, wood, and very soft aluminum, this angle is steepened to approximately 90 degrees. Hard materials such as stainless steel or heat-treated steel require a flat angle of about 150 degrees. Furthermore, the point is ground with a lip or heel relief angle ranging from about 12 to 15 degrees. However, for very soft materials, this angle is usually increased to approximately 18 to 20 degrees.

A common drill used in aircraft sheet metal fabrication and repair is the 135 degree split point. Commonly called aircraft drills, these drills work best on aluminum alloy material. They provide a good starting point with very little burring around

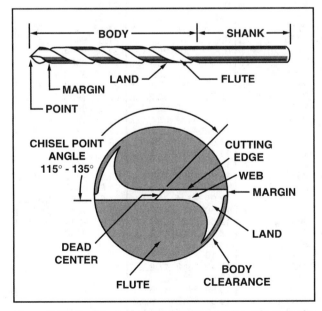

Figure 9-38. On a typical metal drill, the cutting edges on the point perform the actual cutting. The lands and flutes remove the cut material and carry lubricant to the cutting edges.

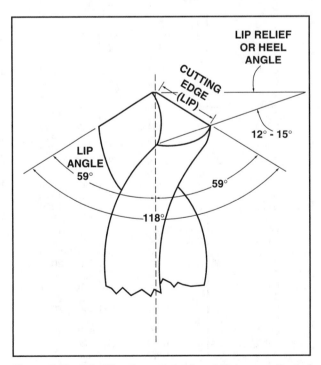

Figure 9-39. Grinding the point of a drill is complicated because of the many angles involved. Because of this, it is best to use a grinding fixture whenever possible.

the exit hole. The 135 degree split point drill works very well for drilling out old rivet heads in repair or replacement work.

DRILL SIZES

There are three methods used to indicate drill sizes: the number, fractional, and letter. **Number drills** range in size from 0.0135 inch for the number 80 to 0.2280 inch for the number 1 drill. **Fractional drills** are available in sets from 1/64 inch (0.0156) to 1/2 inch (0.500). Drill sizes larger than 1/2 inch are typically available individually and are not normally available in sets. **Letter drill** sizes are all larger than number sizes and range from A (0.2340) to the Z (0.4130). The only drill size available in two sets is the 0.2500 inch drill which is the letter E drill and the 1/4 inch drill. [Figure 9-40]

Milli-meter	Dec. Equiv.	Frac-tional	Num-ber	Milli-meter	Dec. Equiv.	Frac-tional	Num-ber	Milli-meter	Dec. Equiv.	Frac-tional	Num-ber	Milli-meter	Dec. Equiv.	Frac-tional	Num-ber	Milli-meter	Dec. Equiv.	Frac-tional	Num-ber	Milli-meter	Dec. Equiv.	Frac-tional	Num-ber
.1	.0039			1.45	.0570			3.2	.1260			5.4	.2126				.3230		P	14.5	.5709		
.15	.0059			1.5	.0591			3.25	.1279				.2130		3	8.25	.3248			14.68	.5781	37/64	
.2	.0079				.0595		53		.1285		30	5.5	.2165			8.3	.3268			15.0	.5906		
.25	.0098			1.55	.0610			3.3	.1299			5.56	.2187	7/32		8.33	.3281	21/64		15.08	.5937	19/32	
.3	.0118			1.59	.0625	1/16		3.4	.1338			5.6	.2205			8.4	.3307			15.48	.6094	39/64	
	.0135		80	1.6	.0629				.1360		29		.2210		2		.3320		Q	15.5	.6102		
.35	.0138				.0635		52	3.5	.1378			5.7	.2244			8.5	.3346			15.88	.6250	5/8	
	.0145		79	1.65	.0649				.1405		28	5.75	.2263			8.6	.3386			16.0	.6299		
.39	.0156	1/64		1.7	.0669			3.57	.1406	9/64			.2280		1		.3390		R	16.27	.6406	41/64	
.4	.0157				.0670		51	3.6	.1417			5.8	.2283			8.7	.3425			16.5	.6496		
	.0160		78	1.75	.0689				.1440		27	5.9	.2323			8.73	.3437	11/32		16.67	.6562	21/32	
.45	.0177				.0700		50	3.7	.1457				.2340		A	8.75	.3445			17.0	.6693		
	.0180		77	1.8	.0709				.1470		26	5.95	.2344	15/64		8.8	.3465			17.06	.6719	43/64	
.5	.0197			1.85	.0728			3.75	.1476			6.0	.2362				.3480		S	17.46	.6875	11/16	
	.0200		76		.0730		49		.1495		25		.2380		B	8.9	.3504			17.5	.6890		
	.0210		75	1.9	.0748			3.8	.1496			6.1	.2401			9.0	.3543			17.86	.7031	45/64	
.55	.0217				.0760		48		.1520		24		.2420		C		.3580		T	18.0	.7087		
	.0225		74	1.95	.0767			3.9	.1535			6.2	.2441			9.1	.3583			18.26	.7187	23/32	
.6	.0236			1.98	.0781	5/64			.1540		23	6.25	.2460		D	9.13	.3594	23/64		18.5	.7283		
	.0240		73		.0785		47	3.97	.1562	5/32		6.3	.2480			9.2	.3622			18.65	.7344	47/64	
	.0250		72	2.0	.0787				.1570		22	6.35	.2500	1/4	E	9.25	.3641			19.0	.7480		
.65	.0256			2.05	.0807			4.0	.1575			6.4	.2520			9.3	.3661			19.05	.7500	3/4	
	.0260		71		.0810		46		.1590		21	6.5	.2559				.3680		U	19.45	.7656	49/64	
.7	.0276				.0820		45		.1610		20		.2570		F	9.4	.3701			19.5	.7677		
	.0280		70	2.1	.0827			4.1	.1614			6.6	.2598			9.5	.3740			19.84	.7812	25/32	
	.0292		69	2.15	.0846			4.2	.1654				.2610		G	9.53	.3750	3/8		20.0	.7874		
.75	.0295				.0860		44		.1660		19	6.7	.2638				.3770		V	20.24	.7969	51/64	
	.0310		68	2.2	.0866			4.25	.1673			6.75	.2657	17/64		9.6	.3780			20.5	.8071		
.79	.0312	1/32		2.25	.0885			4.3	.1693			6.75	.2657			9.7	.3819			20.64	.8125	13/16	
.8	.0315				.0890		43		.1695		18		.2660		H	9.75	.3838			21.0	.8268		
	.0320		67	2.3	.0905			4.37	.1719	11/64		6.8	.2677			9.8	.3858			21.03	.8281	53/64	
	.0330		66	2.35	.0925				.1730		17	6.9	.2716				.3860		W	21.43	.8437	27/32	
.85	.0335				.0935		42	4.4	.1732				.2720		I	9.9	.3898			21.5	.8465		
	.0350		65	2.38	.0937	3/32			.1770		16	7.0	.2756			9.92	.3906	25/64		21.83	.8594	55/64	
.9	.0354			2.4	.0945			4.5	.1771				.2770		J	10.0	.3937			22.0	.8661		
	.0360		64		.0960		41		.1800		15	7.1	.2795				.3970		X	22.23	.8750	7/8	
	.0370		63	2.45	.0964			4.6	.1811				.2811		K		.4040		Y	22.5	.8858		
.95	.0374				.0980		40		.1820		14	7.14	.2812	9/32		10.32	.4062	13/32		22.62	.8906	57/64	
	.0380		62	2.5	.0984			4.7	.1850		13	7.2	.2835				.4130		Z	23.0	.9055		
	.0390		61		.0995		39	4.75	.1870			7.25	.2854			10.5	.4134			23.02	.9062	29/32	
1.0	.0394				.1015		38	4.76	.1875	3/16		7.3	.2874			10.72	.4219	27/64		23.42	.9219	59/64	
	.0400		60	2.6	.1024			4.8	.1890		12		.2900		L	11.0	.4330			23.5	.9252		
	.0410		59		.1040		37		.1910		11	7.4	.2913			11.11	.4375	7/16		23.81	.9375	15/16	
1.05	.0413			2.7	.1063			4.9	.1929				.2950		M	11.5	.4528			24.0	.9449		
	.0420		58		.1065		36		.1935		10	7.5	.2953			11.51	.4531	29/64		24.21	.9531	61/64	
	.0430		57	2.75	.1082				.1960		9	7.54	.2968	19/64		11.91	.4687	15/32		24.5	.9646		
1.1	.0433			2.78	.1094	7/64		5.0	.1968			7.6	.2992			12.0	.4724			24.61	.9687	31/32	
1.15	.0452				.1100		35		.1990		8		.3020		N	12.30	.4843	31/64		25.0	.9843		
	.0465		56	2.8	.1102			5.1	.2008			7.7	.3031			12.5	.4921			25.03	.9844	63/64	
1.19	.0469	3/64			.1110		34		.2010		7	7.75	.3051			12.7	.5000	1/2		25.4	1.0000	1	
1.2	.0472				.1130		33	5.16	.2031	13/64		7.8	.3071			13.0	.5118						
1.25	.0492			2.9	.1141				.2040		6	7.9	.3110			13.10	.5156	33/64					
1.3	.0512				.1160		32	5.2	.2047			7.94	.3125	5/16		13.49	.5312	17/32					
	.0520		55	3.0	.1181				.2055		5	8.0	.3150			13.5	.5315						
1.35	.0531				.1200		31	5.25	.2067				.3160		O	13.89	.5469	35/64					
	.0550		54	3.1	.1220			5.3	.2086			8.1	.3189			14.0	.5512						
1.4	.0551			3.18	.1250	1/8			.2090		4	8.2	.3228			14.29	.5625	9/16					

Figure 9-40. Drill sizes are given in number, fractional, and letter form. Notice that the 0.2500 decimal equivalent is the only size that has both a letter and a fractional form.

Since it is often difficult to tell the exact size of a given drill there are several commercially produced gauges available that simplify this task. [Figure 9-41]

USE OF TWIST DRILLS

When drilling metal with either a hand drill or drill press, it is important that eye protection be used at all times. Also, when drilling steel either a soluble oil or lard must be used to lubricate the cutting edge. Lubrication distributes the heat generated when metal is cut. However, aluminum is typically drilled without the use of any lubricant.

One of the most critical aspects of drilling is the use of the correct drill speed. As a general rule, harder material should be drilled at a slower speed whereas softer metal is drilled at a higher speed. For example, when drilling a 1/4 inch hole in a piece of aluminum, the proper drill speed is about 2,500 rpm. However, when drilling this same size hole in stainless steel, a drill speed of around 500 rpm is used so the drill does not overheat and burn. [Figure 9-42]

	ALUMINUM	BRASS	TOOL STEEL	STAINLESS STEEL
HOLE SIZE (INCH)	DRILL SPEED (RPM)			
1/8	5000	2500	2000	1000
1/4	2500	1200	1000	500
1/2	1200	600	500	250
1	600	300	250	125

Figure 9-42. In the drill speed chart above, you can see the relationship between hole size, drill speed, and the type of material used.

HOLE CUTTERS

For cutting holes in sheet metal that are larger than what can be drilled or for countersinking, a hole cutter is used. There are three types of hole cutters that are used in the aviation industry. They are the hole saw, the fly cutter, and the countersink. The **hole saw** or circular saw is used for cutting large holes in sheet metal. This saw is available in diameters from 9/16 inch up to 4 1/8 inch. The arbor for this saw has a round shank that fits into a standard 1/4 inch electric or air drill and has a special pilot drill in the center. The pilot drill has a very short body and a long shank so that the smooth shank is in the metal when the saw is cutting. This helps keep the pilot hole from becoming enlarged and holds the saw centered. [Figure 9-43]

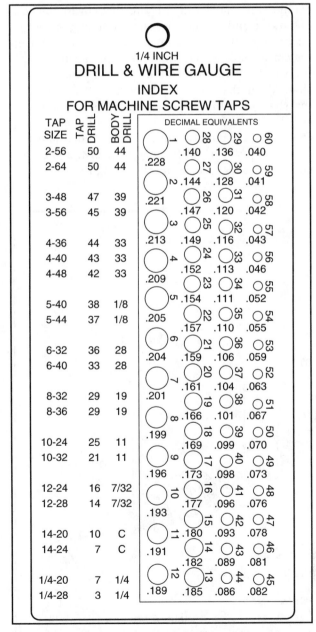

Figure 9-41. Most drill gauges have holes in which you insert a drill in the appropriate hole. The gauge illustrated above also indicates tap sizes so you can drill the proper size hole for a tap.

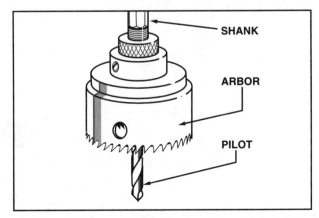

Figure 9-43. Prior to using a hole saw you need to mark the center of the hole so you know where to start the pilot bit.

The **fly cutter** is an adjustable hole cutter typically used with a drill press. This cutting tool is set so that its radius from the center of the pilot drill is exactly the radius of the hole. When using a fly cutter, the pilot drills out the center of the hole and the cutting tool is fed slowly after it. [Figure 9-44]

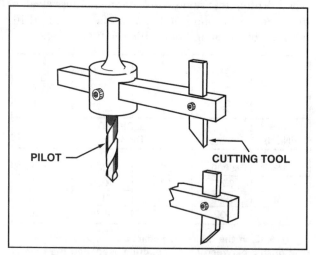

Figure 9-44. On a fly cutter, the cutting tool is reversible in its holder so that the edge of the hole being cut can have either a straight or beveled edge.

When using a fly cutter, it is extremely important to back the metal being cut with scrap wood. This scrap wood is firmly clamped to the table of the drill press so that there is no danger of the cutter grabbing the metal and spinning it. If the cutter is fed into the metal too fast, there is danger of the cutter grabbing the metal instead of cutting it.

A **countersink** is used to cut away small portions of metal so the head of a rivet can sit flush with the surface. Almost all flush rivets used in aircraft construction have a head angle of 100 degrees. Therefore, the most generally used countersink has a 100 degree head with an adjustable stop. Some countersinks have a fiber collar that covers the end of the stop and prevents it from scratching the work. To use the adjustable stop countersink, a cutter of the correct size is placed on the cutter shaft and the proper size pilot is inserted into the center of the shaft. Then, the body and the stop are screwed onto the shaft, and the entire assembly is put into the chuck of a 1/4 inch air or electric drill motor. Next, trial holes are countersunk in scrap metal and the stop is adjusted until the diameter of the countersunk hole is correct for the rivet. [Figure 9-45]

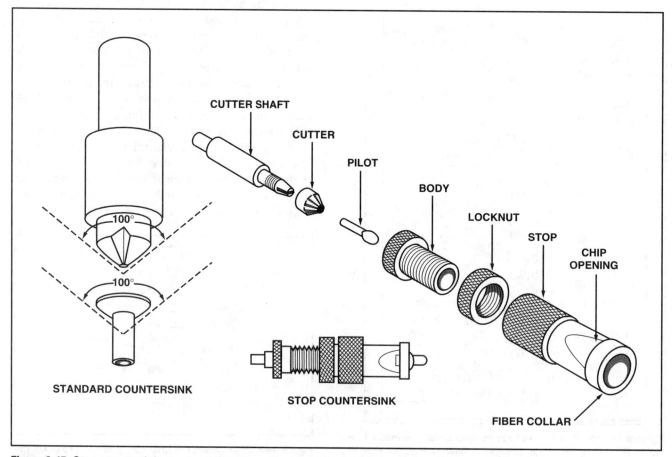

Figure 9-45. Stop countersinks are used to recess a rivet head flush with the material being riveted.

REAMERS

Drills cannot make holes with sufficient accuracy for certain applications. However, if a hole is drilled undersized and then reamed to a finished size, extremely accurate hole sizes are obtained. For example, if a structure is to be held together with both rivets and bolts, no relative motion can exist between the bolt and its hole. For this reason, close tolerance bolts are used in holes that require a light press fit to get the bolt into the hole.

Reamers are delicate tools that have hardened and sharpened cutting edges. When a reamer is fed into a hole it is turned in the direction that allows a cut. After the hole is reamed to the proper size, the reamer must continue to turn in the cutting direction to remove it from the hole. When drilling a pilot hole for a reamer, the hole should be about 0.001 to 0.003 inch smaller than the reamer. This allows the reamer to cut rather than slide through the hole. When a reamer is turned without cutting, it is quickly dulled. There are two basic types of hand reamers used by the aviation technician. They include the **solid reamer** and the **adjustable** or **expansion reamer**. [Figure 9-46]

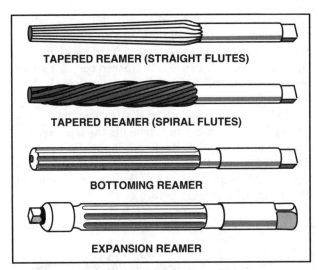

Figure 9-46. With a solid reamer such as the tapered reamer and bottoming reamer only one size hole can be cut. However, with an expansion reamer several different size holes can be cut with the same reamer.

TAPS

Taps are a form of cutting tool used to cut threads on the inside of a hole. Like drills, some are made of high carbon steel or high speed steel, while others contain special alloys. The tap sizes most commonly used are available for cutting both National Fine (NF) and National Coarse (NC) threads. [Figure 9-47]

NATIONAL FINE THREAD SERIES MEDIUM FIT. CLASS 3 (NF)				NATIONAL COARSE THREAD SERIES MEDIUM FIT. CLASS 3 (NC)			
SIZE AND THREADS	DIAMETER OF BODY FOR THREAD	BODY DRILL	TAP DRILL SIZE	SIZE AND THREADS	DIAMETER OF BODY FOR THREADS	BODY DRILL	TAP DRILL SIZE
0-80	.060	52	3/64	1-64	.073	47	#53
1-72	.073	47	#53	2-56	.086	42	#51
2-64	.086	42	#50	3-48	.099	37	5/64
3-56	.099	37	#46	4-40	.112	31	#44
4-48	.112	31	#42	5-40	.125	29	#39
5-44	.125	25	#38	6-32	.138	27	#36
6-40	.138	27	#33	8-32	.164	18	#29
8-36	.164	18	#29	10-24	.190	10	#26
10-32	.190	10	#21	12-24	.216	2	#17
12-28	.216	2	#15	1/4-20	.250	1/4	#7
1/4-28	.250	F	#3	5/16-18	.3125	5/16	#F
5/16-24	.3125	5/16	I	3/8-16	.375	3/8	5/16
3/8-24	.375	3/8	Q	7/16-14	.4375	7/16	U
7/16-20	.4375	7/16	W	1/2-13	.500	1/2	27/64
1/2-20	.500	1/2	7/16	9/16-12	.5625	9/16	31/64
9/16-18	.5625	9/16	1/2	5/8-11	.625	5/8	17/32
5/8-18	.625	5/8	9/16	3/4-10	.750	3/4	41/64
3/4-16	.750	3/4	11/16	7/8-9	.875	7/8	49/64
7/8-14	.875	7/8	51/64	1-8	1.000	1.0	7/8
1-14	1.000	1.0	49/64				

Figure 9-47. A typical chart shows both the body drill and the tap drill sizes for standard screw threads from size 0-80 up through one inch.

The three most commonly used taps in aviation maintenance are the taper tap, the plug tap, and the bottoming tap. The **taper tap** is used to begin the tapping process, because it is tapered back for six to seven threads. This tap cuts a complete thread when it is cutting above the taper and is the only tap needed when tapping holes that extend through a material. The **plug tap** supplements the taper tap for tapping holes in thick stock, but tapers for only the first three to five threads. The **bottoming tap** is not tapered and is used to cut full threads to the bottom of a blind hole. [Figure 9-48]

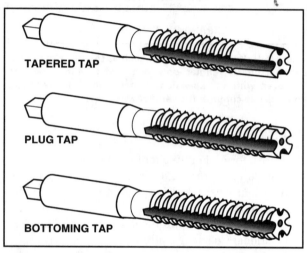

Figure 9-48. The three most common taps include the tapered tap, the plug tap, and the bottoming tap.

Small taps are held with a T-handle tap wrench, while larger taps are typically held with a solid handle tap wrench. Both handles require two hands to turn them. A wrench is usually not used to turn a tap because pressure cannot be applied evenly to both sizes of the tap, making it difficult to tap a straight hole. [Figure 9-49]

DIES

Threading dies are used to cut threads on the outside of rods, bolts, and pipes. A die is typically either round or hexagonal in shape with a round hole through the center that is lined with cutting teeth. Additional holes are provided so that the cut metal chips can fall out.

There are two basic types of dies. The most common is the solid hexagonal rethreading die, which is sometimes called a thread restorer, or thread chaser. This type of die is screwed onto a damaged male thread with the appropriate size wrench. The die straightens any flattened threads and removes a minimum amount of material.

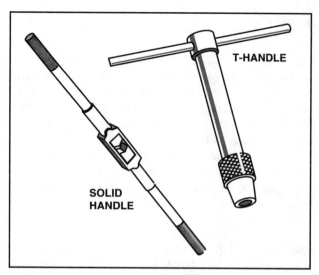

Figure 9-49. When using either the T-handle or the solid handle to turn a tap, make sure to apply even pressure to both sides of the handle.

The second type of die is the **split die** which is round and must be held in a die stock to use. Most round dies have an adjusting screw that allows you to make small adjustments depending on the class of fit desired. For example, when the spring action of the die forces the sides together, a loose fit results. However, when the die is spread apart with the adjusting screw, a tight fit is cut. [Figure 9-50]

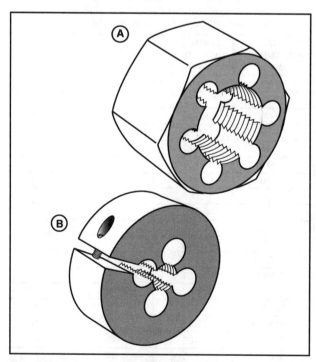

Figure 9-50. (A) — The hexagonal die is typically used to rethread or clean existing threads and, therefore, it is turned with a wrench. (B) — The round split die, on the other hand, is used to cut new threads and must be held in a die stock.

TURNING TOOLS

Turning tools permit the user to install and remove threaded fasteners and include such common tools as screwdrivers, wrenches, and sockets. These tools are probably the most commonly used and, therefore, you should have a good understanding of how to properly use each.

SCREWDRIVERS

A screwdriver is the most familiar and, often times, the most misused tool. The screwdriver's handy shape and wide assortment of sizes makes it tempting to use as a punch, a chisel, or a pry bar. However, screwdrivers should only be used for their intended purpose — to turn screws.

Screwdrivers are available with both wood and high-impact plastic handles. The blades come in a wide variety of shank and blades sizes. Standard screwdrivers are typically available with shanks from about 1 1/2 inches up to lengths of 10 to 12 inches. Special application screwdrivers are available with blades up to 20 inches long. [Figure 9-51]

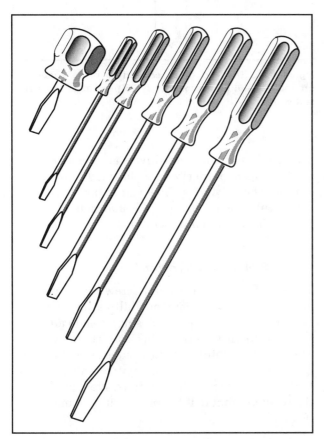

Figure 9-51. Screwdrivers are available with several shank lengths ranging from 1 1/2 inches to about 12 inches. However, some special-purpose screwdrivers may have a shank length of 20 inches and more.

Offset screwdrivers are available for turning screws where there is no clearance for a straight screwdriver. There are two types of offsets, one with two blades at right angles to each other, and one with four blades oriented in 45 ° increments. [Figure 9-52]

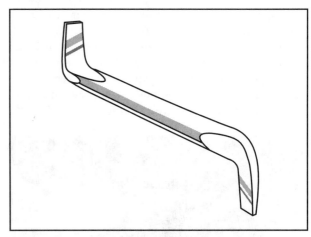

Figure 9-52. An offset screwdriver like the one shown here allows screws to be removed and tightened when there is no clearance for a straight screwdriver.

Two types of screwdrivers that are useful for line technicians working away from their toolboxes are the **reversible-blade** screwdriver and the **interchangeable head** screwdriver. The reversible-blade screwdriver has a regular slotted blade on one end, and a blade that fits a recessed-head screw on the other end. The interchangeable head screwdriver has a hollow magnetized shaft that holds 1/4-inch hex bits. The bits come in many types and sizes and typically fit inside the hollow handle of the screwdriver for storage. [Figure 9-53]

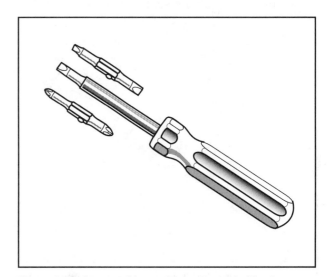

Figure 9-53. A screwdriver with replaceable bits is a very handy tool for aircraft maintenance technicians working away from their tool boxes.

Most airplanes have stressed inspection panels held on with many countersunk recessed-head screws. On each inspection, these screws must be removed and replaced. The time involved to complete this process makes this a major part of an inspection. To help decrease the time spent removing inspection panels, most shops are equipped with air or battery powered screwdrivers. These tools accept a standard 1/4-inch screwdriver bit. Many of these screwdrivers allow you to adjust the amount of torque applied to a screw. Once the preset torque value is reached, a chuck slips inside the screwdriver preventing the screw from being over-torqued. [Figure 9-54]

Figure 9-54. Prior to using an air-driven screwdriver you should make sure you understand how to set the proper torque.

For shops that do not have air or battery powered screwdrivers, a Yankee ratchet-type screwdriver serves as a fair substitute. This screwdriver accepts replaceable bits that are spun by working the handle in and out over its spiral shaft. [Figure 9-55]

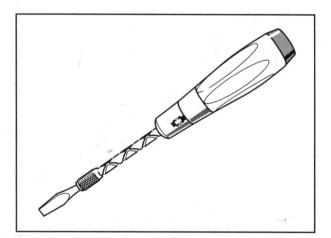

Figure 9-55. A Yankee spiral-type screwdriver operates by pushing the handle down over its shaft. This tightens or loosens the screw depending on the screwdriver setting.

SLOT SCREWDRIVERS

When inserting or removing slotted screws the screwdriver blade should fill at least 75 percent of the screw slot width. For a screwdriver to be effective, it should be kept sharp, with the side of the blade entering the screw slot parallel to the base of the screw slot. A dull screwdriver blade or one that is ground improperly can slip out of the screw slot and damage the work. [Figure 9-56]

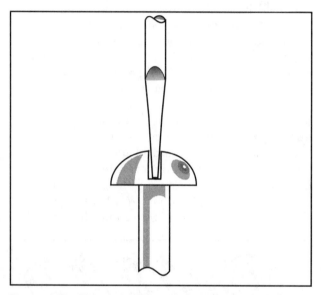

Figure 9-56. The end of a screwdriver blade should be ground flat and fill approximately 75 percent of the screw slot width.

The shanks of some screwdrivers are square or have a hex-shaped bolster where the blade enters the handle. These features allow you to turn the screwdriver with a wrench to provide additional leverage in torquing or loosening screws.

CROSS-POINT SCREWDRIVERS

The two types of cross-point screwdrivers used in aviation maintenance are the Phillips, and the Reed and Prince. These types look similar and care must be taken to ensure use of the correct screwdriver. The cross in a Phillips screw head is cut with a double taper and the sides are not exactly parallel. Furthermore, the Phillips screwdriver has a relatively blunt point that fits into a flat-bottomed hole. [Figure 9-57]

The Reed and Prince screw slot has straight sides and forms a more perfect cross than the Phillips screwdriver. There is also only one taper to the

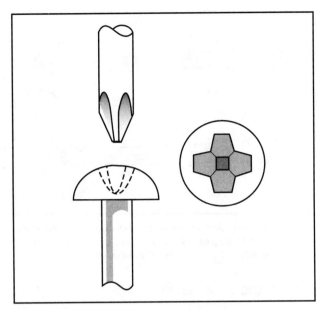

Figure 9-57. A Phillips head screwdriver has a relatively blunt point that fits into a flat-bottomed hole.

point of a Reed and Prince screwdriver and it has a smaller width than the Phillips. [Figure 9-58]

TRI-WING™

Some airlines use screws with a special screw head produced by the Phillips Screw Company. These screws have three slots instead of the four slots found on the regular cross-point screw. The screw head and bit are identified by their registered trade name **TRI-WING**. [Figure 9-59]

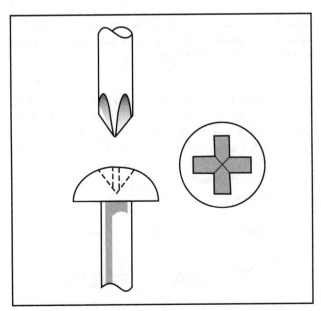

Figure 9-58. The Reed and Prince screwdriver has a perfect cross on its tip instead of the taper a Phillips screwdriver has.

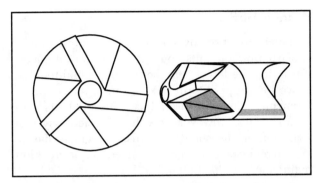

Figure 9-59. The Tri-Wing screw and bit have three wings instead of a straight slot or cross shaped tip.

POSIDRIV™

A Posidriv bit resembles the Phillips bit in that it has a cross shape. However, there is a significant difference in the bottom of the recess that allows the bit to interlock with the screw head. This design is produced by the Phillips Screw Company and provides a tighter and more positive connection between the screwdriver tip and the recess in the screw. [Figure 9-60]

TORX™

The Torx screwdriver is used to remove and install torx-type screws which have a six-pointed slot. The Torx type screwdriver should not be used for an Allen type screw. Using the wrong type of screwdriver can damage both the screw and the screwdriver.

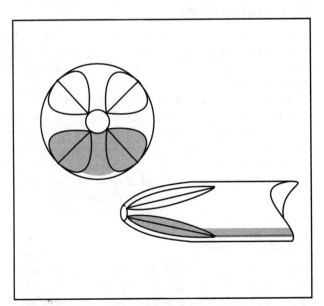

Figure 9-60. The Posidriv point provides a tight connection between the screwdriver tip and the recess in the screw, thereby minimizing the chance of the screwdriver slipping out of the screw head.

WRENCHES

The majority of removable fasteners used in aircraft construction are bolts and nuts. Therefore it makes sense that the largest number of individual tools in a well-equipped tool box is an assortment of wrenches. American aircraft use nuts and bolts with English sizes. However, now that the metric system is becoming the standard across the globe, metric size wrenches may also be needed for aviation maintenance. [Figure 9-61]

ENGLISH (INCHES)	METRIC (mm)
1/4-5/16	6-7
3/8-7/16	8-9
1/2-9/16	10-11
5/8-11/16	12-13
3/4-13/16	15-18
7/8-15/16	21-24
1-1 1/8	

Figure 9-61. Although most of the fasteners used on aircraft are in English sizes, you may come across some metric hardware. The above is a chart of the most common sizes found.

Good quality wrenches are forged of chrome-vanadium steel. This combination of metals is an extremely tough alloy. After being forged, burrs are removed and then the wrenches are plated with cadmium or hard chrome to protect them from rust. The plating process improves the wrench's appearance and makes it easier to clean.

OPEN-END WRENCHES

Open-end wrenches have an opening in each end that fits a bolt head or nut. The openings of an open-end wrench are parallel to each other and are normally angled at 15 degrees to the handle. This angle allows you to turn a nut even when the space for the handle is severely restricted. However, while the 15 degree head angle is standard, there are many other angles available. For example, one type of open-end wrench has the same size opening on both ends with one opening angled at 30 degrees and the opposite opening at 60 degrees. [Figure 9-62]

BOX-END WRENCHES

Nuts that are exceptionally tight can spread the jaws on even the best open-end wrench. To break the torque on tight nuts a box-end wrench is used. Box-end wrenches have a six- or twelve-point opening attached to each end and offset from the axis of the handle by about 15 degrees. [Figure 9-63]

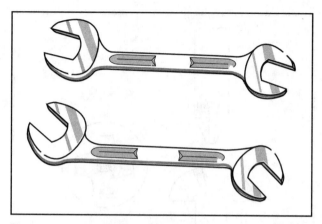

Figure 9-62. The end of an open-end wrench is typically attached at a 15 degree angle. However, on some wrenches the head is attached at a 30 to 60 degree angle.

COMBINATION WRENCHES

The disadvantage of a box-end wrench is the limitation of always having to lift and reposition the wrench in order to continue loosening a fastener. On the other hand, an open-end wrench is much easier to slip off and onto a nut. The combination wrench has the advantage of both a box-end and an open-end wrench. This popular configuration has a box end broached on one end, and an open end of the same size attached to the other end. This allows hard nuts to be broken loose with the box end and

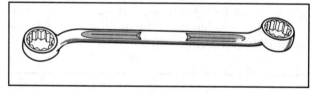

Figure 9-63. Because of their enclosed design, box-end wrenches are less likely to slip.

then removed with the open end. The box end is typically angled 15 degrees to the handle to allow clearance for your hands between the wrench and the work. The open end of the wrench is offset by 15 degrees to the axis of the handle to allow for a new grip on the nut each 15 degrees of handle movement. [Figure 9-64]

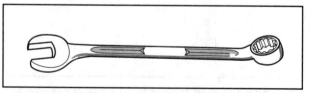

Figure 9-64. Combination wrenches have the advantages of a box-end and an open-end.

FLARE NUT WRENCHES

As you know, aircraft fluid lines are connected to components with flare nuts. While these nuts are typically not tightly torqued, they are often situated in locations where the swinging of the wrench handle is severely restricted. To help remedy this, a special type of box-end wrench is used. A slot is cut into the box end to allow the wrench to slip over a fluid line and then the hex of the nut is engaged each 15 degrees of handle movement. [Figure 9-65]

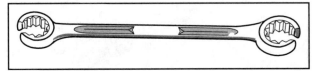

Figure 9-65. Since a flare nut wrench can engage a nut every 15 degrees of handle movement, it is ideal for use in confined areas.

RATCHETING OPEN-END WRENCHES

A special design of open-end wrench allows nuts to be turned without removing the wrench. One of the jaws of an open-end wrench is cut back just enough to engage one edge of a nut. This type of wrench is called a ratcheting open-end wrench and, while it has no ratcheting mechanism, the action it allows is ratcheting. [Figure 9-66]

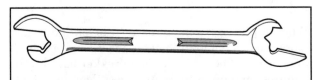

Figure 9-66. A ratcheting open-end wrench is left open on one end to allow a ratcheting action.

RATCHETING BOX-END WRENCHES

Another type of ratcheting wrench is the ratcheting box-end. This type of wrench consists of a box-end wrench set into a handle with a ratcheting mechanism. [Figure 9-67]

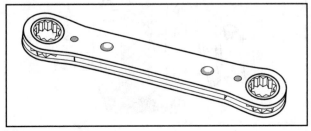

Figure 9-67. Most ratcheting box-end wrenches are locked in one direction, thus to change the direction of movement, the wrench is turned over.

SOCKETS

Sockets are tools that typically have a square hole in one end for a drive handle, and another end with a six- or twelve-point opening designed to fit different sized nuts. Sockets are available in both standard and metric sizes and can be used with a wide variety of handles and extensions.

Socket sets are available in a wide variety of drive sizes. However, in aviation maintenance the 1/4 inch square drive and the 3/8 inch square drive are the most popular. Additional drives that are available include the 1/2 inch, 3/4 inch, and 1 inch square drives. In fact, for extremely large work, socket wrenches are available in square drives as large as 2 1/2 inches.

STANDARD SOCKETS

Standard sockets are available in all of the popular drive sizes and with either four-, six-, eight-, or twelve-point openings. These sockets are also deep enough to fit over a bolt head or a nut if too much shank does not protrude. The six- and twelve-point sockets are used in aviation, whereas the four- and eight-point sockets are available to turn square head pipe plugs. [Figure 9-68]

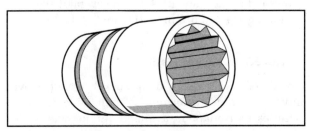

Figure 9-68. Standard sockets are available with four, six, eight, and 12 point openings.

DEEP SOCKETS

There are several applications where a bolt extends through a nut too far for a standard socket to grip the nut. In these cases deep sockets are available to

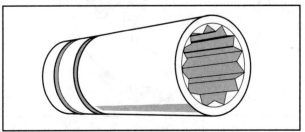

Figure 9-69. Deep sockets aid in the removal of items such as spark plugs.

allow the socket to grip the nut and still allow room for the bolt end. [Figure 9-69]

FLEX SOCKETS

When additional clearance is needed between the socket drive and the socket, a flex socket is used. Flex sockets have a pivot point between the drive handle end of the socket and the nut end. These sockets are made with both six- and twelve-point

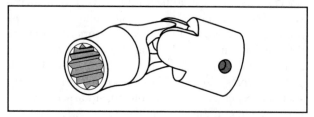

Figure 9-70. Flex sockets have a pivot point that allows for more flexibility.

openings and are available in the drive sizes most used in aviation maintenance. [Figure 9-70]

CROWSFOOT SOCKETS

Nuts are sometimes placed in locations on aircraft where neither a box-end, nor open-end wrench, or standard socket wrench can be used. The crowsfoot socket is designed to reach these nuts and is available with open, box, and flare-nut ends. Furthermore, crowsfoot sockets are available in several drive sizes including 1/4 and 3/8 inch. [Figure 9-71]

HANDLES AND ADAPTERS

The chief advantage for using socket wrenches over any type of nut-turning device is the wide variety of handles and adapters available. Some accessories

Figure 9-71. Crowsfoot sockets are used in locations where a box-end, open-end, or standard socket wrench cannot be used.

include ratchets, breaker bars, speed handles, extensions, universal joints, and adapters. For example, when a socket is snapped onto a ratchet handle, minimum handle movement is required to turn a nut or bolt. Ratchets are available with long or short handles, and with either solid or flexible heads. When selecting a ratchet, choose one that has a

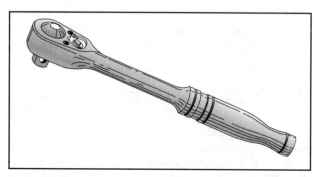

Figure 9-72. Most 3/8 inch ratchet handles are approximately 10 inches in length and their ratcheting action is reversible.

small handle movement between the positions at which the pawl grips the drive gear. [Figure 9-72]

If a socket drive does not have any ratcheting ability, several **ratchet adapters** are available to convert the ratchet into a reversible ratcheting wrench. [Figure 9-73]

When a nut is extremely hard to break loose, and more force is required than a ratchet handle is built to take, a socket is placed on a **breaker bar** and the required amount of force applied. When more force is required than can be applied with a breaker bar, use a socket and handle of the next larger drive size. It is not recommended that you use a piece of pipe over the handle of a breaker bar to increase the leverage. [Figure 9-74]

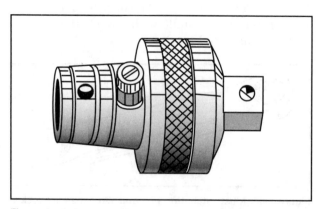

Figure 9-73. Ratchet adapters fit onto nonratcheting drives and extensions to convert them into a ratcheting wrench.

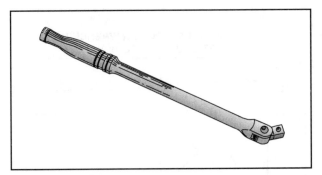

Figure 9-74. A breaker bar is used when additional leverage is needed to break loose a tight nut.

Time is an expensive commodity in aviation maintenance and, therefore, any tool that decreases the time required for an inspection or repair is typically used. One tool that can save a great deal of time is the speed handle. A **speed handle** resembles a bow-type brace and has a socket or a screwdriver bit snapped onto its end. Screws or nuts are turned much faster with a speed handle than they are with

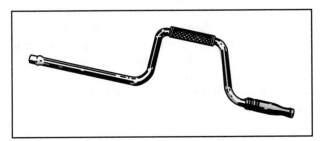

Figure 9-75. Speed handles allow screws and nuts to be turned much faster than when using a conventional screwdriver or ratchet.

a conventional screwdriver or ratchet wrench. [Figure 9-75]

Straight bar-type **extensions** are used to put sockets further away from the wrench handle. These extensions are made of forged steel alloys and are available in lengths from less than two inches up to two or three feet long. [Figure 9-76]

Universal joints have a square opening on one end that fits onto a socket drive or extension and a male socket drive on the other end. The universal joint is

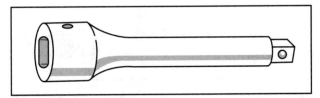

Figure 9-76. Extensions are used with socket wrenches to get to nuts or bolts in hard-to-reach places.

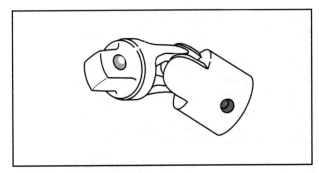

Figure 9-77. Universal joints allow a socket and a drive to pivot independently of each other.

used to tighten or loosen nuts and bolts that cannot be accessed with a straight extension. [Figure 9-77]

Adapters are available to allow different size sockets and drives to fit together. For example, an adapter allows a 1/4 inch drive socket to fit onto a 3/8 inch drive handle or vice-versa. Adapters are also available between 3/8 inch and 1/2 inch drive components. When using an adapter to put a smaller socket on a larger drive, use good judgment because the additional leverage obtained on the drive can break the adapter or the socket. There might also be enough added force to strip the threads of a fastener. [Figure 9-78]

IMPACT TOOLS

Impact tools are turning tools that come in both hand and power types. They are used when corrosion or rust on a fastener causes it to resist any loosening effort. A sharp blow from a **hand-held impact driver** utilizes mechanical advantage to give the fasteners a quick twist.

An impact driver set for aviation maintenance technicians consists of a driver, an assortment of special six-point impact sockets, and bits for the screw sizes and types most often found on airplanes. To use an impact driver, select the proper bit or socket and

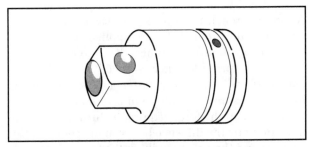

Figure 9-78. Exercise discretion when using adapters since the added force exerted has the potential to break either the adaptor or the socket, or strip threads on a fastener.

insert it onto the driver. Next, place the impact driver on the fastener and strike the driver with a sharp hammer blow. Some stubborn fasteners may need more than one blow before they can be turned with a conventional wrench. An impact driver has both a forward and reverse setting. The reason for this is that it is sometimes necessary to slightly tighten a fastener in order to break it free. However, use care

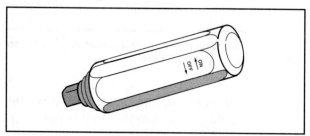

Figure 9-79. A hand-held impact driver and hammer are used to loosen corroded fasteners.

not to over tighten the fastener and further damage the structure. [Figure 9-79]

Hand-held impact drivers typically break loose most stubborn fasteners. However, some fasteners may require an air-driven impact tool. Power impact wrenches apply force in a series of jerks or impacts. This means that an impact wrench set to a specific level of torque actually applies a much higher peak torque than what the wrench is set for. These torque spikes, or peaks, cannot be used on any fastener whose torque is critical, because it over stresses the fastener. [Figure 9-80]

TORQUE WRENCHES

The holding power of a threaded fastener is greatly increased when it is placed under an initial tensile load that is greater than the loads the fastener is subjected to. This task is accomplished by tightening a bolt or nut to a pre-determined torque, or pre-load, with a torque wrench.

A torque wrench is a precision measuring tool that measures the amount of force applied to a fastener. Under controlled conditions, the amount of force required to turn a fastener is directly related to the tensile stress within the fastener. The amount of torque, measured in inch-pounds or foot-pounds, is the product of the force required to turn the fastener multiplied by the distance between the center of the fastener and the point at which the force is applied. For example, a torque wrench has a length permanently established between the center of the drive

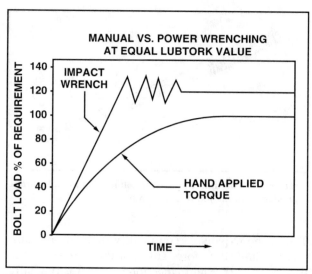

Figure 9-80. The torque produced by a power impact tool is applied in a series of blows or spikes of torque. For this reason, impact wrenches cannot be used on fasteners where torque is critical.

square and a pivot in the handle. The force applied is measured by the amount the beam deflects. If you remember from Chapter 2, Hooke's law states that the amount a beam deflects is directly related to the force applied. Therefore, if the lever is exactly 12 inches long and a force of 30 pounds is applied to the handle, a torque of 360 inch-pounds is produced on the fastener.

$$12 \text{ in.} \times 30 \text{ lbs.} = 360 \text{ in./lbs.}$$

There are three basic types of torque wrenches typically found in aviation maintenance shops. They are the deflecting-beam type, the torsion bar type, and the toggle type.

The **deflecting-beam torque wrench** is one of the simplest. The square drive is on one end of an accurately ground beam with a handle mounted on a pivot at the other end. The pivot ensures that force is always applied at a specific point. A pointer attached to the end of a beam holds the drive square, and a scale is mounted near the handle end. When force is applied to the handle, the beam bends and the pointer moves across the dial measuring the

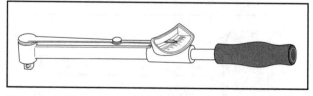

Figure 9-81. The torque measured by a torque wrench is the product of the length between the drive square and the point of force on the handle.

amount the beam bends. The amount of bend is directly proportional to the amount of torque applied. [Figure 9-81]

A bar accurately deflects in torsion as well as bending when a force is applied. This principle is used in the **torsion bar torque wrench**. The drive square of a torsion bar-type wrench is accurately ground and has a rack gear on one end. When the bar is twisted, the

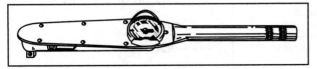

Figure 9-82. A torsion bar-type torque wrench uses deflection as well as a bending force to indicate torque.

rack moves across a pinion gear in a dial indicator which shows the amount of bar defection. The deflection is calibrated in inch-pounds, foot-pounds, or in meter or centimeter-kilograms. [Figure 9-82]

A **toggle torque wrench** is pre-set to the desired

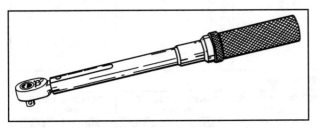

Figure 9-83. A toggle torque wrench releases or snaps over when a pre-set torque is reached.

torque before it is put on a fastener. When this pre-set torque is reached a sound is heard and the handle releases a few degrees. The release indicates that the desired torque is reached. Once the release is reached all force should be removed. [Figure 9-83]

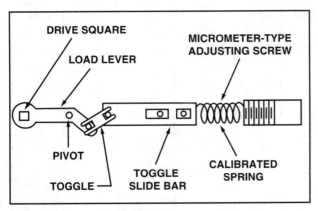

Figure 9-84. The force required to cause a toggle-type torque wrench to snap is determined by the amount of compression on the spring in the wrench handle.

Rather than measuring the deflection of a beam, the toggle-type torque wrench uses a calibrated compression or spring to apply a force to a load lever. When the torque applied to the drive square reaches the pre-set value, the toggle forces the toggle slide bar back enough for the toggle to snap over. [Figure 9-84]

When an adapter is used to reach a particular fastener, the indication of torque on a torque wrench has to be modified to find the actual torque being applied. Remember, when the length of a torque bar changes, the scale used on the torque wrench is no longer accurate. For example, a torque wrench has a length between the drive square and the handle pivot of 20 inches, and a five-inch extension. To find the torque applied to a fastener with an indication of 120 inch-pounds on the wrench, use the following formula:

$$T_A = \frac{T_W (L + A)}{L}$$

Where:

T_A = Actual (desired) torque

T_W = Apparent (indicated) torque

L = Length of torque wrench

A = Added length

When this formula is used and the torque wrench reads 120 inch-pounds, the amount of torque actually applied on the fastener is 150 inch-pounds. [Figure 9-85]

By shifting the variables the same formula can be used to determine what a torque wrench will indicate for a given torque on a fastener. The formula to do this is:

$$T_W = \frac{T_A \times L}{(L + A)}$$

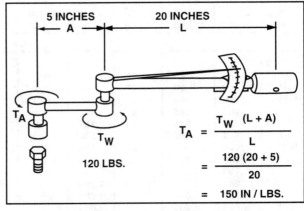

Figure 9-85. The actual amount of torque applied to a fastener when a five inch extension is used with 120 pounds of indicated torque is 150 inch-pounds.

Using the same figures as before, it is found that in order to apply 150 inch-pounds of torque on a fastener with a five-inch extension, the torque handle scale needs to read 120 inch-pounds.

USING A TORQUE WRENCH

If a fastener is under torqued, there is danger of the joint being subjected to unnecessary loads resulting in premature failure. On the other hand, when a fastener is over torqued, the threads are over stressed

	BOLTS STEEL TENSION		BOLTS STEEL TENSION		BOLTS ALUMINUM	
	AN3 THRU AN20 AN42 THRU AN49 AN73 THRU AN81 AN173 THRU AN186 MS20033 THRU MS20046 MS20073 MS20074 AN509 NK9 MS24694 AN525 NK525 MS27039		MS20004 THRU MS20024 NAS144 THRU NAS158 NAS333 THRU NAS340 NAS583 THRU NAS590 NAS624 THRU NAS644 NAS1303 THRU NAS1320 NAS172 NAS174 NAS517 **STEEL SHEAR BOLT** NAS464		AN3DD THRU AN20DD AN173DD THRU AN186DD AN509DD AN525D MS27039D MS24694DD	

	NUTS		NUTS		NUTS	
	STEEL TENSION	STEEL SHEAR	STEEL TENSION	STEEL SHEAR	ALUMINUM TENSION	ALUMINUM SHEAR
	AN310 AN315 AN363 AN365 NAS1021 MS17825 MS21045 MS20365 MS20050 NAS679	AN320 AN364 NAS1022 MS17826 MS20364	AN310 AN315 AN363 AN365 MS17825 MS20365 MS21045 NAS1021 NAS679 NAS1291	AN320 AN364 NAS1022 MS17826 MS20364	AN365D AN310D NAS1021D	AN320D AN364D NAS1022D

FINE THREAD SERIES

NUT-BOLT SIZE	TORQUE LIMITS IN-LBS		TORQUE LIMITS IN-LBS		TORQUE LIMITS IN-LBS		TORQUE LIMITS IN-LBS		TORQUE LIMITS IN-LBS		TORQUE LIMITS IN-LBS	
	MIN.	MAX.	MIN.	MAX.	MIN.	MAX.	MIN.	MAX.	MIN.	MAX.	MIN.	MAX.
8-36	12	15	7	9	—	—	—	—	5	10	3	6
10-32	20	25	12	15	25	30	15	20	10	15	5	10
1/4-28	50	70	30	40	80	100	50	60	30	45	15	30
5/16-24	100	140	60	85	120	145	70	90	40	65	25	40
3/8-24	160	190	95	110	200	250	120	150	75	110	45	70
7/16-20	450	500	270	300	520	630	300	400	180	280	110	170
1/2-20	480	690	290	410	770	950	450	550	280	410	160	260
9/16-18	800	1,000	480	600	1,100	1,300	650	800	380	580	230	360
5/8-18	1,100	1,300	660	780	1,250	1,550	750	950	550	670	270	420
3/4-16	2,300	2,500	1,300	1,500	2,650	3,200	1,600	1,900	950	1,250	560	880
7/8-14	2,500	3,000	1.500	1,800	3,550	4,350	2,100	2,600	1,250	1,900	750	1,200
1-14	3,700	4,500	2,200	3,300	4,500	5,500	2,700	3,300	1,600	2,400	950	1,500
1 1/8-12	5,000	7,000	3,000	4,200	6,000	7,300	3,600	4,400	2,100	3,200	1,250	2,000
1 1/4-12	9,000	11,000	5,400	6,600	11,000	13,400	6,600	8,000	3,900	5,600	2,300	3,650

COARSE THREAD SERIES

NUT-BOLT SIZE	MIN.	MAX.	MIN.	MAX.	MIN.	MAX.	MIN.	MAX.	MIN.	MAX.	MIN.	MAX.
8-32	12	15	7	9	—	—	—	—	—	—	—	—
10-24	20	25	12	15	—	—	—	—	—	—	—	—
1/4-20	40	50	25	30	—	—	—	—	—	—	—	—
5/16-18	80	90	48	55	—	—	—	—	—	—	—	—
3/8-16	160	185	95	110	—	—	—	—	—	—	—	—
7/16-14	235	255	140	155	—	—	—	—	—	—	—	—
1/2-13	400	480	240	290	—	—	—	—	—	—	—	—
9/16-12	500	700	300	420	—	—	—	—	—	—	—	—
5/8-11	700	900	420	540	—	—	—	—	—	—	—	—
3/4-10	1,150	1,600	700	950	—	—	—	—	—	—	—	—
7/8-9	2,200	3,000	1,300	1,800	—	—	—	—	—	—	—	—
1-8	3,700	5,000	2,200	3,000	—	—	—	—	—	—	—	—
1 1/8-8	5,500	6,500	3,300	4,000	—	—	—	—	—	—	—	—
1 1/4-8	6,500	8,000	4,000	5,000	—	—	—	—	—	—	—	—

Figure 9-86. Most torque values are provided by the manufacturer. However, if a value isn't provided, a table of torque values is used.

and can fail. Therefore, it is very important that fasteners be torqued to their specified value. Furthermore, it is important to realize that, unless otherwise specified, all torque values given are for clean dry threads. [Figure 9-86]

When a self-locking nut is torqued, the nut should be run down on the threads until it nearly contacts the washer. The amount of torque required to run the nut down should be measured and this value added to the amount of torque needed for the fastener. The torque needed to turn the nut down is called friction drag torque.

The accuracy of a torque measurement is assured only when torque is applied with a smooth and even motion. Remember that impact-type wrenches should never be used on any fastener whose torque

is critical. If a nut is accidentally over torqued, it should be loosened and then retorqued to the proper value. Never back off a nut or a bolt and leave it untorqued.

When installing a castle nut, start alignment with the cotter pin hole at minimum recommended torque, plus friction drag. If the hole and nut castellation do not align, change washers and try again. Exceeding the maximum recommended torque is not recommended.

Torque wrenches are precision measuring tools and should be periodically checked for accuracy. They should also be properly handled to ensure the torque indicated is actually what is being applied. [Figure 9-87]

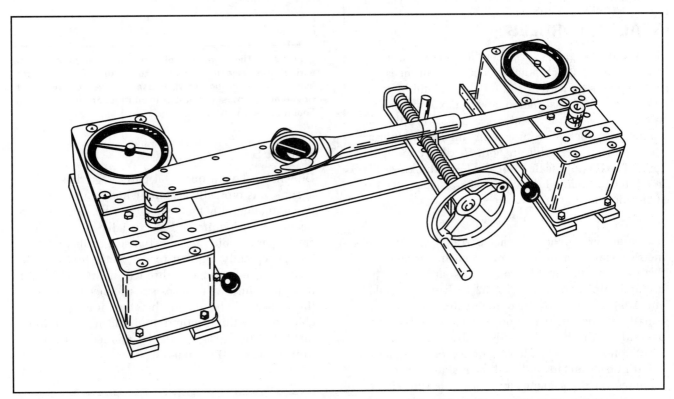

Figure 9-87. A torque analyzer is available in many shops so that wrenches can be calibrated by certificated technicians.

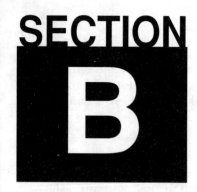
MEASURING AND LAYOUT TOOLS

In aviation maintenance there are many jobs that require precise measuring for close tolerance fits. Because of this, you must be familiar with several measuring devices. Some of the more common measuring tools include protractors, calipers, and telescoping gauges. In this section we will discuss each of these layout tools in detail.

SCALES OR RULES

Steel scales or rulers are essential to have in both six inch and 12 inch lengths. This type of measuring device is typically used for sheet metal layout, and for taking measurements where extreme precision is not required. Scales are made of either a tempered carbon steel or a satin-finished stainless steel, and are available in both flexible and rigid form. The flexible scale typically has a thickness of about 0.015 inch, while a rigid scale is about 0.040 inch thick.

Scales are graduated in exact portions of either a metric measurement, or a fractional measurement. Scales with fractional graduations are typically divided into increments of 1/32 inch on one side and 1/64 inch on the other side. A decimal scale is usually divided in tenths or fiftieths of an inch on one side while the other side is divided in increments of 1/100 inch. Metric graduations are measured in centimeters and millimeters, and are often included on the same scale. Since it is sometimes necessary to convert from metric to fractional or decimal form and vice versa, it is a good idea to keep a conversion chart with your layout tools.

Because the end of a metal scale is not precisely cut, the cut, or factory end should not be used as a measuring guide. Instead, you should always begin measuring somewhere after the first few markings on a scale to ensure a correct measurement. The one inch mark is typically used as the starting point because it is easily subtracted from the final measurement. [Figure 9-88]

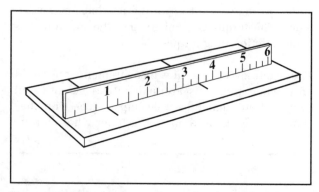

Figure 9-88. The proper use of a steel scale requires that the end of the scale not be used to make a measurement. Instead, the one inch mark is used as the starting point. However, you must remember to subtract one inch from the final measurement.

SCRIBES

Dimension layout on metal parts, regardless of the accuracy, is typically accomplished by using layout dye and a marking tool called a scribe. Scribes have needle-sharp points and are usually made of hard steel or are carbide tipped. To use a scribe, a layout dye is typically applied to the metal first and the scribe is used to scratch through the dye. However, this procedure will causes stress concentrations on the surface of a bend and, therefore, it is not acceptable to use this method to indicate bend lines. Instead bend lines should be marked with a soft tipped marker. [Figure 9-89]

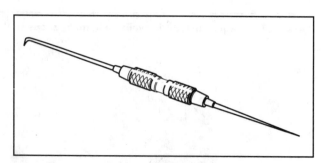

Figure 9-89. A scribe is typically used with layout dye to mark reference points on a material. However, scribe marks should never be used for bend lines as they cause stress concentrations that can lead to component failure.

COMBINATION SET

An elaboration of the steel scale is the combination set. This measuring device consists of a 12 inch steel scale and three heads that move up and down along the scale. The heads of a combination set are removable to allow use of any combination of the individual components. [Figure 9-90]

STOCK HEAD

The stock head of a combination set is normally used as a square. One side of the head is exactly perpendicular to the scale, while the other side meets the scale at a 45 degree angle. To determine if something is level, the stock head can be removed from the scale and the built-in bubble level used.

PROTRACTOR HEAD

A bubble level and a 360 degree protractor are built into the protractor head of a combination set. This head has a lock that releases the protractor and allows it to rotate within its frame. This feature makes the head particularly useful for measuring control surface travel. For example, the surface of a protractor head is locked in a streamline, or neutral position, and its head is placed on a control surface that is in its neutral position. Next, the protractor is turned until its bubble level is centered. Now, when the control surface is fully deflected the degree of travel is represented by the angular difference between the level and the scale.

To determine the total amount of control surface movement, both extreme measurements must be taken. To accomplish this, the control surface is moved to its other extreme of travel and the bubble level is centered. The sum of the two extreme readings is the amount of control surface movement.

CENTER HEAD

The center head of a combination set is used for finding the center of circular objects such as a piece of round bar stock. To use a center head, put the two flat blades of the center head over the end of a piece of circular material that is held perpendicular to the scale. Since the scale bisects the center head angle into two equal parts, the scale also bisects the cir-

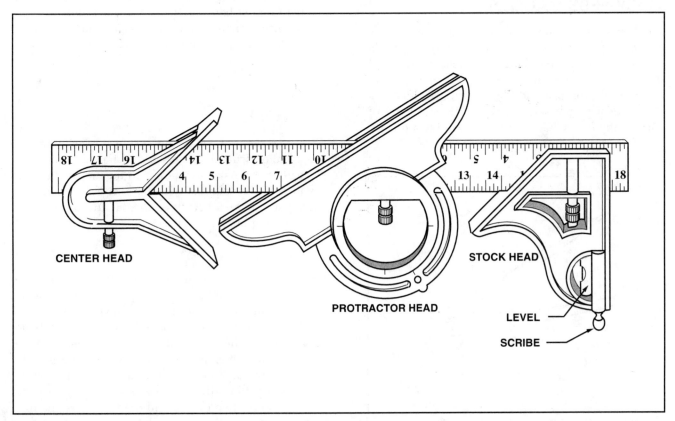

Figure 9-90. The combination set typically includes a center head, a protractor head, and a stock head. The stock head usually has a level and a scribe built into it.

cular material. Next, scribe a line along the edge of the scale that passes through the center of the circular material. Now rotate the center head 90 degrees and scribe another mark along the edge of the scale blade. The point where the two scribe marks intersect is the center of the circular material.

DIVIDERS

Dividers are layout tools that are used to accurately lay out circles and arcs and for transferring dimensions on metal. A divider has two legs with needle-sharp points that are joined in the center with a pivot. Since a divider does not have a built-in scale, it does not provide a measurement when used as a measuring device. To begin laying out dimensions, first cover the metal with a layout dye and mark a base line with a very sharp scribe. Next, make a very light punch mark on the base line to use as a reference for all measurements. After this is complete, the divider is set to the desired distance with the use of a steel machinist scale. Then one divider leg is placed on the punch mark and a scratch is made with the other divider leg through the transfer dye. This procedure is far more accurate than measuring a dimension directly with a steel scale. [Figure 9-91]

THICKNESS GAUGES

Thickness gauges are used to measure clearances between two surfaces, such as a piston ring's fit in a ring groove or its end gap clearance. In addition, they can be used with a surface plate and arbor to check a part for twist or warp. A typical thickness gauge consists of a stack of steel or brass blades of varying thickness. Each blade is ground to a precise

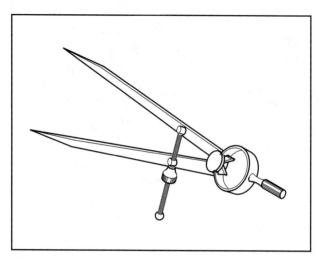

Figure 9-91. A divider is used to lay out circles and arcs on your work.

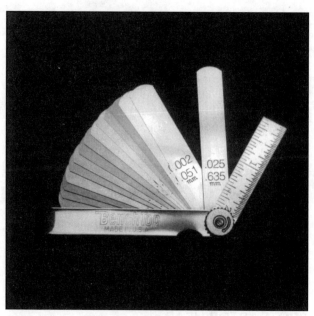

Figure 9-92. Thickness gauges allow quick measurement of the clearance between two surfaces.

dimension, accurate to ten thousandth of an inch. To measure a clearance between two surfaces, select a blade approximately the size of the clearance to be measured and insert it between the two surfaces. If the blade is too loose, try the next larger size. When a blade slides between the two surfaces with a slight amount of friction, the size designation printed on the blade indicates the clearance. [Figure 9-92]

CALIPERS

Calipers are a type of measuring device typically used to measure diameters and distances or for comparing sizes. As an aviation maintenance technician you must be familiar with three types of calipers. They are the inside caliper, the outside caliper, and the hermaphrodite caliper.

Calipers are very similar to dividers in that they have two legs with some type of pivot. **Inside calipers** are used to measure the inside diameter of a hole, and have legs that point outward. **Outside calipers**, on the other hand, are used to measure the outside diameter of an object and have legs that point inward. When using either type of calipers you adjust the caliper until it fits snugly across the widest part of an object, and then measure the distance between the caliper leg points with a steel scale. [Figure 9-93]

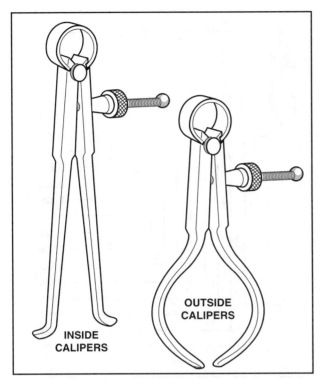

Figure 9-93. Inside calipers have legs with ends that curve to the outside, whereas outside calipers have legs with ends that curve to the inside.

Hermaphrodite calipers are used to scribe marks that are a specific distance from a radius edge. These calipers have one sharp-pointed leg, and one leg that curves to the inside. To use hermaphrodite calipers, the material being worked on is first covered with layout dye, and the distance required is adjusted between the two caliper legs. The caliper point is then moved along the radius edge as the sharp point is drawn across the surface. [Figure 9-94]

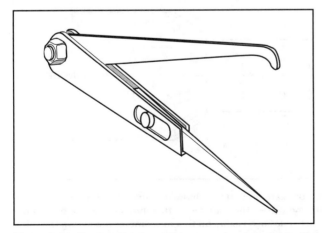

Figure 9-94. Hermaphrodite calipers are used to mark distances away from edges. These calipers have the straight leg of a divider and a leg that curves inward similar to an outside caliper.

MICROMETER CALIPERS

The precision measuring tool most universally used by the aviation maintenance technician is the micrometer caliper. These instruments are used to measure the thickness of sheet metal, the out-of-roundness of cylindrical objects such as piston pins, and the degree of stretch of valve stems. Furthermore, when used in conjunction with other devices, micrometers can measure the fit of many engine parts. The basic parts of a micrometer caliper include a **frame**, which resembles a C-clamp, an **anvil**, a **spindle**, a **sleeve**, and a **thimble**. [Figure 9-95]

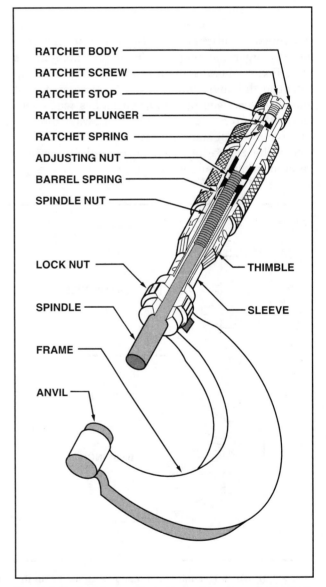

Figure 9-95. Attached to the bottom of the frame is the anvil, while the spindle is attached to the top of the frame. The sleeve covers the spindle body and is marked in increments of .025 inch. The thimble spins around the spindle and sleeve and is used to adjust the distance between the anvil and spindle.

The internal threads on the spindle nut have a pitch of 40 threads per inch and are extremely accurate. The micrometer sleeve is marked with a one-inch longitudinal line that is divided into ten equal segments, or one-tenth-inch increments. Each section between the tenths divisions is further divided into four equal parts of 0.025 inch each. [Figure 9-96]

In addition to the marks on the sleeve, the thimble is marked in 25 equal sections with each mark representing 0.001 inch. Therefore, when the thimble is rotated one complete revolution, from zero around to zero, one line, or 0.025 inch is exposed. [Figure 9-97]

Micrometer calipers are also made to measure objects larger than one inch. To do this, the frame is modified so there is a one-, two-, or three-inch gap between the anvil and spindle. A micrometer that measures between one and two inches reads zero when the anvil is exactly one inch away from the spindle. Special standard gauge blocks are used to check the zero adjustment on more-than-one-inch micrometers. [Figure 9-98]

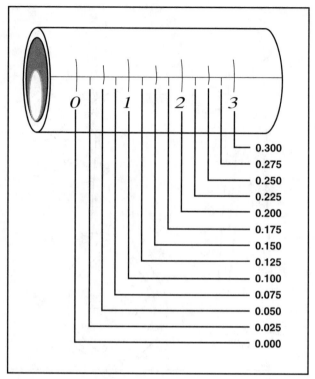

Figure 9-96. Each mark on the sleeve of the micrometer caliper represents 0.025 inch.

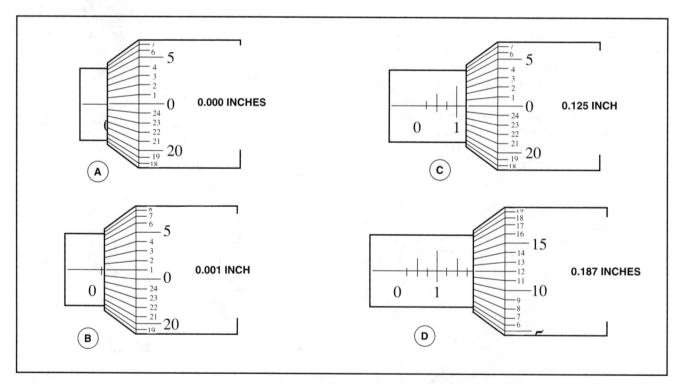

Figure 9-97. (A) — This micrometer shows no marks on the sleeve and the zero line on the thimble in line with the reference line on the sleeve. When set this way, the spindle and anvil should be touching. **(B)** — The thimble on this micrometer has moved out 0.001 inch and the zero mark on the sleeve is just visible. **(C)** — There are four lines visible on this micrometer indicating the thimble has moved out at least 0.100 inch. The thimble has also moved out another complete revolution or 0.025 inch. When added together, the thimble has moved out 0.125 inch. **(D)** — There are seven lines visible on the sleeve indicating the thimble has moved out at least 0.175 inch. The 12 on the thimble is lined up with the reference line on the sleeve indicating the thimble has moved out an additional 0.012 inch. When added together, the thimble has moved out 0.187 inch.

It is possible for a micrometer to slip out of calibration. However, most micrometers can be recalibrated. To do this, a very precisely-sized piece of metal called a **standard** or **gauge block** is inserted between the anvil and the spindle. The spindle is then closed on the block to create a pressed fit. A pressed fit means that they come together with just enough force to meet each other squarely. A special spanner wrench is then slipped into a hole in the

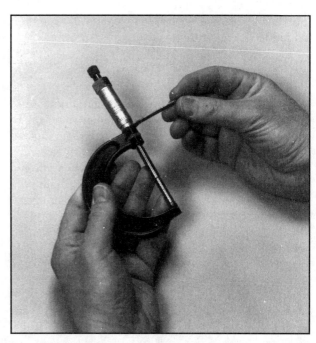

Figure 9-99. A micrometer caliper is calibrated by carefully rotating the sleeve until its longitudinal line exactly aligns with the zero mark on the thimble.

micrometer sleeve and carefully rotated until the longitudinal line is in exact alignment with the zero mark on the thimble. This calibration should be accomplished by a certified technician because temperature variation must be taken into account. [Figure 9-99]

VERNIER SCALES

At times it is necessary to make measurements that are smaller than one-thousandth of an inch. In these situations, you must use a micrometer with a vernier scale. A vernier scale is a secondary scale found on some precision measuring tools that further divides the primary scale into even smaller increments. On a micrometer, the vernier scale divides each one-thousandth of an inch into ten equal increments with each increment being equal to one ten thousandth of an inch.

The vernier micrometer is like the standard micrometer except that it has eleven extra lines on the sleeve. These lines are marked longitudinally, just above the longitudinal reference line. To read a vernier micrometer, begin by reading the sleeve and thimble just like a standard micrometer. Then look at the vernier scale and identify the horizontal line that lines up with an increment on the thimble. The

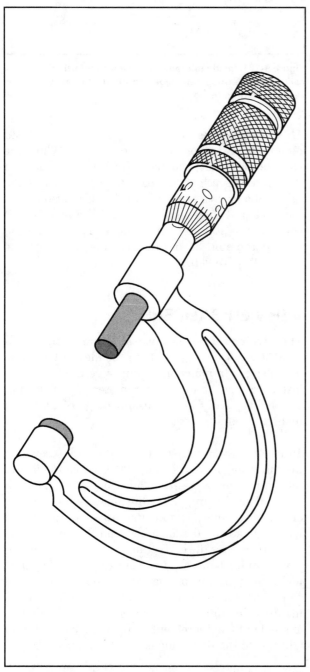

Figure 9-98. Micrometer calipers for measuring between one and two inches have a one-inch space between the anvil and the spindle when the micrometer reads zero.

number associated with the horizontal line is then added to the measurement. [Figure 9-100]

INSIDE MICROMETERS

An inside micrometer uses a single micrometer head consisting of a basic micrometer mechanism and various extensions to cover a wide range of measurements. An inside micrometer caliper is read in exactly the same manner as an outside micrometer. To use an inside micrometer place it inside the opening being measured and adjust the micrometer until it is the same size as the opening. Once this is done, the micrometer is removed from the opening and read. [Figure 9-101]

MICROMETER DEPTH GAUGE

The depth of grooves or recesses in a part are accurately measured with a micrometer depth gauge.

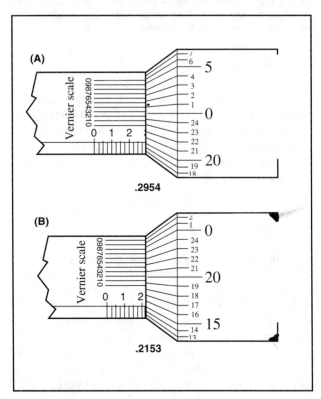

Figure 9-100. (A) — The sleeve indicates a measurement of at least 0.275 inch. The thimble indicates an additional 0.020 inch. On the vernier scale, the line representing 0.0004 lines up with an increment on the thimble and, therefore, is added to the measurement. The total measurement is 0.2954 (0.275 + 0.020 + 0.0004 = 0.2954). (B) — The sleeve indicates a measurement of at least 0.2 inch and the thimble an additional 0.015 inch. On the vernier scale, the horizontal line representing 0.0003 lines up with an increment on the thimble and, therefore, is added to the total measurement. The resulting dimension is 0.2153 inch (0.2 + 0.015 + 0.0003 = 0.2153).

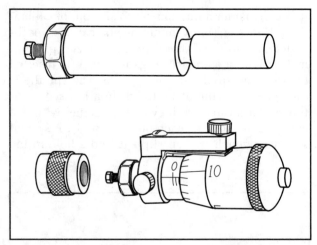

Figure 9-101. Inside micrometers use a standard micrometer caliper and various extensions to obtain the inside diameters and widths of slots.

This device has a standard micrometer head mounted onto a precisely ground bar. When the spindle of the micrometer is flush with the face of the bar, the depth gauge reads zero. To measure the depth of a groove or recess, the bar is placed across a groove and the spindle is screwed down until it contacts the bottom of the groove. The reading on the micrometer head indicates the depth of the groove. [Figure 9-102]

VERNIER CALIPERS

The vernier caliper is a versatile precision instrument used to measure both inside and outside dimensions. In many situations, a vernier caliper is faster to use than a micrometer. Furthermore, calipers typically have a useful range of up to six inches. [Figure 9-103]

To use a vernier caliper, loosen the lock screws on the movable jaw and clamp. Next, move the jaw to the approximate position and lock the clamp in place. Then, using the adjustment screw on the clasp, move the jaw into its correct position and lock it. Read the distance between the jaws on the vernier scale for the correct inside or outside dimension as indicated. To read a vernier caliper, begin by identifying the approximate dimension opposite the zero mark on the sliding scale. Once this is done, identify the number on the sliding scale that is aligned with an increment on the fixed bar scale and add this to the first dimension. [Figure 9-104]

The English vernier caliper has 25 spaces on the vernier scale that occupy exactly the same distance as 24 spaces on the bar scale. Each graduation on the bar

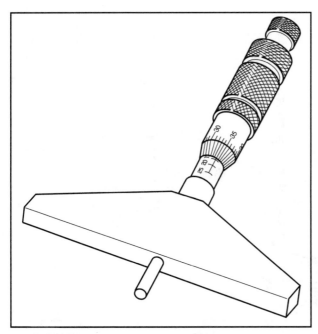

Figure 9-102. To allow micrometer depth gauges to measure several different grooves, recesses, and bores, most have interchangeable spindles of various lengths.

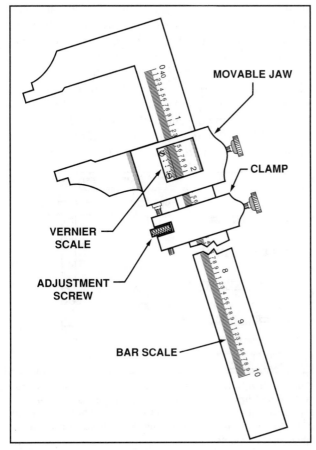

Figure 9-103. A vernier caliper consists of a bar scale with one fixed jaw and one movable jaw. The movable jaw typically has a clamp to lock the jaw in place, as well as an adjustment screw for making small jaw movements.

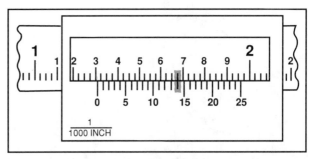

Figure 9-104. The caliper indicates a measurement of at least 1.3 inches. Also notice that the increment representing 0.014 on the movable scale is aligned with an increment on the fixed scale indicating that it must be added to the first measurement. The total measurement is 1.314 inches (1.3 + 0.014 = 1.314).

represents 25-thousandths of an inch (0.025). The line on the vernier scale that is aligned with one of the bar scale marks indicates the number of thousandths of an inch to add to the measurement indicated on the bar before the zero on the vernier scale.

A metric vernier caliper is read in the same way as an English caliper in that the scale on the movable jaw aligns with a line on the bar. The difference is that each number on the metric bar scale represents one centimeter, or 10 millimeters, with each graduation representing one-half millimeter. The graduations on a metric vernier scale represent divisions of 1/50 millimeter. [Figure 9-105]

DIAL INDICATORS

Dial indicators are precision measuring instruments used to determine the amount of movement existing between certain engine and airframe parts. These indicators are also used to determine an out-of-round condition on a shaft as well as the plane of rotation of a disk.

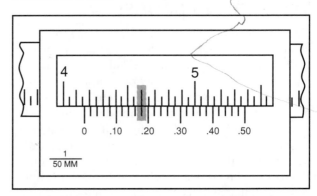

Figure 9-105. By referencing the zero indication on the vernier scale you can see that the dimension is larger than 41.5 millimeters. By looking down the vernier scale you can also see that only the 18 is aligned with the bar scale. Therefore, 0.18 must be added to the final dimension. The total is 41.68 millimeters or 4.168 centimeters.

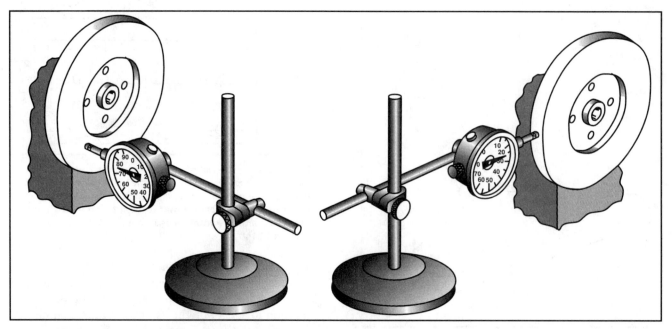

Figure 9-106. Dial indicators are used to check shafts for out-of-round and for bends, and they are also useful for checking back-lash in gears and for measuring axle end play.

Engine crankshaft runout is checked with a dial indicator by mounting the dial indicator to its furnished arm and then clamping the assembly to its stand. The dial indicator is then put in contact with a smooth part of the crankshaft. A **preload** is then applied to the dial indicator by pressing it against the shaft until the pointer deflects a few thousandths of an inch. Next, the indicator is zeroed by rotating the dial or turning the bezel on the outside of the instrument case until the needle is opposite the zero mark. When the crankshaft is turned, the needle indicates the maximum deviation in both directions. [Figure 9-106]

SMALL HOLE GAUGES AND TELESCOPING GAUGES

The inside diameter of holes are sometimes too small to be measured with an inside micrometer caliper. In these situations the holes are measured with small hole gauges. For example, to accurately measure the diameter of a 1/4 inch hole, the ball end of a hole gauge is placed inside the hole to be measured and the end of the shaft is twisted to expand the ball until it fits the hole. The gauge is then locked and removed so that the dimension of the ball can be measured with a micrometer caliper. A small hole gauge is also used to measure the fit between a shaft and its bushing.

Holes with diameters between 5/16 inch and six inches are accurately measured with T-shaped **telescoping gauges**. A telescoping gauge typically has one or two spring loaded arms. These arms are collapsed and placed inside the hole to be measured. The lock on the end of the shaft is then released, allowing the arms of the telescoping gauge to spring outward against the walls of the hole. The gauge is maneuvered until it is square inside the hole, then, it is locked by twisting the end of the handle. The gauge is removed from the hole, and the length of the arm is measured with a micrometer caliper. [Figure 9-107]

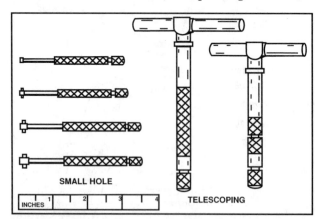

Figure 9-107. Small hole gauges that measure inside diameters range from 1/8 to 1/2 inch. Telescoping gauges are used for measuring the inside diameter of holes that range from 3/16 to six inches. Both are used in conjunction with a micrometer caliper.

CHAPTER 10

FLUID LINES AND FITTINGS

INTRODUCTION

All aircraft, from the smallest trainers to the largest transports, have systems to direct the flow of fluids from their source to the units requiring them. These systems consist of hoses, tubing, fittings, and connectors and are often referred to as an aircraft's "plumbing." Even though aircraft fluid lines and related hardware are very reliable and require little maintenance, they cannot be overlooked. For example, because of the variety of fluids used in aircraft, the requirements for fluid lines differ greatly. Therefore, it is very important that you, as a maintenance technician, understand the different types of fluid lines used, their applications, and the inspection and maintenance requirements of each. For example, replacement lines must be of the same size and material as the original line, and the correct fittings must be selected and properly installed. An error in the selection or installation of a component could result in damage to a unit, loss of fluid, or complete system failure. Regular inspection and time-specified replacements ensure continuous and safe operation.

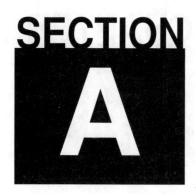

SECTION A

RIGID FLUID LINES

A single aircraft typically contains several different types of rigid fluid lines. Each type of line has a specific application. However, as a rule, rigid tubing is used in stationary applications and where long, relatively straight runs are possible. Systems that typically utilize rigid tubing include fuel, oil, oxygen, and instrument systems.

MATERIALS

Many fluid lines used in early aircraft were made of copper tubing. However, copper tubing proved troublesome because it became hard and brittle from the vibration encountered during flight, and eventually failed. To help prevent failures and extend the life of copper tubing, it must be periodically annealed to restore it to a soft condition. Annealing is accomplished by heating the tube until it is red hot and then quenching in cold water. When working on an aircraft that has copper tubing, the tubing should be annealed each time it is removed. Furthermore, copper lines must be regularly inspected for cracks, hardness, and general condition.

Today, aluminum-alloy and corrosion-resistant steel lines have replaced copper in most applications. Aluminum tubing comes in a variety of alloys. For example, in low pressure systems (below 1,000 psi) such as those used for instrument air or ventilating air, commercially pure aluminum tubing made from 1100-H14 (half-hard), or aluminum alloy 3003-H14 (half-hard) is used. Low pressure fuel and oil and medium pressure (1,000 to 1,500 psi) hydraulic and pneumatic systems often use lines made of 5052-O aluminum alloy. This alloy, even in its annealed state, is about one and three-quarters times stronger than half-hard, commercially pure aluminum. Occasionally, 2024-T aluminum alloy is used for fluid lines because of its higher strength. However, it is not as flexible and, therefore, is more difficult to bend and flare without cracking.

Aluminum alloy tubes are identified in a number of ways. For example, on larger tubes, the alloy designation is stamped directly on the tube's surface. However, on small tubing, the alloy designation is typically identified by a colored band. These color bands are no more than 4 inches wide and are painted on the tube's ends

and mid section. When a band consists of two colors, one-half the width is used for each color. [Figure 10-1]

Corrosion-resistant steel tubing, either annealed or 1/4 hard, is used in high pressure systems (3,000 psi). Applications include high pressure hydraulic, pneumatic and oxygen systems. Corrosion-resistant steel is also used in areas that are subject to physical damage from dirt, debris, and corrosion caused by moisture, exhaust fumes, and salt air. Such areas include flap wells and external brake lines. Another benefit of corrosion-resistant steel tubing is that it has a higher tensile strength which permits the use of tubing with thinner walls. As a result, the installation weight is similar to that of thicker-walled aluminum alloy tubing.

Repairs to aircraft tubing must be made with materials that are the same as the original, or are an approved substitute. One way to ensure that a replacement is made of the same material is to compare the code markings on the replacement tube to those on the original. If the manufacturer's markings are difficult or impossible to read, an alternative method of identification must be used. While aluminum alloy and steel tubing are readily distinguished from one another, it is often difficult to determine whether a material is carbon steel or stainless steel, or whether it is 1100, 2024, 3003, or 5052 aluminum alloy. For this reason, samples of tubing can be tested for hardness, magnetic properties, and reaction to concentrated nitric acid. Hardness is tested by filing or scratching a material

ALUMINUM ALLOY NUMBER	COLOR OF BAND
1100	WHITE
3003	GREEN
2014	GRAY
2024	RED
5052	PURPLE
6053	BLACK
6061	BLUE AND YELLOW
7075	BROWN AND YELLOW

Figure 10-1. Small-diameter aluminum tubing carries a color code to indicate its alloy.

with a scriber, whereas a magnet distinguishes between annealed austenitic and ferritic stainless steels. Typically, the austenitic types are nonmagnetic, whereas the straight chromium carbon and low-alloy steels are strongly magnetic.

Since stainless steel resists corrosion caused by nitric acid, a sample of tubing that corrodes when acid is applied is a carbon, nickel, or copper alloy steel. Nitric acid attacks these alloys at different rates and yields different colors in the corrosion product. [Figure 10-2]

SIZE DESIGNATIONS

The size of rigid tubing is determined by its outside diameter in increments of 1/16 inch. Therefore, a -4 "B" nut tubing is 4/16 or 1/4 inch in diameter. A tube diameter is typically printed on all rigid tubing.

Another important size designation is wall thickness, since this determines a tube's strength. Like the outside diameter, wall thickness is generally printed on the tube in thousandths of an inch.

One dimension that is not printed on rigid tubing is the inside diameter. However, since the outside diameter and wall thickness are indicated, the inside diameter is determined by subtracting twice the wall thickness from the outside diameter. For example, if you have a piece of -8 tubing with a wall thickness of 0.072 inches, you know the inside diameter is .356 inches, $0.5 - (2 \times .072) = 0.356$.

FABRICATING RIGID TUBING

When it is necessary to replace a rigid fluid line, you may obtain a replacement tube assembly from the aircraft manufacturer or fabricate a replacement in the shop. Most shops have the necessary tools to fabricate replacement lines, and as a technician you must be familiar with their operation and limitations.

MATERIAL	MAGNET TEST	NITRIC ACID TEST
CARBON STEEL	STRONGLY MAGNETIC	SLOW CHEMICAL ACTION BROWN
18-8	NON-MAGNETIC	NO ACTION
MONEL	SLIGHTLY MAGNETIC	RAPID CHEMICAL ACTION GREENISH BLUE
NICKEL STEEL	NON-MAGNETIC	RAPID CHEMICAL ACTION GREENISH BLUE

Figure 10-2. Magnet and acid tests help identify various ferrous alloys.

TUBE CUTTING

When cutting a new piece of tubing, you should always cut it approximately 10 percent longer than the tube being replaced. This provides a margin of safety for minor variations in bending. After determining the correct length, cut the tubing with either a fine-tooth hacksaw or a roller-type tube cutter. A tube cutter is most often used on soft metal tubing such as copper, aluminum, or aluminum alloy. However, they are not suitable for stainless-steel tubing because they tend to work harden the tube.

To use a tube cutter, begin by marking the tube with a felt-tip pen or scriber. Next place the tubing in the cutting tool and align the cutting wheel with the cutting mark. Once aligned, gently tighten the cutting wheel onto the tube using the thumbscrew. When the cutting wheel is snug, rotate the cutter around the tube and gradually increase pressure on the cutting wheel every 1 to 2 revolutions. Be careful not to apply too much pressure at one time, as it could deform the tube or cause excessive burring. Once cut, remove all burrs with a special deburring tool. The end of the tube must be smooth and polished so that no sharp edges can produce stress concentrations and cracks when the tube is flared. After the tube has been cut and deburred, blow it out with compressed air to remove metal chips that could become imbedded in the tube. [Figure 10-3]

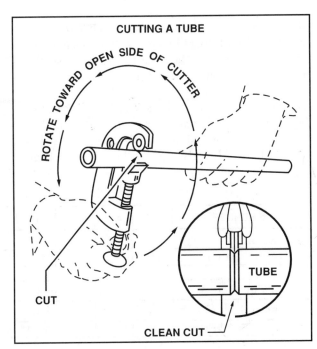

Figure 10-3. To use a tube cutter, gradually increase pressure on the cutting wheel while rotating the cutter toward its open side.

TUBE BENDING

Some applications require rigid lines with complex bends and curves. When duplicating these lines, you must be able to produce bends that are 75 percent of the original tube diameter and free of kinks. Any deformation in a bend affects the flow of fluid. [Figure 10-4]

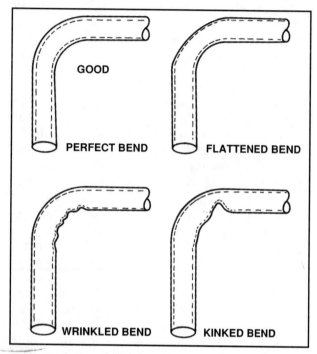

Figure 10-4. A good bend maintains at least 75 percent of the original tube diameter and is free of wrinkles and kinks.

To help reduce the chance of making a bad bend, there are several charts that illustrate standard bend radii for different size tubes. The information on these charts should be adhered to closely. [Figure 10-5]

Tubing under 1/4 inch made of soft metal and having a thin wall can usually be bent by hand. This is accomplished by using a tightly wound steel coil spring that fits snugly around the tubing to keep it

from collapsing. In an emergency, a tube can be bent by first packing it full of clean, dry sand, sealing the ends, and then making the bends. However, when using this method, it is extremely important that every particle of sand be removed from the tube before it is installed.

A variation of the sand method of bending is used in some factories for making complex bends. This process involves filling a tube with an extremely low melting point metal alloy such as Wood's metal or Cerrobend. These alloys are melted in boiling water and then poured into the tube. Once the alloy sets, the tube is bent. When the bending operation is completed, the tube is placed back in a vat of boiling water where the alloy melts and drains out of the tube.

Tubing larger than 1/4 inch in diameter typically requires bending tools to minimize flattening and distortion. Small diameter tubing of between 1/4 inch and 1/2 inch can be bent with a hand bending tool. When using a hand bender, the tube is inserted between the radius block and the slide bar and held in place by a clip. The slide bar handle is then moved down to bend the tubing to the angle needed. The number of degrees the tube is bent is read on the scale of the radius block, opposite the incidence mark on the slide bar. [Figure 10-6]

Production bending is done with a tube bender similar to the one illustrated in Figure 10-7. A tube is clamped between the radius block and the clamp bar, and the radius block is turned by a gear which is driven by the handle. As the radius block turns, the tube is bent between the radius block and a guide bar. The degree of bend is determined by the amount the radius block is turned. Thin-wall tubing is kept from collapsing by using a mandrel, which is a smooth round-end bar that fits into the tube at the bend point. It holds the tube so it cannot collapse.

*TYPE BENDER	A B	A B	B	B	B	B C	B	B C	B	B C	C	B C	C
TUBE OD	1/8"	3/16"	1/4"	5/16"	3/8"	3/8"	7/16"	1/2"	1/2"	5/8"	5/8"	3/4"	3/4"
STANDARD BEND	3/8"	7/16"	9/16"	11/16"	11/16"	15/16"	1 3/8"	1 1/2"	1 1/4"	2"	1 1/2"	2 1/2"	1 3/4"

*TYPE BENDER	C	B	C	C	C	C	C	C	C	C	C	C	C
TUBE OD	7/8"	1"	1"	1 1/8"	1 1/4"	1 3/8"	1 3/8"	1 1/2"	1 1/2"	1 3/4"	2"	2 1/2"	3"
STANDARD BEND	2"	3 1/2"	3"	3 1/2"	3 3/4"	5"	6"	5"	6"	7"	8"	10"	12"

* A—HAND   B—PORTABLE HAND BENDERS   C—PRODUCTION BENDER

Figure 10-5. To avoid making bad tubing bends, refer to a chart for standard bend radii.

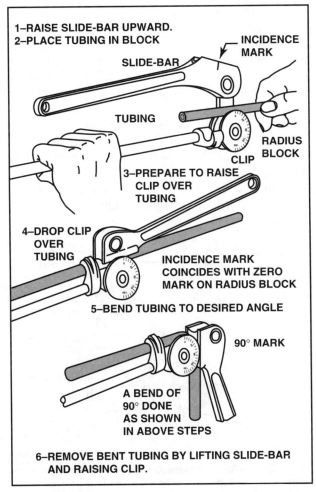

Figure 10-6. Use a hand-held tubing bender to form tubing 1/2 inch or smaller in diameter.

JOINING RIGID TUBING

Sections of rigid tubing can be joined to another tube or to a fitting by several methods. These include single and double-flare connectors, flareless

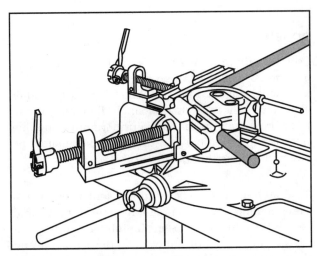

Figure 10-7. A production-type tube bender is used to bend tubing larger than 1/2 inch in diameter.

connectors, or a hose and clamps over a beaded tube. The type of fittings used is determined by the pressure range, the routing, and the material being used for the lines. Whenever you must replace a fitting, make sure you select a fitting made of the same material as the original.

TUBE FLARING

Much of the rigid tubing used in modern aircraft is connected to components by flaring the tube ends and using flare-type fittings. A flared-tube fitting consists of a sleeve and a B-nut. Using this type connector eliminates damage to the flare caused by the wiping or ironing action as the nut is tightened. The sleeve provides added strength and supports the tube so that vibration does not concentrate at the flare. The nut fits over the sleeve and, when tightened, draws the sleeve and flare tightly against a male fitting to form a seal. The close fit between the inside of the flared tube and the flare cone of the male fitting provides the actual seal. Therefore, these two surfaces must be absolutely clean and free of cracks, nicks, and scratches. Aircraft fittings have a flare angle of 37 degrees and are not interchangeable with automotive-type fittings, which have a flare angle of 45 degrees. [Figure 10-8]

There are two types of flares used in aircraft plumbing systems, the **single flare** and the **double flare**. As discussed, the flare provides the sealing surface and is subject to extremely high pressures. Because of

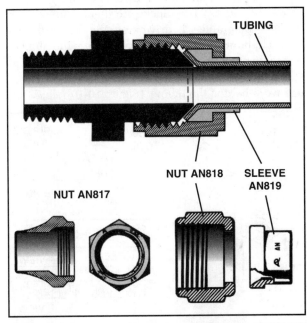

Figure 10-8. The combination of an AN 818 nut and AN 819 sleeve provides a tight, leak-free attachment that does not damage the flare.

this, flares must be properly formed to prevent leaks or failures. [Figure 10-9]

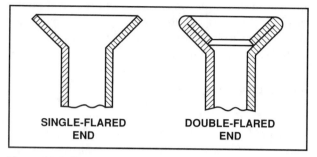

Figure 10-9. Single- and double-flared ends are used on rigid tubing.

A flare which is made too small produces a weak joint, and may leak or pull apart. On the other hand, if a flare is too large it may interfere with the installation of the nut and result in leakage. In either case, if a fitting leaks when properly torqued, you should inspect the flare and fitting components for proper manufacture and assembly. Do not overtighten a leaky fitting.

SINGLE FLARE

A single flare is formed with either an impact-type flaring tool, or one having a flaring cone with a rolling action. To form a flare using an impact-type flaring tool, the tube must be cut squarely and the ends polished. Before the tube is flared, a B-nut and sleeve are slipped on the tube. The tube is then placed in the proper size hole between the halves of the flaring blocks and the plunger is centered over the tube. Once centered, project the end of the tube about 1/16 inch above the blocks. The blocks are then clamped in a vice and the plunger is driven into the tube with several light blows of a hammer, making sure the plunger is rotated one-half turn after each blow. It is important to use as few blows as possible, since too many blows can work-harden the tubing. [Figure 10-10]

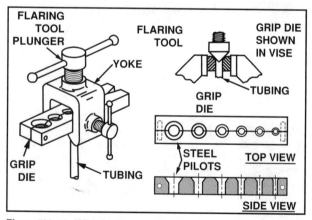

Figure 10-10. With an impact-type flaring tool, the tubing is clamped in flaring blocks while the plunger is driven into the tube to produce the flare.

Roll-type flaring tools are quite popular in aviation maintenance shops because they are entirely self-contained and produce a good flare. A typical roll-type tool can flare tubing from 1/8 to 3/4 inch outside diameter. To use this tool, rotate the dies until the two halves of the correct size are aligned and then insert the tube against the stop. Once you have clamped the dies together, lubricate the flaring cone. The flaring cone is then turned into the end of the tube and rollers in the cone burnish the metal as it expands into the die. When the flare is formed, the handle is reversed to release the dies, and the tube is removed from the tool. [Figure 10-11]

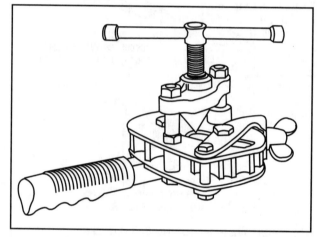

Figure 10-11. With a roll-type flaring tool the flaring cone is turned into the tube eliminating the need for hammering.

Single flares must be made to certain tolerances. Specifically, both the diameter and the radius of the flare must be within specified ranges to ensure a durable, leak-free connection. As previously discussed, a flare that is too small or too large can leak or lead to failure. [Figure 10-12]

DOUBLE FLARE

Soft aluminum tubing with an outside diameter of 3/8 inch or smaller can be double-flared to provide a stronger connection. A double flare is smoother and more concentric than a single flare and, therefore, provides a better seal. Furthermore, a double flare is more durable and resistant to the shearing effect of torque. [Figure 10-13]

To double-flare a piece of tubing, cut it off in the same manner as a single-flare, remove all burrs, and polish the end. Next, insert the tubing into the flaring die to the depth allowed by the stop pin and then clamp the dies. Insert the upsetting tool into the die and, with as few blows of a hammer as possible, upset the tubing. Once the flare is started, insert the flaring tool and strike it with a hammer to fold the metal down into the tubing and form the double flare. [Figures 10-14]

FLARED TUBE FITTINGS

Flared fittings are identified by either an AN or MS number. However, prior to World War II fittings were made to an AC standard. Since AC fittings are still used in some older aircraft, it is important that you be able to identify the differences in fittings. For example, an AN fitting has a shoulder between the

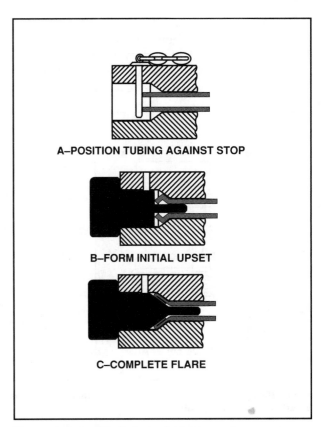

A–POSITION TUBING AGAINST STOP

B–FORM INITIAL UPSET

C–COMPLETE FLARE

Figure 10-13. Double flaring a rigid tube produces a stronger flare that is more resistant to leakage.

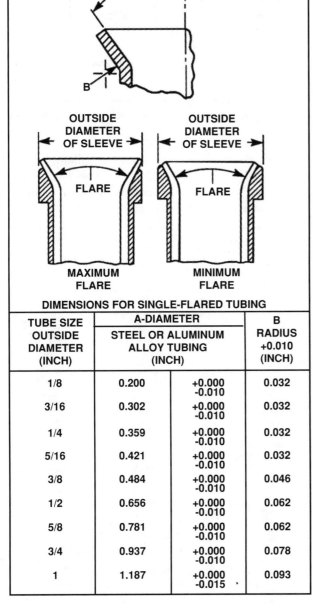

DIMENSIONS FOR SINGLE-FLARED TUBING

TUBE SIZE OUTSIDE DIAMETER (INCH)	A-DIAMETER		B RADIUS +0.010 (INCH)
	STEEL OR ALUMINUM ALLOY TUBING (INCH)		
1/8	0.200	+0.000 -0.010	0.032
3/16	0.302	+0.000 -0.010	0.032
1/4	0.359	+0.000 -0.010	0.032
5/16	0.421	+0.000 -0.010	0.032
3/8	0.484	+0.000 -0.010	0.046
1/2	0.656	+0.000 -0.010	0.062
5/8	0.781	+0.000 -0.010	0.062
3/4	0.937	+0.000 -0.010	0.078
1	1.187	+0.000 -0.015	0.093

Figure 10-12. To prevent leakage, be sure that single flares fall within proper dimensional tolerances.

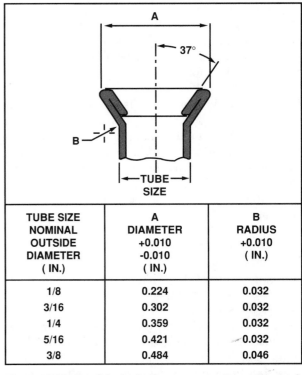

TUBE SIZE NOMINAL OUTSIDE DIAMETER (IN.)	A DIAMETER +0.010 -0.010 (IN.)	B RADIUS +0.010 (IN.)
1/8	0.224	0.032
3/16	0.302	0.032
1/4	0.359	0.032
5/16	0.421	0.032
3/8	0.484	0.046

Figure 10-14. As with single flares, proper dimensional tolerances must be observed for double flares to produce a leak-free seal.

end of the threads and the flare cone. The AC fitting does not have this shoulder. Another difference between AC and AN fittings inlcudes the sleeve design. The AN sleeve is noticeably longer than the AC sleeve of the same size. [Figure 10-15]

When joining tapered pipe thread fittings a small amount of thread lubricant or a thin strip of Teflon™ tape on the male threads ensures that fittings screw together tightly enough to form a complete seal. However, some manufacturers do not rec-

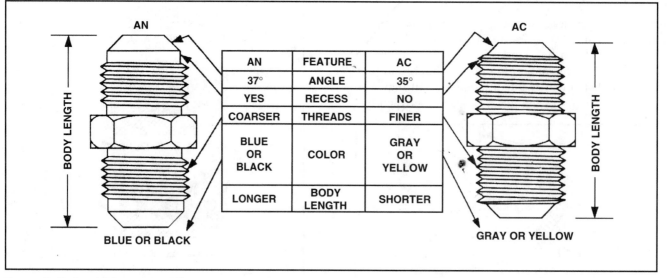

AN	FEATURE	AC
37°	ANGLE	35°
YES	RECESS	NO
COARSER	THREADS	FINER
BLUE OR BLACK	COLOR	GRAY OR YELLOW
LONGER	BODY LENGTH	SHORTER

Figure 10-15. In addition to the differences mentioned, there are also differences in the flare angle, the type of threads, the body length, and the fitting color.

Flared-tube fittings are made of aluminum alloy, steel, or copper base alloys. For identification purposes, all AN steel fittings are colored black, and all AN aluminum fittings are colored blue. The AN 819 aluminum bronze sleeves are cadmium plated and are not colored. AN fittings come in a variety of shapes and sizes, each with a specific use. As an aircraft technician, you must be familiar with the most common fittings used on aircraft. [Figure 10-16]

One specific type of fitting is the **universal bulkhead fitting**. As the name implies, a bulkhead fitting is used to support a line that passes through a bulkhead. Bulkhead fittings have straight machine threads, similar to those on common nuts and bolts. Therefore, flared tube connections, crush washers, or synthetic seals must be used to make these connections fluid-tight.

Fluid lines are commonly attached to components by tapered pipe thread fittings. **Tapered pipe thread fittings** create a seal by wedging the tapered external male thread and the tapered internal female threads. This is the same type of thread used in household plumbing and automotive applications. These threads taper 1/16 inch to the inch. When working with these fittings, care must be exercised when screwing them into cast aluminum or magnesium housings so that the fitting is not screwed tight enough to crack the casting. [Figure 10-17]

ommend Teflon tape because tape fragments can be introduced into a system if the tape is not applied correctly. For this reason you should always follow the manufacturer's directions.

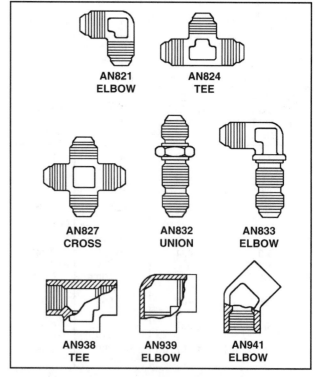

Figure 10-16. AN fittings are available in many shapes and sizes to suit a variety of installation requirements.

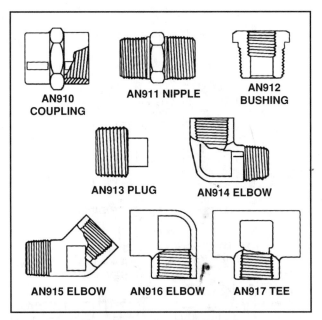

Figure 10-17. Pipe fittings use tapered threads to provide a seal. They are not interchangeable with AN fittings.

FLARELESS FITTINGS

The heavy wall tubing used in some high-pressure systems is difficult to flare. For these applications, the flareless fitting is designed to provide leak-free attachments without flares. Although the use of flareless fittings eliminates the need to flare the tube, a step referred to as **presetting** is necessary prior to installation of a new flareless tube assembly. Presetting is the process of applying enough pressure to the sleeve to cause it to cut into the outside of the tube.

To preset a flareless fitting, lubricate a nut and sleeve, sometimes called a ferrule, and slip them over the end of a tube. Next, screw the nut onto the presetting tool, making sure the tube is square against the bottom of the tool. Now, screw the nut down by hand until it tightens the ferrule against the presetting tool. The final tightening depends upon the tubing. For example, for aluminum alloy tubing up to and including 1/2 inch outside diameter, tighten the nut from one to one and one-sixth turns. For steel tubing and aluminum alloy tubing over 1/2 inch, tighten from one and one-sixth to one and one-half turns. Once this is done, remove the tube from the presetting tool and examine the ferrule. The tube should have a uniform indentation, or "bite," around its end, indicating that the tube is square and has bottomed evenly into the tool. In addition, the bite should be even and the tube material should raise to a height of at least 50 percent of the ferrule thickness. The ferrule may rotate on the tube, but it should not move back and forth. Furthermore, the ferrule should be slightly bowed to provide the proper sealing against the edge of the fitting. [Figure 10-18]

BEADING

Large diameter lines carrying low-pressure fluids such as engine return oil and cooling air are typically joined by a rubber hose that is slipped over the tube ends and held in place with screw-type hose clamps. However, for this to be effective the tube must be **beaded** first. This can be accomplished with either a power beader or a hand beading tool. The diameter and wall thickness of the tube being

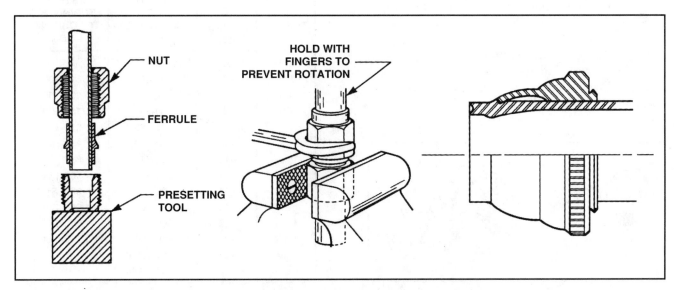

Figure 10-18. Flareless fittings must be preset before assembly to provide a tight seal. A properly preset fitting produces a slight, uniform indentation in the rigid tube, whereas, excessive presetting cuts deeply into the tube and weakens it.

beaded determine which is used. For example, a hand-beading tool is used with tubing having 1/4 inch to 1 inch outside diameter. When using a hand beading tool, the bead is formed by a beader frame with the proper rollers. The sizes, which are marked on the rollers in sixteenths of an inch, correspond with the outside tube diameter. Separate rollers are required for the inside of different sized tubing so care must be taken to use the correct parts when beading.

The beading tool operates somewhat like a tube cutter in that the roller is screwed down while rotating the beading tool around the tubing. However, the inside and outside of the tube must be lubricated with a light oil to reduce friction during beading. [Figure 10-19]

When joining two beaded tubes, begin by slipping the hose over the beads and centering the hose clamps between the ends of the hose and the beads. Next, tighten the clamps finger-tight followed by one and one-half to two complete turns using a wrench or pliers. When doing this, be careful not to overtighten the clamps or you could cause excessive "cold-flow," which is indicated by deep, permanent impressions in the hose. [Figure 10-20]

RIGID TUBING INSTALLATION

Before installing a tube assembly in an aircraft, inspect the tube carefully for nicks, scratches, and dents. In addition, you should inspect the fittings and tube for cleanliness. Never apply sealing compound or anti-seize to a fitting's sealing surfaces since these surfaces depend on metal-to-metal contact to seal. If any foreign matter is present it should be removed so the seal is not compromised.

Before securing a line assembly in place, be sure that it is properly aligned. Furthermore, since rigid tubing expands and shifts when pressurized, an installation that is under tension is undesirable.

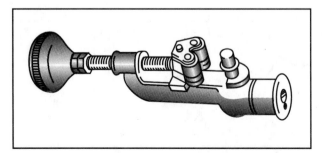

Figure 10-19. Tube beading provides a convenient sealing method for low-pressure applications.

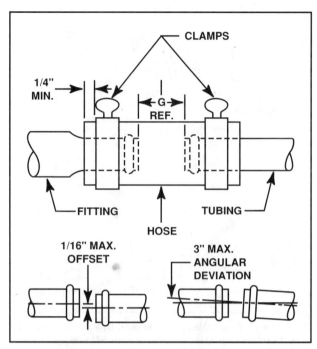

Figure 10-20. If hose clamps are tightened excessively, the hose "cold flows" through the clamp and becomes susceptible to leakage. It is also important that the beaded tubes are not offset or misaligned.

Never pull an assembly into alignment by tightening the nut. Also, since overtightening a fitting may damage the sealing surface or cut off a flare, fittings should always be installed to the specified torque using a torque wrench. [Figure 10-21]

TUBING O.D.	FITTING SIZE	ALUMINUM ALLOY TUBING, NUT TORQUE INCH-LBS.	STEEL TUBING, NUT TORQUE INCH-LBS.
1/8	–2	20-30	
3/16	–3	30-40	90-100
1/4	–4	40-65	135-150
5/16	–5	60-85	180-200
3/8	–6	75-125	270-300
1/2	–8	150-250	450-500
5/8	–10	200-350	650-700
3/4	–12	300-500	900-1,000
7/8	–14	500-600	1,000-1,100
1	–16	500-700	1,200-1400,
1-1/4	–20	600-900	1,200-1,400
1-1/2	–24	600-900	1,500-1,800
1-3/4	–28	850-1,050	
2	–32	950-1,150	

Figure 10-21. The amount of torque applied to a fitting varies with the type of tubing and the fitting size.

Installation of a properly preset flareless fitting is made by tightening the fitting by hand until it bottoms. If this is not possible, a wrench may be used, however, be alert to the first signs of bottoming. Final tightening, on the other hand, is completed with a wrench by turning the nut 1/6 of a turn. When doing this, use a wrench on the male fitting to prevent it from turning. Some manufacturers specify torque limits for this type of fitting. [Figure 10-22]

After all connections are made, the system should be pressure tested. If a connection leaks, some manufacturers allow the nut to be tightened an additional 1/6 turn.

WRENCH TORQUE FOR 304 1/8 H STEEL TUBES		
TUBE O.D.	WALL THICKNESS	WRENCH TORQUE (IN.-LBS.)
3/16	0.016	90-110
3/16	0.020	90-110
1/4	0.016	110-140
1/4	0.020	110-140
5/16	0.020	100-120
3/8	0.020	170-230
3/8	0.028	200-250
1/2	0.020	300-400
1/2	0.028	400-500
1/2	0.035	500-600
5/8	0.020	300-400
5/8	0.035	600-700
5/8	0.042	700-850
3/4	0.028	650-800
3/4	0.049	800-960
1	0.020	800-950
1	0.065	1,600-1,750
WRENCH TORQUE FOR 304-1A OR347-1A STEEL TUBES		
3/8	0.042	145-175
1/2	0.028	300-400
1/2	0.049	500-600
1	0.035	750-900
WRENCH TORQUE FOR 6061-T6 OR T4 ALLOY TUBES		
1/4	0.035	110-140
3/8	0.035	145-175
1/2	0.035	270-330
1/2	0.049	320-380
5/8	0.035	360-440
5/8	0.049	425-525
3/4	0.035	380-470
1	0.035	750-900
1-1/4	0.035	900-1,100

Figure 10-22. When a manufacturer does not specify torques for flareless fittings, a table of standard torque values should be used.

Since rigid tubing flexes and expands under pressure, all runs of tubing must have at least one bend between the fittings to absorb these strains. Furthermore, all bends must be located in such a way that they can be supported by clamps so that the stress from vibrations and from expansion and contraction is not placed on a fitting. [Figure 10-23]

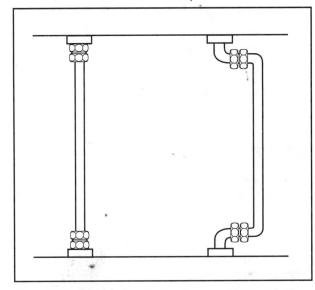

Figure 10-23. To help absorb the stress caused by vibration, tubing should have at least one bend.

ROUTING AND SECURING

All fluid lines should be routed through the aircraft in such a way that they have the shortest practical length. Furthermore, for the sake of appearance and ease of attachment, all fluid lines should follow the structural members of the aircraft and should be secured with appropriate clamps. For example, all fuel lines must be bonded to the structure with integrally bonded line support clamps. [Figure 10-24]

It is important that no fluid line be allowed to chafe against any control cable or aircraft structure. Furthermore, every effort should be made to prevent

TUBE O.D.	DISTANCE BETWEEN SUPPORTS
1/8 – 3/16	9"
1/4 – 5/16	12"
3/8 – 1/2	16"
5/8 – 3/4	22"
1 – 1-1/4	30"
1-1/2 – 2	40"

Figure 10-24. As you can see, smaller diameter fuel lines require more support than larger fuel lines. The primary reason for this is because small diameter tubing is not as strong as larger tubing.

fluid lines from coming in contact with electrical wiring bundles or conduit carrying electrical wires. However, if it is impossible to separate fluid lines from electrical wire bundles, the wire bundle must be routed above the fluid line, and it must be clamped securely to the structure. Under no circumstances should a wire bundle be supported by a fuel line or any line carrying flammable fluid.

By the same token, you should avoid routing fluid lines through passenger compartments. However, if a fluid line must be routed through a passenger, crew, or baggage compartment, it must be supported and protected against damage or installed in such a way that it cannot be used as a hand hold.

SUPPORT CLAMPS

Support clamps are used to secure fluid lines to the aircraft structure, or to assemblies in the engine nacelle. In addition to providing support, these clamps prevent chafing and reduce stress. The two clamps most commonly used are the rubber-cushioned clamp and the plain clamp. The **rubber cushioned clamp** secures lines which are subject to vibration. The clamp's rubber cushion reduces the transmission of vibrations to the line and prevents chafing. In areas subject to contamination by fuel or phosphate ester type hydraulic fluid, cushioned clamps utilizing Teflon are used. Although these do not provide the same level of cushion, they are highly resistant to deterioration. The **plain clamp** is used in areas that are not subject to vibration and typically consists of a metal band formed into a circle. [Figure 10-25]

A third type of clamp used to secure metal fuel, oil, or hydraulic lines is the bonded clamp. **Bonded clamps** have an electrical lead that is connected to the aircraft structure to ground a tube. When installing a bonded clamp, be sure to remove any paint or anodizing from the tube where the bonding clamp is fastened. Unbonded clamps should be used only to secure wiring.

REPAIRING RIGID TUBING

The most common problem encountered with fluid lines is leakage. If a fluid line leaks at a fitting, pressure should be removed from the line and the fitting checked for proper torque. It is never proper to over-torque a fitting in an attempt to stop a leak. If a fitting leaks after it is determined to be properly torqued, the nut should be unscrewed and the sealing surfaces carefully examined. If there is any sign

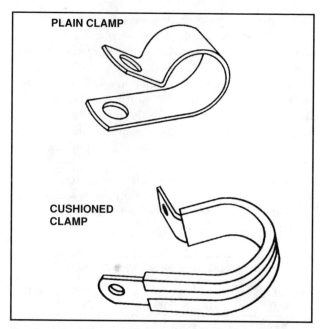

Figure 10-25. The most commonly used types of clamps for aircraft are the rubber cushioned clamp and the plain clamp.

of damage, the fitting must be replaced. Any crack or deformity in a flare is cause for rejection.

Minor dents and scratches in tubing may be repaired. Scratches or nicks in aluminum alloy tubing that are no deeper than 10 percent of the wall thickness and not in the heel of a bend can be repaired by burnishing. However, tubing with severe die marks, seams, or splits must be replaced. A dent less than 20 percent of the tube diameter is permitted if it is not in the heel of a bend.

To remove dents, the tubing must be removed from the aircraft and a bullet of the proper size drawn through the tube using a short length of cable. The bullet can also be pushed through the tube using a dowel rod. The "bullet" used for this operation may be a ball bearing or a slug made of steel or some other hard metal. In soft aluminum tubing, a bullet made of hardwood can be used. [Figure 10-26]

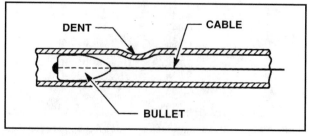

Figure 10-26. One way to remove small dents in metal tubing is to draw a small metal slug, or bullet, through a tube.

Damaged sections of tubing are replaced by cutting out the damaged area and splicing in a new section. This is accomplished by carefully removing the damaged section and inserting a replacement section of the same size and material. When doing this, both ends of the undamaged tube and replacement tube are flared and secured using standard unions, sleeves, and nuts. [Figure 10-27]

IDENTIFICATION OF FLUID LINES

Large aircraft contain plumbing systems for many different types of fluids. Because of this, it is important that each line be clearly identified. This is generally accomplished by marking tubing with color bands, symbols, or writing. The symbols are generally printed on one-inch wide tape or decals and secured at regular intervals along a line. On lines four inches or larger in diameter, or those subject to extreme temperatures, steel tags are used instead of marking tape. In areas where there is the possibility that tape, decals, or tags may be drawn into the induction system, paint is used.

In addition to color bands, some lines carrying fuel are marked with the word "FLAM." This identifies the lines as carrying a flammable fluid. Lines carrying fluids that are physically dangerous such as oxygen, nitrogen, or Freon are marked "PHDAN."

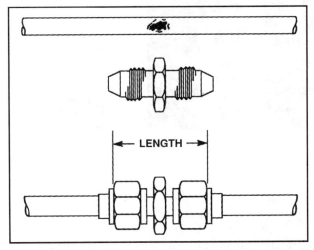

Figure 10-27. Remove damaged tubing and replace it with a new piece of tubing. The cut ends are flared and connected with standard hardware.

Additional markings are sometimes provided to identify a line's function. These include PRESSURE, RETURN, DRAIN, and VENT. [Figure 10-28]

Generally, tapes and decals are placed on both ends of a line and at least once in each compartment through which the line runs. In addition, identification markers are placed immediately adjacent to each valve, regulator, filter, or other accessory within a line. Where paint or tags are used, location requirements are the same as for tapes and decals.

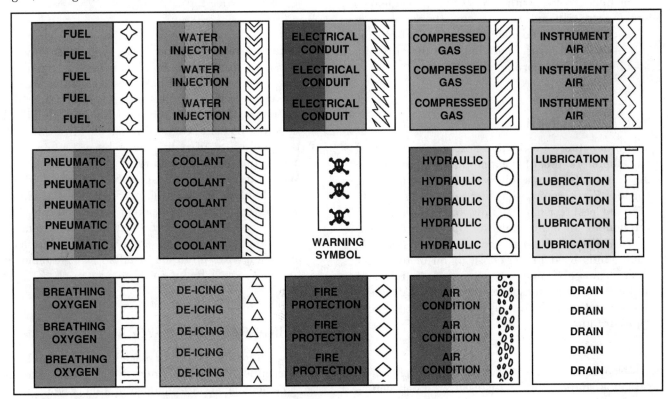

Figure 10-28. Color codes identify the fluid a line carries and warn of potential hazards.

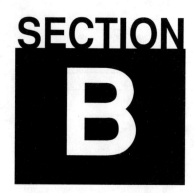

SECTION B

FLEXIBLE FLUID LINES

Flexible fluid lines are used extensively on aircraft to connect stationary parts to moving parts and in areas of high vibration. Therefore, as a maintenance technician you must be familiar with the various types of flexible hoses available, including those designed for special applications. Furthermore, since aircraft systems operate with different fluids under a wide range of pressures, it is imperative that you be able to identify the type of hose that is compatible with each fluid and strong enough to contain its pressure. In addition, a number of fluid line fittings have been used in aircraft over the years, and they are often not compatible with each other. If different types are mixed, the line is likely to leak.

FLEXIBLE HOSE CONSTRUCTION

Flexible hose construction generally consists of an inner liner covered with layers of reinforcement to provide strength, and an outer cover to protect from physical damage. The materials and manufacturing process of each layer determine the suitability of a specific hose for a particular application.

THE INNER LINER

The inner liner of a flexible hose carries the fluid and, therefore, must have a minimum porosity and be chemically compatible with the material being carried. Furthermore, the liner must be smooth to offer the least resistance to flow, and remain flexible throughout an entire range of operating temperatures. There are basically four different synthetic compounds used in the construction of the inner liner. They are neoprene, Buna-N®, butyl, and Teflon®. Each of these compounds has different characteristics and is compatible with different types of fluid.

1. **Neoprene** is a form of synthetic rubber that is abrasion resistant and is used with petroleum-based fluids.

2. **Buna-N®** is a synthetic rubber compound that is also used to carry petroleum-based products. In fact, Buna-N is better suited to carry petroleum products than neoprene.

3. **Butyl** is a synthetic rubber compound made from petroleum raw materials and, therefore, breaks down if used with petroleum products. However, butyl is excellent as an inner liner for fluid lines carrying phosphate ester-base hydraulic fluids such as Skydrol®.

4. **Teflon®** is the DuPont trade name for Tetrafluoroethylene resin. Teflon® has an extremely broad operating temperature range (–65°F to +450°F) and is compatible with nearly every liquid used. Furthermore, its unique wax-like surface offers minimum resistance to fluid flow. Because of its unique chemical structure, Teflon experiences less volumetric expansion than rubber and has an almost limitless shelf and service life.

When installing Teflon hose, you must always observe minimum bend radius restrictions. Furthermore, after the hose has been in service, it "takes a set," or becomes somewhat rigid. Therefore, if a hose is removed from the aircraft, it must not be bent against this set. To further reduce the possibility of damaging a hose, certain hose assemblies are manufactured with a pre-set shape. When this is done, they are usually shipped with a wire holding the assembly in its pre-set shape. When working with a hose that is shipped this way, the wire should remain in place until the assembly is installed. If a pre-set assembly is removed from an aircraft, a wire should be installed prior to removal to help the hose maintain its shape.

REINFORCEMENT LAYERS

The reinforcement layers placed over an inner liner determine the strength of a hose. Common reinforcement layers are made of cotton, rayon, polyester fabric, carbon-steel wire, or a stainless steel wire braid. Since hose has a tendency to increase in diameter and decrease in length when pressure is applied, the design of the reinforcement is critical. The proper design of the reinforcement layers can minimize these dimensional changes.

OUTER COVER

A protective outer cover, usually made of rubber-impregnated fabric or stainless steel braid, is put over the reinforcement to protect the hose from physical damage. In areas of high heat the outer cover is often designed as an integral fire-sleeve to provide extra protection.

The outer cover of almost all aircraft flexible hose is marked with a **lay line**, which consists of a yellow, red, or white stripe running the length of the hose. In addition to a stripe, the information needed to identify the hose, such as the MIL-SPEC number, the manufacturer's name or symbol, the dash number representing the hose size, and in some cases, the manufacturer's part number along with the year and quarter the hose was manufactured. In addition to identifying a hose, the lay line shows if a hose is twisted when it is installed. When a hose is installed properly, the lay line runs straight with no twists. [Figure 10-29]

TYPES OF FLEXIBLE HOSE

While aircraft hose is manufactured to meet a variety of applications, the types of hose are normally classified by the amount of pressure they are designed to withstand. These include low-pressure, medium-pressure, and high-pressure.

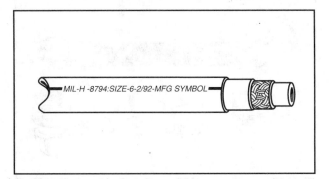

MIL-H-8794:SIZE-6-2/92-MFG SYMBOL

Figure 10-29. A lay line shows if a hose is twisted when it is installed.

LOW-PRESSURE HOSE

Most air or vacuum hoses and some aircraft instrument lines are not required to carry high pressures. Therefore, low pressure rubber hose is typically used with these types of installations. These hoses have a seamless inner tube and a reinforcement made of a single layer of cotton braid. An outer cover of ribbed or smooth rubber is used to protect the reinforcement from physical abrasion. [Figure 10-30]

MEDIUM-PRESSURE HOSE

Medium-pressure hose is used with fluid pressures up to 3,000 psi. However, its maximum operating pressure varies with its diameter. For example, smaller sizes carry pressure up to 3,000 psi while larger sizes are often restricted to lower pressures. Medium-pressure hose has a seamless inner liner with one layer of cotton braid and one layer of stainless-steel reinforcement. A braid of rough oil-resistant rubber-impregnated cotton is usually used as an outer cover. If the hose is used with a petroleum-based fluid, its inner liner is made of synthetic rubber and its outer braid is gray-black. However, if the hose is used with Skydrol or any phosphate-ester

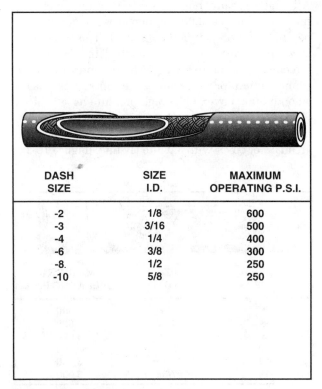

DASH SIZE	SIZE I.D.	MAXIMUM OPERATING P.S.I.
-2	1/8	600
-3	3/16	500
-4	1/4	400
-6	3/8	300
-8	1/2	250
-10	5/8	250

Figure 10-30. The maximum operating pressure of low pressure hose varies with the size.

based hydraulic fluid, the inner liner is made of synthetic Butyl rubber and the outer braid is colored green with SKYDROL written on it. [Figure 10-31]

DASH SIZE	SIZE I.D.	MAXIMUM OPERATING P.S.I.
-2	1/8	3,000
-3	3/16	3,000
-4	1/4	3,000
-5	5/16	2,000
-6	13/32	2,000
-8	1/2	1,750
-10	5/8	1,500
-12	7/8	800
-16	1-1/8	600
-20	1-3/8	500
-24	1-13/16	350
-32	2-3/8	250
-40	3	200

Figure 10-31. Depending on diameter, medium-pressure hose can carry up to 3,000 psi.

HIGH-PRESSURE HOSE

All high-pressure hose has a maximum operating pressure of at least 3,000 psi and uses a synthetic rubber liner to carry petroleum products. This inner liner is wrapped with two or more steel braids as reinforcement. To help distinguish high-pressure hose from medium-pressure hose, the entire hose has a smooth outer cover. Most high-pressure hose is black with a yellow lay line. However, a hose designed to carry Skydrol has a Butyl rubber inner liner and a green outer cover with a white lay line. [Figure 10-32]

DASH SIZE	SIZE I.D.	MAXIMUM OPERATING P.S.I.
-4	7/32	3,000
-6	11/32	3,000
-8	7/16	3,000
-10	9/16	3,000
-12	11/16	3,000
-16	7/8	3,000

Figure 10-32. All high-pressure hose withstands at least 3,000 psi of pressure, regardless of the hose diameter.

SIZE DESIGNATION

The size of a flexible hose is determined by its inside diameter and is measured in increments of 1/16 inch. Like rigid tubing, a dash number indicates the tube diameter. For example, a -10 identifies a 10/16 or 5/8 inch hose.

REPLACING FLEXIBLE HOSE

At each regular inspection interval, hose assemblies should be checked for deterioration. Indications of leakage, mechanical damage, separation of the braid from the covering, or broken wire braids indicate replacement may be necessary. Replacement hose assemblies are obtained from the manufacturer, a hose assembly shop, or they can be fabricated in the shop. The decision on where to obtain the replacement assembly is usually based on the complexity of the hose assembly and the type of fittings used.

FITTINGS

Fittings provide a convenient method of connecting flexible hoses to components. Flexible hose fittings are typically classified by the way they are attached to a hose and by the amount of pressure they can withstand.

SWAGED FITTINGS

Hoses using swaged end fittings are assembled on special machinery that is typically not found in the shop. These fittings cannot be removed and reused. Therefore, replacement lines with swaged fittings must be obtained from the manufacturer or a properly equipped hose assembly shop. [Figure 10-33]

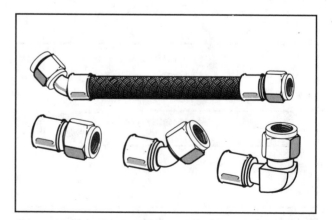

Figure 10-33. Swaged fittings require special machinery for assembly and cannot be reused.

REUSABLE FITTINGS

Some hose assemblies incorporate reusable fittings consisting of a socket, a nipple, and a nut. When a failure occurs in a hose with this type of fitting, a replacement assembly can generally be fabricated. However, prior to reusing any fittings, they must be removed from the damaged hose assembly and carefully inspected. Any damage to the sealing surface or the threads is cause for rejection. Furthermore, the nut should be inspected for signs of cracking and damage caused by wrenches. Damaged components should not be reused. While all fittings used on aircraft must conform to MIL-SPEC, they are manufactured by several companies. Therefore, it is a good practice not to mix components manufactured by different companies. [Figure 10-34]

To install a reusable fitting, begin by determining the length of the hose required. One way to do this is to use the damaged hose assembly with the fittings installed as a pattern. The length of hose required extends from the inside of one socket to the inside of the opposite socket. [Figure 10-35]

One thing to keep in mind is that flexible fluid lines must have between five and eight percent slack to allow for the change in dimensions caused by fluid pressure. Under pressure, flexible hose contracts in length and expands in diameter.

Once you know the proper hose length, mark the cut length on an identical type of new hose and cut both ends off square. The cuts may be made with a cutting tool designed especially for hose, or with a fine tooth hacksaw. After the hose is cut, place a socket in a vise taking care to protect the socket from the vise jaws. For low-pressure and some other types of fittings, a wooden vise block can be fabricated to hold the socket securely with minimal chance for damage. With the socket secured in the vise, screw the hose into the socket until it bottoms. Then, back off the hose just enough to prevent the rubber in the inner liner from obstructing the hole.

With the socket securely in place, lubricate the end of the assembly tool and force it into the hose to open the inner liner enough for the nipple to be inserted. The nipple screws into the socket with the assembly tool and squeezes the tube tight between the outside of the nipple and the inside of the socket. This squeezing action provides a strong physical attachment between the hose and the fitting, and forms a leak-proof seal. Once the nipple is screwed completely into the socket, back it off from 1/32 to 1/16 inch to allow the nut to turn freely on

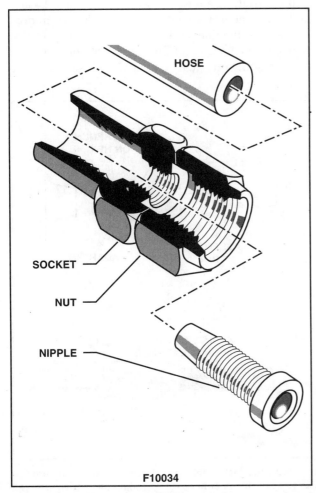

F10034

Figure 10-34. The reusable fitting can be removed and installed with basic hand tools in the shop.

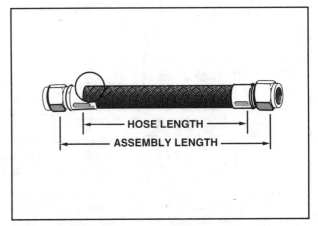

Figure 10-35. To ensure that the final assembly is the correct length, use the old hose as a pattern.

the tube assembly. The inside of the nipple forms the sealing surface which mates with the flare cone of the fitting, and the nut pulls these two sealing surfaces together. When the assembly tool is screwed out of the hose, the inside of the hose should be blown out with compressed air and the entire hose inspected for physical condition. [Figure 10-36]

Once fabricated, test the hose assembly by forcing fluid through it in both directions. This helps verify that the inner liner is secure. If it is not secure, the liner can create a sort of check valve allowing flow in only one direction. The hose should then be "proof-tested" by capping one end of the hose and applying pressure to the inside of the hose assembly. This test is normally conducted at twice the working pressure of the hose, and is held for one to five minutes.

With some high-pressure fluid lines, proof testing can create a safety hazard. In this situation, a proof test must be conducted so the operator is protected in case the hose or fitting fails to pass the pressure test.

The final step in the fabrication of a fluid line should always be to clean the new hose assembly. It is very important to remove any debris that may be in the assembly. Clean all traces of lubricants used during the assembly operation, and remove all traces of fluid used for proof-testing. A mild soap and water solution is generally acceptable. Once clean, dry the hose assembly and place protective caps in both fittings to prevent damage prior to installation.

When installing reusable fittings on some high-pressure hoses the outer cover may have to be stripped from the hose to the depth of the socket. However, before attempting this you should consult the manufacturer's service information. By the same token, you should consult with the manufacturer's service information before attempting to field assemble a hose used on any oxygen system. These assemblies may require special types of hose, fittings, and have special cleaning requirements. For example, because of the fire hazard you should never use petroleum based lubricants on an oxygen line or fitting.

FLEXIBLE HOSE INSTALLATION

Before installing a hose assembly, verify that the aircraft manufacturer specifies a flexible hose is appropriate. If a flexible hose is permitted, inspect the replacement hose carefully. Check for proper type and length, physical damage and cleanliness.

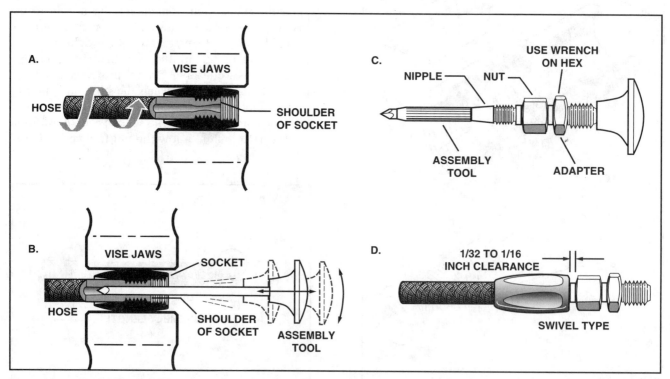

Figure 10-36. Flexible hose fabrication consists of four basic steps. (A) — With the socket held in a vise, screw the hose into the socket. (B) — Next, lubricate the assembly tool and force it into the hose to open the inner liner. (C) — The nipple is then screwed into the socket with the assembly tool. (D) — Once installed, there should be approximately 1/32 to 1/16 inch clearance between the nut and socket to allow the nut to turn freely.

Furthermore, you must ensure that the hose cure date and assembly date are within limits for that type of material. Part number, cure date, and assembly date of hose assemblies are found on the hose identification tag. It is important that the lay line be straight when the hose is installed. Any spiraling is an indication that the hose is twisted and is under an undue amount of strain when there is pressure in the line. Flexible hose should be installed so that it is subject to a minimum of flexing during operation. Although hose must be supported at least every 24 inches, closer supports are desirable. A flexible hose must never be stretched tightly between two fittings. [Figure 10-37.]

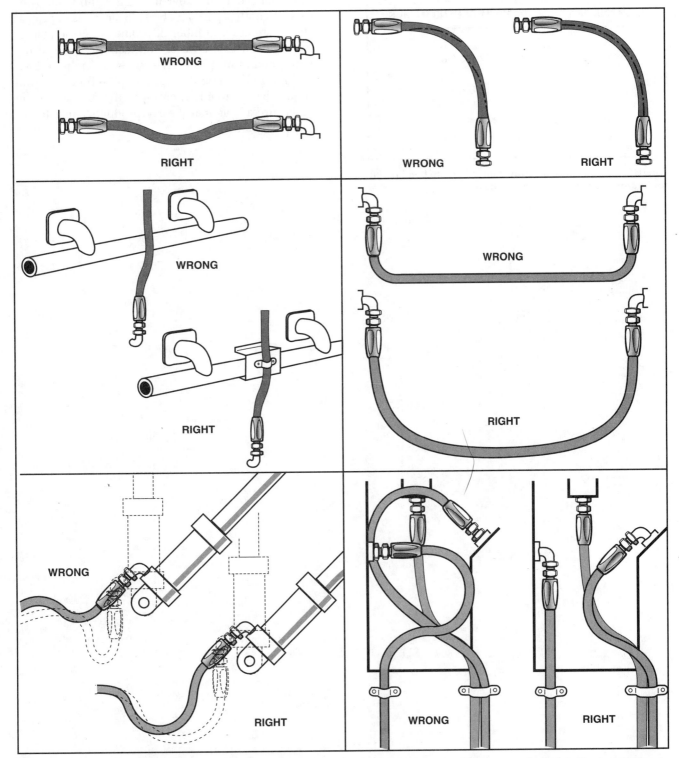

Figure 10-37. Never install a flexible hose tightly between two fittings. Always ensure there is enough slack in the installation to allow for expansion and contraction.

The minimum bend radius for flexible hose is determined by the type of hose being used and its size. Bends that are too sharp reduce the bursting pressure of flexible hose below its rated value. [Figure 10-38.]

DASH NUMBER	MINIMUM BEND RADIUS
-4	3"
-5	3 3/8"
-6	5"
-8	5 3/4"
-10	6 1/2"
-12	7 3/4"
-16	9 5/8"

Figure 10-38. This table illustrates the minimum bend radius for intermediate-pressure flexible hose.

PROTECTIVE SLEEVES

In certain areas, flexible hose must be protected from wear caused by abrasion or extreme heat. For example, if a fluid line must pass near a hot exhaust manifold, the line must be protected with a suitable fire shield. On the other hand, if a fluid line rubs against another part, an abrasion sleeve is appropriate. There are a number of products on the market designed for this type of application. Some of the more common protective sleeves include heat shrink, nylon spiral wrap, and Teflon. Caution should be observed when replacing fire sleeves on older aircraft, as many early products consisted of asbestos braid.

NONDESTRUCTIVE TESTING

CHAPTER 11

INTRODUCTION

As aircraft increased in complexity, the rising cost of down time became a growing concern to aircraft operators. Airlines could not afford to ground an aircraft for long periods to conduct maintenance or inspections. Furthermore, as components became more expensive, new methods of inspection had to be developed to allow inspection without disassembly or destruction of the part. The inspection techniques developed are known as **nondestructive testing** (NDT). This chapter discusses the various methods of NDT including the fundamentals of visual, liquid penetrant, magnetic particle, eddy current, ultrasonic, and radiographic inspections.

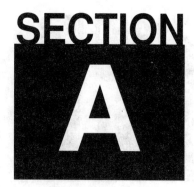
BASIC INSPECTIONS

The most fundamental method of inspecting aircraft structures and components is through visual inspection. This method is irreplaceable in certain circumstances and limited in others. In any case, nothing can be inspected visually unless it is uncovered and made visible.

The basic tools required to conduct a visual inspection include a good light, a mirror, and some form of magnifying glass. Flashlights are typically used to give spot-type illumination to the inspection area. Some flashlights have flexible extensions that allow you to illuminate inaccessible areas. [Figure 11-1]

A **borescope** is an optical device similar in principle to a telescope in that it enlarges objects like a magnifying glass. However, a borescope has a small lens mounted on a shaft with a built-in light source that illuminates the area being inspected. Borescopes are typically used to inspect inside engines using the spark plug hole for access. This optical device allows inspection without disassembly. [Figure 11-2]

FIBEROPTIC SCOPE

A fiberoptic borescope is similar to a standard borescope, but has a flexible, articulated probe that can bend around corners. This allows you to view areas deep inside an assembly that previously required disassembly to inspect. In a typical borescope, a bundle of optical glass fibers transmit light from a light source to the scope's end, or probe. The probe is then inserted into the structure being inspected. The maximum length available for fiberoptic borescopes is four feet. [Figure 11-3]

Special attachments allow mounting of a camera to the borescope to photographically record what is seen through the scope. The pictures taken aid in describing a situation to an inspector for airworthiness determination.

VIDEO SCOPE

Another borescope used for inspecting inaccessible locations is the video borescope. The video scope is similar to a fiberoptic scope, except that the image is recorded by a tiny light-sensitive chip in the end of the probe and transmitted electronically to a video monitor. The video scope provides a high quality image of the area being viewed and can easily be adapted to video recording equipment for inspection records and review. [Figure 11-4]

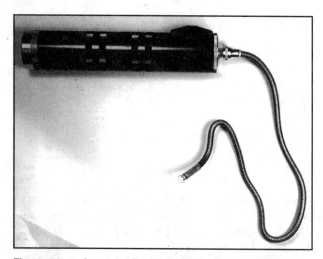

Figure 11- 1. A special flashlight with an extension for the bulb allows inspection of hard-to-reach locations.

Figure 11-2. A borescope allows inspection of internal engine components without disassembling the engine.

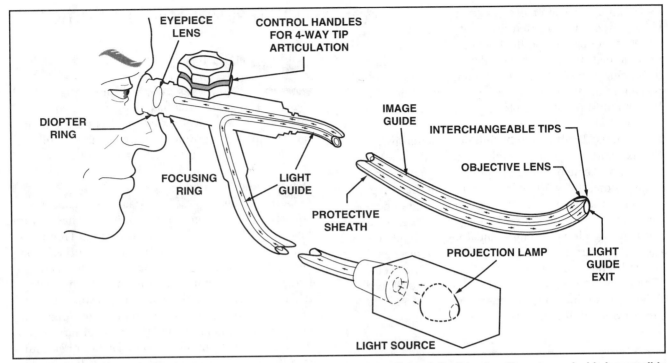

Figure 11-3. A fiberoptic borescope uses the principle of light transmission through flexible glass fibers to see inside inaccessible areas.

WELD INSPECTION

While several means of nondestructive testing are used to inspect welds, visual inspection is the most practical and thus most common. A good weld is uniform in width, with even ripples that taper off smoothly into the base metal. There should be no burn marks or signs of overheating, and no oxide should form on the base metal more than 1/2 inch from the weld. Furthermore, a good weld must be free of gas pockets, porosity, and inclusions. [Figure 11-5]

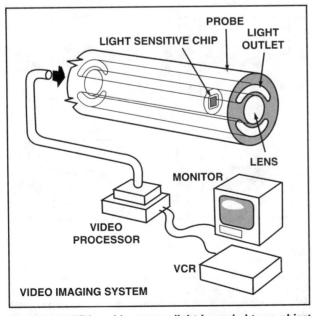

Figure 11-4. With a video scope, light is carried to an object by fiberoptics or light-emitting diodes. The image is viewed through a lens by a light sensitive chip and transmitted to a video processor where the electronic signal from the chip is assembled and output to a monitor and VCR as appropriate.

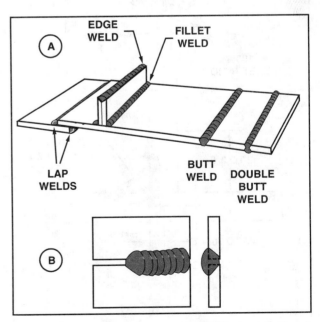

Figure 11-5. (A) — The most common welds encountered in aviation are the butt weld, double butt weld, lap weld, fillet weld, and edge weld. (B) — A good weld tapers evenly into the base metal and shows good penetration free of gas pockets.

Penetration is the depth of fusion in a weld, and is the most important characteristic of a good weld. Penetration depends on the thickness of the material to be joined, the size of the filler rod, and welding technique. A typical butt weld should penetrate 100 percent of the thickness of the base metal, while a fillet weld must penetrate 25 to 50 percent. [Figure 11-6]

Poor welds display certain telltale characteristics. For example, too much acetylene makes the molten metal boil, causing bumps along the center and craters along the weld's edge. A cold weld has irregular edges and considerable variation in depth of penetration, while excessive heat produces a weld with pitting along its edges and long, pointed ripples. If a part is cooled too quickly after being welded, cracks often appear adjacent to the weld. Whenever a welded joint displays any of these defects, all of the old weld must be removed and the joint rewelded. [Figure 11-7]

LIQUID PENETRANT INSPECTION

Liquid penetrant inspection is a method of nondestructive inspection suitable for locating cracks, porosity, or other types of faults open to the surface. Penetrant inspection is usable on ferrous and nonferrous metals, as well as nonporous plastic material. The primary limitation of dye penetrant inspection is that a defect must be open to the surface.

Dye penetrant inspection is based on the principle of **capillary attraction**. The area being inspected is covered with a penetrating liquid that has a very low viscosity and low surface tension. This penetrant is allowed to remain on the surface long enough to allow the capillary action to draw the penetrant into any fault that extends to the surface. After sufficient time, the excess penetrant is washed off and the surface is covered with a developer. The developer, by the process of reverse capillary action, blots the penetrant out of cracks or other faults forming a visible line in the developer. If an indication is fuzzy instead of sharp and clear, the probable cause is that the part was not thoroughly washed before the developer was applied. [Figure 11-8]

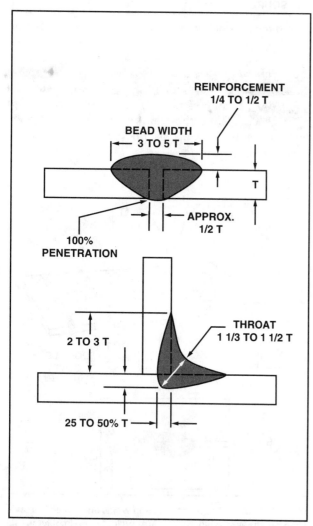

Figure 11-6. Proper penetration is the single most important characteristic of a good weld.

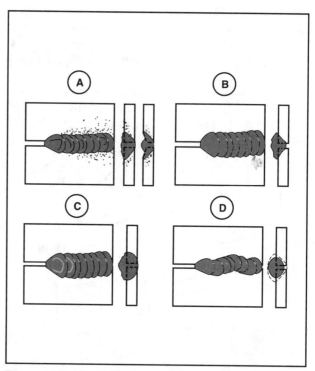

Figure 11-7. (A) — Excessive heat causes pitting and a pointed appearance on the ripples. (B) — Insufficient heat causes a weld with rough, irregular edges that are not feathered into the base metal. Penetration is poor. (C) — Too much acetylene produces a weld with craters along the edges. (D) — A cold weld displays irregular edges and considerable variation in the depth of penetration.

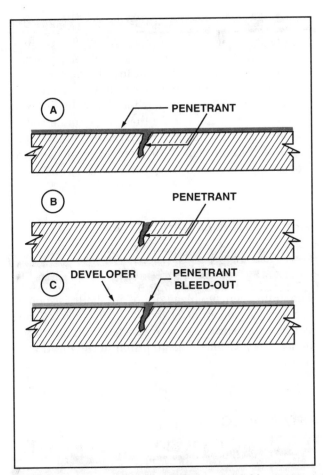

Figure 11-8. (A) — When performing a liquid penetrant inspection, the penetrant is spread over the surface of the material being examined, and allowed sufficient time for capillary action to take place. (B) — The excess penetrant is then washed from the surface, leaving any cracks and surface flaws filled. (C) — An absorbent developer is sprayed over the surface where it blots out any penetrant. The crack then shows up as a bright line against the white developer.

There are two types of dyes used in liquid penetrant inspection: fluorescent and colored. An ultraviolet light is used with the fluorescent penetrant and any flaw shows up as a green line. With the colored dye method, faults show up as red lines against the white developer. [Figure 11-9]

PREPARATION

When using liquid penetrant it is important that the surface be free of grease, dirt, and oil. Only when the surface is perfectly clean can the penetrant be assured of getting into cracks or faults. The best method of cleaning a surface is with a volatile petroleum-based solvent, which effectively removes all traces of oil and grease. However, some materials are damaged by these solvents, so care must taken to ensure the proper cleaner is used. If vapor degreasing is not practical, the part is cleaned by scrubbing with a solvent or a strong detergent solution. Parts to be inspected with liquid penetrant should not be cleaned by abrasive blasting, scraping, or heavy brushing. These methods tend to close any discontinuities on the surface and hide defects that could otherwise be detected. After the part is clean, rinse and dry it thoroughly.

PENETRANT APPLICATION

Penetrant is typically applied to a surface by immersing the part in the liquid or by swabbing or brushing a penetrant solution onto the part's sur-

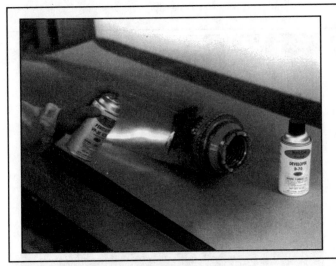
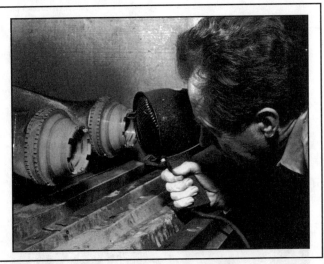

Figure 11-9. (A) left — Colored dye is used in this penetrating liquid so that examination under white light can be accomplished. (B) right— A fluorescent dye is used in this penetrant inspection and then the part is examined under black or ultraviolet light where any fault appears as a vivid green mark.

face. However, some manufacturers do offer dye penetrant in spray cans to allow application in small areas for localized inspection. Whichever system is used, the area inspected is completely covered with the penetrating liquid which is then allowed to remain on the surface for the manufacturer's recommended length of time. [Figure 11-10]

The amount of time required for a penetrant to cure is called its **dwell time** and is determined by the size and shape of the discontinuities being looked for. For example, small, thin cracks require a longer dwell time than large and more open cracks. Dwell time is decreased if a part is heated. However, if the part gets too hot the penetrant evaporates.

REMOVAL OF SURFACE PENETRANT

Liquid penetrants are typically removed using either water, an emulsifying agent, or a solvent. **Water-soluble penetrants** are the easiest to remove. Typically this type of penetrant is flushed away with water that is sprayed at a pressure of 30 to 40 psi, with an adjustable spray nozzle. The spray nozzle is held at a 45 degree angle to the surface to avoid washing the penetrant out of cracks or faults.

Post-emulsifying penetrants are not water soluble. They must be treated with an emulsifying agent before they can be washed from a part's surface. This allows you to control the amount of penetrant that is removed prior to cleaning. By varying the emulsifier dwell time, surface penetrant can be emulsified while the penetrant absorbed into cracks or other defects is left untouched. As a result, the surface penetrant is rinsed off but the absorbed penetrant remains to expose the defect.

Some penetrants are neither water soluble nor emulsifiable, but instead are **solvent-removeable**. When using this type of penetrant, excess penetrant is removed with an absorbent towel, and the part's surface is then wiped with clean towels dampened with solvent. The solvent should not be sprayed onto the surface nor should the part be immersed in the solvent, since this will wash the penetrant out of faults or dilute it enough to prevent proper indication in the developer.

APPLICATION OF DEVELOPER

There are three kinds of developers used to draw penetrants from faults. While all three types do the same job, the methods of their application differs. Penetrant begins to bleed out of any fault as soon as the surface penetrant is removed. Because of this, covering the surface to be inspected with developer as soon as possible helps to pinpoint the location of any fault.

DRY DEVELOPER

Dry developer is a loose powder material such as talcum that adheres to the penetrating liquid and acts as a blotter to draw the penetrant out of any surface faults. When using a dry developer, the part is typically placed in a bin of loose developer. For larger components, dry powder is applied with a soft brush, or blown over the surface with a powder gun. After the powder remains on the surface for the recommended time, the excess is removed with low-pressure air flow.

Figure 11-10. Spray cans of cleaner, penetrant, and developer make dye penetrant inspection an effective and handy method of field inspection. However, large parts inspected by dye penetrant are typically dipped in vats of penetrant.

The penetrant used with a dry developer is often treated with a fluorescent dye, or with a colored dye. These parts are typically examined under black light so faults appear as a green indication as the light causes the dye to fluoresce, or glow. Colored dye penetrants are usually red and any faults appear as red marks, clearly visible on the surface. [Figure 11-11]

WET DEVELOPER

A wet developer is similar to a dry developer in that it is applied as soon as the surface penetrant is rinsed off the part. Wet developer typically consists of a white powder mixed with water that is either flowed over a surface, or a part is immersed in it. The part is then air-dried and inspected in the same way as a part on which dry developer was used. Wet developers are typically used with penetrants that are treated with either fluorescent or colored dyes.

NONAQUEOUS DEVELOPER

The most commonly used developer for field maintenance is the nonaqueous type. Nonaqueous developer consists of a white chalk-like powder suspended in a solvent that is normally applied from a pressure spray can, or sprayed onto a surface with a paint gun. The part being inspected must be thoroughly dry before a thin, moist coat of developer is applied. The developer dries rapidly and pulls out any penetrant that exists within a fault. The penetrant stains the developer and is easily seen with a black light when a fluorescent penetrant is used. If a white light is desired use a colored dye.

Figure 11-11. This part was removed from a dry developer and examined under a black light. The fault is circled using a felt-tip pen so it is visible during the repair procedure. Never use a pencil to mark faults since the lead in pencils cause dissimilar metal corrosion.

MAGNETIC PARTICLE INSPECTION

The nondestructive inspection method most often used for parts made of iron or iron alloys is magnetic particle inspection. In this method of inspection, a part is magnetized and an oxide containing magnetic particles is poured or sprayed over the part's surface. Any discontinuities in the material, either on or near the surface, create disruptions in the magnetic field around the part. A discontinuity is a disruption in a part's normal physical structure that may or may not affect the usefulness of the part. The magnetic particles in the oxide align with these disruptions.

Magnetic particle inspection is useful for detecting cracks, splits, seams, and voids that form when a metal ruptures. It is also useful for detecting cold shuts and inclusions of foreign matter that occurred when the metal was cast or rolled. However, some types of subsurface discontinuities do not produce sharp enough magnetic poles to attract the oxide and form a good indication of the fault.

PRINCIPLES OF MAGNETIC INSPECTION

If you recall from Chapter 3, when a material containing large amounts of iron is subjected to a strong magnetic field, the magnetic domains within the material align themselves and the part becomes magnetized. When this happens, the part develops both a north and south pole and lines of flux flow in a continuous stream from the north pole to the south pole. If a break occurs within the part another set of magnetic poles appears, one on either side of the break. Therefore, when conducting magnetic particle inspection, these poles attract the magnetic particles in the oxide thereby giving you an indication of the break.

MAGNETIC ORIENTATION

In order to detect a crack with magnetic particle inspection, the part must be magnetized in such a way that the lines of flux are perpendicular to the fault. This is because a flaw that is parallel to the lines of flux causes a minimal disruption in the magnetic field. On the other hand, a defect that is perpendicular to the field creates a large disruption, and is easy for an inspector to detect. To ensure that the flux lines are nearly perpendicular to a flaw, a part should be magnetized both longitudinally and circularly.

LONGITUDINAL MAGNETISM

In longitudinal magnetization, the magnetizing current flows either through a coil in which the part is placed, or through a coil around a soft iron yoke. In either method, the magnetic field is oriented along the material so that magnetic fields form on either side of faults located across the material. [Figure 11-12]

CIRCULAR MAGNETIZATION

When magnetizing current flows through a part being inspected, the lines of magnetic flux created encircle the part. When this occurs, flaws or faults located along the material are magnetized and, therefore, attract magnetic particles. Current is sent through the part by placing it between the heads of magnetizing equipment. However, if the part is tubular, it is slipped over a conductive rod that is then placed between the heads of a magnetizing

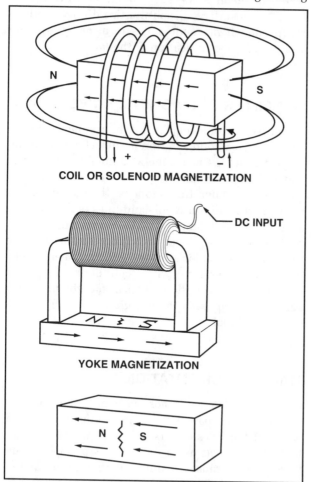

COIL OR SOLENOID MAGNETIZATION

YOKE MAGNETIZATION

Figure 11-12. When a part is magnetized in a coil, or solenoid, the lines of flux pass through the material longitudinally. The same holds true if a part is magnetized using a coil wrapped around a soft iron yoke. In both cases, as the flux lines pass through the part longitudinally, faults that run across the part are detected.

machine. Nuts and similar small parts are magnetized in this way. [Figure 11-13]

Large flat objects are circularly magnetized by using test probes that are held firmly against the surface with current passed through them. The magnetic field is oriented perpendicular to current that flows between the probes. [Figure 11-14]

Either circular or longitudinal magnetization can reveal defects that are 45 degrees to the magnetic field. However, to ensure that all flaws are identified, a part should be magnetized both longitudinally and circularly.

METHODS OF MAGNETIZATION

Ferrous materials can be magnetized in a variety of ways. For example, simply striking a piece of iron can induce a weak magnetic field. However, for the purposes of magnetic particle inspection, magnetic fields must be precisely controlled. Therefore, magnetic particle inspection employs direct current magnetization, half-wave rectified DC magnetization, or alternating current magnetization.

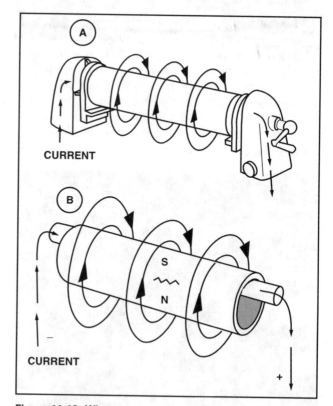

Figure 11-13. When current passes through a part, lines of flux encircle the part making it circularly magnetized. The same holds true when a circular or tubular part is placed over a current-carrying conductor. This circular magnetization allows for the detection of faults extending lengthwise along the part.

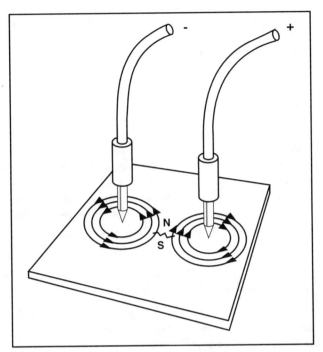

Figure 11-14. Current flowing through a flat object from high-current probes magnetizes the part circularly and detects cracks or faults that are in line with the probes.

DIRECT CURRENT

Pure direct current at voltages from 110 to 440 has excellent penetrating qualities and is suitable for magnetizing parts in coils and with yokes. However, DC has the disadvantage of being difficult to change its value as required for inspecting objects of different sizes.

HALF-WAVE RECTIFIED DC

The problem with using pure DC to magnetize parts is that pure DC is not readily available in most shops. However, commercial frequency alternating current is available and can be rectified to DC with a half-wave rectifier. In addition, by controlling the AC input the DC output can be adjusted to any value. Half-wave DC has the identical penetrating qualities as straight DC, and its pulsating nature helps distribute the magnetic particles so they arrange themselves over any fault.

ALTERNATING CURRENT

The principle of magnetization is based on the magnetic domains of a material aligning with the external magnetizing force. If the magnetizing force is produced by alternating current, the domain alignment reverses each cycle, thus changing the magnetic polarity. Therefore, the magnetic field pro-

duced by AC differs from that produced by DC in that the field strength is almost totally concentrated on the surface of the material.

If a part is subjected to an AC field and the flow of current is suddenly interrupted, the part becomes magnetized. However, the penetration is very shallow. This limits the usefulness of AC magnetization to locating surface faults only.

TESTING MEDIUM

The medium used to indicate the presence of a fault by magnetic particle inspection is ferromagnetic. In other words, the material is finely divided, has a high permeability, and a low retentivity. Furthermore, for operator safety it is nontoxic. There is no one medium that is best for all applications. However, in general, these materials are extremely fine iron oxides that are dyed gray, black, red, or treated with a dye that causes them to fluoresce when illuminated with ultraviolet light.

The iron oxides are often used dry, but can be mixed with kerosene or some other light oil and sprayed over a surface. Dry particles require no special preparation, making them well suited for field applications where portable equipment is used. Dry particles are typically applied with hand shakers, spray bulbs, or powder guns. [Figure 11-15]

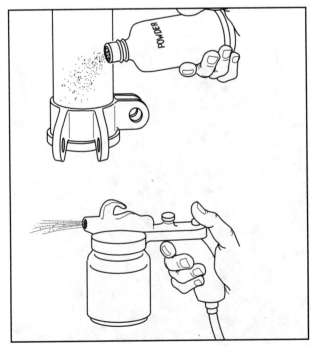

Figure 11-15. Hand shakers, spray bulbs, or powder guns are typically used to apply dry iron oxides.

Wet particles are flowed over a part as a bath. The wet method is typically used with stationary equipment that continuously agitates the bath to keep the particles in suspension. Particles are either mixed in the vehicle with proportions recommended by the manufacturer or they come pre-mixed. The particle concentration of the bath requires close monitoring with adjustments made as necessary each time the system is used. Measuring particle concentration is accomplished by collecting a sample of the agitated bath in a centrifuge tube. A volume of particles settles to the bottom of the tube allowing measurement and comparison to be made against the manufacturers standardization guide.

Since the bath is continuously recycled, it often becomes contaminated and discolored. When this happens, you must drain and clean the equipment, then refill with a fresh bath.

TESTING METHODS

Different types of magnetizing procedures must be used for different applications. The two methods you must be familiar with are the residual magnetism method and the continuous magnetism method.

RESIDUAL MAGNETISM

When a part is magnetized and the magnetizing force is removed before the testing medium is applied, the part is tested by the residual method. This procedure relies on a part's residual or permanent magnetism. The residual procedure is only used with steels that have been heat-treated for stressed applications. [Figure 11-16]

CONTINUOUS MAGNETISM

Continuous magnetization requires that a part be subjected to the magnetizing force when the testing medium is applied. The continuous process of magnetization is most often used to locate invisible defects since it provides a greater sensitivity in locating subsurface discontinuities than does residual magnetism. [Figure 11-17]

INSPECTION

The color of dye used with the magnetic particles determines the type of light used for the inspection. If gray, black, or red dye is used, the inspection is made in white light. However, if a fluorescent dye is used, the part is inspected using a black light in a dark booth.

The skill and experience of an inspector is a critical factor in determining the effectiveness of a magnetic particle inspection. For example, while faults that are open to the surface are easy to identify, some subsurface faults can be extremely difficult to detect due to the fact that there are no strong north and south poles.

Figure 11-16. This propeller blade was magnetized and is being inspected by the residual method.

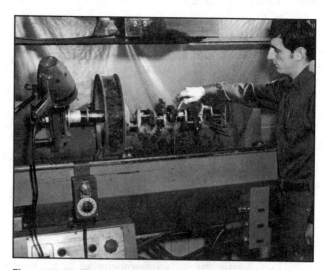

Figure 11-17. This crankshaft is magnetized by flowing current through the two heads of the magnetizing machine. Oxide is suspended in a light oil and is flowed over the magnetized part to inspect it.

FATIGUE CRACKS

Fatigue cracks give sharp, clear patterns, generally uniform and unbroken throughout their length. These cracks are often jagged in appearance, as compared with the straight indications of a seam, and they often change direction slightly in localized areas. [Figure 11-18]

Fatigue cracks are only found in parts that were in service. These cracks are usually in highly stressed areas of a part where a stress concentration exists. It is important to recognize that even a small fatigue crack indicates that failure of the involved part(s) is in progress.

HEAT-TREAT CRACKS

Heat-treat cracks have a smooth outline, and are usually less clear with less buildup than fatigue cracks. On thin sections, such as cylinder barrel walls, heat-treat cracks may give very heavy patterns. These heat-treat cracks have a characteristic form, consisting of short jagged lines grouped together. [Figure 11-19]

SHRINK CRACKS

Shrink cracks give a sharp, clear pattern and the line is usually very jagged. Since the walls of shrink cracks are close together, their indications generally build up to less extent than indications of fatigue cracks.

GRINDING CRACKS

Grinding cracks are fine, sharp, and seldom have a buildup because of their limited depth. Grinding cracks vary from single-line indications to a heavy network of lines. Grinding cracks are generally related to the direction of grinding. For example, the crack typically begins and continues at right angles to the motion of a grinding wheel, resulting in a generally symmetrical pattern. Indications of grinding cracks frequently are identified by means of this symmetrical relationship.

SEAMS

Indications of seams are typically straight, sharp, and fine. They are often intermittent and sometimes have very little buildup.

HAIRLINE CRACKS

Hairline cracks are very fine seams in which the faces are forced very close together during fabrication. Hairline indications are very fine and sharp, with very little buildup. Discontinuities of this type are normally considered detrimental only in highly stressed parts.

INCLUSIONS

Inclusions are nonmetallic materials that have been trapped in the solidifying metal during the manufacturing process. Such examples include, slag materials and chemical compounds. They are usually elongated and strung out as the metal is worked in subsequent processing operations. Nonmetallic contaminants that get into a metal when it is cast show up as a broad and fuzzy indication.

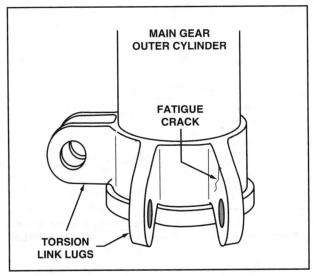

Figure 11-18. The fatigue crack in this landing gear outer cylinder changes direction, is jagged and should not be confused with a seam.

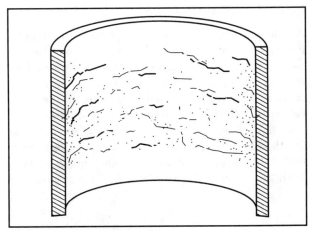

Figure 11-19. The heat-treat cracks on cylinder barrel walls typically have very heavy patterns.

Inclusions appear in varying sizes and shapes, from stringers that are easily visible to the eye, to particles only visible under magnification. In a finished part, inclusions occur as either surface or subsurface discontinuities. Indications of subsurface inclusions are usually broad and fuzzy, seldom continuous, and are typically even in width or density throughout their length. Larger inclusions appear more clearly defined especially if they are near the surface. However, close examination typically reveals their lack of definition and the indication of several parallel lines, rather than a single line. These characteristics distinguish a heavy inclusion from a crack.

DEMAGNETIZATION

Magnetization of a part after it is inspected is often detrimental to its operation in an aircraft. Therefore, before a part is returned to service, it is required to be thoroughly demagnetized. This is accomplished through either AC or DC demagnetization.

AC DEMAGNETIZATION

In order to demagnetize a part, the magnetic domains must be disorganized. To accomplish this, the part is subjected to a magnetizing force opposite the force used to magnetize it. If the magnetizing force was AC, the domains alternate in polarity, and if the part is slowly removed from the field while current is still flowing, the reversing action progressively becomes weaker. Thus, the domains are left with random orientation and the part is demagnetized.

DC DEMAGNETIZATION

AC current does not penetrate a surface very deeply. For this reason, complete demagnetization of some parts require DC demagnetization. To accomplish this, a part is placed in a coil and subjected to more current than initially used to magnetize the part. Current is flowed through the coil and then the direction of current flow is reversed while decreasing the amount. The direction of current flow continues to be reversed in direction and decreased until the lowest value of current flow is reached.

The presence of any residual magnetism is checked with a magnet strength indicator. Parts are demagnetized to within the limits specified in the appropriate overhaul manual before they are returned to service.

TEST SENSITIVITY AND STANDARDS

Test sensitivity is the effect many factors have on the ability of a test system. Nondestructive tests are designed for specific applications to find flaws and determine a part's serviceability. Some factors affecting sensitivity include the method of magnetization, the magnetizing amperage, the current type (AC or DC), the type of particles used, and the method of particle application. A test with high sensitivity locates small defects, while a test with low sensitivity reveals only coarse flaws.

Once these factors are determined, the system requires testing with a test bar that has a known flaw. This test bar is the standard for the parts being tested. Therefore, if defects on the test bar become visible, begin the process on the parts to be tested. [Figure 11-20]

Designing magnetic particle inspections for various applications takes experience and special training. However, the airframe and powerplant technician can successfully perform inspections when manufacturers' requirements are adhered to. Manufacturer's requirements for testing engine parts are detailed and specific. Any flaw found on tested parts, such as a hairline crack, is cause for part rejection. These specifications determine the sensitivity of the test for the technician.

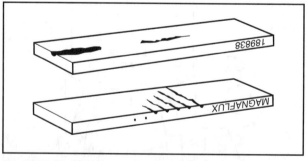

Figure 11-20. The types of defects found on a test bar are indicative of what you could find on an object being tested.

ELECTRONIC INSPECTION

The different types of visual inspection that were discussed in Section A are effective for identifying surface defects and some subsurface flaws. In fact, simple inspection techniques are often the most effective. However, many aircraft components are made of nonferrous materials that must be checked for internal imperfections. To accomplish this, inspection personnel use several electronic inspection methods that provide an electronic view of a component's internal structure.

EDDY CURRENT INSPECTION

Eddy current inspection is a testing method that requires little or no part preparation and can detect surface and subsurface flaws in most metals. Furthermore, it can differentiate among metals and alloys, as well as a metal's heat treat condition. Eddy current inspection is based on the principle of current acceptance. In other words, it determines the ease with which a material accepts induced current. As AC is induced into a material being tested, the AC is measured to determine the material's characteristics.

Eddy currents are electrical currents that flow through electrically conductive material under the influence of an induced electromagnetic field. The ease with which a material accepts the induced eddy currents is determined by four properties: its conductivity, permeability, mass, and by the presence of any voids or faults.

The **conductivity** of a metal varies with alloy type, grain size, degree of heat treatment, and tensile strength. Therefore, an eddy current test can differentiate among any of these characteristics. To perform a conductivity test, a comparison probe and a test probe are held on a reference, or sample material and the meter is balanced to a **null indication**. Then, when the test probe is placed on an unknown test material, the meter deflects from the null position if the conductivity of the test material is different from that of the reference.

The **permeability** of a material is the measure of its ability to accept lines of magnetic flux. If a bridge-type eddy current tester is zeroed with the test probe on a reference material, it indicates an off-zero reading when the test probe is placed on a material whose permeability is different from that of the reference. [Figure 11-21]

The eddy current meter indirectly measures current flowing in the test probe, and the probe current is proportional to the current induced into a test specimen. The mass of the material being tested determines the ease with which eddy currents flow. When the probe is placed on the surface of a test material and then zeroed, the meter moves off zero when the probe is moved over any part with a different mass. For example, if the probe passes over an area containing corrosion or some sort of discontinuity, the meter needle deflects, indicating a decrease in mass. However, as the probe is moved over a surface free from faults, it remains steady.

ABSOLUTE METHOD OF INSPECTION

In the absolute method of inspection, bridge-type eddy current equipment is used to identify a mater-

Figure 11-21. The eddy current probe is being zeroed by placing it inside a piece of material with the same characteristics as the propeller blade to be inspected.

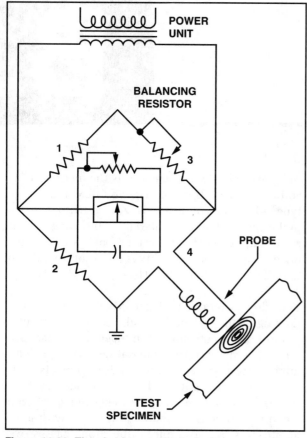

Figure 11-22. The absolute method of eddy current inspection measures the amount of current required to induce a given amount of eddy current into the test specimen.

ial's characteristics by measuring the amount of probe current that flows when current is induced into a test specimen. The meter is zeroed by balancing the bridge when the probe is held in contact with a sound piece of test material. When the probe is placed on the test specimen, it causes an off-zero meter reading if the characteristics of the test specimen differ from those of the reference material. [Figure 11-22]

COMPARISON METHOD OF INSPECTION

The comparison method of inspection uses a double-coiled probe. Instead of zeroing to a standard piece of material, the comparison method indicates differences in characteristics between the material under the reference probe and that under the test probe. The meter type eddy current instrument is very effective for the inspection of conductive materials and is well suited for field applications. While relatively simple to operate, it is limited in that only part of the output information is represented by the meter. [Figure 11-23]

Several eddy current instruments which use a two dimensional display, such as an X-Y oscilloscope, are available without the limitations associated with the meter type instrument. The two dimen-

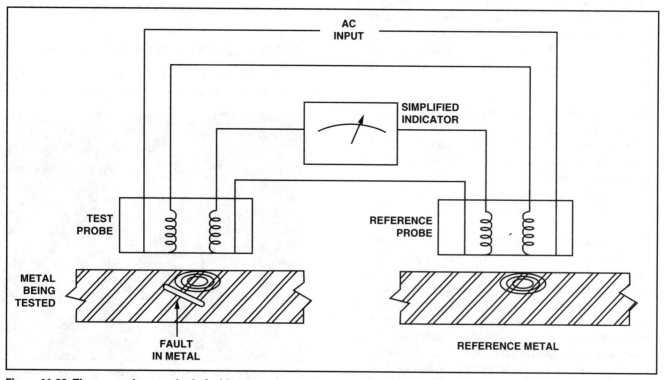

Figure 11-23. The comparison method of eddy current inspection compares the eddy current induced into a reference material with the eddy current flowing in the material being tested.

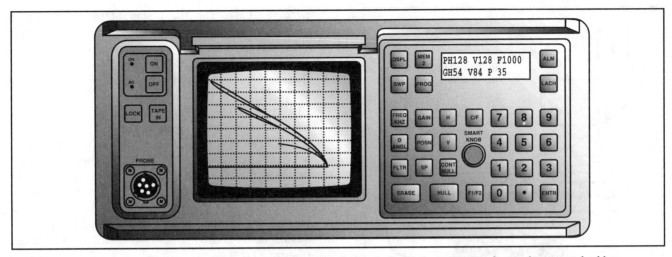

Figure 11-24. Eddy current instruments with a two-dimensional display indicate the presence of a crack as stretched loops.

sional display allows the operator to analyze all the output information from an eddy current test. [Figure 11-24]

ULTRASONIC INSPECTION

Ultrasonic inspection is the only form of nondestructive testing that can be used on plastics, ceramics, and most metals. To understand how ultrasonic testing works, you must first understand how sound is produced and transmitted.

If you recall from Chapter 2, all sound is a product of vibration. When a body vibrates, it produces sound waves that are transmitted by the air surrounding the body. Under normal conditions, these sound waves propagate longitudinally from the source of vibration and are called **longitudinal waves**. However, a second type of wave propagation occurs at right angles to the direction of the sound. This type of wave propagation occurs only in materials made of tightly bonded molecules, such as solids, and are called **transverse**, or **shear waves**. Shear waves that travel along the surface of a material, and do not appreciably extend into the material, are known as **surface**, or **Rayleigh**, waves.

Ultrasonic waves used for nondestructive inspection vary in frequency from 200 kilohertz to 25 megahertz, and are either reflected, focused, or refracted. This sound energy propagates through a solid or liquid material with little loss in wave energy in much the same manner as a focused beam of light travels through air. This property makes sound energy usable, not only in nondestructive inspection, but for sonar operation and for ultrasonic cleaning.

PIEZOELECTRICITY

Some materials produce electricity when they are struck, pressed, bent, or otherwise distorted. Materials that possess this property are called piezoelectric materials. In addition to producing current, piezoelectric materials vibrate when subjected to alternating current. This makes these types of materials useful as **transducers** for introducing physical vibrations into other materials. [Figure 11-25]

Ultrasonic testing equipment is based on an electronic oscillator that produces AC of the proper frequency, which is amplified to the proper strength and sent to a transducer that is touching the material being tested. The transducer causes the test material to vibrate at the oscillator's frequency. When the vibrations reach the other side of the

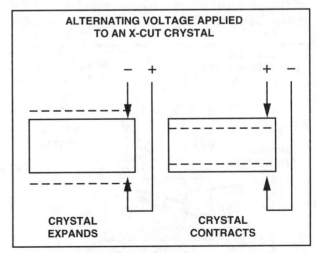

Figure 11-25. Generation of ultrasonic vibration is created by piezoelectric materials that physically distort when subjected to an electrical potential.

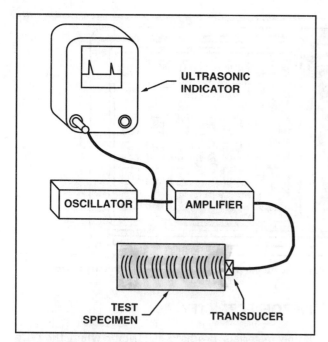

Figure 11-26. An oscillator generates AC energy which is amplified and sent to a transducer that creates mechanical vibrations within the material.

material and bounce back, they create an electrical impulse at the transducer that is seen on the CRT display. [Figure 11-26]

ENERGY INTRODUCED INTO TEST MATERIAL

There are three basic ways in which ultrasonic energy is introduced into the test specimen. The first is by direct contact on only one side of the material. The energy is transmitted from this point and the return echo is received from the same side. The second method uses a transducer on both sides of the material; one introduces a pulse into the material, and the other receives the signal and sends

it to the CRT. The third way of inducing sound energy into a material is the immersion method. With this method, the test specimen is immersed in water and the transducer beams its energy through the water to the test material's surface.

The type of sound wave drected at a material is determined by the orientation of the transducer to the material's surface. The desired orientation is achieved through the use of acrylic wedges inserted between the transducer and the material's surface. However, when this is done, a film of water or oil should be used between the transducer and the wedge, and the wedge and the material's surface to ensure better contact. This film is called a **coupling**. [Figure 11-27]

FAULT INDICATIONS

Two basic systems are used in ultrasonic inspection. They are the pulse-echo system and resonance system. With the **pulse-echo system** a cathode ray oscilloscope is used in conjunction with a CRT as a fault indicator. A time based signal produces a straight line across a CRT screen and when a pulse of energy is sent into the material, a pip, or peak, occurs on this horizontal line. The energy pulse travels through the material until it reaches its opposite side, which reflects it back to the transducer. The energy received by the transducer causes a second peak on the time based line. There is a definite relationship between the physical distance between the

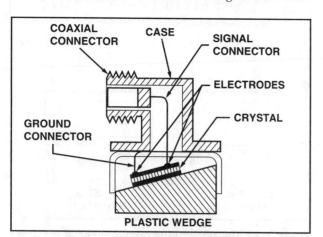

Figure 11-27. (A) — The ultrasonic transducer is mounted on an acrylic adapter and then it beams ultrasonic wave energy into the material being tested. (B) — The transducer is adjusted with reference material before use.

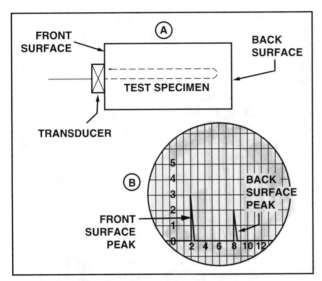

Figure 11-28. (A) — Vibrations are passed through a material and bounce off of its back surface. (B) — A peak is recorded on the base line to represent both the front and back surfaces.

front and back surface of the test piece and the distance between the peaks on the time based line. [Figure 11-28]

A calibrated grid is placed in front of the CRT, and the equipment is adjusted to place the front surface peak on a calibration line to serve as reference. A second adjustment allows the distance between the front and the rear surface peaks to vary so the rear peak can also be placed on a reference line. Once adjusted, a distance between the peaks is established that is representative of the material thickness. Any change in material thickness is indicated by a change in distance between peaks. Furthermore, if a fault or crack exists between the two surfaces a portion of a sound wave will be reflected back to the transducer sooner than the original wave. This is displayed on the CRT as a peak in between the front and rear surface peaks. [Figure 11-29]

Because ultrasonic test equipment indicates the thickness of a material, it is an efficient means of inspecting for corrosion on the inside of a structure. To do this, the transducer is held against a piece of test skin of the same material and thickness, and one that is known to be free of corrosion. The distance between the front and back surface peaks is adjusted to a specific distance as indicated by the marks on the grid. Now, when the transducer is moved over the skin being tested, any movement of the back surface peak closer to the front surface peak indicates a decrease in skin thickness that is possibly caused by corrosion.

In the previous example, a piece of skin of a known thickness is used as a test specimen, or block. In most cases, the manufacturer of the part being tested already has a test block with which to calibrate the instrument, or states the method of ultrasonic calibration. However, no matter what method is used, a calibration must be performed before any ultrasonic inspection has validity.

The second system used in ultrasonic testing is the **resonance system**. Like the pulse-echo system, the resonance system is also used to measure the thickness of material with a consistent thickness and smooth surfaces. However, its principle of operation differs from that of the pulse-echo system in that the resonance system depends on matching the oscillator's frequency to the resonance point of the material being tested. In a typical test, resonance is displayed vertically on a CRT, with material thickness determined by comparing the test information to a transparent scale that is overlaid on the CRT display. This method of inspection is effective in locating wear or corrosion which results in a reduction of thickness of material. It is also used to ensure that plexiglass windows that have been polished to remove surface defects maintain their minimum thickness.

RADIOGRAPHIC INSPECTION

One of the most important methods of nondestructive inspection available is radiographic inspection. Radiographic inspection allows a photographic view inside a structure. In other words, this method uses certain sections of the electromagnetic spectrum to photograph an object's interior.

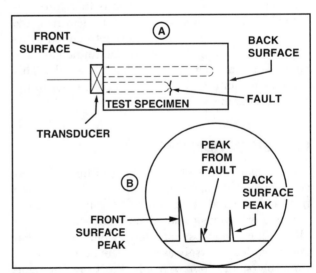

Figure 11-29. Ultrasonic energy bounces off a fault and is reflected on the CRT screen as a peak on the base line between the peaks representing the front and back surfaces.

X-ray and gamma ray radiation are forms of high energy, short wavelength electromagnetic waves. The amount of energy these rays contain is related inversely to their wavelength. In other words, the shorter the wavelength, the greater the energy. They have no electrical charge or mass, travel in straight lines at the speed of light, and are able to penetrate matter. The depth of penetration is dependent upon the ray's energy.

There are certain characteristics that make x-rays and gamma rays especially useful in nondestructive inspection. For example, both types of rays are absorbed by the matter through which they pass. The amount of absorption is proportional to the density of the material. Furthermore, x-rays and gamma rays ionize certain materials, making it possible for them to expose photographic film and cause certain materials to fluoresce, or glow.

GENERATION OF X-RAYS

An x-ray generator consists of a tube containing a heavy insulating envelope. A coil at one end of the tube serves as a cathode that emits electrons when it is heated with electrical current. At the other end of the tube is an anode on which a target is mounted. The target is made of a material that has a high atomic number, and therefore a high density and a high number of electrons. When a high positive voltage is connected to the anode, it draws electrons from the cathode at a high velocity. These electrons strike the target and dislodge electrons from the target. [Figure 11-30]

The intensity of an x-ray beam is determined by the number of electrons available and is therefore controlled by the amount of current used to heat the cathode. The energy of an x-ray is determined by its wavelength, which is in turn determined by the velocity of the electrons striking the anode. This velocity is controlled by the voltage applied to the anode and is measured in kilovolts.

CLASSIFICATION OF X-RAYS

The intensity of the x-rays striking a film or fluorescent screen is related to the distance between the film and the radiation source. If you recall from your study of physics, all electromagnetic radiation obeys the inverse square law, which states that energy varies inversely as the square of the distance from the energy source. Applied to x-ray photography, this means that if the distance between the x-ray source and the film is doubled, the amount of radiation falling on each unit of area decreases to one-fourth. If the distance is tripled, the amount of energy decreases to one-ninth.

In order to get the required amount of radiated energy to the surface of the film, the amount of current supplied to the cathode is controlled. Low current gives low intensity while high current gives high intensity. By the same token, the voltage supplied to the anode determines the amount of energy in the x-ray. The higher the voltage, the more energy it contains. Low-energy x-rays are called soft x-rays and those with high energy are called hard x-rays.

GAMMA RAYS

Gamma rays consist of radiation energy produced by the disintegration of very specific chemical elements known as isotopes. An **isotope** is a form of chemical element that has the same number of protons as a normal atom, but a different number of neutrons. This causes an isotope to have different physical properties since it has a nucleus that is said to be in an excited, or metastable, state.

When the nucleus of an isotope emits gamma radiation energy, the nucleus returns to its **ground state**, the lowest energy state a given atom can have. An atom at its ground state has the same atomic number and the same atomic weight as an excited nucleus. The only difference is that the amount of energy contained in an excited nucleus is more than what is contained in a ground state nucleus. Isotopes are classified according to how long it takes them to return to their ground state.

Gamma radiation is the most damaging type of radiation encountered in aircraft maintenance since it penetrates deep into the human body. Unlike x-rays that are produced only when electrons are flowing within a tube, gamma rays cannot be shut off, controlled, or directed. Therefore, the equipment in which isotopes are kept must provide safe radia-

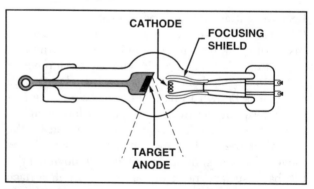

Figure 11-30. X-rays are produced when electrons from the cathode bombard the target anode and dislodge electrons.

tion-proof storage. A camera encased in heavy lead shielding holds the isotope until it is needed to expose the film. [Figure 11-31]

SET-UP AND EXPOSURE

For a permanent record of a radiographic inspection, a sheet of photographic film is placed on one side of the object being inspected, and the radiation source on the other. The film is placed as close to the specimen as possible and the source is oriented so that the radiation penetrates and passes an amount of radiation proportional to the specimen's density. The denser the specimen, the less radiation passes through, and the less the film is exposed. The specimen is then exposed to the radiation source. [Figure 11-32]

Factors that determine the proper exposure include, but are not limited, to the following:

1. Material thickness and density
2. Shape and size of the object
3. Type of defect to be detected
4. Characteristics of the equipment used
5. Exposure distance
6. Exposure angle
7. Film characteristics
8. Types of intensifying screen, if used

FILM

Photographic film is composed of flexible transparent plastic sheets coated with a thin layer of gelatin. This gelatin contains an emulsion of extremely fine silver bromide grains. When the film is exposed to photons of energy, either visible light or radiation energy photons, a latent image is formed on the film. The exposed film is then treated with developer which reduces only the silver bromide grains that were touched by the radiation into clumps of black metallic silver. After all of the affected emulsion is converted, the developing action is stopped with an acid stop bath. The film is then treated with a chemical fixative which removes all unexposed silver bromides in the emulsion and makes the film insensitive to light. All of the chemicals are then washed out of the emulsion, and the film is dried and ready for examination.

The film displays a negative, or reverse, image of the part that was photographed. Because of this, less-dense areas or places with the most radiation exposure are dark. Those places where the density is the greatest get the least radiation and are the clearest.

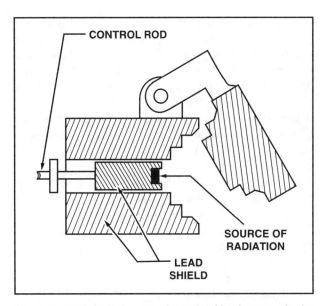

Figure 11-31. A typical camera is made of lead to contain the isotope's gamma radiation. To expose a film and obtain an x-ray, the cover is raised and the control rod is extended to expose the source and provide a wider angle of coverage.

The end result is an x-ray film of the specimen being inspected. Because internal parts or flaws in the specimen have different densities, their images are produced on the film. As a result, internal flaws can be located by careful examination and evaluation. Frequently, these flaws are undiscoverable by any other means.

FLUOROSCOPY

For high-speed radiographic inspection where no permanent record is required, a fluoroscope is used. This system is very similar to the photographic

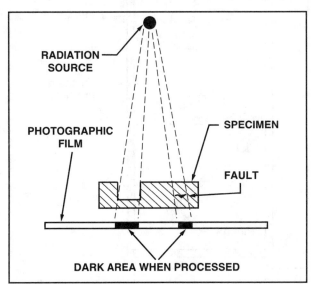

Figure 11-32. Radiation from x-rays or gamma rays penetrates a specimen and exposes a photographic film. The denser the material, the less the film is exposed.

process, except that all equipment is enclosed in a lead box, and instead of using photographic film to record the results, the radiation passes through a specimen and causes a fluorescent screen to glow. The indication on the screen is such that the screen portion receiving the most radiation is the least dense portion of the specimen, and glows the brightest. The image is viewed through a lead glass viewing window. [Figure 11-33]

The main advantage of fluoroscopy is that objects are viewed in real time. Furthermore, the test piece can be moved or rotated in front of the screen by handling devices. Moving the object closer to the x-ray tube for magnification is also possible.

Although fluoroscopy allows you to inspect parts quickly and effectively, it does have some disadvantages. For example, the equipment is bulky and, therefore, is usually not movable. Furthermore, a fluoroscope produces a dim image that is sometimes hard to read as well as less capable of detecting small discontinuities. And finally, the equipment must be designed and constructed with special attention to operator safety.

The initial investment in fluoroscopic equipment is comparatively high. However, once this cost is amortized, the cost per unit inspected is typically lower than for other radiographic inspection processes.

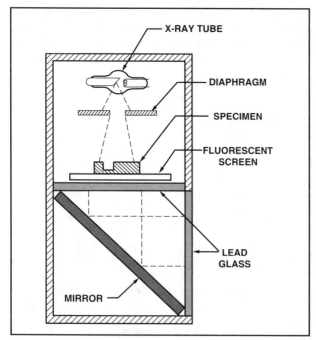

Figure 11-33. If a permanent record is not required, a fluoroscope can give a low-resolution image of a part's interior. The materials viewed on a fluorescent screen give an indication similar to those seen on x-ray film.

RADIOGRAPHIC SAFETY PROCEDURES

Any form of radiation is harmful to the human body. Therefore, care is required when this method of inspection is utilized. The United States Nuclear Regulatory Commission and the Environmental Protection Agency set up safety procedures regarding handling of radioactive materials and their recommendations should be adhered to. As you have just learned, the main reason for using radiographic inspection is its ability to penetrate almost anything. This is also the reason it is difficult to protect personnel from its effects.

Radiation produces changes in all matter through which it passes. This is true of living tissue. When ionizing radiation strikes molecules in the body, the effect is sometimes no more than to dislodge a few electrons. However, an excess of these effects causes irreparable harm. When a complex organism is exposed to ionizing radiation, the degree of damage depends on which of its body cells are changed. Since vital organs are located in the center of the body, penetrating radiation is likely to be the most harmful in these areas. The skin usually absorbs most of the radiation and therefore reacts earliest to radiation.

If the whole body is exposed to a very large dose of radiation, death can result. In general, the type and severity of the pathological effects of radiation depend on the amount of radiation received at one time, and the percentage of the total body exposed. For example, smaller doses of radiation cause blood and intestinal disorders in a short period of time. The more delayed effects are leukemia and cancer. Skin damage and loss of hair are also possible results of exposure to radiation.

The best shielding against radiation is a layer of lead. Care is required to ensure that no holes are in the shielding to allow radiation to escape. Persons working around x-ray or gamma ray equipment should wear **radiation monitoring film badges**, or **dosimeters**. These special devices are worn while in the vicinity of radioactive sources. At specified intervals, dosimeters are checked to see if the wearer has been exposed to an abnormally high amount of radiation.

Access to areas where x-ray or gamma ray equipment is being used should be controlled to prevent personnel from being exposed to radiation energy. Radiographic inspection rooms require some type of visible indication outside the door to show when an inspection is in progress. A system notifying personnel of danger areas should be observed when portable equipment is used in the shop or hangar. Furthermore, any container carrying radioactive material must display the symbol for radioactive material.

INSPECTING COMPOSITES

Composite structures contain materials that sometimes make nondestructive testing difficult. For example, many honeycomb structures are metal-backed. This renders x-ray inspection ineffective, since an x-ray that will penetrate the metal backing will be too powerful to provide resolution on composite material. In addition, dye penetrant is generally ineffective on laminated composite structure, since the laminations and weave of composite materials absorbs and retains the dye, much like a crack or other defect. However, a number of methods of NDT are effective on composites.

COIN TAP TEST

Although it is one of the most simple tests available, the coin tap test is also one of the most effective on laminated, bonded, and honeycomb materials. To detect structural flaws with this method, you merely tap the edge of a coin lightly along an area you suspect is damaged. Undamaged material produces a solid ringing sound, while a damaged area makes a hollow thud. Impact damage to laminated structure, such as ice and rain impingement on radomes, is quickly and easily found using the coin tap test.

THERMOGRAPHY

Thermography locates flaws in a part by measuring temperature variations at the part's surface. A part is heated, and temperature differences are then measured with an infrared camera or film. Thermography requires a knowledge of the test material's thermal conductivity, which is then compared to a reference standard.

RADIOGRAPHY

Although x-rays are not effective on certain bonded structures, other types of radiographic inspection can detect surface cracks and internal damage on many composite structures. In particular, radiography can detect water inside honeycomb core cells.

LASER HOLOGRAPHY

To inspect a part using laser holography, the part is heated and then photographed using a laser light source and a special camera system. Laser holography can detect disbonds, entrapped water, and impact damage in a variety of composites.

CHAPTER 12

CLEANING AND CORROSION

INTRODUCTION

Corrosion is the inevitable result when metals are exposed to water and air. Since all aircraft contain some metal, they require constant inspection and cleaning to minimize the destructive effects of corrosion. Dirt and grease are visually unappealing, and can hide corrosion and structural damage. Furthermore, dirt can trap moisture and corrosive materials next to aircraft structures, hastening the formation of corrosion. Therefore, you must be aware of proper cleaning procedures that remove built up contamination without damaging the materials used in structural components. In addition, you must learn appropriate methods of corrosion control to minimize corrosion's effects once it has become established.

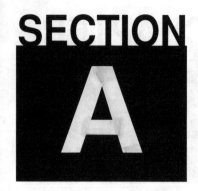

SECTION A

AIRCRAFT CLEANING

While a clean aircraft is more visually appealing than a dirty one, appearance is a secondary consideration in aircraft cleaning. Appendix D of FAR Part 43 requires that the airframe and engine be cleaned before performing an annual or 100 hour inspection. In addition, common sense and good maintenance practice require that an aircraft be kept clean. Dirt can cover up cracked or damaged components as well as trap moisture and solvents that lead to corrosion. If allowed to accumulate over time, dirt and debris can build to a considerable weight and decrease the useful load of the aircraft. Therefore, the cleaning of the aircraft exterior and interior should not be taken lightly.

The materials and procedures presented in this section are intended as guidelines only. You should always follow the manufacturer's recommendations for cleaners, solvents, and cleaning procedures.

EXTERIOR CLEANING

The first and most important step in corrosion control is to keep the exterior of an aircraft thoroughly clean. However, before you start washing an aircraft there are certain areas and components that must be protected from cleaners and water sprays. For example, pitot tubes and static openings should always be plugged or taped prior to cleaning an aircraft to prevent water ingestion. Furthermore, wheel and brake assemblies should be covered to keep out cleaning agents. Although not readily apparent, it is extremely important to use the cleaning compounds and other chemicals that are recommended by the aircraft manufacturer, or are MIL SPEC approved for the particular application. The use of nonapproved or ordinary commercial compounds can result in a condition known as **hydrogen embrittlement**. Hydrogen embrittlement results when a chemical reaction produces hydrogen gas that is absorbed into a metal. This process subsequently reduces a metal's ductility and allows the formation of cracks and stress corrosion.

When washing an aircraft the aircraft should be parked on a wash rack or in an area where it can be hosed down. Avoid washing an aircraft in the sun to help prevent the surface from drying before the cleaner has time to penetrate the film and dirt. For the main part of the aircraft exterior, use a 1:5 or a 1:3 mixture of water and an emulsion-type cleaner that meets MIL-C-15769 specifications. Brush or spray the mixture onto the surface and allow it to stand for a few minutes, then rinse it off with a high-pressure stream of warm water.

The engine cowling and wheel well area usually have grease or oil deposits that require special treatment. Typically, these areas must be soaked with a 1:2 mixture of emulsion cleaner and water. After allowing the cleaner to remain on the surface for a few minutes, scrub the heavily soiled areas with a soft bristle brush to completely loosen the dirt, and rinse it with a high-pressure stream of warm water. [Figure 12-1]

Figure 12-1. Wheel wells accumulate deposits of dirt, oil, and hydraulic fluid. An emulsifying cleaner should be sprayed into the wheel well and allowed to soak, then washed off with water.

Stubborn exhaust stains may require a 1:2 mixture of cleaner with Varsol or kerosene. Mix these ingredients into a creamy emulsion and apply it to the surface. Let it stand for a few minutes, then work all of the loosened residue with a bristle brush and hose it off with a high-pressure stream of warm water. This treatment may be repeated if the first application does not remove all of the stain.

The type of materials to be used in cleaning depends on the nature of the elements that need to be removed. For example, to remove oil, grease, or soft preservative compounds, dry-cleaning solvent, or **naphtha**, is often used. The two most common types of naphtha are aliphatic naphtha and aromatic naphtha. **Aliphatic naphtha** is a hydrocarbon solvent that dissolves oil and grease but does not harm rubber or acrylic components. **Aromatic naphtha**, on the other hand, attacks rubber and acrylic compounds. Because it is safe on most materials, aliphatic naphtha is frequently used to wipe down cleaned surfaces before painting.

Chemical cleaners must be used with great care in cleaning assembled aircraft. The danger of entrapping a potentially corrosive solvent in faying surfaces and crevices counteracts any advantages in their speed and effectiveness. For example, caustic cleaners can cause corrosion on aluminum or magnesium alloys and, therefore, should not be used. Magnesium engine parts should be washed with a commercial solvent and decarbonized, and then scraped or grit blasted. Before they are painted, magnesium engine parts should be wiped down with a dichromate solution to improve paint adhesion. When cleaning aluminum, you should always use cleaners which are relatively neutral and easy to remove. If you must use an abrasive to remove corrosion products from aluminum structure, use aluminum wool or aluminum oxide sandpaper. Carborundum paper, crocus cloth, and steel wool must be avoided, since they can lead to the formation of dissimilar corrosion in aluminum.

When using high-pressure water spray you must exercise caution. This is especially true around the engine components. For example, if high-pressure water is used to degrease an engine, you must avoid spraying electrical components such as magnetos and wiring harnesses. Furthermore, high-pressure water spray can rinse the lubricant from bearings and grease fittings. Therefore, after washing wheel wells, flap tracks, or hinges, you should lubricate them to force out any water and prevent corrosion.

EXTERIOR FINISH MAINTENANCE

A clean, polished surface denies corrosion a place to start. Therefore, aircraft should be kept clean and waxed. All drain openings must be kept open, and deposits which have formed from engine exhausts must be removed before they build up excessively. Avoid damage to aircraft by not using harmful cleaning, polishing, brightening or paint-removing materials. Use only those compounds which conform to existing government or established industry specifications, or products that have been specifically recommended by the aircraft manufacturer as being satisfactory for the intended application. Observe the product manufacturer's recommendations concerning use of their agent.

NONMETAL CLEANING

Nonmetalic aircraft components sometimes require different cleaning techniques than metal parts. For example, the slightest amount of dust on a plastic or plexiglass surface can scratch the finish if rubbed with a dry cloth. Furthermore, the use of dry cloths also builds up static charges on the window that attracts more dust. Therefore, before washing a plastic window, rinse the area with water first. Once clean, dry the window with a soft cloth to prevent streaking.

Oil and hydraulic fluid attack and rapidly destroy the rubber in aircraft tires. Therefore, whenever these fluids are spilled on a tire, they should be immediately wiped off with a dry towel. The tire should then be washed with soap and water. Because most cleaning solvents are petroleum-based, soap and water are the only approved solution for cleaning tires.

Rubber deice boots have a conductive coating to help dissipate static charges. Furthermore, some composite structures such as radomes are painted with special materials that are transparent to radio signals. These areas should be cleaned gently and never subjected to abrasives or stiff brushes. If you have a question as to what type of cleaning solution to use, consult the manufacturer's specifications.

POWERPLANT CLEANING

Accumulated dirt on powerplants can cover defects and lead to overheating. Therefore, aircraft engines should be cleaned on a regular basis. However, when doing this all electrical components in the engine compartment must be protected from solvent and soap. This includes wrapping the magnetos so no water can get in the vents. If the powerplant is located over a landing gear, the gear's brake and tire assemblies should be covered in plastic. If you use a high-pressure water spray, avoid spraying the starter, alternator, and air intakes with solvent or water rinse.

Prior to washing an engine, inspect it for excessive oil leakage. Then, apply a soap or solvent solution to the engine and allow it to set for several minutes. Heavy accumulations of dirt and grease can be scrubbed with a bristle brush to loosen them. The engine is then rinsed and allowed to dry. If volatile solvent was used, be sure it has dried before starting the engine to minimize the risk of fire. Lubricate all controls and rod ends in the engine compartment, and remove protective covers that were installed on electrical components.

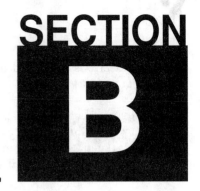

TYPES OF CORROSION

CORROSION

Corrosion and its control are of primary importance to all aircraft operators. Corrosion weakens primary structural members, which must then be replaced or reinforced in order to sustain flight loads. Such replacements or reinforcements are costly, time-consuming, and result in unscheduled delays.

Most metals exist in nature as chemical compounds such as oxides or chlorides. For example, aluminum is never found in nature in its pure state, but must be refined from an ore such as alumina (Al_2O_3). However, when pure aluminum is exposed to the elements, it combines with oxygen and eventually changes back to alumina. Corrosion then, is simply a process wherein metals return to a natural state.

Corrosion is a natural phenomenon which attacks metal by chemical or electrochemical action and converts it into a metallic compound, such as an oxide, hydroxide, or sulfate. Substances that cause corrosion are called **corrosive agents**. Water or water vapor containing salt combine with oxygen in the atmosphere to produce the most prominent corrosive agents. Additional corrosive agents include acids, alkalis, and salts.

The appearance of corrosion varies with various metals. For example, on aluminum alloys and magnesium it appears as surface pitting and etching, often combined with a grey or white powdery deposit. However, on copper and copper alloys corrosion forms a greenish film and on steel a reddish rust. When corrosion deposits are removed, the metal's surface may appear etched and pitted, depending upon the length and severity of attack. If deep enough, these pits may become sites for crack development. Some types of corrosion can travel beneath surface coatings and can spread until the part fails.

There are two general classifications of corrosion, chemical and electrochemical; however, both types involve two simultaneous changes. The metal that is attacked or oxidized suffers an anodic change, and the corrosive agent is reduced and suffers a cathodic change.

CHEMICAL CORROSION

Pure chemical corrosion results from direct exposure of a bare surface to caustic liquid or gaseous agents. The most common agents causing direct chemical corrosion include:

1. Spilled battery acid or fumes from batteries.
2. Residual flux deposits resulting from inadequately cleaned, welded, brazed, or soldered joints.
3. Entrapped caustic cleaning solutions.

ELECTROCHEMICAL CORROSION

Electrochemical corrosion is similar to the electrolytic reaction that takes place in a dry cell battery. To understand how this happens, recall the structure of the atom from Chapter 2. When the number of electrons matches the number of protons in an atom, the atom is said to be electrically balanced. However, if there are more or fewer electrons than protons, the atom is said to be charged and is called an ion. If there are more electrons than protons, it is a negative ion, but if there are more protons than electrons, it is a positive ion. An ion is unstable, always seeking to lose or gain electrons so it can change back into a balanced, or neutral, atom.

Metals are arranged to show the relative ease with which they ionize in what is called the **electrochemical series**. The earlier a metal appears in the series, the more easily it gives up electrons. In other words, a metal that gives up electrons is known as an anodic metal and corrodes easily. On the other hand, metals that appear later in the series do not give up electrons easily and are called cathodic metals. [Figure 12-2]

Many metals become ionized due to galvanic action when brought into contact with dilute acids, salts, or alkalis, such as those found in industrially contaminated air. For example, if an aluminum structure

is in contact with moisture having a trace of hydrochloric acid, a chemical reaction takes place between the acid and the aluminum to form aluminum chloride and hydrogen.

$$2Al + 6HCl \rightarrow 2AlCl_3 + 3H_2$$

The hydrogen is released as a gas, and aluminum chloride, which is a salt, forms as a white powder on the surface of the metal. This powder is the visible evidence of corrosion.

Corrosion is an electrochemical action in which one metal is changed into a chemical salt. When two dissimilar metals are in contact with each other in the presence of some electrolyte such as hydrochloric acid or plain water, the less active metal acts as the cathode and attracts electrons from the anode. As the electrons are pulled away from the anode the metal corrodes.

Perhaps the easiest way to visualize what is actually taking place is to consider the action of a battery. If two metals are immersed in an electrolyte of acid, saline, or alkaline solution, a battery is formed and produces a flow of electrons between the two metals. The process continues as long as there are active materials in the metal and electrolyte and the cathode and anode are connected by a conductive path. [Figure 12-3]

ELECTROCHEMICAL SERIES FOR METALS

(NOBILITY) MOST ANODIC – WILL GIVE UP ELECTRONS MOST EASILY.

MAGNESIUM
ZINC
CLAD 7075 ALUMINUM ALLOY
COMMERCIALLY PURE ALUMINUM (1100)
CLAD 2024 ALUMINUM ALLOY
CADMIUM
7075-T6 ALUMINUM ALLOY
2024-T3 ALUMINUM ALLOY
MILD STEEL
LEAD
TIN
COPPER
STAINLESS STEEL
SILVER
NICKEL
CHROMIUM
GOLD

Figure 12-2. The metals listed above are arranged in order of electrode potential. Any metal appearing in this series is anodic to any metal which follows it and will corrode if subjected to galvanic action.

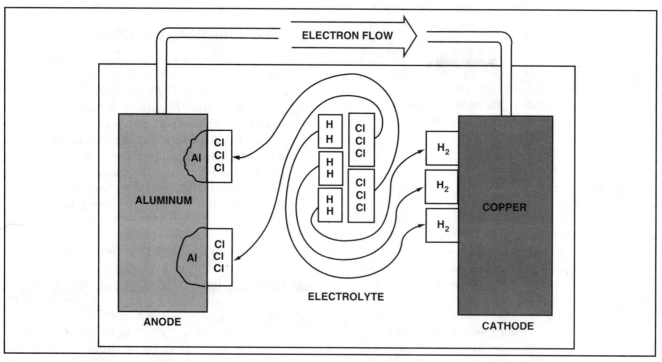

Figure 12-3. A simple battery explains the formation of corrosion, as electrons leaving the anode attract chlorine ions from the electrolyte to form aluminum chloride, which is the visible evidence of corrosion.

This example shows a piece of copper and a piece of aluminum in a weak solution of hydrochloric acid and water. In the electrochemical series, aluminum is considerably more active than copper. When electrons flow from the aluminum through the conductor to the copper, positive aluminum ions are left. Two of these ions attract six negative chlorine atoms from the acid and form two molecules of aluminum chloride ($ALCL_3$) on the surface of the aluminum. This eats away some of the base metal. The six positive hydrogen ions remaining in the acid are attracted to the copper by the electrons which came from the aluminum. These electrons neutralize the hydrogen ions and the six atoms form three hydrogen molecules ($3H_2$) and leave the surface as free hydrogen gas.

One of the basic characteristics of metals is their electrode potential. In other words, when two dissimilar metals are placed in an electrolyte, an electrical potential exists. This potential forces electrons in the more negative material, the anode, to flow to the less negative material, the cathode, when a conductive path is provided. As discussed earlier, corrosion occurs when electrons leave an element.

If all of the aluminum used in the construction of aircraft were pure aluminum, corrosion would not be a problem. However, as discussed in Chapter 7, aluminum must be alloyed with other metals to increase its strength. The most common alloying agent is copper. In a structural piece of alloyed aluminum the microscopic grains of copper and aluminum serve as the cathode and anode of a galvanic cell. Aluminum is more negative than copper and acts as the anode in the electrochemical action. There is no flow of electrons between the two alloying agents within the metal until an external path is provided to form a complete circuit. This path is furnished by the electrolyte, which can be a surface film of moisture containing such pollutants as acids, salts, or other industrial contaminants. [Figure 12-4]

The electrode potential difference between the aluminum and copper grains causes positive ions to exist within the aluminum. When the electrolyte film covers the surface, the aluminum ions attract chlorine ions from the hydrochloric acid and form aluminum chloride, the salt of corrosion. Hydrogen ions are attracted to the copper by the electrons from the aluminum. These hydrogen ions become neutralized and form molecules (H_2) which leave the surface as a free gas. Corrosion forms on the anodic aluminum, but no corrosion products are evident on the copper cathode.

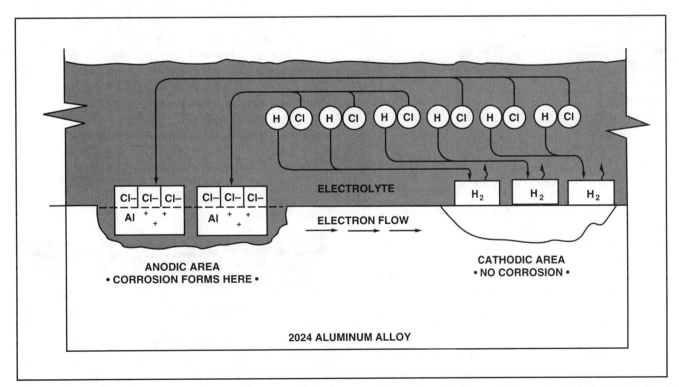

Figure 12-4. Aluminum alloys have both anodic and cathodic areas. Corrosion appears at the anodic area when chlorine atoms from the electrolyte join the aluminum chloride.

This type of electrochemical attack produces pits filled with corrosion salts, and is usually rather localized. However, if the entire surface is covered with a strong electrolyte, corrosion can develop uniformly over an extensive area. This type of corrosion is called a **direct chemical attack**.

This basic introduction shows the four requirements for the formation of corrosion:

1. Presence of a metal that will corrode (anode).
2. Presence of a dissimilar conductive material (cathode) which has less tendency to corrode.
3. Presence of a conductive liquid (electrolyte).
4. Electrical contact between the anode and cathode (usually metal-to-metal contact, or a fastener).

Corrosion control, therefore, consists of preventing the chemical action by eliminating one or more of these basic requirements.

TYPES OF CORROSION

Corrosion is a very general term and may appear in a variety of forms, depending on the metal involved and the corrosion-producing agents present. As an A&P technician, you must be familiar with the different types of corrosion as well as how to identify each.

OXIDATION

One of the simpler forms of corrosion is "**dry**" corrosion or, as it is most generally known, oxidation. When a metal such as aluminum is exposed to a gas containing oxygen, a chemical reaction takes place on the surface between the metal and the gas. Two aluminum atoms join three oxygen atoms to form aluminum oxide (AL_2O_3). If the metal is iron or steel, two atoms of iron join three atoms of oxygen to form iron oxide, or rust (Fe_2O_3).

There is one big difference between iron oxide and aluminum oxide. The film of aluminum oxide is unbroken and, therefore, once it has formed, further reaction with oxygen slows dramatically. Iron oxide, on the other hand, forms a porous, interrupted film. Since the film is not air tight, the metal continues to react with the oxygen in the air until the metal is completely eaten away. [Figure 12-5]

The best way to protect iron from dry corrosion is to keep oxygen from coming into contact with its surface. This is done temporarily by covering the surface with oil or grease, or permanently with a coat of paint.

Figure 12-5. Iron oxide, more commonly known as rust, forms a porous scale on the surface of a material. Therefore, it will continue to convert the iron oxide as long as it remains on the surface.

Aluminum alloy can be protected from oxidation by the formation of an oxide film on its surface. This film insulates the aluminum from any electrolyte, and prevents further reaction with oxygen. The protection afforded by an aluminum oxide coating is the principal reason for cladding (Alclad) aluminum alloy used in structural applications.

UNIFORM SURFACE CORROSION

Where an area of unprotected metal is exposed to an atmosphere containing battery fumes, exhaust gases, or industrial contaminants, a uniform attack over the entire surface occurs. This dulling of the surface is caused by microscopic amounts of the metal being converted into corrosion salts. If these deposits are not removed and the surface protected against further action, the surface becomes so rough that corrosion pits form.

Corrosion sometimes spreads under the surface and cannot be recognized by either roughening of the surface or by a powdery deposit. Instead, the paint or plating lifts off the surface in small blisters due to the pressure of the underlying accumulation of corrosion products.

A common type of **uniform surface corrosion** is caused by the reaction of metallic surfaces with atmospheric contaminants. These include airborne chlorine or sulphur compounds, oxygen, or moisture in the atmosphere. Reactive compounds from exhaust gases, as well as fumes from storage batteries, frequently cause uniform surface corrosion. The amount of damage caused by uniform surface corrosion is ordinarily determined by comparing the thickness of the corroded metal with that of an undamaged specimen.

PITTING CORROSION

Pitting is a likely result of uniform surface corrosion left untreated. Pits form as localized anodic areas, and corrosive action continues until an appreciable percentage of the metal thickness is converted into salts. In extreme cases, this can eat completely through the metal. Pitting corrosion is usually detected by the appearance of clumps of white powder on the surface. [Figure 12-6]

Figure 12-6. Pitting corrosion forms clumps of powdery salts of corrosion on the surface of the metal as it is corroded.

The pits found in this type of corrosion usually have a rather short, well-defined edge with walls that run almost perpendicular to the surface of the metal. All forms of pits have one thing in common, regardless of their shape. They penetrate deeply into the metal and cause damage completely out of proportion to the amount of metal consumed. [Figure 12-7]

A metal's protective coating must be removed or penetrated before the destructive chemical action leading to pitting can occur. However, once pitting

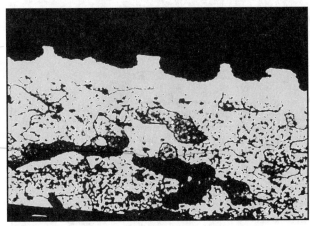

Figure 12-7. This photomicrograph of the surface of an aluminum sheet shows pitting that is a result of corrosion.

begins, it is propagated by means of concentration cells or galvanic action.

GALVANIC CORROSION

This common type of corrosion occurs any time two dissimilar metals make electrical contact in the presence of an electrolyte. For example, galvanic corrosion can take place where dissimilar metal skins are riveted together, or where aluminum inspection plates are attached to the structure with steel screws.

When metals of the same galvanic grouping are joined together, they show little tendency for galvanic corrosion. But metals of one group corrode when they are held in contact with those in another group. The further apart the groups, the more active the corrosion. In addition, corrosion is much more rapid when the anodic metal is smaller than that of the cathodic metal. The reason for this is the greater area of the cathode allows a higher rate of electron flow, accelerating the speed of the reaction. On the other hand, if the corroding metal (anode) is larger than the less active metal (cathode) corrosion is slow and superficial. [Figure 12-8]

CONCENTRATION CELL CORROSION

Concentration cell corrosion, or crevice corrosion, is corrosion of metals in a metal-to-metal joint, corrosion at the edge of a joint even though the metals are identical, or corrosion of a spot on a metal's surface covered by a foreign material. Oxygen concen-

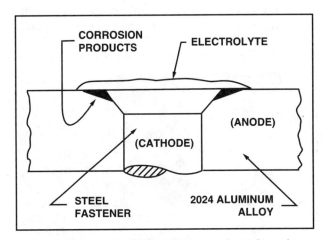

Figure 12-8. A potential for galvanic corrosion exists when a steel fastener is used to hold a 2024 aluminum alloy inspection plate in place. If contaminated moisture gets between the two metals, the aluminum is corroded rather than the steel screw. However, if 2024 aluminum alloy and magnesium are riveted together and covered by an electrolyte, the magnesium corrodes instead of the aluminum.

tration cells, metal ion concentration cells, and active-passive cells are the three general types of concentration cell corrosion.

OXYGEN CONCENTRATION CELL CORROSION

When water covers the surface of an aluminum aircraft skin and seeps into the cracks between lap joints, oxygen concentration cell corrosion can form. Since water in an open area readily absorbs oxygen from the air, it attracts electrons from the metal to form negative hydroxide ions:

$$2H_2O + O_2 + 4\,\text{Electrons} \rightarrow 4(OH)$$

The electrons required to form these negative ions come from the metal itself. The area between the skins does not give up electrons to the water on its surface because there is not enough oxygen there to form hydroxide ions. Instead, its electrons flow to the cathodic surface — the open area. The area between the skins, having lost electrons, now contains positive aluminum ions, and this area becomes the anode.

Electrons flow within the metal from the anode to the cathode, leaving positive metal ions in the area between the sheets. These positive aluminum ions attract negative hydroxide ions from the open water and the aluminum corrodes, forming aluminum hydroxide. The unusual characteristic of this type of corrosion is that it forms in the areas where there is a deficiency of oxygen. [Figure 12-9]

Oxygen concentration cell corrosion can occur on aluminum, magnesium, or on ferrous metals. It forms under marking tape of ferrules on aluminum tubing, beneath sealer that has loosened, and under bolt or screw heads. [Figure 12-10]

When dirt or other oxygen-excluding contamination forms on an anodized surface and the oxide film is scratched, oxygen concentration cell corrosion can prevent the protective film from re-forming.

METAL ION CONCENTRATION CELL CORROSION

The electrode potential within a metal is dependent on the different metals that make up the alloy. However, a potential difference can occur if an electrolyte having a nonuniform concentration of metal ions covers the surface. For example, some metal-to-metal joints rub against each other creating a high concentration of metal ions adjacent to a low concentration of metal ions.

Figure 12-10. Oxygen concentration cell corrosion has formed under the ferrule on this tube.

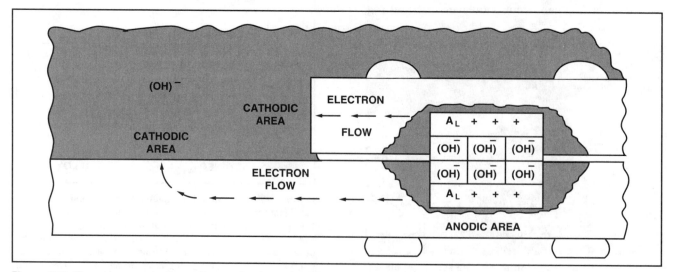

Figure 12-9. Oxygen concentration cell corrosion forms in the areas where there is a deficiency of oxygen.

As with oxygen concentration cell corrosion, when water absorbs oxygen, it attracts electrons. When these electrons leave the aluminum to form negative hydroxide ions, positive aluminum ions are formed. The water can move freely on the surface of the metal, and this movement continually carries the aluminum ions away. However, the water between the two pieces of metal does not move, and it therefore has a higher concentration of metal ions than there is in the open. The metal area between the faying surfaces, where the concentration of positive metal ions is the highest, becomes the cathode and attracts electrons from the skin in the open area, which is the anode.

As electrons flow from the anode to the cathode, they leave positive aluminum ions on the surface near the supply of negative hydroxide ions. These hydroxide ions join the aluminum ions to form aluminum hydroxide, a corrosion salt. [Figure 12-11]

Note the difference between the two types of concentration cell corrosion. The metallic ion concentration cell corrosion forms on the open surface, while oxygen concentration cell corrosion forms in the closed areas between the faying surfaces. Corrosion on aircraft is usually complex and is generally composed of more than one type.

ACTIVE-PASSIVE CELLS

Metals which depend on a tightly adhering passive film for corrosion protection, such as corrosion resistant steel, are prone to rapid corrosive attack by active-passive cells. The corrosive action usually starts as an oxygen concentration cell. As an example, salt deposits on a metal surface in the presence of water containing oxygen can create the oxygen cell. If the passive film is broken beneath a salt deposit, the active metal beneath the film will be exposed to corrosive attack. An electrical potential develops between the large area of the cathode (passive film) and the small area of the anode (active metal).

FILIFORM CORROSION

Filiform corrosion is a special form of oxygen concentration cell corrosion or crevice corrosion which occurs on metal surfaces having an organic coating system. It is recognized by its fine threadlike lines under a polyurethane enamel finish. Filiform corrosion often results when the **wash primer** used on a metal has not been properly cured. A wash primer is a two-part metal preparation material in which phosphoric acid converts the surface of the metal into a phosphate film that protects the metal from

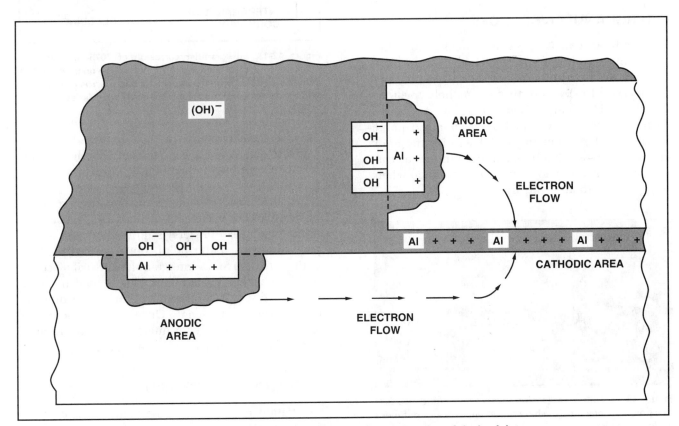

Figure 12-11. Metallic ion concentration cell corrosion forms in the open at the edge of the lap joint.

corrosion, and provides an excellent bond for paint. This conversion process relies on moisture in the air and, if there is not enough moisture to convert all of the acid, some acid remains on the metal. If a dense polyurethane finish is then applied, the acid becomes trapped and reacts with the aluminum alloy to form corrosion. [Figure 12-12]

Filiform corrosion shows itself as a puffiness under the paint film and is first noticed around rivet heads and along the lap joints of skins. When the paint film is broken, you will notice that the puffiness was caused by the growth of the powdery salts of corrosion. There is no cure for filiform corrosion short of stripping all of the paint, removing the corrosion, treating the metal's surface, and refinishing the aircraft. Filiform corrosion does not usually form under acrylic lacquer, because it is porous enough to allow moisture from the air to penetrate the film and complete the conversion of any excess acid. Filiform corrosion can be prevented by storing aircraft in an environment with a relative humidity below 70 percent, using coating systems having a low rate of diffusion for oxygen and water vapors, and by washing aircraft to remove acidic contaminants from the surface, such as those created by pollutants in the air.

INTERGRANULAR CORROSION

Intergranular corrosion is an attack along the grain boundaries of a material. Micro-photographs of aluminum alloys show that they are comprised of extremely tiny grains held together by chemical bonds. Each grain has a clearly defined boundary which, from a chemical point of view, differs from the metal within the grain center. The grain boundary and grain center can react with each other as anode and cathode when in contact with an electrolyte.

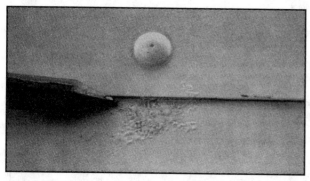

Figure 12-12. Filiform corrosion forms under the dense film of a polyurethane finish, where it first appears as a puffiness under the finish.

Recall from Chapter 7 that heat treatment diffuses an alloying agent into the parent metal. Also recall that after the heat treatment process, the alloy must be quenched promptly after it is removed from the heat bath. If quenching is delayed for even a few seconds, particles of the alloying agent precipitate out of the metal matrix and can become quite large. If quenching is delayed too long, these metal grains can reach a size that produces areas of dissimilar metals large enough to form effective cathodes and anodes so that intergranular corrosion can form. [Figure 12-13]

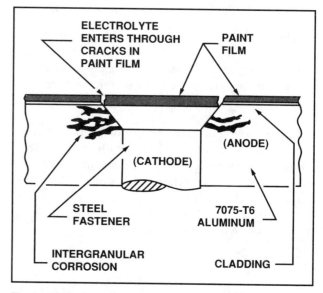

Figure 12-13. Intergranular corrosion of 7075-T6 aluminum adjacent to a steel fastener. The electrolyte needed for this action is supplied from the surface through corrosion deposits, and the attack continues along the grain boundaries.

Spot or seam welding, through localized heating, can also cause grain enlargement that leaves the metal susceptible to intergranular corrosion.

With some forms of intergranular corrosion small blisters can occur beneath the surface. The surface metal over these blisters is quite thin and, when pricked with a knife point, opens a cavity filled with corrosion. Since intergranular corrosion occurs within the metal itself, rather than on the surface, it is quite difficult to detect without ultrasonic or eddy-current equipment. Once found, about the only practical remedy for intergranular corrosion is replacement of the part. [Figure 12-14]

EXFOLIATION CORROSION

Exfoliation corrosion is an extreme case of intergranular corrosion. It occurs chiefly in extruded

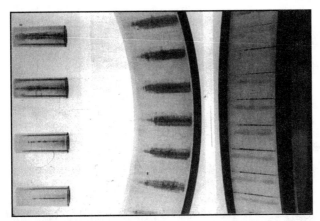

Figure 12-14. The water marks on this hardened steel bearing race are indications of intergranular corrosion in the hardened metal.

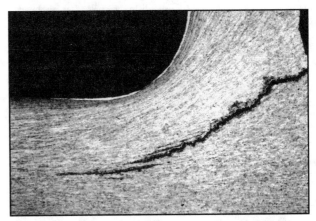

Figure 12-16. Stress corrosion forms at the end of a crack and rapidly enlarges the crack until the material fails.

materials, such as channels or angles, where the grain structure is more laminar (layer-like) than in rolled sheets or castings. This type of corrosion occurs along the grain boundaries and causes the material to separate, or delaminate. As with other types of intergranular corrosion, by the time it is evident on the surface, the metal has been damaged beyond salvage. [Figure 12-15]

STRESS CORROSION

Stress corrosion occurs when metal is subjected to a tensile stress in the presence of a corrosive environment. The stresses in the metal can come from improper quenching after heat treatment, or from an interference fit of a fastener. Stress corrosion can be transgranular or intergranular in nature. Cracks caused by stress corrosion grow rapidly as the corrosive attack concentrates at the end of the crack rather than along its sides. [Figure 12-16]

Since stress corrosion can occur only in the presence of tensile stresses, one method for preventing this type of corrosion in some heat-treated aluminum alloy parts is to **shot-peen** the surface to provide a uniform compressive stress on the surface. By doing this, the compressive stresses must be overcome by tensile forces before stress corrosion can form.

Common locations for stress corrosion to form are between rivets in a stressed skin, around pressed-in bushings, and tapered pipe fittings. If stress corrosion is severe enough, it may be visible through careful visual inspection. However, dye penetrant inspection is required to find the actual extent of the crack.

FRETTING CORROSION

When two surfaces fit tightly together but can move relative to one another, corrosion occurs. This type of corrosion is the result of the abrasive wear caused by the two surfaces rubbing against each other. This rubbing, known as fretting, prevents the formation of a protective oxide film, exposing active metal to the atmosphere. By the time this type of corrosion makes its appearance on the surface, the damage is usually done and the parts must be replaced.

If the contact areas are small and sharp, deep grooves resembling brinell markings or pressure indentations can be worn in the rubbing surface. As a result, this type of corrosion has also been called **false brinelling.**

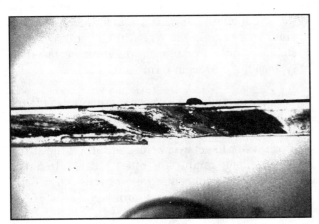

Figure 12-15. Exfoliation corrosion is an example of extreme intergranular corrosion in an extruded material.

Fretting corrosion occurs around rivets in a skin and is indicated by dark deposits around the rivet heads streaming out behind, giving the appearance of **rivet smoking**. Rivets showing this sign of fretting must be drilled out and replaced. [Figure 12-17]

CORROSIVE AGENTS

Corrosive agents are substances that are capable of causing a corrosive reaction. Most corrosive agents fall into one of two categories, acids and alkalis. However, care must be taken not to overlook other less obvious corrosive agents such as the atmosphere which contains moisture, salts, or corrosive industrial agents.

ACIDS AND ALKALIS

Almost all acids and alkalis form effective electrolytes as they react with metals to form metallic salts, but some electrolytes are more active than others. For example, the sulfuric acid found in aircraft batteries is especially active in corroding aluminum. However, a weak solution of chromic or phosphoric acid is often used as a surface treatment to prepare metal for painting.

Ferrous metals are subject to damage from both acids and alkalis, but aluminum is more vulnerable to strong alkaline solutions than it is to acids. For example, an aluminum structure can be severely corroded if it is allowed to remain in contact with a concrete floor. Water draws out enough lime from the cement to form an alkaline solution that corrodes the aluminum.

SALTS

It is very important to remember that many compounds other than sodium chloride fall into the category of salts. In general, salts are the result of a metallic element combining with a nonmetal. The resulting compound is almost always a good electrolyte, and can promote corrosive attack. Magnesium is particularly vulnerable to corrosive attack from an electrolyte formed by salt solutions.

MERCURY

Although it is not commonly found in any quantity around aircraft, there is a definite possibility that mercury could be spilled in an aircraft. Hazardous cargos are often carried in aircraft, and damage from a shifting load can result in damaged containers and hazardous spills.

Mercury attacks aluminum by a chemical reaction known as **amalgamation**. In this process, the mercury attacks along the grain boundaries within the alloy, and in a very short time completely destroys it.

Extreme care must be exercised when removing spilled mercury, as it is "slippery" and flows through tiny cracks to get to the lowest part of the structure where it causes extensive damage. In addition, mercury and its vapors are poisonous to humans and precautions must be taken to avoid exposure to it.

If mercury is spilled, remove every particle with a vacuum cleaner having a mercury trap in the suction line, or with a rubber suction bulb or medicine dropper. Never attempt to remove mercury by blowing it with compressed air. This only scatters it and spreads the damage.

WATER

Pure water reacts with metals to form corrosion or oxidation, but water holding a concentration of salts or other contaminants causes much more rapid corrosion. Seaplanes are in a continual battle with the elements, and every precaution must be taken to stay ahead of corrosion formation.

Seaplanes operating in salt water are especially vulnerable to attack, and when one is taken out of salt water it should be hosed down with large volumes of fresh water to get every trace of salt off the structure. Seaplane ramps are often located in areas where there is a concentration of industrial wastes, making the water even more corrosive.

Float bottoms are subject to the abrasive effect of high-velocity water on takeoff and landings. Since

Figure 12-17. Fretting corrosion causes black residue to trail out behind the rivet, causing it to appear to smoke.

this abrasion tends to damage the natural protective oxide film, seaplanes must be carefully inspected to detect any damage which would allow water to get to the base metal of the structure.

AIR

It is obviously impossible to isolate a structure from the air in which it exists, but the very presence of air is a factor in the deterioration of metal. Marine atmosphere and air above industrial areas hold large concentrations of salts. The chemicals precipitate out of the air and collect on the surface of an aircraft where they attract moisture from the air.

ORGANIC GROWTHS

For years, water which condensed in fuel tanks produced relatively minor corrosion problems. Small perforated metal containers of potassium dichromate crystals protected the fuel tanks by changing any water into a mild chromic acid solution, which inhibited corrosion.

Jet aircraft, however, use a high viscosity fuel which holds more water in suspension than other aviation fuels. Jet aircraft also fly higher than reciprocating engine aircraft and the low-temperature flight conditions associated with these altitudes cause water that is entrained in the fuel to condense out and collect in the bottom of the tanks. This water contains microscopic animal and plant life called microbes. These organic bodies live in the water and feed on the hydrocarbon fuel. Furthermore, the dark insides of the fuel tank promote their growth, and in very short periods of time these tiny creatures multiply and form a scum inside the tank. This scum can grow to cover the entire bottom of a tank and hold water in contact with the tank structure. This provides a place for concentration cell corrosion to form. If the scum forms along the edge of the sealant in an integral fuel tank, the sealant can pull away from the structure, causing a leak and an expensive resealing operation. [Figure 12-18]

It is virtually impossible to prevent the formation of this scum as long as microbes are allowed to live in fuel. The most successful solution to the problem has been to use an additive in fuel which kills these organic growths and prevents the formation of the corrosion-forming scum.

In addition to preventing corrosion by its biocidal action, this same fuel additive also serves as an antifreeze to prevent entrained water from freezing. Should entrained water freeze, the resulting slush could plug the fuel screens and cause fuel starvation.

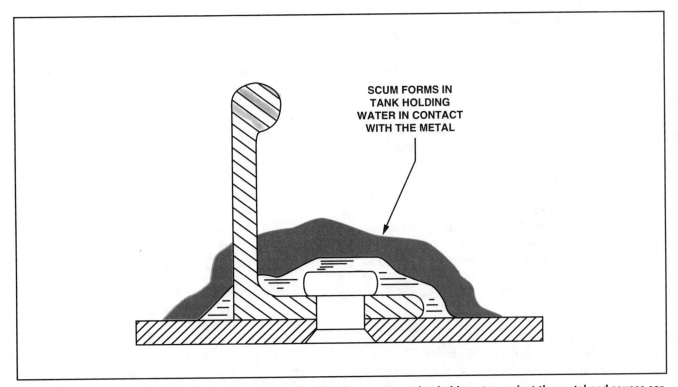

Figure 12-18. Microbial growth within a turbine aircraft fuel tank forms a scum that holds water against the metal and causes corrosion to form.

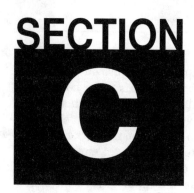

SECTION C

CORROSION DETECTION

Exotic inspection equipment is often needed for certain parts of an aircraft. However, corrosion can often be detected by careful visual inspection of the airplane structure.

For example, corrosion of aluminum or magnesium appears as a white or gray powder along the edges of skins and around rivet heads. Furthermore, since corrosion salts have more volume than sound aluminum they tend to push out against the skin. Therefore, small blisters appearing under the finish on painted surfaces or at lap joints are an indication of corrosion.

DETECTION METHODS

To aid in the detection of corrosion you should use every inspection aid that is available to you. For example, the complex structure of modern aircraft makes the use of magnifying glasses, mirrors, borescopes, fiber optics, and other optical inspection tools imperative for a good visual inspection.

Stress corrosion cracks are sometimes difficult, if not impossible, to detect by visual inspection alone. However, as discussed in Chapter 11, any fault that is open to the surface may be found through a dye penetrant inspection. The main limitation of dye penetrant inspection is the fact that it can fail to find cracks that are so full of corrosion products that the dye cannot penetrate. Also, penetrant can not get in a crack that is filled with oil or grease. Porous or rough surfaces are almost impossible to clean of all penetrant, so materials with rough surfaces do not lend themselves to this type of inspection.

Another means of corrosion detection is through the use of ultrasonic equipment. As discussed in the previous chapter, there are two types of ultrasonic indications used for corrosion detection: the **pulse-echo** and the **resonance** method.

In the pulse-echo method, a pulse of ultrasonic energy is directed into the structure. This energy travels through the material to its opposite side and then bounces back. When the return pulse is received, it is displayed on a CRT screen as a spike, which establishes a time base representing the material's thickness. Any change in a material's thickness, such as that caused by corrosion, causes the return to occupy a shorter space and thus indicate the extent of damage.

The second method of inspection using ultrasonic energy is the resonance method. This method operates on the principle that for any given thickness of material, there is a specific frequency of ultrasonic energy that resonates, or produces the greatest amount of return. In other words, if metal has been eaten away by corrosion, its resonant frequency is different from that of sound metal.

Like ultrasonic inspection, radiological inspection such as x-ray is used to determine if there is any corrosion on the inside of a structure. However, x-ray inspection requires extensive training and experience for proper interpretation of the results. Furthermore, the use of x-ray involves some danger, because exposure to the radiation energy used in this process can cause burns, damage to the blood, and possibly death.

CORROSION-PRONE AREAS

Modern airplanes are made of thin, reactive metals which can tolerate very little loss of strength. Therefore, one of the most important functions you perform as an A&P technician is to check the entire aircraft for indications of corrosion which could degrade the strength of the structure. Almost all parts of an airplane are subject to this type of damage, but certain areas are more prone than others.

ENGINE EXHAUST AREA

Reciprocating and turbine engines generate power by converting chemical energy from a hydrocarbon fuel into heat energy. Because of the inefficiency of the engine, much of this heat along with energy-rich gases pass out of the engine through the exhaust. The gases contain all of the constituents for a potent electrolyte, and because of their elevated temperature, corrosion forms extremely rapidly.

Exhaust areas must be carefully inspected and all exhaust residue removed before corrosion has a chance to start. Cracks and seams in the exhaust track are prime areas for corrosion. Fairings on the nacelles, hinges, and inspection door fasteners all contain crevices which invite the formation of corrosion.

BATTERY COMPARTMENTS AND VENTS

Batteries store electrical energy by converting it into chemical energy and are therefore active chemical plants, complete with environment-polluting exhausts. Airplanes having lead-acid batteries must have their boxes protected by a material that resists corrosion from the sulfuric acid fumes, and airplanes with nickel-cadmium batteries must have their battery areas protected with an alkaline-resistant finish. These finishes can have a bitumastic (tar) base, a rubber base, or can be polyurethane finishes.

During an inspection, the areas around a battery must be carefully checked, especially under the battery, and all traces of corrosion must be removed and the area refinished. Furthermore, you should verify that the battery has a sump vent jar containing an absorbent pad moist with a neutralizing agent. Lead-acid batteries require bicarbonate of soda, and boric acid is used for a nickel-cadmium battery. These sump vent jars should be checked to see that the pads are moist and that there is no leakage. [Figure 12-19]

Figure 12-19. The strong acid or alkaline fumes around a battery box can cause corrosion on the adjacent structure.

All battery box vent openings should be clear and the intake and exhaust lines free and open. If battery electrolyte is spilled during servicing, it must be cleaned up immediately and the area neutralized. To verify an area is completely neutralized you can blot the water on the surface with a piece of **litmus paper**. If the area is acidic, the paper turns pink, and if it is alkaline, it turns blue. The entire area should be neutral and it should not change the color of the paper.

LAVATORIES AND FOOD SERVICE AREAS

Organic materials such as food and human waste are highly corrosive to aluminum surfaces. Therefore, areas in proximity to these materials such as lavatories and food service areas must be inspected with extreme care. Food service areas can be troublesome if there is a possibility of food debris getting into cracks under or behind the galley where it cannot be removed. While this material in itself may not be corrosive, it can hold water which can cause the structure to corrode. [Figure 12-20]

The lavatory area is an especially important area to check for corrosion. Human wastes are usually acidic, and rapidly promote corrosion if allowed to remain on the skin of the airplane or to get into cracks and seams in the structure. Furthermore, some disinfectants used in lavatory areas can cause further damage to the aircraft. Because of this, you should check the disinfectant carried in the airplane to be sure that it is not of a type that is harmful to aluminum.

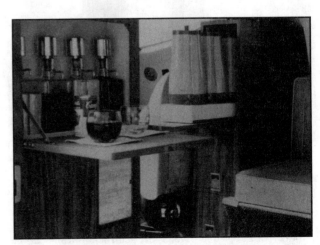

Figure 12-20. The food service area in an aircraft is subject to corrosion damage if spilled food traps water against the structure where it cannot be removed.

In addition to lavatories, the area around and behind the discharge of relief tubes must be inspected carefully for indications of corrosion. If corrosion is present, it should be removed and the area painted with an acid-proof paint.

WHEEL WELLS AND LANDING GEAR

Probably no one area of an airplane is subjected to as much hard service as the wheel well area. On takeoff and landing, debris from the runway surface is thrown up into this area. This can be especially troublesome in the winter when chemicals are used on runways for ice control. Furthermore, abrasion can remove protective lubricants and coatings, and water and mud can freeze and cause damage. [Figure 12-21]

When inspecting for corrosion in and around wheel wells it is important to remember that corrosion can take place in any one of several components including the electrical components, such as antiskid sensors, squat switches, and limit switches. Bolt heads and nuts on magnesium wheels are susceptible to galvanic corrosion, and concentration cell corrosion can form under the marking tape or ferrules on aluminum tubing. Special care must be taken to search out any area where water can be entrapped.

Because of the many complicated shapes, assemblies, and fittings found in the wheel well and landing gear areas, complete coverage of the area with paint film is difficult to attain and a partially applied preservative tends to mask corrosion rather than prevent it. Furthermore, due to heat generated by braking action, preservatives cannot be used on

some main landing gear wheels. During inspection of this area, pay particular attention to the following trouble spots:

1. Magnesium wheels, especially around bolt-heads, lugs, and wheel-web areas, particularly for the presence of entrapped water or its effects.
2. Exposed rigid tubing, especially at B-nuts and ferrules, under clamps and tubing identification tapes.
3. Exposed position indicator switches and other electrical equipment.
4. Crevices between stiffeners, ribs, and lower skin surfaces, which are typical water and debris traps.

Manufacturer's maintenance manuals give detailed information on corrosion inspections and areas that are prone to corrosion. [Figure 12-22]

EXTERNAL SKIN AREAS

One of the first places corrosion appears on the surface of an aircraft is along seams and lap joints. It is here that concentration cell corrosion frequently appears and, in clad skins, it is here that the sheared edges of Alclad material are exposed without the protection of the pure aluminum. There is also a danger of water or cleaning solvents becoming entrapped in the lap joints and providing an effective electrolyte.

Corrosion may start in a spot welded seam if the spot welding process has left an enlarged grain structure in the metal. The area is then susceptible to corrosion when moisture seeps between the skins. To check for bulging along spot welds hold a straightedge along the row of spots, and if there is corrosion in the seam, the surface bulges between the spots and shows up as a wavy skin. Corrosion in a seam can progress to such a degree that the spots actually pull apart. [Figure 12-23]

Figure 12-21. Wheel wells are corrosion-prone areas because abrasion from water and dirt removes the protective finish, and dirt and grease can trap and hold moisture against the aluminum.

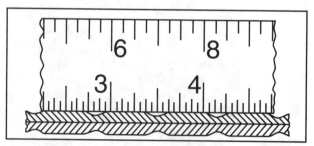

Figure 12-23. Corrosion in a spot-welded seam can be detected by laying a straightedge over the seam. Corrosion causes the seam to have a wavy surface.

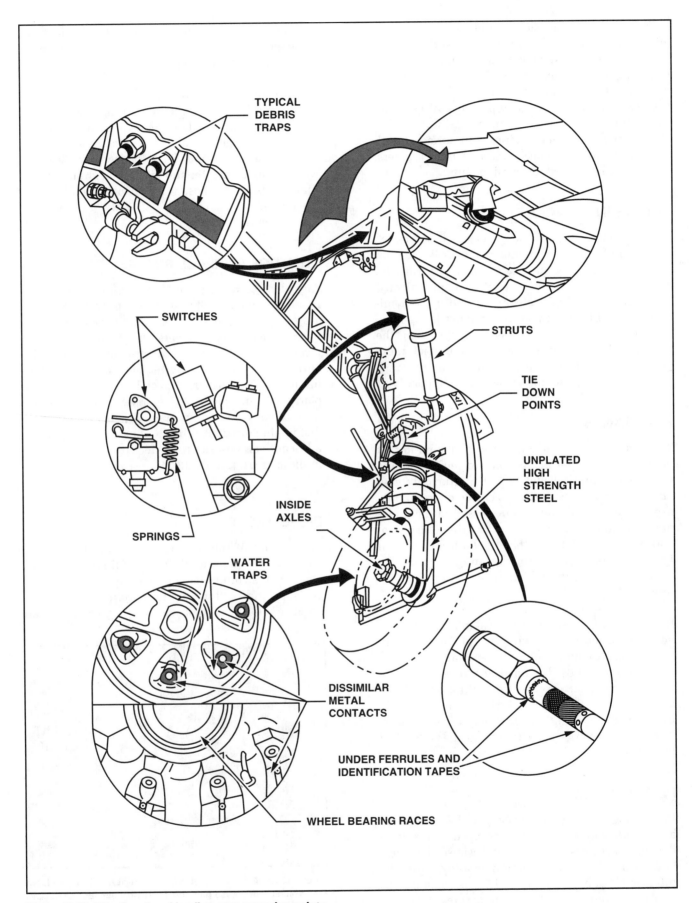

Figure 12-22. Wheel well and landing gear corrosion points.

Relatively little corrosion trouble is experienced with magnesium skins if the original surface finish and insulation are adequately maintained. However, trimming, drilling, or riveting destroys some of the original surface treatment, which can never be completely restored by touchup procedures. Any inspection for corrosion on magnesium skins should include all magnesium skin surfaces, with special attention to edges, areas around fasteners, and cracked, chipped, or missing paint.

ENGINE INLET AREAS

One of the most vital parts of an aircraft is the area directly in front of the engine where air is taken in. This area is usually quite large on jet aircraft, and air rushes into the engine at a high velocity. Abrasion by this high-velocity air and by contaminants carried in the air, tends to remove any protective coating. Therefore, **abrasion strips** along the leading edge of intake ducts help protect these areas. However, careful inspection and maintenance of the finish in these inlet areas is still imperative.

FUEL TANKS

There is probably no single place in an aircraft that is less accessible for inspection and repair than the inside of an integral fuel tank. Unfortunately, fuel tanks are highly susceptible to corrosion formation, especially in jet aircraft. As discussed in the previous section, organic growth is the primary cause of corrosion in fuel tanks that hold turbine fuel. If ignored, this organic growth can grow into the water-holding scum which attaches to the aluminum alloy skin.

In addition to organic growth, sealants used to convert the structure of a wing into a fuel tank are impervious to fuel, but only resistant to water. Therefore, it is possible for water to seep through the sealer and cause oxygen concentration cell corrosion. Corrosion under the sealant is extremely difficult to detect and must usually be found with x-ray or ultrasonic inspection from the outside of the wing.

PIANO HINGES

Piano hinges on control surfaces and access doors are ideal locations for dissimilar metal corrosion to develop. The reason for this is because the hinge body is usually made of aluminum alloy, while the pin is made of hard carbon steel. Furthermore, since it is almost impossible to keep hinge crevices clean,

dirt and dust accumulate and hold moisture between the pin and the hinge body. Dissimilar metal corrosion then develops.

If corrosion within a piano hinge goes undetected the pin may rust or freeze in the hinge, or break off and become impossible to remove. To help prevent this, piano hinges should be kept as clean and dry as possible and should be lubricated with a spray which displaces water and leaves an extremely thin film of lubricant. These water-displacing lubricants are manufactured to meet MIL-C-16713 specifications. [Figure 12-24]

CONTROL SURFACE RECESSES

Any place on an airplane that is difficult to inspect is an area where corrosion has an opportunity to grow. For example, some airplanes have areas in the wing or empennage where the movable surfaces recess back into the fixed structure. Hinges are buried back in these cavities and are difficult to lubricate. Furthermore, special attention must be paid when inspecting these areas to remove every trace of corrosion and provide drains for any water that might collect. A thin film of a water-displacing lubricant can be used to protect the skin lap joints in these areas. [Figure 12-25]

BILGE AREAS

The bottom of the fuselage below the floor is an area where water and all forms of liquid and solid debris can accumulate and cause corrosion. These areas are ideal for the formation of corrosion because of the almost constant exposure to an

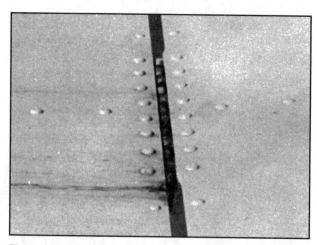

Figure 12-24. A piano hinge must be kept dry and free from dirt which could cause galvanic corrosion because of the different metals used in the hinge.

Figure 12-25. Control surface recesses should be kept clean and dry to prevent the formation of corrosion where it is difficult to inspect.

electrolyte. Furthermore, because of their inaccessibility, corrosion often goes undetected until it has caused major damage.

Airplanes having areas prone to accumulate water are typically provided with **drain holes**. However, dirt and other debris also collect here, and drain holes often become clogged. Therefore, on every inspection, you should carefully inspect any area where water might accumulate and make sure all drain holes are clear. [Figure 12-26]

Air-powered vacuum cleaners can be used to remove dirt or water collected in these areas, and a water-displacing liquid spray that forms a thin film on the surface of the metal can be used to prevent further contact with moisture.

LANDING GEAR BOXES

Few areas in a modern fixed-gear aircraft are as highly stressed, yet as difficult to inspect, as the landing gear box structure. The landing gear attaches into the fuselage by a strong, heavy-gauge aluminum alloy box structure which is under the floor and is accessible for inspection only through a relatively small access hole. Although this area is well-protected, water can collect if the drain holes become plugged. Therefore, during inspections all drains should be opened, and the entire enclosed area generously sprayed with a water-displacing lubricant film. [Figure 12-27]

ENGINE MOUNT STRUCTURE

When a reciprocating aircraft engine is started, the heavy current from the starter must return to the battery through the engine mount. This current flows through joints in the mount and creates the potential difference required for corrosion to form in these areas. To protect welded steel tube mounts from internal corrosion the tubing should be periodically filled with hot linseed oil or other type of tubing oil. The oil is then drained and the drain hole plugged with a drive screw or a self-tapping metal screw.

CONTROL CABLES

The cables used in an aircraft control system are made of either carbon steel or corrosion-resistant steel. If carbon steel cable is left unprotected and water is allowed to get between the cable strands, the cable will corrode. The corrosion that forms on the inside of the cable is difficult to detect. If corrosion is suspected, release the tension from the cable and open the strands by twisting them against the

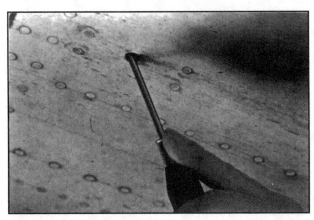

Figure 12-26. The drain holes in the bottom of a structure must be kept open to prevent moisture from collecting and causing corrosion.

Figure 12-27. The landing gear box is one of the most highly stressed areas in an aircraft structure, and is one of the least accesible for adequate inspection.

lay, allowing you to see between the strands. Cable showing any indication of corrosion should be replaced. To prevent this type of corrosion you should spray the cable with a water-displacing type of lubricant. For cables in a seaplane or those exposed to agricultural chemicals, coat them with a waxy grease such as Par-Al-Ketone. [Figure 12-28]

Figure 12-28. A steel control cable must be carefully inspected to be sure there is no corrosion in the strands. If there is any sign of corrosion, the cable must be replaced.

WELDED AREAS

Aluminum torch welding requires the use of a **flux** to exclude oxygen from the weld. This flux may contain lithium chloride, potassium chloride, potassium bisulphide, or potassium fluoride. All of these compounds are extremely corrosive to aluminum and, therefore, all traces of flux must be removed after welding is completed. Welding flux is soluble in water and can be removed with hot water and a nonmetallic bristle brush.

ELECTRONIC EQUIPMENT

The use of copper, lead, tin, and other metals in electronic wiring and printed circuit boards makes them a target for corrosion. Therefore, circuit boards are typically protected by sealing the wiring and circuit boards with a transparent film which excludes all oxygen and moisture from the surface. Detection and repair of corrosion in these areas is a highly specialized field and is not usually the job of an A&P technician.

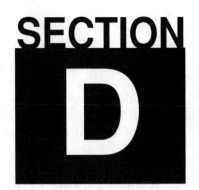

TREATMENT OF CORROSION

Regardless of the type of corrosion or the metal involved, corrosion treatment requires three basic steps:

1. Remove as much of the corrosion as possible.
2. Neutralize any residual material.
3. Restore the protective surface film.

CORROSION REMOVAL

As previously discussed, the first step in corrosion control must be cleaning the surface. After the surface is completely clean and it is determined that corrosion does exist, the damage must be carefully assessed and a decision must be made as to what action should be taken. All corrosion products must be removed as soon as they are discovered, because corrosion continues as long as the deposits remain on the surface.

Corrosion under a paint film cannot be thoroughly inspected without first removing all of the paint. However, before using an unfamiliar paint remover, first test it on a piece of metal similar to that of the structure to be worked on. One thing to keep in mind is never use a caustic paint remover.

Prior to applying a paint remover, all areas not to be stripped should be masked with heavy aluminum foil to keep the stripper from accidently coming into contact with these areas. Water-rinsable paint remover having a syrupy consistency is usually best for aircraft surfaces. This type of remover is applied with a brush by daubing it on the surface rather than brushing it on. Cover the surface with a heavy coating of remover, and allow it to stand until the paint swells and wrinkles up. This breaks the bond between the finish and the metal.

It may be necessary to reapply the remover. If so, scrape the old paint away with a plastic or aluminum scraper, and apply the second coat of remover. This allows the active chemicals to get to the lower layers of finish.

When stripping large areas, spread a sheet of polyethylene plastic over the wet paint remover to slow its drying time. After all of the finish has swelled up and broken away from the surface, it should be rinsed off with hot water or with live steam. A stiff nylon bristle brush may be required around rivet heads and along seams to get all of the stubborn paint that adheres to these places. [Figure 12-29]

Observe caution when using any paint remover. Many solvents used in paint removers attack rubber and synthetic rubber products. Therefore, tires, hoses, and seals must be protected from contact with paint removers.

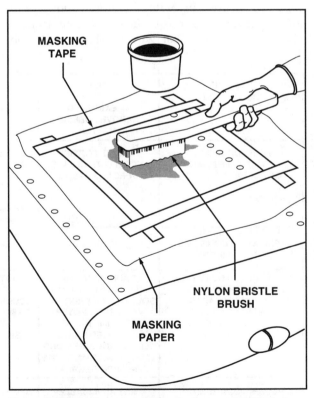

Figure 12-29. Surface preparation is the most critical phase of aircraft refinishing. Cleaning and paint stripping must be performed properly to produce a quality finish.

Most paint removers are usually highly toxic; therefore, special care must be exercised to avoid contact with the skin and eyes. If paint remover is spilled or splashed on your skin, flush the area with water immediately. If any gets into your eyes, flood them with water, and get to a doctor as soon as possible.

TREATMENT OF ALUMINUM ALLOYS

In general, corrosion of aluminum can be more effectively treated on the aircraft than corrosion occurring on other structural materials. Treatment includes the mechanical removal of as much of the corrosion as practicable, the neutralization of residual materials by chemical means, and, finally, the restoration of the permanent surface coating. [Figure 12-30]

MECHANICAL CORROSION REMOVAL

After the paint is removed from a corroded area, all traces of corrosion must be removed from the surface. Very mild corrosion may be removed by using a neutral household abrasive cleaner, such as Bon-Ami™, but be sure that the abrasive does not contain chlorine. Nylon scrubbers, such as "ScotchBrite™" pads, can also be used to remove mild corrosion. More severe corrosion can be removed by brushing with aluminum wool or with an aluminum wire brush. Under no circumstances should you use a steel wire brush or steel wool

since traces of the steel can become embedded in the aluminum and lead to severe corrosion. Blasting the surface with glass beads smaller than 500 mesh can be used to remove corrosion from pits. After using abrasives or brushing, examine the metal with a five- to ten-power magnifying glass to ensure that all traces of the corrosion have been removed. [Figure 12-31]

Severely corroded aluminum alloys must be given more drastic treatment to remove all corrosion. In these situations rotary files or power grinders using rubber wheels impregnated with aluminum oxide are used to grind out corrosion damage. However, when using either of these tools watch carefully to be sure that the minimum amount of material is removed.

After an examination with a five- or ten-power magnifying glass shows no trace of corrosion remaining, remove about two thousandths of an inch more material to be sure that the ends of the intergranular cracks have been reached. Finish by sanding the area smooth with 280-grit, then 400-grit, sandpaper. Clean the area with solvent or an emulsion cleaner, and neutralize the surface with an inhibitor.

CHEMICAL NEUTRALIZATION

After removing all corrosion, treat the surface with a five percent chromic acid solution to neutralize

TYPE OF CORROSION	STEP 1 CLEANING TO REMOVE FOREIGN MATTER	STEP 2 PAINT STRIPPING (WHEN APPLICABLE)	STEP 3 CORROSION REMOVAL	STEP 4 SURFACE TREATMENT (WHEN APPLICABLE)
LIGHT OR HEAVY PITTING OR ETCHING OF ALUMINUM (CLAD)	REMOVE FOREIGN MATTER WITH CLEANER SPEC MIL-C-25789	READILY ACCESSIBLE AREAS: STRIP WITH STRIPPER, SPEC MIL-R-25134 CONFINED AREAS: STRIP WITH SOLVENT	REMOVE CORROSION WITH BRIGHTNER, SPEC MIL C-25378 OR BY MECHANICAL METHOD	CHROMATE CONVERSION COATING, SPEC MIL-C-5541
INTERGRANULAR OR EXFOLIATION CORROSION OF ALUMINUM	AS ABOVE	AS ABOVE	REMOVE CORROSION BY MECHANICAL METHOD	AS ABOVE
LIGHT OR HEAVY CORROSION ON SMALL ALUMINUM PARTS WHICH CAN BE REMOVED FOR TREATMENT	PAINTED PARTS: CLEAN AND STRIP IN SOLUTION OF PAINT AND VARNISH REMOVER, SPEC MIL-R-7751. UNPAINTED PARTS: CLEAN WITH COMPOUND, SPEC P-C-425,MIL C-5543 OR VAPOR DEGREASE	NOT REQUIRED: IF CLEANING ACCOMPLISHED WITH PAINT AND VARNISH REMOVER SPEC MIL R-7751	REMOVE CORROSION AND OXIDE FILM BY IMMERSION OF PARTS IN PHOSPHORIC-CHROMATE ACID SOLUTION	IMMERSION CHROMATE CONVERSION COATING, SPEC MIL-C-5541
STRESS CORROSION CRACKING OF ALUMINUM	NOT APPLICABLE	SEE STEP 1	SEE STEP 1	SEE STEP 1

Figure 12-30. Typical corrosion removal and treatment procedures for aluminum alloys.

METALS OR MATERIALS TO BE PROCESSED	RESTRICTIONS	OPERATION	ABRASIVE PAPER OR CLOTH			ABRASIVE FABRIC OR PAD	ALUMINUM	STAINLESS STEEL	PUMICE 350 MESH OR FINER	ABRASIVE WHEEL
			ALUMINUM OXIDE	SILICON CARBIDE	GARNET					
FERROUS ALLOYS	DOES NOT APPLY TO STEEL HEAT TREATED TO STRENGTHS TO 220,000 PSI AND ABOVE	CORROSION REMOVAL OF FAIRING	150 GRIT OR FINER	180 GRIT OR FINER		FINE TO ULTRA FINE	X	X	X	X
		FINISHING	400				X	X	X	
ALUMINUM ALLOYS EXCEPT CLAD ALUMINUM	DO NOT USE SILICON CARBIDE ABRASIVE	CORROSION REMOVAL OF FAIRING	150 GRIT OR FINER		7/0 GRIT OR FINER	VERY FINE AND ULTRA FINE	X		X	X
		FINISHING	400				X		X	
CLAD ALUMINUM	SANDING LIMITED TO THE REMOVAL OF MINOR SCRATCHES	CORROSION REMOVAL OF FAIRING	240 GRIT OR FINER		7/0 GRIT OR FINER	VERY FINE AND ULTRA FINE			X	X
		FINISHING	400						X	
MAGNESIUM ALLOYS		CORROSION REMOVAL OF FAIRING	240 GRIT OR FINER			VERY FINE AND ULTRA FINE	X		X	X
		FINISHING	400				X		X	
TITANIUM		CLEANING AND FINISHING	150 GRIT OR FINER	180 GRIT OR FINER				X	X	X

Figure 12-31. Abrasives for corrosion removal.

any remaining corrosion salts. After the acid has been on the surface for at least five minutes, it should be washed off with water and allowed to dry. Alodine treatment conforming to MIL-C-5541 will also neutralize corrosion, as well as form a protective film on the metal's surface. Application of alodine treatment is discussed later.

PROTECTIVE COATING

Although some operators prefer a bare aluminum finish to save weight, most aircraft owners utilize a painted finish. Paint is not only attractive and distinctive, but also provides additional protection from impact damage and corrosion.

CLADDING

Pure aluminum is considered to be noncorrosive. However, this is not altogether true because aluminum readily combines with oxygen to form an oxide film. This film is so dense that it excludes air from the metal's surface thereby preventing additional corrosion from forming. The disadvantage of using pure aluminum is that it is not strong enough for aircraft structural components and, therefore, must be alloyed with other metals.

As discussed earlier, once aluminum is alloyed, the alloying agent creates the possibility of dissimilar metal corrosion. However, aluminum alloys can be protected from corrosion and at the same time made attractive in appearance by coating them with a layer of pure aluminum. This is known as cladding. In the manufacture of clad aluminum, pure aluminum is rolled onto the surface of an aluminum alloy and accounts for five to ten percent of the total sheet thickness. The cladding material is anodic as compared to the core material, and any corrosion that takes place attacks the cladding rather than the core.

SURFACE OXIDE FILM

The characteristic of aluminum cladding to form an oxide film on its surface is of real value in protecting aluminum from corrosion. However, in areas where cladding is not practical, metallurgists have found other ways of forming films on metal surfaces that are hard, decorative, waterproof, and airtight. Furthermore, these films typically have the added benefit of acting as a base for paint finishes to adhere to. The process of applying an oxide film is performed in the factories by an electrolytic process known as **anodizing**.

The anodizing process is an electrolytic treatment in which a part is bathed in a lead vat containing a solution of chromic acid and water. This process forms an oxide film on the part that protects the alloy from further corrosion.

After the oxide film has formed, the part is washed in hot water and air-dried. Aluminum treated by this process is not appreciably affected with regard to its tensile strength, its weight, or its dimensions. The anodic film on aluminum alloy is normally a light gray color, varying to a darker gray for some of the alloys. However, some aluminum alloy parts, such as fluid line fittings, are dyed for identification.

In addition to preventing corrosion, the anodic film produced by the anodizing process also acts as an electrical insulator. Therefore, the film must be removed before any electrical connection can be made. Bonding straps are often connected directly to an aluminum alloy part, and for this attachment, the anodized film must be removed by sanding or scraping.

When small parts are fabricated in the field, or when the protective anodizing film has been damaged or removed, the part can have a protective film applied through chemical rather than an electrolytic process. This process is known as **alodizing** and uses a chemical that meets specification MIL-C-5541 and is available under several proprietary names, such as Alodine 1201.

Prior to alodizing a component, all traces of corrosion must be removed. The surface should then be cleaned with a metal cleaner until it supports an unbroken water film. In this **water break test**, any break in the film of rinse water indicates the presence of wax, grease, or oil on the surface, and further cleaning must be done.

While the surface is still wet, brush or spray on a liberal coating of the chemical, allow it to stand for two to five minutes, then rinse it off. If the surface is not kept wet while the chemical is working, streaks may appear and the film will not adequately protect the metal. Therefore, you should work an area that you can keep wet.

After the chemical has had its full working time, flush it from the surface with a spray of fresh water. If a swab or a brush is used, exercise care not to damage the film. The surface is ready to paint after the alodine solution dries. A satisfactory application produces a uniform yellowish-brown iridescent film or an invisible film, depending on the chemical used.

If a powder appears on the surface after the material is dried, it is an indication of poor rinsing or failure to keep the surface wet during the time the chemical was working. If the powder shows up, the part must be re-treated.

WARNING: Rags or sponges used in the application of conversion coating chemicals must be kept wet or thoroughly washed out before they are discarded. Rags drying with the chemical in them constitute a fire hazard.

ORGANIC FILM

One of the most universally used corrosion control devices for metal surfaces is a good coat of paint. Paint adherence is not a problem on porous surfaces, but on smooth surfaces, such as those found on aluminum, the surface must be prepared in order for the paint to have a rough surface to which it can adhere. An aluminum surface is typically roughened with a mild **chromic acid etch**, or by the formation of an oxide film through anodizing or alodizing. The surface can also be mechanically roughened by carefully sanding it with 400-grit sandpaper. When sandpaper is used, it is absolutely imperative that every bit of sanding dust be removed with a damp rag before the primer is applied. Shop rags or hand towels obtained from a commercial service do not normally make good rags for washing surfaces prior to painting. These rags, though clean, frequently contain silicone or other surface contaminants that are incompatible with finishing materials. After removing dust and contaminants, perform a final cleaning of the surface with aliphatic naphtha or an approved prep solvent.

Zinc chromate primer has been used for years with laquer and enamel. It is an inhibiting primer, meaning that the film is slightly porous and water can enter it causing chromate ions to be released and held on the surface of the metal. This ionized surface prevents electrolytic action and inhibits the formation of corrosion. Zinc chromate conforms to specification MIL-P-8585A and can have either a yellow-green or a dark green color. It is thinned with toluol or some of the proprietary reducers made especially for zinc chromate. Prior to applying zinc chromate, the surface to be painted is cleaned of all fingerprints and traces of oil. Then a thin, wet coat of zinc chromate is applied with a spray gun. Because zinc chromate is toxic, always use an appropriate filter mask or respirator when spraying zinc chromate or other paint products.

The synthetic resin base of a zinc chromate primer provides a good bond between the finish and the metal. It also has the property of being dope-proof, which means aircraft dope does not cause it to lift. When a repair is made or a part is fabricated, zinc chromate is often applied to stop corrosion before it ever gets a chance to start.

A **wash primer** is used in aircraft factories for priming new aircraft before they are painted. This two-part primer consists of a resin and an alcohol-phosphoric acid catalyst. The material is mixed and allowed to stand for a short time. It is then sprayed onto the surface with a very light tack coat, followed by a full-bodied wet coat. It cures quickly enough that the topcoats can be sprayed on within less than an hour after the surface is primed.

Epoxy primers are one of the most popular primers for use under polyurethane finishes because they provide maximum corrosion protection. A typical epoxy primer consists of two component materials that produce a tough, dope-proof sandwich coat between the finish and the surface. Epoxy primers can be used on aluminum, magnesium, or steel. For maximum corrosion protection, they can be applied over a wash primer.

The new finishes that are available for aircraft use give beauty and protection far greater than the older, more familiar materials. However, the critical nature of their mixing and application makes it imperative that the manufacturer's instructions be followed in detail. Most finishing materials are classified as "systems" and as such, each has a specific set of recommended materials with which it is compatible.

TREATMENT OF FERROUS METALS

As with other materials, ferrous metal structures and components require some form of protection from corrosive agents to maintain their strength. However, because of their different chemical composition, ferrous materials must be treated differently to achieve the same results.

MECHANICAL CORROSION REMOVAL

Unlike aluminum, the oxide film that forms on ferrous metals is porous and attracts moisture. Therefore, if any trace of iron oxide remains on an iron alloy, it continues to convert the metal into corrosion. The most effective method of removing rust is by mechanical means. Abrasive paper and wire brushes can be used, but the most thorough means

of removing all corrosion from unplated steel parts is by **abrasive blasting**. Abrasive blasting is typically done using sand, aluminum oxide, or glass beads. If a part has been plated, either with cadmium or with chromium, exercise care to protect the plating, since it is usually impossible to restore it in the field.

Highly stressed steel parts, such as those used in landing gear and engines, must be cleaned with extreme care. If corrosion is found on these parts, it should be eliminated immediately by removing the absolute minimum amount of material. A fine stone, fine abrasive paper, or even pumice typically works well. Wire brushes should not be used since they cause minute scratches which can produce stress concentrations that potentially weaken a part. If abrasive blasting is used, it must be done with caution, using a very fine-grit abrasive or glass beads.

After all corrosion has been removed, any rough edges caused by pitting must be faired with a fine stone, or with 400-grit abrasive paper. The surface should then be primed as soon as possible. A dry, clean surface is an ideal setting for corrosion, and if not protected immediately, new damage rapidly sets in. Zinc chromate primer is used to protect most freshly cleaned steel surfaces. [Figure 12-32]

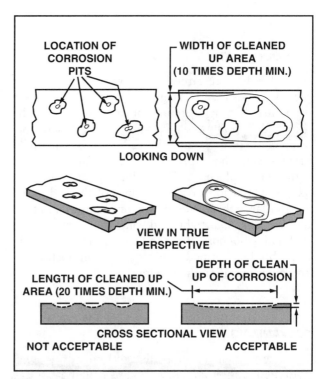

Figure 12-32. Clean the area around corrosion pits to a width that is 10 times the depth of the corrosion, and a length that is 20 times the depth of corrosion. This ensures that all traces of corrosion are eliminated.

SURFACE TREATMENT

A number of options exist for treating ferrous metal surfaces. These options include plating, galvanizing, painting, and metal spraying.

NICKEL OR CHROME PLATING

One way to protect ferrous metals from corrosion is through chrome plating. This plating process produces an airtight coating over the surface that excludes moisture from the base metal. There are two types of chrome plating used in aircraft construction: decorative and hard chrome.

Decorative chrome is used primarily for its appearance and surface protection, while **hard** chrome is used to form a wear-resistant surface on piston rods, cylinder walls, and other parts which are subject to abrasion. Parts to be plated with hard chrome are normally ground undersize and plated back to their proper dimension. Engine cylinder walls are often plated with a porous chrome, whose surface has thousands of tiny cracks which hold oil to aid in lubrication.

CADMIUM PLATING

Almost all steel aircraft hardware is cadmium-plated. This soft, silvery-gray metal is electroplated onto the steel to a minimum thickness of 0.005 inch. It provides an attractive finish as well as protection against corrosion. When the cadmium plating on a part is scratched through to the steel, galvanic action takes place and the cadmium corrodes. The oxides which form on the surface of the cadmium are similar to those which form on aluminum in that they are dense, airtight, and watertight. This means that no further corrosion can take place once the initial film has formed. This type of protection is known as **sacrificial corrosion.**

GALVANIZING

Steel parts such as firewalls are typically treated with a coating of zinc in a process called galvanizing. The protection afforded by this process is similar to that provided by cadmium plating in that when penetrated, the zinc corrodes and forms an airtight oxide film. Steel is galvanized by passing it through vats of molten zinc and then rolling it smooth through a series of rollers.

METAL SPRAYING

Aircraft engine cylinders are sometimes protected from corrosion by spraying molten aluminum on their surface. To accomplish this process, a steel cylinder barrel is sand-blasted absolutely clean, then aluminum wire is fed into an acetylene flame where the wire is melted and blown onto the surface by high-pressure compressed air. Corrosion protection afforded by this treatment is sacrificial corrosion, similar to that provided by cadmium and zinc coating.

ORGANIC COATINGS

The most common organic coating used to protect ferrous metals is paint. However, like aluminum, the surface must be properly prepared to ensure a good bond. Dry abrasive blasting typically removes all of the surface oxides and roughens the surface enough to provide a good bond for the paint. However, parts which have been cadmium-plated must normally have their surface etched with a five percent solution of chromic acid before the primer adheres.

After surface preparation, a thin, wet coat of zinc chromate primer is sprayed on and allowed to dry. The final finish can usually be applied after about an hour.

TREATMENT OF MAGNESIUM ALLOYS

Magnesium is one of the most active metals used in aircraft construction. However, because of its excellent weight-to-strength ratio, designers accept its corrosive tendencies. Magnesium alloys do not naturally form a protective film on their surfaces the way aluminum does, so special care must be taken so that the chemical or electrolytic film applied during the manufacturing process is not destroyed.

MECHANICAL REMOVAL OF CORROSION

When magnesium corrodes, the corrosion products occupy more space than the metal. Therefore, magnesium corrosion typically raises paint or, if it forms between lap joints, it swells the joints. When corrosion is found on a magnesium structure, all traces must be removed and the surface treated to inhibit further corrosion.

Since magnesium is anodic to almost all of the commonly used aircraft structural metals, corrosion should not be removed with metallic tools. Any metallic tool can leave contaminants embedded in the metal that cause further damage. Therefore, stiff nonmetallic bristle brushes or nylon scrubbers are used to remove corrosion. If corrosion exists in the form of deep pits the corrosion must be cut out with sharp carbide-tipped cutting tools or scrapers. If abrasive blasting is used to remove corrosion from magnesium, use only glass beads which have been used for nothing but magnesium.

Many engine parts are made of magnesium, and these parts require special cleaning procedures. Because of the high temperatures and contaminants in engine compartments, carbon deposits build up on engine cases and become baked on. These contaminants are removed through a process called **decarbonization**.

A decarbonizing unit consists of a heated tank and a decarbonizing agent, either water soluble or hydrocarbon based. Parts are immersed in the heated liquid which loosens the accumulated carbon. Complete removal, however, sometimes requires brushing, scraping, or grit blasting. Magnesium parts must not be placed in the decarbonizing tank with steel parts, and metallic cleaning materials such as brushes or abrasives are not to be used.

SURFACE TREATMENT

After all of the corrosion has been removed, a chromic acid pickling solution, which conforms to MIL-M-3171A Type 1 (Dow No. 1), is applied. A satisfactory substitute for this solution may be made by adding about 50 drops of sulfuric acid to a gallon of 10 percent chromic acid solution. Apply this to the surface with rags and allow it to stand for about ten or fifteen minutes, then rinse the part thoroughly with hot water.

A treatment which forms a more protective film is a dichromate conversion treatment such as Dow No. 7, which conforms to MIL-M-3171A Type IV. This solution is applied to the metal and allowed to stand until a golden brown oxide film forms uniformly on the surface. Once this occurs, rinse the surface with cold water and dry it with compressed air. The oxide film is extremely soft when it is wet, and it must be protected from excessive wiping or touching until it dries and hardens. This film is continuous and protects the magnesium surface from corrosion by excluding all electrolyte from its surface.

Like aluminum alloys, magnesium can also have a film deposited on its surface by electrolytic methods. Anodizing magnesium with the Dow No. 17 process produces a hard, surface oxide film which serves as a good base for further protection by a coat of paint.

Magnesium is such an active metal, and magnesium skins are usually so thin that it is absolutely essential that only the proper solutions and proper procedures be used for corrosion treatment. Rather than mixing your own pickling and conversion coating solutions, use prepared chemicals meeting the appropriate MIL specifications, and follow the manufacturer's recommendations in detail.

CORROSION PREVENTION

As stated earlier, the best way to prevent the formation of corrosion is to eliminate one or more of its basic requirements. This is typically done by removing the electrode potential difference within the metal or preventing the introduction of an electrolyte.

DISSIMILAR METAL INSULATION

It is often necessary for different metals be held in contact with each other. When this is the case, dissimilar metal or galvanic corrosion can take place. In order to minimize this danger, the areas to be joined are sprayed with two coats of zinc chromate primer, and a strip of pressure-sensitive vinyl tape is placed between the surfaces before they are assembled. [Figure 12-33]

POWERPLANT PRESERVATION

Reciprocating engines present special problems for controlling corrosion. Oil absorbs contaminants such as carbon, sulphur, and hydrocarbons as an engine operates. If the oil remains in the engine for any length of time, these contaminants can form powerful corrosives that attack the internal parts of the engine. Therefore, engines that are to be stored must first be preserved.

When preparing an engine for storage, the engine must be drained of oil and refilled with a suitable preservative. Some manufacturers require that the engine be run for a short period to heat the preservative oil and coat all internal surfaces. When the engine is stopped, the spark plugs are removed and the propeller is rotated until a cylinder is at bottom center. The cylinder is then sprayed with preservative oil. When all cylinders are coated, desiccant plugs are installed in the spark plug holes to absorb any moisture. It is important not to move the propeller after applying the preservative oil, since the pistons will break the preservative seal and allow corrosion to form.

In addition to the engine, the propeller should be wiped with an oily cloth. This leaves a protective film on the propeller's surface which inhibits the formation of corrosion. Never use acid or alkaline cleaners on metal propeller blades.

FASTENERS

When steel fasteners are used in an aluminum structure and are subjected to a corrosive environment, the holes should be drilled and countersunk, treated with a conversion coating material such as Alodine, and then primed with zinc chromate. Furthermore, the fasteners should be coated with the primer and installed wet. This treatment does not ensure a completely insulated joint, but it does exclude moisture.

CONTACTING METALS	ALUMINUM ALLOY	CADMIUM PLATE	ZINC PLATE	CARBON AND ALLOY STEELS	LEAD	TIN COATING	COPPER AND ALLOYS	NICKEL AND ALLOYS	TITANIUM AND ALLOYS	CHROMIUM PLATE	CORROSION RESISTING STEEL	MAGNESIUM ALLOYS
ALUMINUM ALLOY				▓	▓	▓	▓	▓	▓	▓	▓	▓
CADMIUM PLATE				▓	▓	▓	▓	▓	▓	▓	▓	▓
ZINC PLATE				▓	▓	▓	▓	▓	▓	▓	▓	▓
CARBON AND ALLOY STEELS	▓	▓	▓				▓	▓	▓	▓	▓	▓
LEAD	▓	▓	▓				▓	▓	▓	▓	▓	▓
TIN COATING	▓	▓	▓				▓	▓	▓	▓	▓	▓
COPPER AND ALLOYS	▓	▓	▓	▓	▓	▓						▓
NICKEL AND ALLOYS	▓	▓	▓	▓	▓	▓						▓
TITANIUM AND ALLOYS	▓	▓	▓	▓	▓	▓						▓
CHROMIUM PLATE	▓	▓	▓	▓	▓	▓						▓
CORROSION RESISTING STEEL	▓	▓	▓	▓	▓	▓						▓
MAGNESIUM ALLOYS	▓	▓	▓	▓	▓	▓	▓	▓	▓	▓	▓	

Figure 12-33. This chart lists the metal combinations requiring a protective separator. The separating materials can be metal primer, aluminum tape, washers, grease, or sealant, depending on the metals involved.

SEALERS AND SEALANTS

The higher speeds of modern aircraft have brought requirements for aerodynamic sealers used to fair the edges of wing skins. Airplanes flying at high altitudes require pressurized cabins and as a consequence, every place that plumbing or electrical wiring passes through a bulkhead must be sealed with some form of plastic sealer. Integral fuel tanks are actually a part of the wing structure, and every rivet joint must be sealed to prevent fuel leakage.

Electrical connectors are potted, or sealed, with a special sealing compound to exclude water which would corrode the terminals and lead to intermittent electrical contact.

All of these sealers have specific requirements and are used for unique operating conditions. Therefore, when replacing any sealer, be very sure to use exactly the same material that was originally used, or a substitute specifically approved by the airframe manufacturer. These materials are usually of the two-component type, having one container of resin and one of accelerator or catalyst. The mixing sequence, proportions, time, and temperature must be rigidly adhered to. Otherwise, there is likely to be a poor bond between the sealer and the metal. [Figure 12-34]

When a repair is made, every trace of the old sealer must be removed and the surface prepared for the new sealer. Clean the surface with the recommended cleaner and remove every trace of fuel, oil, and fingerprints. Apply the new sealer and feather the edges down exactly as the instructions specify. This type of material is critical and has no tolerance for careless work or methods of application which are not recommended.

Figure 12-34. Fuel tanks are sealed with a special resilient sealer in all of the seams.

GROUND HANDLING AND SERVICING

INTRODUCTION

While the goal of an aviation technician is to maintain aircraft in such a manner as to assure safe flight, you must also be concerned with creating a safe environment while an aircraft is on the ground. For example, the fuel tanks of transport aircraft contain large amounts of highly flammable fuel and, therefore, can pose a considerable risk of fire. In addition, rotating propellers and operating turbojet engines present a serious risk of injury or death to ground personnel. Therefore, you must make every effort to prevent injury to personnel and damage to aircraft while maintenance and servicing are being performed.

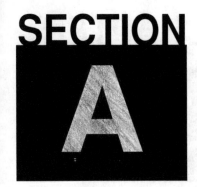
Accidents in the workplace are one of the leading causes of death and disability in the United States. One reason for this is that after working at a job for a period of time, many people become complacent and do not give workplace safety the attention it requires. Aircraft operation areas contain many dangers to personnel, but a sound safety program and an aware workforce can reduce these dangers dramatically. Make workplace safety one of your primary job duties.

ELECTRICAL SAFETY

Every aircraft maintenance shop uses electrical power for day to day activities. While electricity performs many useful functions, you must remember that it can injure or kill if mishandled. Therefore, it is the responsibility of everyone that uses electrical power to be aware of the safety procedures regarding it.

The human body conducts electricity. Furthermore, electrical current passing through the body disrupts the nervous system and causes burns at the entry and exit points. Common 110/120-volt AC house current is particularly dangerous because it affects nerves in such a way that a person holding a current-carrying wire is unable to release it. Since water conducts electricity, you must avoid handling electrical equipment while standing on a wet surface or wearing wet shoes. The water provides a path to ground and heightens the possibility of electric shock.

To understand how common hand tools can create an electrical hazard, consider a typical electric drill that has an AC motor inside a metal housing. One wire is connected to the power terminal of the motor, and the other terminal connects to ground through a white wire. If there are only two wires in

the cord and the power lead becomes shorted to the housing, the return current flows to ground through the operator's body. However, if the drill motor is wired with a three-conductor cord, return current flows through the third (green) wire to ground.

Most shop equipment operating on 110/120-volt single-phase alternating current utilizes a three-conductor cord. The black insulated wire carries the power, while the white wire is the ground and is connected to the earth ground where the power enters the building. The green wire is the equipment ground and connects the housing of the equipment to the earth ground through the long, round pin in the male plug. [Figure 13-1]

To minimize the risk of shock, make sure that all electrical equipment is connected with three-wire extension cords of adequate capacity. Furthermore, do not use cords that are frayed, or that have any of the wires exposed, and be sure to replace any plugs that are cracked.

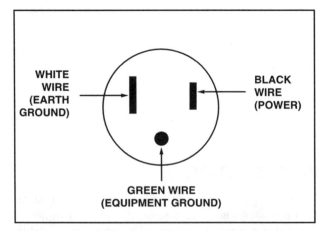

Figure 13-1. The black wire in a three-conductor extension cord carries the electrical power while the white wire serves as the earth ground, and the green wire serves as the equipment ground.

ELECTRICAL FIRE SAFETY

As you recall from your study of electricity, electrical current flowing through a conductor produces heat. The higher the resistance, the greater the heat. Because of this, it is possible for a plug on an extension cord to make such a poor connection that its resistance causes enough heat to start a fire. Therefore, never overload extension cords or wall outlets. Furthermore, always make sure electrical cords are in good shape before using them.

RADIATION HAZARDS

Many large transport category aircraft contain potentially dangerous radioactive materials. For example, the balance weights on the flight controls of many large aircraft contain depleted uranium 238. In aircraft cabin areas, radioactive tritium is used in luminescent devices such as exit signs. Some optical instruments and electronic equipment contain radioactive thorium. In addition, many smoke detectors contain radioactive americium 241.

All of these materials emit low-level radiation and can be dangerous if handled or disposed of improperly. For example, if particles of radioactive material are accidently swallowed or inhaled, serious long-term health consequences can result. As an aircraft maintenance technician working around these radiation sources, you must treat all radioactive materials with the respect they deserve.

HAZARD COMMUNICATION PROGRAM

The Occupational Safety and Health Administration (OSHA) Communication Standard 29, CFR 1910.1200, as well as most states' Right-To-Know laws, require maintenance shops to develop a formal Hazard Communication Program. The program's purpose is to make all personnel aware of shop materials that are considered hazardous or potentially hazardous, and train them in the proper handling and disposal of these materials. The shop program consists of:

1. A formal written program stating compliance and training procedures.
2. A complete inventory of all hazardous materials on the premises.

3. A Material Safety Data Sheet (MSDS) for each item listed on the inventory.
4. Labeling of all pertinent containers and equipment.

MATERIAL SAFETY DATA SHEETS

OSHA regulations require an employer to have copies of relevant Material Safety Data Sheets that are readily available to all shop personnel at all times. These data sheets allow for quick reference in case of a chemical spill or injury. In the case of a chemical injury, a copy of the pertinent data sheet(s) should be sent along to the emergency room to ensure proper medical attention. [Figure 13-2]

A Material Safety Data Sheet consists of nine basic sections:

1. Product identification including trade name, and the address and emergency phone number of the manufacturer/supplier.
2. Principal ingredients including percentages of mixture by weight.
3. Physical data describing the substance's appearance, odor, and specific technical information such as boiling point, vapor pressure, solubility, etc.
4. Fire and explosion hazard potential.
5. Reactivity data including stability and incompatibility with other substances.
6. First aid and health hazard data.
7. Ventilation and personal protection—gloves, goggles, respirator, etc.
8. Storage and handling precautions.
9. Spill, leak, and disposal procedures.

CONTAINER LABELING

Chemical hazard labels vary in size, style, and the amount of information they convey. However, all hazardous materials utilize the same color coding and hazard indexing information. A typical hazard label consists of four color-coded diamonds arranged into one large diamond. The colors used in the table are red, blue, yellow, and white. Within three of the colored areas a number from zero to four appears. The label area colored red indicates a material's flammability hazard. A zero indicates materials that are normally stable and that do not burn unless heated. A rating of four, however,

Material Safety Data Sheet

Identity (Trade Name As Used On Label)

Manufacturer

MSDS Number*

Address

CAS Number*

Date Prepared

Phone Number (For Information)

Prepared By*

Emergency Phone Number   Telex*

Note: Blank spaces are not permitted. If any item is not applicable, or no information is available, the space must be marked to indicate that.

SECTION 1 - MATERIAL IDENTIFICATION AND INFORMATION

COMPONENTS — Chemical Name & Common Names (Hazardous Components 1% or greater; Carcinogens 0.1% or greater)	%*	OSHA PEL	ACGIH TLV	OTHER LIMITS RECOMMENDED
Non-Hazardous Ingredients				
TOTAL	100			

SECTION 2 - PHYSICAL / CHEMICAL CHARACTERISTICS

Boiling Point	Specific Gravity (H_2O = 1)
Vapor Pressure (mm Hg and Temperature)	Melting Point
Vapor Density (Air = 1)	Evaporation Rate (_____ = 1)
Solubility in Water	Water Reactive

Appearance and Odor

SECTION 3 - FIRE AND EXPLOSION HAZARD DATA

Flash Point and Method Used	Auto-Ignition Temperature	Flammability Limits in Air % by Volume	LEL	UEL

Extinguisher Media

Special Fire Fighting Procedures

Unusual Fire and Explosion Hazards

*Optional

Figure 13-2. Material Safety Data Sheets (MSDS) provide information on hazardous materials that are present in the workplace. Furthermore, all employers must maintain current copies of all Material Safety Data Sheets for reference at any time.

applies to highly combustible gases and volatile liquids with flash points below 73° F and boiling points below 100° F (NFPA Class IA). The blue area of the label rates a substance's health hazard from no significant risk (0), to life threatening or permanently damaging with single or repeated exposures (4). The yellow area of the label rates a substance's reactivity. A zero rating applies to materials which are normally stable, even under fire conditions, and which do not react with water. On the other hand, materials rated at four are readily capable of detonation or explosive decomposition at normal temperatures and pressures.

The white area of the label indicates a personal protection index. Unlike the other three ratings given, the personal protection index incorporates an alphabetical rating system using the letters A through K. Each letter indicates different combinations of protective equipment to be worn when working with hazardous materials. For example, the letter "A" indicates the minimum required equipment, including safety glasses or goggles. A "K" rat-

ing requires the use of a full body suit, boots and head mask with independent air supply. You should always use the recommended safety equipment when handling hazardous materials. [Figure 13-3]

SAFETY AROUND COMPRESSED GASES

Compressed gases are found in all aircraft maintenance shops. For example, compressed air powers pneumatic drill motors, rivet guns, paint spray guns, and cleaning guns. In addition, compressed nitrogen is used to inflate tires and shock struts while compressed acetylene is used in welding.

Most shop compressed air is held in storage tanks and routed throughout the shop in high pressure lines. This high pressure presents a serious threat of injury. For example, if a concentrated stream of compressed air is blown across a cut in the skin, it is possible for the air to enter the bloodstream and cause severe injury or death. For this reason, air

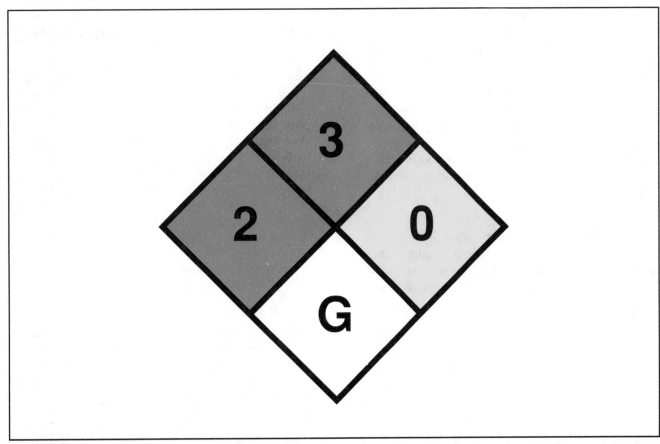

Figure 13-3. The material described by this sample hazard label presents a serious flammability hazard (red), a moderate hazard to health (blue), and a minimum stability hazard (yellow). When handling this material, the use of goggles, gloves and respirator (G) is required for personal protection.

dusting guns are usually equipped with a restrictor that reduces the pressure at their discharge to 30 psi or less.

Be very careful when using compressed air not to blow dirt or chips into the face of anyone standing nearby. To prevent eye injury, you should wear eye protection when using pneumatic tools. To prevent injury from a ruptured hose, always keep air hoses and fittings in good condition.

Far too many accidents occur when inflating or deflating tires. Therefore, wheel assemblies being worked on should be placed in a safety cage to minimize injury if the wheel or tire fails during inflation. Always use calibrated tire gauges, and make certain to use a regulator that is in good working condition.

High-pressure compressed gases are especially dangerous if they are mishandled. Oxygen and nitrogen are often found in aviation maintenance shops stored in steel cylinders under a pressure of around 2,000 psi. These cylinders have brass valves screwed into them. If a cylinder should be knocked over and the valve broken off, the escaping high-pressure gas would propel the tank like a rocket. Because this would create a substantial hazard, you should make sure that all gas cylinders are properly supported. A common method of securing high pressure cylinders in storage is by chaining them to a building. Furthermore, a cap should be securely installed on any tank that is not connected into a system. This protects the valve from damage. [Figure 13-4]

It is extremely important that oxygen cylinders be treated with special care. In addition to having all of the dangers inherent with other high-pressure gases, oxygen always possesses the risk of explosion and combustion. For example, you must never allow oxygen to come in contact with petroleum products such as oil or grease, since oxygen causes these materials to ignite spontaneously and burn. Furthermore, never use an oily rag, or tools that are oily or greasy, to install a fitting or a regulator on an oxygen cylinder.

To minimize the risk of fire, use only an approved MIL Specification thread lubricant when assembling oxygen system components. When checking

Figure 13-4. Be sure the protective cap is screwed on a cylinder containing high-pressure gas to prevent damage to the valve when the cylinder is not connected into a system.

oxygen systems for leaks, use only an approved leak check solution that contains no oil.

SAFETY AROUND MACHINE TOOLS

Many kinds of high-speed cutting tools are commonly found in aviation maintenance shops and can be dangerous if misused. However, these tools pose little threat when used for their intended purpose and reasonable safety precautions are observed. For example, do not use any machine tools with which you are not familiar, or any tool whose safety features you are unfamiliar with. The guards and safety covers found on many tools have been put there to protect the operator. Some of these guards may appear to interfere with the operation of the equipment. However, they must never be removed or disabled. The slight inconvenience they cause is more then compensated for by the added safety they provide. [Figure 13-5]

Dull cutting tools present a greater threat of injury than sharp tools since a dull or improperly sharpened tool requires excessive forces to do its job. As a result, the work can be grabbed or thrown out of the machine. Therefore, always make sure a cutting tool is sharp and serviceable before you use it.

When using a drill press be sure that the material being worked is securely clamped to the drill press

Figure 13-5. Be sure the eye-protection shields are in place when using a bench grinder.

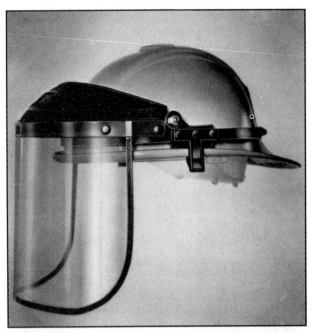

Figure 13-6. Face shields or goggles should be used when drilling, grinding, or sawing.

table before you begin drilling a hole. If this is not done, it is possible for the drill or the cutter to grab the metal and spin it around, effectively slicing anything in its way. Furthermore, never leave a chuck key in a drill motor or a drill press. If the switch is accidentally turned on, the key will be thrown out with considerable force.

The use of eye protection can not be overstressed. Chips coming off metalworking tools can easily penetrate deeply into your eyes. In addition, if someone working near you is using compressed air, a blast of air can easily pick up dirt or dust and spray into your face. To prevent eye injuries, always wear eye protection when using power tools, or when you must enter areas where they are being used. [Figure 13-6]

In addition to eye protection, you should always wear the appropriate clothing when in the shop. For example, you should never wear ties or other clothing that could get caught in a spinning tool. Furthermore, if you wear your hair long, tie it back to keep it out of the way.

When adjusting or changing the blade or bit on a power tool, disconnect the tool from its power source. When maintenance is performed on a power tool that cannot be disconnected from its power source, the electrical junction box for that tool should be turned off and "locked out," to prevent someone from accidentally turning the power back on.

Power tools are one of the greatest timesavers found in a maintenance shop, but you must not allow their convenience to cause you to misuse the tool. In other words, never be in a hurry around a power tool and never use a tool for a purpose for which it is not intended. Most important of all, think before using any tool.

WELDING

Welded repairs are common in aircraft maintenance and shops should provide a means of safely accomplishing the task. Welding should be performed only in areas that are designated for the purpose. If a part needs to be welded, remove it and take it to the welding area. Welding areas should be equipped with proper tables, ventilation, tool storage, and fire extinguishing equipment. If welding is to be accomplished in a hangar, no other aircraft should be within 35 feet of the hanger, and the area should be roped off and clearly marked.

FIRE SAFETY

Aviation maintenance shops harbor all of the requirements for fires, so fire prevention is a vital concern. All combustible materials should be stored in proper containers in areas where spontaneous combustion cannot occur. Since dope and paint solvents are so volatile, they should be stored in a cool, ventilated area outside of the shop.

Spilled gasoline, sanding dust, and dried paint overspray should never be swept with a dry broom, since static electricity can cause a spark and ignite them. Always flush these combustible products with water before sweeping them.

Always be aware of the possibility of fire and provide for exits when putting aircraft in the hangar. Be sure that fire extinguishers are properly serviced, clearly marked and never obstructed. The key to fire safety is a knowledge of what causes fire, how to prevent them, and how to put them out. [Figure 13-7]

FIRE PROTECTION

Since aircraft fuels, paints, and solvents are highly flammable, you must take every precaution to prevent fires where these materials are present. However, you should also be aware of the proper procedures to observe if a fire does start.

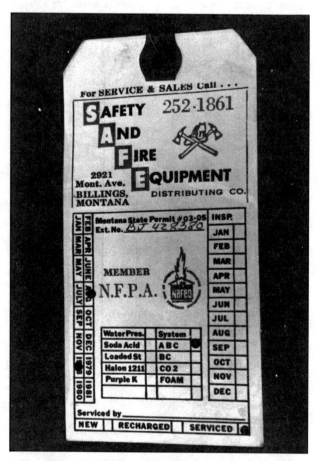

Figure 13-7. Be sure that all fire extinguishers are properly serviced, and are clearly marked to indicate the type of fire they are suited for.

A fire is a chemical reaction between a material and oxygen, in which the material is reduced to its elements with the release of a great deal of heat. Three conditions must be met for a fire to occur. First, there must be fuel, which is any material that combines with oxygen. Second, there must be a supply of oxygen. Third, the temperature of the fuel must be raised to its **kindling point**, which is the temperature at which combustion occurs.

Different types of fuel have different kindling temperatures. For example, gasoline combines with oxygen at a relatively low temperature. On the other hand, materials such as wood must reach a considerably higher temperature before they ignite.

The concentration of available oxygen also affects a material's combustibility. A petroleum product such as oil or grease ignites at room temperature if it is blanketed with pure oxygen. Steel alloy is normally not combustible, but it burns when it is heated red-hot and a stream of pure oxygen is fed into it.

The very nature of aircraft makes them highly susceptible to fire. They carry large amounts of highly flammable fuel, as well as oxygen under high pressure. Because of this, aviation technicians must take proper precautions to prevent fires in aircraft, and have the knowledge and tools to deal with fire when it happens.

CLASSIFICATION OF FIRES

Fire protection begins with a knowledge of the type of fires, what materials are involved, and which extinguishing materials work best for each type. The National Fire Protection Association defines four classes of fires.

Class A fires are those in which solid combustible materials such as wood, paper, or cloth burn. Aircraft cabin fires are usually of this class. **Class B fires** involve combustible liquids such as gasoline, oil, turbine fuel and many of the paint thinners and solvents. Since many of these flammable liquids float, water is sometimes not the best extinguishing agent to use. **Class C fires** are those in which energized electrical equipment is involved. Special care must be exercised in the selection of a fire extinguisher for class C fires, since some fire suppressants conduct electricity. A **Class D fire** is one in which some metal, such as magnesium, is burning. Since Class D fires burn with intense heat, the use of the improper type of extinguisher can intensify these fires, rather than extinguish them. [Figure 13-8]

TYPES OF FIRE EXTINGUISHERS

A fire is extinguished by either cooling the fuel below its kindling temperature or by depriving it of oxygen. All fire extinguishers work on one of these principles.

WATER EXTINGUISHERS

Water can only be used for Class A fires, such as aircraft cabin fires, where electricity is not involved. Most modern water-type extinguishers consist of a container of water in which an antifreeze material

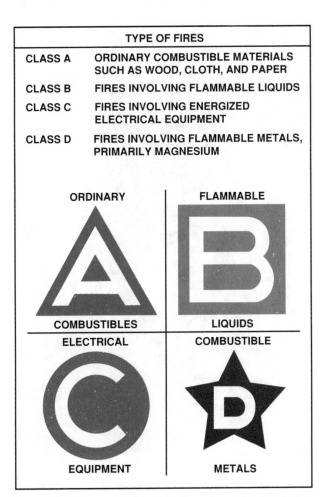

TYPE OF FIRES	
CLASS A	ORDINARY COMBUSTIBLE MATERIALS SUCH AS WOOD, CLOTH, AND PAPER
CLASS B	FIRES INVOLVING FLAMMABLE LIQUIDS
CLASS C	FIRES INVOLVING ENERGIZED ELECTRICAL EQUIPMENT
CLASS D	FIRES INVOLVING FLAMMABLE METALS, PRIMARILY MAGNESIUM

ORDINARY — A COMBUSTIBLES

FLAMMABLE — B LIQUIDS

ELECTRICAL — C EQUIPMENT

COMBUSTIBLE — D METALS

Figure 13-8. Fires are divided into four classes to allow selection of the proper extinguishing agent.

has been mixed. The water is propelled from the extinguisher by a charge of carbon dioxide. Once the extinguisher is activated, all of the propellant is discharged and a new cartridge must be installed when the extinguisher is serviced.

CARBON DIOXIDE EXTINGUISHERS

Carbon dioxide fire extinguishers consist of a steel cylinder filled with the gas under pressure. When the gas is released, it expands to many times its compressed volume, its temperature drops, and it blankets a fire in the form of a white snow. The carbon dioxide gas excludes oxygen from the fire and the fire dies out. CO_2 fire extinguishers are available in sizes ranging from a small two-pound unit that can be mounted in a bracket in the cockpits of small aircraft, to large units that are mounted on wheels.

To operate a carbon dioxide extinguisher, the valve is opened and the CO_2 is directed at the base of the fire from a horn attached to the valve. Since carbon dioxide is heavier than air and is electrically nonconductive, it is effective on both Class B and Class C fires. Furthermore, carbon dioxide extinguishers are particularly well-suited for engine intake and carburetor fires, since they leave no residue. However, never use CO_2 fire extinguishers on Class D fires. The cooling effect of the carbon dioxide on the metal can cause an explosive reaction of the metal.

HALOGENATED HYDROCARBON EXTINGUISHERS

A **halogen** element is one of the group that contains chlorine, fluorine, bromine, or iodine. Some hydrocarbons combine with halogens to produce very effective fire extinguishing agents. Halogenated hydrocarbons are numbered according to chemical formulas with Halon numbers. Halogenated hydrocarbon fire extinguishers are most effective on Class B and C fires, but can be used on Class A and D fires as well. However, their effectiveness on Class A and D fires is somewhat limited.

The most common fire extinguishing agent for cabin fires in modern aircraft is Halon™ 1301 (also known as Freon™ 13), which is most useful as a fire suppressor. It is not harmful to humans in moderate concentrations. In addition to its use for cabin fires,

Halon 1301 is extremely effective for extinguishing fires in engine compartments of both piston and turbine powered aircraft. In engine compartment installations, the Halon 1301 container is pressurized by compressed nitrogen and is discharged through spray nozzles. [Figure 13-9]

DRY-POWDER EXTINGUISHERS

Bicarbonate of soda, ammonium phosphate, or potassium bicarbonate are used as **dry-powder extinguishants**. They are most effective on Class B and C fires where liquids and live electric circuits are involved. In addition, they are the most effective extinguishant for Class D metal fires.

In a typical unit, dry powder is expelled from the container by compressed nitrogen and blankets the fire, excluding oxygen from the fuel. It also prevents a reflash that would re-ignite the fuel after it has been extinguished. Dry powder extinguishers are not recommended for aircraft because of the potential damage to system components from the loose powder. It is also difficult to remove the residue after a dry extinguisher is discharged. [Figure 13-10]

CHECKING FIRE EXTINGUISHERS

The three most common types of fire extinguishers used on aircraft are the carbon dioxide, the nitrogen-pressurized dry powder, and the halogenated hydrocarbon extinguishers. All fire extinguishers should have seals over their operating handles to indicate if a unit has been discharged. The content

of a CO_2 extinguisher is determined by its weight, with the weight of the empty container stamped on the filler valve. CO_2 extinguishers installed on aircraft require periodic weight checks to verify they contain the proper quantity of agent. Nitrogen-pressurized extinguishers have pressure gauges on them with red and green arcs on their dial. As long as the gauge is indicating in the green arc, the gas pressure is sufficient. [Figure 13-11]

JACKING AND HOISTING

Aircraft must often be raised from a hanger floor for weighing, maintenance, or repair. There are several methods of doing this, however, and you should follow the aircraft manufacturer's instructions. If an aircraft slips out of a hoist or falls off a jack, the cost to repair the aircraft is usually quite high.

It is often necessary to lift only one wheel from the floor to change a tire or to service a wheel or brake. For this type of jacking, some manufacturers have

Figure 13-9. Halogenated hydrocarbon fire extinguishers provide effective fire suppression in aircraft engine compartments.

Figure 13-10. A dry-powder fire extinguisher blankets a fire to exclude oxygen from the fuel.

Figure 13-11. The content of the propellant of a dry-powder extinguisher is indicated by the pressure gauge.

made provisions on the strut for the placement of a short hydraulic jack. When using this method, never place the jack under the brake housing or in

any location that is not specifically approved by the manufacturer. On aircraft with spring steel landing gear legs, manufacturers typically provide a special jack pad that clamps to the gear leg, providing a jack point. It is usually recommended that both wheels NOT be lifted off the floor at the same time when jacking from the landing gear struts. [Figure 13-12]

When jacked from the struts, some aircraft have a tendency to move sideways and tilt the jack as the weight is removed from the tire. If this should occur, lower the jack and straighten it, and then raise the wheel again. To keep the aircraft from moving while it is on the jack, the wheels that are not jacked should be securely chocked.

Most modern aircraft have jack pads located on their main wing spars. In addition, many nose-wheel-type aircraft have an attach point on the tail where a jack stand is placed.

The most important consideration when jacking an aircraft is to follow the manufacturer's instructions in detail. Be sure to use the proper jacks so that the aircraft remains level with no tendency for it to slip off of the jacks. Most higher-capacity jacks have screw-type safety collars to prevent the jack from inadvertently retracting. Be sure that these collars are screwed down as the airplane is raised. Jacks that do not have the screw-type safety usually have holes

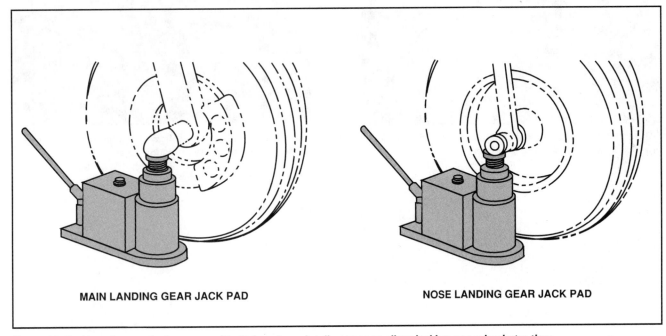

MAIN LANDING GEAR JACK PAD   NOSE LANDING GEAR JACK PAD

Figure 13-12. Aircraft manufacturers provide jack pads on landing gear to allow jacking one wheel at a time.

drilled in the shaft so lock pins can be inserted to guard against the jack retracting. [Figure 13-13]

Many light aircraft can be jacked from only the main spar position by securing a weighted stand to the tail tiedown ring. When using this method, make sure to place enough weight in the stand, or tie the tail to a tiedown ring embedded in the hangar floor. Some aircraft can have their tail held down by weights placed on the main spar of the horizontal stabilizer. However, make sure this procedure is approved by the aircraft manufacturer before attempting it.

Guard against any movement within the aircraft when it is on jacks, since shifting the weight behind the jack could cause the aircraft to tilt enough to fall off the jack.

Before lowering the aircraft, be sure to remove work stands, ladders and other equipment. Items placed under the aircraft while it was on jacks could cause damage when the aircraft is lowered. Furthermore, be sure that the landing gear is down and locked before the aircraft is lowered evenly.

It is possible for some landing gear to produce a side load on the jacks as the weight is taken by the tires, and this must be watched to prevent this side load from causing the jack to tip. Be sure that the oleo struts do not bind and hold the aircraft. If they do bind enough to allow the jack to be lowered away from the wing and the strut should suddenly collapse, it can drop the airplane back onto the jack and cause serious damage.

Always use only the equipment and jacking methods approved by the manufacturer. To do otherwise can cause serious personal injury or major damage to the aircraft.

HOISTING

At times an aircraft must be hoisted, rather than jacked. When this is done, follow the manufacturer's recommendations in detail. Use a hoist of sufficient capacity and, where necessary, place spreader bars between the cables to prevent side loads on the attachment points. [Figure 13-14]

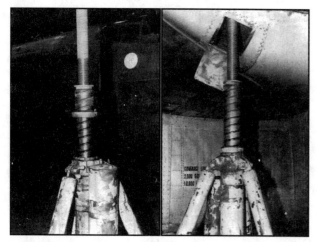

Figure 13-13. A jack's safety collar should be used to keep the jack from collapsing if a seal fails.

Figure 13-14. Be sure to follow the aircraft manufacturer's recommendations when hoisting an aircraft.

SAFETY ON THE FLIGHT LINE

HEARING PROTECTION

The extreme amount of energy released by turbine engines has made the flight line of a modern airport a high-noise area. Continued exposure to this noise can permanently damage your ears and impair your hearing. Because of this, all personnel on the flight line, as well as those in a shop where the noise level is high, should wear some type of hearing protection. This can be either external protectors similar to a pair of large earphones, or internal protectors that fit into the auditory canal of the ear. Either type prevents sound energy from damaging the inner ear mechanism. [Figure 13-15]

FOREIGN OBJECT DAMAGE (FOD)

If foreign objects such as nuts, bolts, and safety wire are drawn into the inlet of a turbine engine, or through the arc of a rotating propeller blade, they can easily cause damage that can lead to catastrophic failure. Furthermore, if these objects are caught in an engine's high-velocity exhaust and strike another person, these objects can cause serious injury. Therefore, it is extremely important that an airport flight line be kept clean. Furthermore, you, as an aircraft maintenance technician, should develop the habit of picking up all loose hardware and rags you find on the ramp and deposit them in a suitable container.

Your personal safety is much more important than damage to equipment. Therefore, be aware that propellers and jet intakes make aircraft operation areas extremely dangerous places. Sound maintenance practice requires that aircraft operating their engines must illuminate their navigation lights and rotating beacons. Furthermore, a second person should be stationed on the ground to warn anyone in the area. If it is necessary to approach an aircraft whose engine is running, be sure that you have the attention of the person in the cockpit so they know where you are at all times. The danger of a propeller is so obvious that it hardly needs mention, but sadly enough, many people have been killed by walking into a spinning propeller. Always be cautious of rotating propellers.

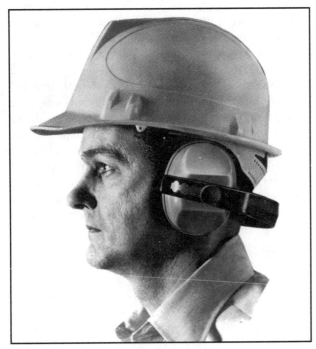

Figure 13-15. The high noise levels encountered in aircraft operation areas can seriously damage your hearing. Hearing protectors such as these can prevent hearing loss.

Turbojet aircraft are dangerous from both the front and the rear. When an engine is running, it moves a large volume of air and produces a low-pressure area in front of the intake. This low-pressure area can draw a person into the engine, and can certainly ingest such items as hats, clipboards, and loose items of clothing. Behind the aircraft in the exhaust area, the high-velocity exhaust can cause severe damage to both people and equipment. [Figure 13-16]

SAFETY AROUND HELICOPTERS

Helicopters are a unique type of aircraft often operated in terrain in which an airplane could not be used. Personnel must often approach and depart

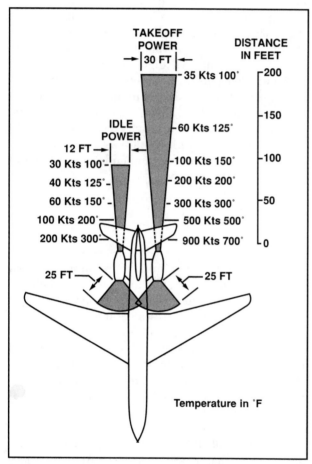

Figure 13-16. When you are in the vicinity of jet aircraft that are running their engines, stay clear of the danger areas shown.

from them when the engine is operating and the rotor turning. When doing this, be sure to stay in the pilot's field of vision. By doing this you will remain clear of the tail rotor. [Figure 13-17]

One thing to keep in mind when working around helicopters is that many of the passengers carried on helicopters are not familiar with this type of aircraft. Therefore, it is your responsibility to watch that they do not endanger either themselves or the helicopter.

TIEDOWN PROCEDURES

An aircraft's lightweight construction coupled with its airfoil-shaped wings and tail surfaces or rotors makes it highly susceptible to damage from wind. The most violent windstorms, the ones that do the most severe damage to aircraft in the United States, are tornados that occur during the early summer months in the central states, and hurricanes which occur along the Gulf and East coastal regions during the months of August through October. In addition, localized thunderstorms can occur throughout the year and appear in almost every part of the country. Regardless of the type of storm, damage can be severe for aircraft that are not protected. However, a great deal of this damage can be minimized if proper protective measures are taken.

The best protective measure you can take to help ensure an aircraft's safety is to put it in a hanger. However, at times a hanger may not be available. If this is the case, an aircraft should be securely tied down and its controls firmly locked in place. For example, most aircraft are equipped with internal control locks that hold the control surfaces in a streamlined position. However, since these locks secure the cockpit control, there is still a possibility that if severe forces were exerted on an aircraft's control surfaces, damage to the control actuating system could result. To prevent this, control surface **battens** are often used to hold a control surface in a streamline position. These battens are clamped

SAFETY AROUND HELICOPTERS

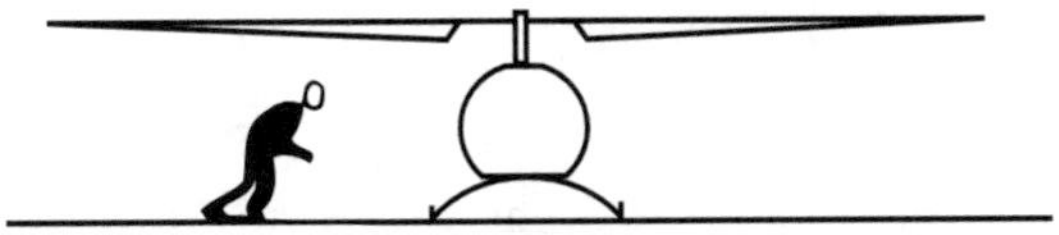

1 — Approach or leave machine in a crouching manner (for extra clearance from main rotor)

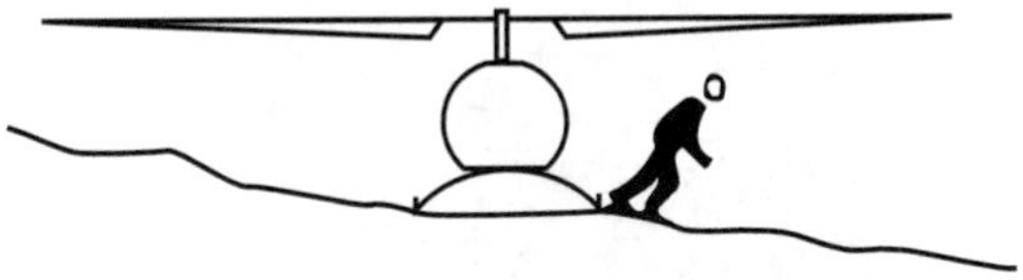

2 — Approach or leave on the down slope side (to avoid main rotor).

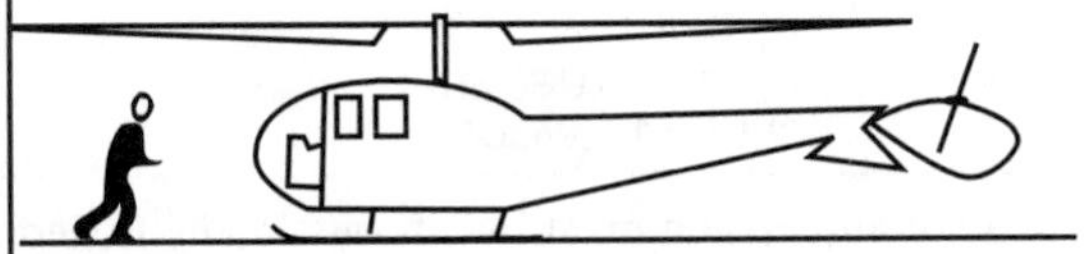

3 — Approach or leave in pilot's field of vision (to avoid tail rotor).

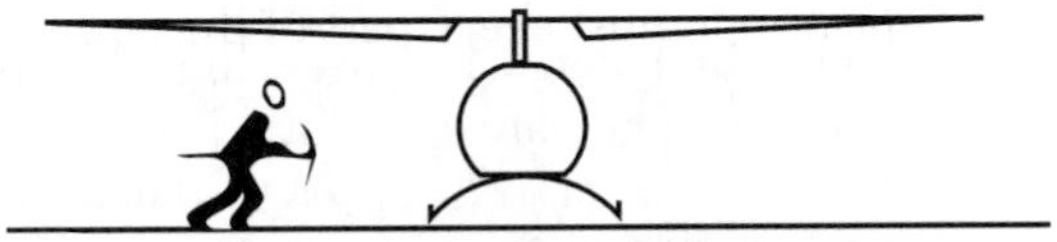

4 — Carry tools horizontally, below waist level (never upright or over shoulder).

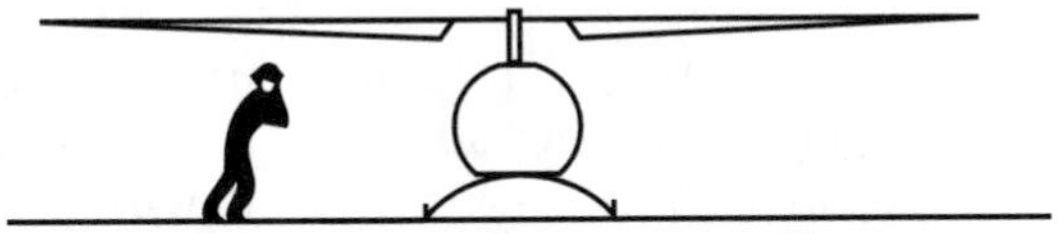

5 — Hold on to hard hat when approaching or leaving machine, unless chin straps are used.

6 — Fasten seat belt on entering helicopter and leave it buckled until pilot signals you to get out.

7 —If leaving machine at hover, get out and off in one smooth, unhurried motion.

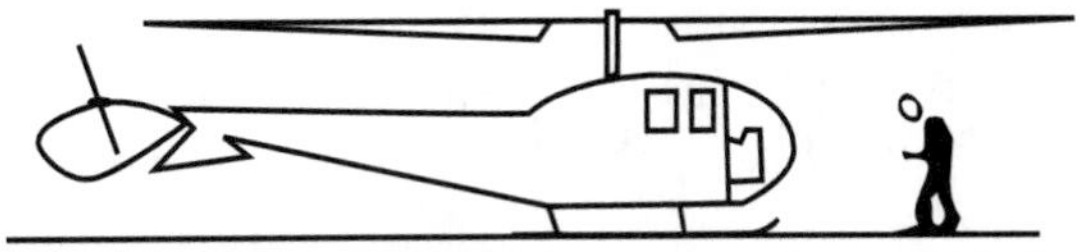

8 — Do not touch bubble or any moving parts (tail rotor linkage, etc.).

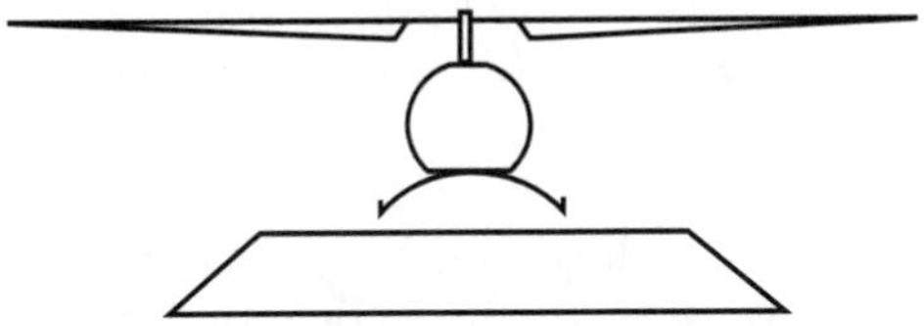

9 — Keep landing pad clear of loose articles.

10 — Loading assistants should always use plastic eye shields.

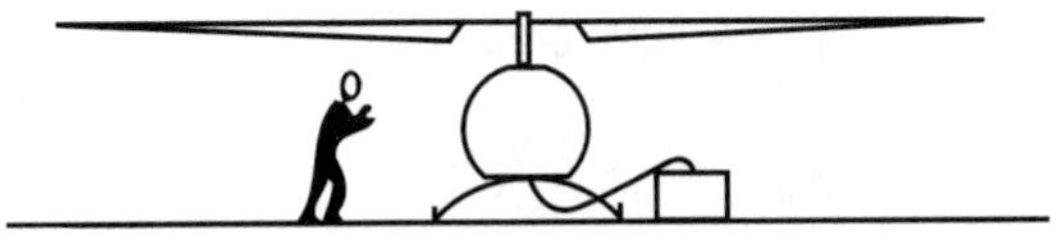

11 — After hooking up cargo sling, move forward and to side to signal pilot

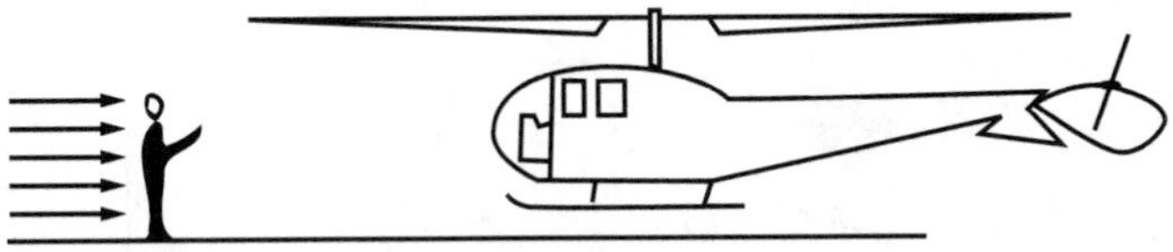

12 — When directing machine for landing, stand with back to wind and arms outstretched toward landing pad.

13 — When directing pilot by radio, give no landing instructions that require acknowledgement as pilot will have both hands busy.

Figure 13-17. Illustrated are some of the more commonly accepted safety practices to use around helicopters.

against a fixed surface and should be lined with one-inch foam rubber. Furthermore, battens should be painted red and have a long red streamer attached so they are easy to see. This helps prevent a pilot from inadvertently leaving them on the controls prior to a flight. [Figure 13-18]

If a tail wheel aircraft is tied down facing into the wind, its elevator should be locked in the full up position so the wind forces the tail down. On the other hand, if a tailwheel-type aircraft is tied down facing away from the wind, the elevator should be locked in the full down position.

If a severe wind is expected, spoiler boards can be secured to the top surface of a wing to destroy lift. These spoilers are often made of 2 × 2 boards on which a one-inch strip of foam rubber is attached. Holes are drilled through the boards so they can be secured with nylon rope. The nylon rope is tied around the wing to hold the spoiler parallel with the wing span approximately one-quarter of the wing's width back from the leading edge. [Figure 13-19]

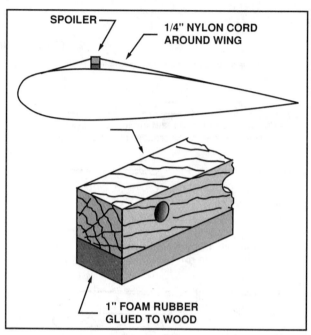

Figure 13-19. If an airplane must be left outside in a windstorm, spoilers can be lashed to the top of the wing to prevent generation of lift.

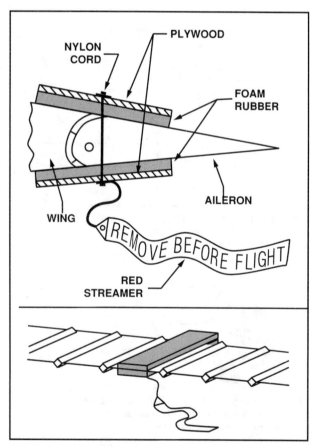

Figure 13-18. Control battens are used to securely lock an aircraft's control surfaces in place. They should have red streamers attached to help prevent their inadvertently being left on the aircraft when it is prepared for flight.

Special care must be taken when securing a set of spoilers so that the ropes are not pulled too tight and the wing's leading or trailing edge damaged. Furthermore, scraps of carpet or foam rubber should be placed under the rope where it contacts the wing to prevent damage to the aircraft's finish.

In addition to securing the control surfaces and attaching spoilers to the wing, all doors and windows should be secured so they cannot be blown open. Furthermore, all engine openings should be covered to keep blowing dirt from entering the engine compartment and the engine itself. Pitot heads should also be covered to exclude water and dirt.

When it comes to tying down an aircraft, most airports have a tiedown area with anchors permanently embedded into a hard-surfaced ramp. However, at some airports, aircraft are tied down to cables that run the length of the flight line. With either method, an aircraft should be secured so that it is headed as nearly into the wind as is practical with as much separation between it and adjacent aircraft as possible. [Figure 13-20]

When parked, an aircraft's nose wheel should be locked in a straight ahead position so the aircraft cannot move from side to side, or weathervane in the wind. In addition, tiedowns should be secured to each wing and to the tail. Although aircraft can be tied down with either rope or

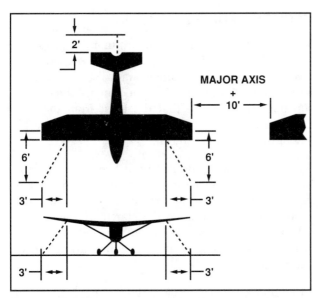

Figure 13-20. When securing an aircraft on a crowded ramp, observe the minimum clearances shown in this diagram.

chain, ropes normally provide the strongest attachment. Nylon is the strongest material for rope, although Dacron and yellow polypropylene also provide sufficient strength. Manila rope should be avoided, if possible, since it has a tendency to shrink when it gets wet, as well as mildew and rot from exposure to weather.

When inserting rope through aircraft tiedown rings the rope should be pulled snug and secured with a bowline knot. The rope should not be pulled tight enough to put a strain on the wing, but must keep the aircraft from rocking back and forth excessively. Proper tension allows about one inch of movement. However, if manila rope is used, a little extra slack must be allowed in the event the rope shrinks. [Figure 13-21]

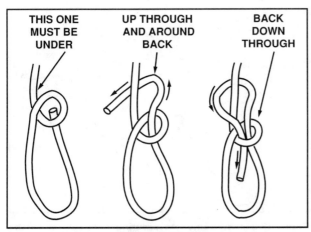

Figure 13-21. The bowline is the most generally used knot for tying an aircraft with ropes.

Chains are used at many airports. Although they have a much longer life than rope and are easier to use, but they are not as strong as the proper size rope. If chains are used, they must be secured to an aircraft by passing the chain through the tiedown ring, then sticking one link through a link in the standing chain and fastening it in place with a snap. Do not allow the snap to take any of the strain, since it is not made for this purpose. [Figure 13-22]

In addition to securing the wings, airplanes with nose wheels should be tied down with one rope through the nose gear tiedown ring and two ropes through the tail tiedown ring. The ropes from the tail should pull away at a 45 degree angle to each side of the tail. Furthermore, make sure that the wheels are blocked with properly fitting chocks in front of and behind the wheels.

SEAPLANES

Seaplanes can be secured by towing them into shallow water or onto a beach or by securing them to a dock or tree. If left in the water, some of the float compartments should be flooded to add weight and assist the tiedown ropes in holding the aircraft.

SKIPLANES

Ski-equipped aircraft are often caught out in storms with no protection from high winds. When this happens, loose snow can be packed around the skis, then doused with water so they freeze in. The tiedown ropes can also be frozen into ice to secure

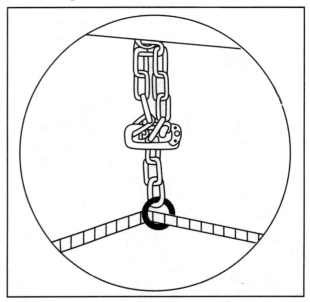

Figure 13-22. When securing an aircraft with a chain, do not depend upon the snap to carry any strain. The snap should be used only to secure a free link through a standing link.

them. One method of doing this is through the use of a deadman anchor. [Figure 13-23]

HELICOPTERS

The fragile nature of a helicopter rotor system makes it extremely important that the manufacturer's operational manual be followed in detail when securing a helicopter. However, a few general considerations should be noted. The helicopter should be headed into the anticipated wind, clear of any other helicopters, airplanes, or buildings by at least a rotor span. The skids should be securely tied to ground tiedown facilities by attaching ropes to the points specified by the helicopter manufacturer. If the helicopter is mounted on wheels, the brakes should be set and chocks placed in front of and behind the wheels. Position the main and tail rotor blades and secure them to the helicopter structure by the method described in the operations manual. [Figure 13-24]

ENGINE STARTING PROCEDURES

Starting an aircraft engine is a specialized procedure and varies with an individual engine and aircraft. Therefore, before starting any aircraft engine, be sure to study the procedures in the appropriate airplane flight manual and get instruction from an experienced operator. However, certain general guidelines apply to all reciprocating and turbine engine powered aircraft.

RECIPROCATING ENGINES

Before starting an engine, insure that the propeller area is clear. The aircraft should not be parked in an area of loose gravel, since small stones can be sucked up into the propeller and damage the blades. Furthermore, be sure that the blast from the propeller does not blow dirt into any hangar or building or onto another aircraft.

There is always a possibility of fire when starting an engine. Because of this, you should always have a carbon dioxide fire extinguisher of adequate capacity available. For starting large aircraft where it is not possible to see the engine when it is being started, a fire guard must be stationed near the engine.

Induction system fires are the type which occur most frequently in reciprocating engines. The reason for this is if an engine is over-primed and then fires back through the carburetor, the gasoline in the induction system can ignite. If this occurs, the best procedure is to continue cranking and start the

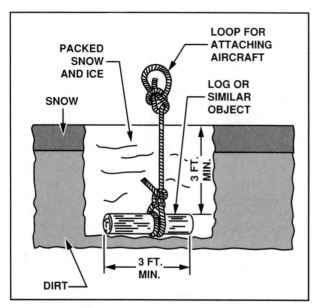

Figure 13-23. A deadman anchor can provide an effective means of securing an aircraft that is caught in a storm away from an airport.

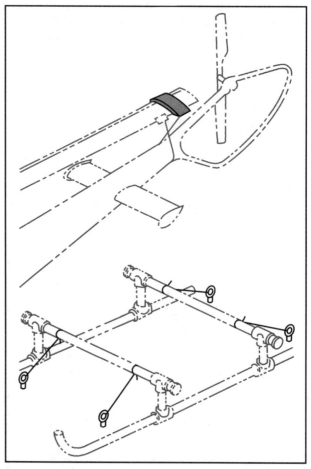

Figure 13-24. In addition to securing a helicopter to the ground, you must also secure the helicopter's rotor blades to its fuselage.

engine if possible. Once started, the engine will suck the fire into the cylinders, and no further damage can occur.

Large radial engines have a problem of oil seeping past the piston rings and getting into the lower cylinders. In fact, in as little as 30 minutes, enough oil can seep by the piston rings creating a problem known as **hydraulic lock**. With hydraulic lock, oil fills the lower cylinders to a point that the piston cannot make its compression stroke. This can cause serious damage if some of the cylinders should fire and drive the piston into this oil. To be sure that none of the cylinders have a hydraulic lock, the engine should be turned through by hand, with the ignition off, until all of the cylinders have passed through their compression strokes.

To start an engine with a typical float-type carburetor, place the mixture control in the full rich position. Almost all reciprocating engines are equipped with either a **carburetor-heat** or an **alternate-air** position on the carburetor air inlet system. For starting and ground operation, these controls should be in the cold position, because heated air is not filtered. Prime the engine as required, and open the throttle about 1/2 inch. Turn the master switch on, and turn the engine over with the starter switch. When the engine starts, check for positive oil pressure and adjust the throttle to produce about 1,000 rpm. To shut the engine down, pull the mixture control to the idle cut-off position. When the engine stops, turn the ignition and master switch off.

A flooded reciprocating engine can be cleared of excessive fuel by placing the mixture control in the idle cutoff position. This shuts off all fuel flow to the cylinders. With the mixture in the cutoff position, place the ignition switch in the off position, open the throttle all the way, and crank the engine with the starter until the fuel charge in the cylinders has been cleared.

Fuel-injected engines have several different starting requirements. For example, once the mixture is placed in the full rich position and the throttle is opened about one inch, both the master switch and fuel pump are turned on until adequate fuel flow is observed. This procedure is required to prime the engine. Once primed, the fuel pump is turned off, the magneto switches are turned on, and the starter is engaged. When the engine starts, check for positive oil pressure.

HAND PROPPING

Hand propping of aircraft engines is a procedure with which all technicians should be familiar, as it is possible for a battery to become discharged when no auxiliary power is available. It is important that no person ever attempt to prop an aircraft until being thoroughly checked out in the procedure. Although hand propping is something that cannot be learned from a book, there are a few general guidelines that are helpful.

One important aspect of propping an aircraft is to be sure that the person in the cockpit is thoroughly familiar with the aircraft and that clear communication is established between you and the person in the cockpit. Check to be sure the brakes are actually holding by attempting to move the aircraft. If the propeller must be rotated to check for a hydraulic lock, or turned backwards to clear an over-prime condition, call out SWITCH OFF. Don't turn the propeller until you hear the answer, SWITCH OFF. When ready to pull the propeller to start the engine, call out the word CONTACT. If the person in the cockpit has all of the controls properly set, they reply CONTACT and then turn the ignition switch on.

When actually pulling on the propeller, be sure that your footing is secure. Do not stand on wet grass or sandy ramp surfaces. Stand close enough to the propeller that you are slightly overbalanced away from the engine rather than into it. Grasp the propeller by laying your palm over the blade. Do not grip your finger around the trailing edge of the blade, as there is danger of the engine kicking back and pulling you into the propeller. As an added safety precaution, even though the control operator has answered your SWITCH OFF, treat the propeller with caution.

As soon as the engine starts, check the oil pressure gauge. If there is no indication of oil pressure within the time limit specified in the aircraft operator's manual (usually 30 seconds), the engine must be shut down and the cause determined. If the oil pressure is good, the engine rpm should be adjusted to the point that the engine runs smoothest and is allowed to warm up at this speed.

TURBINE ENGINES

Turbine engine aircraft usually have enough automatic sequencing by the fuel controls so that starting is less involved than starting a reciprocating engine of comparable power. But the extremely high cost of turbine aircraft, and the possibility of severe damage in the event of an improper start, make it doubly important that only qualified personnel start these engines.

One of the problems that can occur when starting a turbine engine is the lack of sufficient electrical power to get the engine up to proper speed for it to become self-accelerating in its operation. This condition can be minimized through the use of a ground power unit (GPU). Improper starts are usually one of two types, hot starts and hung starts.

HOT STARTS

A hot start is one in which ignition occurs when there is an excessively rich fuel/air mixture. If a hot start is allowed to proceed, the exhaust gas temperature will exceed the allowable limit and the engine will be damaged. To minimize the possibility of a hot start, the exhaust gas temperature, turbine inlet temperature, or interstage turbine temperature gauge must be monitored during a start. If any temperature indication exceeds its allowable limit, fuel to the engine should be shut off immediately. Consult the engine manufacturer's maintenance manual for inspection or overhaul requirements following hot starts.

HUNG STARTS

A hung start is one in which the engine starts, but does not accelerate enough for the compressor to supply sufficient air for the engine to become self-accelerating. Hung starts occur when the starter cuts out too soon, or when the starting power source fails to provide enough energy to rotate the engine to a sufficient speed. The start is normal and the exhaust gas temperature is all right, but the engine does not accelerate to normal operating speed. If a hung start occurs, the engine must be shut down and the cause for insufficient starting speed corrected before another attempt is made.

TURBINE ENGINE START PROCEDURES

Starting procedures for specific turbine engines vary, and the manufacturer's checklist should always be followed. Personnel performing engine run and taxi must be thoroughly checked out on the equipment they are operating. Be certain all safety equipment is in place, and always be aware of emergency procedures in the event of fire. The following procedures are typical for a large transport aircraft, and are intended as general guidelines only.

Be sure the engine inlets are clear and that the area in front of the inlets is clear of loose objects. Check that all doors are closed, and that the area behind the engines is clear. Turn on the aircraft's rotating beacon and signal the fireguard that you are starting an engine.

Turn on the auxiliary hydraulic pump, open the ground interconnect and set the parking brake. Open the bleed valves and verify that the ground power unit or aircraft auxiliary power unit is supplying air. Open the start valve for the engine you are starting, and observe that the start valve light illuminates. You should observe an oil pressure indication and N_2 rotation within two or three seconds. When N_2 reaches 20 percent and N_1 rotation is confirmed, lift the fuel lever out of the cutoff detent into the run position. Immediately check the fuel flow indicator to confirm fuel flow, and then watch the EGT indicator to confirm a light-off. This indicates a start has begun. Continue to observe the EGT indicator and, in case of a hot start, shut the engine down. When engine speed is self-accelerating, release the start switch and monitor the engine instruments for proper indications. [Figure 13-25]

TAXIING AIRCRAFT

Airplanes and helicopters are designed to fly, and movement on the ground is often a rather awkward procedure. Because of this, only qualified persons authorized to taxi aircraft may actually taxi an aircraft. Before starting an engine, be sure that the areas in front and behind the aircraft are clear of people and equipment. A maintenance technician should be checked out by a properly qualified instructor before taxiing a new or different aircraft.

From the cockpit, it is difficult to assure that there is sufficient clearance between the aircraft structure and any buildings or other aircraft. Therefore, it is a good policy to station signalmen where they can watch the wings or rotor and any obstructions. When this is done, it is important that all personnel use the same signals and understand exactly what

Figure 13-25. It is important that the person starting a turbine engine be thoroughly familiar with the procedure.

the signals mean to avoid misunderstanding at a crucial time. The signalman has the responsibility of remaining in a position that is visible from the cockpit. To ensure that you can be seen at all times, make sure that you can see the pilot's eyes while directing him. [Figure 13-26]

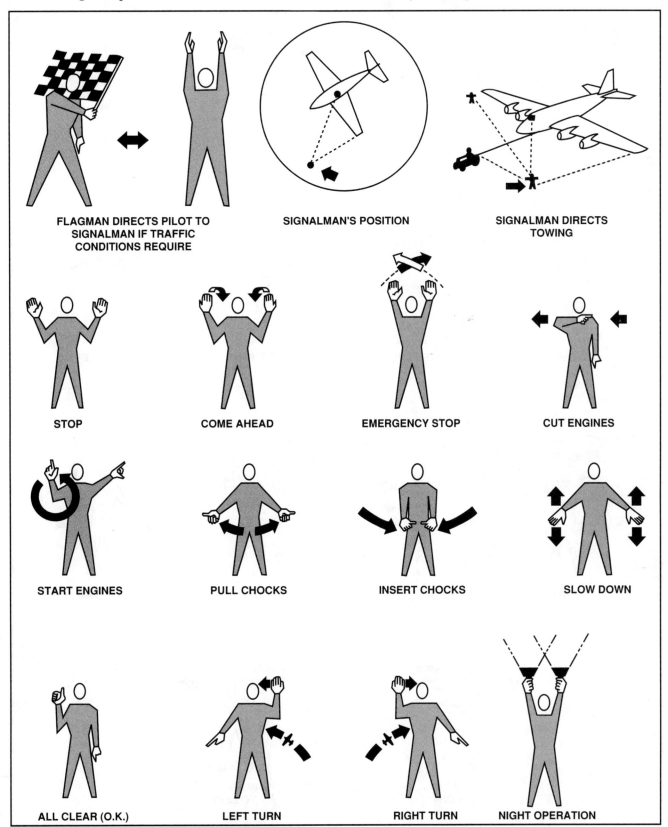

FLAGMAN DIRECTS PILOT TO SIGNALMAN IF TRAFFIC CONDITIONS REQUIRE

SIGNALMAN'S POSITION

SIGNALMAN DIRECTS TOWING

STOP

COME AHEAD

EMERGENCY STOP

CUT ENGINES

START ENGINES

PULL CHOCKS

INSERT CHOCKS

SLOW DOWN

ALL CLEAR (O.K.)

LEFT TURN

RIGHT TURN

NIGHT OPERATION

Figure 13-26. Standard hand signals allow ground personnel to direct the movement of aircraft.

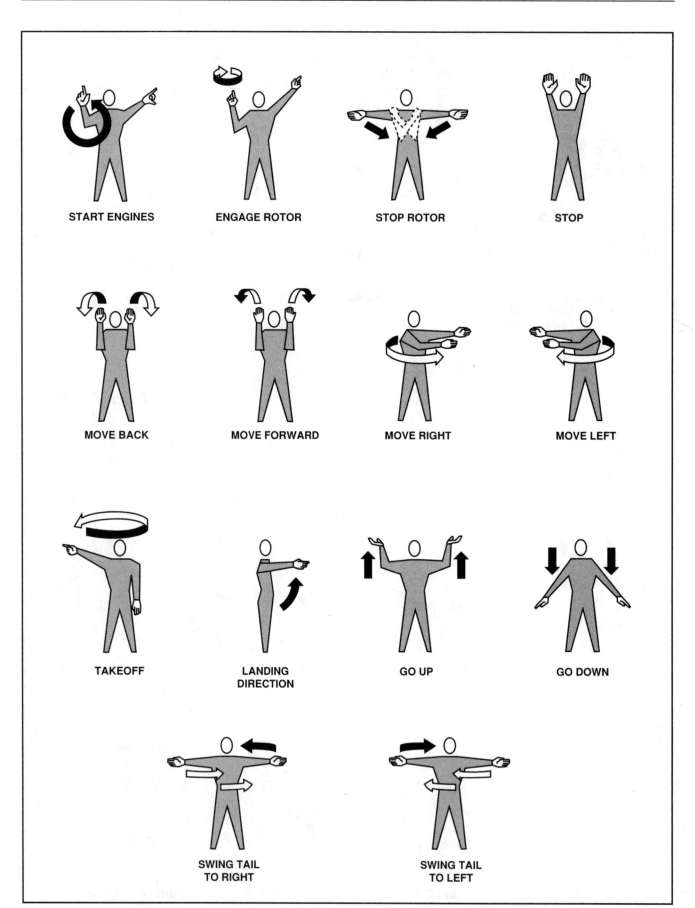

Figure 13-26. Continued.

When taxiing an aircraft at a tower-controlled airport, you typically must receive a clearance from ground control before you begin taxiing. Once the aircraft is in motion, immediately tap the brakes to insure they are working properly. After testing the brakes, test the nose gear steering system to make sure it is operating.

TAILWHEEL AIRCRAFT

Tailwheel aircraft present certain difficulties during taxi because the tail is low leaving the operator's view over the nose obstructed. Because of this, an operator must alternately turn the nose from side to side in a series of **S-turns** during taxi to avoid objects or hazards in the aircraft's path.

The abrupt use of brakes should be avoided on a tailwheel aircraft since a hard brake application could cause the aircraft to nose over. Furthermore, tailwheel aircraft are difficult to taxi in windy conditions, especially crosswinds. The reason for this is that tailwheel aircraft are designed much like a weathervane, with the pivot point at the main landing gear. The vertical stabilizer on the tail and the fuselage behind the main gear present a large surface to a crosswind. Since there is so much more surface area behind the main gear than in front of it, a crosswind creates a powerful turning force that pushes the aircraft's nose into the wind. To avoid losing control of the aircraft, you must exercise extreme caution when taxiing a tailwheel aircraft in crosswind conditions.

LIGHT SIGNALS

Busy airports usually require radio contact between an aircraft and the control tower when the aircraft moves on taxiways or runways. In the event that you must taxi an aircraft that does not have a radio or in the event of radio failure, control towers are equipped with highly directional light guns they can use to signal you. [Figure 13-27]

TOWING

It is often necessary to move an aircraft without using its engines. This can be accomplished by towing the aircraft. Large aircraft are towed with a tractor, or special towing vehicle, and are connected to the vehicle with a special tow bar. Extreme care must be used to avoid towing an aircraft too fast and to be sure that there is always sufficient clearance between the wings and any obstructions.

When an aircraft is being towed, a qualified person should be in the cockpit to operate the aircraft

COLOR AND TYPE OF SIGNAL	MEANING
	ON THE GROUND
STEADY GREEN	CLEARED FOR TAKEOFF
FLASHING GREEN	CLEARED TO TAXI
STEADY RED	STOP
FLASHING RED	TAXI CLEAR OF LANDING AREA (RUNWAY) IN USE
FLASHING WHITE	RETURN TO STARTING POINT ON AIRPORT
ALTERNATING RED AND GREEN	EXERCISE EXTREME CAUTION

Figure 13-27. Located in the cab of the control tower is a powerful light that controllers can use to direct light beams of various colors toward you aircraft. Each color or color combination has a specific meaning for an aircraft on the airport surface.

brakes when needed since the brakes on a towing vehicle are usually insufficient to overcome a large aircraft's momentum. Extra personnel should be assigned to watch the wing tips and tail for clearance between other objects.

The nose gear on most aircraft have a very definite limit to the amount it can be turned and, when towing, it is easy to exceed these limits. If the turning radius is exceeded, the nose gear strut and steering mechanism will be damaged. Damage can be quite extensive, requiring replacement of the nose gear shock strut. Some aircraft have a method of disconnecting a locking device so the nose wheel can be swiveled to facilitate maneuvering. If this is the case, the locking device must always be disconnected when an aircraft is towed. Furthermore, remember to reset the lock after removing the towbar from the aircraft. Persons riding in the aircraft should not attempt to steer the nosewheel when a towbar is attached to the aircraft. [Figure 13-28]

Figure 13-28. When towing an aircraft, be sure that the limits of nose wheel movement are observed since the towing stops can be damaged if these limits are exceeded.

Although small aircraft can be moved by hand, substantial damage can result from careless or improper handling procedures. For example, you should never move an airplane by pulling on its propeller. The propeller is designed to move the aircraft through the air, but the thrust it produces is uniform. Moving the airplane by pulling on one blade puts an asymmetrical load on both the propeller and the engine.

When towing an aircraft, you should always use a tow bar. Most tow bars attach to the nose wheel and are used both to move and steer the airplane. After an aircraft has been towed with a tow bar and parked in the desired position, remove the tow bar from the nose strut and place it beside the nose-wheel, or stow it away. If an engine is started with a tow bar still attached to a nosegear, the tow bar, propeller, and aircraft will typically sustain substantial damage.

When pushing an aircraft, be sure to push only at points that are specified by the aircraft manufacturer. Never push on control surfaces, nor in the center of a strut. NO STEP and NO PUSH decals mean just that.

HELICOPTER GROUND HANDLING

Ground movement of helicopters is different than that for conventional aircraft. The most common landing gear on helicopters consists of a set of skids which allows operation from various surfaces. To move a helicopter on the ground, small wheels are attached to its skids and the helicopter is raised off the ground onto the wheels. [Figure 13-29]

Some helicopters do utilize wheels as landing gear. In this situation, the nose gear is typically free to swivel as the helicopter is taxied. These helicopters have tow bars that attach to the nose gear, and are towed in much the same way as fixed-gear aircraft.

Figure 13-29. The attachment of small wheels to the skids allows for easier ground handling on most light helicopters.

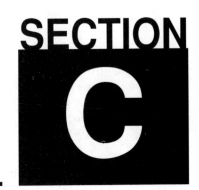
SERVICING AIRCRAFT

GROUND SERVICING EQUIPMENT

From time to time, most large aircraft require some form of auxiliary power to start the engines, provide electricity while the aircraft is on the ground, or provide cabin heating or cooling. For this reason, various types of **ground power units** (GPUs) are available for supplying power when the engines are not running. Some GPUs are mobile units that are driven to the aircraft while others are pulled behind a tug. Some newer airports have power and air outlets built into the tarmac.

ELECTRICAL GROUND POWER UNITS

The batteries used in most smaller aircraft have very limited capacity and, therefore, engine starting requirements may be more than the battery is capable of supplying. For this reason, most airports are equipped with battery carts and cables that can be used to assist an aircraft's battery. For large aircraft, a large self-propelled generator unit is used to assist in starting the aircraft.

It is extremely important when installing a battery on a battery cart that its polarity match the polarity of the aircraft. Reversed polarity can damage the alternator as well as the battery.

Most turbojet transport category aircraft carry an **auxiliary power unit** (APU). An APU is a small turbine engine that supplies compressed air for engine starting and cabin air conditioning, as well as electrical power for various aircraft systems. If the APU is not working, a ground unit is used for these purposes.

HYDRAULIC POWER UNITS

Many aircraft have hydraulically retracted landing gear that must be periodically tested for operation. To do this, the aircraft is lifted off the hangar floor with jacks, and a hydraulic power unit is connected to the aircraft's hydraulic system so the landing gear can be cycled through its retraction and extension cycles to verify that it operates properly.

These hydraulic power supplies, or **mules** as they are commonly called, connect into the aircraft hydraulic system with quick-disconnect fittings. When the fitting is screwed onto the hydraulic power supply, it automatically opens so it can supply hydraulic pressure to the aircraft.

Before connecting a hydraulic power supply make sure that all of the lines are clean so no dirt or contamination gets into the aircraft's system. Normally, these power supplies do not furnish fluid for the aircraft; rather, they use the fluid in the aircraft. Some of the fluid, however, remains in the pump and lines of the unit. [Figure 13-30]

Because of the incompatibility of different types of hydraulic fluid, such as MIL-H-5606 and Skydrol™, a different hydraulic power unit must be used with each type of fluid. If the wrong type of hydraulic fluid is introduced into an aircraft's hydraulic system the entire aircraft hydraulic system would have to be purged and refilled with fresh, uncontaminated fluid. Should any of the aircraft's systems be

Figure 13-30. A hydraulic power unit replaces the engine-driven pump to produce hydraulic pressure for gear retraction tests.

operated with the wrong type of fluid in the system, all components would have to be cleaned, flushed, or possibly disassembled for seal replacement. This would be extremely expensive and may even require the services of a certified repair station.

OXYGEN SERVICING EQUIPMENT

Modern aircraft fly at altitudes where life support systems are needed. Even though most of these aircraft are pressurized, emergency oxygen must be carried in the event the pressurization equipment fails. Most civilian aircraft carry gaseous oxygen in steel cylinders, or bottles, pressurized to approximately 1,800 psi. The cylinders are painted green and labeled "Aviator's Breathing Oxygen."

SERVICE CARTS

Oxygen systems are typically serviced from oxygen carts, which usually carry six high-pressure bottles of oxygen and one bottle of nitrogen. The nitrogen is used for filling hydraulic accumulators and oleo shock struts. To prevent inadvertently mistaking it for oxygen, it is usually laid on the cart opposite the direction of the oxygen cylinders. The oxygen cylinders are all manifolded together and connected to the aircraft service port.

THE "CASCADE" SYSTEM

When servicing an oxygen system always use two people, one to control the flow at the cart and one to monitor the pressure in the aircraft system. To begin servicing, open the valve on the cylinder having the lowest pressure and let the oxygen flow until the pressure stabilizes. Then shut the valve off and open the cylinder having the next lowest pressure. Continue this process until the system is charged from the cylinder having the highest pressure.

This procedure keeps the oxygen cart from having several bottles with pressures too low to charge the system. When done properly, the last two or three bottles increase the pressure without adding significantly to the volume.

Since oxygen presents such a serious fire hazard, you should avoid parking an oxygen cart beside a hydraulic mule, or in any area where petroleum products are likely to come in contact with the oxygen servicing equipment.

AIRCRAFT FUELING

Aircraft fueling is an operational procedure that is conducted more frequently than any other. It must be done under a wide variety of conditions and typically must be completed in a timely manner. The knowledge of aviation fuels and the use of correct procedures are of extreme importance for safety and efficiency. Use of the wrong type, wrong grade, intermixed, or contaminated fuels can lead to engine failure and catastrophe. Therefore, steps must be taken at all levels to assure a clean supply of the proper fuel is delivered to an aircraft.

To better understand what is required to maintain fuel quality as well as the importance of proper fueling procedures, it is first necessary to have a basic knowledge of aviation fuels. This knowledge should include the characteristics of various grades of aviation fuels as well as the importance of controlling contamination.

CHARACTERISTICS OF AVIATION FUELS

Weight is always a primary consideration in aircraft operation. Every extra pound used in the airframe and powerplant subtracts one pound from the aircraft's useful load. For this reason, aviation fuels must have the highest possible energy, or heat value per pound. Typical 100LL aviation gasoline has 18,720 British Thermal Units (BTUs) per pound. Jet A turbine fuel has about 18,401 BTUs per pound. However, Jet A weighs 6.7 pounds per gallon while a gallon of 100LL weighs 6 pounds. In other words, jet fuel is denser than avgas and, as a result, Jet A supplies 123,287 BTUs per gallon whereas 100LL supplies 112,320 BTUs per gallon.

The dynamics of the internal combustion cycle demand certain properties from gasolines. Aircraft engines compound these demands because of the wide range of atmospheric conditions they must operate under. One of the most critical characteristics of aviation gasoline is its **volatility**, which is a measure of a fuel's ability to change from a liquid into a vapor. Volatility is usually expressed in terms of **Reid vapor pressure** which represents the pressure above the liquid required to prevent vapors from escaping from the liquid at a given temperature. The vapor pressure of 100LL aviation gasoline is approximately seven pounds per square inch at 100 degrees F. Jet A, on the other hand, has a vapor pressure of less than 0.1 psi at 100 degrees F and Jet B has a vapor pressure of between two and three pounds per square inch at 100 degrees F.

For obvious reasons, a fuel's volatility is critical to its performance in an aircraft engine. For example, in a piston engine, the fuel must vaporize readily in the carburetor to burn evenly in the cylinder.

Fuel that is only partially atomized leads to hard starting and rough running. On the other hand, fuel which vaporizes too readily can evaporate in the fuel lines and lead to **vapor lock**. Furthermore, in an aircraft carburetor, an excessively volatile fuel causes extreme cooling within the carburetor body when the fuel evaporates. This increases the chances for the formation of carburetor ice, which can cause rough running or a complete loss of engine power. Therefore, the ideal aviation fuel has a high volatility that is not excessive to the point of causing vapor lock.

DETONATION

Reciprocating engine aircraft require high-quality aviation gasolines to insure reliable operation. These fuels are specially formulated to possess certain characteristics that allow them to function reliably in aircraft. To understand the different num-

bers used to designate fuel grades, you must first be familiar with detonation in reciprocating engines.

As you know, when a fuel-air charge enters the cylinder of a piston engine it is ignited by the spark plugs. Ideally, the fuel burns at a rapid but uniform rate. The expanding gases then push the piston downward, turning the crankshaft and creating power.

Detonation is the explosive, uncontrolled burning of the fuel-air charge. It occurs when the fuel burns unevenly or explosively because of excessive cylinder temperature or pressure in the cylinder. Rather than gently pushing the piston down, detonation slams against the cylinder walls and the piston. The pressure wave hits the piston like a hammer, often damaging the piston, connecting rods, and bearings. This is often heard as a knock in the engine. Detonation also causes high cylinder head temperatures and, if allowed to continue, can melt engine components. [Figure 13-31]

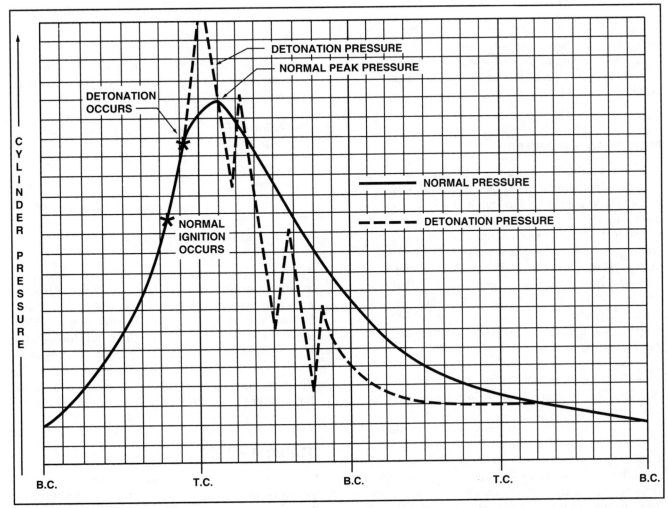

Figure 13-31. This chart illustrates the pressure created in a cylinder as it passes through its various strokes. As you can see, when normal combustion occurs, cylinder pressure builds and dissipates evenly. However, when detonation occurs, cylinder pressure fluctuates dramatically.

Detonation can happen anytime an engine overheats. It also can occur if an improper fuel grade is used. The potential for engine overheating is greatest under the following conditions: use of fuel grade lower than recommended, takeoff with an engine that is already overheated or is very near the maximum allowable temperature, operation at high rpm and low airspeed, and extended operations above 75 percent power with an extremely lean mixture.

PREIGNITION

In a properly functioning ignition system, combustion is precisely timed. In contrast, **preignition** takes place when the fuel/air mixture ignites too soon. Preignition is caused by residual hot spots in the cylinder. A hot spot may be a small carbon deposit on a spark plug, a cracked ceramic spark plug insulator, or almost any damage around the combustion chamber. In extreme cases, preignition can cause serious damage to the engine in a short period of time.

Preignition and detonation often occur simultaneously, and one may cause the other. Inside the aircraft, you will be unable to distinguish between the two, since both are likely to cause engine roughness and high engine temperatures.

PERFORMANCE NUMBERS

Aviation gasoline is formulated to burn smoothly without detonating, or knocking, and fuels are numerically graded according to their ability to resist detonation. The higher the number, the more resistant the fuel is to knocking. The most common grading system used for this purpose is the **octane** rating system. The octane number assigned to a fuel compares the anti-knock properties of that fuel to a mixture of iso-octane and normal heptane. For example, grade 80 fuel has the same anti-knock properties as a mixture of 80 percent iso-octane and 20 percent heptane.

Some fuels have two performance numbers, such as 100/130. The first number is the lean mixture rating, whereas the second number represents the fuel's rich mixture rating. To avoid confusion and to minimize errors in handling different grades of aviation gasolines, it has become common practice to designate the different grades of fuel by the lean mixture performance numbers only. Therefore, aviation gasolines are identified as Avgas 80, 100, and 100LL. Although 100LL performs the same as grade 100 fuel; the "LL" indicates it has a low lead content.

Another way petroleum companies help prevent detonation is to mix **tetraethyl lead** into aviation fuels. However, it has the drawback of forming corrosive compounds in the combustion chamber. For this reason, additional additives such as ethylene bromide are added to the fuel. These bromides actively combine with lead oxides produced by the tetraethyl lead allowing the oxides to be discharged from the cylinder during engine operation.

COLOR CODING OF AVIATION GASOLINES

In the past, there were four grades of aviation gasoline, each identified by color. The old color identifiers were as follows:

80/87—Red

91/96—Blue

100/130—Green

115/145—Purple

The only reason for mentioning the old ratings is because manuals on older airplanes may still contain references to these colors.

The color code for the aviation gasoline currently available is as follows:

80—Red

100—Green

100LL—Blue

Turbine engines can operate for limited periods on aviation gasoline. However, prolonged use of leaded avgas forms tetraethyl lead deposits on turbine blades and decreases engine efficiency. Turbine engine manufacturers specify the conditions under which gasoline can be used in their engines, and these instructions should be strictly followed. Reciprocating engines do not operate on turbine fuel. Never put jet fuel into a piston engined aircraft.

TURBINE FUELS

Aviation turbine fuels are used for powering turbojet, turbo-prop, and turboshaft engines. The types of turbine fuel in use are **JET A** and **JET A-1**, which are kerosene types, and **JET B**, which is a blend of gasoline and kerosene.

The difference between Jet A and Jet A-1 is that Jet A-1 has a freeze point of –47°C (–52.6°F) whereas Jet A has a freeze point of –40°C (–40°F). Jet B, which is similar to JP-4, is normally used by the military, particularly the Air Force. This fuel has an allowable freeze point of –50°C (–58°F).

One thing to keep in mind is that jet fuel designations, unlike those for avgas, are merely numbers that label a particular fuel and do not describe any performance characteristics.

COLOR CODING OF TURBINE FUELS

Unlike the various grades of aviation gasoline that are dyed different colors to aid in recognition, all turbine fuels are colorless or have a light straw color. Be aware that off-color fuel may not meet specifications and, therefore, should not be used in aircraft.

ADDITIONAL MARKINGS

In addition to coloring fuels, a marking and coding system has been adopted to identify the various airport fuel handling facilities and equipment, according to the kind and grade of fuel they contain. For example, all aviation gasolines are identified by name, using white letters on a red background. In contrast, turbine fuels are identified by white letters on a black background.

Valves, loading and unloading connections, switches, and other control equipment are color-coded to the grade or type of fuel they dispense. The fuel in piping is identified by name and by colored bands painted or decaled around the pipe at intervals along its length. [Figure 13-32]

Fuel trucks and hydrant carts are marked with large fuel identification decals on each side of the tank or

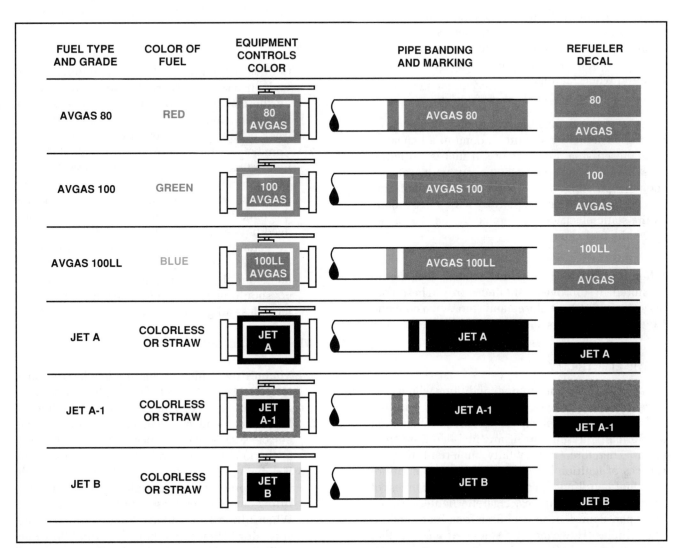

Figure 13-32. This illustration depicts the colors of the various grades and types of aviation fuels as well as the markings and color codes used on fuel conduits and controls.

body and have a small decal on the dashboard in the cab. These decals utilize the same color code. The fixed ring around the fueller dome covers and hydrant box lids are also painted in accordance with the color code. In short, all parts of the fueling facility and equipment are identified and keyed into the same marking and color code.

FIRE HAZARDS

Any facility that is storing or handling fuel represents a major fire hazard. This also holds true for facilities that store or handle aviation fuels. Therefore, all personnel should be aware of the danger and be trained on how to handle fuel.

VOLATILITY

When an aircraft is fueled, vapors rise from the tank. The more volatile the fuel (the higher its vapor pressure) and the higher the outside temperature, the more vapors are released and the more caution is required when fueling.

Because of the flammable nature of fuel vapors, no fueling or defueling should be done in a hangar or an enclosed area. Furthermore, if fuel is spilled, it must be wiped up or washed away with water as soon as possible. It is extremely important that spilled fuel never be swept away with a dry broom, as the static electricity generated by the broom can ignite the fuel vapors.

When it comes to storing fuel, aviation fuel should be stored in approved containers only. These containers must be kept closed and stored in a cool and isolated area that has been approved for fuel storage.

STATIC ELECTRICITY

All aviation fuels burn under conditions where they have sufficient oxygen and a source of ignition. Sufficient air and fuel vapors to support combustion are normally present during any fuel-handling operation. Therefore, it is vitally important that all sources of ignition be eliminated in the vicinity of any fuel-handling operation. Obvious sources of ignition include matches, cigarette lighters, smoking, open flames, even backfires from malfunctioning vehicles. However, one source of ignition that may not be so obvious is the sparks created by static electricity.

Static electrical charges are generated in various degrees whenever one body passes through or against another. For example, an aircraft in flight through the air, a fuel truck driving on a roadway, the rapid flow of fuel through a pipe or filter, and even the splashing of fuel into a fuel truck or aircraft during fueling operations all generate static electricity.

To minimize this hazard, it is necessary to eliminate static electrical charges before they can build up to create a static spark. This is accomplished by bonding and grounding all components of the fueling system together with static wires and allowing sufficient time for the charge to dissipate before performing any act which could draw a spark. Contrary to popular belief, the bleeding off of an electrical charge from a body of fuel is not always an instantaneous act. In fact, it can take several seconds to bleed off all static charges from some fuels. Because of this, it is absolutely essential that the following procedures be followed to bleed off static charges. [Figure 13-33]

When handling aviation fuels:

1. Connect a grounding cable (static wire) from the fuel truck or hydrant cart to ground. Furthermore, when loading a fuel truck connect the static wire from the loading rack to the fuel truck before operating the dome cover.
2. Connect a static wire from the fuel truck, hydrant cart, pit or cabinet to the aircraft.
3. When conducting overwing fueling, connect the fuel nozzle static wire to the aircraft before the tank cover is opened. Underwing nozzles need not be bonded to the aircraft.
4. In general, the dispensing unit should be grounded first, and should ultimately be bonded to the receiving unit. Dome or tank covers should never be opened during a fuel transfer unless all grounds and bonds are in place.

When handling turbine fuels:

1. Minimize splashing during the loading of a fuel truck by placing the end of the loading spout at, or as near as possible to the compartment bottom.
2. Do not suspend or lower metal or conductive objects such as gauge tapes, sample containers, or thermometers into a tank or fuel truck while it is being filled. Give any static charge which may be present a few minutes to bleed off after filling before using these devices.
3. When filling large storage tanks, minimize the splashing action by slowing the initial flow rate until the end of the tank inlet line is covered with at least two feet of fuel.

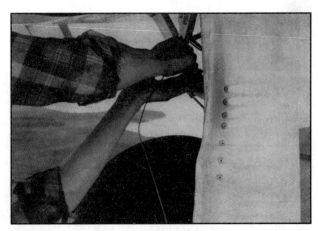

Figure 13-33. Before opening the fuel tanks or connecting the fuel hose, be sure that the aircraft and the fuel truck are connected together by an electrical bonding cable.

CONTAMINATION CONTROL

Aviation fuels are subject to several types of contaminants. The more common forms of aviation fuel contaminants include solids, water, surfactants, micro-organisms, and intermixing of grades or types of fuel. Surfactants and micro-organisms contaminants have become serious problems with the advent of turbine fuels.

As a general rule, the ease with which a fuel can hold contaminants varies with the fuel's viscosity. **Viscosity** is the property of a fluid that describes its thickness or resistance to flow. Fluids that have a high viscosity tend to retain contaminants in suspension. This is a disadvantage with turbine fuels since their high viscosity allows them to hold liquid and solid material that does not easily settle out. Furthermore, since the quantity of fuel passing through a turbine engine per hour is considerably greater than with piston engines, any contamination in the fuel can accumulate in the fuel control unit rapidly. As a result, turbine engines are much more sensitive to fuel cleanliness.

Operational differences between piston and turbine powered aircraft also affect requirements for clean fuel. For example, in turbine powered aircraft, it is not uncommon to record in-flight fuel temperatures of −25 degrees F. These temperatures can cause water molecules to precipitate out of the fuel and freeze. This produces ice crystals that can then accumulate and interrupt fuel flow.

WATER

Water has always been one of the major contamination problems with aviation fuel. It condenses out of

the air in storage tanks, fuel trucks, and even in aircraft fuel tanks. Water exists in aviation fuels in one of two forms, dissolved and free.

All aviation fuels dissolve water in varying amounts depending upon the fuel composition and temperature. This can be likened to humidity in the air. Any water in excess of that which dissolves is called free water. Lowering the fuel temperature causes dissolved water to precipitate out as free water, somewhat similar to the way fog is created. Typically, dissolved water does not pose a problem to aircraft and cannot be removed by practical means.

Free water can appear as water slugs or as entrained water. A **water slug** is a relatively large amount of water appearing in one body or layer. A water slug can be less than a pint and is sometimes measured in hundreds of gallons. **Entrained water** is water which is suspended in tiny droplets. Individual droplets may or may not be visible to the naked eye, but they can give the fuel a cloudy or hazy appearance, depending upon their size and number. Entrained water usually results when a water slug and fuel are violently agitated, as when they pass through a pump, and usually settles out in time. Entrained water may also be formed by lowering the temperature of a fuel saturated with dissolved water. Because of its high viscosity, entrained water is often visible in turbine fuel as a water haze.

Most aircraft engines can tolerate dissolved water. However, large slugs of free water can cause engine failure, and ice from slugs and entrained water can severely restrict fuel flow by plugging aircraft fuel filters and other mechanisms.

Most fuel systems that are subject to ice crystals are protected by filter-heating devices. These devices can satisfactorily deal with dissolved and even entrained water, however, there is little margin for handling large amounts of free water.

Water can enter an airport fuel system though leaks in the seals of equipment, or it may be brought into a system when fuel is delivered. The best means of minimizing the amount of water entering a system is through inspection and maintenance of equipment, and by making certain that only dry fuel is received.

Water can be detected in many ways. Free water, lying in the bottom of underground storage tanks, can be discovered by applying a water finding paste to the end of a gauge stick and placing the gauge in the storage tank. When doing this, you must always allow at least 30 seconds for the paste to react, as its reaction time can be slowed by other contaminants. In the case of above ground tanks and equipment, a sample can be drawn into a container and the free water actually observed. A small amount of liquid vegetable dye (cake coloring) is sometimes helpful to outline the free water in a sample. It mixes with and colors the water but is insoluble in the fuel.

Water is removed from fuel by providing adequate filtration or separation equipment. With turbine fuels, floating suction devices, and allowing the fuel to settle for at least two hours after filling minimizes water contamination. All storage and fuel truck tank bottoms and filter/separator sumps should be checked for water, and any accumulations removed on a daily basis.

SOLID CONTAMINANTS

Solid contaminants are those which do not dissolve in fuel. The most common contaminants are iron rust and scale, sand, and dirt. However, other debris such as metal particles, dust, lint, particles of filter media, rubber, valve lubricants, and even bacterial sludge can be encountered. Solid contaminants are typically introduced into fuel at every stage of its movement from the refinery to the aircraft.

The maximum amount of solids that an aircraft can tolerate depends on the type of aircraft and fuel system, and the number and size of the solid contaminants. Close tolerance mechanisms in modern turbine engines can be damaged by particles as small as 1/20 the diameter of a human hair.

The best method of controlling solids is to limit their introduction into the fuel. Rusty lines, tanks, and containers obviously should not be used. Furthermore, covers and caps should be kept tightly closed until you are ready to begin pumping fuel. Exercise care to keep lint from wiping rags or wind-blown sand, dirt, and dust from entering the system during filling or fueling operations. Fueling nozzles and loading spouts should be cleaned before use, and dust caps and other protective devices replaced after they are used. Furthermore, filters should be regularly inspected and maintained in accordance with the operating specifications.

Because solid contaminants generally appear in relatively small numbers and sizes in relation to the volume of fuel, their detection can be difficult. Aviation gasoline is generally considered "clean" if a one-quart sample is clear of any sediment when viewed in a clean and dry clear glass container. It may be helpful to swirl the container so that a vortex is created. The solid contaminants, if present, tend to collect at the bottom beneath the vortex.

Because the fuel control units used on turbine powered aircraft are extremely sensitive, turbine fuels must be cleaner that aviation gasoline. While a visual inspection is adequate for operational checks, a millipore test should be performed from time to time. The **millipore** test is a filter-type test capable of detecting microscopic solid contaminants down to .8 of a micron in size, or about 1/120 the diameter of a human hair.

MICROBIAL GROWTHS

Microbial growths have become a critical problem in some turbine fuel systems. There are over 100 different varieties of micro-organisms which can live in the free water which accumulates in sumps and on the bottom of storage and aircraft tanks. Many of these micro-organisms are airborne and, therefore, fuel is constantly exposed to this type of contaminant.

The principle effects of micro-organisms are:

1. Formulation of a sludge or slime which can foul filter/separators and fueling mechanisms.
2. Emulsification of the fuel.
3. Creation of corrosive compounds and offensive odors.

Severe corrosion of aircraft fuel tanks has been attributed to micro-organisms and considerable expense has been incurred removing these growths and repairing the damage they cause. The actual determination of microbial content, or number of colonies, is reserved for the laboratory. Any evidence of black sludge or slime, or even a vegetative-like mat growth should be removed. Growths also appear as dark brown spots on some filter/separator element socks. The socks should be replaced whenever this condition is discovered.

Because microbes thrive in water, a simple and effective method to prevent or retard their growth is to eliminate the water. A common way of doing this is by introducing a fuel additive during the fueling process. [Figure 13-34]

Figure 13-34. A biocidal agent is added to turbine fuel to prevent microbial growths that cause corrosion in the fuel tanks.

SURFACTANTS

The term surfactants is a contraction of the words SURFace ACTive AgeNTS. Surfactants consist of soap or detergent-like materials that occur naturally in fuel, or can be introduced during refining or handling. Surfactants are usually more soluble in water than in fuel and reduce the surface tension between water and fuel. This stabilizes suspended water droplets and contaminants in the fuel. They are attracted to the elements of filter/separators and can make them ineffective. Surfactants, in large concentrated quantities, usually appear as a tan to dark brown liquid with a sudsy-like consistency.

Surfactants alone do not constitute a great threat to aircraft. However, because of their ability to suspend water and dirt in the fuel and inhibit filter action, they allow these contaminants to get into an aircraft's fuel system. Surfactants have become one of the major contaminants in aviation turbine fuels, and can cause fuel gauge problems. There is no established maximum limit on the level of surfactants which can be safely contained in a fuel, and there are no simple tests for determining their concentration in fuel. The common danger signals of a surface contaminated facility are:

1. Excess quantities of dirt and/or free water going through the system.
2. Discovery of sudsy-like liquid in tank and filter/separator sumps.
3. Malfunctioning of filter/separators.
4. Slow effective settling rates in storage tanks.

MISCELLANEOUS CONTAMINANTS

Miscellaneous contaminants can include either soluble or insoluble materials or both. Fuel can be con-taminated by mixing with other grades or types of fuels, by picking up compounds from concentrations in rust and sludge deposits, by additives, or by any other of a number of soluble materials.

The greatest single danger to aircraft safety from contaminated fuels cannot be attributed to solids, exotic micro-organisms, surfactants, or even water. It is contamination resulting from human error. It is the placing of the wrong grade or type of fuel into an aircraft, the mixing of grades, or any other type of human error that allows off-specification fuels to be placed aboard an aircraft. The possibility of human error can never be eliminated, but it can be minimized through careful design of fueling facilities, good operating procedures, and adequate training.

FUELING PROCEDURES

The fueling process begins with the delivery of fuel to the airport fueling facility, usually by tank truck. Quality control begins by checking the bill of lading for the proper amount and grade of fuel.

Fuel testing should begin with the tank truck. The personnel receiving the fuel delivery must determine that the proper type of fuel is in the truck, and samples taken and checked for visible contamination. Once all of these checks are completed, the truck is connected to the correct unloading point, and unloading can proceed.

Turbine fuel should be allowed to settle a minimum of two hours after any disturbance. Therefore, once a quantity of turbine fuel is delivered, it should be allowed to sit in its storage tank for at least two hours before it is pumped into an aircraft. Aviation gasolines do not need time to settle before being withdrawn for use; however, no withdrawals may be made from a tank while it is receiving fuel from a transport truck.

FROM A FUEL TRUCK

Aircraft can have fuel pumped directly into their tanks from over the wing tank openings, or from a single point source under the wing. Typically, overwing fueling is done with a fuel truck whereas underwing fueling is done from a pit through single-point fueling.

Before driving a fuel truck to an aircraft, be sure that the sumps have been drained and that the sight gauges show that the fuel is bright and clear. Furthermore, fire extinguishers must be in place and fully charged. Approach the aircraft with a fuel truck parallel to the wings and stop the truck in front of the aircraft. Set the parking brake on the truck and connect the static bonding wire between the truck and the aircraft.

Prior to removing the aircraft's fuel tank cap verify that you have the proper grade of fuel. This is done by reading the placard near the filler cap. [Figure 13-35]

Put a mat over the wing so the fuel hose can not scratch the finish, connect the static bonding wire between the nozzle and the aircraft and remove the fuel tank cap. Remove the dust cap from the nozzle, and when inserting the nozzle into the tank be sure that the end of the nozzle does not contact the bottom of the tank, as it could dent the thin metal. Should the fuel tank be a fuel cell, contact with the nozzle could puncture the cell and cause a serious leak.

Misfueling is a constant danger that can frequently result in a complete engine failure. To help prevent misfueling accidents, the nozzles used to pump turbine fuel are larger than the nozzles used to pump aviation gasoline. Furthermore, FAR 23.973 specifies that all general aviation aircraft utilizing aviation gasoline have restricted fuel tank openings that will not allow the nozzle used to pump Jet A to fit in the tank opening. While it is possible for a jet or turbine engine to run on gasoline, a piston engine will not run on Jet A.

UNDERGROUND STORAGE SYSTEM

Most of the large airports that service transport category aircraft have underground storage tanks and buried fuel lines. This arrangement allows the aircraft to be fueled without having to carry the fuel to the aircraft in tank trucks. Since most aircraft that are fueled from this type of system use under-wing fueling, the method is discussed here.

A service truck having filters, water separators, and a pump is driven to the aircraft and its inlet hose is connected to the underground hydrant valve. The discharge hose or hoses from the servicer are attached to the fueling ports on the aircraft and, with a properly qualified maintenance person in the aircraft monitoring the fuel controls, the valves are opened and the pumps started. The person monitoring the fuel controls can determine the sequence in which the tanks are filled and can shut off the fuel when the correct load has been taken on board.

Some large corporate aircraft also have single point refueling systems. However, in most cases, control of the fueling sequence is from an outside control panel located under an access cover. A service technician must be checked out on these systems before operating them. Should there by any questions about the operation, ask for assistance from the pilot-in-command of the aircraft. [Figure 13-36]

DEFUELING

It is sometimes necessary to remove fuel from an aircraft, either for maintenance reasons or because of a change in flight plans after the aircraft was serviced. Defueling is accomplished in much the same manner as fueling, with many of the same safety precautions used.

Figure 13-35. An aircraft fuel tank must be clearly marked with the proper grade of fuel required.

Figure 13-36. High rates of fuel flow can be put into an aircraft system from the underwing fueling ports.

Never defuel an aircraft inside a hangar, or in any area with inadequate ventilation. Be sure that all of the proper safeguards are taken with regard to neutralizing any static electricity that builds up when the fuel flows through the lines.

If only a small quantity of fuel is off-loaded and there is no reason to suspect contamination, the fuel may be taken back to stock. On the other hand, the quality of the off-loaded fuel could be suspect if an engine failed and a large quantity of fuel was removed. This fuel should be segregated, preferably in a fuel truck, and quarantined until its quality is assured.

In no event should suspected fuel be returned to storage or placed aboard another aircraft. If acceptable fuel is returned to storage, make sure it is taken back into a tank containing the same grade of fuel and that complete quality control procedures are followed.

If an aircraft is defueled into drums, be sure that the drums are clean and that the bungs are replaced and tightened immediately after the drums are filled. Some companies, and some aircraft operations manuals, do not allow fuel that has been stored in drums to be reused in an aircraft. Frequently this fuel is used in ramp vehicles, space heaters, and GPUs.

REVIEW OF SAFETY PROCEDURES

Review the safety procedures that must be observed when fueling or defueling an aircraft:

1. Be sure that only the correct grade of fuel is put into an aircraft. Remember that aviation gasoline comes in various grades and the wrong grade can cause severe damage to the engine. Turbine fuel in a reciprocating engine can cause severe detonation and engine failure, and the improper use of aviation gasoline in a turbine engine can also be harmful.

2. Be sure that the fuel truck, or servicer, is properly bonded to the aircraft and the fuel nozzle is bonded to the structure before the cover is taken from the fuel tank.

3. Wipe up spilled fuel or flood it with water. Do not sweep spilled fuel with a dry broom.

4. Be sure that there are no open fires in the vicinity of the fueling or defueling operations.

5. Be sure that fire extinguishers suitable for a Class B fire are available. Either CO_2 or dry powder units are generally used.

6. Protect the aircraft structure from damage from the fuel hose and from the nozzle.

7. Be sure that the radio or radar are not used during fueling or defueling, and that no electrical equipment is turned on or off, except for the equipment needed for the fueling operation.

8. When defueling, be sure that the fuel is not contaminated if it is to be used again.

9. Be sure that the filters in the tank truck or servicer remove all traces of water and contamination and that the fuel pumped into an aircraft is bright and clear.

10. If a biocidal additive is required, be sure that it is mixed with the fuel in the proper concentration.

11. If the aircraft is being fueled in the rain, be sure that the tank opening is covered to exclude water from the tank.

12. Be sure that dust covers and caps are placed over the end of the fuel nozzles and any open fuel lines when they are not in use.

13. Drive the tank truck parallel to the wing of the aircraft and be sure that the parking brake is set so the truck cannot roll into an aircraft.

14. When conducting underwing pressure fueling, be sure that the pressure used and the delivery rate are those specified by the manufacturer of the aircraft.

15. If any fuel is spilled onto your body, wash it off with soap and water as soon as possible. Do not wear any clothing on which fuel has been spilled.

MAINTENANCE PUBLICATIONS, FORMS, AND RECORDS

CHAPTER 14

INTRODUCTION

Screwdrivers, wrenches, and sockets are some of the tools you use to maintain and repair aircraft. However, another tool that must not be neglected consists of the publications and forms that serve as a guide and record for your maintenance activities. For example, a typical transport aircraft contains many miles of electric wire for indicating, lighting, and control systems. Attempting to troubleshoot a maintenance problem without access to the manufacturer's wiring manual would be a frustrating experience. A thorough knowledge of regulatory publications such as the Federal Aviation Regulations, and nonregulatory material such as manufacturer's manuals, is every bit as important as the tools in your toolbox.

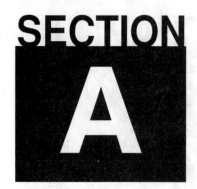

THE FEDERAL AVIATION ADMINISTRATION

For more than 20 years after the Wright brothers first flight, aviation was virtually an unregulated industry. Anyone who wanted could design, build, and fly an aircraft without obtaining approval from the government. Predictably, the safety record of early air transport was not very good, and the government responded by passing the Air Commerce Act of 1926. The 1926 Act required registration of aircraft and the licensing of pilots and mechanics. In 1938, the Air Commerce Act was repealed and replaced by the Civil Aeronautics Act. This law created the Civil Aeronautics Administration and Civil Aeronautics Board which established and enforced all regulations pertaining to civil aviation.

The **Federal Aviation Act of 1958** repealed the Civil Aeronautics Act and created the **Federal Aviation Agency**. The agency was brought into the Department of Transportation in 1967 and renamed the **Federal Aviation Administration** (FAA). Currently, the FAA is charged with the responsibility of regulating and promoting civil aviation in the United States. The FAA's powers and responsibilities are quite broad, governing virtually every aspect of aircraft manufacture, operation, and maintenance.

The primary regulatory tools of the FAA are the **Federal Aviation Regulations (FARs)**. When the FAA wants to adopt a new rule or regulation, it formulates a proposal called a **Notice of Proposed Rule Making** and publishes it in the *Federal Register*. Comments and suggestions are then solicited from the general public and interested parties. After a prescribed comment period, the proposals are adopted and printed in *Title 14 of the Code of Federal Regulations*, thereby becoming federal law. To help organize, the FARs are broken down into separate sections, or **parts**. For example, FAR Part 65 prescribes the requirements, privileges, and limitations for certification of airmen other than flight crewmembers, which includes aviation maintenance technicians.

Copies of the FARs are available from the Government Printing Office or from a number of private suppliers, including the publisher of this textbook. Since the regulations change frequently, all copies must be periodically updated. Repair stations, aviation maintenance training schools, and others affected by the FARs are required to keep their FARs updated. For your course of study, you may use an FAR textbook produced by the government or some other publisher. These publications serve as a good general guide to the content of the regulations. However, when you take your practical test, you must have access to a current set of updated FARs.

MAINTENANCE FARS

Many regulations do not affect the maintenance technician and require no discussion. Others, however, are of vital importance to technicians in the performance of their duties. It is imperative for all technicians to be familiar with these regulations, and to follow them when exercising the privileges of an A&P certificate. Some of the regulations concerning aircraft maintenance and inspection are listed below:

1. FAR Part 01, Definitions and Abbreviations
2. FAR Part 13, Investigation and Enforcement Procedures
3. FAR Part 21, Certification Procedures for Products and Parts
4. FAR Part 23, Airworthiness Standards. Normal, Utility and Acrobatic aircraft
5. FAR Part 25, Airworthiness Standards, Transport Category Airplanes
6. FAR Part 27, Airworthiness Standards, Normal Category Rotorcraft
7. FAR Part 33, Airworthiness Standards: Aircraft Engines
8. FAR Part 35, Airworthiness Standards: Propellers
9. FAR Part 39, Airworthiness Directives
10. FAR Part 43, Maintenance, Preventive Maintenance, Rebuilding and Alterations

11. FAR Part 45, Identification and Registration Markings
12. FAR Part 47, Aircraft Registration
13. FAR Part 65, Certification: Airmen other than Flight Crewmembers
14. FAR Part 91, General Operating and Flight Rules
15. FAR Part 121, Certification and Operations: Domestic, Flag, and Supplemental Air Carriers and Commercial Operators of Large Aircraft
16. FAR Part 125, Certification and Operations: Airplanes having a seating capacity of 20 or more passengers, or a maximum pay load capacity of 6,000 lbs or more
17. FAR Part 127, Certification and Operation of Scheduled Air Carriers with Helicopters
18. FAR Part 135, Air Taxi Operators and Commercial Operators
19. FAR Part 137, Agricultural Aircraft Operators
20. FAR Part 145, Repair Stations
21. FAR Part 147, Aviation Maintenance Technician Schools
22. FAR Part 183, Representatives of the Administrator

The regulations most important to maintenance technicians are discussed in detail.

FAR PART 21

FAR Part 21 lists the requirements for establishing and maintaining the certification of aircraft and components. When an aircraft is manufactured, an inspector determines if it conforms to that model's **type certificate**. The type certificate lists all pertinent information on an aircraft or accessory design. If the aircraft is in conformity, it is issued an **airworthiness certificate**, signifying that it meets the standards for service. The airworthiness certificate stays with the aircraft throughout its service life and is transferred when ownership changes. However, an aircraft must be maintained properly in order for the airworthiness certificate to remain valid.

To ensure that aircraft perform reliably the FAA requires that all installed parts and appliances conform to **technical standard orders** (TSOs). TSOs are a set of specifications that call for parts to meet certain quality standards. A TSO stamp on a part means that the part meets FAA requirements for manufacturing quality, and is approved for installation on aircraft.

If a TSO part is called for by the aircraft manufacturer, installation of a noncertified item can invalidate the aircraft's airworthiness certificate. The FARs specify that it is the responsibility of the person or agency installing a part to verify that it conforms to the proper standard. Therefore, you must make sure the hardware and appliances you install on aircraft are certified.

Companies can obtain a **parts manufacturing approval** (PMA) from the FAA to produce replacement parts. However, the manufacturer must prove to the FAA that their product meets performance and quality standards. Again, the installation of nonapproved parts can affect the airworthiness of the aircraft and lead to catastrophic failure. Always check parts for TSO and PMA conformity.

FAR PART 23

FAR Part 23 describes in detail the performance characteristics various aircraft must demonstrate to be airworthy. It specifies requirements for every component and system installed on an aircraft, often down to the smallest detail. Maintenance technicians can use Part 23 to verify that a particular aircraft or component is in conformity with its type certificate. For example, when cockpit instruments are repaired or replaced, the technician installing the instrument must check that the range markings painted on the instrument face are correct. The aircraft's approved flight manual gives the correct operating speeds and ranges for that particular model. Part 23 specifies color codes and instrument face markings that must be on all flight and engine instruments.

FAR PART 39 — AIRWORTHINESS DIRECTIVES

When an unsafe condition exists with an aircraft, engine, propeller, or accessory, the FAA issues an **Airworthiness Directive** (AD) to notify concerned parties of the condition and to describe the appropriate corrective action. No person may operate an aircraft to which an AD applies, except in accordance with the requirements of that AD. AD compliance is mandatory, and the time in which the compliance must take place is listed within the AD. Information provided in an Airworthiness Directive is considered approved data for the purpose of the AD. The compliance record for ADs must be entered into the aircraft's permanent records.

Airworthiness Directives are issued biweekly. The biweekly listings are published for small general aviation aircraft and accessories in one volume, while the larger aircraft and their accessories are published in a separate volume. This separation of different aircraft categories provides operators with a much simpler means of filing ADs.

ADs are listed by a six digit numerical number. The first two digits denote the year an AD is issued. For example, all ADs issued during the year 1996 begin with the number 96-. The third and fourth digits of the AD number denote the biweekly issue in which the AD was first published. There are twenty six issues of the biweekly AD listing issued each year, and the issues are numbered beginning with number 01. The last two digits indicate the number of the AD in the specified biweekly listing. For example, the fourth AD issued in the first biweekly publication in May 1996 would be issued the number: 96-10-04.

FAR PART 43

FAR Part 43 — Maintenance, Preventive Maintenance, Rebuilding, and Alteration is one of the most critical sections for the aviation technician to study. As its title indicates, Part 43 outlines the fundamental standards for aircraft inspection, maintenance, and repair, as well as all record keeping requirements.

REPAIRS AND ALTERATIONS

A **repair** is an operation that restores an item to a condition of practical operation or to original condition, whereas an **alteration** is any change in the configuration or design of an aircraft. The FAA divides aircraft repairs and alterations into two categories: major and minor. A **major repair** is one that, if improperly done, might appreciably affect weight, balance, structural strength, performance, powerplant operation, flight characteristics, or other airworthiness factors. It is also a repair that cannot be performed using elementary operations. A **major alteration** is an alteration not listed in the product's specifications that might affect the product's performance in a similar fashion to a major repair.

Appendix A of FAR Part 43 lists examples of major repairs and alterations to airframes, engines, propellers and appliances. Some examples are listed below.

1. Airframe major alterations: alterations to wings, fuselage, engine mounts or control systems. Changes to the wing or to fixed or movable control surfaces which affect flutter and vibration characteristics.

2. Powerplant major alterations: conversion of an aircraft engine from one approved model to another, replacing engine structural parts with parts not supplied by the original manufacturer.

3. Propeller major alterations: changes in blade, hub, or governor design. Installation of a propeller deicing system.

4. Appliance major alterations: alterations of the basic design not made in accordance with recommendations of the appliance manufacturer or in accordance with an FAA Airworthiness Directive.

5. Airframe major repairs: airframe repairs involving reinforcing, splicing, and manufacturing of primary structural members or their replacement, when their replacement is by fabrication such as riveting or welding.

6. Powerplant major repairs: separation or disassembly of the crankcase or crankshaft of certain reciprocating powerplants. Special repairs to structural engine parts by plating, welding, or other methods.

7. Propeller major repairs: any repairs to, or straightening of, steel blades, shortening of blades, overhaul of controllable pitch propellers.

8. Appliance major repairs: calibration of instruments or radios, overhaul of pressure carburetors, pressure fuel cells, and oil and hydraulic pumps.

This is not a complete listing of major repairs and alterations. Always refer to the appropriate section of Appendix A to find out precisely what classification a job falls under before starting. If there is any question as to whether a repair or alteration is major or minor, contact the local FAA office.

The FAA defines **minor repairs and alterations** as those that are not major repairs and alterations. Since this definition is not very specific, it is sometimes difficult to distinguish which category a repair or alteration falls into. However, as a general rule, the complexity of the work being done is a good indication of whether a repair or alteration is major or minor. For example, Appendix A of FAR Part 43 specifically states that replacement of an engine mount by riveting or welding is an airframe major repair. But if the same engine mount is attached by bolts and is replaced, it is considered a minor repair. The same holds true of other airframe and engine components. Generally, if a part is replaced with one exactly the same as the original, and elementary operations are used in the installation of a replacement part, the procedure is considered a minor repair or alteration. Records of minor repairs and alterations need only be entered into the aircraft's permanent maintenance records.

PREVENTIVE MAINTENANCE

Preventive maintenance consists of preservation, upkeep, and the simple replacement of small parts. Under some circumstances, the FARs allow licensed airmen other than maintenance personnel to perform preventive maintenance. For example, if an aircraft owner holds at least a private pilot license, the owner can, among other things, change an aircraft's oil and replace or repair a landing gear tire. A complete listing of those items that are classified as preventive maintenance is given in Appendix A of FAR Part 43.

INSPECTION CHECKLISTS

FAR 43.15 lists the performance criteria for performing inspections and specifically states that a checklist which meets the minimum requirements listed in FAR 43 Appendix D must be used for all annual and 100-hour inspections. This, however, does not preclude you from developing a more extensive checklist or using one prepared by a repair station or manufacturer. As long as the checklist covers the items listed in Appendix D it may be used. Most major aircraft manufacturers provide inspection checklists for their aircraft by type and model number. These forms are readily available through the manufacturer's representatives and are highly recommended. They meet the minimum requirements of Appendix D and contain many details covering specific items of equipment installed on a particular aircraft. In addition, they often include references to service bulletins and letters which could otherwise be overlooked.

FAR PART 65

FAR Part 65 discusses the certification requirements as well as the privileges and limitations for aviation maintenance technicians. Because of the amount of pertinent information contained in Part 65, Chapter 15 is dedicated to discussing Part 65 in detail.

FAR PART 91

FAR Part 91 is entitled General Operating and Flight Rules of which Subpart E contains the regulations governing the maintenance, preventive maintenance, and alteration of U.S. registered aircraft. Unlike Part 43, Subpart E of Part 91 outlines the inspections that must be performed on all aircraft.

Inspection of airframes, powerplants, and appliances is the single most effective way to identify potential problems and ensure safe operation. As a result, inspections are one of an aircraft maintenance technician's primary duties. The FAA requires aircraft and their associated components to be inspected regularly. The frequency of these inspections depends on the type and use of the aircraft or component.

ANNUAL INSPECTIONS

FAR Part 91 states that all general aviation aircraft must go through an annual inspection to remain airworthy. All annual inspections are based on calendar months and, therefore, are due on the last day of the 12th month after the last annual was completed. For example, if a previous annual was completed on June 11, 1995, the next annual inspection is due on June 30, 1996.

Annual inspections must be performed regardless of the number of hours flown in the previous year. Furthermore, they may only be performed by airframe and powerplant mechanics holding an **inspection authorization** (IA). The IA can not delegate the inspection duties to an airframe and powerplant mechanic, nor may an IA merely supervise an annual inspection.

If the person performing the annual inspection finds a discrepancy that renders the aircraft unairworthy, they must provide the aircraft owner with a written notice of the defect. Furthermore, the aircraft may not be operated until the defect is corrected. However, if the owner wants to fly the aircraft to a different location to have the repairs performed, a special flight permit may be obtained to ferry the aircraft to the place where repairs are to be made.

100-HOUR INSPECTION

All general aviation aircraft that are operated for hire must be inspected every 100 flight hours. This inspection is in addition to the annual inspection requirement, and covers the same items as the annual inspection. The major difference is that an A&P technician may perform a 100-hour inspection. As in the case of an annual inspection, the person conducting a 100-hour inspection cannot delegate inspection duties.

The operating hours are the primary consideration for determining when the next 100-hour inspection is due. As the name implies, a 100-hour inspection is due 100 hours after the last 100-hour inspection was completed, regardless of the date. However, there is a provision for extending the 100-hour interval, up to a maximum of 10 hours, to permit the aircraft to fly to a place where the inspection can be accomplished. However, when this is done, the number of hours in excess of the 100-hour interval are deducted from the next inspection interval. For example, if a flight to a place where a 100-hour inspection can be conducted takes the aircraft six hours beyond the 100-hour inspection interval, the next 100-hour inspection would be due in 94 hours. In other words, the next inspection interval is shortened by the same amount of time the previous inspection was extended.

PROGRESSIVE INSPECTIONS

The progressive inspection is designed for aircraft operators who do not wish to have their aircraft grounded for several days while an annual or 100-hour inspection is being accomplished. Instead, the inspection may be performed in segments each 90 days, or at each 25 hour interval. The procedures for establishing a progressive inspection program are covered in FAR Part 91.409(d) and the owner must request this program. Furthermore, approval must be received from the FAA prior to beginning a progressive inspection program.

CONTINUOUS INSPECTION PROGRAMS

The continuous inspection program is designed for operators of large commercial carrier aircraft operating under FAR Parts 121, 127, and 135. Like a progressive inspection program, a continuous inspection program must be approved by the FAA. These inspection programs are very comprehensive and require complex maintenance facilities with large numbers of technical personnel. Most large airlines operate under the continuous inspection programs of FAR Part 121.

ALTIMETER AND STATIC SYSTEM CHECKS

FAR 91.411 requires periodic altimeter and static system checks for aircraft that operate in controlled airspace under instrument flight rules. These checks must be made in accordance with the guidelines stated in FAR Part 43, Appendix E, and performed each 24 months. Furthermore, the test must be done whenever the static system is opened,

excluding those times when the system is drained using a preinstalled drain and when the alternate static source is opened.

TRANSPONDER CHECKS

A **transponder** is an electronic device aboard an aircraft that enhances the aircraft's identity on an air traffic control (ATC) radar screen. Because of the important role transponders play in safety, they must be checked every 24 calendar months. This requirement is spelled out in FAR 91.413 and testing of transponder equipment must be in accordance with FAR Part 43, Appendix E

EMERGENCY LOCATOR CHECK (ELT)

An **emergency locator transmitter** (ELT) emits a radio distress beacon at 121.5 megahertz when it is triggered by the pilot or by abrupt deceleration of the aircraft. To enhance the chances of rescue in case of an accident, FAR 91.207 requires that all U.S. registered aircraft be equipped with an ELT. ELTs must be inspected every 12 calendar months for proper installation, battery condition and expiration date, operation of the controls and crash sensor, and radio signal strength.

SPECIAL INSPECTIONS

At times, unusual incidents dictate that a special inspection be performed. While not specifically required by the FARs, most manufacturers have several categories of special inspections that must be performed after an aircraft has been subjected to an unusual flight condition.

Overweight or Hard Landing

This inspection is performed any time an aircraft has experienced an overweight, or unusually hard landing. The structural damage which can occur during this type landing can lead to catastrophic failure and, therefore, manufacturers typically publish a set of instructions outlining what should be inspected as well as the method of inspection used.

Severe Turbulence Inspection

Severe turbulence exerts substantial stress on all areas of an aircraft structure. Therefore, when an aircraft flies through severe turbulence, a detailed inspection of the airframe should be conducted.

Hot Starts, Stackfires, Sudden Stoppage Inspections

Inspection of an engine after a hot start, stack fire, or sudden stoppage should be in accordance with the manufacturer's instructions. Chapter 14 of AC 43.13-1B contains instructions for sudden stoppage inspections. At times, complete disassembly of the engine is required to check for hidden damage.

RECURRING INSPECTIONS

At times, an airframe component, engine, or accessory requires a recurring inspection at specified intervals. For example, a manufacturer may require that a mechanic check the torque of cylinder base nuts at every second 100 hour inspection. This means that the actual interval between inspections is 200 hours.

Airworthiness directives often specify recurring inspections after a certain time in service. For example, assume a new AD is issued that requires a wing spar cap inspection every 150 hours on aircraft with 7,500 hours in service or less, and every 50 hours on aircraft with more than 7,500 hours. Suppose the technician performing a 100-hour inspection on an aircraft found it had 5,257 hours in service, and the AD was complied with 28 hours ago. To determine the next inspection time, subtract 28 hours from the 150 hour inspection interval to get a figure of 122 hours. Add this to the total airframe time in service to find that the inspection is due next at 5,379 hours.

ADVISORY CIRCULARS (ACs)

Many of the technical publications and regulations issued by the FAA are complex in nature and often require additional explanation. As a result, the FAA issues Advisory Circulars (ACs) to inform, explain, and provide further guidance. Advisory circulars are informative only and cannot be used as approved data unless incorporated into a regulation or an airworthiness directive. Advisory circulars are listed in a numerical sequence closely following the same subject areas covered by the FARs. Some of the subject areas are:

00	General
10	Procedural Rules
20	Aircraft
60	Airmen
120	Aircarriers, Air Travel Clubs, and Operators for Compensation and Hire: Certification and Operations

Within the general subject areas are more specific subjects that also have a subject number. For example, within the general subject of Aircraft, the specific subject of maintenance, preventive maintenance, and rebuilding and alterations is assigned the number 43, which is also the number of the FAR Part that covers maintenance.

One of the most popular Advisory Circulars for maintenance technicians is in the AC43 series. **AC 43.13-1B and -2A**, Acceptable Methods, Techniques and Practices is a highly technical publication covering most of the aircraft maintenance areas which the A&P technician must service. It contains information on standard hardware and torque values, acceptable repair methods, and inspection techniques. [Figure 14-1]

Another popular AC within the 43 series is **AC43.9-1E** which gives information on how to complete of a Major Repair and Alteration Form (FAA Form 337). In particular, it lists FAA approved data sources that can be used as a reference for major repairs and alterations. Data obtained from Airworthiness Directives, approved manufacturer's instructions, Type Certificate Data Sheets, Supplemental Type Certificates, and Technical Standard Orders are considered approved.

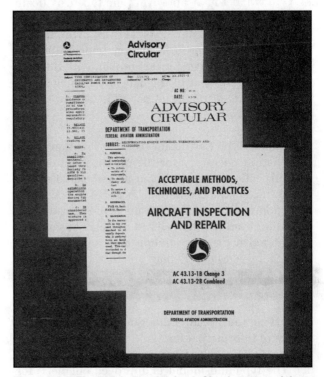

Figure 14-1. The FAA issues Advisory Circulars to explain or clarify the Federal Aviation Regulations. Advisory Circulars are nonregulatory in nature unless incorporated into an FAR and generally do not contain approved data.

To improve aviation safety, the FAA gathers information on mechanical problems and difficulties discovered by aviation maintenance technicians working in the field. When encountering a new or unusual maintenance problem, the technician is requested to fill out and mail an **FAA Form 8010-4 Malfunction or Defect Report** detailing the problem. When the FAA detects a trend forming with a particular aircraft or appliance, it publishes this information in **AC43-16**, General Aviation Airworthiness Alerts. Alerts are issued monthly to distribute the information gathered with the goal of improving service reliability.

For a complete listing of ACs, the FAA publishes an advisory circular checklist. Periodically, the Advisory Circular Checklist (AC 00-2) is revised and reissued to inform you of the current status of ACs. The checklist also provides you with pricing and ordering information. Some ACs are free while others are available at cost. You can order either type through the Department of Transportation, U.S. Government Printing Office.

TYPE CERTIFICATE DATA SHEETS

As previously discussed, new airframes, powerplants, and appliances are issued a type certificate once they meet FAA approval. The type certificate number assigned to the product is also used on the Type Certificate Data Sheet (TCDS), which lists technical and other information concerning the product. The type certificate number, with the date and revision number of the data sheet, is enclosed in a box located in the upper right corner of the data sheet.

An aircraft Type Certificate Data Sheet contains the information necessary for the proper maintenance and inspection of an aircraft or its associated equipment. Information contained in a TCDS is considered approved data for use on a 337 Form. Any deviation from the items listed in the data sheets is considered a major alteration and must be documented by a Major Repair or Alteration Form, a **Supplemental Type Certificate** (STC), or by compliance with an Airworthiness Directive. [Figure 14-2]

The Type Certificate Data Sheet is the primary source of information for:

1. The type and model of approved engine(s) for the model aircraft.
2. The minimum fuel grade for the approved engine(s).

3. The maximum approved rpm and the horsepower rating of the engine(s).
4. Propellers approved for use, rpm limits, and operating restrictions, if any.
5. Airspeed limits for the aircraft in knots and mph.
6. Center of Gravity range, in inches from the datum.
7. Empty weight center of gravity range may be listed, if it has been established by the manufacturer.
8. Location of the reference datum line.
9. Means to level the aircraft for weighing purposes.
10. All maximum weights allowed for various compartments and locations within the aircraft.
11. Oil and fuel capacity and fuel tank moment arms.
12. Control surface movements in degrees.
13. Required equipment necessary for operation of the aircraft.
14. Any additional equipment found necessary for certification of the aircraft.
15. Any placards which must be displayed in full view of the pilot will be listed in Note #2 of the data sheet.

Type Certificate Data Sheets are used during annual inspections to ensure that an aircraft conforms to its type certificate. For example, a TCDS lists an aircraft's various airspeed limits. These limits must be marked on the face of the aircraft's airspeed indicator in accordance with FAR Part 23.1545 and the technician performing the inspection must check these instrument markings to be sure they agree with those in the TCDS.

The TCDS also contains information required to perform a weight and balance check on an aircraft. For example, the leveling means, location of the datum, and the center of gravity range are listed in the data sheets and must be used in weight and balance calculations.

AIRCRAFT SPECIFICATIONS

Prior to 1958, aircraft were certificated under the Civil Air Regulations. These regulations specified that information on certificated aircraft, engines, and powerplants be listed in aircraft specifications. The specifications were similar to Type Certificate Data Sheets with the addition of an equipment list. Many of the older specifications for one type aircraft often required up to sixty or seventy pages of information in the equipment list. The necessity of

DEPARTMENT OF TRANSPORTATION
FEDERAL AVIATION ADMINISTRATION

A12CE
Revision 11
BEECH
60
A60
May 7, 1973

TYPE CERTIFICATE DATA SHEET NO. A12CE

This data sheet which is part of type certificate No. A12CE prescribes conditions and limitations
under which the product for which the type certificate was issued meets the airworthiness requirements
of the Federal Aviation Regulations.

Type Certificate Holder   Beech Aircraft Corporation
                                 Wichita, Kansas 67201

I - Model 60, 4 or 6 PCLM (Normal Category), Approved February 1, 1968
    Model A60, 4 or 6 PCLM (Normal Category), Approved January 30, 1970

Engines	Lycoming TIO-541-E1A4 or TIO-541-E1C4 (2 of either or 1 of each)
Fuel	100/130 minimum grade aviation gasoline
Engine limits	For all operations, 2900 r.p.m. (380 b. hp.)
Propeller and propeller limits	(a) Two (in any combination) Hartzell three-blade propellers
	Diameter: 74 in., (Normal) Minimum allowable for repair 73 1/2 in.
	(No further reduction permitted)
	Pitch settings at 30 in. sta.:
	low 14°, high 81.7°
	HC-F 3 YR-2/C7479-2R
	or HC-F-3 YR-2/C7497B-2R
	or HC-F 3 YR-2F/FC 7479B-2R
	or HC-F 3 YR-2F/FC 7479B-2R
	(b) Beech 60-389000-3 governor

Airspeed limits		
	Never exceed	235 knots
	Maximum structural cruising	208 knots
	Maneuvering	161 knots
	Maximum flap extension speed	
	Approach position 15°	175 knots
	Full down position 30°	135 knots
	Landing gear extended	174 knots
	Landing gear operating	174 knots

C.G. range (landing
  gear extended)

    (+134.2) to (+139.2) at 6725 lb.
    (+128.0) to (+139.2) at 5150 lb. or less
    Straight line variation between points given
    Moment change due to retracting landing gear (+857 in.-lb.)

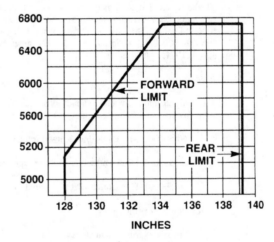

Empty wt. C.G. range	None
Maximum weight	Takeoff and landing 6725 lb.
Ramp weight	6819 lb.

Figure 14-2. A Type Certificate Data Sheet is issued to an aircraft when the FAA approves its design. The TCDS contains important information for use in aircraft maintenance.

- 2 -   A12CE

No. of seats	4 (2 at +141, 2 at +173) (add 2 at +205)
Maximum baggage	500 lb. at +75 (nose compartment
(structural limit)	655 lb. at +212 (aft area of cabin)
Fuel capacity	142 gal. (+138) comprising two interconnected cells in each wing
	or
	204 gal. (+139) comprising three cells in each wing and one cell
	in each nacelle (four cells interconnected)
	See NOTE 1 for data on system fuel
Oil capacity	26 qt. (+88)
(wet sump)	
Max. operating limit	30,000 ft. pressure altitude

Control surface movements					
Wing flaps			Maximum	30°	
Aileron	Up	25°	Down	15°	
Aileron tab (L.H. only)	Up	10°	Down	10°	
Aileron tab anti-servo	Up	12°	Down	7°	
Elevator	Up	17°	Down	15°	
Elevator tab (L.H. only)	Up	10°	Down	30°	
Elevator tab servo	Up	6°	Down	7°	
Rudder	Right	33°	Left	28°	
Rudder tab	Right	26°	Left	26°	

Serial Nos. eligible	Model 60: P-3 thru P-126 (except P-123)
	Model A60: P-123, P-127 and up (see NOTE 3)
Datum	Located 100 in. forward of front pressure bulkhead
Leveling means	Drop plumb line between leveling screws in cabin door frame rear edge
Certification basis	Part 23 of the Federal Aviation Regulations effective February 1, 1965
	as amended by 1, 2, 3, and 12; and Special Conditions dated May 16, 1967,
	forwarded with FAA letter dated June 1, 1967; approved for flight into
	known icing conditions when equipped as specified in the approved airplane
	flight manual.
	Application for Type Certificate dated December 22, 1965.
	Type Certificate No. A12CE issued February 1, 1968, obtained by the
	manufacturer under delegation option procedures.
Production basis	Production Certificate No. 8 issued and Delegation Option Manufacturer
	No. CE-2 authorized to issue airworthiness certificates under delegation
	option provisions of Part 21 of the Federal Aviation Regulations.

Equipment   The basic required equipment as prescribed in applicable airworthiness
regulations (see Certification basis) must be installed in the aircraft
for certification. This equipment must include, for all operations, Airplane Flight Manual
P/N 60-590000-5D dated January 15, 1971, amended
July 1, 1971, or later issue.

In addition:

1. For flights into known icing conditions, these flight manual
supplements and the   equipment noted therein:

   60-590001-17   Flight Into Known Icing Conditions.

   60-590001-11   Continuous Pressure Operated Surface Deice System.

   60-590001-13 Goodrich Electrothermal Propeller Deice System.

2. For all other operations:

   Pre-stall warning indicator P/N 151-6, 151-7, or 190-2 (Safe
   Flight Corporation).

Figure 14-2. A Type Certificate Data Sheet is issued to an aircraft when the FAA approves its design. The TCDS contains important information for use in aircraft maintenance. (continued)

- 3 -   A12CE

NOTE 1.   Current weight and balance data including list of equipment included in certificated empty
weight and loading instructions when necessary must be provided for each aircraft at the
time of original certification.

The certificated empty weight and corresponding center of gravity locations must include
unusable fuel of 24 lb. at (+135).

NOTE 2.   The following placard must be displayed in front of and in clear view of the pilot:

"This airplane must be operated in the normal category in compliance with the operation
limitations stated in the form of placards, markings and manuals."

NOTE 3.   Fuselage pressure vessel structural life limit — refer to the latest revision of the
Airplane Flight Manual for mandatory retirement time.

NOTE 4.   Model 60 (S/N P-3 thru P-126 except P-123) when modified to Beech dwg. 60-5008 and
Model A60 (S/N P-123, P-127 and up) eligible for a masximum weight of 6775 lb.

NOTE 5.   A landing weight of 6435 lb. must be observed if 10 PR tires are installed on aircraft
not equipped with 60-810012-15 (LH) or 60-810012-16 (RH) shock struts.

. . . END . . .

Figure 14-2. A Type Certificate Data Sheet is issued to an aircraft when the FAA approves its design. The TCDS contains important information for use in aircraft maintenance. (continued)

updating and revising the specifications became more costly each year. Because of this, the equipment list was removed from the specifications and Type Certificate Data Sheets were issued for new aircraft and equipment under the provisions of FAR Part 23. Newer aircraft are provided with an equipment list which is delivered to the aircraft owner at time of delivery. Many of the newer aircraft models have this information included in a section of the aircraft flight operating handbook.

Aircraft that were originally certified with aircraft specifications do have the option of changing to the TCDS. Therefore, when conducting a conformity inspection or weight and balance check on an older aircraft that was originally certificated under the CARs, it may be necessary to look in both the aircraft specifications and the Type Certificate Data Sheets.

AIRCRAFT LISTINGS

When the total number of any type aircraft, engine, and propeller still on the aircraft registry falls below fifty, its specifications and Type Certificate Data Sheets are no longer published. Instead, their information is transferred to an aircraft, engine, or propeller listing as appropriate. Type approvals which have expired, or for which the manufacturer no longer holds a production certificate, are also transferred to the Listing Section.

SUPPLEMENTAL TYPE CERTIFICATES

The FAA allows a product to deviate from the original configuration detailed on the Type Certificate Data Sheet if it is modified according to data provided on a Supplemental Type Certificate. Supplemental Type Certificates (STCs) are issued in accordance with FAR Part 21, Subpart E, and are a common method for approving the replacement of an original engine with another model, modifying an aircraft for a specific use such as short take off and landing, or installing equipment not originally certified on an aircraft.

Any individual or organization may apply for a Supplemental Type Certificate, and an STC may be issued to more than one applicant for the same design change, providing each applicant shows compliance with the applicable airworthiness requirement. However, the applicant must show sufficient proof that the alteration meets applicable airworthiness requirements. This is normally

accomplished using engineering data and static and flight testing information. When a Supplemental Type Certificate is issued, the holder may alter aircraft to meet the specifications of the certificate, offer kits for the modification, or offer the plans and use of the STC as approved data. STCs which have been issued are published in the **Summary of Supplemental Type Certificates**, and listed by aircraft make and model. [Figure 14-3]

MANUFACTURER'S PUBLICATIONS

Aircraft manufacturers provide various manuals with their products to assist technicians in inspection, maintenance, and repair. With few exceptions, manufacturer's manuals are acceptable data. The technician must use manufacturer's maintenance manuals when performing maintenance.

ATA SPECIFICATION 100

At one time, the organization of data in manufacturer's publications was left up to the individual producing the manual. As a result, there was little uniformity among different publications, and much time was wasted as technicians had to learn each manufacturer's particular system. To correct this, the Air Transport Association of America (ATA) issued specifications for the organization of Manufacturers Technical Data. The ATA specification calls for the organization of an aircraft's technical data into individual systems which are numbered. Each system also has provisions for subsystem numbering. For example, all of the technical information on the Fire Protection system has been designated as Chapter 26 under the ATA 100 specifications, with fire detection equipment further identified by the sub-chapter number 2610, and fire extinguishing equipment as 2620. Because of this specification, maintenance information for all transport aircraft is arranged in the same way. [Figure 14-4]

General aviation aircraft manufacturers are in the process of standardizing their maintenance information and ATA Specification 100 will be used as the format for this standardization.

MAINTENANCE MANUALS

A manufacturer's maintenance manual is the primary reference tool for the aviation maintenance technician working on aircraft. Airframe maintenance manuals generally cover an aircraft and all of the equipment installed on it when it is in service. Powerplant maintenance manuals, on the other

United States of America
Department of Transportation — Federal Aviation Administration

Supplemental Type Certificate

Number AB123CD

This certificate, issued to John Doe Aircraft Services
1234 Airport Road
Anywhere, USA 12345-1000

certifies that the change in the type design for the following product with the limitations and conditions therefor as specified hereon meets the airworthiness requirements of Part 3 *of the* Civil Air Regulations.

Original Product — Type Certificate Number: 1A2
　　　　Make: Piper
　　　　Model: PA-18-135, PA-18A-135, PA-18S-135, PA-18AS-135, PA-18-150.
　　　　　　　　PA-18A-150, PA-18S-150, PA-18AS-150.

Description of Type Design Change.

Installation of McCauley 1A175/GM 8241 propeller on the -135 models listed above, and installation of McCauley 1A175/GM8241 through GM8244 on the -150 models listed above per page 2 of Joe Doe Aircraft Services Instruction Sheet dated 20 October 1968 and amended 15 May 1969 and 7 January 1970.

Limitations and Conditions:

The approval of this change in type design applies basically to Piper PA-18 models only. This approval should not be extended to other aircraft of this model on which other previously approved modifications are incorporated unless it is determined by the installer that the interrelationship between this change and any of those other previously approved modifications will introduce no adverse effect on the airworthiness of that aircraft. This determination should include consideration of significant changes in weight distribution such as an increase in the fixed disposable weight in the fuselage.

This certificate and the supporting data which is the basis for approval shall remain in effect until surrendered, suspended, revoked, or a termination date is otherwise established by the Administrator of the Federal Aviation Administration.

Date of application: 1 September 1967

Date of issuance: 27 May 1968

Date amended: 9-20-68, 10-15-68, 5-12-69,
6-18-69, 1-8-70, and
8-4-71

By direction of the Administrator

Robert J. Smith

ROBERT J. SMITH, Chief
Engineering and Manufacturing Branch
Northern Region

(Title)

Any alteration of this certificate is punishable by a fine of not exceeding $1,000, or imprisonment not exceeding 3 years, or both.

FAA Form 8110-2 (10-68)

This certificate may be transferred in accordance with FAR 21.47.

Figure 14-3. Supplemental type certificates are available for aircraft owners who want to install an engine, propeller, or appliance that is not on the original Type Certificate.

hand, cover areas of the engines that are not dealt with in the airframe manual.

Maintenance manuals provide information on routine servicing, system descriptions and functions, handling procedures, and component removal and installation. In addition, these manuals contain basic repair procedures and troubleshooting guides for common malfunctions. Maintenance information presented in these manuals is considered acceptable data by the FAA, and may be approved data for the purpose of major repairs and alterations.

OVERHAUL MANUAL

Overhaul manuals contain information on the repair and rebuilding of components that can be removed from an aircraft. These manuals contain multiple illustrations showing how individual components are assembled as well as list individual part numbers.

ILLUSTRATED PARTS CATALOG

Parts catalogs show the location and part numbers of items installed on an aircraft. They contain detailed exploded views of all areas of an aircraft to assist the technician in locating parts. Illustrated parts catalogs are generally not considered acceptable data for maintenance and repair by the FAA.

WIRING MANUALS

The majority of aircraft electrical systems and their components are illustrated in individual wiring manuals. Wiring manuals contain schematic diagrams to aid in electrical system troubleshooting. They also list part numbers and locations of electrical system components.

STRUCTURAL REPAIR MANUALS

For repair of serious damage, structural repair manuals are used. These manuals contain detailed information for repair of an aircraft's primary and secondary structure. The repairs described in a structural repair manual are developed by the manufacturer's engineering staff, and thus are usually considered approved data by the FAA.

SERVICE BULLETINS AND NOTES

One way manufacturers communicate with aircraft owners and operators is through service bulletins and service notes. Service bulletins are issued to inform aircraft owners and technicians of possible design defects, modifications, servicing changes, or other information that may be useful in maintaining an aircraft or component. On occasion, service bulletins are made mandatory and are incorporated into airworthiness directives to correct an unsafe condition.

System	Sub	Title	System	Sub	Title	System	Sub	Title
21	00	AIR CONDITIONING		50	Steering	57	00	WINGS
	10	Compression		60	Position & Warning		10	Main Frame
	20	Distribution		70	Supplementary Gear		20	Auxiliary
	30	Pressurization Control	33	00	LIGHTS		30	Plates/Skin
	40	Heating		10	Flight Compartment		40	Attach Fittings
	50	Cooling		20	Passenger Compartment		50	Flight Surfaces
	60	Temperature Control		30	Cargo & Service Compartment	61	00	PROPELLERS-General
	70	Humidity Regulation		40	Exterior		10	Propeller Assembly
22	00	AUTO PILOT		50	Emergency Lighting		20	Controlling
	10	Amplification	34	00	NAVIGATION		30	Braking
	20	Actuation		10	Air Data Instrumentation		40	Indicating
	30	Controlling		20	Altitude & Direction Inst.	65	00	ROTORS
	40	Indicating		30	Radio Navigation		10	Main rotor
	50	Sensing		40	Radar Navigation		20	Anti-Torque Rotor Assy.
	60	Coupling		50	Proximity Warning		30	Accessory Driving
23	00	COMMUNICATIONS		60	Position Computing		40	Controlling
	10	HF	35	00	OXYGEN		50	Braking
	20	VHF		10	Crew		60	Indicating
	30	PA & Pass. Entertainment		20	Passenger	71	00	POWER PLANT
	40	Interphone		30	Portable		10	Cowling
	50	Audio Integrating	36	00	PNEUMATIC		20	Mounts
	60	Static Discharging		10	Distribution		30	Fireseals
	70	Voice Recorders		20	Indicating		40	Attach Fitting
24	00	ELECTRICAL POWER	37	00	VACUUM		50	Electrical Harness
	10	Generator Drive		10	Distribution	72	00	ENGINE TURBINE
	20	AC Generation		20	Indicating		10	Reduction Gear & Shaft Sect.
	30	DC Generation	38	00	WATER/WASTE		20	Air Inlet Section
	40	External Power		10	Potable		30	Compressor Section
	50	Elect. Load Distribution		20	Wash		40	Combustion Section
25	00	EQUIP./FURNISHINGS		30	Waste Disposal		50	Turbine Section
	10	Flight Compartment		40	Air Supply		60	Accessory Drives
	20	Passenger Compartment	49	00	AIRBORNE AUX. POWER		70	By-Pass Section
	30	Buffet/Galley		10	Power Plant	73	00	ENGINE FUEL & CONTROL
	40	Lavatory		20	Engine		10	Distribution
	50	Cargo & Accessory Compartment		30	Engine Fuel & Control		20	Controlling
	60	Emergency		40	Ignition Starting		30	Indicating
26	00	FIRE PROTECTION		50	Air	74	00	IGNITION
	10	Detection		60	Engine Controls		10	Electrical Power Supply
	20	Extinguishing		70	Indicating		20	Distribution
27	00	FLIGHT CONTROLS		80	Exhaust		30	Switching
	10	Aileron & Tab		90	Oil	75	00	AIR
	20	Rudder & Tab	51	00	STRUCTURES		10	Engine Anti-Icing
	30	Elevator & Tab	52	00	DOORS		20	Accessory Cooling
	40	Horiz. Stabilizer Control		10	Passenger/Crew		30	Compressor Control
	50	Flaps		20	Emergency Exit		40	Indication
	60	Spoiler & Drag		30	Cargo	76	00	ENGINE CONTROLS
	70	Gust Lock & Dampener		40	Service		10	Power Control
	80	Lift Augmenting		50	Fixed Interior		20	Emergency Shutdown
28	00	FUEL		60	Entrance Stairs	77	00	ENGINE INDICATING
	10	Storage		70	Door Warning		10	Power
	20	Distribution		80	Landing Gear		20	Temperature
	30	Dump	53	00	FUSELAGE		30	Analyzers
	40	Indications		10	Main Frame	78	00	EXHAUST
29	00	HYDRAULIC POWER		20	Auxiliary Structure		10	Collector
	10	Main		30	Plates, Skins		20	Noise Suppressor
	20	Auxiliary		40	Attach Fittings		30	Thrust Reverser
	30	Indicating		50	Cones & Fillets/Fairings	79	00	OIL
30	00	ICE & RAIN PROTECTION	54	00	NACELLES/PYLONS		10	Storage
	10	Airfoil		10	Main Frame		20	Distribution
	20	Air Intakes		20	Auxiliary Structure		30	Indicating
	30	Pitot and Static		30	Plates/Skin	80	00	STARTING
	40	Windows & Windshields		40	Attach Fittings		10	Cranking
	50	Antennas & Radomes		50	Fillets/Fairings		20	Igniting
	60	Propellers/Rotors	55	00	STABILIZERS	81	00	TURBINES
	70	Water Lines		10	Horizontal		10	Power Recovery
	80	Detection		20	Elevator		20	Turbo-Supercharger
31	00	INSTRUMENTS		30	Vertical	82	00	WATER INJECTION
	10	Panels		40	Rudder		10	Storage
	20	Independent Instruments		50	Attach Fittings		20	Distribution
32	00	LANDING GEAR	56	00	WINDOWS		30	Dumping & Purging
	10	Main & Doors		10	Flight Compartment		40	Indicating
	20	Nose & Doors		20	Cabin	83	00	ACCESSORY GEAR BOXES
	30	Extension & Retraction		30	Door		10	Drive Shaft Section
	40	Wheels and Brakes		40	Inspection & Observation		20	Gearbox Section

Figure 14-4. To ensure uniformity in maintenance documentation, ATA codes are assigned to all aircraft systems and subsystems. For example, all brake systems fall under the ATA 32-40 code.

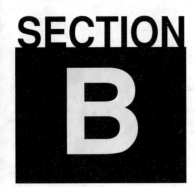

MAINTENANCE FORMS

As you recall, safety is the primary goal for all aviation professionals. To achieve this goal, the FAA requires thorough documentation of all work performed on aircraft. Any time an aircraft is maintained or repaired, an entry must be made in a logbook or on a special form. Therefore you, as an aviation maintenance technician, must become familiar with the forms, certificates, and records that document your maintenance activities.

INSPECTION FORMS

FAR Part 43, Appendix D is the authority covering the scope and detail of items to be inspected during an annual and 100-hour inspection. However, the material listed in Appendix D is not in a format easily adaptable by the technician performing an inspection. Furthermore, Appendix D is more a directive than an inspection form, because it is designed to cover all makes and models of general aviation aircraft. Therefore, the scope of Appendix D is very broad and cannot effectively cover all the specific areas in detail.

As mentioned earlier, FAR Part 43 requires all persons performing an annual or 100-hour inspection to use a checklist that includes at least those items presented in Part 43, Appendix D. The checklist you use may be of your own design or it may be provided by the aircraft manufacturer. All major aircraft manufacturers provide inspection checklists for their aircraft by type and model number. These forms are readily available through the manufacturer's representative and are highly recommended. They meet the minimum requirements of FAR Part 43, Appendix D, and contain many details covering specific items of equipment installed on a particular aircraft as well as references to service bulletins and letters which could otherwise be overlooked. [Figure 14-5]

FAA FORM 337, MAJOR REPAIR AND ALTERATION

All major repairs and alterations to aircraft, powerplants, or appliances require strict compliance with

FAA directives and the manufacturer's recommendations. In order to maintain control of the type of repairs performed and the structural integrity of an aircraft and its components, the FAA requires that a record of all major repairs and alterations be reported on FAA Form 337. Once completed, one copy of the 337 form should be given to the aircraft owner and a second copy should be sent to the local Flight Standards District Office within 48 hours after the aircraft is approved for return to service.

The front of a 337 form contains space for complete identification of the aircraft, powerplant, or appliance, as well as the information concerning the aircraft owner or operator, the person making the repairs, and the person approving the aircraft for return to service. The form should be filled out in a manner similar to the following recommendations.

1. AIRCRAFT: In this section you must fill in the name of the aircraft manufacturer: (Piper Aircraft Co., Cessna Aircraft Co., Beechcraft), the aircraft's model designation, (Cherokee 140, Cessna 150, 172), the aircraft's serial number as it appears on the aircraft data plate, and the N-number appearing on the Owner's Aircraft Registration Certificate.
2. OWNER: In this section the aircraft owner's name and address must be inserted. This information is typically obtained from the aircraft registration certificate.
3. FOR FAA USE ONLY: Do not write in this space.
4. UNIT IDENTIFICATION: If a repair or alteration was performed on an airworthy aircraft, no entry is necessary. If the powerplant, propeller or appliance is a serviceable item, not installed on an aircraft at the present time, the identifying information must be entered in block #4, and the form filed with the component until such time as the item is installed on an airworthy aircraft. At the time of installation on an aircraft, blocks 1 and 2 can be filled in and the form processed at that time.

INSPECTION REPORT

This form meets requirements of FAR Part 43   Work Order No. _____

Make	Model	Serial No.	Registration No.

Owner	Date

Type of Inspection	Tach Time

A. PROPELLER GROUP

	L	R	100	500	insp.

1. Inspect spinner and back plate
2. Inspect blades for nicks and cracks
3. Inspect hub for cracks and corrosion
4. Check for grease and oil leaks
5. Check mounting bolts and safety
6. Constant speed — check blades for tightness in hub pitot tube
* 7. Constant speed — remove prop, remove sludge
* 8. Lubricate as per manual
9. Inspect complete assembly
10. Replace spinner

B. ENGINE GROUP

1. Remove engine cowls
2. Clean cowling, check for cracks, missing fasteners, etc.
3. Compression check: /80

   L.   #1   #2   #3   #4   #5   #6
   R.   #1   #2   #3   #4   #5   #6

* 4. Drain oil
5. Check oil screens and clean
6. Replace oil filter element
7. Check oil temp sender unit for leaks and security
8. Clean and check oil radiator fins
* 9. Remove and flush oil radiator
10. Check and clean fuel screens
11. Drain carburetor
*12. Service fuel injector nozzles
13. Check fuel system for leaks
14. Check oil lines for leaks and security
15. Check fuel lines for leaks and security
*16. Service air cleaner
17. Check induction air and heat ducts
18. Check condition of carb heat box
19. Check mag points for proper clearance
20. Check mags for oil seal leakage
21. Check breaker felts for lubrication
22. Check distributor block for cracks, burned areas, corrosion, height of contact springs
23. Check ignition harness and insulators

Figure 14-5. When performing an inspection, you must use a checklist that follows the requirements of FAR Part 43, Appendix D.

5. TYPE: Mark the proper column to identify the type of repair, i.e., a major repair or a major alteration.

6. CONFORMITY STATEMENT: Within item number 6 you should enter your name and address where it asks for the agency name and address. Under the kind of agency, enter a check mark in the proper block followed by your A&P license number under the box asking for the certificate number. Once the repair or alteration is complete read the statement in section D, and if the information is true and correct, date the form and sign your name in the space provided.

If you are employed by a certified repair station (CRS) the conformity statement would be filled out differently. For example, in the box asking for the agency's name and address you would insert the repair station's name and address, followed by the applicable box being checked and the station's certificate number. The signature of the person performing the work would not change.

7. APPROVAL FOR RETURN TO SERVICE: This section of the FAA Form 337 is filled in by a person authorized by FAR Part 65 to return an aircraft to service after a major repair or alteration.

8. DESCRIPTION OF THE WORK PERFORMED: The reverse side of the FAA Form 337 contains space for the technician to completely describe the maintenance or alteration performed. When doing this, reference should be made to manufacturer's drawings, aircraft station numbers, approved data for the type of repairs performed and all information necessary to ensure that the aircraft has been returned to its original condition, or has been altered in accordance with approved engineering data.

When filling out this section, use as many sheets or pages as necessary to completely describe the nature and extent of maintenance performed. Do not show weight and balance calculations on the FAA Form 337. Weight and balance information should be entered in the aircraft weight and balance data, not on the repair form. Do not leave any blank spaces where someone else may enter additional information on the form. Line out all unused portions of the form.

If you have any reservations as to the technical data pertaining to the repair, you should contact the local FAA office prior to commencing the repairs. In many instances, field approval of the repair may be obtained, based on recommendations in AC 43.13-1B, or the maintenance manual supplied by the aircraft, powerplant, or appliance manufacturer. In many instances, these publications are acceptable to the Administrator, but are not officially approved data. By contacting the local FAA office in advance, the proposed repair procedures can be approved prior to beginning the repairs. [Figure 14-6]

FAA FORM 8010-4, MALFUNCTION OR DEFECT REPORT

Information concerning malfunctions and maintenance problems encountered on all types of aircraft in daily operations is processed through the Aviation Standards National Field Office in Oklahoma City. This information is published in AC 43-16, "General Aviation Airworthiness Alerts," and disseminated throughout the country. This information makes a significant contribution to the continued safety of aircraft operations and maintenance by notifying operators of trends and maintenance problems before they become serious.

The source of this information is the A&P technician performing the daily maintenance and inspection of aircraft. A simple postage-paid postcard form

AC 43.9-1E
Appendix 1

5/21/87

APPENDIX 1. FAA FORM 337 (FRONT), MAJOR REPAIR AND ALTERATION (AIRFRAME, POWERPLANT, PROPELLER, OR APPLIANCE)

	MAJOR REPAIR AND ALTERATION (Airframe, Powerplant, Propeller, or Appliance)	Form Approved OMB No. 2120-0020
US Department of Transportation Federal Aviation Administration		**For FAA Use Only** Office Identification

INSTRUCTIONS: Print or type all entries. See FAR 43.9, FAR 43 Appendix B, and AC 43.9-1 (or subsequent revision thereof) for instructions and disposition of this form. This report is required by law (49 U.S.C. 1421). Failure to report can result in a civil penalty not to exceed $1,000 for each such violation (Section 901 Federal Aviation Act of 1958).

	Make Cessna	Model 182
1. Aircraft	Serial No. 15-10521	Nationality and Registration Mark N-3763
2. Owner	Name *(As shown on registration certificate)* William Taylor	Address *(As shown on registration certificate)* 36 Main Street Cambria, Pennsylvania 15946

3. For FAA Use Only

The data identified herein complies with the applicable airworthiness requirements and is approved for the above described aircraft, subject to conformity inspection by a person authorized by FAR Part 43.
AEA-GADO-19   April 15, 1991   *Ralph Burlingame*
District Office   Date   Ralph Burlingame
                                                              Signature of FAR Inspector

4. Unit Identification				5. Type	
Unit	Make	Model	Serial No.	Repair	Alteration
AIRFRAME	~~~~~~~~~~~~~~~~~ *(As described in Item 1 above)* ~~~~~~~~~~~~~~~~~			X	
POWERPLANT					
PROPELLER					
APPLIANCE	Type				
	Manufacturer				

6. Conformity Statement

A. Agency's Name and Address	B. Kind of Agency		C. Certificate No.
George Morris High Street Johnstown, Pennsylvania 15236	X	U.S. Certificated Mechanic	1305888
		Foreign Certificated Mechanic	
		Certificated Repair Station	
		Manufacturer	

D. I certify that the repair and/or alteration made to the unit(s) identified in item 4 above and described on the reverse or attachments hereto have been made in accordance with the requirements of Part 43 of the U.S. Federal Aviation Regulations and that the information furnished herein is true and correct to the best of my knowledge.

Date March 19, 1991	Signature of Authorized Individual *George Morris* George Morris

7. Approval for Return To Service

Pursuant to the authority given persons specified below, the unit identified in item 4 was inspected in the manner prescribed by the Administrator of the Federal Aviation Administration and is   ☒ APPROVED   ☐ REJECTED

BY	FAA Flt. Standards Inspector	Manufacturer	Inspection Authorization	Other *(Specify)*
	FAA Designee	Repair Station	Person Approved by Transport Canada Airworthiness Group	
Date of Approval or Rejection April 9, 1991	Certificate or Designation No. 237412	Signature of Authorized Individual Donald Pauley	*Donald Pauley*

FAA Form 337 (4-87)

Figure 14-6. An FAA Form 337 is used to document major repairs and alterations to airframes, powerplants, propellers, and appliances.

is provided by the FAA for technicians to enter the basic information as to the type of malfunction or defective component as well as the type and model of aircraft. The form is then forwarded to the National Field Office through the local FAA office. Submission of these forms is voluntary, but they are very important to the continued safe operation of general aviation aircraft. If the information is serious enough, an Airworthiness Directive (AD) could be issued as a result of the information submitted. [Figure 14-7]

verify it meets the requirements of FAR Part 21. This certificate may also be referred to as an FAA Form 8100-2. Airworthiness Certificates are issued without an expiration date and, therefore, remain valid as long as all maintenance and airworthiness directives are complied with, and the aircraft is properly registered in the United States. The airworthiness certificate must be displayed in the aircraft so that it is legible to passengers and crewmembers. [Figure 14-8]

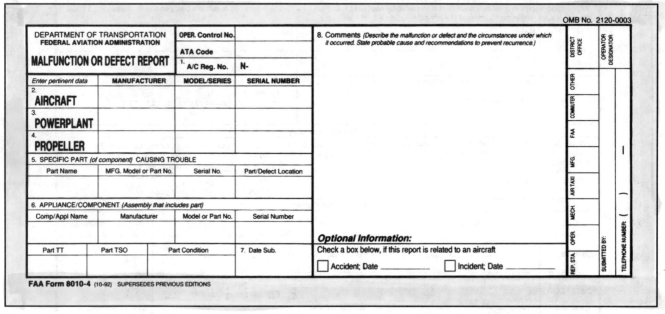

Figure 14-7. Aviation maintenance technicians perform a valuable service by using FAA Form 8010-4 Malfunction or Defect Reports. This voluntary program gathers information on general aviation maintenance problems.

AIRCRAFT FORMS

To be considered airworthy, specific documents must be carried on board an aircraft. While it is the aircraft operator's responsibility to verify that these items are present on the aircraft before each flight, it is considered good practice for maintenance technicians to check for their presence during an annual or 100-hour inspection. These documents can be remembered by the acronym ARROW, which stands for:

A — Airworthiness certificate.
R — Registration certificate.
R — Radio station license (required only for operations outside the U.S.).
O — Operating instructions.
W — Weight and balance information.

AIRWORTHINESS CERTIFICATE

The FAA issues an airworthiness certificate to an aircraft after it is manufactured and inspected to

Aircraft licensed in the experimental, restricted, and agricultural categories may be issued a special airworthiness certificate called an FAA Form 8130-7. These special airworthiness certificates are effective only for the time period listed on the certificate.

Figure 14-8. A Standard Airworthiness Certificate is issued to an aircraft and must remain on the aircraft during operation.

REGISTRATION CERTIFICATE

A Certificate of Aircraft Registration, AC Form 8050-3, is issued by the FAA to an aircraft owner when an aircraft is purchased. This registration remains valid as long as the ownership remains the same. Furthermore, the certificate of registration must be present in the aircraft whenever the aircraft is operated. [Figure 14-9]

RADIO STATION LICENSE

The Federal Communications Commission requires all radio transmitters installed in aircraft to be registered and licensed if flight operations are conducted outside the United States. This radio station license must be kept with the aircraft anytime a radio transmitter is installed and operated during international flights.

OPERATING INSTRUCTIONS

The operating limitations and instructions for a particular aircraft are located in an FAA-Approved Airplane Flight Manual or Pilot's Operating Handbook. The flight manual must be on board the aircraft during flight. However, aircraft built prior to March 1, 1979, were certified without an approved flight manual and must have the operating limitations visible on placards, instrument markings, or approved flight material.

WEIGHT AND BALANCE DATA

Since no two aircraft weigh the same, weight and balance data must be kept with every aircraft. This data must contain an equipment list that identifies the weight and moment of each accessory added to an aircraft. As discussed in Chapter 6, any modification or change to an aircraft or its installed equipment requires a new set of weight and balance figures to be calculated.

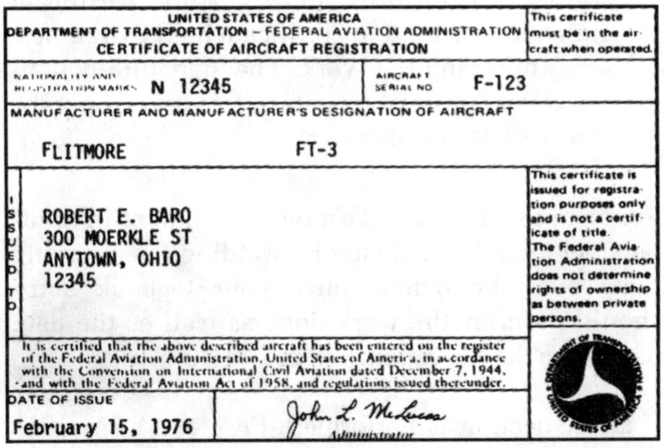

ment requires a new set of weight and balance figures to be calculated.

MAINTENANCE RECORDS

In the past, the historical record of aircraft and engine operating hours, maintenance actions performed, and inspections accomplished on an aircraft and its powerplants were entered in the aircraft and powerplant logbooks in chronological order. These logbooks typically consisted of bound volumes of simple forms. Entries were made by maintenance personnel and by the pilot/owner of the aircraft. These records were, in many cases, intermixed making record research time consuming and tedious.

The present requirements for aircraft records are defined in FAR Part 91.417 and clearly state that the aircraft owner is primarily responsible for maintaining an aircraft's required maintenance records. However, as an aircraft technician you are required to document any maintenance you perform. Therefore, aircraft records are more of a shared responsibility. Maintenance records are divided into two categories: **permanent records**, which must be kept with the aircraft as long as it is in service, and **temporary records**, which can be disposed of after a specified period of time.

PERMANENT RECORDS

Permanent records, as defined by FAR Part 91, are those records which must be retained by the aircraft owner until the aircraft is destroyed or permanently removed from service. If the aircraft is sold, the permanent records must be transferred with the aircraft. The six types of records in this category are:

1. The total time in service of the airframe, each engine, and each propeller.
2. The current status of life-limited parts of each airframe, engine, propeller, and appliance.
3. The time since the last overhaul of all items on the aircraft which are required to be overhauled at a specified time interval.
4. The current inspection status of the aircraft and the time since last inspection.
5. The current status of applicable Airworthiness Directives (ADs), including the method of compliance, the AD number and revision date, and the time and date when the next action is required, if any.
6. Copies of any FAA Form 337 for each major repair or alteration to the airframe and the currently installed powerplants, rotors, propellers, and appliances.

UNITED STATES OF AMERICA
DEPARTMENT OF TRANSPORTATION – FEDERAL AVIATION ADMINISTRATION
CERTIFICATE OF AIRCRAFT REGISTRATION

This certificate must be in the aircraft when operated.

NATIONALITY AND REGISTRATION MARKS N 12345
AIRCRAFT SERIAL NO. F-123

MANUFACTURER AND MANUFACTURER'S DESIGNATION OF AIRCRAFT
FLITMORE   FT-3

ISSUED TO
ROBERT E. BARO
300 MOERKLE ST
ANYTOWN, OHIO
12345

This certificate is issued for registration purposes only and is not a certificate of title. The Federal Aviation Administration does not determine rights of ownership as between private persons.

It is certified that the above described aircraft has been entered on the register of the Federal Aviation Administration, United States of America, in accordance with the Convention on International Civil Aviation dated December 7, 1944, and with the Federal Aviation Act of 1958, and regulations issued thereunder.

DATE OF ISSUE
February 15, 1976

John L. McLucas
Administrator

Figure 14-9. A Certificate of Aircraft Registration is issued to an aircraft owner when the aircraft is purchased.

These permanent records are maintained in several different ways, depending on the size and complexity of the aircraft. For example, for small single engine aircraft used by private individuals or for flight training purposes, simple bound paper logbooks meet the record keeping requirements of FAR Part 91. A separate logbook for the airframe, the engine(s), and the propeller(s) must be maintained in order to comply with the regulations.

The aircraft operating time in service requirements are tracked through the use of a recording tachometer or electrically operated hour meter (Hobbs meter) and the current operating time should be periodically entered in the aircraft logbooks. By doing this, the inspection status of the aircraft, as well as the time accumulated since the the last inspection, are easily computed by reference to the time recorded at the last inspection. The current status of life-limited parts installed on the airframe, engine(s), propeller(s), rotor, and appliances are entered in the appropriate logbook, with the date and hours in service at the time of installation. In the event there is previous operating time on a component or appliance you are installing, the time should be noted and the replacement times corrected as necessary.

The current status of applicable airworthiness directives and the other required data may be entered on a separate AD record and attached to the aircraft logbooks. A separate AD record should be maintained for the airframe, engines, propellers, and appliances. Recurring ADs, and the necessary actions required, should be clearly marked and recorded in the applicable logbooks.

TEMPORARY RECORDS

Temporary records are those records which may be disposed of after the work is repeated or superseded by other work or for one year after the work is completed. There are two categories of temporary records.

1. Records of maintenance to an airframe, engine, propeller, rotor, or appliance. This refers to maintenance actions of a routine nature, such as repacking wheel bearings and other minor maintenance actions which are periodically repeated and are not major repairs or major alterations.

2. Records of the 100-hour, annual, or progressive inspections. Since these inspections are repeated at prescribed intervals, the old inspection records may be disposed of when the latest entry is entered in the aircraft records.

The option to dispose of temporary records may not be an easy task for the small general aviation aircraft owner. Since the record keeping requirements for these aircraft are easily met by data entries in bound paper logbooks, removal of these entries would probably destroy the logbook or at least produce records that appear incomplete. For this reason, it is suggested that the temporary maintenance records for small aircraft be retained and not removed from the aircraft logbooks. However, for an executive or corporate aircraft operator, the aircraft maintenance records are typically maintained by a record keeping staff. Furthermore, the aircraft logbooks often consist of loose leaf binders making removal and disposal of aircraft maintenance records easy.

MAINTENANCE RECORD ENTRIES

FAR 43.9 requires that certain information be entered into an aircraft's maintenance records after a repair or alteration is performed and the aircraft is returned to service. Upon completion of the work, the person releasing the aircraft to service must make a logbook entry containing at least the following information.

1. A description of work performed, or reference to acceptable data.
2. The date the work was completed.
3. The name of the person performing the work.
4. If the work performed on the item was performed satisfactorily, the signature, certificate number, and kind of certificate held by the person approving the work. The signature constitutes the approval for return to service only for the work performed.

For example, if you make a repair to a dent in tubular steel joined at a cluster by welding a reinforcing plate over the dented area, your logbook entry should describe the work done as well as the date used to make the repair. [Figure 14-10]

If an inspection is performed, FAR 43.11 specifies that the person releasing the aircraft to service must make a logbook entry containing:

MAINTENANCE RECORD
F.A.R. 43.9

DATE OF COMPLETION	AIRCRAFT TIME IN	DESCRIPTION OF WORK PERFORMED OR APPROVED DATA USED	AGENCY & CERTIFICATE NO. WORK PERFORMED/RETURNED TO SERVICE
12/2/95	1836:27	REPAIRED DENT AT FUSE-LAGE STATION 112.5 BY WELDING REINFORCING PLATE OVER DENTED AREA. REPAIR PERFORMED I.A.W. AC-43-13-1A PAGE 35 & CESSNA SRM PAGE 212.	O.D. Cleaner A&P 554221910

26

Figure 14-10. A maintenance logbook entry should briefly describe the work performed and reference an approved maintenance manual source. In this example, AC 43-13.1A states that a dent a tubular steel cluster is repaired by welding a specially formed steel plate over the dented area and surrounding tubes.

1. The inspection type and a brief description of its extent.
2. The date of the inspection and the aircraft total time in service.
3. The signature, certificate number, and type of certificate held by the person approving or disapproving the item for return to service.
4. Except for progressive inspections, an airworthy aircraft release must contain a statement similar to "I certify that this aircraft has been inspected in accordance with (insert type) inspection and was determined to be in airworthy condition."

If the aircraft is not approved for return to service because of needed maintenance, noncompliance with applicable specifications, airworthiness directives or other approved data, the statement "I certify that this aircraft has been inspected in accordance with (insert type) inspection and a list of discrepancies and unairworthy items dated (date) has been provided for the aircraft owner or operator." [Figure 14-11]

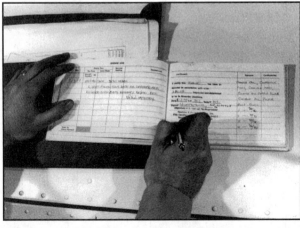

Figure 14-11. Aviation maintenance technicians must document all inspection, maintenance, and repair actions in an appropriate format such as a logbook.

CHAPTER 15

MECHANIC PRIVILEGES AND LIMITATIONS

INTRODUCTION

When you have successfully completed your studies and passed your exams, you will receive a Mechanic's Certificate with airframe and powerplant ratings. This certificate entitles you to certain privileges and limitations. This chapter discusses many of the regulations governing certified Airframe and Powerplant technicians. However, this chapter does not look at every aspect of FAR Part 65 in detail. Therefore, as a responsible and competent technician, you must take the time to familiarize yourself with the regulations governing, as well as the privileges and limitations of, your ratings.

SECTION A

THE MECHANIC CERTIFICATE

As discussed in Chapter 14, FAR Part 65 entitled Certification: Airmen Other Than Flight Crewmembers defines the requirements for licensing maintenance personnel and lists the privileges and limitations of the various technician ratings. Under the current regulations, there are two certificates for maintenance personnel described in Part 65, each with different privileges and limitations. They are the mechanic certificate and the repairman certificate. In addition, there are two ratings issued to certificated mechanics — the airframe rating and the powerplant rating. [Figure 15-1]

ELIGIBILITY REQUIREMENTS

An applicant for a mechanic certificate must be able to read, write and understand the English language, and be at least 18 years old. Furthermore, the applicant must have completed an FAA approved maintenance technician school, or have documented evidence of a minimum of 18 months practical experience for either the airframe or powerplant rating or a total of 30 months experience for both the Airframe and Powerplant ratings.

After satisfying these requirements, an applicant must successfully pass a general, airframe, and powerplant computerized examination covering 45

subject areas. The General exam consists of 50 questions whereas the Airframe exam consists of 100 questions on airframe structures, systems, and components. The Powerplant exam also has 100 questions. However, the subject matter you are tested on includes powerplant theory, maintenance, systems, and components.

As a test applicant you will receive three test scores for each examination. You must receive a minimum score of 70 percent on all three exams. After receiving your computerized test scores, you may then apply for the oral and practical examination, which is administered by an FAA examiner or a **Designated Mechanic Examiner** (DME). The oral and practical examination covers the same 45 technical subject areas as the computerized exam. For the oral exam, you must correctly answer at least three of four questions on each of the 45 technical subject areas. For the practical exam, however, you must demonstrate the ability to perform the practical examination projects assigned by the DME.

MECHANIC PRIVILEGES AND LIMITATIONS

A certificated mechanic may perform or supervise the maintenance, preventive maintenance, or alteration of an aircraft, appliance, or part for which the mechanic is appropriately rated. However, the mechanic must have performed this work at an earlier date, or be able to show the ability to perform the work to the satisfaction of the FAA. This can be accomplished by simply performing the work under the supervision of a certificated and rated mechanic who has had previous experience in the operation being done.

A mechanic certificate with an airframe rating allows a technician to approve and return to service an airframe, or any related part or appliance, after performing, supervising, or inspecting its maintenance or alteration. Furthermore, the technician can perform 100-hour inspections on airframes and related parts or appliances and approve them for

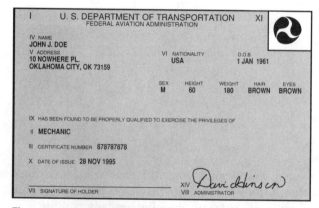

Figure 15-1. A Mechanic's Certificate with airframe and powerplant ratings allows you to perform maintenance on aircraft. However, you must understand and comply with the Federal Aviation Regulations governing your activities.

return to service. However, a technician with an airframe rating may not inspect or return to service an airframe or related part or appliance that has undergone a major repair or alteration.

A technician holding a powerplant rating can approve and return to service powerplants, propellers, and accessories after performing, supervising, or inspecting their maintenance or alteration. Furthermore, a powerplant rating allows a technician to perform 100-hour inspections on powerplants and propellers and approve them for return to service. However, like the airframe rating, a powerplant rating does not permit a technician to inspect or return to service a powerplant, propeller, or accessory that has undergone a major repair or alteration.

Holders of both airframe and powerplant ratings can perform minor repairs and alterations to airframes, powerplants, propellers, and components, and approve these items for return to service. In addition, an A&P can perform major repairs and alterations to airframes, powerplants, and components. However, an A&P may not approve a major repair or alteration for return to service. Furthermore, an A&P cannot perform major repairs or alterations to propellers, or perform any repairs or alterations to instruments. An A&P can, however, perform a 100-hour inspection on airframes, powerplants, propellers, accessories, and instruments and approve and return them to service. It is important to note, however, that an A&P cannot delegate his inspection duties in a 100-hour inspection. In addition, an A&P cannot perform annual inspections. However, they can correct discrepancies an authorized inspector finds on an annual inspection.

RECENT EXPERIENCE REQUIREMENTS

An Airframe and Powerplant Mechanic Certificate remains in effect until it is surrendered, suspended, or revoked. However, to exercise the privileges of a certificate and rating, a technician must have served as a mechanic in the capacity their certificate and rating allows, or have supervised other mechanics, for at least 6 months out of the previous 24 months.

THE INSPECTION AUTHORIZATION

Technicians who have held an A&P rating for a minimum of three years and who have been actively engaged in maintaining general aviation aircraft for at least two years can apply for an Inspection Authorization (IA). The IA applicant must have the equipment, tools, facilities, and inspection data necessary to perform annual inspections and approve major repairs and alterations to aircraft. Furthermore, all IA applicants must have a fixed base of operations where they may be contacted during a normal work week. If an IA applicant meets these qualifications, the FAA office having jurisdiction over the applicant's geographical location will schedule a date for the inspection authorization examination. The examination consists of three tests that normally require at least five hours to complete. Furthermore, IA applicants must provide their own technical data for the examination.

The first test consists of 10 multiple choice questions on FAR Part 65, paragraphs 91 through 95, and AC 65-19, the FAA Inspection Authorization Study Guide. The applicant is allowed 20 minutes to complete this test. The second test, administered only after the applicant has successfully passed the first test, consists of an open book examination in which the applicant must provide all the technical data required to properly answer the questions. The FAA examiner assigns the applicant an aircraft by type, model, and serial number, as well as an engine type, propeller, and any component or appliance desired. The test questions are then answered using the assigned aircraft for reference when necessary. The 20 questions in this exam must be answered in full and include the technical data where the answer was located. The publication number, as well as the page and paragraph number must be referenced as part of the answer. A time limit of two hours is allowed for completion of the second test.

The third test is given after the second test has been graded and a passing score achieved. The content of the third test is similar to the second test, but with different questions. The third test also consists of 20 questions and has a two hour time limit.

The Inspection Authorization rating is awarded upon successful completion of the three exams. Failure of any test terminates the testing, and a waiting period of 90 days is required before an applicant may reapply for the IA examinations. There are no exceptions to the 90 day waiting period.

IA PRIVILEGES

In addition to all the privileges of an Airframe and Powerplant rating, an Inspection Authorization permits a technician to perform an annual inspection on an aircraft and approve it for return to service. Furthermore, an IA can perform major repairs and alterations made on airframes and powerplants and approve the work for return to service. However, an IA cannot perform major repairs and alterations to propellers, or make any repairs or alterations to instruments. These tasks must be performed by an appropriately rated repair station. However, once a repair station has conducted the work, a technician holding an IA rating can inspect and approve those items for return to service.

It is important to note that an authorized inspector cannot delegate his inspection duties for annual and 100-hour inspections to another person. In other words, the IA must actually perform the inspection. However, the IA can allow other technicians to perform a progressive inspection as long as the IA supervises their work.

INSPECTION AUTHORIZATION DURATION

Unlike an A&P technician, an IA's privileges can expire. To prevent this from happening, an IA must maintain the recent experience requirements of their Airframe and Powerplant rating as well as keep a fixed base of operation, and maintain the equipment, facilities, and inspection data necessary to perform their work.

The IA rating expires on March 31 of each year. To renew the rating, the FAA requires evidence that the IA has performed at least one of the following:

1. Conducted an annual inspection every 90 days.
2. Inspected at least two major repairs or alterations every 90 days.
3. Performed or supervised at least one progressive inspection.
4. Attended and completed an approved IA refresher course.
5. Pass an oral test given by an FAA inspector to demonstrate the IA's knowledge of applicable regulations and standards.

THE REPAIRMAN CERTIFICATE

Many aircraft repair facilities work on components and subassemblies. Technicians working in these facilities performing maintenance activities such as component overhaul and rebuilding do not require the broad training an Airframe and Powerplant technician does. However, they do require training on the specific duties they must perform. Once a person receives the appropriate training they are issued a repairman's certificate. FAR Parts 65 and 145 specify

the requirements for repairmen's certificates for technicians performing specialized maintenance functions at certificated repair stations. The holder of a repairman's certificate can perform and supervise the maintenance, preventive maintenance, and alteration of an aircraft or its components for which his employer is certified. [Figure 15-2]

To be eligible for a repairman certificate, a technician must be qualified to perform maintenance on aircraft or components appropriate to the job for which the technician is employed. The repairman certificate is issued to a technician for the repair facility in which they are employed. Therefore, when a repairman leaves the employment of the designated repair station, the certificate must be surrendered.

DRUG AND ALCOHOL TESTING

As a response to several drug and alcohol-related transportation accidents, the Department of Transportation began requiring drug testing of employees in safety and security related positions. Aircraft maintenance is obviously safety related, and aviation maintenance technicians are today subject to drug and alcohol testing under FAR Parts 65, 121, and 135. They are also subject to suspension or revocation of their licenses if they are convicted in state or federal courts of drug-related offenses.

FAR 65.23 requires employees of Part 121 and Part 135 certificate holders to submit to testing for drugs when requested by an employer. Refusal to take a drug test is grounds for denial of an application for any FAA license or rating, and suspension or revocation of any certificate or rating currently held.

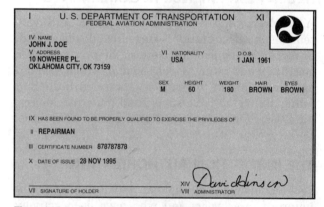

Figure 15-2. A repairman's certificate allows a technician to perform many of the same maintenance activities as an A&P mechanic. However, the repairman's certificate must be surrendered when the technician leaves the place of employment.

Under Appendix I of Part 121, employees can be tested for evidence of drug use for the following reasons.

1. Pre-employment testing. This means an employer must test prospective employees for evidence of drug use prior to hiring them.
2. Periodic testing. All employees performing safety-sensitive functions must submit to periodic drug tests.
3. Random testing. A certain percentage of an employer's workers must be tested randomly each year. Employees to be tested are selected at random by use of a random-number table or computer-based random number generator.
4. Post-accident testing. Any employee performing a safety-sensitive function will be tested for evidence of drug and alcohol use when the employee is involved in an accident.
5. Reasonable cause testing. Employers must test employees who are reasonably suspected of using a prohibited drug, or being under the influence of alcohol.
6. Return to duty and follow-up testing. If an employee refuses to take a drug test, or takes the test and fails to pass, the employee must be tested for drugs before being allowed to return to work. The employee must also submit to periodic unannounced drug tests if the employee returns to work after failing or refusing to take a drug test.

INDEX

D

T

X

Y

Z

SPECIAL INTEREST REFERENCE BOOKS

A&P MECHANICS CERTIFICATION GUIDE

Outlines all FAA established qualifications to become a licensed A&P technician. This is an FAA reprint. ISBN: #0-89100-082-8. 72 pages.

ITEM NUMBER JS312615

AIRCRAFT INSPECTION & MAINTENANCE RECORDS

Describes the various inspections technicians are required to perform and the proper way to record them in the maintenance records. A real time-saver for the A&P. ISBN: #0-89100-094-1. 92 pages.

ITEM NUMBER JS312677

ACCEPTABLE METHODS, TECHNIQUES AND PRACTICES/AIRCRAFT ALTERATIONS

A newly revised FAA reprint, this is the "bible" of aircraft maintenance. It is a combined edition of both the AC 43.13-1B and -2A series handbooks that establish the standards for inspection, repair and alterations. ISBN# 0-89100-306-1. 768 pages.

ITEM NUMBER JS312617

BEST OF AMJ MAINTENANCE TIPS

The best of the Aviation Mechanics Journal's popular "Maintenance Tips". These "tips" from the "pros" in the field will help you to do things better, faster and easier. ISBN# 0-89100-341-X. 184 pages.

ITEM NUMBER JS312629

STANDARD AVIATION MAINTENANCE HANDBOOK

Hundreds of the most commonly-used charts, diagrams, scales and measurements for aviation maintenance personnel in a "tool box"-sized, spiral-bound reference guide. ISBN# 0000-89100-282-0. 240 pages.

ITEM NUMBER JS312624

AIRCRAFT TECHNICAL DICTIONARY (3RD EDITION)

Updated and expanded. The best way to learn the aviation maintenance language. ISBN# 0-89100-410-6.
504 pages.

ITEM NUMBER JS312625

VISIT YOUR JEPPESEN DEALER OR CALL 1-800-621-JEPP
PLEASE CALL FOR PRICING

AVIATION MAINTENANCE TRAINING

A&P TECHNICIAN GENERAL WORKBOOK

Contains questions covering the general section of aviation technician training. It is designed to be used with the A&P technician general textbook. Answer key bound into back of workbook. ISBN# 0-88487-212-2. 204 pages.

ITEM NUMBER JS322710

A&P TECHNICIAN GENERAL STUDY GUIDE

An excellent, self-contained study program. Provides coverage of key points required to answer the FAA's computerized test questions. Also includes: a reprint of the FAA's computerized test questions, answers, explanations and references. Each question is page referenced for easy study. ISBN# 0-88487-204-1. 168 pages.

ITEM NUMBER JS312691

FAR HANDBOOK FOR AVIATION MAINTENANCE TECHNICIANS

Contains the most current and easy-to-use FAR information available. Updated annually to include all changes published by the FAA in the previous year, the changes are identified to make them easy to find. The handbook contains only those FAR parts which are pertinent to an aviation maintenance technician: 1, 13, 21, 23, 27, 33, 34, 35, 39, 43, 45, 47, 65, 91, 119, 125, 135, 145, 147, 183 as well as applicable SFARS.

As an added bonus, each book contains a copy of applicable Advisory Circulars including: AC 20-62, AC 20-109, AC 21-12, AC 39-7, AC 43-9, AC 43.9-1, AC 65-11, and AC 65-19G, Inspection Authorization Study Guide. ISBN# 0-88487-200-9.

ITEM NUMBER JS312616

VISIT YOUR JEPPESEN DEALER OR CALL 1-800-621-JEPP
PLEASE CALL FOR PRICING

AVIATION MAINTENANCE TRAINING

Jeppesen's General, Airframe and Powerplant training materials also come in complete kit form. Comprehensive, developed by respected experts in the field, these kits give you the information needed to succeed in obtaining the Airframe and Powerplant license.

AIRFRAME

Airframe Kit Includes the following: A&P Technician Airframe Textbook • A&P Technician Airframe Workbook • A&P Technician Airframe Study Guide • Student Kit Bag

ITEM NUMBER JS302128

KIT ITEMS ARE ALSO SOLD SEPARATELY.

POWERPLANT

Powerplant Kit Includes the following: A&P Technician Powerplant Textbook • A&P Technician Powerplant Workbook • A&P Technician Powerplant Study Guide • Powerplant Exam Package • Student Kit Bag

ITEM NUMBER JS302184

KIT ITEMS ARE ALSO SOLD SEPARATELY.

TECHSTAR PRO

An innovative handheld computer and personal organizer for maintenance technicians and students. The Techstar Pro combines the latest technology and ease of use to give you a Scientific Calculator, 7-function aviation computer and 8-function personal organizer. Great for the Maintenance student and approved for FAA Knowledge Exams.

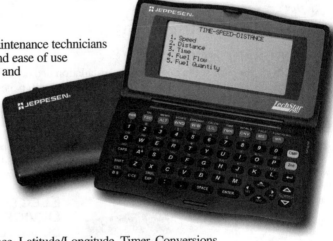

SCIENTIFIC CALCULATOR

In addition to performing basic mathematic functions, the Techstar Pro will calculate square roots, powers, roots, scientific notation, and all trigonometric and algebraic functions.

FLIGHT COMPUTER

Time/Speed/Distance, Altitude/Airspeed, Wind, Weight & Balance, Latitude/Longitude, Timer, Conversions

PERSONAL ORGANIZER

Telephone/address, Personal Memo/To Do, Daily Scheduler, Trip Expense Log, Monthly Calender, Local and UTC Time, Calculator

ITEM NUMBER JS505000

VISIT YOUR JEPPESEN DEALER OR CALL 1-800-621-JEPP
PLEASE CALL FOR PRICING

TOPICAL MAINTENANCE BOOKS SUPPORTING THE GENERAL TEXTBOOK

MATHEMATICS & PHYSICS FOR AVIATION PERSONNEL

Reviews basic physics principles and minimum mathematics required to solve simple equations involved in the further study of aerodynamics, aircraft material factors and related subjects. ISBN# 0-89100-399-1. 100 pages.

ITEM NUMBER JS312619

PHYSICS FOR AVIATION

A complete physics course for aviation-related trades. Not a review, it's the only physics book geared for aviation training that uses "every day" examples to explain every facet of the subject. ISBN# 0-89100-411-4. 184 pages.

ITEM NUMBER JS312620

NONDESTRUCTIVE TESTING FOR AIRCRAFT

Provides the student technician with a working knowledge of the nondestructive test methods used for today. Explores the advantages and disadvantages of each, and the problems which might be encountered. Methods covered are visual, liquid penetrant, magnetic particle, eddy current, ultrasonic and radiographic. Glossary included. By Douglas C. Latia. ISBN# 0-89100-415-7. 122 pages.

ITEM NUMBER JS312640

AIRCRAFT WEIGHT & BALANCE

Includes important weight and balance information related to types of aircraft, positioning of jacks, weighing the craft and more. ISBN# 0-89100-096-8. 112 pages.

ITEM NUMBER JS312634

AIRCRAFT SYSTEMS & COMPONENTS

This book by Don Garrett is written in a direct, easy-to-understand style. Designed principally for A&P and Airway Science programs, it works equally well as a basic text for any course of study that requires a knowledge of aircraft systems. Starting with a background in electricity, the content brings the reader through complex systems and helps develop a high degree of understanding.

ITEM NUMBER JS312685

AIRCRAFT BATTERIES

An in-depth look at the construction, installation and servicing of lead-acid and nickel-cadmium batteries. ISBN# 0-89100-052-6. 36 pages.

ITEM NUMBER JS312644

AIRCRAFT CORROSION CONTROL

Discusses corrosion's causes and effects; its removal and prevention. Complete with glossary, study questions and answers. ISBN# 0-89100-111-5. 56 pages.

ITEM NUMBER JS312630

ELECTRONIC CIRCUIT DEVICES

Discusses basic Electronic Circuit theory, AC power controls, transistors, Sine wave oscillators and much more. ISBN# 0-89100-192-1. 280 pages.

ITEM NUMBER JS312663

DC CIRCUITS

DC electrical circuits explained step-by-step. Useful addition to any technical library. ISBN# 0-89100-121-2. 168 pages.

ITEM NUMBER JS312635

VISIT YOUR JEPPESEN DEALER OR CALL 1-800-621-JEPP

PLEASE CALL FOR PRICING